10. $\dfrac{d}{dx} e^x = e^x, \qquad \dfrac{d}{dx} e^{f(x)} = f'(x)e^{f(x)}$

11. $\dfrac{d}{dx} \ln x = \dfrac{1}{x}, \quad x > 0;$

$\dfrac{d}{dx} \ln f(x) = f'(x) \cdot \dfrac{1}{f(x)}, \quad f(x) > 0$

12. $\dfrac{d}{dx} \ln |x| = \dfrac{1}{x}, \quad x < 0;$

$\dfrac{d}{dx} \ln |f(x)| = f'(x) \cdot \dfrac{1}{f(x)}, \quad f(x) < 0$

13. If $\dfrac{dP}{dt} = kP$, then $P(t) = P_0 e^{kt}$.

14. If $\dfrac{dP}{dt} = -kP$, then $P(t) = P_0 e^{-kt}$.

15. $\dfrac{d}{dx} a^x = (\ln a)a^x$

16. $\dfrac{d}{dx} \log_a x = \dfrac{1}{\ln a} \cdot \dfrac{1}{x}, \quad x > 0;$

$\dfrac{d}{dx} \log_a |x| = \dfrac{1}{\ln a} \cdot \dfrac{1}{x}, \quad x < 0$

17. $a^x = e^{x(\ln a)}$

APPLIED CALCULUS

THIRD EDITION

APPLIED CALCULUS

THIRD EDITION

MARVIN L. BITTINGER
BERNARD B. MORREL

Indiana University—Purdue University at Indianapolis

ADDISON-WESLEY PUBLISHING COMPANY

Reading, Massachusetts · Menlo Park, California · New York
Don Mills, Ontario · Wokingham, England · Amsterdam
Bonn · Sydney · Singapore · Tokyo · Madrid
San Juan · Milan · Paris

Football is like calculus.
Someone has to show you how to do it,
but once you've got er' figgered, why, she's easy.

BERT JONES, Quarterback, *Baltimore Colts*
(Bert Jones majored in mathematics in college.)

Sponsoring Editor	Jerome Grant
Managing Editor	Karen Guardino
Design, Editorial, and Production Services	Quadrata, Inc.
Technical Art Supervisor	Joseph Vetere
Technical Art Consultant	Connie Hulse
Illustrator	Tech-Graphics
Manufacturing Supervisor	Roy Logan
Cover Photograph	Marshall Henrichs
Cover Designer	Eileen R. Hoff

Library of Congress Cataloging-in-Publication Data
Bittinger, Marvin L.
    Applied calculus/Marvin L. Bittinger, Bernard B. Morrel—3rd ed.
        p.   cm.
    Includes index.
    ISBN 0-201-52984-X
    1. Calculus.   I. Morrel, Bernard B.   II. Title.
    QA303.B6447   1993
    515—dc20

92-1729
CIP

Photo credits appear on page 822.

2 3 4 5 6 7 8 9 10-DO-959493

Appropriate for a two-term course, this text is an introduction to calculus as applied to business, economics, the life and physical sciences, the social sciences, and many general areas of interest to all students. A course in intermediate algebra is a prerequisite for the text, although Chapter 1 provides sufficient review to unify the diverse backgrounds of most students.

What's New in the Third Edition?

The style, format, and approach of the second edition have been retained in this new edition, but the text has been improved in the following ways.

The text is now printed in an extremely functional use of four colors. The use of four colors is not just window dressing, but has been carried out in a methodical and precise manner so that its use enhances the readability of the text for the student and the instructor. The following illustrates.

For example, when two curves are graphed using the same set of axes, one is usually red and the other blue, with the red graph the curve of major importance. This is exemplified in the following graph from Chapter 1 (p. 57). Note that the equation labels are the same color as the curve. When the instructions say "Graph," the dots match the color of the curve.

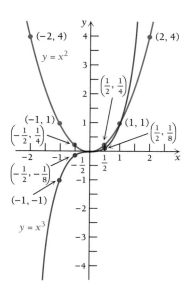

In the following figure from Chapter 2 (p. 118), we see how color is used to distinguish between secant and tangent lines. Throughout the text, blue is used for secant lines and red for tangent lines.

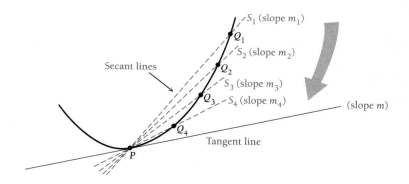

In the following graph and text development from Chapter 3 (p. 195), the color red denotes substitution in equations and blue highlights maximum and minimum values (second coordinates). This specific use of color is carried out further, as shown in the figure that follows. Note also that when dots are used for emphasis other than just mere plotting, they are black.

We then find second coordinates by substituting in the original function:

$$f(-3) = (-3)^3 + 3(-3)^2 - 9(-3) - 13 = 14;$$
$$f(1) = (1)^3 + 3(1)^2 - 9(1) - 13 = -18.$$

Thus there is a relative maximum at $(-3, 14)$ and a relative minimum at $(1, -18)$. The relative extrema are shown in the graph in Fig. 5.

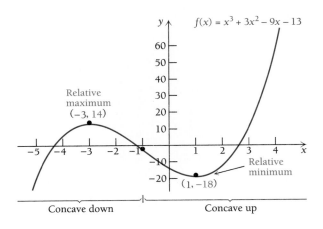

Beginning with integration in Chapter 5, the color amber is used to identify the area, unless certain special concepts are considered in the initial learning process. The following figure from Chapter 5 (p. 396) illustrates the vivid use of blue and red for the curves and labels and amber for the area.

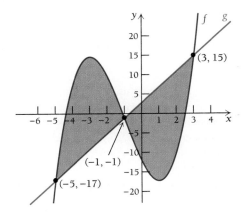

In Chapter 7 (p. 515), Figure 12 shows the use of extra colors for clarity. The top of the surface is blue and the bottom of the surface is rose. This is carried out in all the examples in Section 7.2.

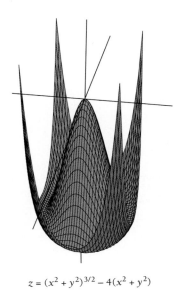

$$z = (x^2 + y^2)^{3/2} - 4(x^2 + y^2)$$

The number of applications has been significantly expanded in this edition, with those relating to business and economics increased by about 40%.   This is in response to user preference to have more applications in the fields in which most of the students are majoring.

The applications in the exercise sets are now grouped under headings that identify them as business and economics, life and physical sciences, social science, and general interest.   This allows the instructor to gear the assigned exercises to a particular student and also allows the student to know whether a particular exercise applies to his or her major.

Section 3.3 in the second edition, on graphing using calculus techniques, has been expanded to three sections, 3.1–3.3 in the third edition.   This is also in response to user feedback to have more development that includes greater use of relative extrema, absolute extrema, and asymptotes of all types.

The text now contains a greater emphasis on graphing.   This is carried out not only by the expansion of graphing topics in Chapter 3, but also by the location of many more graphs in the exercises as well as the inclusion of many *Computer–Graphing Calculator Exercises*. These exercises lend themselves to the use of the many types of graphing calculators and software packages now on the market. In particular, the student can use THE CALCULUS EXPLORER software published by Addison-Wesley Publishing Company. These exercises have a special indication at the ends of exercise sets.

Content Features

Intuitive Approach. Although the word "intuitive" has many meanings and interpretations, its use here means "experience based." Throughout the text, when a particular concept is discussed, its presentation is designed so that the students' learning process is based on their earlier mathematical experience or a new experience presented by the author before the concept is formalized. This is illustrated by the following situations.

- The definition of the derivative in Chapter 2 is presented in the context of a discussion of average rates of change (see p. 107). This presentation is more accessible and realistic than the strictly geometric idea of slope.

- When maximum problems involving volume are introduced (see p. 238), a function is derived that is to be maximized. Instead of forging ahead with the standard calculus solution, the student is first asked to stop, make a table of function values, graph the function, and then estimate the maximum value. This experience provides students with more insight into the problem. They recognize that not only do different dimensions yield different volumes, but also the dimensions yielding the maximum volume may be conjectured or estimated as a result of the calculations.

- The understanding behind the definition of the number e in Chapter 4 is explained both graphically and through a discussion of continuously compounded interest (see pp. 281–283).

Applications. Relevant and factual applications drawn from a broad background of fields are integrated throughout the text as applied problems and exercises, and are also featured in separate application sections. These have been updated and expanded in this edition; for example, there are 40% more applications to business and economics.

- An early discussion in Chapter 1 of compound interest, total revenue, cost and profit, as well as supply and demand functions sets the stage for subsequent applications that permeate the book (see pp. 46, 144, and 243).

- All techniques of differentiation, including the Extended Power Rule, the Chain Rule, and material on higher derivatives, are covered in Chapter 2. This is prior to the introduction of applications of differentiation in Chapter 3, which includes maximum–minimum applications (see pp. 236–255) and implicit differentiation and related rates (see pp. 263–270).

- When the exponential model is studied in Chapter 4, other applications, such as continuously compounded interest and the demand for natural resources, are also considered (see pp. 309–336). Growth and decay are covered in separate sections in Chapter 4 to allow room for the many worthwhile applications that relate to these concepts. The material on elasticity has been enhanced.

- Applications of integration are covered in a separate chapter and include a thorough coverage of continuous probability (see pp. 435–505).

- The treatment of double integrals over (nonrectangular) regions has been improved and enhanced.

Coverage.   Many aspects of the treatment and coverage of topics in this text have been commended by reviewers. The following are a few examples.

- Trouble areas in algebra are integrated into coverage of traditional precalculus topics in Chapter 1.

- Relative maxima and minima, Sections 3.1–3.3, and absolute maxima and minima, Section 3.4, are covered in separate sections in Chapter 3, so that students obtain a gradual buildup of these topics as they consider graphing using calculus concepts (see pp. 176–235).

- The text also includes a section on differential equations, Section 6.7, coverage of probability, Sections 6.4 and 6.5, and a chapter on the trigonometric functions, Chapter 8.

- For the convenience of the instructor who wishes to give a unified treatment of differential equations, portions of Sections 4.3 and 6.7 are repeated in Chapter 9.

Pedagogical Features

Interactive approach.   As each new section begins, its objectives are stated in the margin. These can be spotted easily by the student, and they provide the answer to the typical question, "What material am I responsible for?" Each section contains margin exercises that are for practice, development, or exploration. As students work through the material, they are encouraged to do the margin exercises. This involves them actively in the development of each topic, enhances the readability of the text, and gives them a deeper understanding of the material they are studying. All margin exercises have answers in the back of the text. It is recommended that students work out all these problems, stopping to do them when the text so indicates (see pp. 150 and 383).

Variety of Exercises.   Not only have the exercise sets been enhanced by the inclusion of many more business–economics applications, detailed art pieces, and extra graphs, but by the following additional features:

- *Computer–Graphing Calculator Exercises.* These exercises lend themselves to the use of the many types of graphing calculators and software packages now on the market. In particular, the student can use THE CALCULUS EXPLORER software published by Addison-Wesley Publishing Company. These exercises have a special indication at the ends of exercise sets (see pp. 223 and 388).

- *Calculator Exercises.* Exercises and examples geared to the use of a regular scientific calculator are also included throughout the text and are highlighted by the symbol ▦ (see pp. 69 and 289).

- *Synthesis Exercises.* Synthesis exercises are included at the ends of most exercise sets and all chapter tests. They require students to go beyond the immediate objectives of the section or chapter and are designed both to challenge students and to make them think about what they are learning (see pp. 165 and 404–405).

- *Exploratory Exercises.* These exercises emphasize analysis of data produced from functions, home experiments, or theoretical situations, and are designed to build students' mathematical experience and understanding of concepts (see p. 368).

- *Applications.* A section of applied problems is included in most exercise sets. The problems are grouped under headings that identify them as business and economics, life and physical sciences, social sciences, or general interest. Each problem is accompanied by a brief description of its subject matter (see pp. 251 and 320).

Tests and Reviews

- *Summary and Review.* At the end of each chapter is a summary and review. These are designed to provide students with all the material they need for successful review. Answers are at the back of the book, together with section references so that students can easily find the correct material to restudy if they have difficulty with a particular exercise. New to this edition in each summary and review is a list of *Terms to Know,* accompanied by page references to help students key in on important concepts (see p. 349).

- *Tests.* Each chapter ends with a chapter test that includes synthesis questions. There is also a Cumulative Review at the end of the text that can serve as a final examination as well. The Answers, with section references to the chapter tests and the Cumulative Review, are at the back of the book. Six additional forms of each of the chapter tests and the final examination with answer keys appear, ready for classroom use, in the Instructor's Resource Manual.

Supplements for the Instructor

Instructor's Manual and Printed Test Bank.   The Instructor's Manual and Printed Test Bank (IM & PTB) contains six alternative test forms with answers for each chapter test and six comprehensive final examinations. The tests are completely different from those for the fourth edition. The IM & PTB also has answers for the even-numbered exercises in the exercise sets. These can be easily copied and handed out to students. The answers to the odd-numbered exercises are at the back of the text. The IM & PTB additionally contains some extra application topics, namely *Maximum Sustainable Harvest* and *Minimizing Travel Time in a Building.*

Instructor's Solutions Manual.   This manual by Judith A. Penna contains worked-out solutions to all exercises in the exercise sets. A *Student's Solutions Manual* contains solutions to all odd-numbered exercises in the exercise sets. Either or both of these could be made available to the student.

Software

Computerized Testing

- *OmniTest II (IBM PC).*   This algorithm-driven testing system for the IBM allows you to create up to 99 variations of a customized test with just a few keystrokes, choosing from over 300 open-ended and multiple-choice test items. Instructors can also select and print tests by chapter, level of difficulty, type of problem or their own coding.

The IBM testing program, OmniTest II, also allows users to enter their own test items and edit existing items in an easy-to-use WYSI-WYG format with variable spacing.

- *OmniTest Macintosh.* This algorithm-driven testing system for the Macintosh allows you to create up to 99 variations of a customized test with just a few keystrokes, choosing from over 300 open-ended and multiple-choice test items.

Supplements for the Student

Student's Solutions Manual by Judith A. Penna. Complete worked-out solutions are provided in this booklet for all odd-numbered exercises in the exercise sets. This supplement is available to instructors and is for sale to students.

Options for Integrating Technology

Exploring Business Calculus with a Graphing Calculator by Charlene Beckmann. Containing laboratory explorations and problems to be completed with a graphing calculator, this supplement may be used with any graphing calculator. A Solutions Manual is also available.

The Calculus Explorer. Consisting of 27 programs ranging from functions to vector fields, this software enables the instructor and student to use the computer as an "electronic chalkboard." *The Explorer* is highly interactive and allows for manipulation of variables and equations to provide graphical visualization of mathematical relationships that are not intuitively obvious. *The Explorer* provides user-friendly operation through an easy-to-use completely menu-driven system, extensive on-line documentation, superior graphics capability, and fast operation. An accompanying manual includes sections covering each program, including examples and exercises. Available for IBM PC/compatibles.

References to where this software can be applied are marked in a special taupe-colored box under the heading *Computer-Graphing Calculator Exercises* at the ends of exercise sets in the text.

Related Addison-Wesley Software Titles. There are numerous other software packages distributed by Addison-Wesley. *MathCAD, Student Edition* is an electronic scratchpad for the IBM-PC that allows you to write equations on the PC exactly as you would on paper. MathCAD automatically calculates and displays results as numbers or graphs. *Lotus*

1-2-3, Student Version is a fully functional version of Lotus 1-2-3 with a 64-column–by–256-row spreadsheet that includes all the database, calculating, and graphics capabilities of Lotus 1-2-3 with teaching materials and enhancements particularly suited for educational use. Contact your Addison-Wesley sales representative for more information about these titles.

Acknowledgments

The authors have taken many steps to ensure the accuracy of the manuscript. The graphic art pieces have been computer-generated. A number of devoted individuals comprised the team that was responsible for monitoring each step of the production process in such careful detail. In particular, we would like to thank Judith A. Beecher and Judith A. Penna for their helpful suggestions, proofreading, and checking of the art. Their efforts above and beyond the call of duty deserve infinite praise. The authors also wish to express their appreciation to their students for providing suggestions and criticisms so willingly during the preceding editions; to John Jobe of Oklahoma State University for preparing the videotapes; to Mike Penna of IUPUI for his help with the computer graphics; and to Barbara Johnson, John K. Baumgart, and Stuart Ball for their precise checking and proofreading of the manuscript for mathematical accuracy. We would also like to thank the users of the text who have sent us corrections, comments, and constructive criticism—in particular, David Kast, University of Wisconsin (Marathon Center), John McCarthy, Washington University, and Russell E. Wright, Napa Valley College.

In addition, the following reviewers provided thoughtful and insightful comments that helped immeasurably in the revision of this text.

Barbara M. Brook, Camden County College
Peter Herron, Suffolk Community College
William Keils, San Antonio College
Richard Nadel, Florida International University
Richard Semmler, Northern Virginia Community College
Frank Smith, Kent State University—Main Campus
Sherman Wong, CUNY–Baruch College
Cathleen M. Zucco, LeMoyne College

M.L.B.
B.B.M.

INDEX OF APPLICATIONS

CONTENTS

3

APPLICATIONS OF DIFFERENTIATION

4

EXPONENTIAL AND LOGARITHMIC FUNCTIONS

INTEGRATION

APPLICATIONS OF INTEGRATION

FUNCTIONS OF SEVERAL VARIABLES

8

TRIGONOMETRIC FUNCTIONS

9

DIFFERENTIAL EQUATIONS

10

TAYLOR SERIES AND l'HÔPITAL'S RULE

11

NUMERICAL TECHNIQUES

1

ALGEBRA REVIEW, FUNCTIONS, AND MODELING

The number of oranges in a pile of the type shown in the photograph is approximated by the function

$$f(x) = \tfrac{1}{6}x^3 + \tfrac{1}{2}x^2 + \tfrac{1}{3}x,$$

where $f(x)$ is the number of oranges and x is the number of layers. Find the number of oranges when the number of layers is 7, 10, and 12.

THE MATHEMATICS

To solve the problem, we substitute 7 for x, 10 for x, and 12 for x in the function

$$\underbrace{f(x) = \tfrac{1}{6}x^3 + \tfrac{1}{2}x^2 + \tfrac{1}{3}x.}$$

↑

This is a formula for a *function*.

Although this chapter is mainly a review of basic concepts of algebra used in calculus, it does introduce many topics that we will consider several times throughout the text. These topics are functions and their graphs, supply and demand, total cost, revenue, and profit, and the concept of a mathematical model.

1.1

Rename without exponents.

1. 3^4 2. $(-3)^2$

3. $(1.02)^3$ 4. $\left(\dfrac{1}{4}\right)^2$

5. $(-5)^4$ 6. -5^4

EXPONENTS, MULTIPLYING, AND FACTORING

Exponential Notation

Let us review the meaning of an expression

$$a^n,$$

where a is any real number and n is an integer; that is, n is a number in the set $\{\ldots, -3, -2, -1, 0, 1, 2, 3, \ldots\}$. The number a is called the **base** and n is called the **exponent**. When n is greater than 1, then

$$a^n = \underbrace{a \cdot a \cdot a \cdots a}_{n \text{ factors}}.$$

In other words, a^n is the product of n **factors**, each of which is a.

In later sections of the text, we will consider a^n when n is any real number.

EXAMPLE 1   Rename without exponents.

a) $4^3 = 4 \cdot 4 \cdot 4 = 64$

b) $(-2)^5 = (-2)(-2)(-2)(-2)(-2) = -32$

c) $(-2)^4 = (-2)(-2)(-2)(-2) = 16$

d) $-2^4 = -(2^4) = -(2)(2)(2)(2) = -16$

e) $(1.08)^2 = 1.08 \times 1.08 = 1.1664$

f) $\left(\dfrac{1}{2}\right)^3 = \dfrac{1}{2} \cdot \dfrac{1}{2} \cdot \dfrac{1}{2} = \dfrac{1}{8}$ ❖

DO EXERCISES 1–6. (EXERCISES ARE IN THE MARGIN.)

We define an exponent of 1 as follows:

$$a^1 = a, \quad \text{for any real number } a.$$

In other words, any real number to the first power is that number itself.

We define an exponent of 0 as follows:

$$a^0 = 1, \quad \text{for any nonzero real number } a.$$

That is, any nonzero real number a to the zero power is 1.

EXAMPLE 2   Rename without exponents.

a) $(-2x)^0 = 1$ b) $(-2x)^1 = -2x$ c) $\left(\dfrac{1}{2}\right)^0 = 1$

d) $e^0 = 1$ e) $e^1 = e$ f) $\left(\dfrac{1}{2}\right)^1 = \dfrac{1}{2}$ ❖

DO EXERCISES 7–12.

The meaning of a negative integer as an exponent is as follows:

$$a^{-n} = \dfrac{1}{a^n}, \quad \text{for any nonzero real number } a.$$

In other words, any nonzero real number a to the $-n$ power is the reciprocal of a^n.

EXAMPLE 3   Rename without negative exponents.

a) $2^{-5} = \dfrac{1}{2^5} = \dfrac{1}{2 \cdot 2 \cdot 2 \cdot 2 \cdot 2} = \dfrac{1}{32}$

b) $10^{-3} = \dfrac{1}{10^3} = \dfrac{1}{10 \cdot 10 \cdot 10} = \dfrac{1}{1000}$, or 0.001

c) $\left(\dfrac{1}{4}\right)^{-2} = \dfrac{1}{\left(\dfrac{1}{4}\right)^2} = \dfrac{1}{\dfrac{1}{4} \cdot \dfrac{1}{4}} = \dfrac{1}{\dfrac{1}{16}} = 1 \cdot \dfrac{16}{1} = 16$

d) $x^{-5} = \dfrac{1}{x^5}$

e) $e^{-k} = \dfrac{1}{e^k}$

f) $t^{-1} = \dfrac{1}{t^1} = \dfrac{1}{t}$ ❖

DO EXERCISES 13–19.

Properties of Exponents

Note the following:

$$b^5 \cdot b^{-3} = (b \cdot b \cdot b \cdot b \cdot b) \cdot \dfrac{1}{b \cdot b \cdot b} = \dfrac{b \cdot b \cdot b}{b \cdot b \cdot b} \cdot b \cdot b = 1 \cdot b \cdot b = b^2.$$

Rename without exponents.

7. $(5t)^0$ 8. $(5t)^1$

9. k^0 10. m^1

11. $\left(\dfrac{1}{4}\right)^1$ 12. $\left(\dfrac{1}{4}\right)^0$

Rename without negative exponents.

13. 2^{-4} 14. 10^{-2}

15. $\left(\dfrac{1}{4}\right)^{-3}$ 16. t^{-7}

17. e^{-t} 18. M^{-1}

19. $(x + 1)^{-2}$

Multiply.

20. $t^4 \cdot t^5$

21. $t^{-4} \cdot t$

22. $10e^{-4} \cdot 5e^{-9}$

23. $t^{-3} \cdot t^{-4} \cdot t$

24. $4b^5 \cdot 6b^{-2}$

We could have obtained the same result by adding the exponents. This is true in general.

THEOREM 1

For any nonzero real number a and any integers n and m,

$$a^n \cdot a^m = a^{n+m}.$$

(To multiply when the bases are the same, add the exponents.)

EXAMPLE 4 Multiply.

a) $x^5 \cdot x^6 = x^{5+6} = x^{11}$

b) $x^{-5} \cdot x^6 = x^{-5+6} = x$

c) $2x^{-3} \cdot 5x^{-4} = 10x^{-3+(-4)} = 10x^{-7}$, or $\dfrac{10}{x^7}$

d) $r^2 \cdot r = r^{2+1} = r^3$ ❖

DO EXERCISES 20–24.

Note the following:

$$b^5 \div b^2 = \frac{b^5}{b^2} = \frac{b \cdot b \cdot b \cdot b \cdot b}{b \cdot b} = \frac{b \cdot b}{b \cdot b} \cdot b \cdot b \cdot b = 1 \cdot b \cdot b \cdot b = b^3.$$

We could have obtained the same result by subtracting the exponents. This is true in general.

THEOREM 2

For any nonzero real number a and any integers n and m,

$$\frac{a^n}{a^m} = a^{n-m}.$$

(To divide when the bases are the same, subtract the exponent in the denominator from the exponent in the numerator.)

EXAMPLE 5 Divide.

a) $\dfrac{a^3}{a^2} = a^{3-2} = a^1 = a$

b) $\dfrac{x^7}{x^7} = x^{7-7} = x^0 = 1$

c) $\dfrac{e^3}{e^{-4}} = e^{3-(-4)} = e^{3+4} = e^7$

d) $\dfrac{e^{-4}}{e^{-1}} = e^{-4-(-1)} = e^{-4+1} = e^{-3}$, or $\dfrac{1}{e^3}$ ❖

DO EXERCISES 25–30.

Note the following:
$$(b^2)^3 = b^2 \cdot b^2 \cdot b^2 = b^{2+2+2} = b^6.$$

We could have obtained the same result by multiplying the exponents. This is true in general.

THEOREM 3

For any nonzero real number a and any integers n and m,
$$(a^n)^m = a^{nm}.$$
(To raise a power to a power, multiply the exponents.)

EXAMPLE 6   Simplify.

a) $(x^{-2})^3 = x^{-2 \cdot 3} = x^{-6}$, or $\dfrac{1}{x^6}$

b) $(e^x)^2 = e^{2x}$

c) $(2x^4y^{-5}z^3)^{-3} = 2^{-3}(x^4)^{-3}(y^{-5})^{-3}(z^3)^{-3}$
$$= \frac{1}{2^3}x^{-12}y^{15}z^{-9}, \text{ or } \frac{y^{15}}{8x^{12}z^9}$$ ❖

DO EXERCISES 31–35.

Multiplication

The distributive laws are important in multiplying. The laws are as follows:

The Distributive Laws

For any numbers A, B, and C,
$$A(B+C) = AB + AC \quad \text{and} \quad A(B-C) = AB - AC.$$

Divide.

25. $\dfrac{x^6}{x^2}$

26. $\dfrac{x^2}{x^6}$

27. $\dfrac{e^t}{e^t}$

28. $\dfrac{e^2}{e^k}$

29. $\dfrac{e^5}{e^{-7}}$

30. $\dfrac{e^{-5}}{e^{-7}}$

Simplify.

31. $(x^{-4})^3$

32. $(e^2)^2$

33. $(e^x)^3$

34. $(5x^3y^5)^2$

35. $(4x^{-5}y^{-6}z^2)^{-4}$

Multiply.

36. $2(x + 7)$

37. $P(1 - i)$

38. $(x - 4)(x + 7)$

39. $(a - b)(a - b)$

40. $(a - b)(a + b)$

Multiply.

41. $(x - h)^2$

42. $(3x + t)^2$

43. $(5t - m)(5t + m)$

EXAMPLE 7 Multiply.

a) $3(x - 5) = 3 \cdot x - 3 \cdot 5 = 3x - 15$

b) $P(1 + i) = P \cdot 1 + P \cdot i = P + Pi$

c) $(x - 5)(x + 3) = (x - 5)x + (x - 5)3$
$$= x \cdot x - 5x + 3x - 5 \cdot 3$$
$$= x^2 - 2x - 15$$

d) $(a + b)(a + b) = (a + b)a + (a + b)b$
$$= a \cdot a + ba + ab + b \cdot b$$
$$= a^2 + 2ab + b^2$$ ❖

DO EXERCISES 36–40.

The following formulas, which are obtained using the distributive laws, are also useful in multiplying.

$$(A + B)^2 = A^2 + 2AB + B^2, \quad (1)$$
$$(A - B)^2 = A^2 - 2AB + B^2 \quad (2)$$
$$(A - B)(A + B) = A^2 - B^2 \quad (3)$$

EXAMPLE 8 Multiply.

a) $(x + h)^2 = x^2 + 2xh + h^2$

b) $(2x - t)^2 = (2x)^2 - 2(2x)t + t^2 = 4x^2 - 4xt + t^2$

c) $(3c + d)(3c - d) = (3c)^2 - d^2 = 9c^2 - d^2$ ❖

DO EXERCISES 41–43.

Factoring

Factoring is the reverse of multiplication. That is, to factor an expression, we find an equivalent expression that is a product. Always remember to look first for a common factor.

EXAMPLE 9 Factor.

a) $P + Pi = P \cdot 1 + P \cdot i = P(1 + i)$ We used a distributive law.

b) $2xh + h^2 = h(2x + h)$

c) $x^2 - 6xy + 9y^2 = (x - 3y)^2$

d) $x^2 - 5x - 14 = (x - 7)(x + 2)$ We looked for factors of -14 whose sum is -5.

e) $6x^2 + 7x - 5 = (2x - 1)(3x + 5)$ We first considered ways of factoring the first coefficient—for example, $(2x \quad)(3x \quad)$. Then we looked for factors of -5 such that when we multiply, we obtain the given expression.

Factor.

44. $P - Pi$

f) $x^2 - 9t^2 = (x - 3t)(x + 3t)$ We used the formula $(A - B)(A + B) = A^2 - B^2$.   ❖

DO EXERCISES 44–48.

45. $x^2 + 10xy + 25y^2$

In later work, we will consider expressions like
$$(x + h)^2 - x^2.$$
To simplify this, first note that

46. $4x^2 + 28x + 40$

$$(x + h)^2 = x^2 + 2xh + h^2.$$
Subtracting x^2 on both sides of this equation gives us
$$(x + h)^2 - x^2 = 2xh + h^2.$$
Factoring out an h on the right side, we get

47. $25c^2 - d^2$

$$(x + h)^2 - x^2 = h(2x + h). \qquad (4)$$
Let us now use this result to compare two squares.

EXAMPLE 10   How close is $(3.1)^2$ to 3^2?

Solution   Substituting $x = 3$ and $h = 0.1$ in Eq. (4), we get

48. $12y^2 + 13y - 14$

$$(3.1)^2 - 3^2 = 0.1(2 \cdot 3 + 0.1) = 0.1(6.1) = 0.61.$$
Thus, $(3.1)^2$ differs from 3^2 by 0.61.   ❖

DO EXERCISE 49.

Compound Interest

49. How close is $(5.1)^2$ to 5^2?

Suppose we invest P dollars at interest rate i, compounded annually. The amount A_1 in the account at the end of one year is given by
$$A_1 = P + Pi = P(1 + i) = Pr,$$
where, for convenience,
$$r = 1 + i.$$

There is a formula for finding the amount in a savings account after a certain period of time.

50. *Business: Compound interest.* Suppose $1000 is invested at 9%, compounded annually. How much is in the account at the end of 2 years?

Going into the second year, we have Pr dollars, so by the end of the second year, we would have the amount A_2 given by

$$A_2 = A_1 \cdot r = (Pr)r = Pr^2 = P(1 + i)^2.$$

Going into the third year, we have Pr^2 dollars, so by the end of the third year, we would have the amount A_3 given by

$$A_3 = A_2 \cdot r = (Pr^2)r = Pr^3 = P(1 + i)^3.$$

In general, we have the following.

THEOREM 4

If an amount P is invested at interest rate i, compounded annually, in t years it will grow to the amount A given by

$$A = P(1 + i)^t.$$

EXAMPLE 11 *Business: Compound interest.* Suppose $1000 is invested at 8%, compounded annually. How much is in the account at the end of 2 years?

Solution We substitute 1000 for P, 0.08 for i, and 2 for t into the equation $A = P(1 + i)^t$ and get

$$A = 1000(1 + 0.08)^2 = 1000(1.08)^2 = 1000(1.1664) = \$1166.40. \quad \diamond$$

DO EXERCISE 50.

For interest that is compounded quarterly, we can find a formula like the one above (see Fig. 1).

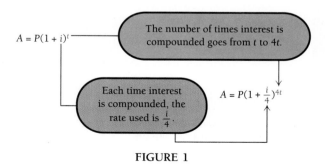

FIGURE 1

In general, the following theorem applies.

THEOREM 5

If a principal P is invested at interest rate i, compounded n times a year, in t years it will grow to an amount A given by

$$A = P\left(1 + \frac{i}{n}\right)^{nt}.$$

EXAMPLE 12   *Business: Compound interest.* Suppose $1000 is invested at 8%, compounded quarterly. How much is in the account at the end of 2 years?

Solution   We use the equation $A = P(1 + i/n)^{nt}$, substituting 1000 for P, 0.08 for i, 4 for n (compounding quarterly), and 2 for t. Then we get

$$A = 1000\left(1 + \frac{0.08}{4}\right)^{4 \times 2}$$

$$= 1000(1 + 0.02)^8$$

$$= 1000(1.02)^8$$

$$= 1000(1.171659381) \qquad \text{Using a calculator to approximate } (1.02)^8$$

$$= 1171.659381$$

$$\approx \$1171.66.$$

A calculator note: A calculator with a $\boxed{y^x}$ key and a ten-digit readout was used to find $(1.02)^8$ in Example 12. The number of places on a calculator may affect the accuracy of the answer. Thus you may occasionally find that your answers do not agree with those at the back of the book, which were found on a calculator with a ten-digit readout. In general, when using a calculator, do all your computations, and round only at the end, as in Example 12. Usually, your answer will agree to at least four digits. It might be wise to consult with your instructor on the accuracy required.

❖

DO EXERCISE 51.

51. *Business: Compound interest.* Suppose $1000 is invested at 11%, compounded semiannually ($n = 2$). How much is in the account at the end of 3 years?

EXERCISE SET 1.1

Rename without exponents.

1. 5^3

2. 7^2

3. $(-7)^2$

4. $(-5)^3$

5. $(1.01)^2$

6. $(1.01)^3$

7. $\left(\dfrac{1}{2}\right)^4$

8. $\left(\dfrac{1}{4}\right)^3$

9. $(6x)^0$

10. $(6x)^1$

11. t^1

12. t^0

13. $\left(\dfrac{1}{3}\right)^0$

14. $\left(\dfrac{1}{3}\right)^1$

Rename without negative exponents.

15. 3^{-2}

16. 4^{-2}

17. $\left(\dfrac{1}{2}\right)^{-3}$

18. $\left(\dfrac{1}{2}\right)^{-2}$

19. 10^{-1}

20. 10^{-4}

21. e^{-b}

22. t^{-k}

23. b^{-1}

24. h^{-1}

Multiply.

25. $x^2 \cdot x^3$

26. $t^3 \cdot t^4$

27. $x^{-7} \cdot x$

28. $x^5 \cdot x$

29. $5x^2 \cdot 7x^3$

30. $4t^3 \cdot 2t^4$

31. $x^{-4} \cdot x^7 \cdot x$

32. $x^{-3} \cdot x \cdot x^3$

33. $e^{-t} \cdot e^t$

34. $e^k \cdot e^{-k}$

Divide.

35. $\dfrac{x^5}{x^2}$

36. $\dfrac{x^7}{x^3}$

37. $\dfrac{x^2}{x^5}$

38. $\dfrac{x^3}{x^7}$

39. $\dfrac{e^k}{e^k}$

40. $\dfrac{t^k}{t^k}$

41. $\dfrac{e^t}{e^4}$

42. $\dfrac{e^k}{e^3}$

43. $\dfrac{t^6}{t^{-8}}$

44. $\dfrac{t^5}{t^{-7}}$

45. $\dfrac{t^{-9}}{t^{-11}}$

46. $\dfrac{t^{-11}}{t^{-7}}$

47. $\dfrac{ab(a^2b)^3}{ab^{-1}}$

48. $\dfrac{x^2y^3(xy^3)^2}{x^{-3}y^2}$

Simplify.

49. $(t^{-2})^3$

50. $(t^{-3})^4$

51. $(e^x)^4$

52. $(e^x)^5$

53. $(2x^2y^4)^3$

54. $(2x^2y^4)^5$

55. $(3x^{-2}y^{-5}z^4)^{-4}$

56. $(5x^3y^{-7}z^{-5})^{-3}$

57. $(-3x^{-8}y^7z^2)^2$

58. $(-5x^4y^{-5}z^{-3})^4$

Multiply.

59. $5(x - 7)$

60. $4(x - 3)$

61. $x(1 - t)$

62. $x(1 + t)$

63. $(x - 5)(x - 2)$

64. $(x - 4)(x - 3)$

65. $(a - b)(a^2 + ab + b^2)$

66. $(x^2 - xy + y^2)(x + y)$

67. $(2x + 5)(x - 1)$

68. $(3x + 4)(x - 1)$

69. $(a - 2)(a + 2)$

70. $(3x - 1)(3x + 1)$

71. $(5x + 2)(5x - 2)$

72. $(t - 1)(t + 1)$

73. $(a - h)^2$

74. $(a + h)^2$

75. $(5x + t)^2$

76. $(7a - c)^2$

77. $5x(x^2 + 3)^2$

78. $-3x^2(x^2 - 4)(x^2 + 4)$

Use the following equation (Eq. 1) for Exercises 79–81.

$$\begin{aligned}
(x + h)^3 &= (x + h)(x + h)^2 \\
&= (x + h)(x^2 + 2xh + h^2) \\
&= (x + h)x^2 + (x + h)2xh + (x + h)h^2 \\
&= x^3 + x^2h + 2x^2h + 2xh^2 + xh^2 + h^3 \\
&= x^3 + 3x^2h + 3xh^2 + h^3 \qquad\qquad (1)
\end{aligned}$$

79. $(a + b)^3$

80. $(a - b)^3$

81. $(x - 5)^3$

82. $(2x + 3)^3$

Factor.

83. $x - xt$

84. $x + xh$

85. $x^2 + 6xy + 9y^2$

86. $x^2 - 10xy + 25y^2$

87. $x^2 - 2x - 15$

88. $x^2 + 8x + 15$

89. $x^2 - x - 20$

90. $x^2 - 9x - 10$

91. $49x^2 - t^2$

92. $9x^2 - b^2$

93. $36t^2 - 16m^2$

94. $25y^2 - 9z^2$

95. $a^3b - 16ab^3$

96. $2x^4 - 32$

97. $a^8 - b^8$

98. $36y^2 + 12y - 35$

99. $10a^2x - 40b^2x$

100. $x^3y - 25xy^3$

101. $2 - 32x^4$

102. $2xy^2 - 50x$

103. $9x^2 + 17x - 2$

104. $6x^2 - 23x + 20$

105. $x^3 + 8$
    (*Hint:* See Exercise 66.)

106. $a^3 - 27b^3$
    (*Hint:* See Exercise 65.)

107. $y^3 - 64t^3$

108. $m^3 + 1000p^3$

109. Use the equation

$$(x + h)^2 - x^2 = h(2x + h)$$

to answer the following.

a) How close is $(4.1)^2$ to 4^2?

b) How close is $(4.01)^2$ to 4^2?

c) How close is $(4.001)^2$ to 4^2?

110. From Eq. (1) above, it follows that

$$(x + h)^3 - x^3 = h(3x^2 + 3xh + h^2).$$

Use the equation to answer the following.

a) How close is $(4.1)^3$ to 4^3?

b) How close is $(4.01)^3$ to 4^3?

c) How close is $(4.001)^3$ to 4^3?

APPLICATIONS

❖ Business and Economics

The symbol ▦ indicates an exercise designed to be done using a calculator.

111. *Compound interest.* Suppose $1000 is invested at 16%. How much is in the account at the end of 1 year, if interest is compounded:

a) annually?

b) semiannually?

c) quarterly?

d) daily? (▦ with $\boxed{y^x}$ key)

e) hourly? (▦ with $\boxed{y^x}$ key)

112. *Compound interest.* Suppose $1000 is invested at 10%. How much is in the account at the end of 1 year, if interest is compounded:

a) annually?

b) semiannually?

c) quarterly?

d) daily? (▦ with $\boxed{y^x}$ key)

e) hourly? (▦ with $\boxed{y^x}$ key)

Determining monthly payments on a loan. If P dollars are borrowed for a home mortgage, the monthly payment M, made at the end of each month for n months, is given by

$$M = P\left[\frac{\frac{i}{12}\left(1 + \frac{i}{12}\right)^n}{\left(1 + \frac{i}{12}\right)^n - 1}\right],$$

where i is the annual interest rate and n is the total number of monthly payments.

113. ▦ The mortgage on a house is $43,000, the interest rate is $8\frac{3}{4}\%$, and the loan period is 25 years. What is the monthly payment?

114. ▦ The mortgage on a house is $100,000, the interest rate is $9\frac{1}{2}\%$, and the loan period is 30 years. What is the monthly payment?

1.2

EQUATIONS, INEQUALITIES, AND INTERVAL NOTATION

Equations

Basic to the solution of many equations are two simple principles. We can add the same number on both sides of a true equation and still obtain a true equation. We can also multiply by any number on both sides of a true equation and obtain a true equation.

THE ADDITION PRINCIPLE

If an equation $a = b$ is true, then the equation $a + c = b + c$ is true for any number c.

THE MULTIPLICATION PRINCIPLE

If an equation $a = b$ is true, then the equation $ac = bc$ is true for any number c.

When solving an equation, we use these equation-solving principles and other properties of real numbers to get the variable alone on one side. Then it is easy to determine the solution.

EXAMPLE 1 Solve: $-\frac{5}{6}x + 10 = \frac{1}{2}x + 2$.

Solution We first multiply on both sides by 6 to clear of fractions:

$$6(-\tfrac{5}{6}x + 10) = 6(\tfrac{1}{2}x + 2) \qquad \text{Using the Multiplication Principle}$$

$$6(-\tfrac{5}{6}x) + 6 \cdot 10 = 6(\tfrac{1}{2}x) + 6 \cdot 2 \qquad \text{Using the Distributive Law}$$

$$-5x + 60 = 3x + 12 \qquad \text{Simplifying}$$

$$60 = 8x + 12 \qquad \begin{array}{l}\text{Using the Addition Principle:}\\ \text{We add } 5x \text{ on both sides.}\end{array}$$

$$48 = 8x \qquad \text{Adding } -12 \text{ on both sides}$$

$$\tfrac{1}{8} \cdot 48 = \tfrac{1}{8} \cdot 8x \qquad \text{Multiplying by } \tfrac{1}{8} \text{ on both sides}$$

$$6 = x.$$

The variable is now alone, and we see that 6 is the solution. We can check by substituting 6 into the original equation. ❖

DO EXERCISE 1.

To solve applied problems, we first translate to mathematical language, usually an equation. Then we solve the equation and check to see whether the solution to the equation is a solution to the problem.

EXAMPLE 2 *Life science: Weight gain.* After a 5% gain in weight, an animal weighs 693 lb. What was its original weight?

Solution We first translate to an equation:

$$\underbrace{(Original\ weight)}_{w} + 5\% \underbrace{(Original\ weight)}_{w} = 693$$

$$w + 5\% \qquad\qquad w \qquad = 693.$$

Now we solve the equation:

$$w + 5\%w = 693$$
$$1 \cdot w + 0.05w = 693$$
$$(1 + 0.05)w = 693$$
$$1.05w = 693$$
$$w = \frac{693}{1.05} = 660.$$

Check: $660 + 5\% \times 660 = 660 + 0.05 \times 660 = 660 + 33 = 693.$ ❖

DO EXERCISE 2.

The third principle for solving equations is the *Principle of Zero Products*.

THE PRINCIPLE OF ZERO PRODUCTS

For any numbers a and b, if $ab = 0$, then $a = 0$ or $b = 0$; and if $a = 0$ or $b = 0$, then $ab = 0$.

An equation being solved by this principle must have a 0 on one side and a product on the other. The solutions are then obtained by setting each factor equal to 0 and solving the resulting equations.

EXAMPLE 3 Solve: $3x(x - 2)(5x + 4) = 0.$

1. Solve: $-\frac{7}{8}x + 5 = \frac{1}{4}x - 2.$

2. *Business: Investment increase.* An investment is made at 14%, compounded annually. It grows to $826.50 at the end of 1 year. How much was invested originally?

3. Solve: $5x(x + 2)(2x - 3) = 0$.

Solution

$$3x(x - 2)(5x + 4) = 0$$

$$3x = 0 \quad \text{or} \quad x - 2 = 0 \quad \text{or} \quad 5x + 4 = 0 \qquad \text{Using the Principle of Zero Products}$$

$$\tfrac{1}{3} \cdot 3x = \tfrac{1}{3} \cdot 0 \quad \text{or} \qquad\quad x = 2 \quad \text{or} \qquad\quad 5x = -4 \qquad \text{Solving each separately}$$

$$x = 0 \qquad \text{or} \qquad\quad x = 2 \quad \text{or} \qquad\quad x = -\tfrac{4}{5}$$

The solutions are 0, 2, and $-\tfrac{4}{5}$. ❖

Note that the Principle of Zero Products can be applied *only* when a product is 0. For example, although we may know that $ab = 8$, we *do not know* that $a = 8$ or $b = 8$.

Solve.

4. $x^2 + x = 12$

DO EXERCISE 3.

EXAMPLE 4 Solve: $4x^3 = x$.

Solution

$$4x^3 = x$$

$$4x^3 - x = 0 \qquad \text{Adding } -x$$

$$x(4x^2 - 1) = 0$$

$$x(2x - 1)(2x + 1) = 0 \qquad \text{Factoring}$$

$$x = 0 \quad \text{or} \quad 2x - 1 = 0 \quad \text{or} \quad 2x + 1 = 0 \qquad \text{Using the Principle of Zero Products}$$

$$x = 0 \quad \text{or} \qquad\quad 2x = 1 \quad \text{or} \qquad\quad 2x = -1$$

$$x = 0 \quad \text{or} \qquad\quad x = \tfrac{1}{2} \quad \text{or} \qquad\quad x = -\tfrac{1}{2}.$$

5. $x^3 = x$

The solutions are 0, $\tfrac{1}{2}$, and $-\tfrac{1}{2}$. ❖

DO EXERCISES 4 AND 5.

Inequalities

The principles for solving inequalities are similar to those for solving equations. We can add the same number on both sides of an inequality. We can also multiply on both sides by the same nonzero number, but if that number is negative, we must reverse the inequality sign. The following are the inequality-solving principles.

THE INEQUALITY-SOLVING PRINCIPLES

If the inequality $a < b$ is true, then:

I 1. $a + c < b + c$ is true, for any c.

I 2. $a \cdot c < b \cdot c$, for any positive c.

I 3. $a \cdot c > b \cdot c$, for any negative c.

Similar principles hold when $<$ is replaced by $\leq$ and $>$ is replaced by $\geq$.

EXAMPLE 5   Solve: $17 - 8x \geq 5x - 4$.

Solution

$$17 - 8x \geq 5x - 4$$
$$-8x \geq 5x - 21 \qquad \text{Adding } -17$$
$$-13x \geq -21 \qquad \text{Adding } -5x$$
$$-\tfrac{1}{13}(-13x) \leq -\tfrac{1}{13}(-21) \qquad \text{Multiplying by } -\tfrac{1}{13} \text{ and}$$
$$\qquad\qquad\qquad\qquad\qquad \textit{reversing} \text{ the inequality sign}$$
$$x \leq \tfrac{21}{13}$$

Any number less than or equal to $\frac{21}{13}$ is a solution.   ❖

DO EXERCISES 6 AND 7.

EXAMPLE 6   *Business: Total sales.* Raggs, Ltd., a clothing firm, determines that its total revenue, in dollars, from the sale of x suits is given by

$$200x + 50.$$

Determine the number of suits that the firm must sell so that its total revenue will be more than $70,050.

Solution   We translate to an inequality and solve:

$$200x + 50 > 70{,}050$$
$$200x > 70{,}000 \qquad \text{Adding } -50$$
$$x > 350. \qquad \text{Multiplying by } \tfrac{1}{200}$$

Thus the company's total revenue will exceed $70,050 when it sells more than 350 suits.   ❖

DO EXERCISE 8.

Solve.

6. $3x < 11 - 2x$

7. $16 - 7x \leq 10x - 4$

8. *Business: Total revenue.* In Example 6, determine the number of suits that the firm must sell so that its total revenue will be more than $95,050.

Intervals between houses are analogous to intervals on a number line.

9. Write interval notation for each graph.

a)

-3 -2 -1 0 1 2 3 4

b)

-3 -2 -1 0 1 2 3 4

10. Write interval notation for each of the following.

a) The set of all numbers x such that $-1 < x < 4$

b) The set of all numbers x such that $-\frac{1}{4} < x < \frac{1}{4}$

Interval Notation

The set of real numbers corresponds to the set of points on a line.

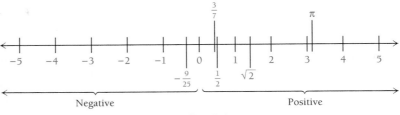

FIGURE 1

For real numbers a and b such that $a < b$ (a is to the left of b on a number line), we define the **open interval** (a, b) to be the set of numbers between, but not including, a and b. That is,

$$(a, b) = \text{the set of all numbers } x \text{ such that } a < x < b.$$

The inequality $a < x < b$ is true when both $a < x$ is true *and* $x < b$ is true.

FIGURE 2

The graph of (a, b) is shown in color in Fig. 2. The open circles and the parentheses indicate that a and b are not included. The numbers a and b are called **endpoints**.

DO EXERCISES 9 AND 10.

The notation (a, b) is an example of what is called *interval notation*. We now define other kinds of *interval notation*.

The **closed interval** $[a, b]$ is the set of numbers between and including a and b. That is,

$$[a, b] = \text{the set of all numbers } x \text{ such that } a \le x \le b.$$

FIGURE 3

The graph of $[a, b]$ is shown in color in Fig. 3. The solid circles and the brackets indicate that a and b are included.

There are two kinds of **half-open intervals** defined as follows:

$(a, b]$ = the set of all numbers x such that $a < x \leq b$.

FIGURE 4

Note in Fig. 4 that the open circle and the parenthesis in $(a, b]$ indicate that a is not included. The solid circle and the bracket indicate that b is included. Also,

$[a, b)$ = the set of all numbers x such that $a \leq x < b$.

FIGURE 5

Note in Fig. 5 that the solid circle and the bracket in $[a, b)$ indicate that a is included. The open circle and the parenthesis indicate that b is not included.

DO EXERCISES 11 AND 12.

Some intervals are of unlimited extent in one or both directions. In such cases, we use the infinity symbol ∞. For example,

$[a, \infty)$ = the set of all numbers x such that $x \geq a$.

FIGURE 6

Note that ∞ is not a number and this is indicated with a parenthesis by the infinity symbol.

(a, ∞) = the set of all numbers x such that $x > a$.

FIGURE 7

$(-\infty, b]$ = the set of all numbers x such that $x \leq b$.

FIGURE 8

11. Write interval notation for each graph.

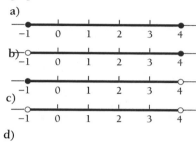

a)

b)

c)

d)

12. Write interval notation for each of the following.

a)   The set of all numbers x such that $-\sqrt{2} < x < \sqrt{2}$

b)   The set of all numbers x such that $0 \leq x < 1$

c)   The set of all numbers x such that $-6.7 < x \leq -4.2$

d)   The set of all numbers x such that $3 \leq x \leq 7\frac{1}{2}$

13. Write interval notation for each graph.

a)

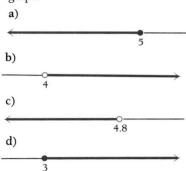

5

b)

4

c)

4.8

d)

3

14. Write interval notation for each of the following.

a) The set of all numbers x such that $x \geqslant 8$

b) The set of all numbers x such that $x < -7$

c) The set of all numbers x such that $x > 10$

d) The set of all numbers x such that $x \leqslant -0.78$

$(-\infty, b) =$ the set of all numbers x such that $x < b$.

b

FIGURE 9

We can name the entire set of real numbers using $(-\infty, \infty)$.

$(-\infty, \infty)$

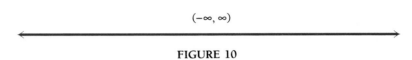

FIGURE 10

DO EXERCISES 13 AND 14.

Any point in an interval that is not an endpoint is an **interior point**.

Interior point
x

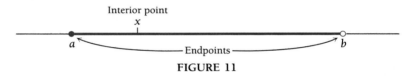

a — Endpoints — b

FIGURE 11

Note that all the points in an open interval are interior points.

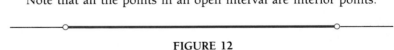

FIGURE 12

EXERCISE SET 1.2.

Solve.

1. $-7x + 10 = 5x - 11$

2. $-8x + 9 = 4x - 70$

3. $5x - 17 - 2x = 6x - 1 - x$

4. $5x - 2 + 3x = 2x + 6 - 4x$

5. $x + 0.8x = 216$

6. $x + 0.5x = 210$

7. $x + 0.08x = 216$

8. $x + 0.05x = 210$

9. $2x(x + 3)(5x - 4) = 0$

10. $7x(x - 2)(2x + 3) = 0$

11. $x^2 + 1 = 2x + 1$

12. $2t^2 = 9 + t^2$

13. $t^2 - 2t = t$

14. $6x - x^2 = x$

15. $6x - x^2 = -x$

16. $2x - x^2 = -x$

17. $9x^3 = x$

18. $16x^3 = x$

19. $(x - 3)^2 = x^2 + 2x + 1$

20. $(x - 5)^2 = x^2 + x + 3$

21. $3 - x \leqslant 4x + 7$

22. $x + 6 \leqslant 5x - 6$

23. $5x - 5 + x > 2 - 6x - 8$

24. $3x - 3 + 3x > 1 - 7x - 9$

25. $-7x < 4$ **26.** $-5x \geqslant 6$

27. $5x + 2x \leqslant -21$ **28.** $9x + 3x \geqslant -24$

29. $2x - 7 < 5x - 9$ **30.** $10x - 3 \geqslant 13x - 8$

31. $8x - 9 < 3x - 11$ **32.** $11x - 2 \geqslant 15x - 7$

33. $8 < 3x + 2 < 14$ **34.** $2 < 5x - 8 \leqslant 12$

35. $3 \leqslant 4x - 3 \leqslant 19$ **36.** $9 \leqslant 5x + 3 < 19$

37. $-7 \leqslant 5x - 2 \leqslant 12$ **38.** $-11 \leqslant 2x - 1 < -5$

Write interval notation for each graph in Exercises 39–46.

39.

40.

41.

42.

43.

44.

45.

46.

Write interval notation for each of the following.

47. The set of all numbers x such that $-3 \leqslant x \leqslant 3$

48. The set of all numbers x such that $-4 < x < 4$

49. The set of all numbers x such that $-14 \leqslant x < -11$

50. The set of all numbers x such that $6 < x \leqslant 20$

51. The set of all numbers x such that $x \leqslant -4$

52. The set of all numbers x such that $x > -5$

APPLICATIONS

❖ **Business and Economics**

53. *Investment increase.* An investment is made at 11%, compounded annually. It grows to $721.50 at the end of 1 year. How much was invested originally?

54. *Investment increase.* An investment is made at 13%, compounded annually. It grows to $904 at the end of 1 year. How much was invested originally?

55. *Total revenue.* A firm determines that the total revenue, in dollars, from the sale of x units of a product is

$$3x + 1000.$$

Determine the number of units that must be sold so that its total revenue will be more than $22,000.

56. *Total revenue.* A firm determines that the total revenue, in dollars, from the sale of x units of a product is

$$5x + 1000.$$

Determine the number of units that must be sold so that its total revenue will be more than $22,000.

❖ **Life and Physical Sciences**

57. *Weight gain.* After a 6% gain in weight, an animal weighs 508.8 lb. What was its original weight?

58. *Weight gain.* After a 7% gain in weight, an animal weighs 363.8 lb. What was its original weight?

❖ **Social Sciences**

59. *Population increase.* After a 2% increase, the population of a city is 826,200. What was the former population?

60. *Population increase.* After a 3% increase, the population of a city is 741,600. What was the former population?

❖ **General Interest**

61. *Grade average.* To get a B in a course, a student's average must be greater than or equal to 80% (at least 80%) and less than 90%. On the first three tests, the student scores 78%, 90%, and 92%. Determine the scores on the fourth test that will guarantee a B.

62. *Grade average.* To get a C in a course, a student's average must be greater than or equal to 70% and less than 80%. On the first three tests, the student scores 65%, 83%, and 82%. Determine the scores on the fourth test that will guarantee a C.

1.3

GRAPHS AND FUNCTIONS

OBJECTIVES

a) Given a function and several inputs, find the outputs.

b) Graph a given equation or function.

c) Decide whether a graph is that of a function.

A graph offers the opportunity to visualize relationships. We see graphs of all kinds in magazines and newspapers. Examples are shown in Fig. 1. The type of graph shown in the figure on selected interest rates is the kind we will study most in calculus. It shows changes over a period of time. One topic we might be interested in is how the change of time affects the change in interest rates.

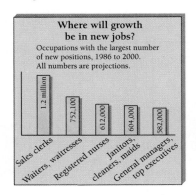

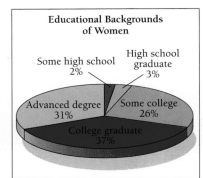

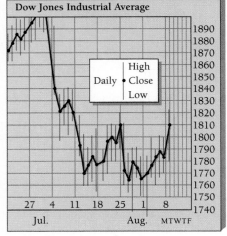

A graph of the Dow Jones Industrial Average.

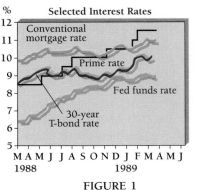

FIGURE 1

Graphs

Each point in the plane corresponds to an ordered pair of numbers. Note that the pair (2, 5) is different from the pair (5, 2) (see Fig. 2). This is why we call (2, 5) an **ordered pair**. The first member, 2, is called the **first coordinate** of the point, and the second member, 5, is called the **second coordinate**. Together these are called the *coordinates of the point*. The vertical line is often called the *y-axis,* and the horizontal line is often called the *x-axis.*

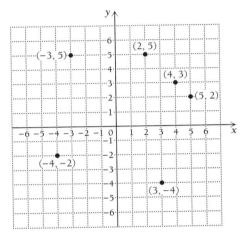

FIGURE 2

DO EXERCISE 1.

Graphs of Equations

A **solution** of an equation in two variables is an ordered pair of numbers that, when substituted for the variables, gives a true sentence. If not directed otherwise, we usually take the variables in alphabetical *order.* For example, $(-1, 2)$ is a solution to the equation $3x^2 + y = 5$, because when we substitute -1 for x and 2 for y, we get a true sentence:

$$\frac{\begin{array}{c|c} 3x^2 + y \ = \ 5 \\ \hline 3(-1)^2 + 2 & 5 \\ 3 + 2 & \\ 5 & \text{True} \end{array}}{}$$

DO EXERCISE 2.

---

1. Plot the ordered pairs (2, 0), (0, 2), $(-1, 3)$, (4, 3), and $(-2, -3)$.

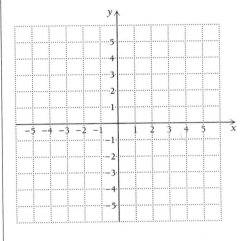

2. Decide whether each pair is a solution of $x^2 - 2y = 6$.

   a) $(-2, -1)$

   b) $(3, 0)$

3. Graph $y = -2x + 1$.

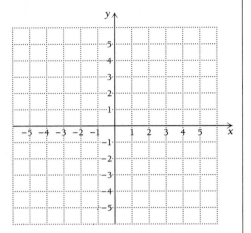

DEFINITION

The *graph* of an equation is a drawing that represents all the solutions of the equation.

We obtain the graph of an equation by plotting enough ordered pairs (that are solutions) to see a pattern. The graph could be a line, a curve (or curves), or some other configuration.

EXAMPLE 1   Graph: $y = 2x + 1$.

Solution   We find some ordered pairs that are solutions and arrange them in a table. To find an ordered pair, we can choose *any* number for x and then determine y. For example, if we choose -2 for x, then $y = 2(-2) + 1 = -4 + 1 = -3$. We substituted -2 for x in the equation $y = 2x + 1$. For balance, we make some negative choices for x, as well as some positive choices. If a number takes us off the graph paper, we usually omit the pair from the graph.

x	y	(x, y)
-2	-3	$(-2, -3)$
-1	-1	$(-1, -1)$
0	1	$(0, 1)$
1	3	$(1, 3)$
2	5	$(2, 5)$

(1) Choose any x.
(2) Compute y.
(3) Form the pair (x, y).
(4) Plot the points.

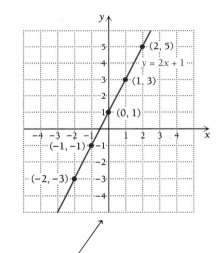

FIGURE 3

After we plot the points, we look for a pattern in the graph. If we had enough of the points, they would make a solid line. We can draw the line with a ruler and label it $y = 2x + 1$ (Fig. 3). ❖

DO EXERCISE 3.

EXAMPLE 2   Graph: $y = x^2 - 1$.

Solution

x	y	(x, y)
-2	3	$(-2, 3)$
-1	0	$(-1, 0)$
0	-1	$(0, -1)$
1	0	$(1, 0)$
2	3	$(2, 3)$

(1)   Choose any x.
(2)   Compute y.
(3)   Form the pair (x, y).
(4)   Plot the points.

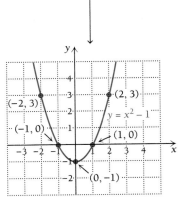

FIGURE 4

This time the pattern of the points is a curve called a *parabola*. We fill in the pattern and obtain the graph (Fig. 4). Note that we must plot enough points to see a pattern.   ❖

DO EXERCISE 4.

EXAMPLE 3   Graph: $x = y^2$.

4. Graph $y = x^2 - 3$.

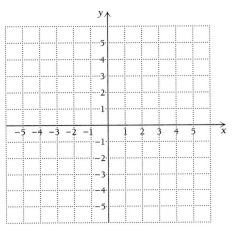

5. Graph $x = y^2 + 1$.

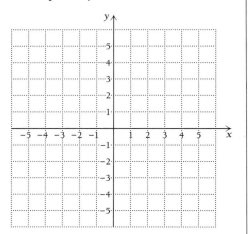

Solution This time x is expressed in terms of the variable y. Thus we first choose numbers for y and then compute x.

x	y	(x, y)
4	-2	$(4, -2)$
1	-1	$(1, -1)$
0	0	$(0, 0)$
1	1	$(1, 1)$
4	2	$(4, 2)$

(1) Choose any y.
(2) Compute x.
(3) Form the pair (x, y).
(4) Plot the points.

FIGURE 5

We plot these points (Fig. 5), keeping in mind that x is still the first coordinate and y the second. We look for a pattern and complete the graph. ❖

DO EXERCISE 5.

Functions

A **relation** is any set of ordered pairs. Thus the solutions of an equation in two variables form a relation.

A **function** is a special kind of relation. Such relations are of fundamental importance in calculus.

A Function as an Input–Output Relation

DEFINITION

A *function* is a relation that assigns to each "input," or first coordinate, a unique "output," or second coordinate. The set of all input numbers is called the *domain.* The set of all output numbers is called the *range.*

EXAMPLE 4 Squaring numbers is a function. We can take any number x as an input. We square that number to find the output, x^2.

Input	Output
-3	9
1.73	2.9929
k	k^2
$\sqrt{a}$	a
$1 + t$	$(1 + t)^2$, or $1 + 2t + t^2$

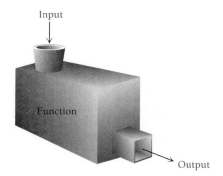

Input

Function

Output

FIGURE 6

Think of a function as a machine (Fig. 6). Think of putting a member of the domain (an input) into the machine. The machine squares the input and gives you the output, a member of the range. The domain of this function is the set of all real numbers, because any real number can be squared.   ❖

DO EXERCISE 6.

It is customary to use letters such as f and g to represent functions. Suppose that f is a function and x is a number in its domain. For the input x, we can name the output as

$$f(x), \text{ read "} f \text{ of } x \text{," or "the value of } f \text{ at } x.\text{"}$$

6. The operation of "taking the reciprocal" is a function. That is, the operation of going from x to $1/x$ is a function defined for all numbers except 0. Thus the domain is the set of all nonzero real numbers.

Complete this table for the reciprocal function.

Input	Output
5	
$-\frac{2}{3}$	
$\frac{1}{4}$	
$\frac{1}{a}$	
k	
$1 + t$	

7. The reciprocal function is given by
$$f(x) = \frac{1}{x}.$$

Find $f(5)$, $f(-2)$, $f(\frac{1}{4})$, $f(1/a)$, $f(k)$, $f(1 + t)$, and $f(x + h)$.

If f is the squaring function, then $f(3)$ is the output for the input 3. Thus, $f(3) = 3^2 = 9$.

EXAMPLE 5   The squaring function is given by
$$f(x) = x^2.$$
Find $f(-3)$, $f(1)$, $f(k)$, $f(\sqrt{k})$, $f(1 + t)$, and $f(x + h)$.

Solution

$$f(-3) = (-3)^2 = 9,$$
$$f(1) = 1^2 = 1,$$
$$f(k) = k^2,$$
$$f(\sqrt{k}) = (\sqrt{k})^2 = k,$$
$$f(1 + t) = (1 + t)^2 = 1 + 2t + t^2,$$
$$f(x + h) = (x + h)^2 = x^2 + 2xh + h^2$$

To find $f(x + h)$, remember what the function does: It squares the input. Thus, $f(x + h) = (x + h)^2 = x^2 + 2xh + h^2$. This amounts to replacing x on both sides of $f(x) = x^2$ by $x + h$.   ❖

DO EXERCISE 7.

8. A function t is given by
$$t(x) = x + x^2.$$
Find $t(5)$, $t(-5)$, and $t(x + h)$.

EXAMPLE 6   A function f subtracts the square of an input from the input. A description of f is given by
$$f(x) = x - x^2.$$
Find $f(4)$ and $f(x + h)$.

Solution   We replace the x's on both sides by the inputs. Thus,

$$f(4) = 4 - 4^2 = 4 - 16 = -12;$$
$$f(x + h) = (x + h) - (x + h)^2 = x + h - (x^2 + 2xh + h^2)$$
$$= x + h - x^2 - 2xh - h^2.$$ ❖

DO EXERCISE 8.

Taking square roots is *not* a function, because an input can have more than one output. For example, the input 4 has two outputs, 2 and -2.

When a function is given by a formula, and nothing is said about the domain, its domain is understood to be the set of all numbers that can be substituted into the formula. For example, consider the reciprocal function

$$f(x) = \frac{1}{x}.$$

The only number that cannot be substituted into the formula is 0. We say that f is *not defined at* 0, or $f(0)$ *does not exist*. The domain consists of all nonzero real numbers.

EXAMPLE 7   Taking principal square roots (nonnegative roots) is a function. Let g be this function. Then g can be described as

$$g(x) = \sqrt{x}.$$

Recall from algebra that the symbol $\sqrt{a}$ represents the nonnegative square root of a for $a \geq 0$. There is only one such real-number root.

a)   Find the domain of this function.

b)   Find $g(0)$, $g(2)$, $g(a)$, $g(16)$, and $g(t + h)$.

Solution

a)   The domain consists of numbers that can be substituted into the formula. We can take the principal square root of any nonnegative number. The principal square root of a negative number is not a real number. (Taking square roots of negative numbers would require us to consider complex numbers, which we will not cover in this text.) Thus the domain consists of all nonnegative numbers.

b)
$$g(0) = \sqrt{0} = 0,$$
$$g(2) = \sqrt{2},$$
$$g(a) = \sqrt{a},$$
$$g(16) = \sqrt{16} = 4,$$
$$g(t + h) = \sqrt{t + h}$$ ❖

DO EXERCISE 9.

A Function as a Mapping

We can also think of a function as a "mapping" of one set to another (Fig. 7).

DEFINITION

A *function* is a mapping that associates with each number x in one set (called the *domain*) exactly one number y in another set (called the *range*).
---

9. Subtracting 3 from a number and then taking the reciprocal is a function f given by

$$f(x) = \frac{1}{x - 3}.$$

a)   What is the domain of this function? Explain.

b)   Find $f(5)$, $f(4)$, $f(2.5)$, and $f(x + h)$.

The number of people at the beach at a given time is a function of the temperature, although a formula may not be readily available.

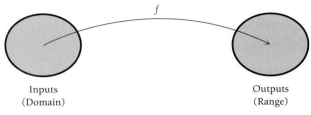

FIGURE 7

For example, the squaring function maps members of the set of real numbers to members of the set of nonnegative numbers (Fig. 8).

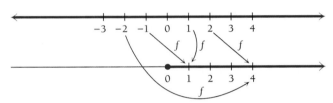

FIGURE 8

The statement

$$y = f(x)$$

means that the number x is mapped to the number y by the function f. Functions are often implicit in certain equations. For example, consider

$$xy = 2.$$

For any nonzero x, there is a unique number y that satisfies the equation. This yields a function that is given explicitly by

$$y = f(x) = \frac{2}{x}.$$

On the other hand, consider the equation

$$x = y^2.$$

A positive number x would be related to two values of y, namely $\sqrt{x}$ and $-\sqrt{x}$. Thus this equation is not an implicit description of a function that maps inputs x to outputs y.

Graphs of Functions

Consider again the squaring function. The input 3 is associated with the output 9. The input–output pair (3, 9) is one point on the *graph* of this function.

DEFINITION

The *graph* of a function f is a drawing that represents all the input–output pairs $(x, f(x))$. In cases where the function is given by an equation, the graph of a function is the graph of the equation $y = f(x)$.

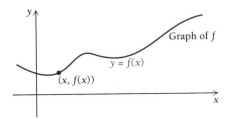

FIGURE 9

It is customary to locate input values (the domain) on the horizontal axis and output values (the range) on the vertical axis.

EXAMPLE 8   Graph: $f(x) = x^2 - 1$.

Solution

x	$f(x)$	$(x, f(x))$
-2	3	$(-2, 3)$
-1	0	$(-1, 0)$
0	-1	$(0, -1)$
1	0	$(1, 0)$
2	3	$(2, 3)$

(1) Choose any x.
(2) Compute y.
(3) Form the pair (x, y).
(4) Plot the points.

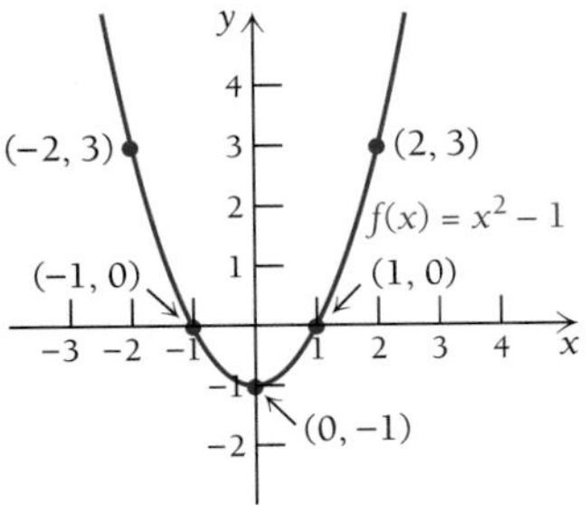

FIGURE 10

We plot the input–output pairs from the table and, in this case, draw a curve to complete the graph (Fig. 10). ❖

10. Graph $f(x) = -2x + 1$.

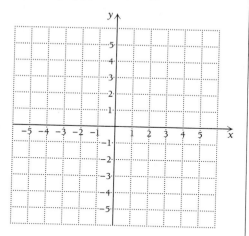

DO EXERCISES 10 AND 11.

Figure 11 illustrates how the idea of a mapping is connected with the graph of a function.

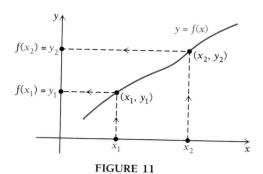

FIGURE 11

The Vertical-Line Test

Let us now determine how we can look at a graph and decide whether it is a graph of a function. We already know that

$$x = y^2$$

does not yield a function that maps a number x to a unique number y. Its graph is shown in Fig. 12. Note that there is a point x_1 that has two outputs. This means that we have a vertical line that meets the graph in more than one place.

11. Graph $g(x) = x^2 - 3$.

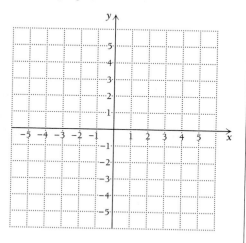

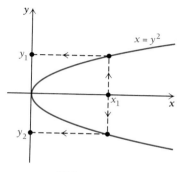

FIGURE 12

The Vertical-Line Test
A graph is the graph of a function provided no vertical line meets the graph more than once.

EXAMPLE 9 Determine whether each of the following is a graph of a function.

a)

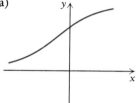

b)

c)

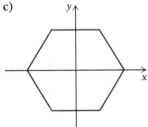

d)

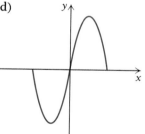

Solution

a) A function. No vertical line meets the graph more than once.

b) Not a function. A vertical line (in fact, many) meets the graph more than once.

c) Not a function.

d) A function.   ❖

DO EXERCISE 12.

Functions Defined Piecewise

Sometimes functions are defined piecewise. That is, there are different output formulas for different parts of the domain.

12. Determine whether each of the following is the graph of a function.

a)

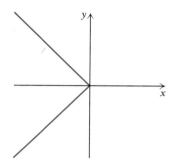

b)

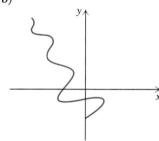

c)

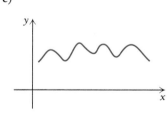

d)

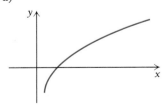

EXAMPLE 10 Graph the function defined as follows.

$$f(x) = \begin{cases} 4, & \text{for } x \leqslant 0 \\ & \text{(This means that for any input } x \text{ less than or equal} \\ & \text{to 0, the output is 4.)} \\ 4 - x^2, & \text{for } 0 < x \leqslant 2 \\ & \text{(This means that for any input } x \text{ greater than 0 and} \\ & \text{less than or equal to 2, the output is } 4 - x^2.) \\ 2x - 6, & \text{for } x > 2 \\ & \text{(This means that for any input } x \text{ greater than 2, the} \\ & \text{output is } 2x - 6.) \end{cases}$$

Solution See the graph in Fig. 13.

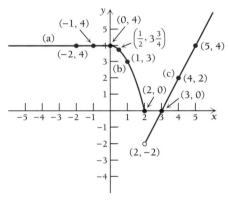

FIGURE 13

a) We graph $f(x) = 4$ for inputs less than or equal to 0 (that is, $x \leqslant 0$).

For $f(x) = 4$,

$f(-2) = 4$,

$f(-1) = 4$, and

$f(0) = 4$.

x	$f(x)$	$(x, f(x))$
-2	4	$(-2, 4)$
-1	4	$(-1, 4)$
0	4	$(0, 4)$

b) We graph $f(x) = 4 - x^2$ for inputs greater than 0 and less than or equal to 2 (that is, $0 < x \leqslant 2$).

For $f(x) = 4 - x^2$,

$f(\frac{1}{2}) = 4 - (\frac{1}{2})^2 = 3\frac{3}{4}$,

$f(1) = 4 - 1^2 = 3$, and

$f(2) = 4 - 2^2 = 0$.

x	$f(x)$	$(x, f(x))$
$\frac{1}{2}$	$3\frac{3}{4}$	$(\frac{1}{2}, 3\frac{3}{4})$
1	3	(1, 3)
2	0	(2, 0)

The solid circle indicates that the point (2, 0) is part of the graph.

c) We graph $f(x) = 2x - 6$ for inputs greater than 2 (that is, $x > 2$).

For $f(x) = 2x - 6$,

$f(3) = 2(3) - 6 = 0$,

$f(4) = 2(4) - 6 = 2$, and

$f(5) = 2(5) - 6 = 4$.

x	$f(x)$	$(x, f(x))$
3	0	(3, 0)
4	2	(4, 2)
5	4	(5, 4)

The open circle at (2, −2) indicates that the point is not part of the graph. ❖

DO EXERCISE 13.

13. Graph the function defined as follows:

$$f(x) = \begin{cases} x + 3, & \text{for } x \le -2, \\ 1, & \text{for } -2 < x \le 3, \\ x^2 - 10, & \text{for } x > 3. \end{cases}$$

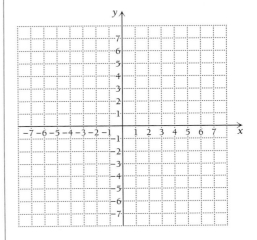

Some Final Remarks

Almost all the functions in this text can be described by equations. Some functions, however, cannot. For example, there will be a function that assigns grades to students in this course, but that function will most likely not have a formula.

We sometimes use the terminology y *is a function of* x. This means that x is an input and y is an output. We often refer to x as the **independent variable** when it represents inputs and y as the **dependent variable** when it represents outputs. We may refer to "a function $y = x^2$," without naming it with a letter f. We may simply refer to x^2 (alone) as a function.

In calculus we will be studying how outputs of a function change when the inputs change.

1. A function f is given by

$$f(x) = 2x + 3.$$

This function takes a number x, multiplies it by 2, and adds 3.

a) Complete this table.

Input	Output
4.1	
4.01	
4.001	
4	

b) Find $f(5)$, $f(-1)$, $f(k)$, $f(1 + t)$, and $f(x + h)$.

2. A function f is given by

$$f(x) = 3x - 1.$$

This function takes a number x, multiplies it by 3, and subtracts 1.

a) Complete this table.

Input	Output
5.1	
5.01	
5.001	
5	

b) Find $f(4)$, $f(-2)$, $f(k)$, $f(1 + t)$, and $f(x + h)$.

3. A function g is given by

$$g(x) = x^2 - 3.$$

This function takes a number x, squares it, and subtracts 3. Find $g(-1)$, $g(0)$, $g(1)$, $g(5)$, $g(u)$, $g(a + h)$, and $g(1 - h)$.

4. A function g is given by

$$g(x) = x^2 + 4.$$

This function takes a number x, squares it, and adds 4. Find $g(-3)$, $g(0)$, $g(-1)$, $g(7)$, $g(v)$, $g(a + h)$, and $g(1 - t)$.

5. A function f is given by

$$f(x) = (x - 3)^2.$$

This function takes a number x, subtracts 3 from it, and squares the result.

a) Find $f(4)$, $f(-2)$, $f(0)$, $f(a)$, $f(t + 1)$, $f(t + 3)$, and $f(x + h)$.

b) Note that f could also be given by

$$f(x) = x^2 - 6x + 9.$$

Explain what this does to an input number x.

6. A function f is given by

$$f(x) = (x + 4)^2.$$

This function takes a number x, adds 4 to it, and squares the result.

a) Find $f(3)$, $f(-6)$, $f(0)$, $f(k)$, $f(t - 1)$, $f(t - 4)$, and $f(x + h)$.

b) Note that f could also be given by

$$f(x) = x^2 + 8x + 16.$$

Explain what this does to an input number x.

Graph the function.

7. $f(x) = 2x + 3$

8. $f(x) = 3x - 1$

9. $g(x) = -4x$

10. $g(x) = -2x$

11. $f(x) = x^2 - 1$

12. $f(x) = x^2 + 4$

13. $g(x) = x^3$

14. $g(x) = \frac{1}{2}x^3$

Determine whether the graph is that of a function.

15.

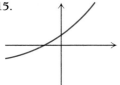

16.

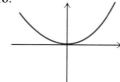

17.

18.

19.

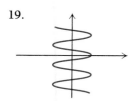

20.

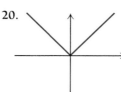

21.

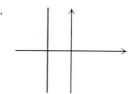

22.

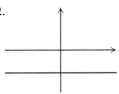

In Exercises 23–26, the vertical dashed lines are not part of the graph.

23.

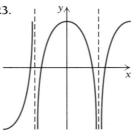

24.

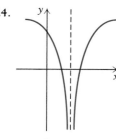

25.

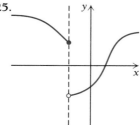

26.

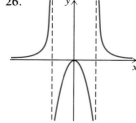

27. **a)** Graph $x = y^2 - 1$.

   b) Is this a function?

28. **a)** Graph $x = y^2 - 3$.

   b) Is this a function?

29. For $f(x) = x^2 - 3x$, find $f(x + h)$.

30. For $f(x) = x^2 + 4x$, find $f(x + h)$.

Graph.

31. $f(x) = \begin{cases} 1, & \text{for } x < 0, \\ -1, & \text{for } x \geqslant 0 \end{cases}$

32. $f(x) = \begin{cases} 2, & \text{for } x \text{ an integer}, \\ -2, & \text{for } x \text{ not an integer} \end{cases}$

33. $f(x) = \begin{cases} -3, & \text{for } x = -2, \\ x^2, & \text{for } x \neq -2 \end{cases}$

34. $f(x) = \begin{cases} -2x - 6, & \text{for } x \leqslant -2, \\ 2 - x^2, & \text{for } -2 < x < 2, \\ 2x - 6, & \text{for } x \geqslant 2 \end{cases}$

APPLICATIONS

❖ **Business and Economics**

35. *Total revenue.* Raggs, Ltd., a clothing firm, determines that its total revenue from the sale of x suits is given by the function

$$R(x) = 200x + 50,$$

where $R(x)$ = the revenue, in dollars, from the sale of x suits. Find $R(10)$ and $R(100)$.

36. *Compound interest.* The amount of money in a savings account at 14%, compounded annually, depends on the initial investment x and is given by the function

$$A(x) = x + 14\%x,$$

where $A(x)$ = the amount in the account at the end of 1 year. Find $A(100)$ and $A(1000)$.

SYNTHESIS EXERCISES

Solve for y in terms of x. Decide whether the resulting equation represents a function.

37. $2x + y - 16 = 4 - 3y + 2x$

38. $2y^2 + 3x = 4x + 5$

39. $(4y^{2/3})^3 = 64x$

40. $(3y^{3/2})^2 = 72x$

COMPUTER-GRAPHING CALCULATOR EXERCISES

Exercises such as these are designed to be done with the use of a computer software graphing package, such as The Calculus Explorer, or with a graphing calculator. Sketch graphs of the following functions. More complicated graphing will be considered in Section 1.5. Graph sketching, using calculus, will be considered in detail in Chapter 3.

41. $f(x) = 4 - x^2$

42. $f(x) = (x + 2)^2 - 3$

43. $f(x) = -(x - 3)^2 + 1$

44. $f(x) = -2x^2 + 10x - 7$

THE CALCULUS EXPLORER:
PowerGrapher

Find function values of some of the functions in this exercise set. See how this program can help to make graph sketching much easier.

1.4
SLOPE AND LINEAR FUNCTIONS

OBJECTIVES

a) Graph equations of the type $y = b$ and $x = a$.

b) Graph linear functions.

c) Find an equation of a line given its slope and one point on the line.

d) Find the slope of the line containing a given pair of points.

(continued)

Horizontal and Vertical Lines

Let us consider graphs of equations $y = b$ and $x = a$.

EXAMPLE 1

a) Graph $y = 4$.

b) Decide whether the relation is a function.

Solution

a) The graph consists of all ordered pairs whose second coordinate is 4 (Fig. 1). To see how a pair such as $(-2, 4)$ could be a solution of $y = 4$, we can consider the equation above in the form

$$0x + y = 4.$$

Then $(-2, 4)$ is a solution because

$$0(-2) + 4 = 4$$

is true.

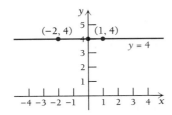

FIGURE 1

b) The vertical-line test holds. Thus this is a function.   ❖

DO EXERCISE 1.

EXAMPLE 2

a) Graph $x = -3$.

b) Decide whether it is a function.

Solution

a) The graph consists of all ordered pairs whose first coordinate is -3 (Fig. 2). To see how a pair such as $(-3, 4)$ could be a solution of

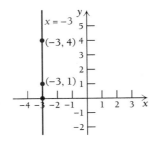

FIGURE 2

e) Find an equation of the line containing a given pair of points.

f) Given an equation of a line, find the slope and the y-intercept.

g) Solve problems involving linear functions.

1. a) Graph $y = 3$.

b) Decide whether it is a function.

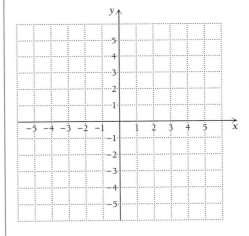

2. a) Graph $x = 1$.

b) Decide whether it is a function.

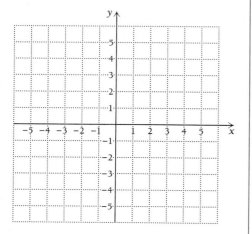

$x = -3$, we can consider the equation above in the form

$$x + 0y = -3.$$

Then $(-3, 4)$ is a solution because

$$(-3) + 0(4) = -3$$

is true.

b) This is *not* a function. It fails the vertical-line test. The line itself meets the graph more than once—in fact, infinitely many times. ❖

DO EXERCISE 2.

In general, we have the following.

THEOREM 6

The graph of $y = b$, a horizontal line, is the graph of a function.
The graph of $x = a$, a vertical line, is not the graph of a function.

The Equation $y = mx$

Consider the following table of numbers and look for a pattern.

x	1	-1	$-\frac{1}{2}$	2	-2	3	-7	5
y	3	-3	$-\frac{3}{2}$	6	-6	9	-21	15

Note that the ratio of the bottom number to the top one is 3. That is,

$$\frac{y}{x} = 3, \quad \text{or} \quad y = 3x.$$

Ordered pairs from the table can be used to graph the equation $y = 3x$ (Fig. 3). Note that this is a function.

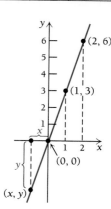

FIGURE 3

THEOREM 7

The graph of the function given by

$$y = mx \quad \text{or} \quad f(x) = mx$$

**is the straight line through the origin (0, 0) and the point (1, m).
The constant m is called the *slope* of the line.**

DO EXERCISE 3.

Various graphs of $y = mx$ for positive m are shown in Fig. 4. Note that such graphs slant up from left to right. A line with large positive slope rises faster than a line with smaller positive slope.

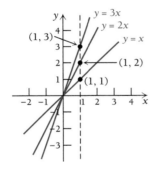

FIGURE 4

When $m = 0$, $y = 0x$, or $y = 0$. Figure 5 is a graph of $y = 0$. Note that this is the x-axis and is a horizontal line.

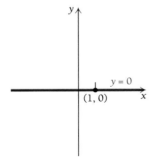

FIGURE 5

Graphs of $y = mx$ for negative m are shown in Fig. 6. Note that such graphs slant down from left to right.

3. a) Graph $y = -2x$.

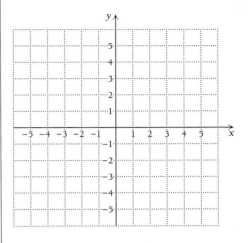

b) Is this a function?

c) What is the slope?

Lines of various slopes.

4. Consider these lines.

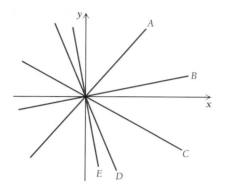

a)   Which lines have positive slope?

b)   Which lines have negative slope?

c)   Which line has the largest slope?

d)   Which line has the smallest slope?

Hair will grow 6 inches in 12 months.

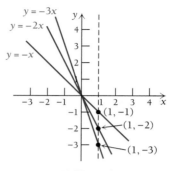

FIGURE 6

DO EXERCISE 4.

Direct Variation

There are many applications involving equations like $y = mx$, where m is some positive number. In such situations we say that we have **direct variation**, and m (the slope) is called the **variation constant**, or **constant of proportionality**. Generally, only positive values of x and y are considered.

DEFINITION

The variable y *varies directly* as x if there is some positive constant m such that $y = mx$. We also say that y is *directly proportional* to x.

EXAMPLE 3   *Life science: Hair growth.* The number N of inches that human hair will grow is directly proportional to the time t in months. Hair will grow 6 in. in 12 months.

a)   Find an equation of variation.

b)   How many months does it take for hair to grow 10 in.?

Solution

a)   Since $N = mt$, then $6 = m(12)$ and $\frac{1}{2} = m$. Thus, $N = \frac{1}{2}t$ (Fig. 7).

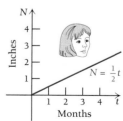

FIGURE 7

b) To find how many months it takes for hair to grow 10 in., we solve

$$10 = \tfrac{1}{2}t$$

and get

$$20 = t.$$

Thus it takes 20 months for hair to grow 10 in.   ❖

DO EXERCISE 5.

The Equation $y = mx + b$

Compare the graphs of the equations

$$y = 3x \quad \text{and} \quad y = 3x - 2$$

(Fig. 8). Note that the graph of $y = 3x - 2$ is a shift 2 units downward of the graph of $y = 3x$, and that $y = 3x - 2$ has y-intercept $(0, -2)$. Note also that the graph of $y = 3x - 2$ is a graph of a function.

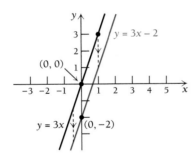

FIGURE 8

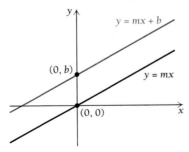

FIGURE 9

DO EXERCISE 6.

5. *Life science: Newspaper recycling.* The number T of trees saved by recycling is directly proportional to the height h, in inches, of a stack of recycled newspaper.

    a) It is known that a stack of newspaper 36 in. high will save 1 tree. Find an equation of variation expressing T as a function of h.

    b) How many trees are saved by a stack of paper that is 162 in. (13.5 ft) high?

6. a) Using the same set of axes, graph

$$y = 3x$$

    and

$$y = 3x + 1.$$

    b) How can the graph of $y = 3x + 1$ be obtained from the graph of $y = 3x$?

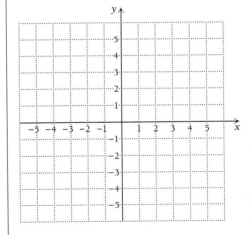

7. Find the slope and the y-intercept of $2x + 3y - 6 = 0$.

DEFINITION

A *linear function* is given by

$$y = mx + b \quad \text{or} \quad f(x) = mx + b$$

and has a graph (Fig. 9) that is the straight line parallel to $y = mx$ with y-intercept $(0, b)$. The constant m is called the *slope*.

When $m = 0$, $y = 0x + b = b$, and we have what is known as a **constant function**. The graph of such a function is a horizontal line.

The Slope-Intercept Equation

Any nonvertical line l is uniquely determined by its slope m and its y-intercept $(0, b)$. In other words, the slope describes the "slant" of the line, and the y-intercept is the point at which the line crosses the y-axis. Thus we have the following definition.

DEFINITION

$y = mx + b$ is called the *slope–intercept equation* of a line.

EXAMPLE 4 Find the slope and the y-intercept of $2x - 4y - 7 = 0$.

Solution We solve for y:

$$-4y = -2x + 7$$

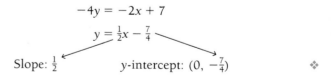

$$y = \tfrac{1}{2}x - \tfrac{7}{4}$$

Slope: $\tfrac{1}{2}$ y-intercept: $(0, -\tfrac{7}{4})$ ❖

DO EXERCISE 7.

The Point–Slope Equation

Suppose we know the slope of a line and some point on the line other than the y-intercept. We can still find an equation of the line.

EXAMPLE 5 Find an equation of the line with slope 3 containing the point $(-1, -5)$.

Solution From the slope–intercept equation, we have

$$y = 3x + b,$$

so we must determine b. Since $(-1, -5)$ is on the line, it follows that

$$-5 = 3(-1) + b,$$

so

$$-2 = b \quad \text{and} \quad y = 3x - 2. \qquad \qquad ❖$$

DO EXERCISE 8.

If a point (x_1, y_1) is on the line

$$y = mx + b, \qquad \qquad (1)$$

it must follow that

$$y_1 = mx_1 + b. \qquad \qquad (2)$$

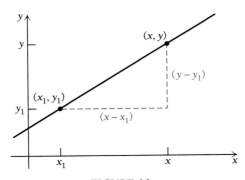

FIGURE 10

Subtracting Eq. (2) from Eq. (1) eliminates the b's, and we have

$$\begin{aligned}
y - y_1 &= (mx + b) - (mx_1 + b) \\
&= mx + b - mx_1 - b \\
&= mx - mx_1 \\
&= m(x - x_1).
\end{aligned}$$

8. Find an equation of the line with slope -4 containing the point $(2, -7)$.

Lines of the same slope.

9. Find an equation of the line with slope -4 containing the point $(2, -7)$.

DEFINITION

$y - y_1 = m(x - x_1)$ is called the *point–slope equation* of a line.

This definition allows us to write an equation of a line given its slope and the coordinates of *any* point on it.

EXAMPLE 6 Find an equation of the line with slope 3 containing the point $(-1, -5)$.

Solution Substituting in

$$y - y_1 = m(x - x_1),$$

we get

$$y - (-5) = 3[x - (-1)].$$

Simplifying and solving for y, we get the slope–intercept equation, as found in Example 5:

$$y + 5 = 3(x + 1)$$
$$y + 5 = 3x + 3$$
$$y = 3x + 3 - 5$$
$$y = 3x - 2. \qquad ❖$$

DO EXERCISE 9.

Computing Slope

We now determine a method of computing the slope of a line when we know the coordinates of two of its points. Suppose that (x_1, y_1) and (x_2, y_2) are the coordinates of two different points, P_1 and P_2, respectively, on a line that is not vertical. Consider a right triangle with legs parallel to the axes, as shown in Fig. 11.

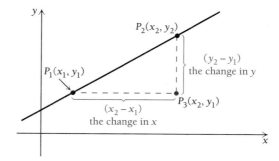

FIGURE 11

The point P_3 with coordinates (x_2, y_1) is the third vertex of the triangle. As we move from P_1 to P_2, y changes from y_1 to y_2. The change in y is $y_2 - y_1$. Similarly, the change in x is $x_2 - x_1$. The ratio of these changes is the slope. To see this, consider the point–slope equation,

$$y - y_1 = m(x - x_1).$$

Since (x_2, y_2) is on the line, it must follow that

$$y_2 - y_1 = m(x_2 - x_1).$$

Since the line is not vertical, the two x-coordinates must be different, so $x_2 - x_1$ is nonzero and we can divide by it to get the following theorem.

THEOREM 8

$$m = \frac{y_2 - y_1}{x_2 - x_1} = \frac{\text{change in } y}{\text{change in } x} = \text{slope of line containing points} \atop (x_1, y_1) \text{ and } (x_2, y_2)$$

EXAMPLE 7   Find the slope of the line containing the points $(-2, 6)$ and $(-4, 9)$.

Solution   We have

$$m = \frac{y_2 - y_1}{x_2 - x_1} = \frac{6 - 9}{-2 - (-4)}$$

$$= \frac{-3}{2} = -\frac{3}{2}.$$

Note that it does not matter which point is taken first, so long as we subtract the coordinates in the same order. In this example, we can also find m as follows:

$$m = \frac{9 - 6}{-4 - (-2)}$$

$$= \frac{3}{-2} = -\frac{3}{2}. \qquad ❖$$

DO EXERCISES 10–13.

If a line is horizontal, the change in y for any two points is 0 (Fig. 12a). Thus a horizontal line has slope 0. If a line is vertical, the change in x for any two points is 0 (Fig. 12b). Thus the slope is *not defined* because we cannot divide by 0. A vertical line has no slope. Thus "0 slope" and "no slope" are two very distinct concepts.

Find the slope of the line containing the given pair of points.

10. (1, 3) and (2, 5)

11. $(-6, 4)$ and (2, 5)

12. (4, 7) and (6, -10)

13. (3, 5) and $(-1, 5)$

Find the slope, if it exists, of the line containing the given pair of points.

14. $(4, -7)$ and $(-2, -7)$

(a)

(b)

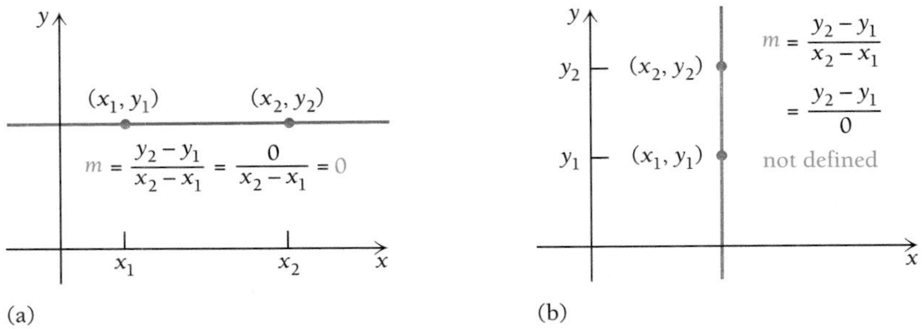

FIGURE 12

DO EXERCISES 14 AND 15.

Applications of Linear Functions

Many applications are modeled by linear functions.

15. $(4, -7)$ and $(4, -9)$

EXAMPLE 8   *Business: Total cost.* Raggs, Ltd., a clothing firm, has **fixed costs** of $10,000 per year. These costs, such as rent, maintenance, and so on, must be paid no matter how much the company produces. To produce x units of a certain kind of suit, it costs $20 per unit in addition to the fixed costs. That is, the **variable costs** for producing x of these units is $20x$ dollars. These are costs that are directly related to production, such as material, wages, fuel, and so on. Then the **total cost** $C(x)$ of producing x suits in a year is given by a function C:

$$C(x) = (\text{Variable costs}) + (\text{Fixed costs}) = 20x + 10{,}000.$$

a)   Graph the variable-cost, the fixed-cost, and the total-cost functions.

b)   What is the total cost of producing 100 suits? 400 suits?

c)   How much more does it cost to produce 400 suits than 100 suits?

Solution

a)   The variable-cost and fixed-cost functions appear in Fig. 13(a). The total-cost function is shown in Fig. 13(b). From a practical standpoint, the domains of these functions are nonnegative integers 0, 1, 2, 3, and so on, since it does not make sense to make a negative number of suits or a fractional number of suits. It is common practice to draw the graphs as though the domains were the entire set of nonnegative real numbers.

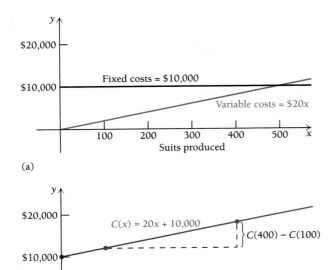

(a)

(b)

FIGURE 13

16. *Business: Total cost.* Rework Example 8, given that variable costs = $30x$, fixed costs = $15,000$, and total costs = $C(x) = 30x + 15,000$.

b)   The total cost of producing 100 suits is

$$C(100) = 20 \cdot 100 + 10,000 = \$12,000.$$

The total cost of producing 400 suits is

$$C(400) = 20 \cdot 400 + 10,000$$
$$= \$18,000.$$

c)   The extra cost of producing 400 suits rather than 100 suits is given by

$$C(400) - C(100) = \$18,000 - \$12,000$$
$$= \$6000.$$ ❖

DO EXERCISE 16.

EXAMPLE 9   *Business: Profit-and-loss analysis.* In reference to Example 8, Raggs, Ltd., determines that its total revenue from the sale of x suits is $80 per suit. That is, the total revenue $R(x)$ is given by the function

$$R(x) = 80x.$$

a) Graph $R(x)$ and $C(x)$ using the same set of axes.

b) The total profit $P(x)$ is given by a function P:

$$P(x) = (\text{Total revenue}) - (\text{Total costs}) = R(x) - C(x).$$

Determine $P(x)$ and draw its graph using the same set of axes.

c) The company will *break even* at that value of x for which $P(x) = 0$ (that is, no profit and no loss). This is where $R(x) = C(x)$. Find the **break-even value** of x.

Solution

a) The graphs of $R(x) = 80x$ and $C(x) = 20x + 10,000$ are shown in Fig. 14. When $C(x)$ is above $R(x)$, a loss will occur. This is shown by the color-shaded region. When $R(x)$ is above $C(x)$, a gain will occur. This is shown by the gray-shaded region.

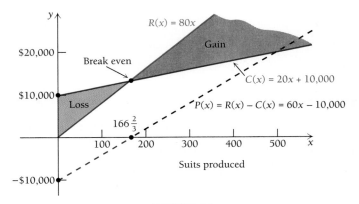

FIGURE 14

b) We see that

$$P(x) = R(x) - C(x) = 80x - (20x + 10,000) = 60x - 10,000.$$

The graph of $P(x)$ is shown by the dashed line. The color dashed line shows a "negative" profit, or loss. The black dashed line shows a "positive" profit, or gain.

c) To find the break-even value, we solve $R(x) = C(x)$:

$$R(x) = C(x)$$
$$80x = 20x + 10{,}000$$
$$60x = 10{,}000$$
$$x = 166\tfrac{2}{3}.$$

How do we interpret the fractional answer, since it is not possible to produce $\tfrac{2}{3}$ of a suit? We simply round to 167. Estimates of break-even values are usually sufficient since companies want to operate well away from break-even values in order to maximize profit.   ❖

DO EXERCISE 17.

17. *Business: Profit-and-loss analysis.* Rework Example 9, given that

$$C(x) = 30x + 15{,}000$$

and

$$R(x) = 90x.$$

EXERCISE SET 1.4

Graph.
1. $y = -4$
2. $y = -3.5$
3. $x = 4.5$
4. $x = 10$

Graph. Find the slope and the y-intercept.
5. $y = -3x$
6. $y = -0.5x$
7. $y = 0.5x$
8. $y = 3x$
9. $y = -2x + 3$
10. $y = -x + 4$
11. $y = -x - 2$
12. $y = -3x + 2$

Find the slope and the y-intercept.
13. $2x + y - 2 = 0$
14. $2x - y + 3 = 0$
15. $2x + 2y + 5 = 0$
16. $3x - 3y + 6 = 0$

Find an equation of the line:
17. with $m = -5$, containing $(1, -5)$.

18. with $m = 7$, containing $(1, 7)$.
19. with $m = -2$, containing $(2, 3)$.
20. with $m = -3$, containing $(5, -2)$.
21. with y-intercept $(0, -6)$ and slope $\tfrac{1}{2}$.
22. with y-intercept $(0, 7)$ and slope $\tfrac{4}{3}$.
23. with slope 0, containing $(2, 3)$.
24. with slope 0, containing $(4, 8)$.

Find the slope of the line containing the given pair of points, if it exists.
25. $(-4, -2)$ and $(-2, 1)$
26. $(-2, 1)$ and $(6, 3)$
27. $(2, -4)$ and $(4, -3)$
28. $(-5, 8)$ and $(5, -3)$
29. $(3, -7)$ and $(3, -9)$
30. $(-4, 2)$ and $(-4, 10)$
31. $(2, 3)$ and $(-1, 3)$
32. $(-6, \tfrac{1}{2})$ and $(-7, \tfrac{1}{2})$
33. $(x, 3x)$ and $(x + h, 3(x + h))$
34. $(x, 4x)$ and $(x + h, 4(x + h))$

35. $(x, 2x + 3)$ and $(x + h, 2(x + h) + 3)$

36. $(x, 3x - 1)$ and $(x + h, 3(x + h) - 1)$

37.–48. Find an equation of the line containing each pair of points in Exercises 25–36.

APPLICATIONS

❖ **Business and Economics**

49. *Investment.* A person makes an investment of P dollars at 14%. After 1 year, it grows to an amount A.

  a) Show that A is directly proportional to P.

  b) Find A when $P = \$100$.

  c) Find P when $A = \$273.60$.

50. *Profit-and-loss analysis.* A ski manufacturer is planning a new line of skis. For the first year, the fixed costs for setting up the new production line are $22,500. The variable costs for producing each pair of skis are estimated at $40. The sales department projects that 3000 pairs can be sold during the first year at a price of $85 per pair.

  a) Formulate a function $C(x)$ for the total cost of producing x pairs of skis.

  b) Formulate a function $R(x)$ for the total revenue from the sale of x pairs of skis.

  c) Formulate a function $P(x)$ for the total profit from the production and sale of x pairs of skis.

  d) What profit or loss will the company realize if the expected sales of 3000 pairs occurs?

  e) How many pairs must the company sell in order to break even?

51. *Profit-and-loss analysis.* Boxowitz, Inc., a computer firm, is planning to sell a new minicalculator. For the first year, the fixed costs for setting up the new production line are $100,000. The variable costs for producing each calculator are estimated at $20. The sales department projects that 150,000 calculators can be sold during the first year at a price of $45 each.

  a) Formulate a function $C(x)$ for the total cost of producing x calculators.

  b) Formulate a function $R(x)$ for the total revenue from the sale of x calculators.

  c) Formulate a function $P(x)$ for the total profit from the production and sale of x calculators.

  d) What profit or loss will the firm realize if the expected sales of 150,000 calculators occurs?

  e) How many calculators must the firm sell in order to break even?

52. *Straight-line depreciation.* A company buys an office machine for $5200 on January 1 of a given year. The machine is expected to last for 8 years, at the end of which time its *trade-in value*, or *salvage value* will be $1100. If the company figures the decline in value to be the same each year, then the *book value*, or *salvage value*, after t years, $0 \leqslant t \leqslant 8$, is given by the linear function

$$V(t) = C - t\left(\frac{C - S}{N}\right),$$

where C = the original cost of the item ($5200), N = the number of years of expected life (8), and S = the salvage value ($1100).

  a) Find the linear function for the straight-line depreciation of the office machine.

  b) Find the salvage value after 0 years, 1 year, 2 years, 3 years, 4 years, 7 years, and 8 years.

53. *Profit-and-loss analysis.* A college student decides to mow lawns in the summer. The initial cost of the lawnmower is $250. Gasoline and maintenance costs are $1 per lawn.

  a) Formulate a function $C(x)$ for the total cost of mowing x lawns.

  b) The student determines that the total-profit function for the lawnmowing business is given by $P(x) = 9x - 250$. Find a function for the total revenue from mowing x lawns. How much does the student charge per lawn?

  c) How many lawns must the student mow before making a profit?

54. *Sales commissions.* A person applying for a sales position is offered alternative salary plans:

Plan A: A base salary of $600 per month plus a commission of 4% of the gross sales for the month.

Plan B: A base salary of $700 per month plus a commission of 6% of the gross sales for the month in excess of $10,000.

a) For each plan, formulate a function that expresses monthly earnings as a function of gross sales x.

b) For what values of gross sales is plan B preferable?

❖ **Life and Physical Sciences**

55. *Energy conservation.* The R-factor of home insulation is directly proportional to its thickness T.

a) Find an equation of variation if $R = 12.51$ when $T = 3$ in.

b) What is the R-factor for insulation that is 6 in. thick?

56. *Nerve impulse speed.* Impulses in nerve fibers travel at a speed of 293 ft/sec. The distance D traveled in t sec is given by $D = 293t$. How long would it take an impulse to travel from the brain to the toes of a person who is 6 ft tall?

57. *Brain weight.* The weight B of a human's brain is directly proportional to his or her body weight W.

a) It is known that a person who weighs 200 lb has a brain that weighs 5 lb. Find an equation of variation expressing B as a function of W.

b) Express the variation constant as a percent and interpret the resulting equation.

c) What is the weight of the brain of a person who weighs 120 lb?

58. *Muscle weight.* The weight M of the muscles in a human is directly proportional to his or her body weight W.

a) It is known that a person who weighs 200 lb has 80 lb of muscles. Find an equation of variation expressing M as a function of W.

b) Express the variation constant as a percent and interpret the resulting equation.

c) What is the muscle weight of a person who weighs 120 lb?

59. *Stopping distance on glare ice.* The stopping distance (at some fixed speed) of regular tires on glare ice is given by a linear function of the air temperature F,

$$D(F) = 2F + 115,$$

where $D(F)$ = the stopping distance, in feet, when the air temperature is F, in degrees Fahrenheit.

a) Find $D(0°)$, $D(-20°)$, $D(10°)$, and $D(32°)$.

b) Graph $D(F)$.

c) Explain why the domain should be restricted to the interval $[-57.5°, 32°]$.

60. *Spread of an organism.* A certain kind of organism is released over an area of 2 mi^2. It grows and spreads over more area. The area covered by the organism after time t is given by a linear function

$$A(t) = 1.1t + 2,$$

where $A(t)$ = the area covered, in square miles, after time t, in years.

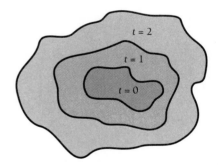

The muscle weight is directly proportional to body weight.

a) Find $A(0)$, $A(1)$, $A(4)$, $A(10)$.

b) Graph $A(t)$.

c) Why should the domain be restricted to the interval $[0, \infty)$?

61. *Estimating heights.* An anthropologist can use certain linear functions to estimate the height of a male or female, given the length of certain bones. The *humerus* is the bone from the elbow to the shoulder. Let $x =$ the length of the humerus in centimeters. Then the height, in centimeters, of a male with a humerus of length x is given by

$$M(x) = 2.89x + 70.64.$$

The height, in centimeters, of a female with a humerus of length x is given by

$$F(x) = 2.75x + 71.48.$$

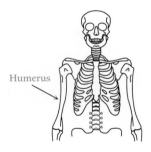

Humerus

A 45-cm humerus was uncovered in a ruins.

a) If we assume it was from a male, how tall was he?

b) If we assume it was from a female, how tall was she?

❖ **Social Sciences**

62. *Urban population.* The population of a town is P. After a growth of 2%, its new population is N.

a) Assuming that N is directly proportional to P, find an equation of variation.

b) Find N when $P = 200,000$.

c) Find P when $N = 367,200$.

63. *Median age of women at first marriage.* Our society is marrying at a later age. The median age of women at first marriage can be approximated by the linear function

$$A(t) = 0.08t + 19.7,$$

where $A(t) =$ the median age of women at first marriage the tth year after 1950. Thus, $A(0)$ is the median age of women at first marriage in the year 1950, $A(30)$ is the median age in 1980, and so on.

a) Find $A(0)$, $A(1)$, $A(10)$, $A(30)$, and $A(40)$.

b) What will be the median age of women at first marriage in 1996?

c) Graph $A(t)$.

COMPUTER-GRAPHING CALCULATOR EXERCISES

THE CALCULUS EXPLORER:
PowerGrapher

Graph some of the total-revenue, total-cost, and total-profit functions in this exercise set using the same set of axes. Identify regions of profit and loss.

1.5

OTHER TYPES OF FUNCTIONS

Quadratic Functions

OBJECTIVES

a)   Graph a given function.

b)   Convert from radical notation to fractional exponents and from fractional exponents to radical notation.

c)   Simplify expressions with fractional exponents.

d)   Determine the domain of a rational function and graph certain rational functions.

e)   Find the equilibrium point, given a supply function and a demand function.

DEFINITION

A *quadratic function* f is given by

$$f(x) = ax^2 + bx + c, \quad \text{where } a \neq 0.$$

We have already considered some such functions—for example, $f(x) = x^2$ and $g(x) = x^2 - 1$. We graph quadratic functions using the following information.

The graph of a quadratic function $f(x) = ax^2 + bx + c$ is called a *parabola*.

a)   It is always a cup-shaped curve, like those in Examples 1 and 2.

b)   It has a turning point, or *vertex*, at a point whose first coordinate is given by

$$x = -\frac{b}{2a}.$$

c)   It has the vertical line $x = -b/2a$ as a line of symmetry (not part of the graph).

d)   It opens up if $a > 0$ or opens down if $a < 0$.

EXAMPLE 1   Graph: $y = x^2 - 2x - 3$.

Solution   Let us first find the vertex, or turning point. The x-coordinate of the vertex is

$$x = -\frac{b}{2a} = -\frac{-2}{2(1)} = 1.$$

Substituting 1 for x in the equation, we find the second coordinate of the vertex:

$$y = x^2 - 2x - 3 = 1^2 - 2(1) - 3 = 1 - 2 - 3 = -4.$$

The vertex is $(1, -4)$. The vertical line $x = 1$ is the line of symmetry of the graph. We choose some x-values on each side of the vertex, compute y-values, plot the points, and graph the parabola (Fig. 1).

$$y = x^2 - 2x - 3$$

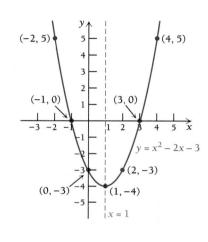

x	y	
1	−4	← Vertex
0	−3	
2	−3	
3	0	
4	5	
−1	0	
−2	5	

FIGURE 1

EXAMPLE 2   Graph: $y = -2x^2 + 10x - 7$.

Solution   Let us first find the vertex, or turning point. The x-coordinate of the vertex is

$$x = -\frac{b}{2a}$$

$$= -\frac{10}{2(-2)} = \frac{5}{2}.$$

Substituting $\frac{5}{2}$ for x in the equation, we find the second coordinate of the vertex:

$$y = -2x^2 + 10x - 7 = -2(\tfrac{5}{2})^2 + 10(\tfrac{5}{2}) - 7$$
$$= -2(\tfrac{25}{4}) + 25 - 7 = \tfrac{11}{2}.$$

The vertex is $(\frac{5}{2}, \frac{11}{2})$, and the line of symmetry is $x = \frac{5}{2}$. We choose some x-values on each side of the vertex, compute y-values, plot the points, and graph the parabola (Fig. 2).

$y = -2x^2 + 10x - 7$

x	y
$\frac{5}{2}$	$\frac{11}{2}$	$\longleftarrow$ Vertex
0	-7
1	1
2	5
3	5
4	1
5	-7

FIGURE 2   ❖

DO EXERCISES 1 AND 2.

First coordinates of points at which a quadratic function intersects the x-axis (x-intercepts), if they exist, can be found by solving the quadratic equation $ax^2 + bx + c = 0$. If real-number solutions exist, they can be found using the *quadratic formula.*

THEOREM 9

The Quadratic Formula

The solutions of any quadratic equation $ax^2 + bx + c = 0$, $a \neq 0$, are given by

$$x = \frac{-b \pm \sqrt{b^2 - 4ac}}{2a}.$$

When solving a quadratic equation, first try to factor and then use the Principle of Zero Products, as we did in Section 1.2. When factoring is not possible or seems difficult, try the quadratic formula. It will always give the solutions. When $b^2 - 4ac < 0$, there are no real-number solutions, but there are solutions in an expanded number system called the

1. Graph $y = x^2 - 6x + 4$.

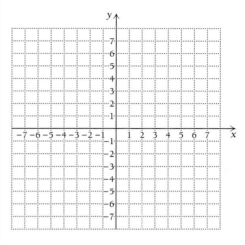

2. Graph $y = -2x^2 + 4x + 1$.

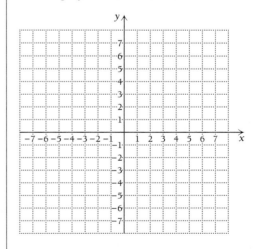

3. Solve $3x^2 = 7 - 2x$.

complex numbers. In this text we will be considering only real-number solutions.

EXAMPLE 3   Solve: $3x^2 - 4x = 2$.

Solution   We first find standard form $ax^2 + bx + c = 0$, and then determine a, b, and c:

$$3x^2 - 4x - 2 = 0,$$
$$a = 3, \quad b = -4, \quad c = -2.$$

We then use the quadratic formula:

$$x = \frac{-b \pm \sqrt{b^2 - 4ac}}{2a}$$

$$= \frac{-(-4) \pm \sqrt{(-4)^2 - 4(3)(-2)}}{2 \cdot 3}$$

$$= \frac{4 \pm \sqrt{16 + 24}}{6} = \frac{4 \pm \sqrt{40}}{6}$$

$$= \frac{4 \pm \sqrt{4 \cdot 10}}{6} = \frac{4 \pm 2\sqrt{10}}{6}$$

$$= \frac{2(2 \pm \sqrt{10})}{2 \cdot 3} = \frac{2 \pm \sqrt{10}}{3}.$$

The solutions are $(2 + \sqrt{10})/3$ and $(2 - \sqrt{10})/3$.   ❖

DO EXERCISE 3.

Polynomial Functions

Linear and quadratic functions are part of a general class of *polynomial functions.*

DEFINITION

A *polynomial function* f is given by

$$f(x) = a_n x^n + a_{n-1} x^{n-1} + \cdots + a_2 x^2 + a_1 x^1 + a_0,$$

where n is a nonnegative integer and $a_n, a_{n-1}, \ldots, a_1, a_0$ are real numbers, called the *coefficients* of the polynomial.

The following are examples of polynomial functions:

$f(x) = -5,$ (A constant function)
$f(x) = 4x + 3,$ (A linear function)
$f(x) = -x^2 + 2x + 3,$ (A quadratic function)
$f(x) = 2x^3 - 4x^2 + x + 1.$ (A cubic function)

In general, graphing polynomial functions other than linear and quadratic functions is difficult. We use calculus to sketch such graphs in Chapter 3. Some **power functions**, such as

$$y = ax^n,$$

are relatively easy to graph.

EXAMPLE 4   Using the same set of axes, graph $y = x^2$ and $y = x^3$.

Solution   We set up a table of values, plot the points, and then draw the graph (Fig. 3).

x	x^2	x^3
-2	4	-8
-1	1	-1
$-\frac{1}{2}$	$\frac{1}{4}$	$-\frac{1}{8}$
0	0	0
$\frac{1}{2}$	$\frac{1}{4}$	$\frac{1}{8}$
1	1	1
2	4	8

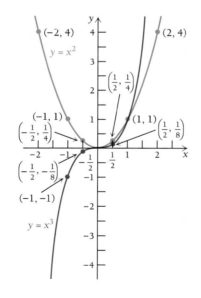

FIGURE 3   ❖

4. Using the same set of axes, graph $y = x^2$ and $y = x^4$.

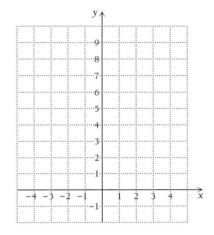

DO EXERCISE 4.

5. Determine the domain of each function.

a) $g(x) = \dfrac{x^2 - 4}{x + 2}$

b) $t(x) = \dfrac{x - 7}{x^2 + 4x - 5}$

c) $k(x) = \dfrac{1}{x - 5}$

Rational Functions

DEFINITION

Functions given by the quotient, or ratio, of two polynomials are called *rational functions*.

The following are examples of rational functions:

$$f(x) = \frac{x^2 - 9}{x - 3},$$

$$g(x) = \frac{x^2 - 16}{x + 4},$$

$$h(x) = \frac{x - 3}{x^2 - x - 2}.$$

The domain of a rational function is restricted to those input values that do not result in division by zero. Thus for f above, the domain consists of all real numbers except 3. To determine the domain of h, we set the denominator equal to 0 and solve:

$$x^2 - x - 2 = 0$$
$$(x + 1)(x - 2) = 0$$
$$x = -1 \quad \text{or} \quad x = 2.$$

Therefore, -1 and 2 are not in the domain. The domain of h consists of all real numbers except -1 and 2.

DO EXERCISE 5.

The graphing of most rational functions is rather complicated and is best dealt with using the tools of calculus that we will develop in Chapters 2 and 3. At this point we will consider graphs that are fairly easy and leave the more complicated graphs for Chapter 3.

EXAMPLE 5   Graph: $f(x) = \dfrac{x^2 - 9}{x - 3}$.

Solution   This particular function can be simplified before we graph the function. We do so by factoring the numerator and removing a factor of 1 as follows:

$$f(x) = \frac{x^2 - 9}{x - 3} = \frac{(x-3)(x+3)}{x-3}$$

$$= \frac{x-3}{x-3} \cdot \frac{x+3}{1} = x + 3, \quad x \neq 3.$$

This simplification assumes that x is not 3. The number 3 is not in the domain because it would result in division by zero. Thus we can express the function as follows:

$$y = f(x) = x + 3, \quad x \neq 3.$$

To find function values, we substitute any value for x other than 3. We make calculations as in the following table and draw the graph (Fig. 4). The open circle at the point (3, 6) indicates that it is not part of the graph.

x	y
1	4
0	3
2	5
-3	0
4	7
-1	2
-2	1

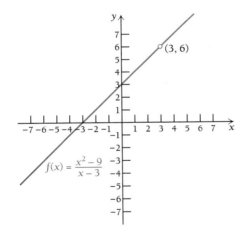

FIGURE 4

DO EXERCISE 6.

One important class of rational functions is given by $y = k/x$.

EXAMPLE 6   Graph: $y = 1/x$.

Solution   We make a table of values, plot the points, and then draw the graph (Fig. 5).

6. Graph $g(x) = \dfrac{x^2 - 4}{x + 2}$.

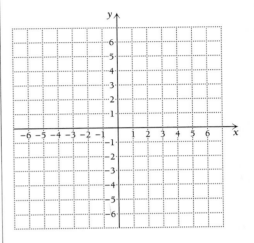

7. Graph $y = \dfrac{-1}{x}$.

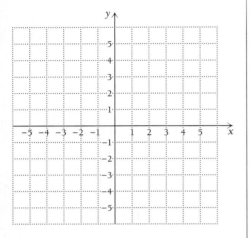

x	-3	-2	-1	$-\frac{1}{2}$	$-\frac{1}{4}$	$\frac{1}{4}$	$\frac{1}{2}$	1	2	3
y	$-\frac{1}{3}$	$-\frac{1}{2}$	-1	-2	-4	4	2	1	$\frac{1}{2}$	$\frac{1}{3}$

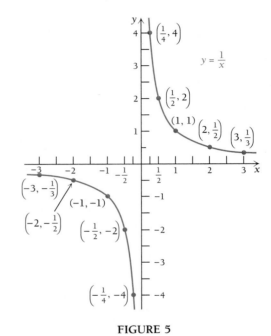

FIGURE 5

In Example 6, note that 0 is not in the domain of the function since it would yield a denominator of 0. The function is decreasing over the intervals $(-\infty, 0)$ and $(0, \infty)$. It is an example of **inverse variation**.

DEFINITION

> y *varies inversely* as x if there is some positive number k such that $y = k/x$. We also say that y is *inversely proportional* to x.

DO EXERCISE 7.

EXAMPLE 7   *Business: Stocks and gold.* Certain economists theorize that stock prices are inversely proportional to the price of gold. That is, when the price of gold goes up, the prices of stock go down; and when the price of gold goes down, the prices of stock go up. Let us assume that the

Dow-Jones Industrial Average, D, an index of the overall price of stock, is inversely proportional to the price of gold, G, in dollars per ounce. One day the Dow-Jones was 2639 and the price of gold was $364 per ounce. What will the Dow-Jones be if the price of gold drops to $300?

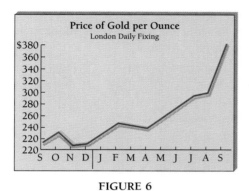

FIGURE 6

8. *Business: Demand.* The quantity sold, x, of a certain kind of radio is inversely proportional to the price, p. It was found that 240,000 radios will be sold when the price per radio is $12.50. How many will be sold if the price is $18.75?

Solution

a)   We know that $D = k/G$, so $2639 = k/364$ and $k = 960{,}596$. Thus, $D = 960{,}596/G$.

b)   We substitute 300 for G and compute D:

$$D = \frac{960{,}596}{300} \approx 3202.$$

Warning! Do not put too much "stock" in the equation of Example 7. It is meant to give us an idea of economic relationships. An equation to predict the stock market accurately has not been found!   ❖

DO EXERCISE 8.

Absolute-Value Functions

The following is an example of an absolute-value function and its graph. The absolute value of a number is its distance from 0 on a number line. We denote the absolute value of a number x as $|x|$.

EXAMPLE 8   Graph: $f(x) = |x|$.

Solution   We make a table of values, plot the points, and then draw the graph (Fig. 7).

9. Graph $f(x) = |x - 1|$. To find an output, take an input, subtract 1 from it, and then take the absolute value.

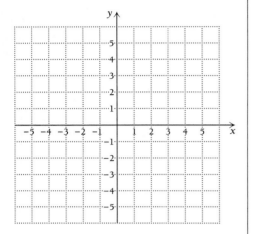

x	-3	-2	-1	0	1	2	3
$f(x)$	3	2	1	0	1	2	3

We can think of this function as being defined piecewise by considering the definition of absolute value:

$$f(x) = |x| = \begin{cases} x, & \text{if } x \geq 0, \\ -x, & \text{if } x < 0. \end{cases}$$

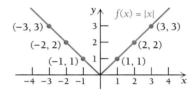

FIGURE 7 ❖

DO EXERCISE 9.

Square-Root Functions

The following is an example of a square-root function and its graph.

EXAMPLE 9 Graph: $f(x) = -\sqrt{x}$.

Solution The domain of this function is just the nonnegative numbers—the interval $[0, \infty)$. You can find approximate values of square roots on your calculator.

We set up a table of values, plot the points, and then draw the graph (Fig. 8).

x	0	1	2	3	4	5	10
$f(x)$, or $-\sqrt{x}$	0	-1	-1.4	-1.7	-2	-2.2	-3.2

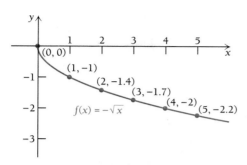

FIGURE 8

DO EXERCISES 10 AND 11.

Power Functions with Fractional Exponents

We are motivated to define fractional exponents so that the laws we discussed in Section 1.1 still hold. For example, if the laws of exponents are to hold, we would have

$$a^{1/2} \cdot a^{1/2} = a^{1/2 + 1/2} = a^1 = a.$$

Thus we are led to define $a^{1/2}$ as $\sqrt{a}$. Similarly, we are led to define $a^{1/3}$ as the cube root of a, $\sqrt[3]{a}$. In general,

$$a^{1/n} = \sqrt[n]{a}, \quad \text{provided } \sqrt[n]{a} \text{ is defined.}$$

Again, if the laws of exponents are to hold, we would have

$$\sqrt[n]{a^m} = (a^m)^{1/n} = (a^{1/n})^m = a^{m/n}.$$

An expression $a^{-m/n}$ is defined by

$$a^{-m/n} = \frac{1}{a^{m/n}} = \frac{1}{\sqrt[n]{a^m}}.$$

EXAMPLE 10   Convert to fractional exponents.

a) $\sqrt[3]{x^2} = x^{2/3}$

b) $\sqrt[4]{y} = y^{1/4}$

c) $\dfrac{1}{\sqrt[3]{b^5}} = \dfrac{1}{b^{5/3}} = b^{-5/3}$

d) $\dfrac{1}{\sqrt{x}} = \dfrac{1}{x^{1/2}} = x^{-1/2}$

e) $\sqrt{x^8} = x^{8/2}$, or x^4

DO EXERCISES 12–17.

---

10. Graph $f(x) = \sqrt{x}$.

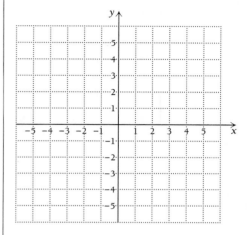

11. Find the domain of $y = \sqrt{2x + 3}$. (*Hint:* Solve $2x + 3 \geqslant 0$.)

Convert to fractional exponents.

12. $\sqrt[4]{t^3}$

13. $\sqrt[5]{y}$

14. $\dfrac{1}{\sqrt[5]{x^2}}$

15. $\dfrac{1}{\sqrt[3]{t}}$

16. $\sqrt{x^6}$

17. $\sqrt{x^7}$

Convert to radical notation.

18. $y^{1/7}$

19. $x^{3/2}$

20. $t^{-3/2}$

21. $b^{-1/2}$

Simplify.

22. $4^{5/2}$

23. $27^{2/3}$

Caribou in their territorial area.

EXAMPLE 11 Convert to radical notation.

a) $x^{1/3} = \sqrt[3]{x}$

b) $t^{6/7} = \sqrt[7]{t^6}$

c) $x^{-2/3} = \dfrac{1}{x^{2/3}} = \dfrac{1}{\sqrt[3]{x^2}}$

d) $e^{-1/4} = \dfrac{1}{e^{1/4}} = \dfrac{1}{\sqrt[4]{e}}$ ❖

DO EXERCISES 18–21.

EXAMPLE 12 Simplify.

a) $8^{5/3} = (8^{1/3})^5 = (\sqrt[3]{8})^5 = 2^5 = 32$

b) $81^{3/4} = (81^{1/4})^3 = (\sqrt[4]{81})^3 = 3^3 = 27$ ❖

DO EXERCISES 22 AND 23.

Earlier when we graphed $f(x) = \sqrt{x}$, we were also graphing $f(x) = x^{1/2}$, or $f(x) = x^{0.5}$. The power functions

$$f(x) = ax^k, \quad k \text{ fractional,}$$

do arise in application. For example, the *home range* of an animal is defined as the region to which the animal confines its movements. It has been hypothesized in statistical studies* that the area H of that region can be approximated using the body weight W of an animal by the function

$$H = W^{1.41}.$$

We can approximate function values using a power key $\boxed{y^x}$ on a calculator.

W	0	10	20	30	40	50
H	0	26	68	121	182	249

Note that

$$H = W^{1.41} = W^{141/100} = \sqrt[100]{W^{141}}.$$

The graph is shown in Fig. 9. Note that this is an increasing function. As body weight increases, the area over which the animal moves increases.

*See J. M. Emlen, *Ecology: An Evolutionary Approach,* p. 200 (Reading, MA: Addison-Wesley, 1973).

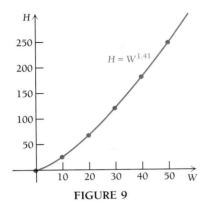

FIGURE 9

Supply and Demand Functions

Supply and demand in economics are modeled by increasing and decreasing functions. Although specific scientific formulas for these concepts are not generally known, the notions of increasing and decreasing help us understand the ideas.

Demand Functions

Look at the following table and graph (Fig. 10). They show the relationship between the price p per bag of sugar and the quantity x of 5-lb bags that the consumer will buy at that price.

Demand schedule

Price (p) per 5-lb bag	Quantity (x) of 5-lb bags, in millions
$5	4
4	5
3	7
2	10
1	15

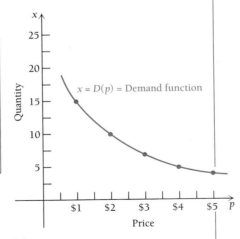

FIGURE 10

Note that as the price per bag increases, the quantity demanded by the consumer decreases; and as the price per bag decreases, the quantity demanded by the consumer increases. Thus it is natural to think of x as a function of p,

$$x = D(p),$$

and to graph the function with the price p on the horizontal axis and the quantity x on the vertical axis. Thus, for a **demand function** D, $D(p)$ is the quantity x of units demanded by the consumer when the price is p.

In some situations it may be appropriate to consider the demand function as $D(x) = p$, where price is a function of the quantity demanded.

Supply Functions

Look at the following table and graph (Fig. 11). They show the relationship between the price p per bag of sugar and the quantity x of 5-lb bags that the seller is willing to supply, or sell, at that price.

Supply schedule

Price (p) per 5-lb bag	Quantity (x) of 5-lb bags, in millions
$1	0
2	10
3	16
4	20
5	22

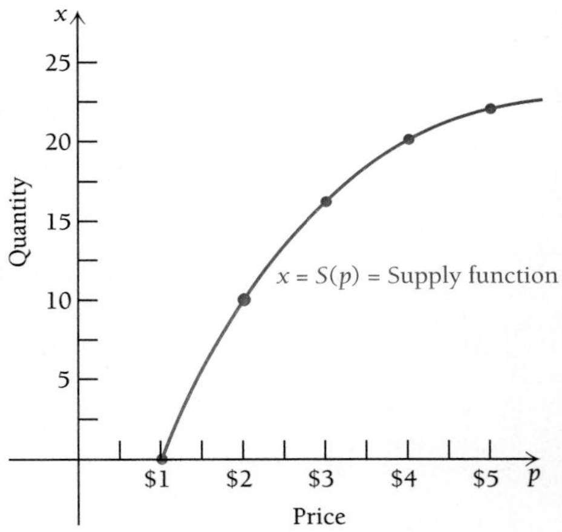

FIGURE 11

Note that as the price per bag increases, the more the seller is willing to supply; and as the price per bag decreases, the less the seller is willing to supply. Thus it is natural to think of x as a function of p,

$$x = S(p),$$

and to graph the function with the price p on the horizontal axis and the quantity x on the vertical axis. Thus, for a **supply function** S, $S(p)$ is the quantity x of items that the seller is willing to supply, or sell, at price p.

In some situations it may be appropriate to consider the supply function as $S(x) = p$, where price is a function of the quantity supplied.

Let us now look at these curves together (Fig. 12). As price increases, supply increases and demand decreases; and as price decreases, supply decreases and demand increases. The point of intersection of the two curves (p_E, x_E) is called the **equilibrium point**. The equilibrium price p_E (in this case, $2 per bag) is the point at which the amount x_E (in this case, 10 million bags) that the seller willingly supplies is the same as the amount that the consumer willingly demands. The situation is analogous to a buyer and seller haggling over the sale of an item. The equilibrium point, or selling price, is what they finally agree on.

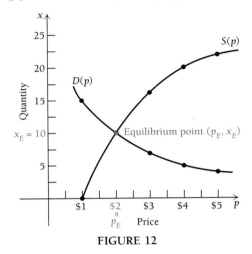

FIGURE 12

EXAMPLE 13   *Economics: Equilibrium point.* Find the equilibrium point for the demand and supply functions

$$D(p) = 300 - 49p \quad \text{and} \quad S(p) = 100 + p.$$

Solution   To find the equilibrium point, we set $D(p) = S(p)$ and solve:

$$300 - 49p = 100 + p$$
$$-50p = -200$$
$$p = \frac{-200}{-50} = 4.$$

Thus, $p_E = 4$. To find x_E, we substitute p_E into either $D(p)$ or $S(p)$. We choose $S(p)$:

$$x_E = S(p_E) = S(4) = 100 + 4 = 104.$$

Thus the equilibrium quantity is 104 units and the equilibrium point is ($4, 104).

DO EXERCISE 24.

24. *Economics: Equilibrium point.* Find the equilibrium point for the demand and supply functions

$$D(p) = 1000 - 60p$$

and

$$S(p) = 200 + 4p.$$

EXERCISE SET 1.5

Using the same set of axes, graph the pair of equations.

1. $y = \frac{1}{2}x^2$ and $y = -\frac{1}{2}x^2$
2. $y = \frac{1}{4}x^2$ and $y = -\frac{1}{4}x^2$
3. $y = x^2$ and $y = (x - 1)^2$
4. $y = x^2$ and $y = (x - 3)^2$
5. $y = x^2$ and $y = (x + 1)^2$
6. $y = x^2$ and $y = (x + 3)^2$
7. $y = |x|$ and $y = |x + 3|$
8. $y = |x|$ and $y = |x + 1|$
9. $y = x^3$ and $y = x^3 + 1$
10. $y = x^3$ and $y = x^3 - 1$
11. $y = \sqrt{x}$ and $y = \sqrt{x + 1}$
12. $y = \sqrt{x}$ and $y = \sqrt{x - 2}$

Graph.

13. $y = x^2 - 4x + 3$
14. $y = x^2 - 6x + 5$
15. $y = -x^2 + 2x - 1$
16. $y = -x^2 - x + 6$
17. $y = \dfrac{2}{x}$
18. $y = \dfrac{3}{x}$
19. $y = \dfrac{-2}{x}$
20. $y = \dfrac{-3}{x}$
21. $y = \dfrac{1}{x^2}$
22. $y = \dfrac{1}{x - 1}$
23. ▣ $y = \sqrt[3]{x}$
24. $y = \dfrac{1}{|x|}$
25. $f(x) = \dfrac{x^2 - 9}{x + 3}$
26. $g(x) = \dfrac{x^2 - 4}{x - 2}$
27. $f(x) = \dfrac{x^2 - 1}{x - 1}$
28. $g(x) = \dfrac{x^2 - 25}{x + 5}$

Solve.

29. $x^2 - 2x = 2$
30. $x^2 - 2x + 1 = 5$
31. $x^2 + 6x = 1$
32. $x^2 + 4x = 3$
33. $4x^2 = 4x + 1$
34. $-4x^2 = 4x - 1$
35. $3y^2 + 8y + 2 = 0$
36. $2p^2 - 5p = 1$

Convert to fractional exponents.

37. $\sqrt{x^3}$
38. $\sqrt{x^5}$
39. $\sqrt[5]{a^3}$
40. $\sqrt[4]{b^2}$
41. $\sqrt[7]{t}$
42. $\sqrt[8]{c}$
43. $\dfrac{1}{\sqrt[3]{t^4}}$
44. $\dfrac{1}{\sqrt[5]{b^6}}$
45. $\dfrac{1}{\sqrt{t}}$
46. $\dfrac{1}{\sqrt{m}}$
47. $\dfrac{1}{\sqrt{x^2 + 7}}$
48. $\sqrt{x^3 + 4}$

Convert to radical notation.

49. $x^{1/5}$
50. $t^{1/7}$
51. $y^{2/3}$
52. $t^{2/5}$
53. $t^{-2/5}$
54. $y^{-2/3}$
55. $b^{-1/3}$
56. $b^{-1/5}$
57. $e^{-17/6}$
58. $m^{-19/6}$
59. $(x^2 - 3)^{-1/2}$
60. $(y^2 + 7)^{-1/4}$

Simplify.

61. $9^{3/2}$
62. $16^{5/2}$
63. $64^{2/3}$
64. $8^{2/3}$
65. $16^{3/4}$
66. $25^{5/2}$

Determine the domain of the function.

67. $f(x) = \dfrac{x^2 - 25}{x - 5}$
68. $f(x) = \dfrac{x^2 - 4}{x + 2}$
69. $f(x) = \dfrac{x^3}{x^2 - 5x + 6}$
70. $f(x) = \dfrac{x^4 + 7}{x^2 + 6x + 5}$
71. $f(x) = \sqrt{5x + 4}$
72. $f(x) = \sqrt{2x - 6}$

APPLICATIONS

❖ **Business and Economics**

Find the equilibrium point for the given demand and supply functions.

73. $D(p) = \dfrac{8 - p}{2}$, $S(p) = p - 2$

74. $D(p) = 800 - 43p$, $S(p) = 210 + 16p$
75. $D(p) = 1000 - 10p$, $S(p) = 250 + 5p$
76. $D(p) = 8800 - 30p$, $S(p) = 7000 + 15p$

77. $D(p) = \dfrac{5}{p}, S(p) = \dfrac{p}{5}$

78. $D(p) = \dfrac{4}{p}, S(p) = \dfrac{p}{4}$

79. $D(p) = (p - 3)^2, 0 \leq p \leq 3;$
$S(p) = p^2 + 2p + 1$

80. $D(p) = (p - 4)^2, 0 \leq p \leq 4;$
$S(p) = p^2 + 2p + 6$

81. $D(p) = 5 - p, 0 \leq p \leq 5; S(p) = \sqrt{p + 7}$

82. $D(p) = 7 - p, 0 \leq p \leq 7; S(p) = 2\sqrt{p + 1}$

83. *Stock dividends and prime rate.* It is theorized that the dividends paid on utilities stocks are inversely proportional to the prime (interest) rate. Recently, the dividends D on the stock of Indianapolis Power and Light were $2.09 per share and the prime rate was 19%. The prime rate R dropped to 17.5%. What dividends would be paid if the assumption of inverse proportionality is correct?

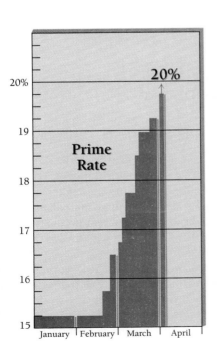

❖ **Life and Physical Sciences**

84. (📱 with $\boxed{y^x}$ key) *Territorial area.* The *territorial area* of an animal is defined to be its defended region, or exclusive region. For example, a lion has a certain region over which it is ruler. It has been hypothesized in statistical studies (see the footnote on p. 64) that the area T of that region can be approximated using the body weight W of the animal by the power function

$$T = W^{1.31}.$$

Complete the table of approximate function values and graph the function.

W	0	10	20	30	40	50	100	150
T	0	20						

85. *Fruit stacking.* The number of oranges in a pile of the type shown in the chapter opening photograph is approximated by the function

$$f(x) = \frac{1}{6}x^3 + \frac{1}{2}x^2 + \frac{1}{3}x,$$

where $f(x)$ is the number of oranges and x is the number of layers. Each layer is an equilateral triangle. Find the number of oranges when the number of layers is 7, 10, and 12.

86. *Life science: Pollution control.* Pollution control has become a very important concern in all countries. If controls are not put in force, it has been predicted that the function

$$P = 1000t^{5/4} + 14{,}000$$

will describe the average pollution, in particles of pollution per cubic centimeter, in most cities at time *t*, in years, where $t = 0$ corresponds to 1970 and $t = 24$ corresponds to 1994.

a) Predict the pollution in 1970, 1994, and 2000.

b) Graph the function over the interval $[0, 30]$.

An input *a* for which $f(a) = 0$ is called a *zero* of the function. Sketch a graph of each of the following functions and estimate the zeros. Most computer software packages include the capability of estimating zeros. If a graphing calculator is to be used, a ZOOM feature, if available, will be most helpful for estimation.

87. $f(x) = x^3 - 4x$ **88.** $f(x) = x - x^3$

89. $f(x) = x^3 - 9x^2 +$ **90.** $f(x) = x^4 + 4x^3 -$
       $27x + 50$ $36x^2 - 160x + 300$

THE CALCULUS EXPLORER:
PowerGrapher, Bisections, Secants, Newton

Graph some of the functions in this exercise set. You may need to adjust the axis scaling in order to see the behavior of the graph. Do this using the XMIN, XMAX, YMIN, and YMAX commands. Use the root finder program to estimate the roots in exercises 87–90.

1.6

MATHEMATICAL MODELING

OBJECTIVE

a) Use curve fitting to find a model for a set of data; then use the model to make predictions.

What Is a Mathematical Model?

When the essential parts of a problem are described in mathematical language, we say that we have a **mathematical model.** For example, the arithmetic of the natural numbers constitutes a mathematical model for situations in which counting is the essential ingredient. Situations in which calculus can be brought to bear often require the use of equations and functions, and typically there is concern with the way a change in one variable affects a change in another.

Mathematical models are abstracted from real-world situations (Fig. 1). Procedures within the mathematical model then give results that allow one to predict what will happen in that real-world situation. To the extent that these predictions are inaccurate or the results of experimentation do not conform to the model, the model is in need of modification.

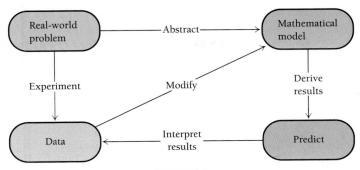

FIGURE 1

Figure 1 shows that mathematical modeling is an ongoing, possibly everchanging, process. This is often the case. For example, finding a mathematical model that will enable accurate prediction of population growth is not a simple problem. Surely any population model one might devise will need to be altered as further relevant information is acquired.

Although models can reveal worthwhile information, one must always be cautious using them. An interesting case in point is a study* showing that world records in *any* running race can be modeled by a linear function. In particular, for the mile run,

$$R = -0.00582x + 15.3476,$$

where R is the world record in minutes and x is the year. Roger Bannister shocked the world in 1954 by breaking the 4-min mile. Had people been aware of this model, they would not have been shocked, for when we substitute 1954 for x, we get

$$R = -0.00582(1954) + 15.3476 = 3.97532 \approx 3{:}58.5.$$

The actual record was 3:59.4. Although this model will continue for 40 to 50 years to be worthwhile in predicting the world record in the mile run, we see that we can't get meaningful answers to some questions. For example, we could use the model to find when the 1-min mile will be broken. We set $R = 1$ and solve for x:

$$1 = -0.00582x + 15.3476$$

$$2465 = x.$$

Most track athletes would assure us that the 1-min mile is beyond human capability. If fact, at the time of this writing, experienced runners think

---

*Ryder, H. W., H. J. Carr, and P. Herget, "Future Performance in Footracing," *Scientific American* **234** (June 1976): 109–119.

Use the model for the world record in the mile run.

1. What will the world record be in 1998?

2. When will the world record be 3:45.0? (*Hint:* 3:45.0 = 3.75.)

the record will never reach 3:40.0, the current world record being 3:46.31. Going to an even further extreme, we see that the model predicts that the 0-min mile will be run in 2637. In conclusion, one must be careful in the use of any model. (You will develop this model in Exercise Set 7.5.)

DO EXERCISES 1 AND 2.

In Section 1.5, we saw an example of a mathematical model using supply and demand functions. In general, the idea is to find a function that fits observations and theoretical reasoning (including common sense) as well as possible. In Section 7.5, we will see how calculus can be used to develop and analyze models. For now we will consider one type of modeling procedure using a somewhat oversimplified procedure that we call *curve fitting*.

Curve Fitting

The four types of function shown in Fig. 2 fit many situations.

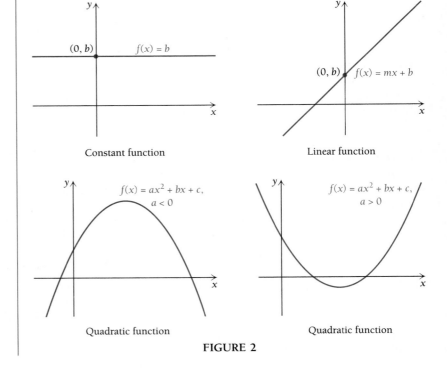

Constant function

Linear function

Quadratic function

Quadratic function

FIGURE 2

The following is a procedure that sometimes works for finding mathematical models.

Curve Fitting

Given a set of data:

1. **Graph the data.**

2. **Look at the data and determine whether a known function seems to fit.**

3. **Find a function that fits the data by using data points to derive the constants.**

The following problem is based on factual data.

EXAMPLE 1 *Business: Average price of a movie ticket.* The table below shows the average price of a movie ticket for various years. As of this writing, the price of most first-run movies ranges from $6 to as much as $7.50, but keep in mind that we are considering average price. The average price becomes lower due to many factors such as senior citizen discounts, children's prices, and special volume discounts on passes, as well as reduced rates at second-run theaters.

From the given set of data, (a) find a model and (b) use the model to find the average price of a ticket in the year 2010.

In the table we let t = the number of years since 1985. This means that for 1985, $t = 0$, and for 1990, $t = 5$.

Year (t)	Average price of a movie ticket
1987 ($t = 2$)	$3.83
1988 ($t = 3$)	3.99
1990 ($t = 5$)	4.29

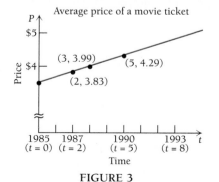

FIGURE 3

3. *Business: Taxes from each dollar earned.* The following table relates the part of each dollar earned that goes for taxes, *A*, to the year *t*, where *t* is measured from 1929. That is, *t* = 0 corresponds to 1929, *t* = 30 corresponds to 1959, and so on.

a) Find a linear model that fits the data using the points (21, 29¢) and (31, 34¢).

b) Use the model to find the part of each dollar earned in 1996 (*t* = 67) that goes for taxes.

Year (*t*)	Part of each dollar earned that goes for taxes, *A*
1929 (*t* = 0)	12¢
1950 (*t* = 21)	29¢
1960 (*t* = 31)	34¢
1977 (*t* = 48)	42¢

Solution

a) From the graph shown in Fig. 3, it looks as though a linear function fits these data fairly well. We consider the linear model

$$P = mt + b, \tag{1}$$

where *P* = the average price of a movie ticket *t* years since 1985. To derive the constants (or parameters) *m* and *b*, we choose two data points. Although this choice is subjective, you should not pick a point that deviates greatly from the general pattern. We choose the points (3, $3.99) and (5, $4.29). Since the points are to be solutions of Eq. (1), it follows that

$$3.99 = m \cdot 3 + b, \quad \text{or} \quad 3.99 = 3m + b, \tag{2}$$
$$4.29 = m \cdot 5 + b, \quad \text{or} \quad 4.29 = 5m + b. \tag{3}$$

This is a system of equations. We can subtract each side of Eq. (2) from Eq. (3) to eliminate *b*:

$$0.30 = 2m.$$

Then we have

$$m = \frac{0.30}{2} = 0.15.$$

Substituting 0.15 for *m* in Eq. (3), we get

$$4.29 = 5(0.15) + b$$
$$4.29 = 0.75 + b$$
$$3.54 = b.$$

Substituting these values of *m* and *b* into Eq. (1), we get the function (model) given by

$$P = 0.15t + 3.54. \tag{4}$$

b) The average price of a movie ticket in the year 2010 is found by letting *t* = 25 in Eq. (4):

$$P = 0.15(25) + 3.54 = \$7.29. \qquad \text{❖}$$

DO EXERCISE 3.

To repeat, the technique of curve fitting illustrated here is an over-simplified method of finding models. Other techniques such as the least-squares method (Section 7.5) and computer simulations are used for more thorough research.

EXAMPLE 2   *Life science: Hours of sleep and death rate.* The table below and the graph in Fig. 4 relate death rate y, in deaths per 100,000 males, to the average number of hours of sleep, x. For the given set of factual data, (a) find a model and (b) use the model to find the death rate for those who sleep 2 hr, 8 hr, 10 hr.

Average number of hours of sleep (x)	Death rate per 100,000 males (y)
5	1121
7	626
9	967

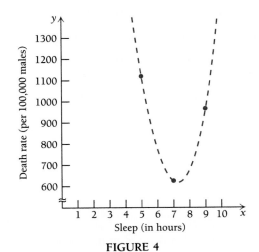

FIGURE 4

Solution

a) From the graph in Fig. 4, it looks as though a quadratic function might fit:

$$y = ax^2 + bx + c. \qquad (1)$$

To derive the constants (or parameters) a, b, and c, we use the three data points $(5, 1121)$, $(7, 626)$, and $(9, 967)$. Since these points are to

4. *General interest: Number of daytime accidents.*

Age (x) of driver, in years	Number of daytime accidents (A) committed by driver of age x
20	420
40	150
50	146
70	400

a)   For the given set of data, graph the data and find a quadratic function that fits the data. Use the data points (20, 420), (40, 150), and (70, 400). First give exact fractional values for *a, b,* and *c.* Then give decimal values rounded to the nearest thousandth.

b)   How many daytime accidents are committed by a driver of age 16?

be solutions to Eq. (1), it follows that

$$1121 = a \cdot 5^2 + b \cdot 5 + c, \quad \text{or} \quad 1121 = 25a + 5b + c,$$
$$626 = a \cdot 7^2 + b \cdot 7 + c, \quad \text{or} \quad 626 = 49a + 7b + c,$$
$$967 = a \cdot 9^2 + b \cdot 9 + c, \quad \text{or} \quad 967 = 81a + 9b + c.$$

We solve this system of three equations in three variables using procedures of algebra and get

$$a = 104.5, \quad b = -1501.5, \quad \text{and} \quad c = 6016.$$

Substituting these values of *a, b,* and *c* into Eq. (1), we get the function (model) given by

$$y = 104.5x^2 - 1501.5x + 6016.$$

b)   The death rate for 2 hr is given by

$$y = 104.5(2)^2 - 1501.5(2) + 6016 = 3431.$$

The death rate for 8 hr is given by

$$y = 104.5(8)^2 - 1501.5(8) + 6016 = 692.$$

The death rate for 10 hr is given by

$$y = 104.5(10)^2 - 1501.5(10) + 6016 = 1451.$$

DO EXERCISE 4.

EXERCISE SET 1.6

For each exercise, parts (a) and (b) are as follows:

a)   Graph the data given.

b)   Look at the data and determine whether one of the four models discussed in this section seems to fit.

APPLICATIONS

❖ **Business and Economics**

1. *Total cost in the clothing business.* Raggs, Ltd., keeps track of its total cost of producing *x* items of a certain suit. These data are shown to the right.

c)   Use the data points (1, $10,030) and (3, $10,094) to find a linear function that fits the data.

Number of suits (x)	Total cost (C) of producing x suits
0	$10,000
1	10,030
2	10,059
3	10,094

d) Predict the total cost of producing 4 suits and 10 suits.

e) Use the data points (0, $10,000) and (2, $10,059) to find a linear function that fits the data.

f) Use the model of part (e) to predict the total cost of producing 4 suits and 10 suits.

2. *Total cost in the pizza business.* Pizza, Unltd., keeps track of its total cost of producing x pizzas. These data are shown below.

Number of pizzas (x)	Total cost (C) of producing x pizzas
0	$1000.00
1	1001.00
2	1001.80
3	1002.50

c) Use the data points (1, $1001.00) and (3, $1002.50) to find a linear function that fits the data.

d) Predict the total cost of producing 4 pizzas and 100 pizzas.

e) Use the data points (0, $1000.00) and (2, $1001.80) to find a linear function that fits the data.

f) Use the model of part (e) to predict the total cost of producing 4 pizzas and 100 pizzas.

3. *Total sales.*

Year (t)	Total sales (S), in dollars
1	$100,310
2	100,290
3	100,305
4	100,280

c) Use the data points (1, $100,310) and (2, $100,290) to find a linear function that fits the data.

d) This data set approximates a constant function. What procedure, apart from that of part (c), could you use to find the constant?

4. *Total sales.*

Year (t)	Sales (S), in dollars
1	$10,000
2	21,000
3	27,000
4	37,000

c) Use the data points (1, $10,000) and (4, $37,000) to find a linear function that fits the data.

d) Use the model to predict the sales of the company in the fifth year.

❖ **General Interest**

The problems in Exercises 5 and 6 are based on factual data.

5. *VCR counter readings.* The instruction booklet for a video cassette recorder (VCR) includes a table relating the counter readings and the time, in hours, that the tape has run.

Counter readings (N)	Time (t), that the tape has run
000	0
300	1
500	2
675	3
800	4

c) Use the data points (0, 0), (300, 1), and (500, 2) to find a quadratic equation that fits the data.

d) The counter readings do not give times for programs on the half hour. Use the function to find the counter reading after the tape has run for $\frac{1}{2}$ hr, $1\frac{1}{2}$ hr, and $2\frac{1}{2}$ hr.

6. *Accidents at certain speeds.*

Travel speed (x), in miles per hour	Number (N) of vehicles involved in an accident at nighttime (for every 100 million miles of travel)
20	10,000
30	2,000
40	400
50	250
60	250
70	350
80	1,500

c) Use the data points (20, 10,000), (50, 250), and (80, 1500) to find a quadratic function that fits the data.

d) Use the model to find the number of vehicles involved in an accident at 30 mph.

CHAPTER SUMMARY AND REVIEW

1

TERMS TO KNOW

Base, p. 2
Exponent, p. 2
Factor, p. 2
Distributive law, p. 5
Factoring, p. 6
Compound interest, p. 7
Addition Principle, p. 12
Multiplication Principle, p. 12
Principle of Zero Products, p. 13
Inequality, p. 14
Open interval, p. 16
Closed interval, p. 16
Half–open interval, p. 17
Endpoint, p. 16
Interior point, p. 18
Ordered pair, p. 21
First coordinate, p. 21
Second coordinate, p. 21
Solution, p. 21

Graph, p. 22, 29
Relation, p. 24
Function, p. 24, 27
Input, p. 24
Output, p. 24
Domain, p. 24
Range, p. 24
Vertical-line test, p. 31
Functions defined piecewise, p. 31
Independent variable, p. 33
Dependent variable, p. 33
Slope, p. 39, 41, 45
Direct variation, p. 40
Variation constant, p. 40
Linear function, p. 41
Constant function, p. 42
Slope–intercept equation, p. 42
Point–slope equation, p. 44
Fixed costs, p. 46

Variable costs, p. 46
Total cost, p. 46
Break-even value, p. 48
Quadratic function, p. 53
Parabola, p. 53
Vertex, p. 53
Quadratic formula, p. 55
Polynomial function, p. 56
Power function, p. 57
Rational function, p. 58
Inverse variation, p. 60
Absolute-value function, p. 61
Square-root function, p. 62
Demand function, p. 66
Supply function, p. 66
Equilibrium point, p. 67
Mathematical model, p. 70
Curve fitting, p. 73

REVIEW EXERCISES

These review exercises are for test preparation. They can also be used as a lengthened practice test. Answers are at the back of the book. The answers also contain bracketed section references, which tell you where to restudy if your answer is incorrect.

1. Rename without an exponent: $(2.01)^2$.

2. Divide: $\dfrac{y^{-2}}{y^{-5}}$.

3. Multiply: $y^{-2} \cdot y^{-5}$.

4. Simplify: $(3t^{-2}m^4)^5$.

Multiply.

5. $(3x - t)^2$

6. $(3x - t)(3x + t)$

7. $(3x - t)(4x + 5t)$

Factor.

8. $x^2 - 2x - 48$

9. $25x^2 - 16t^2$

10. $a^2 - 8ab + 16b^2$

11. $21x^2 - 19x + 4$

12. *Business: Compound interest.* Suppose $1100 is invested at 10%, compounded semiannually. How much is in the account at the end of 1 year?

13. *Business: Compound interest.* Suppose $4000 is invested at 12%, compounded annually. How much is in the account at the end of 2 years?

Solve.

14. $3 - 6x \leqslant 5x - 8$

15. $3 - 6x = 5x - 8$

16. $16x^2 - 25 = 0$

17. $x(x - 3)(2x + 5) = 0$

18. Write interval notation for the set of all numbers x such that $-6 < x \leqslant 1$.

19. Write interval notation for the set of all positive numbers.

A function f is given by $f(x) = 2x^2 - x + 3$. Find each of the following.

20. $f(-2)$

21. $f(1 + h)$

22. $f(0)$

A function f is given by $f(x) = (1 - x)^2$. Find each of the following.

23. $f(-5)$

24. $f(2 - h)$

25. $f(4)$

Graph.

26. $y = |x + 1|$

27. $f(x) = (x - 2)^2$

28. $f(x) = \dfrac{x^2 - 16}{x + 4}$

Determine whether each of the following is a graph of a function.

29.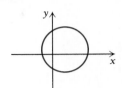

30.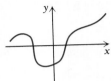

31.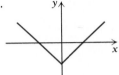

Graph.

32. $x = -2$

33. $y = 4 - 2x$

34. Write an equation of the line containing the points $(4, -2)$ and $(-7, 5)$.

35. Write an equation of the line with slope 8, containing the point $(\frac{1}{2}, 11)$.

36. For the linear equation $y = 3 - \frac{1}{6}x$, find the slope and the y-intercept.

37. *Life science: Muscle weight and body weight.* The weight M of muscles in a human is directly proportional to its body weight W. It is known that a person who weighs 150 lb has 60 lb of muscles. What is the muscle weight of a person who weighs 180 lb?

38. *Business: Profit-and-loss analysis.* A furniture manufacturer has fixed costs of $80,000 for producing a certain type of classroom chair for universities. The variable cost for producing x chairs is $16x$ dollars. The chairs will be sold at a price of $28 each.

   a) Formulate a function $R(x)$ for the total revenue from the sale of x chairs.

   b) Formulate a function $C(x)$ for the total cost of producing x chairs.

   c) Formulate a function $P(x)$ for the total profit from the production and sale of x chairs.

   d) What profit or loss will the company realize from expected sales of 8000 chairs?

e) How many chairs must the company sell in order to break even?

39. Convert to radical notation: $y^{1/6}$.

40. Convert to fractional exponents: $\sqrt[20]{x^3}$.

41. Simplify: $25^{3/2}$.

Find the domain of the function.

42. $f(x) = \sqrt{18 - 6x}$

43. $g(x) = \dfrac{1}{x^2 - 1}$

44. *Economics: Equilibrium point.* Find the equilibrium point if $D(p) = p^2 - 11p + 51$, $0 \leqslant p \leqslant 5$, and $S(p) = p^2 + 18$.

45. Derive a linear function that fits the data points $(6, -1)$ and $(0, 10)$.

46. Derive a quadratic function that fits the data points $(1, 4)$, $(3, 8)$, and $(-1, 8)$.

47. Graph:
$$f(x) = \begin{cases} x - 5, & \text{for } x > 2, \\ x + 3, & \text{for } x \leqslant 2. \end{cases}$$

SYNTHESIS EXERCISE

48. Simplify: $(64^{5/3})^{-1/2}$.

EXERCISES FOR THINKING AND WRITING

49. Explain the idea of a function in as many ways as possible.

50. Explain why the slope of a vertical line is not defined but the slope of a horizontal line is 0.

51. Explain and compare the situations in which one would use the slope-intercept equation rather than the point–slope equation.

52. Discuss and relate the concepts of fixed cost, total cost, total revenue, and total profit.

CHAPTER TEST

1. Rename without a negative exponent: e^{-k}.

2. Divide: $\dfrac{e^{-5}}{e^8}$.

3. Multiply: $(x + h)^2$.

4. Factor: $25x^2 - t^2$.

5. *Business: Compound interest.* A person makes an investment at 13%, compounded annually. It has grown to $1039.60 at the end of 1 year. How much was originally invested?

6. Solve: $-3x < 12$.

7. A function is given by $f(x) = x^2 - 4$. Find (a) $f(-3)$ and (b) $f(x + h)$.

8. What are the slope and the y-intercept of $y = -3x + 2$?

9. Find an equation of the line with slope $\frac{1}{4}$, containing the point $(8, -5)$.

10. Find the slope of the line containing the points $(-2, 3)$ and $(-4, -9)$.

11. *Life science: Body fluids.* The weight F of fluids in a human is directly proportional to body weight W. It is known that a person who weighs 180 lb has 120 lb of fluids. Find an equation of variation expressing F as a function of W.

12. *Business: Profit-and-loss analysis.* A record company has fixed costs of $10,000 for producing a record master. Thereafter, the variable costs are $0.50 per record for

duplicating from the record master. Revenue from each record is expected to be $1.30.

a) Formulate a function $C(x)$ for the total cost of producing x records.

b) Formulate a function $R(x)$ for the total revenue from the sale of x records.

c) Formulate a function $P(x)$ for the total profit from the production and sale of x records.

d) How many records must the company sell in order to break even?

13. *Economics: Equilibrium point.* Find the equilibrium point for the demand and supply functions

$$D(p) = (p - 7)^2, 0 \leqslant p \leqslant 7, \quad \text{and} \quad S(p) = p^2 + p + 4.$$

14. Graph: $y = 4/x$.

15. Convert to fractional exponents: $1/\sqrt{t}$.

16. Convert to radical notation: $t^{-3/5}$.

17. Graph: $f(x) = \dfrac{x^2 - 1}{x + 1}$.

Determine the domain of the function.

18. $f(x) = \dfrac{x^2 + 20}{(x - 2)(x + 7)}$

19. $f(x) = \sqrt{5x + 10}$

20. Find a linear function that fits the data points $(1, 3)$ and $(2, 7)$.

21. Find a quadratic function that fits the data points $(1, 5)$, $(2, 9)$, and $(3, 4)$.

22. Write interval notation for this graph.

23. Graph:

$$f(x) = \begin{cases} x^2 + 2, & \text{for } x \geqslant 0, \\ x^2 - 2, & \text{for } x < 0. \end{cases}$$

SYNTHESIS EXERCISE

24. *Economics: Demand.* The demand function for a product is given by

$$x = D(p) = 800 - p^3, \$0 \leqslant p \leqslant \$9.28.$$

a) Find the number of units sold when the price per unit is $6.50.

b) Find the price per unit when 720 units are sold.

2

DIFFERENTIATION

AN APPLICATION

A firm estimates that it will sell N units of a product after spending a dollars on advertising, where

$$N(a) = -a^2 + 300a + 6$$

and a is measured in thousands of dollars. What is the rate of change of the number of units sold with respect to the amount spent on advertising?

THE MATHEMATICS

The rate of change $N'(a)$ with respect to the amount spent on advertising is given by

$$N'(a) = -2a + 300.$$

This function is a *derivative*.

With this chapter we begin our study of calculus. The first concepts we consider are those of limits and continuity. Then we apply those concepts to establishing the first of the two building blocks of calculus: differentiation. *Differentiation* is a process that takes a formula for a function and derives a formula for another function, called a *derivative,* that allows us to find the slope of the tangent line to a curve at a point. We also find that a derivative can represent an instantaneous rate of change. Throughout the chapter we will learn various techniques for finding derivatives.

2.1

OBJECTIVES

a) Find

$$\lim_{x \to a} f(x),$$

if it exists.

b) Determine whether a graph is that of a continuous function.

c) Determine whether a function is continuous at a given point a.

LIMITS AND CONTINUITY

In this section we give an intuitive (meaning "based on prior and present experience") treatment of two important concepts: *limits* and *continuity*.

Limits

Suppose a defensive football team is on its own 5-yd line and is given a penalty. Such a penalty is half the distance to the goal. Suppose it keeps getting such a penalty. Then the offense moves from the 5-yd line to the $2\frac{1}{2}$-yd line to the $1\frac{1}{4}$-yd line, and so on. Note that it can never score under this premise, but its distance to the goal continues to get closer and closer to 0. We say that the *limit* is 0.

One important aspect of the study of calculus is the analysis of how function values, or outputs, change when the inputs change. Basic to this study is the notion of limit. Suppose the inputs get closer and closer to some number. If the corresponding outputs get closer and closer to a number, that number is called a *limit*.

Consider the function f given by

$$f(x) = 2x + 3.$$

Suppose we select input numbers x closer and closer to the number 4, and look at the output numbers $2x + 3$. Study the following input–output table and the graph shown in Fig. 1.

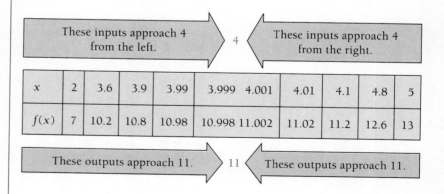

	These inputs approach 4 from the left.				4		These inputs approach 4 from the right.			
x	2	3.6	3.9	3.99	3.999	4.001	4.01	4.1	4.8	5
$f(x)$	7	10.2	10.8	10.98	10.998	11.002	11.02	11.2	12.6	13

These outputs approach 11.	11	These outputs approach 11.

In the table and the graph, we see that as input numbers approach 4 from the left, output numbers approach 11. As input numbers approach 4

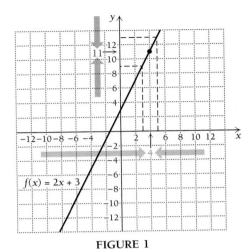

FIGURE 1

from the right, output numbers approach 11. Thus we say:

As *x approaches* 4,   2*x* + 3 *approaches* 11.

An arrow, →, is often used for the word "approaches." Thus the statement above can be written:

As *x* → 4,   2*x* + 3 → 11.

The number 11 is said to be the *limit* of 2*x* + 3 as *x* approaches 4. We can abbreviate this statement as follows:

$$\lim_{x \to 4} (2x + 3) = 11.$$

This is read, "The limit, as *x* approaches 4, of 2*x* + 3 is 11."

DEFINITION

A function *f* has the *limit L* as *x* approaches *a*, written

$$\lim_{x \to a} f(x) = L,$$

if we can get *f*(*x*) as close to *L* as we wish by taking *x* sufficiently close to *a*, but not equal to *a*.

DO EXERCISE 1.

1. Consider $f(x) = 3x - 1$.

   a) Complete this table. (▦ helpful, though not necessary)

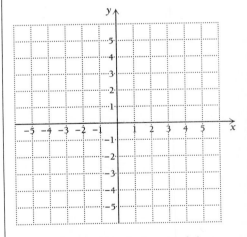

x	f(x)
5	
5.8	
5.9	
5.99	
5.999	
6.001	
6.01	
6.1	
6.4	
7	

   b) Find $\lim_{x \to 6} f(x)$.

   c) Graph $f(x) = 3x - 1$.

Use the graph to find each of the following limits.

   d) $\lim_{x \to -1} f(x)$ e) $\lim_{x \to 2} f(x)$

   f) $\lim_{x \to 0} f(x)$

2. Consider the function $g(x) = \dfrac{1}{x-3}$.

a) Complete this table. (▦ helpful)

Inputs, x	Outputs, $g(x)$
1	
2	
2.9	
2.99	
2.999	
3.001	
3.01	
3.1	
3.8	
4	

3 ← → ?

b) Find $\lim\limits_{x \to 3} g(x)$.

c) Graph $g(x) = \dfrac{1}{x-3}$.

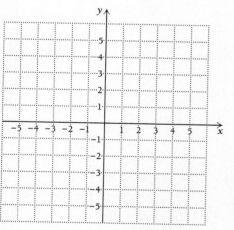

Use the graph to find each of the following limits, if they exist.

d) $\lim\limits_{x \to 3} g(x)$ **e)** $\lim\limits_{x \to 1} g(x)$

f) $\lim\limits_{x \to 4} g(x)$

EXAMPLE 1 Consider the function

$$F(x) = \frac{1}{x}.$$

Find each of the following limits, if they exist.

a) $\lim\limits_{x \to 3} F(x)$ **b)** $\lim\limits_{x \to 0} F(x)$

Solution

a) From the graph shown in Fig. 2, we see that as inputs x approach 3 from either the left or the right, outputs $1/x$ approach $\frac{1}{3}$. We can also check this using an input–output table. Thus we have

$$\lim_{x \to 3} F(x) = \frac{1}{3}.$$

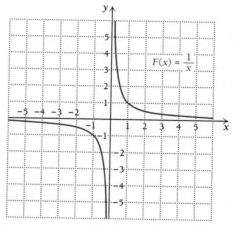

FIGURE 2

b) From either the graph or an input–output table, we see that as inputs x approach 0 from the left, outputs become more and more negative without bound. Similarly, as inputs x approach 0 from the right, outputs get larger and larger without bound. We say that

$$\lim_{x \to 0} \frac{1}{x} \quad \text{does not exist.}$$ ❖

DO EXERCISE 2.

The following is very important to keep in mind when determining whether a limit exists.

> **In order for a limit to exist, both of the limits from the left and the right must exist and be the same.**

We use the notation

$$\lim_{x \to a^+} f(x) \quad \text{to indicate the limit from the right}$$

and

$$\lim_{x \to a^-} f(x) \quad \text{to indicate the limit from the left.}$$

Then in order for a limit to exist, both of the limits above must exist and be the same.

EXAMPLE 2   Consider the function H defined as follows:

$$H(x) = \begin{cases} 2x + 2, & \text{for } x < 1, \\ 2x - 2, & \text{for } x \geq 1. \end{cases}$$

Graph the function and find each of the following limits, if they exist.

a) $\lim_{x \to 1} H(x)$ b) $\lim_{x \to -3} H(x)$

Solution   The graph is shown in Fig. 3.

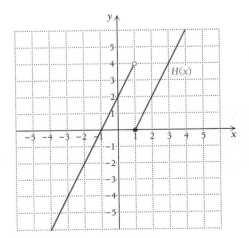

FIGURE 3

3. Consider this graph

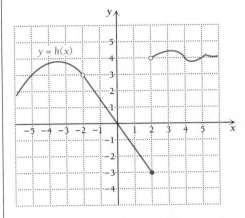

Find each of the following limits, if they exist.

a) $\lim_{x \to 2^+} h(x)$

b) $\lim_{x \to 2^-} h(x)$

c) $\lim_{x \to 2} h(x)$

d) $\lim_{x \to 0^+} h(x)$

e) $\lim_{x \to 0^-} h(x)$

f) $\lim_{x \to 0} h(x)$

g) $\lim_{x \to -2^+} h(x)$

h) $\lim_{x \to -2^-} h(x)$

i) $\lim_{x \to -2} h(x)$

4. Consider this function:

$$f(x) = \begin{cases} 2 - x^2, & \text{for } x \geqslant 0, \\ x^2 - 2, & \text{for } x < 0. \end{cases}$$

a) Graph $y = f(x)$.

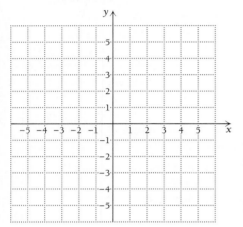

Find each of the following limits, if they exist.

b) $\lim\limits_{x \to 0} f(x)$

c) $\lim\limits_{x \to -2} f(x)$

a) As inputs x approach 1 from the left, outputs $H(x)$ approach 4. Thus the limit from the left is 4. That is,

$$\lim_{x \to 1^-} H(x) = 4.$$

But as inputs x approach 1 from the right, outputs $H(x)$ approach 0. Thus the limit from the right is 0. That is,

$$\lim_{x \to 1^+} H(x) = 0.$$

Since the limit from the left, 4, is not the same as the limit from the right, 0, we say that

$$\lim_{x \to 1} H(x) \quad \text{does not exist.}$$

b) As inputs x approach -3 from the left, outputs $H(x)$ approach -4, so the limit from the left is -4. That is,

$$\lim_{x \to -3^-} H(x) = -4.$$

As inputs x approach -3 from the right, outputs $H(x)$ approach -4, so the limit from the right is -4. That is,

$$\lim_{x \to -3^+} H(x) = -4.$$

Since the limits from the left and from the right exist and are the same, we have

$$\lim_{x \to -3} H(x) = -4. \qquad \qquad ❖$$

DO EXERCISES 3 AND 4. (EXERCISE 3 IS ON THE PREVIOUS PAGE.)

The following is also important in finding limits.

> **The limit at a number *a* *does not depend* on the function value at *a*** even if that function value, $f(a)$, exists. That is, whether or not a limit exists at *a* has nothing to do with the function value $f(a)$.

EXAMPLE 3   Consider the function G defined as follows:

$$G(x) = \begin{cases} 5, & \text{for } x = 1, \\ x + 1, & \text{for } x \neq 1. \end{cases}$$

Graph the function and find each of the following limits, if they exist.

a) $\lim\limits_{x \to 1} G(x)$

b) $\lim\limits_{x \to 3} G(x)$

Solution The graph is shown in Fig. 4.

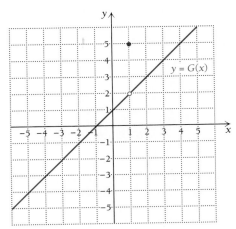

FIGURE 4

a) As inputs x approach 1 from the left, outputs $G(x)$ approach 2, so the limit from the left is 2. As inputs x approach 1 from the right, outputs $G(x)$ approach 2, so the limit from the right is 2. Since the limit from the left, 2, is the same as the limit from the right, 2, we have

$$\lim\limits_{x \to 1} G(x) = 2.$$

Note that the limit, 2, is not the same as the function value at 1, which is $G(1) = 5$.

b) We have $\lim\limits_{x \to 3} G(x) = 4$. We also know that $G(3) = 4$. In this case, the function value and the limit are the same. ❖

DO EXERCISES 5 AND 6. (EXERCISE 6 IS ON THE FOLLOWING PAGE.)

Continuity

When the limit of a function is the same as its function value, it satisfies a condition called **continuity at a point**. We now consider the concept of continuity.

5. Consider this function:

$$f(x) = \begin{cases} -3, & \text{for } x = -2, \\ x^2, & \text{for } x \neq -2. \end{cases}$$

a) Graph $y = f(x)$.

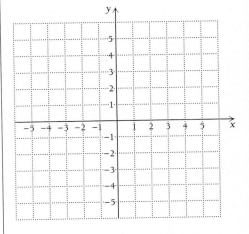

Find each of the following limits, if they exist.

b) $\lim\limits_{x \to -2^+} f(x)$

c) $\lim\limits_{x \to -2^-} f(x)$

d) $\lim\limits_{x \to -2} f(x)$

e) $\lim\limits_{x \to 2^+} f(x)$

f) $\lim\limits_{x \to 2^-} f(x)$

g) Does $\lim\limits_{x \to -2} f(x) = f(-2)$?

h) Does $\lim\limits_{x \to 2} f(x) = f(2)$?

6. Consider this graph.

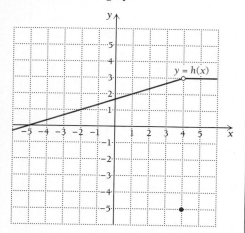

Find each of the following limits, if they exist.

a) $\lim\limits_{h \to 1} h(x)$

b) $\lim\limits_{h \to 4} h(x)$

c)   Does $\lim\limits_{h \to 1} h(x) = h(1)$?

d)   Does $\lim\limits_{h \to 4} h(x) = h(4)$?

Figure 5 shows the graphs of functions that are *continuous* over the whole real line $(-\infty, \infty)$.

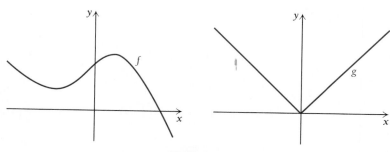

FIGURE 5

Note that there are no "jumps" or holes in the graphs. For now we will use a somewhat intuitive definition of continuity, which we will refine. We say that a function is **continuous over, or on, some interval** of the real line if its graph can be traced without lifting the pencil from the paper. If a function has one point at which it fails to be continuous, then we say that it is *not continuous* over the whole real line. Figures 6–8 show graphs of functions that are *not* continuous over the whole real line. For functions G and H (Figs. 7 and 8, respectively), the open circle indicates that the point is not part of the graph.

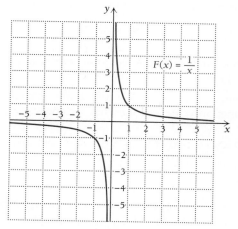

FIGURE 6

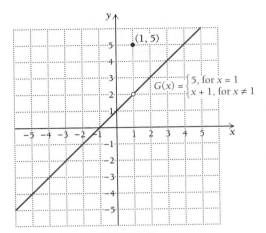

$$G(x) = \begin{cases} 5, \text{ for } x = 1 \\ x + 1, \text{ for } x \neq 1 \end{cases}$$

FIGURE 7

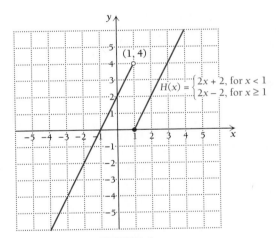

$$H(x) = \begin{cases} 2x + 2, \text{ for } x < 1 \\ 2x - 2, \text{ for } x \geq 1 \end{cases}$$

FIGURE 8

A continuous curve.

In each case, the graph *cannot* be traced without lifting the pencil from the paper. However, each case represents a different situation. Let us discuss why each case fails to be continuous over the whole real line.

The function F (Fig. 6) fails to be continuous over the *whole* real line $(-\infty, \infty)$. Since F is not defined at $x = 0$, the point $x = 0$ is not part of the domain, so $F(0)$ does not exist and there is no point $(0, F(0))$ on the graph. Thus there is no point to trace at $x = 0$. However, F is continuous on the intervals $(-\infty, 0)$ and $(0, \infty)$.

7. Determine whether each of the following functions is continuous.

a)

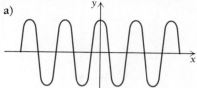

b)

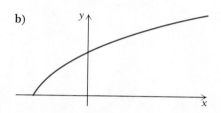

c)

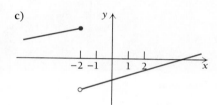

d)

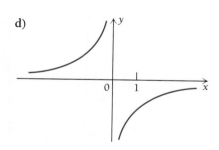

8. a) Decide whether the function in Margin Exercise 7(c) is continuous at −2; at 1.

b) Decide whether the function in Margin Exercise 7(d) is continuous at 1; at 0.

The function G (Fig. 7) is not continuous over the whole real line since it is not continuous at $x = 1$. Let us trace the graph of G starting to the left of $x = 1$. As x approaches 1, $G(x)$ seems to approach 2. However, at $x = 1$, $G(x)$ *jumps* up to 5, whereas to the right of $x = 1$, $G(x)$ *drops* back to some value close to 2. Thus G is discontinuous at $x = 1$.

The function H (Fig. 8) is not continuous over the whole real line since it is not continuous at $x = 1$. Let us trace the graph of H starting to the left of $x = 1$. As x approaches 1, $H(x)$ approaches 4. However, at $x = 1$, $H(x)$ *drops* down to 0; and just to the right of $x = 1$, $H(x)$ is close to 0. Since the limit from the left is 4 and the limit from the right is 0, we see that $\lim_{x \to 1} f(x)$ does not exist. Thus, $H(x)$ is discontinuous at $x = 1$.

DO EXERCISES 7 AND 8.

Limits and Continuity

After working through Example 3, we might ask, "When can we substitute to find a limit?" The answer lies in the following definition, which also provides a more formal definition of continuity.

DEFINITION

A function f is *continuous at $x = a$* if:

a) $f(a)$ exists,   (The output $f(a)$ exists.)

b) $\lim_{x \to a} f(x)$ exists, and   (The limit as $x \to a$ exists.)

c) $\lim_{x \to a} f(x) = f(a)$.   (The limit is the same as the output.)

A function is *continous over an interval I* if it is continuous at each point in I.

Note that we will also use the notation "$\lim_{x \to a} f(x)$" for a limit.

EXAMPLE 4   Determine whether the function given by

$$f(x) = 2x + 3$$

is continuous at $x = 4$.

Solution   This function is continuous at 4 because

a) $f(4)$ exists,   ($f(4) = 11$)

b) $\lim_{x \to 4} f(x)$ exists,   [$\lim_{x \to 4} f(x) = 11$ (as shown earlier)],

c) $\lim_{x \to 4} f(x) = 11 = f(4)$.

In fact, $f(x) = 2x + 3$ is continuous at any point on the real line. ❖

DO EXERCISE 9.

EXAMPLE 5 Determine whether the function in Example 1 is continuous at $x = 0$.

Solution The function is *not* continuous at $x = 0$ because $f(0)$ does not exist. ❖

EXAMPLE 6 Determine whether the function in Example 2 is continuous at $x = 1$.

Solution The function is not continuous at $x = 1$ because $\lim_{x \to 1} H(x)$ does not exist. ❖

EXAMPLE 7 Determine whether the function in Example 3 is continuous at $x = 1$.

Solution The function is not continuous at $x = 1$ because $G(1) = 5$, but $\lim_{x \to 1} G(x) = 2$. ❖

DO EXERCISES 10–12.

CONTINUITY PRINCIPLES

The following continuity principles, which we will not prove, allow us to determine whether a function is continuous.

C1. Any constant function is continuous (such a function never varies).

C2. For any positive integer n and any continuous function f, $[f(x)]^n$ and $\sqrt[n]{f(x)}$ are continuous. When n is even, the inputs of f in $\sqrt[n]{f(x)}$ are restricted to inputs x for which $f(x) \geq 0$.

C3. If $f(x)$ and $g(x)$ are continuous, then so are $f(x) + g(x)$, $f(x) - g(x)$, and $f(x) \cdot g(x)$.

C4. If $f(x)$ and $g(x)$ are continuous, so is $g(x)/f(x)$, so long as the inputs x do not yield outputs $f(x) = 0$.

EXAMPLE 8 Provide an argument to show that

$$f(x) = x^2 - 3x + 2$$

is continuous.

9. Consider

$$f(x) = 3x - 1$$

(see Margin Exercise 1).

a) Does $f(6)$ exist? If so, what is it?

b) Does $\lim_{x \to 6} f(x)$ exist? If so, what is it?

c) Does $\lim_{x \to 6} f(x) = f(6)$?

d) Is f continuous at 6?

10. Determine whether the function in Margin Exercise 2 is continuous at each of the following.

a) $x = 3$

b) $x = -2$

11. Determine whether the function in Margin Exercise 4 is continuous at $x = 0$.

12. Determine whether the function in Margin Exercise 5 is continuous at $x = -2$.

13. Provide an argument to show that the function given by

$$f(x) = \frac{\sqrt[3]{x} - 7x^2}{x - 2}$$

is continuous so long as $x \neq 2$.

Solution First we note that x is a continuous function. We determine this by the definition of continuity. We know that x^2 is continuous by C2. The constant function 3 is continuous by C1 and the function x is continuous, so the product $3x$ is continuous by C3. Thus, $x^2 - 3x$ is continuous by C3, and since the constant 2 is continuous, we can apply C3 again to show that $x^2 - 3x + 2$ is continuous.   ❖

In similar fashion, we can show that any polynomial, such as

$$f(x) = x^4 - 5x^3 + x^2 - 7,$$

is continuous.

A rational function is a quotient of two polynomials

$$r(x) = \frac{g(x)}{f(x)}.$$

Thus by C4, a rational function is continuous so long as the inputs x are not such that $f(x) = 0$. Thus a function like

$$r(x) = \frac{x^4 - 5x^3 + x^2 - 7}{x - 3}$$

is continuous for all values of x except $x = 3$.

DO EXERCISE 13.

EXERCISE SET 2.1

Determine whether each of the following is continuous.

1.

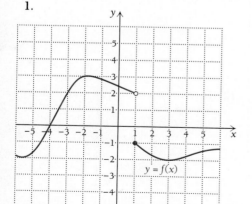

2.

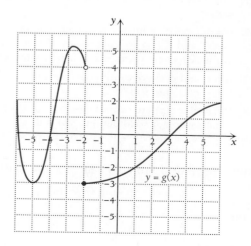

3.

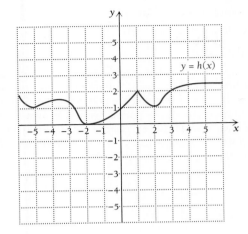

4.

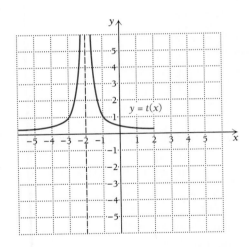

Use the graphs and functions in Exercises 1–4 to answer each of the following.

5. **a)** Find $\lim\limits_{x \to 1^+} f(x)$, $\lim\limits_{x \to 1^-} f(x)$, and $\lim\limits_{x \to 1} f(x)$.

   b) Find $f(1)$.

   c) Is f continuous at $x = 1$?

   d) Find $\lim\limits_{x \to -2} f(x)$.   **e)** Find $f(-2)$.

   f) Is f continuous at $x = -2$?

6. **a)** Find $\lim\limits_{x \to 1^+} g(x)$, $\lim\limits_{x \to 1^-} g(x)$, and $\lim\limits_{x \to 1} g(x)$.

   b) Find $g(1)$.

   c) Is g continuous at $x = 1$?

   d) Find $\lim\limits_{x \to -2} g(x)$.   **e)** Find $g(-2)$.

   f) Is g continuous at $x = -2$?

7. **a)** Find $\lim\limits_{x \to 1} h(x)$.

   b) Find $h(1)$.

   c) Is h continuous at $x = 1$?

   d) Find $\lim\limits_{x \to -2} h(x)$.   **e)** Find $h(-2)$.

   f) Is h continuous at $x = -2$?

8. **a)** Find $\lim\limits_{x \to 1} t(x)$.

   b) Find $t(1)$.

   c) Is t continuous at $x = 1$?

   d) Find $\lim\limits_{x \to -2} t(x)$.   **e)** Find $t(-2)$.

   f) Is t continuous at $x = -2$?

APPLICATIONS

❖ Business and Economics

The postage function. Postal rates are 29¢ for the first ounce and 23¢ for each additional ounce or fraction thereof. Formally speaking, if x is the weight of a letter in ounces, then $p(x)$ is the cost of mailing the letter, where

$$p(x) = 29¢, \quad \text{if } 0 < x \le 1,$$
$$p(x) = 52¢, \quad \text{if } 1 < x \le 2,$$
$$p(x) = 75¢, \quad \text{if } 2 < x \le 3,$$

and so on, up to 11 ounces (at which point postal cost also

depends on distance).
The graph of p is shown here.

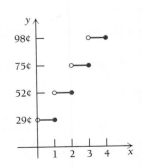

9. Is p continuous at 1? at $1\frac{1}{2}$? at 2? at 2.53?

10. Is p continuous at 3? at $3\frac{1}{4}$? at 4? at 3.98?

Using the graph of the postage function, find each of the following limits, if it exists.

11. $\lim\limits_{x \to 1^-} p(x)$, $\lim\limits_{x \to 1^+} p(x)$, $\lim\limits_{x \to 1} p(x)$

12. $\lim\limits_{x \to 2^-} p(x)$, $\lim\limits_{x \to 2^+} p(x)$, $\lim\limits_{x \to 2} p(x)$

13. $\lim\limits_{x \to 2.3} p(x)$ **14.** $\lim\limits_{x \to 1/2} p(x)$

Total cost. The total cost $C(x)$ of producing x units of a product is shown in the graph below. Note that the curve increases gradually, but jumps suddenly at a certain value. This is probably the point at which new machines must be purchased and extra employees must be employed.

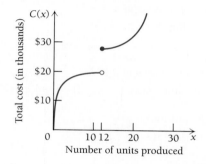

Using the graph, find each of the following limits, if it exists.

15. $\lim\limits_{x \to 12^-} C(x)$, $\lim\limits_{x \to 12^+} C(x)$, $\lim\limits_{x \to 12} C(x)$

16. $\lim\limits_{x \to 20^-} C(x)$, $\lim\limits_{x \to 20^+} C(x)$, $\lim\limits_{x \to 20} C(x)$

17. Is C continuous at 12? at 20?

18. Is C continuous at 6? at 18?

❖ **Social Sciences**

A learning curve. In psychology one often takes a certain amount of time t to learn a task. Suppose that the goal is to do a task perfectly and that you are practicing the ability to master it. After a certain time period, what is known to psychologists as an "I've got it!" experience occurs, and you are able to perform the task perfectly. Where do you think this happens on the learning curve below?

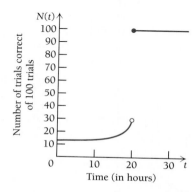

Using the graph, find each of the following limits, if it exists.

19. $\lim\limits_{x \to 20^+} N(t)$, $\lim\limits_{x \to 20^-} N(t)$, $\lim\limits_{x \to 20} N(t)$

20. $\lim\limits_{x \to 30^-} N(t)$, $\lim\limits_{x \to 30^+} N(t)$, $\lim\limits_{x \to 30} N(t)$

21. Is N continuous at 20? at 30?

22. Is N continuous at 10? at 26?

COMPUTER-GRAPHING CALCULATOR EXERCISES

23. Analyze the following function:

$$f(x) = \frac{\sqrt{x-2} - 3}{x - 11}.$$

a) Graph the function over smaller and smaller intervals that contain the number 11. Does the "hole" in the graph become evident?

b) Estimate

$$\lim\limits_{x \to 11} f(x)$$

using input–output tables.

THE CALCULUS EXPLORER
Function Evaluator/Comparer
Use the program to create input–output tables
when evaluating limits.

Continuity at a Point
Use this program to generate problems for
practicing the three-part definition of
continuity.

2.2

MORE ON LIMITS

Using Limit Principles

If a function is continuous at a, we can substitute to find the limit.

EXAMPLE 1   Find $\lim_{x \to 2} (x^4 - 5x^3 + x^2 - 7)$.

Solution   It follows from the Continuity Principles that $x^4 - 5x^3 + x^2 - 7$ is continuous. Thus we can find the limit by substitution:

$$\lim_{x \to 2} (x^4 - 5x^3 + x^2 - 7) = 2^4 - 5 \cdot 2^3 + 2^2 - 7$$
$$= 16 - 40 + 4 - 7$$
$$= -27.$$ ❖

EXAMPLE 2   Find $\lim_{x \to 0} \sqrt{x^2 - 3x + 2}$.

Solution   Using the Continuity Principles, we have shown that $x^2 - 3x + 2$ is continuous so long as x is restricted to values for which $x^2 - 3x + 2$ is nonnegative. Since $x^2 - 3x + 2$ is nonnegative, it follows from Principle C2 that $\sqrt{x^2 - 3x + 2}$ is continuous at $x = 0$. Thus we can substitute to find the limit:

$$\lim_{x \to 0} \sqrt{x^2 - 3x + 2} = \sqrt{0^2 - 3 \cdot 0 + 2}$$
$$= \sqrt{2}.$$ ❖

DO EXERCISES 1 AND 2.

There are Limit Principles that correspond to the Continuity Principles. We can use them to find limits when we are uncertain of the continuity of a function at a given point.

OBJECTIVES

a)   Find

$$\lim_{x \to a} f(x),$$

if it exists, using input–output tables, algebra, or graphs.

b)   Find limits at infinity.

c)   Find a limit like

$$\lim_{h \to 0} (3x^2 + 3xh + h^2).$$

Find the limit, if it exists.

1. $\lim_{x \to -2} (x^4 - 5x^3 + x^2 - 7)$

2. $\lim_{x \to 1} \sqrt{x^2 + 3x + 4}$

LIMIT PRINCIPLES

If $\lim_{x \to a} f(x) = L$ and $\lim_{x \to a} g(x) = M$, then we have the following.

L1. $\lim_{x \to a} c = c.$

(The limit of a constant is the constant.)

L2. $\lim_{x \to a} [f(x)]^n = [\lim_{x \to a} f(a)]^n = L^n,$

$\lim_{x \to a} \sqrt[n]{f(x)} = \sqrt[n]{\lim_{x \to a} f(x)} = \sqrt[n]{L}$, for any positive integer.

When n is even, the inputs of x in $\sqrt[n]{f(x)}$ must be restricted to those inputs x for which $f(x) \geq 0$.

(The limit of a power is the power of the limit, and the limit of a root is the root of the limit.)

L3. $\lim_{x \to a} [f(x) \pm g(x)] = \lim_{x \to a} f(x) \pm \lim_{x \to a} g(x) = L \pm M.$

(The limit of a sum or a difference is the sum or the difference of the limits.)

$\lim_{x \to a} [f(x) \cdot g(x)] = \left[\lim_{x \to a} f(x) \right] \cdot \left[\lim_{x \to a} g(x) \right] = L \cdot M.$

(The limit of a product is the product of the limits.)

L4. $\lim_{x \to a} \dfrac{g(x)}{f(x)} = \dfrac{\lim_{x \to a} g(x)}{\lim_{x \to a} f(x)} = \dfrac{M}{L}$, provided $L \neq 0.$

(The limit of a quotient is the quotient of the limits.)

L5. $\lim_{x \to a} cf(x) = c \cdot \lim_{x \to a} f(x) = c \cdot L.$

(The limit of a constant times a function is the constant times the limit.)

Principle L5 can actually be proved using Principles L1 and L3, but we state it for emphasis.

EXAMPLE 3 Find

$$\lim_{x \to -3} \frac{x^2 - 9}{x + 3}.$$

Solution The graph of the rational function $r(x) = (x^2 - 9)/(x + 3)$ is shown in Fig. 1. We can see from the graph that the limit is -6. But let us also examine the limit using the limit principles.

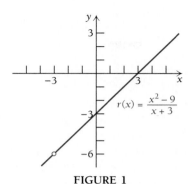

$$r(x) = \frac{x^2 - 9}{x + 3}$$

FIGURE 1

Find the limit, if it exists.

3. $\displaystyle\lim_{x \to -4} \frac{x^2 - 16}{x + 4}$

The function $(x^2 - 9)/(x + 3)$ is not continuous at $x = -3$. We use some algebraic simplification and then some limit principles:

$$\lim_{x \to -3} \frac{x^2 - 9}{x + 3} = \lim_{x \to -3} \frac{(x + 3)(x - 3)}{x + 3}$$

$$= \lim_{x \to -3} (x - 3) \qquad \text{Simplifying, assuming } x \neq -3$$

$$= \lim_{x \to -3} x - \lim_{x \to -3} 3 \qquad \text{By L3}$$

$$= -3 - 3 = -6. \qquad \text{❖}$$

4. $\displaystyle\lim_{x \to 3} \frac{x - 3}{x^2 - 9}$

DO EXERCISES 3 AND 4.

In Section 2.3, we will encounter expressions with two variables, x and h. Our interest is in those limits where x is fixed as a constant and h approaches zero.

EXAMPLE 4   Find $\lim_{h \to 0} (3x^2 + 3xh + h^2)$.

Solution   We treat x as a constant since we are interested only in the way in which the expression varies when $h \to 0$. We use the Limit Principles to find that

$$\lim_{h \to 0} (3x^2 + 3xh + h^2) = 3x^2 + 3x(0) + 0^2 = 3x^2.$$

The reader can check any limit about which there is uncertainty by using an input–output table. On the following page is a table for this limit.

5. a)   Complete the following table.

h	2x + h
1	
0.7	
0.4	
0.1	
0.01	
0.001	

b)   Find

$$\lim_{h \to 0} (2x + h).$$

h	$3x^2 + 3xh + h^2$	
1	$3x^2 + 3x \cdot 1 + 1^2,$	or $3x^2 + 3x + 1$
0.8	$3x^2 + 3x(0.8) + (0.8)^2,$	or $3x^2 + 2.4x + 0.64$
0.5	$3x^2 + 3x(0.5) + (0.5)^2,$	or $3x^2 + 1.5x + 0.25$
0.1	$3x^2 + 3x(0.1) + (0.1)^2,$	or $3x^2 + 0.3x + 0.01$
0.01	$3x^2 + 3x(0.01) + (0.01)^2,$	or $3x^2 + 0.03x + 0.0001$
0.001	$3x^2 + 3x(0.001) + (0.001)^2,$	or $3x^2 + 0.003x + 0.000001$

From the pattern in the table, it appears that

$$\lim_{h \to 0} (3x^2 + 3xh + h^2) = 3x^2. \qquad \maltese$$

DO EXERCISE 5.

Limits and Infinity

In Example 1 of Section 2.1, we discussed that

$$\lim_{x \to 0} \frac{1}{x} \quad \text{does not exist.}$$

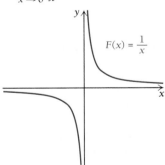

FIGURE 2

Looking at the graph shown in Fig. 2, we note that as x approaches 0 from the right, the outputs increase without bound. These numbers do not approach any real number, though it might be said that the limit from the right is infinity (∞). That is,

$$\lim_{x \to 0^+} \frac{1}{x} = \infty.$$

As x approaches 0 from the left, the outputs become more and more negative without bound. These numbers do not approach any real number, though it might be said that the limit from the left is negative infinity ($-\infty$). That is,

$$\lim_{x \to 0^-} \frac{1}{x} = -\infty.$$

Keep in mind that ∞ and $-\infty$ are not real numbers. We associate ∞ with numbers increasing without bound in a positive direction and $-\infty$ with numbers decreasing without bound in a negative direction.

DO EXERCISE 6.

Limits Involving Infinity

Occasionally we need to determine limits when the inputs get larger and larger without bound, that is, when they approach infinity. In such cases, we are finding *limits at infinity*. Such a limit is expressed as

$$\lim_{x \to \infty} f(x).$$

EXAMPLE 5 Find

$$\lim_{x \to \infty} \left(\frac{3x - 1}{x} \right).$$

Solution One way to find such a limit is to use an input–output table, as follows.

Inputs, x	1	10	50	100	2000
Outputs, $\frac{3x - 1}{x}$	2.0	2.9	2.98	2.99	2.9995

As the inputs get larger and larger without bound, the outputs get closer and closer to 3. Thus,

$$\lim_{x \to \infty} \left(\frac{3x - 1}{x} \right) = 3.$$

6. *General interest: Earned-run average.* A pitcher's *earned-run average* (the average number of runs given up every 9 innings, or 1 game) is given by

$$E = 9 \cdot \frac{n}{i},$$

where n = the number of earned runs allowed and i = the number of innings pitched. Suppose we fix the number of earned runs allowed at 4 and let i vary. We get a function given by

$$E(i) = 9 \cdot \frac{4}{i}.$$

a) Complete the following table, rounding to two decimal places.

Innings pitched (i)	Earned-run average (E)
9	
8	
7	
6	
5	
4	
3	
2	
1	
$\frac{2}{3}$ (2 outs)	
$\frac{1}{3}$ (1 out)	

b) Find

$$\lim_{i \to 0} E(i).$$

c) On the basis of parts (a) and (b), determine a pitcher's earned run average if 4 runs were allowed and there were 0 outs.

7. Consider

$$f(x) = \frac{2x + 5}{x}.$$

a)   Complete this table.
     (📱 helpful)

Inputs, x	Outputs, $\dfrac{2x + 5}{x}$
4	
20	
80	
200	
1,000	
10,000	

b)   Find

$$\lim_{x \to \infty} \frac{2x + 5}{x}.$$

Another way to do this is to use some algebra and the fact that as $x \to \infty$, $b/ax^n \to 0$, for any positive integer n.

We multiply by 1, using $(1/x) \div (1/x)$. This amounts to dividing both the numerator and the denominator by x:

$$\lim_{x \to \infty} \frac{3x - 1}{x} = \lim_{x \to \infty} \frac{3x - 1}{x} \cdot \frac{(1/x)}{(1/x)}$$

$$= \lim_{x \to \infty} \frac{(3x - 1)\dfrac{1}{x}}{x \cdot \dfrac{1}{x}}$$

$$= \lim_{x \to \infty} \frac{3x \cdot \dfrac{1}{x} - 1 \cdot \dfrac{1}{x}}{1}$$

$$= \lim_{x \to \infty} \left(3 - \frac{1}{x} \right) = 3 - 0 = 3. \qquad ❖$$

DO EXERCISE 7.

EXAMPLE 6   Find

$$\lim_{x \to \infty} \frac{3x^2 - 7x + 2}{7x^2 + 5x + 1}.$$

Solution   The function involved is a quotient of two polynomials. The *degree* of a polynomial is its highest power. Note that the degree of each polynomial above is 2.

The highest power of x in the denominator is x^2. We divide both the numerator and the denominator by x^2:

$$\lim_{x \to \infty} \frac{3x^2 - 7x + 2}{7x^2 + 5x + 1} = \lim_{x \to \infty} \frac{3 - \dfrac{7}{x} + \dfrac{2}{x^2}}{7 + \dfrac{5}{x} + \dfrac{1}{x^2}}$$

$$= \frac{3 - 0 + 0}{7 + 0 + 0} = \frac{3}{7}. \qquad ❖$$

EXAMPLE 7   Find

$$\lim_{x \to \infty} \frac{5x^2 + 7x + 9}{3x^3 + 2x - 4}.$$

Solution   The highest power of x in the denominator is x^3. We divide the numerator and the denominator by x^3:

$$\lim_{x \to \infty} \frac{5x^2 + 7x + 9}{3x^3 + 2x - 4} = \lim_{x \to \infty} \frac{\dfrac{5}{x} + \dfrac{7}{x^2} + \dfrac{9}{x^3}}{3 + \dfrac{2}{x^2} - \dfrac{4}{x^3}}$$

$$= \frac{0 + 0 + 0}{3 + 0 - 0} = \frac{0}{3} = 0. \qquad ❖$$

EXAMPLE 8   Find

$$\lim_{x \to \infty} \frac{7x^5 - 8x - 6}{3x^2 + 5x + 2}.$$

Solution   The highest power of x in the denominator is x^2. We divide the numerator and the denominator by x^2:

$$\lim_{x \to \infty} \frac{7x^5 - 8x - 6}{3x^2 + 5x + 2} = \lim_{x \to \infty} \frac{7x^3 - \dfrac{8}{x} - \dfrac{6}{x^2}}{3 + \dfrac{5}{x} + \dfrac{2}{x^2}}$$

$$= \frac{\lim\limits_{x \to \infty} 7x^3 - 0 - 0}{3 + 0 + 0} = \infty.$$

The numerator, in this case, increases without bound positively while the denominator approaches 3. This can be checked with an input–output table. Thus the limit is ∞. $\qquad ❖$

DO EXERCISES 8–10.

Find the limit.

8. $\lim\limits_{x \to \infty} \dfrac{2x^2 + x - 7}{3x^2 - 4x + 1}$

9. $\lim\limits_{x \to \infty} \dfrac{5x + 4}{2x^3 - 3}$

10. $\lim\limits_{x \to \infty} \dfrac{5x^2 - 2}{4x + 5}$

EXERCISE SET   2.2

Find the limit. Use any method: algebra, graphs, or input–output tables.

1. $\lim\limits_{x \to 1} (x^2 - 3)$

2. $\lim\limits_{x \to 1} (x^2 + 4)$

3. $\lim\limits_{x \to 0} \dfrac{3}{x}$

4. $\lim\limits_{x \to 0} \dfrac{-4}{x}$

5. $\lim\limits_{x \to 3} (2x + 5)$

6. $\lim\limits_{x \to 4} (5 - 3x)$

7. $\lim\limits_{x \to -5} \dfrac{x^2 - 25}{x + 5}$

8. $\lim\limits_{x \to -4} \dfrac{x^2 - 16}{x + 4}$

9. $\lim\limits_{x \to -2} \dfrac{5}{x}$

10. $\lim\limits_{x \to -5} \dfrac{-2}{x}$

11. $\lim\limits_{x \to 2} \dfrac{x^2 + x - 6}{x - 2}$

12. $\lim\limits_{x \to -4} \dfrac{x^2 - x - 20}{x + 4}$

2.3

a) Compute an average rate of change of one variable with respect to another.

b) Find a simplified difference quotient.

AVERAGE RATES OF CHANGE

A car travels 110 miles in 2 hours. Its *average speed* is 110 mi/2 hr, or 55 mi/hr. This is the *average rate of change* of distance with respect to time. The concept of average rate of change applies to many situations, as we will see in the following examples.

Figure 1 shows the total production of suits by Raggs, Ltd., during one morning of work. Industrial psychologists have found curves like this typical of the production of factory workers.

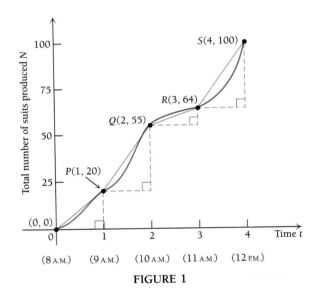

FIGURE 1

EXAMPLE 1 *Business: Production.* What was the number of suits produced from 9 A.M. to 10 A.M.?

Solution At 9 A.M., 20 suits had been produced. At 10 A.M., 55 suits had been produced. In the hour from 9 A.M. to 10 A.M., the number of suits produced was

$$55 \text{ suits} - 20 \text{ suits}, \quad \text{or} \quad 35 \text{ suits}.$$

Note that 35 is the slope of the line from P to Q.

EXAMPLE 2 *Business: Average rate of change.* What was the average number of suits produced per hour from 9 A.M. to 11 A.M.?

Solution   We have

$$\frac{64 \text{ suits} - 20 \text{ suits}}{11 \text{ A.M.} - 9 \text{ A.M.}} = \frac{44 \text{ suits}}{2 \text{ hr}} = 22 \frac{\text{suits}}{\text{hr}} \text{ (suits per hour)}.$$

Note that 22 is the slope of the line from P to R. The line from P to R is not shown in the graph.   ❖

DO EXERCISE 1.

Let us consider a function $y = f(x)$ and two inputs x_1 and x_2. The *change in input*, or the *change in x*, is

$$x_2 - x_1.$$

The *change in output*, or the *change in y*, is

$$y_2 - y_1,$$

where $y_1 = f(x_1)$ and $y_2 = f(x_2)$.

DEFINITION

The *average rate of change of y with respect to x*, as x changes from x_1 to x_2, is the ratio of the change in output to the change in input:

$$\frac{y_2 - y_1}{x_2 - x_1}, \quad \text{where } x_2 \neq x_1.$$

If we look at a graph of the function (Fig. 2), we see that

$$\frac{y_2 - y_1}{x_2 - x_1} = \frac{f(x_2) - f(x_1)}{x_2 - x_1}$$

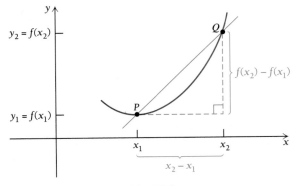

FIGURE 2

1. *Business: Average rate of change.* Refer to the graph in Fig. 1.

   a)   Find the number of suits produced per hour from

   > 8 A.M. to 9 A.M.;
   > 9 A.M. to 10 A.M.;
   > 10 A.M. to 11 A.M.;
   > 11 A.M. to 12 P.M.

   b)   Which interval in part (a) had the highest number?

   c)   Why do you think this happened?

   d)   Which interval in part (a) had the lowest number?

   e)   Why do you think this happened?

   f)   What was the average number of suits produced per hour from 8 A.M. to 12 P.M.?

2. For

$$f(x) = x^3,$$

find the average rates of change and sketch the secant lines as:

a) x changes from 1 to 4;

b) x changes from 1 to 2;

c) x changes from 2 to 4;

d) x changes from -1 to -4.

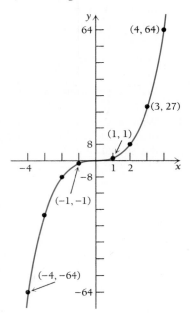

and that this is the slope of the line from $P(x_1, y_1)$ to $Q(x_2, y_2)$. The line $\overleftrightarrow{PQ}$ is called a **secant line**.

EXAMPLE 3   For $y = f(x) = x^2$, find the average rate of change as:

a) x changes from 1 to 3;

b) x changes from 1 to 2;

c) x changes from 2 to 3.

Solution   The graph in Fig. 3 is not necessary to the computations, but gives us a look at two of the secant lines whose slopes are being computed.

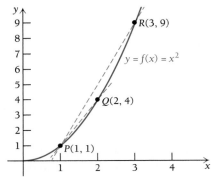

FIGURE 3

a)   When $x_1 = 1$,

$$y_1 = f(x_1) = f(1) = 1^2 = 1;$$

and when $x_2 = 3$,

$$y_2 = f(x_2) = f(3) = 3^2 = 9.$$

The average rate of change is

$$\frac{y_2 - y_1}{x_2 - x_1} = \frac{f(x_2) - f(x_1)}{x_2 - x_1}$$

$$= \frac{9 - 1}{3 - 1} = \frac{8}{2} = 4.$$

b)   When $x_1 = 1$,

$$y_1 = f(x_1) = f(1) = 1^2 = 1;$$

and when $x_2 = 2$,

$$y_2 = f(x_2) = f(2) = 2^2 = 4.$$

The average rate of change is

$$\frac{4 - 1}{2 - 1} = \frac{3}{1} = 3.$$

c) When $x_1 = 2$,

$$y_1 = f(x_1) = f(2)$$
$$= 2^2 = 4;$$

and when $x_2 = 3$,

$$y_2 = f(x_2) = f(3)$$
$$= 3^2 = 9.$$

The average rate of change is

$$\frac{9 - 4}{3 - 2} = \frac{5}{1} = 5.$$

DO EXERCISES 2 AND 3.

For a linear function, the average rates of change are the same for any choice of x_1 and x_2; that is, they are equal to the slope m of the line. As we saw in Example 3 and in Margin Exercise 2, a function that is not linear has average rates of change that vary with the choice of x_1 and x_2.

Difference Quotients

Let us now simplify our notation a bit by eliminating the subscripts. Instead of x_1, we will write simply x.

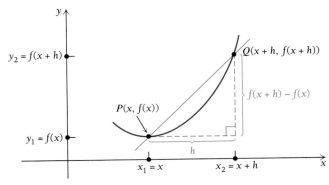

FIGURE 4

3. For

$$f(x) = \tfrac{1}{2}x + 1,$$

find the average rates of change and sketch the secant lines as:

a) x changes from 2 to 4;

b) x changes from 2 to 3;

c) x changes from -1 to 4.

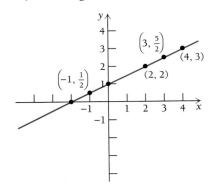

4. For

$$f(x) = 4x^2,$$

complete the following table to find the difference quotients.

$\dfrac{f(x+h)-f(x)}{h}$					
$f(x+h)-f(x)$					
$f(x+h)$					
$f(x)$					
$x+h$					
h	2	1	0.1	0.01	0.001
x	3	3	3	3	3

To get from x_1, or x, to x_2 in Fig. 4, we move a distance h. Thus, $x_2 = x + h$. Then the average rate of change, also called a **difference quotient**, is given by

$$\frac{y_2 - y_1}{x_2 - x_1} = \frac{f(x_2) - f(x_1)}{x_2 - x_1}$$

$$= \frac{f(x + h) - f(x)}{(x + h) - x}$$

$$= \frac{f(x + h) - f(x)}{h}.$$

DEFINITION

The **average rate of change** of f with respect to x is also called the *difference quotient*. It is given by

$$\frac{f(x + h) - f(x)}{h}, \quad \text{where } h \neq 0.$$

The difference quotient is equal to the slope of the line from a point $P(x, f(x))$ to a point $Q(x + h, f(x + h))$.

EXAMPLE 4 For $f(x) = x^2$, find the difference quotient when:

a) $x = 5$ and $h = 3$;

b) $x = 5$ and $h = 0.1$.

Solution

a) We substitute $x = 5$ and $h = 3$ into the formula:

$$\frac{f(x + h) - f(x)}{h} = \frac{f(5 + 3) - f(5)}{3} = \frac{f(8) - f(5)}{3}.$$

Now $f(8) = 8^2 = 64$ and $f(5) = 5^2 = 25$, and we have

$$\frac{f(8) - f(5)}{3} = \frac{64 - 25}{3} = \frac{39}{3} = 13.$$

The difference quotient is 13. It is also the slope of the line from $(5, 25)$ to $(8, 64)$.

b) We substitute $x = 5$ and $h = 0.1$ into the formula:

$$\frac{f(x + h) - f(x)}{h} = \frac{f(5 + 0.1) - f(5)}{0.1} = \frac{f(5.1) - f(5)}{0.1}.$$

Now $f(5.1) = (5.1)^2 = 26.01$ and $f(5) = 25$, and we have

$$\frac{f(5.1) - f(5)}{0.1} = \frac{26.01 - 25}{0.1} = \frac{1.01}{0.1} = 10.1.$$ ❖

DO EXERCISE 4.

For the function in Example 4, let us find a general form of the difference quotient. This will allow more efficient computations.

EXAMPLE 5   For $f(x) = x^2$, find a **simplified** form of the **difference quotient**. Then find the value of the difference quotient when $x = 5$ and $h = 0.1$.

Solution   We have

$$f(x) = x^2,$$

so

$$f(x + h) = (x + h)^2 = x^2 + 2xh + h^2.$$

Then

$$f(x + h) - f(x) = (x^2 + 2xh + h^2) - x^2 = 2xh + h^2.$$

Thus,

$$\frac{f(x + h) - f(x)}{h} = \frac{2xh + h^2}{h} = \frac{h(2x + h)}{h} = 2x + h, \quad h \neq 0.$$

It is important to note that a difference quotient is defined only when $h \neq 0$. The simplification above is valid only for nonzero values of h.
    When $x = 5$ and $h = 0.1$,

$$\frac{f(x + h) - f(x)}{h} = 2x + h = 2 \cdot 5 + 0.1 = 10 + 0.1 = 10.1.$$ ❖

DO EXERCISE 5.

EXAMPLE 6   For $f(x) = x^3$, find a simplified form of the difference quotient.

Solution   Now $f(x) = x^3$, so

$$f(x + h) = (x + h)^3$$
$$= x^3 + 3x^2h + 3xh^2 + h^3.$$

5. For

$$f(x) = 4x^2,$$

find a simplified form of the difference quotient by completing parts (a) through (c). Then complete the table in part (d) using the simplified form.

a)   Find $f(x + h)$.

b)   Find $f(x + h) - f(x)$.

c)   Find $[f(x + h) - f(x)]/h$ and simplify.

d)   Complete the following table.

x	h	$\dfrac{f(x + h) - f(x)}{h}$
6	−3	
6	−2	
6	−1	
6	−0.1	
6	−0.01	
6	−0.001	

6. a)   For
$$f(x) = 4x^3,$$
find a simplified difference quotient.

b)   Complete the following table.

x	h	$\dfrac{f(x+h) - f(x)}{h}$
-2	1	
-2	0.1	
-2	0.01	
-2	0.001	

7. a)   For
$$f(x) = \frac{1}{x},$$
find a simplified difference quotient.

b)   Complete the following table.

x	h	$\dfrac{f(x+h) - f(x)}{h}$
2	3	
2	1	
2	0.1	
2	0.01	
2	0.001	

(This was shown in Exercise Set 1.1.) Then
$$f(x + h) - f(x) = (x^3 + 3x^2h + 3xh^2 + h^3) - x^3$$
$$= 3x^2h + 3xh^2 + h^3.$$

So
$$\frac{f(x+h) - f(x)}{h} = \frac{3x^2h + 3xh^2 + h^3}{h}$$
$$= \frac{h(3x^2 + 3xh + h^2)}{h}$$
$$= 3x^2 + 3xh + h^2, \quad h \neq 0.$$

Again, this is true *only* for $h \neq 0$.   ❖

DO EXERCISE 6.

EXAMPLE 7   For $f(x) = 3/x$, find a simplified form of the difference quotient.

Solution   Now
$$f(x) = \frac{3}{x},$$
so
$$f(x + h) = \frac{3}{x + h}.$$

Then
$$f(x + h) - f(x) = \frac{3}{x + h} - \frac{3}{x}$$
$$= \frac{3}{x + h} \cdot \frac{x}{x} - \frac{3}{x} \cdot \frac{x + h}{x + h}$$
Here we are multiplying by 1 to get a common denominator.
$$= \frac{3x - 3(x + h)}{x(x + h)}$$
$$= \frac{3x - 3x - 3h}{x(x + h)} = \frac{-3h}{x(x + h)}.$$

So
$$\frac{f(x + h) - f(x)}{h} = \frac{\dfrac{-3h}{x(x + h)}}{h} = \frac{-3h}{x(x + h)} \cdot \frac{1}{h} = \frac{-3}{x(x + h)}, \quad h \neq 0.$$

This is true only for $h \neq 0$.   ❖

DO EXERCISE 7.

EXERCISE SET 2.3

For the functions in each of Exercises 1–12, (a) find a simplified difference quotient and (b) complete the following table.

1. $f(x) = 7x^2$
2. $f(x) = 5x^2$
3. $f(x) = -7x^2$
4. $f(x) = -5x^2$
5. $f(x) = 7x^3$
6. $f(x) = 5x^3$
7. $f(x) = \dfrac{5}{x}$
8. $f(x) = \dfrac{4}{x}$
9. $f(x) = -2x + 5$
10. $f(x) = 2x + 3$
11. $f(x) = x^2 - x$
12. $f(x) = x^2 + x$

x	h	$\dfrac{f(x + h) - f(x)}{h}$
4	2	
4	1	
4	0.1	
4	0.01	

APPLICATIONS

❖ **Business and Economics**

13. *Utility.* Utility is a type of function that occurs in economics. When a consumer receives x units of a certain product, a certain amount of pleasure, or utility, U, is derived from them. Below is a typical graph of a utility function.

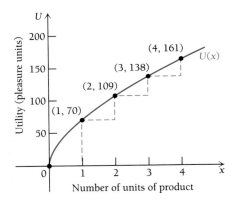

Number of units of product

a) Find the average rate of change of U as x changes from 0 to 1; from 1 to 2; from 2 to 3; and from 3 to 4.

b) Why do you think the average rates of change are decreasing?

14. *Advertising results.* The following graph shows a typical response to advertising. After an amount a is spent on advertising, the company sells $N(a)$ units of a product.

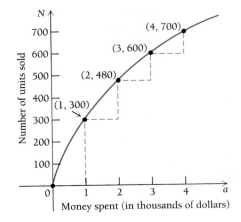

Money spent (in thousands of dollars)

a) Find the average rate of change of N as a changes from 0 to 1; from 1 to 2; from 2 to 3; and from 3 to 4.

b) Why do you think the average rates of change are decreasing?

15. *Total revenue.* A firm determines that the total revenue from the sale of x units of a certain product is given by

$$R(x) = -0.01x^2 + 1000x,$$

where $R(x)$ is in dollars.

a) Find $R(301)$.

b) Find $R(300)$.

c)   Find $R(301) - R(300)$.

d)   Find $\dfrac{R(301) - R(300)}{301 - 300}$.

16.   *Total cost.* A firm determines that the total cost C of producing x units of a certain product is given by

$$C(x) = -0.05x^2 + 50x,$$

where $C(x)$ is in dollars.

a)   Find $C(301)$.   b)   Find $C(300)$.

c)   Find $C(301) - C(300)$.

d)   Find $\dfrac{C(301) - C(300)}{301 - 300}$.

17.   *Total revenue of Pepsico.* Pepsico is a company that owns many fast-food chains, such as Taco Bell and

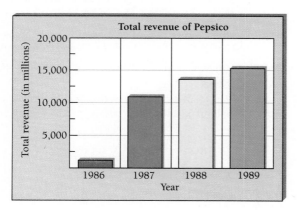

Total revenue of Pepsico

Pizza Hut. It also sells other food items through subsidiaries such as Frito Lay and Pepsi-Cola. It has experienced great sales growth during the past few years. In 1986 the total revenue was $930 million. In 1988 it was $13,007 million and in 1989 it was $15,242 million.

a)   Find the average rate of change of total revenue from 1986 to 1989.

b)   Find the average rate of change of total revenue from 1988 to 1989.

❖   Life and Physical Sciences

18.   *Temperature during an illness.* The temperature T, in degrees Fahrenheit, of a patient during an illness is shown in the graph below, where t = the time in days.

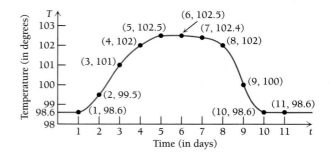

a)   Find the average rate of change of T as t changes from 1 to 10. Using this rate of change, would you know that the person was sick?

b)   Find the average rate of change of T with respect to t, as t changes from 1 to 2; from 2 to 3; from 3 to 4; from 4 to 5; from 5 to 6; from 6 to 7; from 7 to 8; from 8 to 9; from 9 to 10; and from 10 to 11.

c)   When do you think the temperature began to rise?

d)   When do you think the temperature reached its peak?

e)   When do you think the temperature began to subside?

f)   When was the temperature back to normal?

19.   *Memory.* The total number of words $M(t)$ that a person can memorize in time t, in minutes, is shown in the following graph.

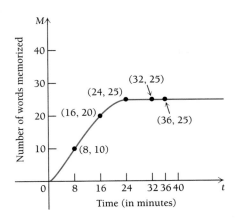

a) Find the average rate of change of M as t changes from 0 to 8; from 8 to 16; from 16 to 24; from 24 to 32; and from 32 to 36.

b) Why do the average rates of change become 0 after 24 min?

20. *Average velocity.* A car is at a distance s, in miles, from its starting point in t hours, given by

$$s(t) = 10t^2.$$

a) Find $s(2)$ and $s(5)$.

b) Find $s(5) - s(2)$. What does this represent?

c) Find the average rate of change of distance with respect to time as t changes from $t_1 = 2$ to $t_2 = 5$. This is known as *average velocity, or speed.*

21. *Average velocity.* An object is dropped from a certain height. It is known that it will fall a distance s, in feet, in t seconds, given by

$$s(t) = 16t^2.$$

a) How far will the object fall in 3 sec?

b) How far will the object fall in 5 sec?

c) What is the average rate of change of distance with respect to time during the period from 3 to 5 sec? This is also *average velocity, or speed.*

22. *Gas mileage.* At the beginning of a trip, the odometer on a car reads 30,680 and the car has a full tank of gas. At the end of the trip, the odometer reads 30,970. It takes 20 gal of gas to fill the tank again.

a) What is the average rate of consumption (that is,

the rate of change of the number of miles with respect to the number of gallons)?

b) What is the average rate of change of the number of gallons with respect to the number of miles?

❖ **Social Sciences**

23. *Population growth.* The two curves shown in the following figure describe the number of people in each of two countries at time t, in years.

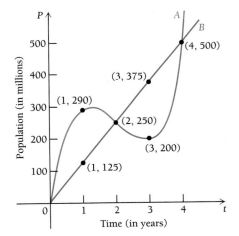

a) Find the average rate of change of each population (the number of people in the population) with respect to time t as t changes from 0 to 4. This is often called an *average growth rate.*

b) If the calculation in part (a) were the only one made, would we detect the fact that the populations were growing differently?

c) Find the average rates of change of each population as t changes from 0 to 1; from 1 to 2; from 2 to 3; and from 3 to 4.

d) For which population does the statement "the population grew by 125 million each year" convey the least information about what really took place?

24. *Divorce rate.* It is known that in 1960 there were 393,000 divorces. In 1982 there were 1,180,000 divorces. Find the average rate of change in the number of divorces with respect to time. This is called an *average divorce rate.*

25. *Marriage rate.* It is known that in 1960 there were 1,523,000 marriages. In 1982 there were 2,495,000 marriages. Find the average rate of change of the number of marriages with respect to time. This is called an *average marriage rate.*

SYNTHESIS EXERCISES

Find the simplified difference quotient.

26. $f(x) = mx + b$

27. $f(x) = ax^2 + bx + c$

28. $f(x) = ax^3 + bx^2$

29. $f(x) = \sqrt{x}$ (*Hint:* Multiply by 1 using $\dfrac{\sqrt{x+h} + \sqrt{x}}{\sqrt{x+h} + \sqrt{x}}.$)

30. $f(x) = x^4$

31. $f(x) = \dfrac{1}{x^2}$

32. $f(x) = \dfrac{1}{1-x}$

33. $f(x) = \dfrac{x}{1+x}$

2.4

DIFFERENTIATION USING LIMITS

OBJECTIVE

a) Given a formula of a function, find a formula for its derivative, find various values of the derivative, and find equations of tangent lines.

The path of the loose ski follows a tangent line to the ski chute at the point where it breaks loose.

Tangent Lines

A line tangent to a circle is a line that touches the circle exactly once (Fig. 1).

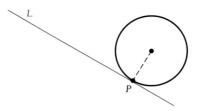

FIGURE 1

This definition becomes unworkable with other curves. For example, consider the curve shown in Fig. 2. Line L touches the curve at point P but meets the curve at other places as well. It is considered a tangent line, but "touching at one point" cannot be its definition.

Note in Fig. 2 that over a small interval containing P, line L does touch the curve exactly once. This is still not a suitable definition of a *tangent line* because it allows a line like M in Fig. 3 to be a tangent line, which we will not accept.

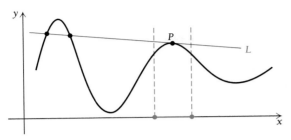

FIGURE 2

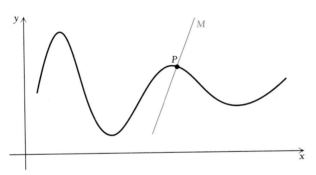

FIGURE 3

Later we will give a definition of a tangent line, but for now we will rely on intuition. In Fig. 4, lines L_1 and L_2 are not tangent lines. All the others are.

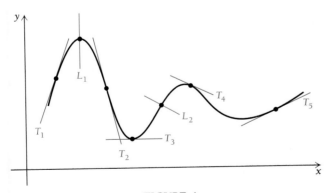

FIGURE 4

DO EXERCISE 1.

1. a) Which of the following appear to be tangent lines?

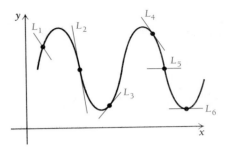

b) Below is a graph of $y = x^2$. Tangent lines are drawn at various points on the graph. Let $m(x)$ = the slope at the point $(x, f(x))$.

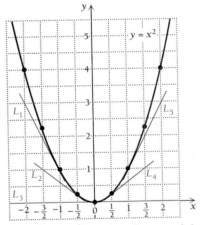

(Exercise 1 is continued at the top of the next page.)

Estimate the slope of each line and complete this table.

Lines	x	$m(x)$
L_1	-1	
L_2	$-\frac{1}{2}$	
L_3	0	
L_4	$\frac{1}{2}$	
L_5	1	

c) Derive a formula for $m(x)$.

Why Do We Study Tangent Lines?

The reason for our study of tangent lines will become evident in Chapter 3. For now, look at the graph of a total-profit function shown in Fig. 5. Note that the largest (or maximum) value of the function occurs at the point where the graph has a horizontal tangent; that is, where the tangent line has slope 0.

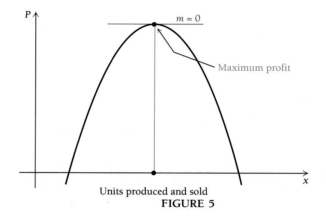

FIGURE 5

Differentiation Using Limits

We will define *tangent line* in such a way that it makes sense for *any* curve. To do this we use the notion of limit.

In Fig. 6, we obtain the line tangent to the curve at point P by considering secant lines through P and neighboring points Q_1, Q_2, and so on.

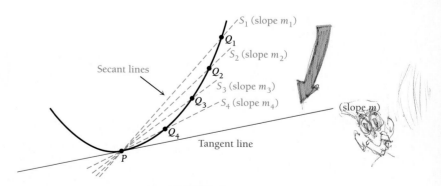

FIGURE 6

As the points Q approach P, the secant lines approach the tangent line. Each secant line has a slope. The slopes of the secant lines approach the slope of the tangent line. In fact, we *define* the **tangent line** as the line that contains the point P and has slope m, where m is the limit of the slopes of the secant lines as the points Q approach P.

DO EXERCISE 2.

How might we calculate the limit m? Suppose P has coordinates $(x, f(x))$ in Fig. 7(a). Then the first coordinate of Q is x plus some number h, or $x + h$. The coordinates of Q are $(x + h, f(x + h))$.

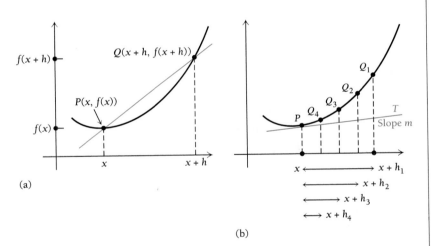

(a)

(b)

FIGURE 7

2. Use a blue colored pencil and draw each secant from point P to the points Q_i. Then use a red colored pencil and draw a tangent line to the curve at point P.

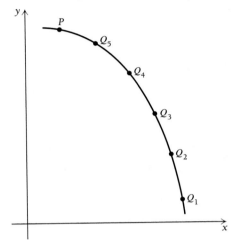

From Section 2.3, we know that the slope of the secant line $\overleftrightarrow{PQ}$ is given by the difference quotient

$$\frac{f(x + h) - f(x)}{h}.$$

Now, as we see in Fig. 7(b), as the points Q approach P, $x + h$ approaches x. That is, h approaches 0. Thus we have the following.

The slope of the tangent line $= m = \lim_{h \to 0} \dfrac{f(x + h) - f(x)}{h}$.

The formal definition of the *derivative of a function f* can now be given. We will designate the derivative at x as $f'(x)$, rather than $m(x)$.

DEFINITION

For a function $y = f(x)$, its *derivative* at x is defined as

$$f'(x) = \lim_{h \to 0} \frac{f(x + h) - f(x)}{h},$$

provided the limit exists. If $f'(x)$ exists, then we say that f is *differentiable* at x.

"Nothing in this world is so powerful as an idea whose time has come."

Victor Hugo

This is the basic definition of *differential calculus*.

Let us now calculate some formulas for derivatives. That is, given a formula for a function f, we will be trying to find a formula for f'.

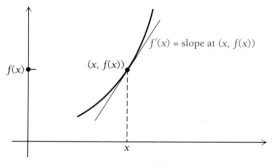

$f'(x)$ = slope at $(x, f(x))$

$(x, f(x))$

$f(x)$

x

FIGURE 8

There are three steps in calculating a derivative.
1. Write down the difference quotient $[f(x + h) - f(x)]/h$.
2. Simplify the difference quotient.
3. Find the limit as h approaches 0.

A formula for the derivative of a linear function

$$f(x) = mx + b$$

is

$$f'(x) = m.$$

Let us verify this using the definition.

EXAMPLE 1 For $f(x) = mx + b$, find $f'(x)$.

Solution We follow the steps in the box above. Thus,

1. $\dfrac{f(x + h) - f(x)}{h} = \dfrac{[m(x + h) + b] - (mx + b)}{h}$

2. $\dfrac{f(x + h) - f(x)}{h} = \dfrac{mx + mh + b - mx - b}{h}$

$= \dfrac{mh}{h} = m, \quad h \neq 0$

3. $\lim\limits_{h \to 0} \dfrac{f(x + h) - f(x)}{h} = \lim\limits_{h \to 0} m = m,$

since m represents a constant. Thus,

$$f'(x) = m.$$

In Margin Exercise 1 you may have conjectured that the function

$$f(x) = x^2$$

has derivative

$$f'(x) = 2x.$$

This would mean that the tangent line at $x = 4$ has slope $f'(4) = 8$. Let us verify this particular case and then the general formula.

EXAMPLE 2 For $f(x) = x^2$, find $f'(4)$.

Solution We have

1. $\dfrac{f(4 + h) - f(4)}{h} = \dfrac{(4 + h)^2 - 4^2}{h}$

2. $\dfrac{f(4 + h) - f(4)}{h} = \dfrac{16 + 8h + h^2 - 16}{h} = \dfrac{8h + h^2}{h}$

$= \dfrac{h(8 + h)}{h} = 8 + h, \quad h \neq 0$

3. $\lim\limits_{h \to 0} \dfrac{f(4 + h) - f(4)}{h} = \lim\limits_{h \to 0} (8 + h) = 8.$

Thus, $f'(4) = 8$.

DO EXERCISE 3.

EXAMPLE 3 For $f(x) = x^2$, find (the general formula) $f'(x)$.

3. For $f(x) = x^2$, find $f'(5)$ using the definition of a derivative.

4. For $f(x) = 4x^2$, find $f'(x)$. Then find $f'(5)$ and interpret the meaning.

Solution

1. We have

$$\frac{f(x + h) - f(x)}{h} = \frac{(x + h)^2 - x^2}{h}.$$

2. In Example 5 of Section 2.3, we showed how this difference quotient can be simplified to

$$\frac{f(x + h) - f(x)}{h} = 2x + h.$$

3. We want to find

$$\lim_{h \to 0} \frac{f(x + h) - f(x)}{h} = \lim_{h \to 0} (2x + h).$$

As $h \to 0$, we see that $2x + h \to 2x$. Thus,

$$\lim_{h \to 0} (2x + h) = 2x,$$

and we have

$$f'(x) = 2x,$$

which tells us, for example, that at $x = -3$, the curve has a tangent line whose slope is

$$f'(-3) = 2(-3), \quad \text{or} \quad -6.$$

We can say, simply, "The curve has slope -6 at the point $(-3, 9)$." ❖

DO EXERCISE 4.

EXAMPLE 4 For $f(x) = x^3$, find $f'(x)$. Then find $f'(-1)$ and $f'(10)$.

Solution

1. We have

$$\frac{f(x + h) - f(x)}{h} = \frac{(x + h)^3 - x^3}{h}.$$

2. In Example 6 of Section 2.3, we showed how this difference quotient can be simplified to

$$\frac{f(x + h) - f(x)}{h} = 3x^2 + 3xh + h^2.$$

3. We then have

$$\lim_{h \to 0} \frac{f(x+h) - f(x)}{h} = \lim_{h \to 0} (3x^2 + 3xh + h^2) = 3x^2.$$

(An input–output table for this is shown in Example 4 of Section 2.2.) Thus, for $f(x) = x^3$, we have $f'(x) = 3x^2$. Then

$$f'(-1) = 3(-1)^2 = 3 \quad \text{and} \quad f'(10) = 3(10)^2 = 300. \qquad \diamondsuit$$

DO EXERCISE 5.

EXAMPLE 5   For $f(x) = 3/x$:

a) Find $f'(x)$.

b) Find $f'(1)$ and $f'(2)$.

c) Find an equation of the tangent line to the curve at the point (1, 3).

d) Find an equation of the tangent line to the curve at the point $(2, \frac{3}{2})$.

Solution

a) 1.   We have

$$\frac{f(x+h) - f(x)}{h} = \frac{[3/(x+h)] - (3/x)}{h}.$$

2.   In Example 7 of Section 2.3, we showed that this difference quotient can be simplified to

$$\frac{f(x+h) - f(x)}{h} = \frac{-3}{x(x+h)}.$$

3.   We want to find

$$\lim_{h \to 0} \frac{f(x+h) - f(x)}{h} = \lim_{h \to 0} \frac{-3}{x(x+h)}.$$

As $h \to 0$, $x + h \to x$, so we have

$$f'(x) = \lim_{h \to 0} \frac{-3}{x(x+h)} = \frac{-3}{x^2}.$$

b) Then

$$f'(1) = \frac{-3}{1^2} = -3 \quad \text{and} \quad f'(2) = \frac{-3}{2^2} = -\frac{3}{4}.$$

5. For $f(x) = 4x^3$, find $f'(x)$. Then find $f'(-5)$ and $f'(0)$.

c) We know that the point $(1, 3)$ is on the graph of the function because $f(1) = 3$. To find an equation of the tangent line to the curve at the point $(1, 3)$, we use the fact that the slope at $x = 1$ is -3, as we found in the preceding work. Now we have

Point: $(1, 3)$,

Slope: -3.

We substitute into the point–slope equation (see Section 1.4):

$$y - y_1 = m(x - x_1)$$
$$y - 3 = -3(x - 1)$$
$$y = -3x + 3 + 3$$
$$y = -3x + 6.$$

The equation of the tangent line to the curve at $x = 1$ is $y = -3x + 6$ (Fig. 9).

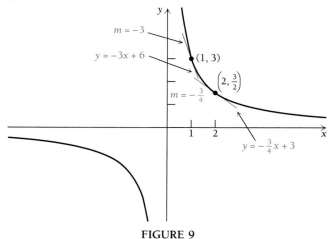

FIGURE 9

d) To find an equation of the tangent line to the curve at $x = 2$, we use the fact that the slope at $x = 2$ is $-\frac{3}{4}$, as we found in part (a). Now we have

Point: $(2, \frac{3}{2})$,

Slope: $-\frac{3}{4}$.

We substitute into the point–slope equation (see Section 1.4):

$$y - y_1 = m(x - x_1)$$
$$y - \tfrac{3}{2} = -\tfrac{3}{4}(x - 2)$$
$$y = -\tfrac{3}{4}x + \tfrac{3}{2} + \tfrac{3}{2}$$
$$y = -\tfrac{3}{4}x + 3.$$

The equation of the tangent line to the curve at $x = 2$ is $y = -\frac{3}{4}x + 3$ (Fig. 9).

Because $f(0)$ does not exist, we cannot evaluate the difference quotient

$$\frac{f(0 + h) - f(0)}{h}.$$

Thus, $f'(0)$ does not exist. We say, "f is not differentiable at 0." When a function is not defined at a point, it is not differentiable at that point. Also, if a function is discontinuous at a point, it is not differentiable at that point. ❖

DO EXERCISE 6.

It can happen that a function f is defined and continuous at a point but that its derivative f' is not. The function f given by

$$f(x) = |x|$$

is an example (Fig. 10). Note that

$$f(0) = |0| = 0,$$

so the function is defined at 0.

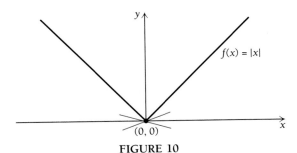

FIGURE 10

Suppose we try to draw a tangent line at $(0, 0)$. A function like this with a corner (not smooth) would seem to have many tangent lines at $(0, 0)$, and thus many slopes. The derivative at such a point would not be unique. Let us try to calculate the derivative at 0.

Since

$$f'(x) = \lim_{h \to 0} \frac{|x + h| - |x|}{h},$$

6. For

$$f(x) = \frac{1}{x}:$$

a)   Find $f'(x)$.

b)   Find $f'(-10)$ and $f'(-2)$.

c)   Find an equation of the tangent line to the curve at the point $(-10, -\frac{1}{10})$.

d)   Find an equation of the tangent line to the curve at the point $(-2, -\frac{1}{2})$.

at $x = 0$, we have

$$f'(0) = \lim_{h \to 0} \frac{|0 + h| - |0|}{h} = \lim_{h \to 0} \frac{|h|}{h}.$$

h	$\dfrac{|h|}{h}$
2	$\dfrac{|2|}{2}$, or $\dfrac{2}{2}$, or 1
1	1
0.1	1
0.01	1
0.001	1

h	$\dfrac{|h|}{h}$
−2	$\dfrac{|-2|}{-2}$, or $\dfrac{2}{-2}$, or −1
−1	−1
−0.1	−1
−0.01	−1
−0.001	−1

Look at the input–output tables. Note that as h approaches 0 from the right, $|h|/h$ approaches 1, but as h approaches 0 from the left, $|h|/h$ approaches −1. Thus,

$$\lim_{h \to 0} \frac{|h|}{h} \quad \text{does not exist,}$$

so

$$f'(0) \text{ does not exist.}$$

If a function has a "sharp point" or "corner," it will not have a derivative at that point (Fig. 11).

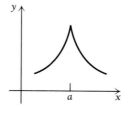 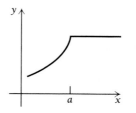 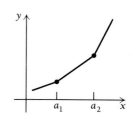

FIGURE 11

DO EXERCISE 7.

A function may also fail to be differentiable at a point by having a vertical tangent at that point. For example, the function shown in Fig. 12 has a vertical tangent at point a. Recall that since the slope of a vertical line is undefined, there is no derivative at such a point.

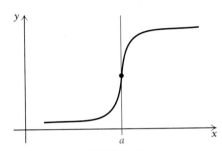

FIGURE 12

The function $f(x) = |x|$ illustrates the fact that although a function may be continuous at each point in an interval I, it may not be differentiable at each point in I. That is, continuity does not imply differentiability. On the other hand, if we know that a function is differentiable at each point in an interval I, then it is continuous over I. That is, if $f'(a)$ exists, then f is continuous at a. The function $f(x) = x^2$ is an example. Also, if a function is discontinuous at some point a, then it is not differentiable at a. Thus when we know that a function is differentiable over an interval, it is *smooth* in the sense that there are no "sharp points," "corners," or "breaks" in the graph.

7. List the points at which the following function is not differentiable.

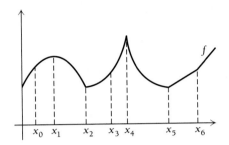

EXERCISE SET 2.4

For each function, find $f'(x)$. Then find $f'(-2)$, $f'(-1)$, $f'(0)$, $f'(1)$, and $f'(2)$, if they exist.

1. $f(x) = 5x^2$
2. $f(x) = 7x^2$
3. $f(x) = -5x^2$
4. $f(x) = -7x^2$
5. $f(x) = 5x^3$
6. $f(x) = 7x^3$
7. $f(x) = 2x + 3$
8. $f(x) = -2x + 5$
9. $f(x) = -4x$
10. $f(x) = \frac{1}{2}x$
11. $f(x) = x^2 + x$
12. $f(x) = x^2 - x$

13. $f(x) = \frac{4}{x}$
14. $f(x) = \frac{5}{x}$
15. $f(x) = mx$
16. $f(x) = ax^2 + bx + c$

17. Find an equation of the tangent line to the graph of $f(x) = x^2$ at the point $(3, 9)$, at $(-1, 1)$, and at $(10, 100)$.

18. Find an equation of the tangent line to the graph of $f(x) = x^3$ at the point $(-2, -8)$, at $(0, 0)$, and at $(4, 64)$.

19. Find an equation of the tangent line to the graph of $f(x) = 5/x$ at the point $(1, 5)$, at $(-1, -5)$, and at $(100, 0.05)$.

20. Find an equation of the tangent line to the graph of $f(x) = 2/x$ at the point $(-1, -2)$, at $(2, 1)$, and at $(10, \frac{1}{5})$.

21. Find an equation of the tangent line to the graph of $f(x) = 4 - x^2$ at the point $(-1, 3)$, at $(0, 4)$, and at $(5, -21)$.

22. Find an equation of the tangent line to the graph of $f(x) = x^2 - 2x$ at the point $(-2, 8)$, at $(1, -1)$, and at $(4, 8)$.

23. List the points in the graph at which the function is not differentiable.

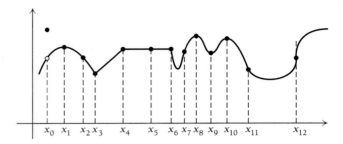

24. Consider the function f given by

$$f(x) = \frac{x^2 - 9}{x + 3}.$$

For what values is this function not differentiable?

APPLICATIONS

❖ **Business and Economics**

25. *The postage function.* Consider the postage function defined in Exercise Set 2.1. At what values is the function not differentiable?

SYNTHESIS EXERCISES

Find $f'(x)$.

26. $f(x) = x^4$

27. $f(x) = \dfrac{1}{x^2}$

28. $f(x) = \dfrac{1}{1 - x}$

29. $f(x) = \dfrac{x}{1 + x}$

30. $f(x) = \sqrt{x}$ $\left(\text{Multiply by 1, using } \dfrac{\sqrt{x + h} + \sqrt{x}}{\sqrt{x + h} + \sqrt{x}}. \right)$

COMPUTER-GRAPHING CALCULATOR EXERCISES

31. *Business: Growth of an investment.* A company determines that the value of an investment is V, in millions of dollars, after time t, in years, where V is given by

$$V(t) = 5t^3 - 30t^2 + 45t + 5\sqrt{t}.$$

Note: Computers and calculators usually only graph functions using the variables y and x, so you may need to change the variables when entering this function.

a) Sketch the graph over the interval $[0, 5]$.

b) Find the equation of the secant line passing through the points $(1, f(1))$ and $(5, f(5))$. Then sketch this secant line using the same axes as in (a).

c) Find the average rate of change of the investment between year 1 and year 5.

d) Repeat steps (b) and (c) for the pairs of points $(1, f(1))$ and $(4, f(4))$; $(1, f(1))$ and $(3, f(3))$; $(1, f(1))$ and $(1.5, f(1.5))$.

e) What appears to be the slope of the tangent line to the graph at the point $(1, f(1))$?

THE CALCULUS EXPLORER
Secant Lines

Use this program to generate secant lines,
show their slope as they change position, and
eventually draw tangent lines.

2.5

DIFFERENTIATION TECHNIQUES: THE POWER AND SUM–DIFFERENCE RULES

OBJECTIVES

a)   Differentiate using the Power Rule, the Sum–Difference Rule, or the rule for differentiating a constant or a constant times a function.

b)   Find the points on the graph of a function at which the tangent line has a given slope.

Leibniz's Notation

When y is a function of x, we will also designate the derivative, $f'(x)$, as*

$$\frac{dy}{dx},$$

which is read "the derivative of y with respect to x." This notation was invented by the German mathematician Leibniz. It does *not* mean dy divided by dx! (That is, we cannot interpret dy/dx as a quotient until meanings are given to dy and dx, which we will not do here.) For example, if $y = x^2$, then

$$\frac{dy}{dx} = 2x.$$

We can also write

$$\frac{d}{dx} f(x)$$

to denote the derivative of f with respect to x. For example,

$$\frac{d}{dx} x^2 = 2x.$$

---

*The notation $D_x y$ is also used.

1. For

$$y = x^3,$$

use the results of previous work to find each of the following.

a) $\dfrac{dy}{dx}$

b) $\dfrac{d}{dx} x^3$

c) $\dfrac{dy}{dx}\Big|_{x=4}$

Historical Note: The German mathematician and philosopher Gottfried Wilhelm von Leibniz (1646–1716) and the English mathematician, philosopher, and physicist Sir Isaac Newton (1642–1727) are both credited with the invention of the calculus, though each made the invention independently of the other. Newton used the dot notation $\dot{y}$ for dy/dt, where y is a function of time, and this notation is still used, though it is not as common as Leibniz's notation.

The value of dy/dx when $x = 5$ can be denoted by

$$\frac{dy}{dx}\bigg|_{x=5}.$$

Thus for $dy/dx = 2x$,

$$\frac{dy}{dx}\bigg|_{x=5} = 2 \cdot 5, \quad \text{or} \quad 10.$$

In general, for $y = f(x)$,

$$\frac{dy}{dx}\bigg|_{x=a} = f'(a).$$

DO EXERCISE 1.

The Power Rule

In the remainder of this section, we will develop rules and techniques for efficient differentiation.

Look for a pattern in the following table, which contains functions and derivatives that we have found in previous work.

Function	Derivative
x^2	$2x^1$
x^3	$3x^2$
x^4	$4x^3$
$\dfrac{1}{x} = x^{-1}$	$-1 \cdot x^{-2} = \dfrac{-1}{x^2}$
$\dfrac{1}{x^2} = x^{-2}$	$-2 \cdot x^{-3} = \dfrac{-2}{x^3}$

Perhaps you have discovered the following theorem.

THEOREM 1

The Power Rule

For any real number k,

$$\frac{d}{dx} x^k = k \cdot x^{k-1}.$$

We proved this theorem for the cases $k = 2, 3$, and -1 in Examples 3 and 4 and Margin Exercise 6, respectively, of Section 2.4. We will not prove the other cases in this text. Note that this rule holds no matter what the exponent. That is, to differentiate x^k, we write the exponent k as the coefficient, followed by x with an exponent 1 less than k.

①   Write the exponent as the coefficient.

②   Subtract 1 from the exponent.

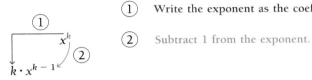

EXAMPLE 1 $\dfrac{d}{dx} x^5 = 5x^4$ ❖

EXAMPLE 2 $\dfrac{d}{dx} x = 1 \cdot x^{1-1}$

          $= 1 \cdot x^0 = 1$ ❖

EXAMPLE 3 $\dfrac{d}{dx} x^{-4} = -4 \cdot x^{-4-1}$

          $= -4x^{-5}, \quad \text{or} \quad -4 \cdot \dfrac{1}{x^5}, \quad \text{or} \quad -\dfrac{4}{x^5}$ ❖

The Power Rule also allows us to differentiate $\sqrt{x}$. To do so, it helps to first convert to an expression with a fractional exponent.

EXAMPLE 4 $\dfrac{d}{dx} \sqrt{x} = \dfrac{d}{dx} x^{1/2} = \dfrac{1}{2} \cdot x^{(1/2)-1}$

          $= \dfrac{1}{2}x^{-1/2}, \quad \text{or} \quad \dfrac{1}{2} \cdot \dfrac{1}{x^{1/2}}, \quad \text{or} \quad \dfrac{1}{2} \cdot \dfrac{1}{\sqrt{x}}, \quad \text{or} \quad \dfrac{1}{2\sqrt{x}}$ ❖

EXAMPLE 5 $\dfrac{d}{dx} x^{-2/3} = -\dfrac{2}{3}x^{(-2/3)-1}$

          $= -\dfrac{2}{3}x^{-5/3}, \quad \text{or} \quad -\dfrac{2}{3} \cdot \dfrac{1}{x^{5/3}}, \quad \text{or} \quad -\dfrac{2}{3\sqrt[3]{x^5}}$ ❖

DO EXERCISES 2–5.

Find $\dfrac{dy}{dx}$ (differentiate).

2. $y = x^6$

3. $y = x^{-7}$

4. $y = \sqrt[3]{x}$

5. $y = x^{-1/4}$

6. Find $g'(x)$ if

$$g(x) = -14.$$

The Derivative of a Constant Function

Look at the graph of the constant function $F(x) = c$, shown in Fig. 1. What is the slope at each point on the graph?

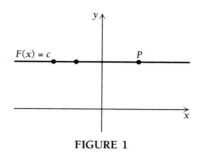

FIGURE 1

We now have the following.

THEOREM 2

The derivative of a constant function is 0. That is, $\dfrac{d}{dx} c = 0$.

Proof.   Let F be the function given by $F(x) = c$. Then

$$\frac{F(x + h) - F(x)}{h} = \frac{c - c}{h}$$

$$= \frac{0}{h} = 0.$$

The difference quotient is always 0. Thus, as h approaches 0, the limit of the difference quotient approaches 0, so $F'(x) = 0$.

DO EXERCISE 6.

The Derivative of a Constant Times a Function

Now let us consider differentiating functions such as

$$f(x) = 5x^2 \quad \text{and} \quad g(x) = -7x^4.$$

Note that we already know how to differentiate x^2 and x^4. Let us again look for a pattern in the results of Exercise Set 2.3.

Function	Derivative
$5x^2$	$10x$
$4x^{-1}$	$-4x^{-2}$
$-7x^2$	$-14x$
$5x^3$	$15x^2$

Perhaps you have discovered the following.

THEOREM 3

The derivative of a constant times a function is the constant times the derivative of the function. Using derivative notation, we can write this as

$$\frac{d}{dx}[c \cdot f(x)] = c \cdot f'(x).$$

Proof.   Let F be the function given by $F(x) = cf(x)$. Then

$$\frac{F(x + h) - F(x)}{h} = \frac{cf(x + h) - cf(x)}{h} = c\left[\frac{f(x + h) - f(x)}{h}\right].$$

As h approaches 0, the limit of the preceding expression is the same as c times $f'(x)$. Thus, $F'(x) = cf'(x)$.

Combining this rule with the Power Rule allows us to find many derivatives.

EXAMPLE 6 $\dfrac{d}{dx} 5x^4 = 5\dfrac{d}{dx} x^4 = 5 \cdot 4 \cdot x^{4 - 1} = 20x^3$ ❖

EXAMPLE 7 $\dfrac{d}{dx}(-9x) = -9\dfrac{d}{dx} x = -9 \cdot 1 = -9$ ❖

With practice you will be able to differentiate many such functions in one step.

EXAMPLE 8 $\dfrac{d}{dx} \dfrac{-4}{x^2} = \dfrac{d}{dx}(-4x^{-2}) = -4 \cdot \dfrac{d}{dx} x^{-2}$

$$= -4(-2)x^{-2 - 1}$$

$$= 8x^{-3}, \quad \text{or} \quad \frac{8}{x^3}$$ ❖

Find $\dfrac{dy}{dx}$ (differentiate).

7. $y = 5x^{20}$

8. $y = -\dfrac{3}{x}$

9. $y = -8\sqrt{x}$

10. $y = 0.16x^{6.25}$

EXAMPLE 9 $\dfrac{d}{dx}(-x^{0.7}) = -1 \cdot \dfrac{d}{dx}x^{0.7}$

$$= -1 \cdot 0.7 \cdot x^{0.7 - 1}$$
$$= -0.7x^{-0.3} \qquad \qquad ❖$$

DO EXERCISES 7–10.

The Derivative of a Sum or a Difference

In Exercise 11 of Exercise Set 2.4, you found that the derivative of

$$f(x) = x^2 + x$$

is

$$f'(x) = 2x + 1.$$

Note that the derivative of x^2 is $2x$, the derivative of x is 1, and the sum of these derivatives is $f'(x)$. This illustrates the following.

THEOREM 4

The Sum–Difference Rule

Sum: The derivative of a sum is the sum of the derivatives:

If $F(x) = f(x) + g(x)$,   then $F'(x) = f'(x) + g'(x)$.

Difference: The derivative of a difference is the difference of the derivatives:

If $F(x) = f(x) - g(x)$,   then $F'(x) = f'(x) - g'(x)$.

Proof.   For the Sum Rule, we have

$$\frac{F(x + h) - F(x)}{h} = \frac{[f(x + h) + g(x + h)] - [f(x) + g(x)]}{h}$$
$$= \frac{f(x + h) - f(x)}{h} + \frac{g(x + h) - g(x)}{h}.$$

As h approaches 0, the two terms on the right approach $f'(x)$ and $g'(x)$, respectively, so their sum approaches $f'(x) + g'(x)$. Thus, $F'(x) = f'(x) + g'(x)$.

The proof of the Difference Rule is similar.

Any function that is a sum or a difference of several terms can be differentiated term by term.

EXAMPLE 10 $\dfrac{d}{dx}(3x + 7) = \dfrac{d}{dx}(3x) + \dfrac{d}{dx}(7)$

$= 3\dfrac{d}{dx}(x) + \dfrac{d}{dx}(7) = 3\cdot 1 + 0 = 3$ ❖

EXAMPLE 11 $\dfrac{d}{dx}(5x^3 - 3x^2) = \dfrac{d}{dx}(5x^3) - \dfrac{d}{dx}(3x^2)$

$= 5\dfrac{d}{dx}x^3 - 3\dfrac{d}{dx}x^2$

$= 5\cdot 3x^2 - 3\cdot 2x$

$= 15x^2 - 6x$ ❖

EXAMPLE 12 $\dfrac{d}{dx}\left(24x - \sqrt{x} + \dfrac{2}{x}\right) = \dfrac{d}{dx}(24x) - \dfrac{d}{dx}(\sqrt{x}) + \dfrac{d}{dx}\left(\dfrac{2}{x}\right)$

$= 24\cdot\dfrac{d}{dx}x - \dfrac{d}{dx}x^{1/2} + 2\cdot\dfrac{d}{dx}x^{-1}$

$= 24\cdot 1 - \dfrac{1}{2}x^{(1/2)-1} + 2(-1)x^{-1-1}$

$= 24 - \dfrac{1}{2}x^{-1/2} - 2x^{-2}$

$= 24 - \dfrac{1}{2\sqrt{x}} - \dfrac{2}{x^2}$ ❖

DO EXERCISES 11–13.

A word of caution! The derivative of

$$f(x) + c,$$

a function plus a constant, is just the derivative of the function,

$$f'(x).$$

The derivative of

$$c\cdot f(x),$$

a function times a constant, is the constant times the derivative

$$c\cdot f'(x).$$

That is, for a product the constant is retained, but for a sum it is not.

Find $\dfrac{dy}{dx}$ (differentiate).

11. $y = -\dfrac{1}{4}x - 9$

12. $y = 7x^4 + 6x^2$

13. $y = 15x^2 + \dfrac{4}{x} + \sqrt{x}$

14. Find the points on the graph of

$$y = \tfrac{1}{3}x^3 - 2x^2 + 4x$$

at which the tangent line is horizontal.

It is important to be able to determine points at which the tangent line to a curve has a certain slope—that is, points at which the derivative attains a certain value.

EXAMPLE 13   Find the points on the graph of $y = -x^3 + 6x^2$ at which the tangent line is horizontal.

Solution   A horizontal tangent has slope 0. Thus we seek the values of x for which $dy/dx = 0$. That is, we want to find x such that

$$-3x^2 + 12x = 0.$$

We factor and solve:

$$-3x(x - 4) = 0$$
$$-3x = 0 \quad \text{or} \quad x - 4 = 0$$
$$x = 0 \quad \text{or} \quad x = 4.$$

We are to find the points *on the graph,* so we must determine the second coordinates from the original equation, $y = -x^3 + 6x^2$.

For $x = 0$, $y = -0^3 + 6 \cdot 0^2 = 0$.
For $x = 4$, $y = -4^3 + 6 \cdot 4^2 = -64 + 96 = 32$.

Thus the points we are seeking are $(0, 0)$ and $(4, 32)$. This is shown on the graph in Fig. 2.

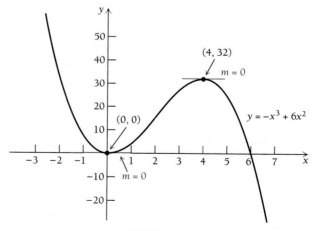

FIGURE 2   ❖

DO EXERCISE 14.

EXAMPLE 14   Find the points on the graph of $y = -x^3 + 6x^2$ at which the tangent has slope 6.

Solution   We want to find values of x for which $dy/dx = 6$. That is, we want to find x such that

$$-3x^2 + 12x = 6.$$

To solve, we add -6 and get

$$-3x^2 + 12x - 6 = 0.$$

We can simplify this equation by multiplying by $-\frac{1}{3}$ since each term has a common factor of -3. This gives us

$$x^2 - 4x + 2 = 0.$$

This is a quadratic equation, not readily factorable, so we use the Quadratic Formula, where $a = 1$, $b = -4$, and $c = 2$:

$$
\begin{aligned}
x &= \frac{-b \pm \sqrt{b^2 - 4ac}}{2a} \\
&= \frac{-(-4) \pm \sqrt{(-4)^2 - 4 \cdot 1 \cdot 2}}{2 \cdot 1} \\
&= \frac{4 \pm \sqrt{8}}{2} \\
&= \frac{2 \cdot 2 \pm 2\sqrt{2}}{2 \cdot 1} \\
&= \frac{2}{2} \cdot \frac{2 \pm \sqrt{2}}{1} \\
&= 2 \pm \sqrt{2}.
\end{aligned}
$$

The solutions are $2 + \sqrt{2}$ and $2 - \sqrt{2}$.

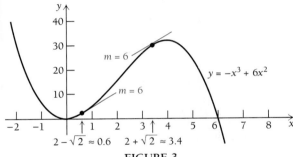

FIGURE 3

15. Find the points on the graph of

$$y = \tfrac{1}{3}x^3 - 2x^2 + 4x$$

at which the tangent line has slope 3.

We determine the second coordinates from the original equation. For $x = 2 + \sqrt{2}$,

$$
\begin{aligned}
y &= -(2 + \sqrt{2})^3 + 6(2 + \sqrt{2})^2 \\
  &= -[(2 + \sqrt{2})^2(2 + \sqrt{2})] + 6(4 + 4\sqrt{2} + 2) \\
  &= -[(6 + 4\sqrt{2})(2 + \sqrt{2})] + 6(6 + 4\sqrt{2}) \\
  &= -[12 + 6\sqrt{2} + 8\sqrt{2} + 8] + 36 + 24\sqrt{2} \\
  &= -[20 + 14\sqrt{2}] + 36 + 24\sqrt{2} \\
  &= -20 - 14\sqrt{2} + 36 + 24\sqrt{2} \\
  &= 16 + 10\sqrt{2}.
\end{aligned}
$$

Similarly, for $x = 2 - \sqrt{2}$,

$$y = 16 - 10\sqrt{2}.$$

Thus the points we are seeking are $(2 + \sqrt{2},\ 16 + 10\sqrt{2})$ and $(2 - \sqrt{2},\ 16 - 10\sqrt{2})$. This is shown on the graph in Fig. 3. ❖

DO EXERCISE 15.

EXERCISE SET 2.5

Find $\dfrac{dy}{dx}$.

1. $y = x^7$

2. $y = x^8$

3. $y = 15$

4. $y = 78$

5. $y = 4x^{150}$

6. $y = 7x^{200}$

7. $y = x^3 + 3x^2$

8. $y = x^4 - 7x$

9. $y = 8\sqrt{x}$

10. $y = 4\sqrt{x}$

11. $y = x^{0.07}$

12. $y = x^{0.78}$

13. $y = \dfrac{1}{2}x^{4/5}$

14. $y = -4.8x^{1/3}$

15. $y = x^{-3}$

16. $y = x^{-4}$

17. $y = 3x^2 - 8x + 7$

18. $y = 4x^2 - 7x + 5$

19. $y = \sqrt[4]{x} - \dfrac{1}{x}$

20. $y = \sqrt[5]{x} - \dfrac{2}{x}$

Find $f'(x)$.

21. $f(x) = 0.64x^{2.5}$

22. $f(x) = 0.32x^{12.5}$

23. $f(x) = \dfrac{5}{x} - x$

24. $f(x) = \dfrac{4}{x} - x$

25. $f(x) = 4x - 7$

26. $f(x) = 7x + 11$

27. $f(x) = 4x + 9$

28. $f(x) = 7x - 14$

29. $f(x) = \dfrac{x^4}{4}$

30. $f(x) = \dfrac{x^3}{3}$

31. $f(x) = -0.01x^2 - 0.5x + 70$

32. $f(x) = -0.01x^2 + 0.4x + 50$

33. $f(x) = 3x^{-2/3} + x^{3/4} + x^{6/5} + \dfrac{8}{x^3}$

34. $f(x) = x^{-3/4} - 3x^{2/3} + x^{5/4} + \dfrac{2}{x^4}$

35. $f(x) = \dfrac{2}{x} - \dfrac{x}{2}$

36. $f(x) = \dfrac{x}{5} + \dfrac{5}{x}$

37. $f(x) = \dfrac{16}{x} - \dfrac{8}{x^3} + \dfrac{1}{x^4}$

38. $f(x) = \dfrac{20}{x^5} + \dfrac{1}{x^3} - \dfrac{2}{x}$

39. $f(x) = \sqrt{x} + \sqrt[3]{x} - \sqrt[4]{x} + \sqrt[5]{x}$

40. $f(x) = \dfrac{x^5 - 3x^4 + 2x^3 - 5x^2 - 8x + 4}{x^2}$

41. Find an equation of the tangent line to the graph of $f(x) = x^3 - 2x + 1$ at the point $(2, 5)$, at $(-1, 2)$, and at $(0, 1)$.

42. Find an equation of the tangent line to the graph of $f(x) = x^2 - \sqrt{x}$ at the point $(1, 0)$, at $(4, 14)$, and at $(9, 78)$.

For each function, find the points on the graph at which the tangent line is horizontal.

43. $y = x^2$

44. $y = -x^2$

45. $y = -x^3$

46. $y = x^3$

47. $y = 3x^2 - 5x + 4$

48. $y = 5x^2 - 3x + 8$

49. $y = -0.01x^2 - 0.5x + 70$

50. $y = -0.01x^2 + 0.4x + 50$

51. $y = 2x + 4$

52. $y = -2x + 5$

53. $y = 4$

54. $y = -3$

55. $y = -x^3 + x^2 + 5x - 1$

56. $y = -\frac{1}{3}x^3 + 6x^2 - 11x - 50$

57. $y = \frac{1}{3}x^3 - 3x + 2$

58. $y = x^3 - 6x + 1$

For each function, find the points on the graph at which the tangent line has slope 1.

59. $y = 20x - x^2$

60. $y = 6x - x^2$

61. $y = -0.025x^2 + 4x$

62. $y = -0.01x^2 + 2x$

63. $y = \frac{1}{3}x^3 + 2x^2 + 2x$

64. $y = \frac{1}{3}x^3 - x^2 - 4x + 1$

SYNTHESIS EXERCISES

65. Find the points on the graph of
$$y = x^4 - \tfrac{4}{3}x^2 - 4$$
at which the tangent line is horizontal.

66. Find the points on the graph of
$$y = 2x^6 - x^4 - 2$$
at which the tangent line is horizontal.

Find dy/dx. Each of the following can be differentiated using the rules developed in this section, but some algebra may be required beforehand.

67. $y = x(x - 1)$

68. $y = (x - 1)(x + 1)$

69. $y = (x - 2)(x + 3)$

70. $y = \dfrac{5x^2 - 8x + 3}{8}$

71. $y = \dfrac{x^5 + x}{x^2}$

72. $y = (5x)^2$

73. $y = (-4x)^3$

74. $y = \sqrt{7x}$

75. $y = \sqrt[3]{8x}$

76. $y = (x - 3)^2$

77. $y = (x + 1)^3$

78. $y = (x - 2)^3(x + 1)$

79. Prove Theorem 4 (Difference Rule).

COMPUTER-GRAPHING CALCULATOR EXERCISES

For each function, sketch the graph of f and f' on the given interval. Then estimate points at which the tangent line is horizontal.

80. $f(x) = 3.8x^5 - 18.6x^3$; $[-3, 3]$

81. $f(x) = 5x^3 - 30x^2 + 45x + 5\sqrt{x}$; $[0, 5]$

THE CALCULUS EXPLORER:
Derivatives

Select several functions $y = f(x)$ in this exercise set and key them into the computer. The program will automatically graph f and f'. Try $f(x) = 20x^3 - 3x^5$.

2.6

a) Given a distance function $s(t)$, find a formula for the velocity $v(t)$ and the acceleration $a(t)$, and evaluate $s(t)$, $v(t)$, and $a(t)$ for given values of t.

b) Given y as a function of x, find the rate of change of y with respect to x, and evaluate this rate of change for values of x.

INSTANTANEOUS RATE OF CHANGE

A car travels 108 miles in 2 hours. Its *average speed* (or *average velocity*) is 108 mi/2 hr, or 54 mi/hr. This is the *average rate of change* of distance with respect to time. At various times during the trip, however, the speedometer did not read 54. Thus we say that 54 is the *average*. A snapshot of the speedometer taken at any instant would indicate *instantaneous* speed, or **instantaneous rate of change**.

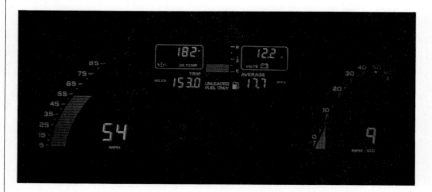

Average rates of change are given by difference quotients. If distance s is a function of time t and if h is the duration of the trip, then average velocity is given by

$$\text{Average velocity} = \frac{s(t + h) - s(t)}{h}.$$

Instantaneous rates of change are found by letting h approach 0. Thus,

$$\text{Instantaneous velocity} = \lim_{h \to 0} \frac{s(t + h) - s(t)}{h} = s'(t).$$

EXAMPLE 1   *Physical science: Velocity.* An object travels in such a way that distance s (in miles) from the starting point is a function of time t (in hours) as follows:

$$s(t) = 10t^2.$$

a) Find the average velocity between the times $t = 2$ and $t = 5$.

b) Find the (instantaneous) velocity when $t = 4$.

An instantaneous velocity.

Solution

a) From $t = 2$ to $t = 5$, $h = 3$, so

$$\frac{s(t + h) - s(t)}{h} = \frac{s(2 + 3) - s(2)}{3}$$

$$= \frac{s(5) - s(2)}{3}$$

$$= \frac{10 \cdot 5^2 - 10 \cdot 2^2}{3}$$

$$= \frac{250 - 40}{3}$$

$$= \frac{210}{3} = 70\frac{\text{mi}}{\text{hr}}.$$

b) The instantaneous velocity is given by $s'(t) = 20t$. Thus,

$$s'(4) = 20 \cdot 4 = 80\frac{\text{mi}}{\text{hr}}.$$ ❖

DO EXERCISE 1.

We generally use the letter v for velocity. Thus we have the following.

DEFINITION

$$\text{Velocity} = v(t) = \lim_{h \to 0} \frac{s(t + h) - s(t)}{h} = s'(t)$$

The rate of change of velocity is called **acceleration.** We generally use the letter a for acceleration. Thus the following definition applies.

DEFINITION

$$\text{Acceleration} = a(t) = v'(t)$$

EXAMPLE 2 *Physical science: Distance, velocity, and acceleration.* For $s(t) = 10t^2$, find $v(t)$ and $a(t)$, where s is in miles and t is in hours. Then find the distance, velocity, and acceleration when $t = 4$ hr.

1. *Physical science: Velocity.* For the function in Example 1:

  a) Find the average velocity between the times $t = 1$ and $t = 6$.

  b) Find the instantaneous velocity when $t = 5$.

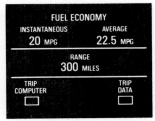

Some of the newer automobiles describe fuel economy by giving *average* miles per gallon and *instantaneous* miles per gallon.

2. *Physical science: Velocity.* An object is dropped from a certain height. It will fall a distance of s feet in t seconds as given by

$$s(t) = 16t^2.$$

a) Find the velocity $v(t)$.

b) Find the object's velocity 2 sec after it has been dropped.

c) Find its velocity 10 sec after it has been dropped.

3. *Physical science: Acceleration.* Referring to Margin Exercise 2, find $a(t)$. In what units should it be expressed?

4. *General interest: Volume.* The volume V of a cubical carton with a side of length s, in feet, is given by

$$V(s) = s^3.$$

a) Find the rate of change of the volume V with respect to the length s of a side.

b) Find the rate of change of the volume when $s = 10$ ft.

Solution We have

$$v(t) = s'(t) = 20t,$$
$$a(t) = v'(t) = 20.$$

Then

$$s(4) = 10(4)^2 = 160 \text{ miles},$$
$$v(4) = 20(4) = 80 \text{ mi/hr}, \quad \text{and}$$
$$a(4) = 20 \text{ mi/hr}^2.$$

If this distance function applies to a vehicle, then at time $t = 4$ hr, the distance is 160 miles, the velocity, or instantaneous speed, is 80 mi/hr, and the acceleration is 20 miles per hour per hour, which we abbreviate as 20 mi/hr^2. ❖

DO EXERCISES 2 AND 3.

In general, derivatives give instantaneous rates of change.

DEFINITION

If **y is a function of x**, then the **(instantaneous)** *rate of change of y with respect to x* is given by the derivative

$$\frac{dy}{dx} = f'(x) = \lim_{h \to 0} \frac{f(x+h) - f(x)}{h}.$$

EXAMPLE 3 *Life science: Volume of cancer tumor.* The spherical volume V of a cancer tumor is given by

$$V(r) = \tfrac{4}{3}\pi r^3,$$

where r is the radius of the tumor, in centimeters.

a) Find the rate of change of the volume with respect to the radius.

b) Find the rate of change of the volume at $r = 1.2$ cm.

Solution

a) $\dfrac{dV}{dr} = V'(r) = 3 \cdot \dfrac{4}{3} \cdot \pi r^2 = 4\pi r^2$

(This turns out to be the surface area.)

b) $V'(1.2) = 4\pi(1.2)^2 = 5.76\pi \approx 18\dfrac{\text{cm}^3}{\text{cm}} = 18 \text{ cm}^2$ ❖

DO EXERCISE 4.

EXAMPLE 4   *Life science: Population growth.* The initial population in a bacteria colony is 10,000. After t hours, the colony has grown to a number $P(t)$ given by

$$P(t) = 10,000(1 + 0.86t + t^2).$$

a)   Find the rate of change of the population P with respect to time t. This is also known as the **growth rate**.

b)   Find the number of bacteria present after 5 hr. Also, find the growth rate when $t = 5$.

Solution

a)   Note that $P(t) = 10,000 + 8600t + 10,000t^2$. Then

$$P'(t) = 8600 + 20,000t.$$

b)   The number of bacteria present when $t = 5$ is given by

$$P(5) = 10,000 + 8600 \cdot 5 + 10,000 \cdot 5^2 = 303,000.$$

The growth rate when $t = 5$ is given by

$$P'(5) = 8600 + 20,000 \cdot 5 = 108,600 \ \frac{\text{bacteria}}{\text{hr}}.$$

Thus at $t = 5$, there are 303,000 bacteria present, and the colony is growing at the rate of 108,600 bacteria per hour.   ❖

DO EXERCISE 5.

Rates of Change in Business and Economics

In the study of business and economics, we are frequently interested in how such quantities as cost, revenue, and profit change with an increase in product quantity. In particular, we are interested in what is called **marginal*** cost or profit (or whatever). This term is used to signify the *rate of change with respect to quantity*. Thus, if

$C(x)$ = the *total cost* of producing x units of a product (usually considered in some time period),

then

$C'(x)$ = the **marginal cost**
 = the rate of change of the total cost with respect to the number of units, x, produced.

---

*The term "marginal" comes from the Marginalist School of Economic Thought, which originated in Austria for the purpose of applying mathematics and statistics to the study of economics.

5.   *Life science: Population growth.* The initial population of a bacteria colony is 10,000. After t hours, the colony grows to a number $P(t)$ given by

$$P(t) = 10,000(1 + 0.97t + t^2).$$

a)   Find the growth rate of the population.

b)   Find the number of bacteria present (the population) when $t = 5$ hr. Find the growth rate when $t = 5$ hr.

c)   Find the number of bacteria present when $t = 6$ hr. Find the growth rate when $t = 6$.

Let us think about these interpretations. The total cost of producing 5 units of a product is $C(5)$. The rate of change $C'(5)$ is the cost per unit at that stage in the production process. That this cost per unit does not include fixed costs is seen in this example:

$$C(x) = \underbrace{(x^2 + 4x)}_{\text{Variable costs}} \quad + \quad \underbrace{\$10{,}000.}_{\text{Fixed costs (constant)}}$$

Because the derivative of a constant is 0,

$$C'(x) = 2x + 4.$$

This verifies an economic principle stating that the fixed costs of a company have no effect on marginal cost.

Following are some other marginal functions. Recall that

$$R(x) = \text{the } \textit{total revenue} \text{ from the sale of } x \text{ units.}$$

Then

$R'(x) = $ the **marginal revenue**

$= $ the rate of change of the total revenue with respect to the number of units, x, sold.

Also,

$P(x) = $ the *total profit* from the production and sale of x units of a product

$= R(x) - C(x)$.

Then

$P'(x) = $ the **marginal profit**

$= $ the rate of change of the total profit with respect to the number of units, x, produced and sold

$= R'(x) - C'(x)$.

EXAMPLE 5   *Business: Marginal revenue, cost, and profit.* Given

$$R(x) = 50x, \quad \text{and}$$
$$C(x) = 2x^3 - 12x^2 + 40x + 10,$$

find each of the following.

a)   Total profit $P(x)$

b)   Total revenue $R(2)$, cost $C(2)$, and profit $P(2)$ from the production and sale of 2 units of the product

c)   Marginal revenue $R'(x)$, cost $C'(x)$, and profit $P'(x)$

d) Marginal revenue $R'(2)$, cost $C'(2)$, and profit $P'(2)$ at that point in the process where 2 units have been produced and sold

Solution

a) The total profit $P(x) = R(x) - C(x)$
$$= 50x - (2x^3 - 12x^2 + 40x + 10)$$
$$= -2x^3 + 12x^2 + 10x - 10$$

b) $R(2) = 50 \cdot 2 = \$100$ (the total revenue from the sale of the first 2 units);

$C(2) = 2 \cdot 2^3 - 12 \cdot 2^2 + 40 \cdot 2 + 10 = \$58$ (the total cost of producing the first 2 units);

$P(2) = R(2) - C(2) = \$100 - \$58 = \$42$ (the total profit from the production and sale of the first 2 units)

c) The marginal revenue $R'(x) = 50$;

The marginal cost $C'(x) = 6x^2 - 24x + 40$;

The marginal profit $P'(x) = R'(x) - C'(x)$
$$= 50 - (6x^2 - 24x + 40)$$
$$= -6x^2 + 24x + 10$$

d) $R'(2) = \$50$ per unit;

$C'(2) = 6 \cdot 2^2 - 24 \cdot 2 + 40 = \$16$ per unit;

$P'(2) = \$50 - \$16 = \$34$ per unit

Note that the marginal revenue in this example is constant. No matter how much is produced and sold, the revenue per unit stays the same. This may not always be the case. Also note that $C'(2)$, or $16 per unit, is not the average cost per unit, which is given by

$$\frac{\text{Total cost of producing 2 units}}{\text{2 units}} = \frac{\$58}{2} = \$29 \text{ per unit.} \qquad ❖$$

In general, we have the following.

DEFINITION

$A(x) = $ the *average cost* of producing x units $= \dfrac{C(x)}{x}$

DO EXERCISE 6.

Let us look at a typical marginal-cost function C' (Fig. 1a) and its associated total-cost function C (Fig. 1b).

6. *Business: Marginal revenue, cost, and profit.* Given

$$R(x) = 50x - 0.5x^2 \quad \text{and}$$
$$C(x) = 10x + 3,$$

find each of the following.

a) $P(x)$

b) $R(40)$, $C(40)$, and $P(40)$

c) $R'(x)$, $C'(x)$, and $P'(x)$

d) $R'(40)$, $C'(40)$, and $P'(40)$

e) Is the marginal revenue constant?

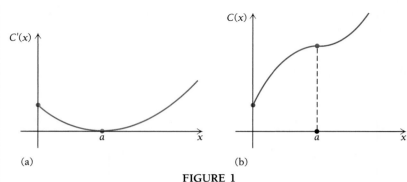

FIGURE 1

Marginal cost normally decreases as more units are produced until it reaches some minimum value at a, and then it increases. (This is probably due to factors such as paying overtime or buying more machinery.) Since $C'(x)$ represents the slope of $C(x)$ and is positive and decreasing up to a, the graph of $C(x)$ turns down as x goes from 0 to a. Then past a, it turns up.

EXERCISE SET **2.6**

1. Given

$$s(t) = t^3 + t,$$

where s is measured in feet and t in seconds, find each of the following.

a) $v(t)$

b) $a(t)$

c) The velocity and acceleration when $t = 4$ sec

2. Given

$$s(t) = 3t + 10,$$

where s is measured in miles and t in hours, find each of the following.

a) $v(t)$

b) $a(t)$

c) The velocity and acceleration when $t = 2$ hr. When the distance function is given by a linear function, we have *uniform motion*.

APPLICATIONS

❖ **Business and Economics**

3. *Marginal revenue, cost, and profit.* Given

$$R(x) = 5x \quad \text{and}$$
$$C(x) = 0.001x^2 + 1.2x + 60,$$

find each of the following.

a) $P(x)$

b) $R(100)$, $C(100)$, and $P(100)$

c) $R'(x)$, $C'(x)$, and $P'(x)$

d) $R'(100)$, $C'(100)$, and $P'(100)$

4. *Marginal revenue, cost, and profit.* Given

$$R(x) = 50x - 0.5x^2 \quad \text{and}$$
$$C(x) = 4x + 10,$$

find each of the following.

a) $P(x)$

b) $R(20)$, $C(20)$, and $P(20)$

c) $R'(x)$, $C'(x)$, and $P'(x)$

d) $R'(20)$, $C'(20)$, and $P'(20)$

5. *Advertising.* A firm estimates that it will sell N units of a product after spending a dollars on advertising, where

$$N(a) = -a^2 + 300a + 6$$

and a is measured in thousands of dollars.

a) What is the rate of change of the number of units sold with respect to the amount spent on advertising?

b) How many units will be sold after spending $10,000 on advertising?

c) What is the rate of change at $a = 10$?

6. *Sales.* A company determines that monthly sales S, in thousands of dollars, after t months of marketing a product is given by

$$S(t) = 2t^3 - 40t^2 + 220t + 160.$$

a) Find the monthly sales after 1 month, 4 months, 6 months, 9 months, and 20 months.

b) Find the rate of change $S'(t)$.

c) Find the rate of change at $t = 1$, $t = 4$, $t = 6$, $t = 9$, and $t = 20$.

Marginal productivity. An employee's monthly productivity, in numbers of units produced, M, is found to be a function of the number of years of service, t. For a certain product, the productivity function is given by

$$M(t) = -2t^2 + 100t + 180.$$

a) Find the productivity of an employee after 5 years, 10 years, 25 years, and 45 years of service.

b) Find the marginal productivity.

c) Find the marginal productivity at $t = 5$, $t = 10$, $t = 25$, and $t = 45$.

8. *Supply.* A supply function for a certain product is given by

$$S(p) = 0.08p^3 + 2p^2 + 10p + 11.$$

a) Find the rate of change of supply with respect to price, dS/dp.

b) How many units will the seller allow to be sold when the price is $3 per unit?

c) What is the rate of change at $p = 3$?

9. *Demand.* A demand function for a certain product is given by

$$D(p) = 100 - \sqrt{p}.$$

a) Find the rate of change of quantity with respect to price, dD/dp.

b) How many units will the consumer want to buy when the price is $25 per unit?

c) What is the rate of change at $p = 25$?

❖ **Life and Physical Sciences**

10. *Stopping distance on glare ice.* The stopping distance on glare ice (at some fixed speed) of regular tires is given by a linear function of the air temperature F,

$$D(F) = 2F + 115,$$

where $D(F) =$ the stopping distance, in feet, when the air temperature is F, in degrees Fahrenheit. Find the rate of change of the stopping distance D with respect to the air temperature F.

11. *Healing wound.* The circumference C, in centimeters, of a healing wound is given by

$$C(r) = 2\pi r,$$

where r is the radius, in centimeters. Find the rate of change of the circumference with respect to the radius.

12. *Healing wound.* The circular area A, in square centimeters, of a healing wound is given by

$$A(r) = \pi r^2,$$

where r is the radius, in centimeters. Find the rate of change of the area with respect to the radius.

13. *Temperature during an illness.* The temperature T of a person during an illness is given by

$$T(t) = -0.1t^2 + 1.2t + 98.6,$$

where T is the temperature, in degrees Fahrenheit, at time t, in days.

a) Find the rate of change of the temperature with respect to time.

b) Find the temperature at $t = 1.5$ days.

c) Find the rate of change at $t = 1.5$ days.

14. *Blood pressure.* For a certain dosage of x cubic centimeters (cc) of a drug, the resulting blood pressure B is approximated by

$$B(x) = 0.05x^2 - 0.3x^3.$$

Find the rate of change of the blood pressure with respect to the dosage. Such a rate of change is often called the *sensitivity*.

15. *Territorial area.* The territorial area T of an animal is defined as its defended, or exclusive, region. The area T of that region can be approximated using the animal's body weight W by

$$T = W^{1.31}$$

(see Section 1.5). Find dT/dW.

16. *Home range.* The home range H of an animal is defined as the region to which the animal confines its movements. The area of that region can be approximated using the animal's body weight W by

$$H = W^{1.41}$$

(see Section 1.5). Find dH/dW.

17. *Sensitivity.* The reaction R of the body to a dose Q of medication is often represented by the general function

$$R(Q) = Q^2\left(\frac{k}{2} - \frac{Q}{3}\right),$$

where k is a constant and R is measured in millimeters, if the reaction is a change in blood pressure, or in degrees of temperature, if the reaction is a change in temperature. The rate of change dR/dQ is defined to be the *sensitivity*. Find a formula for the sensitivity.

❖ Social Sciences

18. *Population growth rate.* The population of a city grows from an initial size of 100,000 to an amount P given by

$$P(t) = 100,000 + 2000t^2,$$

where t is measured in years.

a) Find the growth rate.

b) Find the number of people in the city after 10 years (at $t = 10$ yr).

c) Find the growth rate at $t = 10$.

19. *Median age of women at first marriage.* The median age of women at first marriage can be approximated by the linear function

$$A(t) = 0.08t + 19.7,$$

where $A(t)$ = the median age of women at first marriage the tth year after 1950. Find the rate of change of the median age A with respect to time t.

These bears may be determining territorial area.

❖ **General Interest**

20. *View to the horizon.* The view V, or distance, in miles that one can see to the horizon from a height h, in feet, is given by

$$V = 1.22\sqrt{h}.$$

a) Find the rate of change of V with respect to h.

b) How far can one see to the horizon from an airplane window from a height of 40,000 ft?

c) Find the rate of change at $h = 40{,}000$.

COMPUTER–GRAPHING CALCULATOR EXERCISES

THE CALCULUS EXPLORER:
Derivatives

The program graphs f and f' concurrently. Enter, respectively, various sets of total-revenue, -cost, and -profit functions considered in this section. Then try some other functions.

2.7

DIFFERENTIATION TECHNIQUES: THE PRODUCT AND QUOTIENT RULES

OBJECTIVE

a) Differentiate using the Product and the Quotient Rules.

The Product Rule

The derivative of a sum is the sum of the derivatives, but the derivative of a product is *not* the product of the derivatives. To see this, consider x^2 and x^5. The product is x^7, and the derivative of this product is $7x^6$. The individual derivatives are $2x$ and $5x^4$, and the product of these derivatives is $10x^5$, which is not $7x^6$.

The following is the rule for finding the derivative of a product.

THEOREM 5

The Product Rule

Suppose $F(x) = f(x) \cdot s(x)$, where $f(x)$ is the "first" function and $s(x)$ is the "second" function. Then

$$F'(x) = f(x) \cdot s'(x) + f'(x) \cdot s(x).$$

The derivative of a product is the first function times the derivative of the second function, plus the derivative of the first function times the second function.

---

150 DIFFERENTIATION

Use the Product Rule to find $f'(x)$.

1. $f(x) = 3x^8 \cdot x^{10}$

2. $f(x) = (9x^3 + 4x^2 + 10)(-7x^2 + x^4)$

3. $f(x) = (3x^4 + \sqrt{x})(x^2 - 6x)$

Let us check this for $x^2 \cdot x^5$. There are five steps.

$x^2 \cdot 5x^4 + 2x \cdot x^5$
$= 5x^6 + 2x^6$
$= 7x^6$

1. Write down the first factor.
2. Multiply it by the derivative of the second factor.
3. Write down the derivative of the first factor.
4. Multiply it by the second factor.
5. Add the result of steps (1) and (2) to the result of steps (3) and (4).

EXAMPLE 1 Find $\dfrac{d}{dx}[(x^4 - 2x^3 - 7)(3x^2 - 5x)]$.

Solution We have

$$\frac{d}{dx}(x^4 - 2x^3 - 7)(3x^2 - 5x) = (x^4 - 2x^3 - 7)(6x - 5) + (4x^3 - 6x^2)(3x^2 - 5x).$$

Note that we could have multiplied the polynomials and then differentiated, avoiding the use of the Product Rule, but this would have been more work. ❖

EXAMPLE 2 Given $F(x) = (x^2 + 4x - 11)(7x^3 - \sqrt{x})$, find $F'(x)$.

Solution We rewrite this as

$$F(x) = (x^2 + 4x - 11)(7x^3 - x^{1/2}).$$

Then using the Product Rule, we have

$$F'(x) = (x^2 + 4x - 11)(21x^2 - \tfrac{1}{2}x^{-1/2}) + (2x + 4)(7x^3 - x^{1/2}). ❖$$

DO EXERCISES 1–3.

The Quotient Rule

The derivative of a quotient is *not* the quotient of the derivatives. To see why, consider x^5 and x^2. The quotient x^5/x^2 is x^3, and the derivative of this quotient is $3x^2$. The individual derivatives are $5x^4$ and $2x$, and the quotient of these derivatives $5x^4/2x$ is $(5/2)x^3$, which is not $3x^2$.

The rule for differentiating quotients is as follows.

THEOREM 6

The Quotient Rule

If

$$q(x) = \frac{n(x)}{d(x)},$$

then

$$q'(x) = \frac{d(x) \cdot n'(x) - d'(x) \cdot n(x)}{[d(x)]^2}.$$

The derivative of a quotient is the denominator times the derivative of the numerator, minus the derivative of the denominator times the numerator, all divided by the square of the denominator.

(If we think of the function in the numerator as the first function and the function in the denominator as the second function, then we can reword the Quotient Rule as "the second function times the derivative of the first function minus the derivative of the second function times the first function, all divided by the square of the second function.")

The Quotient Rule is illustrated below.

1. Write down the denominator.

2. Multiply the denominator by the derivative of the numerator.

3. Write a minus sign.

4. Write down the derivative of the denominator.

5. Multiply it by the numerator.

6. Divide by the square of the denominator.

EXAMPLE 3 For $q(x) = x^5/x^3$, find $q'(x)$.

4. For

$$f(x) = \frac{x^9}{x^5},$$

find $f'(x)$ using the Quotient Rule.

Differentiate.

5. $f(x) = \frac{1 - x^2}{x^5}$

6. $f(x) = \frac{x^2 - 1}{x^3 + 1}$

Solution

$$q'(x) = \frac{x^3 \cdot 5x^4 - 3x^2 \cdot x^5}{(x^3)^2}$$

$$= \frac{5x^7 - 3x^7}{x^6}$$

$$= \frac{2x^7}{x^6}$$

$$= 2x$$

DO EXERCISE 4.

EXAMPLE 4   Differentiate $f(x) = \frac{1 + x^2}{x^3}$.

Solution

$$f'(x) = \frac{x^3 \cdot 2x - 3x^2(1 + x^2)}{(x^3)^2}$$

$$= \frac{2x^4 - 3x^2 - 3x^4}{x^6}$$

$$= \frac{-x^4 - 3x^2}{x^6}$$

$$= \frac{-x^2 \cdot x^2 - 3x^2}{x^6}$$

$$= \frac{x^2(-x^2 - 3)}{x^2 \cdot x^4}$$

$$= \frac{-x^2 - 3}{x^4}$$

EXAMPLE 5   Differentiate $f(x) = \frac{x^2 - 3x}{x - 1}$.

Solution

$$f'(x) = \frac{(x - 1)(2x - 3) - 1(x^2 - 3x)}{(x - 1)^2}$$

$$= \frac{2x^2 - 5x + 3 - x^2 + 3x}{(x - 1)^2}$$

$$= \frac{x^2 - 2x + 3}{(x - 1)^2}$$

It is not necessary to multiply out $(x - 1)^2$ to get $x^2 - 2x + 1$.

DO EXERCISES 5 AND 6.

An Application

We discussed earlier that it is typical for a total-revenue function to vary depending on the number of units x sold. Let us see what can determine this. Recall the consumer's demand function, or price function, $x = D(p)$, discussed in Section 1.5. It is the quantity x of units of a product that a seller will purchase when the price is p dollars per unit. This is typically a decreasing function (Fig. 1).

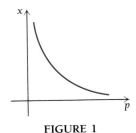

FIGURE 1

The total revenue when the price is p dollars per unit is then

$$R(p) = \text{(Number of units sold)} \cdot \text{(Price charged per unit)},$$

or

$$R(p) = x \cdot p = D(p) \cdot p = p\,D(p).$$

A typical graph of a revenue function is shown in Fig. 2.

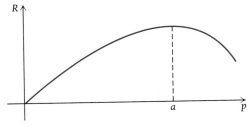

FIGURE 2

As the price increases, $D(p)$ decreases. Because we have a product $p \cdot D(p)$, the revenue typically rises for a while as p increases, but tapers off as $D(p)$ gets smaller and smaller.

Using the Product Rule, we can obtain an expression for the marginal revenue $R'(p)$ in terms of p and $D'(p)$ as

$$R(p) = p\,D(p),$$

7. *Business: Revenue.* A company determines that the demand function for a certain product is given by

$$D(p) = 200 - p.$$

a) Find an expression for the total revenue $R(p)$.

b) Find the marginal revenue $R'(p)$.

so

$$R'(p) = p \cdot D'(p) + 1 \cdot D(p) = pD'(p) + D(p).$$

You need not memorize this. You can merely repeat the Product Rule where necessary.

DO EXERCISE 7.

EXERCISE SET **2.7**

Differentiate.

1. $y = x^3 \cdot x^8$, two ways

2. $y = x^4 \cdot x^9$, two ways

3. $y = \dfrac{-1}{x}$, two ways

4. $y = \dfrac{1}{x}$, two ways

5. $y = \dfrac{x^8}{x^5}$, two ways

6. $y = \dfrac{x^9}{x^5}$, two ways

7. $y = (8x^5 - 3x^2 + 20)(8x^4 - 3\sqrt{x})$

8. $f(x) = (7x^6 + 4x^3 - 50)(9x^{10} - 7\sqrt{x})$

9. $f(x) = x(300 - x)$, two ways

10. $f(x) = x(400 - x)$, two ways

11. $f(x) = (4\sqrt{x} - 6)(x^3 - 2x + 4)$

12. $f(x) = (\sqrt[3]{x} - 5x^2 + 4)(4x^2 + 11x - 5)$

13. $f(x) = (x + 3)^2$ [Hint: $(x + 3)^2 = (x + 3)(x + 3)$.]

14. $f(x) = (5x - 4)^2$

15. $f(x) = (x^3 - 4x)^2$

16. $f(x) = (3x^2 - 4x + 5)^2$

17. $f(x) = 5x^{-3}(x^4 - 5x^3 + 10x - 2)$

18. $f(x) = 6x^{-4}(6x^3 + 10x^2 - 8x + 3)$

19. $f(x) = \left(x + \dfrac{2}{x}\right)(x^2 - 3)$

20. $f(x) = (4x^3 - x^2)\left(x - \dfrac{5}{x}\right)$

21. $f(x) = \dfrac{x}{300 - x}$

22. $f(x) = \dfrac{x}{400 - x}$

23. $f(x) = \dfrac{3x - 1}{2x + 5}$

24. $f(x) = \dfrac{2x + 3}{x - 5}$

25. $y = \dfrac{x^2 + 1}{x^3 - 1}$

26. $y = \dfrac{x^3 - 1}{x^2 + 1}$

27. $y = \dfrac{x}{1 - x}$

28. $y = \dfrac{x}{3 - x}$

29. $y = \dfrac{x - 1}{x + 1}$

30. $y = \dfrac{x + 2}{x - 2}$

31. $f(x) = \dfrac{1}{x - 3}$

32. $f(x) = \dfrac{1}{x + 2}$

33. $f(x) = \dfrac{3x^2 + 2x}{x^2 + 1}$

34. $f(x) = \dfrac{3x^2 - 5x}{x^2 - 1}$

35. $f(x) = \dfrac{3x^2 - 5x}{x^8}$, two ways

36. $f(x) = \dfrac{3x^2 + 2x}{x^5}$, two ways

37. $g(x) = \dfrac{4x + 3}{\sqrt{x}}$

38. $g(x) = \dfrac{6x^2 - 3x}{3\sqrt{x}}$

39. Find an equation of the tangent line to the graph of $y = 8/(x^2 + 4)$ at the point $(0, 2)$ and at $(-2, 1)$.

40. Find an equation of the tangent line to the graph of $y = 4x/(1 + x^2)$ at the point $(0, 0)$ and at $(-1, -2)$.

APPLICATIONS

❖ **Business and Economics**

41. *Marginal demand.* The demand function for a certain product is given by

$$D(p) = \frac{2p + 300}{10p + 11}.$$

a) Find the marginal demand $D'(p)$.
b) Find $D'(4)$.

42. *Marginal revenue.* A company sells fix-it-yourself books. It finds that its total revenue from the sale of x books is given by

$$R(x) = \frac{12{,}000\sqrt{x}}{4 + x^{3/2}}.$$

a) Find the marginal revenue $R'(x)$.
b) Find $R'(4)$.

Marginal revenue. In each of Exercises 43–46, a demand function $x = D(p)$ is given. Find (a) the total revenue $R(p)$ and (b) the marginal revenue $R'(p)$.

43. $D(p) = 400 - p$ **44.** $D(p) = 500 - p$

45. $D(p) = \dfrac{4000}{p} + 3$ **46.** $D(p) = \dfrac{3000}{p} + 5$

47. *Marginal average cost.* In Section 2.6, we defined the average cost of producing x units of a product in terms of the total cost $C(x)$ by

$$A(x) = \frac{C(x)}{x}.$$

Use the Quotient Rule to find a general expression for *marginal average cost* $A'(x)$.

48. *Marginal demand.* In this section we determined that

$$R(p) = p\,D(p).$$

Then

$$D(p) = \frac{R(p)}{p} = \text{the number of units sold when the price is } p \text{ dollars per unit}$$

$$= \text{the demand function.}$$

Use the Quotient Rule to find a general expression for *marginal demand* $D'(p)$.

SYNTHESIS EXERCISES

Differentiate.

49. $f(x) = \dfrac{x^3}{\sqrt{x} - 5}$

50. $g(t) = \dfrac{1 + \sqrt{t}}{t^5 + 3}$

51. $f(v) = \dfrac{3}{1 + v + v^2}$

52. $g(z) = \dfrac{1 + z + z^2}{1 - z + z^2}$

53. $p(t) = \dfrac{t}{1 - t + t^2 - t^3}$

54. $f(x) = \dfrac{\dfrac{2}{3x} - 1}{\dfrac{3}{x^2} + 5}$

55. $h(x) = \dfrac{x^3 + 5x^2 - 2}{\sqrt{x}}$

56. $y(t) = 5t(t - 1)(2t + 3)$

57. $f(x) = x(3x^3 + 6x - 2)(3x^4 + 7)$

58. $g(x) = (x^3 - 8) \cdot \dfrac{x^2 + 1}{x^2 - 1}$

59. $f(t) = (t^5 + 3) \cdot \dfrac{t^3 - 1}{t^3 + 1}$

60. $f(x) = \dfrac{(x^2 + 3x)(x^5 - 7x^2 - 3)}{x^4 - 3x^3 - 5}$

61. $f(x) = \dfrac{(2x^2 + 3)(4x^3 - 7x + 2)}{x^7 - 2x^6 + 9}$

62. $s(t) = \dfrac{5t^8 - 2t^3}{(t^5 - 3)(t^4 + 7)}$

For each function, sketch the graph of f and f' on the given interval. Then estimate points at which the tangent line is horizontal.

63. $f(x) = x(x + 2)(x - 2)$ **64.** $f(x) = x^2(x - 2)(x + 2)$

65. $f(x) = \left(x + \dfrac{2}{x}\right)(x^2 - 3)$ **66.** $f(x) = \dfrac{x^3 - 1}{x^2 + 1}$

67. $f(x) = \dfrac{0.3x}{0.04 + x^2}$ **68.** $f(x) = \dfrac{0.01x^2}{x^4 + 0.0256}$

> **THE CALCULUS EXPLORER:**
> *Derivatives*
>
> Select functions f considered in this exercise set. Enter them to see f and f' concurrently. Try
>
> $$f(x) = \frac{4x}{x^2 + 1}.$$

2.8

OBJECTIVES

THE CHAIN RULE

a) Differentiate using the Extended Power Rule and the Chain Rule.

b) Find the composition of functions.

The Extended Power Rule

How can we differentiate more complicated functions such as

$$y = (1 + x^2)^3, \qquad y = (1 + x^2)^{89}, \quad \text{or} \quad y = (1 + x^2)^{1/3}?$$

For $(1 + x^2)^3$, we can expand and then differentiate. Although this could be done for $(1 + x^2)^{89}$, it would certainly be time-consuming, and such an expansion would not work for $(1 + x^2)^{1/3}$. Not knowing a rule, we might conjecture that the derivative of the function $y = (1 + x^2)^3$ is

$$3(1 + x^2)^2. \tag{1}$$

To check this, we expand $(1 + x^2)^3$ and then differentiate. From Exercise Set 1.1, we know that $(a + h)^3 = a^3 + 3a^2h + 3ah^2 + h^3$, so

$$(1 + x^2)^3 = 1^3 + 3 \cdot 1^2 \cdot (x^2)^1 + 3 \cdot 1 \cdot (x^2)^2 + (x^2)^3$$
$$= 1 + 3x^2 + 3x^4 + x^6.$$

(We could also have done this by finding $(1 + x^2)^2$ and then multiplying again by $1 + x^2$.) It follows that

$$\frac{dy}{dx} = 6x + 12x^3 + 6x^5 = (1 + 2x^2 + x^4)6x$$

$$= 3(1 + x^2)^2 \cdot 2x. \tag{2}$$

Comparing this with Eq. (1), we see that the Power Rule is not sufficient for such a differentiation. Note that the factor $2x$ in the actual derivative, Eq. (2), is the derivative of the "inside" function, $1 + x^2$. This is consistent with the following new rule.

THEOREM 7

The Extended Power Rule

Suppose $g(x)$ is a function of x. Then for any real number k,

$$\frac{d}{dx}[g(x)]^k = k[g(x)]^{k-1} \cdot \frac{d}{dx} g(x).$$

Let us differentiate $(1 + x^3)^5$. There are three steps to carry out.

$(1 + x^3)^5$ **1.** Mentally block out the "inside" function, $1 + x^3$.

$5(1 + x^3)^4$ **2.** Differentiate the "outside" function, $(1 + x^3)^5$.

$5(1 + x^3)^4 \cdot 3x^2$
$= 15x^2(1 + x^3)^4$ **3.** Multiply by the derivative of the "inside" function.

Step (3) is most commonly overlooked. Try not to forget it!

EXAMPLE 1   Differentiate $f(x) = (1 + x^3)^{1/2}$.

Solution

$$\frac{d}{dx}(1 + x^3)^{1/2} = \frac{1}{2}(1 + x^3)^{1/2 - 1} \cdot 3x^2$$

$$= \frac{1}{2}(1 + x^3)^{-1/2} \cdot 3x^2$$

$$= \frac{3x^2}{2\sqrt{1 + x^3}} \qquad \qquad ❖$$

DO EXERCISES 1 AND 2.

EXAMPLE 2   Differentiate $y = (1 - x^2)^3 - (1 - x^2)^2$.

Differentiate.

1. $f(x) = (1 + x^2)^{10}$

2. $y = (1 - x^2)^{1/2}$

3. Differentiate:
$$f(x) = (1 + x^2)^2 - (1 + x^2)^3.$$

4. Differentiate:
$$y = (x - 4)^5(6 - x)^3.$$

Solution Here we combine the Difference Rule and the Extended Power Rule:

$$\frac{dy}{dx} = 3(1 - x^2)^2(-2x) - 2(1 - x^2)(-2x).$$ We differentiate each term using the Extended Power Rule.

Thus,

$$\frac{dy}{dx} = -6x(1 - x^2)^2 + 4x(1 - x^2)$$
$$= x(1 - x^2)[-6(1 - x^2) + 4]$$ Here we factor out $x(1 - x^2)$.
$$= x(1 - x^2)[-6 + 6x^2 + 4]$$
$$= x(1 - x^2)(6x^2 - 2)$$
$$= 2x(1 - x^2)(3x^2 - 1).$$

DO EXERCISE 3.

EXAMPLE 3 Differentiate $f(x) = (x - 5)^4(7 - x)^{10}$.

Solution Here we combine the Product Rule and the Extended Power Rule:

$$f'(x) = (x - 5)^4 \cdot 10(7 - x)^9(-1) + 4(x - 5)^3(1)(7 - x)^{10}$$
$$= -10(x - 5)^4(7 - x)^9 + 4(x - 5)^3(7 - x)^{10}$$
$$= 2(x - 5)^3(7 - x)^9[-5(x - 5) + 2(7 - x)]$$ We factor out $2(x - 5)^3(7 - x)^9$.
$$= 2(x - 5)^3(7 - x)^9[-5x + 25 + 14 - 2x]$$
$$= 2(x - 5)^3(7 - x)^9(39 - 7x).$$

DO EXERCISE 4.

EXAMPLE 4 Differentiate $f(x) = \sqrt[4]{\dfrac{x + 3}{x - 1}}$.

Solution We must use the Quotient Rule to differentiate the inside function, $(x + 3)/(x - 1)$:

$$\frac{d}{dx}\sqrt[4]{\frac{x + 3}{x - 1}} = \frac{d}{dx}\left(\frac{x + 3}{x - 1}\right)^{1/4} = \frac{1}{4}\left(\frac{x + 3}{x - 1}\right)^{1/4 - 1}\left[\frac{(x - 1)1 - 1(x + 3)}{(x - 1)^2}\right]$$
$$= \frac{1}{4}\left(\frac{x + 3}{x - 1}\right)^{-3/4}\left[\frac{x - 1 - x - 3}{(x - 1)^2}\right]$$

then

$$= \frac{1}{4}\left(\frac{x+3}{x-1}\right)^{-3/4} \cdot \frac{-4}{(x-1)^2}$$

$$= \left(\frac{x+3}{x-1}\right)^{-3/4} \cdot \frac{-1}{(x-1)^2}.$$

❖

DO EXERCISE 5.

Composition of Functions and the Chain Rule

The Extended Power Rule is a special case of a more general rule called the *Chain Rule*. Before discussing it, we define the *composition* of functions.

There is a function g that gives a correspondence between women's shoe sizes in the United States and those in Italy. The function is given by $g(x) = 2(x + 12)$, where x is a shoe size in the United States and $g(x)$ is a shoe size in Italy. For example, a shoe size of 4 in the United States corresponds to a shoe size of $g(4) = 2(4 + 12)$, or 32 in Italy.

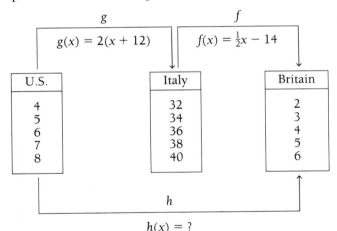

$$h(x) = ?$$

There is also a function f that gives a correspondence between women's shoe sizes in Italy and those in Britain. The function is given by $f(x) = \frac{1}{2}x - 14$, where x is a shoe size in Italy and $f(x)$ is the corresponding shoe size in Britain. For example, a shoe size of 32 in Italy corresponds to a shoe size of $f(32) = \frac{1}{2}(32) - 14$, or 2 in Britain.

It seems reasonable to assume that a shoe size of 4 in the United States corresponds to a shoe size of 2 in Britain and that there is a function h that describes this correspondence. Can we find a formula for h?

5. Differentiate:

$$y = \sqrt[3]{\frac{x+5}{x-4}}.$$

Looking at the tables above, we might guess that such a formula is $h(x) = x - 2$, and that is indeed correct. For more complicated formulas, however, we need to do some algebra.

A shoe size x in the United States corresponds to a shoe size $g(x)$ in Italy, where

$$g(x) = 2(x + 12).$$

Now $2(x + 12)$ is a shoe size in Italy. If we replace x in $f(x)$ by $2(x + 12)$, we can find the corresponding shoe size in Britain:

$$\begin{aligned} f(g(x)) &= \tfrac{1}{2}[2(x + 12)] - 14 \\ &= \tfrac{1}{2}[2x + 24] - 14 \\ &= x + 12 - 14 \\ &= x - 2. \end{aligned}$$

This gives a formula for h: $h(x) = x - 2$. Thus a shoe size of 4 in the United States corresponds to a shoe size of $h(4) = 4 - 2$, or 2 in Britain. The function h is called the **composition** of f and g and is denoted $f \circ g$.

DEFINITION

The *composed* function $f \circ g$, the *composition* of f and g, is defined as

$$f \circ g(x) = f(g(x)).$$

We can visualize the composition of functions as shown in Fig. 1.

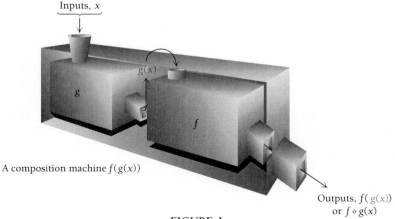

A composition machine $f(g(x))$

FIGURE 1

To find $f \circ g(x)$, we substitute $g(x)$ for x in $f(x)$.

EXAMPLE 5 Given $f(x) = x^3$ and $g(x) = 1 + x^2$, find $f \circ g(x)$ and $g \circ f(x)$.

Solution Consider each function separately:

$$f(x) = x^3 \qquad \text{This function cubes each input.}$$

and

$$g(x) = 1 + x^2. \qquad \text{This function adds 1 to the square of each input.}$$

a) $f \circ g$ first does what g does (adds 1 to the square) and then does what f does (cubes). We find $f(g(x))$ by substituting $g(x)$ for x:

$$f \circ g(x) = f(g(x)) = f(1 + x^2) \qquad \text{Substituting } 1 + x^2 \text{ for } x$$
$$= (1 + x^2)^3$$
$$= 1 + 3x^2 + 3x^4 + x^6.$$

b) $g \circ f$ first does what f does (cubes) and then does what g does (adds 1 to the square). We find $g(f(x))$ by substituting $f(x)$ for x:

$$g \circ f(x) = g(f(x)) = g(x^3) \qquad \text{Substituting } x^3 \text{ for } x$$
$$= 1 + (x^3)^2$$
$$= 1 + x^6. \qquad \text{❖}$$

DO EXERCISE 6.

EXAMPLE 6 Given $f(x) = \sqrt{x}$ and $g(x) = x - 1$, find $f \circ g(x)$ and $g \circ f(x)$.

Solution

$$f \circ g(x) = f(g(x)) = f(x - 1) = \sqrt{x - 1},$$
$$g \circ f(x) = g(f(x)) = g(\sqrt{x}) = \sqrt{x} - 1 \qquad \text{❖}$$

DO EXERCISE 7.

How do we differentiate the composition of functions? The following theorem tells us.

6. Given $f(x) = 3x$ and $g(x) = x^2 - 1$, find $f \circ g(x)$ and $g \circ f(x)$.

7. Given $f(x) = 4x + 5$ and $g(x) = \sqrt[3]{x}$, find $f \circ g(x)$ and $g \circ f(x)$.

8. Given $y = u^2$ and $u = x^3 + 2$, find
$$\frac{dy}{du}, \quad \frac{du}{dx}, \quad \text{and} \quad \frac{dy}{dx}.$$

THEOREM 8

The Chain Rule

The derivative of the composition $f \circ g$ is given by

$$\frac{d}{dx}\,[f \circ g(x)] = \frac{d}{dx}\,[f(g(x))] = f'(g(x)) \cdot \frac{d}{dx}\,g(x).$$

Note that the Extended Power Rule is a special case. Consider the function $f(x) = x^k$. Then for any other function $g(x)$, $f \circ g(x) = [g(x)]^k$ and the derivative of the composition is

$$\frac{d}{dx}\,[g(x)]^k = k[g(x)]^{k-1} \cdot \frac{d}{dx}\,g(x).$$

The Chain Rule often appears in another form. Suppose $y = f(u)$ and $u = g(x)$. Then

$$\frac{dy}{dx} = \frac{dy}{du} \cdot \frac{du}{dx}.$$

EXAMPLE 7   Given $y = 2 + \sqrt{u}$ and $u = x^3 + 1$, find dy/du, du/dx, and dy/dx.

Solution   First we find dy/du and du/dx:

$$\frac{dy}{du} = \frac{1}{2}u^{-1/2} \quad \text{and} \quad \frac{du}{dx} = 3x^2.$$

Then

$$\frac{dy}{dx} = \frac{dy}{du} \cdot \frac{du}{dx}$$

$$= \frac{1}{2\sqrt{u}} \cdot 3x^2$$

$$= \frac{3x^2}{2\sqrt{x^3 + 1}}. \qquad \text{Substituting } x^3 + 1 \text{ for } u \qquad \diamondsuit$$

DO EXERCISE 8.

It is important to be able to recognize how a function can be expressed as a composition.

EXAMPLE 8   Find $f(x)$ and $g(x)$ such that $h(x) = f \circ g(x)$:

$$h(x) = (4x^3 - 7)^6.$$

Solution   This is $4x^3 - 7$ to the 6th power. Two functions that can be used for the composition are $f(x) = x^6$ and $g(x) = 4x^3 - 7$. We can check by forming the composition:

$$h(x) = f \circ g(x) = f(g(x)) = f(4x^3 - 7) = (4x^3 - 7)^6.$$

This is the most "obvious" answer to the question. There can be other less obvious answers—for example, $f(x) = x^3$ and $g(x) = (4x^3 - 7)^2$.   ❖

DO EXERCISE 9.

9. Find $f(x)$ and $g(x)$ such that $h(x) = f \circ g(x)$. Answers may vary.

a) $h(x) = \sqrt[3]{x^2 + 1}$

b) $h(x) = \dfrac{1}{(x + 5)^4}$

EXERCISE SET 2.8

Differentiate.

1. $y = (1 - x)^{55}$

2. $y = (1 - x)^{100}$

3. $y = \sqrt{1 + 8x}$

4. $y = \sqrt{1 - x}$

5. $y = \sqrt{3x^2 - 4}$

6. $y = \sqrt{4x^2 + 1}$

7. $y = (3x^2 - 6)^{-40}$

8. $y = (4x^2 + 1)^{-50}$

9. $y = x\sqrt{2x + 3}$

10. $y = x\sqrt{4x - 7}$

11. $y = x^2\sqrt{x - 1}$

12. $y = x^3\sqrt{x + 1}$

13. $y = \dfrac{1}{(3x + 8)^2}$

14. $y = \dfrac{1}{(4x + 5)^2}$

15. $f(x) = (1 + x^3)^3 - (1 + x^3)^4$

16. $f(x) = (1 + x^3)^5 - (1 + x^3)^4$

17. $f(x) = x^2 + (200 - x)^2$

18. $f(x) = x^2 + (100 - x)^2$

19. $f(x) = (x + 6)^{10}(x - 5)^4$

20. $f(x) = (x - 4)^8(x + 3)^9$

21. $f(x) = (x - 4)^8(3 - x)^4$

22. $f(x) = (x + 6)^{10}(5 - x)^9$

23. $f(x) = -4x(2x - 3)^3$

24. $f(x) = -5x(3x + 5)^6$

25. $f(x) = \sqrt{\dfrac{1 - x}{1 + x}}$

26. $f(x) = \sqrt{\dfrac{3 + x}{2 - x}}$

27. $f(x) = \left(\dfrac{3x - 1}{5x + 2}\right)^4$

28. $f(x) = \left(\dfrac{x}{x^2 + 1}\right)^3$

29. $f(x) = \sqrt[3]{x^4 + 3x^2}$

30. $f(x) = \sqrt{x^2 + 5x}$

31. $f(x) = (2x^3 - 3x^2 + 4x + 1)^{100}$

32. $f(x) = (7x^4$

33. $g(x) = \left(\dfrac{2x}{5x}\right.$

35. $f(x) = \sqrt{\dfrac{x}{x^2}}$

37. $f(x) = \dfrac{(2x}{(3x}$

39. $f(x) = 12(2$

40. $y = 6\sqrt[3]{x^2}$

Find $\dfrac{dy}{du}, \dfrac{du}{dx}$, an

41. $y = \sqrt{u}$ a

42. $y = \dfrac{15}{u^3}$ an

43. $y = u^{50}$ and $u = 4x^3 - 2x^2$

44. $y = \dfrac{u + 1}{u - 1}$ and $u = 1 + \sqrt{x}$

45. $y = u(u + 1)$ and $u = x^3 - 2x$

46. $y = (u + 1)(u - 1)$ and $u = x^3 + 1$

47. Find an equation for the tangent line to the graph of $y = \sqrt{x^2 + 3x}$ at the point $(1, 2)$.

48. Find an equation for the tangent line to the graph of $y = (x^3 - 4x)^{10}$ at the point $(2, 0)$.

49. Consider

$$f(x) = \frac{x^2}{(1 + x)^5}.$$

a) Find $f'(x)$ using the Quotient Rule and the Extended Power Rule.

b) Note that $f(x) = x^2(1 + x)^{-5}$. Find $f'(x)$ using the Product Rule and the Extended Power Rule.

c) Compare your answers to parts (a) and (b).

50. Consider

$$g(x) = (x^3 + 5x)^2.$$

a) Find $g'(x)$ using the Extended Power Rule.

b) Note that $g(x) = x^6 + 10x^4 + 25x^2$. Find $g'(x)$.

c) Compare your answers to parts (a) and (b).

Find $f \circ g(x)$ and $g \circ f(x)$.

51. $f(x) = 3x^2 + 2, \quad g(x) = 2x - 1$

52. $f(x) = 4x + 3, \quad g(x) = 2x^2 - 5$

53. $f(x) = 4x^2 - 1, \quad g(x) = \dfrac{2}{x}$

54. $f(x) = \dfrac{3}{x}, \quad g(x) = 2x^2 + 3$

55. $f(x) = x^2 + 1, \quad g(x) = x^2 - 1$

56. $f(x) = \dfrac{1}{x^2}, \quad g(x) = x + 2$

Find $f(x)$ and $g(x)$ such that $h(x) = f \circ g(x)$. Answers may vary.

57. $h(x) = (3x^2 - 7)^5$

58. $h(x) = \dfrac{1}{\sqrt{7x + 2}}$

59. $h(x) = \dfrac{x^3 + 1}{x^3 - 1}$

60. $h(x) = (\sqrt{x} + 5)^4$

APPLICATIONS

❖ **Business and Economics**

61. *Marginal cost.* A total-cost function is given by

$$C(x) = 1000\sqrt{x^3 + 2}.$$

Find the marginal cost $C'(x)$.

62. *Marginal revenue.* A total-revenue function is given by

$$R(x) = 2000\sqrt{x^2 + 3}.$$

Find the marginal revenue $R'(x)$.

63. *Marginal profit.* Use the total-cost and total-revenue functions in Exercises 61 and 62 to find the marginal profit.

64. *Utility.* Utility is a type of function or concept considered in economics. When a consumer receives x units of a product, a certain amount of pleasure, or utility, U, is derived. Suppose for a new kind of video game device that the following is the utility function related to the number of cartridges obtained:

$$U(x) = 80\sqrt{\frac{2x + 1}{3x + 4}}.$$

Find the marginal utility.

65. *Compound interest.* If $1000 is invested at interest rate i, compounded annually, it will grow in 3 years to an amount A given by

$$A = \$1000(1 + i)^3$$

(see Section 1.1). Find the rate of change, dA/di.

66. *Compound interest.* If $1000 is invested at interest rate i, compounded quarterly, it will grow in 5 years to an amount A given by

$$A = \$1000\left(1 + \frac{i}{4}\right)^{20}.$$

Find the rate of change, dA/di.

67. *Marginal demand.* Suppose that the demand function for a product is given by

$$D(p) = \frac{80,000}{p}$$

and that price p is a function of time given by $p = 1.6t + 9$, where t is in days.

a) Find the demand as a function of time t.

b) Find the marginal demand as a function of time.

c) Find the rate of change of the quantity demanded when $t = 100$ days.

68. *Marginal profit.* A company is selling microcomputers. It determines that its total profit is given by

$$P(x) = 0.08x^2 + 80x + 260,$$

where x is the number of units produced and sold. Suppose that x is a function of time, in months, where $x = 5t + 1$.

a) Find the total profit as a function of time t.

b) Find the marginal profit as a function of time.

c) Find the rate of change of total profit when $t = 48$ months.

SYNTHESIS EXERCISES

Differentiate.

69. $y = \sqrt[3]{x^3 - 6x + 1}$

70. $s = \sqrt[4]{t^4 + 3t^2 + 8}$

71. $y = \dfrac{x}{\sqrt{x - 1}}$

72. $y = \dfrac{(x + 1)^2}{(x^2 + 1)^3}$

73. $u = \dfrac{(1 + 2v)^4}{v^4}$

74. $y = x\sqrt{1 + x^2}$

75. $y = \dfrac{\sqrt{1 - x^2}}{1 - x}$

76. $w = \dfrac{u}{\sqrt{1 + u^2}}$

77. $y = \left(\dfrac{x^2 - x - 1}{x^2 + 1}\right)^3$

78. $y = \sqrt{1 + \sqrt{x}}$

79. $s = \dfrac{\sqrt{t} - 1}{\sqrt{t} + 1}$

80. $y = x^{2/3} \cdot \sqrt[3]{1 + x^2}$

COMPUTER-GRAPHING CALCULATOR EXERCISES

For each function, sketch the graph of f and f' on the given interval. Then estimate points at which the tangent line is horizontal.

81. $f(x) = 1.68x\sqrt{9.2 - x^2}$; $[-3, 3]$

82. $f(x) = \sqrt{6x^3 - 3x^2 - 48x + 45}$; $[-3, 3]$

 THE CALCULUS EXPLORER:
Derivatives, Bisections, Secants, Newton

Use the program to do exercises 81 and 82.

2.9

HIGHER-ORDER DERIVATIVES

Consider the function given by

$$y = f(x) = x^5 - 3x^4 + x.$$

Its derivative f' is given by

$$y' = f'(x) = 5x^4 - 12x^3 + 1.$$

The derivative function f' can also be differentiated. We can think of its

OBJECTIVE

a) Find higher-order derivatives.

1. Find the first six derivatives of
$$f(x) = 2x^6 - x^5 + 10.$$

2. For
$$y = x^7 - x^3,$$
find each of the following.

a) $\dfrac{dy}{dx}$

b) $\dfrac{d^2y}{dx^2}$

c) $\dfrac{d^3y}{dx^3}$

d) $\dfrac{d^4y}{dx^4}$

---

derivative as the rate of change of the slope of the tangent lines. We use the notation f'' for the derivative $(f')'$. We call f'' the *second derivative* of f. It is given by

$$y'' = f''(x) = 20x^3 - 36x^2.$$

Continuing in this manner, we have

$$f'''(x) = 60x^2 - 72x, \qquad \text{The third derivative of } f$$
$$f''''(x) = 120x - 72, \qquad \text{The fourth derivative of } f$$
$$f'''''(x) = 120. \qquad \text{The fifth derivative of } f$$

When notation like $f'''(x)$ gets lengthy, we abbreviate it using a symbol in parentheses. Thus, $f^{(n)}(x)$ is the nth derivative. For the function above,

$$f^{(4)}(x) = 120x - 72,$$
$$f^{(5)}(x) = 120,$$
$$f^{(6)}(x) = 0, \quad \text{and}$$
$$f^{(n)}(x) = 0, \quad \text{for any integer } n \geq 6.$$

DO EXERCISE 1.

Leibniz's notation for the second derivative of a function given by $y = f(x)$ is

$$\frac{d^2y}{dx^2}, \quad \text{or} \quad \frac{d}{dx}\left(\frac{dy}{dx}\right),$$

read "the second derivative of y with respect to x." The 2's in this notation *are not* exponents. If $y = x^5 - 3x^4 + x$, then

$$\frac{d^2y}{dx^2} = 20x^3 - 36x^2.$$

Leibniz's notation for the third derivative is d^3y/dx^3; for the fourth derivative, d^4y/dx^4; and so on:

$$\frac{d^3y}{dx^3} = 60x^2 - 72x, \qquad \frac{d^4y}{dx^4} = 120x - 72, \qquad \frac{d^5y}{dx^5} = 120.$$

DO EXERCISE 2.

EXAMPLE 1   For $y = 1/x$, find d^2y/dx^2.

Solution We have $y = x^{-1}$, so

$$\frac{dy}{dx} = -1 \cdot x^{-1-1} = -x^{-2}, \quad \text{or} \quad -\frac{1}{x^2}.$$

Then

$$\frac{d^2y}{dx^2} = (-2)(-1)x^{-2-1} = 2x^{-3}, \quad \text{or} \quad \frac{2}{x^3}. \qquad ❖$$

DO EXERCISE 3.

EXAMPLE 2 For $y = (x^2 + 10x)^{20}$, find y' and y''.

Solution To find y', we use the Extended Power Rule:

$$y' = 20(x^2 + 10x)^{19}(2x + 10)$$
$$= 20(x^2 + 10x)^{19} \cdot 2(x + 5)$$
$$= 40(x^2 + 10x)^{19}(x + 5).$$

To find y'', we use the Product Rule and the Extended Power Rule:

$$y'' = 40(x^2 + 10x)^{19}(1) + 19 \cdot 40(x^2 + 10x)^{18}(2x + 10)(x + 5)$$
$$= 40(x^2 + 10x)^{19} + 760(x^2 + 10x)^{18} \cdot 2(x + 5)(x + 5)$$
$$= 40(x^2 + 10x)^{19} + 1520(x^2 + 10x)^{18}(x + 5)^2$$
$$= 40(x^2 + 10x)^{18}[(x^2 + 10x) + 38(x + 5)^2]$$
$$= 40(x^2 + 10x)^{18}[x^2 + 10x + 38(x^2 + 10x + 25)]$$
$$= 40(x^2 + 10x)^{18}[39x^2 + 390x + 950]. \qquad ❖$$

DO EXERCISE 4.

Acceleration can be regarded as a second derivative. As an object moves, its distance from a fixed point after time t is some function of the time, say, $s(t)$. Then

$$v(t) = s'(t) = \text{the velocity at time } t$$

and

$$a(t) = v'(t) = s''(t) = \text{the acceleration at time } t.$$

Whenever a quantity is a function of time, the first derivative gives the rate of change with respect to time and the second derivative gives the acceleration. For example, if $y = P(t)$ gives the number of people in a population at time t, then $P'(t)$ represents how fast the size of the pop-

3. For $y = \dfrac{2}{x}$, find $\dfrac{d^2y}{dx^2}$.

4. For $y = (x^2 - 12x)^{30}$, find y' and y''.

5. For $s(t) = 3t + t^4$, find the acceleration $a(t)$.

ulation is changing and $P''(t)$ gives the acceleration in the size of the population.

DO EXERCISE 5.

Find d^2y/dx^2.

1. $y = 3x + 5$

2. $y = -4x + 7$

3. $y = -\dfrac{1}{x}$

4. $y = -\dfrac{3}{x}$

5. $y = x^{1/4}$

6. $y = \sqrt{x}$

7. $y = x^4 + \dfrac{4}{x}$

8. $y = x^3 - \dfrac{3}{x}$

9. $y = x^{-3}$

10. $y = x^{-4}$

11. $y = x^n$

12. $y = x^{-n}$

13. $y = x^4 - x^2$

14. $y = x^4 + x^3$

15. $y = \sqrt{x} - 1$

16. $y = \sqrt{x} + 1$

17. $y = ax^2 + bx + c$

18. $y = \sqrt{a - bx}$

19. $y = (x^2 - 8x)^{43}$

20. $y = (x^3 + 15x)^{20}$

21. $y = (x^4 - 4x^2)^{50}$

22. $y = (x^2 - 3x + 1)^{12}$

23. $y = x^{2/3} + 4x$

24. $y = x^{5/4} - x^{4/5}$

25. $y = (x - 8)^{3/4}$

26. $y = \dfrac{x^5}{40} - \dfrac{x^3}{36}$

27. $y = \dfrac{1}{x^2} + \dfrac{2}{x^3}$

28. $y = \dfrac{4}{x^5} - \dfrac{7}{3x^4}$

29. For $y = x^4$, find d^4y/dx^4.

30. For $y = x^5$, find d^4y/dx^4.

31. For $y = x^6 - x^3 + 2x$, find d^5y/dx^5.

32. For $y = x^7 - 8x^2 + 2$, find d^6y/dx^6.

33. For $y = (x^2 - 5)^{10}$, find d^2y/dx^2.

34. For $y = x^k$, find d^5y/dx^5.

35. If s is a distance given by $s(t) = t^3 + t^2 + 2t$, find the acceleration.

36. If s is a distance given by $s(t) = t^4 + t^2 + 3t$, find the acceleration.

APPLICATIONS

❖ Life and Physical Sciences

37. *Population growth.* A population grows from an initial size of 100,000 to an amount $P(t)$, given by

$$P(t) = 100,000(1 + 0.6t + t^2).$$

What is the acceleration in the size of the population?

38. *Population growth.* A population grows from an initial size of 100,000 to an amount $P(t)$, given by

$$P(t) = 100,000(1 + 0.4t + t^2)$$

What is the acceleration in the size of the population?

SYNTHESIS EXERCISES

Find y', y'', and y'''.

39. $y = x^{-1} + x^{-2}$

40. $y = \dfrac{1}{1 - x}$

41. $y = x\sqrt{1 + x^2}$

42. $y = 3x^5 + 8\sqrt{x}$

43. $y = \dfrac{3x - 1}{2x + 3}$

44. $y = \dfrac{1}{\sqrt{x - 1}}$

45. $y = \dfrac{x}{\sqrt{x - 1}}$

46. $y = \dfrac{\sqrt{x} - 1}{\sqrt{x} + 1}$

Find $f''(x)$.

47. $f(x) = \dfrac{x}{x - 1}$

48. $f(x) = \dfrac{1}{1 + x^2}$

Find the first through the fifth derivative. Be sure to simplify at each stage before continuing.

49. $f(x) = \dfrac{x - 1}{x + 2}$

50. $f(x) = \dfrac{x + 3}{x - 2}$

COMPUTER-GRAPHING CALCULATOR EXERCISES

For each function, sketch the graph of f, f', and f'' on the given interval. Analyze and compare the behavior of each function.

51. $f(x) = 0.1x^4 - x^2 + 0.4$; $[-5, 5]$

52. $f(x) = -x^3 + 3x$; $[-3, 3]$

53. $f(x) = x^4 + x^3 - 4x^2 - 2x + 4$; $[-3, 2]$

54. $f(x) = x^3 - 3x^2 + 2$; $[-2, 4]$

 CALCULUS EXPLORER: *Derivatives*

Use the program to do Exercises 51–54.

CHAPTER SUMMARY AND REVIEW

2

TERMS TO KNOW

Limit, p. 85
Continuity at a point, p. 89, 92
Continuous over an interval, p. 90, 92
Continuous function, p. 92
Limit involving infinity, p. 101
Average rate of change, p. 107
Secant line, p. 108
Difference quotient, p. 110
Simplified difference quotient, p. 111
Tangent line, p. 119
Derivative, p. 120

Differentiation, p. 120
Leibniz notation, p. 129
Power Rule, p. 130
Sum Rule, p. 134
Difference Rule, p. 134
Instantaneous rate of change, p. 140
Velocity, p. 140, 141
Speed, p. 140
Acceleration, p. 141
Growth rate, p. 143

Marginal cost, p. 143
Marginal revenue, p. 144
Marginal profit, p. 144
Average cost, p. 145
Product Rule, p. 149
Quotient Rule, p. 151
Extended Power Rule, p. 157
Composition of functions, p. 160
Chain Rule, p. 162
Higher-order derivative, p. 165

REVIEW EXERCISES

These exercises are for test preparation. They can also be used as a lengthened practice test. Answers are at the back of the book. The answers also contain bracketed section references, which tell you where to restudy if your answer is incorrect.

Find the limit, if it exists.

1. $\lim\limits_{x \to -2} \dfrac{8}{x}$

2. $\lim\limits_{x \to 1} (4x^3 - x^2 + 7x)$

3. $\lim\limits_{x \to -7} \dfrac{x^2 + 4x - 21}{x + 7}$

4. $\lim\limits_{x \to \infty} \dfrac{5x + 6}{x}$

5. $\lim\limits_{x \to \infty} \dfrac{6x^2 + 5x - 7}{3x^4 - 2x + 4}$

Determine whether the graph is continuous.

6.

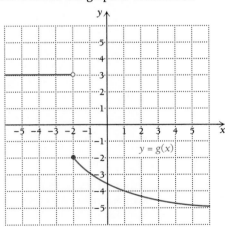

7.

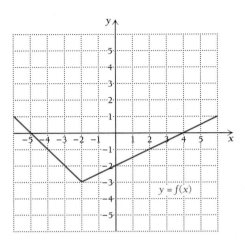

For the function in Exercise 6, answer the following.

8. Find $\lim\limits_{x \to 1} g(x)$

9. Find $g(1)$.

10. Is g continuous at 1?

11. Find $\lim\limits_{x \to -2} g(x)$.

12. Find $g(-2)$.

13. Is g continuous at -2?

14. For $f(x) = x^3 - 2x$, find the average rate of change as x changes from -2 to 1.

15. Find a simplified difference quotient for $g(x) = -3x + 5$.

16. Find a simplified difference quotient for $f(x) = 2x^2 - 3$.

17. Find an equation of the tangent line to the graph of $y = x^2 + 3x$ at the point $(-1, -2)$.

18. Find the points on the graph of $y = -x^2 + 8x - 11$ at which the tangent line is horizontal.

19. Find the points on the graph of $y = 5x^2 - 49x + 12$ at which the tangent line has slope 1.

Find dy/dx.

20. $y = 4x^5$

21. $y = 3\sqrt[3]{x}$

22. $y = \dfrac{-8}{x^8}$

23. $y = 15x^{2/5}$

24. $y = 0.1x^7 - 3x^4 - x^3 + 6$

Differentiate.

25. $f(x) = \dfrac{1}{6}x^6 + 8x^4 - 5x$

26. $y = \dfrac{x^3 + x}{x}$ **27.** $y = \dfrac{x^2 + 8}{8 - x}$

28. $g(x) = (5 - x)^2(2x - 1)^5$

29. $f(x) = (x^5 - 2)^7$ **30.** $f(x) = x^2(4x + 3)^{3/4}$

31. For $y = x^3 - \dfrac{2}{x}$, find $\dfrac{d^5y}{dx^5}$.

32. For $y = x^7 + 3x^2$, find $\dfrac{d^4y}{dx^4}$.

33. Given $s(t) = t + t^4$, find each of the following.

  a) $v(t)$

  b) $a(t)$

  c) The velocity and the acceleration when $t = 2$ sec

34. *Business: Marginal revenue, cost, and profit.* Given $R(x) = 40x$ and $C(x) = 8x^2 - 7x - 10$, find each of the following.

  a) $P(x)$

  b) $R(20)$, $C(20)$, and $P(20)$

  c) $R'(x)$, $C'(x)$, and $P'(x)$

  d) $R'(20)$, $C'(20)$, and $P'(20)$

35. *Life science: Growth rate.* The population of a city grows from an initial size of 10,000 to an amount P, given by $P = 10,000 + 50t^2$, where t is measured in years.

  a) Find the growth rate.

  b) Find the number of people in the city after 20 years (at $t = 20$ yr).

  c) Find the growth rate at $t = 20$.

36. Find $f \circ g(x)$ and $g \circ f(x)$, given that $f(x) = x^2 + 5$ and $g(x) = 1 - 2x$.

SYNTHESIS EXERCISES

37. Find $\lim\limits_{x \to \infty} \dfrac{2 - 5x^5}{4 + 3x^6}$.

38. Differentiate $y = \dfrac{x\sqrt{1 + 3x}}{1 + x^3}$.

EXERCISES FOR THINKING AND WRITING

39. Discuss three different ways in which a function may not be continuous at a point a. Draw graphs to illustrate your discussion.

40. Discuss as many interpretations as you can of the derivative of a function at a point x.

41. Try to prove a rule for finding the derivative of the product of three functions. Describe the rule in words.

42. The following is the beginning of an alternative proof of the Quotient Rule for finding the derivative of a quotient that uses the Product Rule and the Power Rule. Complete the proof giving reasons for each step.

Proof

Let

$$q(x) = \frac{n(x)}{d(x)}.$$

Then

$$q(x) = n(x) \cdot [d(x)]^{-1}.$$

CHAPTER TEST

2

Determine whether the function is continuous.

1.

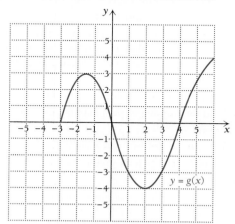

2.

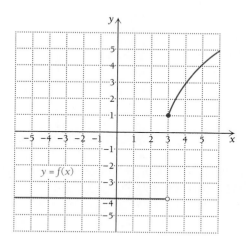

For the function in Question 2, answer the following.

3. Find $\lim\limits_{x \to 3} f(x)$.

4. Find $f(3)$.

5. Is f continuous at 3?

6. Find $\lim\limits_{x \to 4} f(x)$.

7. Find $f(4)$.

8. Is f continuous at 4?

Find the limit, if it exists.

9. $\lim\limits_{x \to 1} (3x^4 - 2x^2 + 5)$

10. $\lim\limits_{x \to 1} \dfrac{x - 1}{x^2 - 1}$

11. $\lim\limits_{x \to 0} \dfrac{7}{x}$

12. $\lim\limits_{x \to \infty} \dfrac{4x - 3}{x}$

13. $\lim\limits_{x \to \infty} \dfrac{2x^5 - 4x + 1}{6x^3 + 5}$

14. Find a simplified difference quotient for
$$f(x) = 3x^2 + 1.$$

15. Find an equation of the tangent line to the graph of $y = x + (4/x)$ at the point $(4, 5)$.

16. Find the points on the graph of $y = x^3 - 3x^2$ at which the tangent line is horizontal.

Find dy/dx.

17. $y = x^{84}$

18. $y = 10\sqrt{x}$

19. $y = \dfrac{-10}{x}$

20. $y = x^{5/4}$

21. $y = -0.5x^2 + 0.61x + 90$

Differentiate.

22. $y = \dfrac{1}{3}x^3 - x^2 + 2x + 4$

23. $y = \dfrac{2x - 5}{x^4}$

24. $f(x) = \dfrac{x}{5 - x}$

25. $f(x) = (x + 3)^4 (7 - x)^5$

26. $y = (x^5 - 4x^3 + x)^{-5}$

27. $f(x) = x\sqrt{x^2 + 5}$

28. For $y = x^4 - 3x^2$, find $\dfrac{d^3y}{dx^3}$.

29. *Business: Marginal revenue, cost, and profit.* Given $R(x) = 50x$ and $C(x) = 0.001x^2 + 1.2x + 60$, find each of the following.

a) $P(x)$

b) $R(10)$, $C(10)$, and $P(10)$

c) $R'(x)$, $C'(x)$, and $P'(x)$

d) $R'(10)$, $C'(10)$, and $P'(10)$

30. *Social sciences: Memory.* In a certain memory experi- ment, a person is able to memorize M words after t minutes, where $M = -0.001t^3 + 0.1t^2$.

a) Find the rate of change of the number of words memorized with respect to time.

b) How many words are memorized during the first 10 min (at $t = 10$)?

c) What is the memory rate at $t = 10$ min?

31. Find $f \circ g(x)$ and $g \circ f(x)$, given that $f(x) = x + x^2$ and $g(x) = x^3$.

SYNTHESIS EXERCISES

32. Find $\lim\limits_{x \to 2} \dfrac{x^3 - 8}{x - 2}$.

33. Differentiate $y = (1 - 3x)^{2/3}(1 + 3x)^{1/3}$.

3.1

OBJECTIVE

a) Find relative extrema of continuous functions using the First-Derivative Test and sketch a graph of the function.

USING FIRST DERIVATIVES TO FIND MAXIMUM AND MINIMUM VALUES AND SKETCH GRAPHS

Figure 1 illustrates a typical life cycle of a retail product. Note that the number of items sold varies with respect to time. Sales begin at a small level and increase to a point of maximum sales after which they decline and taper off to a low level, probably due to the effect of new competitive products. Then the company rejuvenates the product by making product improvements. Think about products such as black and white TVs, color TVs, records, compact discs, and digital audio tape recorders. Where might each be in a typical life cycle and is that curve appropriate?

The graph of a typical life cycle illustrates many of the ideas that we will consider in this chapter.

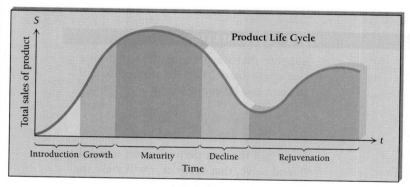

FIGURE 1

Finding the largest and smallest values of a function—that is, the maximum and minimum values—has extensive application. The first and second derivatives of a function are tools of calculus that give us information about the shape of a graph that may be helpful in finding maximum and minimum values of functions and in graphing functions. Throughout this section we will assume that the functions f are continuous, but this does not necessarily imply that f' and f'' are continuous.

Increasing and Decreasing Functions

If the graph of a function rises from left to right on an interval I, it is said to be **increasing** on I. If the graph drops from left to right, it is said to be **decreasing** on I. We can describe this mathematically as follows.

DEFINITION

A function f is *increasing* on I, if for every a and b in I,

$$\text{if } a < b, \text{ then } f(a) < f(b).$$

(If the input a is less than the input b, then the output for a is less than the output for b.)

A function f is *decreasing* on I, if for every a and b in I,

$$\text{if } a < b, \text{ then } f(a) > f(b).$$

(If the input a is less than the input b, then the output for a is greater than the output for b.)

Note that the directions of the inequalities stay the same for an increasing function, but they differ for a decreasing function.

Figures 2–4 show examples of the preceding definition.

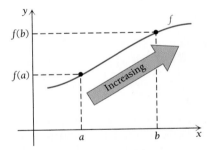

f is an increasing function.
If $a < b$, then $f(a) < f(b)$.
$f'(x) > 0$ for all x.

FIGURE 2

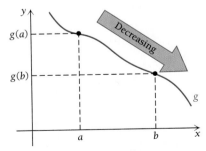

g is a decreasing function.
If $a < b$, then $g(a) > g(b)$.
$g'(x) < 0$ for all x.

FIGURE 3

1. Consider this graph.

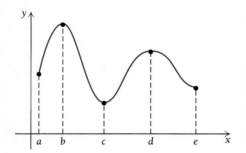

a) Over what intervals is the function increasing?

b) Over what intervals is the derivative positive? To determine this, locate a straightedge at points on the graph and decide whether the slopes of the tangent lines are positive.

c) Over what intervals is the function decreasing?

d) Over what intervals is the derivative negative?

2. Consider this graph.

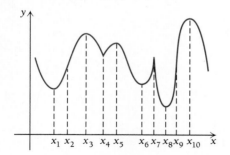

a) At which points are there horizontal tangent lines?

b) At which points does the derivative not exist?

c) Which are critical points?

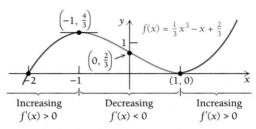

f is increasing on the intervals $(-\infty, -1)$ and $(1, \infty)$ and decreasing on the interval $(-1, 1)$.

FIGURE 4

In Chapter 1 we saw how the slope of a linear function determines whether that function is increasing or decreasing (or neither). For a general function, the derivative yields similar information. Let us investigate how this happens in Margin Exercise 1.

DO EXERCISE 1.

The following theorem shows how we can use derivatives to determine whether a function is increasing or decreasing.

THEOREM 1

If $f'(x) > 0$ for all x in an interval I, then f is increasing on I.
If $f'(x) < 0$ for all x in an interval I, then f is decreasing on I.

Critical Points

Consider the graph of a continuous function shown in Fig. 5.

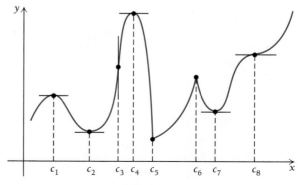

FIGURE 5

Note the following:

1. $f'(c) = 0$ at points c_1, c_2, c_4, c_7, and c_8. That is, the tangent to the graph is horizontal at these points.

2. $f'(c)$ does not exist at points c_3, c_5, and c_6. The tangent line is vertical at c_3 and there is a corner point or sharp point at both c_5 and c_6. (See also the discussion at the end of Section 2.4.)

DEFINITION

A *critical point* of a function is an interior point c of its domain at which the tangent to the graph is horizontal or at which the derivative does not exist. That is, c is a critical point if

$$f'(c) = 0 \quad \text{or} \quad f'(c) \text{ does not exist.}$$

Thus, in Fig. 5:

1. c_1, c_2, c_4, c_7, and c_8 are critical points because $f'(c) = 0$ for each point.

2. c_3, c_5, and c_6 are critical points because $f'(c)$ does not exist at each point.

Note also in Fig. 5 that a function can change from increasing to decreasing *only* at a critical point.

DO EXERCISES 2–4.

Finding Relative Maximum and Minimum Values

Consider the graph shown in Fig. 6. Note the "peaks" and "valleys" at the interior points c_1, c_2, and c_3.

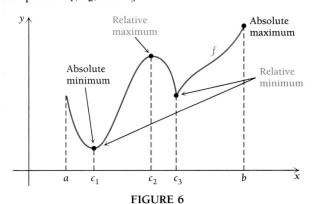

FIGURE 6

3. Try to draw a graph of a continuous function from P to Q that increases on part or parts of $[a, b]$ and decreases on part or parts of $[a, b]$.

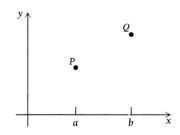

Does the function you drew have any critical points between a and b?

4. Now try to draw a graph of a continuous function from P to Q that increases on part or parts of $[a, b]$ and decreases on part or parts of $[a, b]$, but in such a way that no critical points occur between a and b.

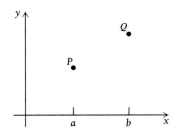

The function value $f(c_2)$ is called a **relative maximum.** Each of the function values $f(c_1)$ and $f(c_3)$ is called a **relative minimum.**

DEFINITION

Suppose that f is a function whose value $f(c)$ exists at input c in an interval I in the domain of f. Then:

$f(c)$ is a *relative minimum* if there exists an open interval I containing c in the domain such that $f(c) \leqslant f(x)$, for all x in I; and

$f(c)$ is a *relative maximum* if there exists an open interval I containing c in the domain such that $f(c) \geqslant f(x)$, for all x in I.

A relative maximum can be thought of as a high point that may or may not be the highest point, or *absolute maximum,* on an interval I. Similarly, a relative minimum can be thought of as a low point that may or may not be the lowest point, or *absolute minimum,* on I. For now we will consider how to find relative maximum or minimum values, stated simply as **relative extrema.**

Look again at Fig. 6. The points at which a continuous function has relative extrema are points where the derivative is 0 or where the derivative does not exist—the critical points.

THEOREM 2

If a function f has a relative extreme value $f(c)$, then c is a critical point, so

$$f'(c) = 0 \quad \text{or} \quad f'(c) \text{ does not exist.}$$

Theorem 2 is very useful, but it is important to understand it precisely. What it says is that when we are looking for points that are relative extrema, the only points we need to consider are those where the derivative is 0 or where the derivative does not exist. We can think of a critical point as a *candidate* for a relative maximum or minimum, and the candidate might or might not provide a relative extremum. That is, Theorem 2 does not say that if a point is a critical point, its function value will necessarily be a relative maximum or minimum. The existence of a critical point does *not* guarantee that a function has a relative maximum or minimum. A useful counterexample is illustrated in Fig. 7, with the graph of

$$f(x) = (x - 1)^3 + 2.$$

Note that

$$f'(x) = 3(x - 1)^2,$$

and

$$f'(1) = 3(1 - 1)^2 = 0.$$

The function has a critical point at $c = 1$, but has no relative maximum or minimum at $c = 1$.

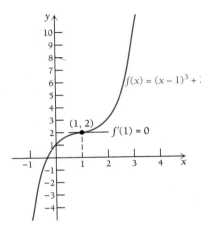

FIGURE 7

Now how can we tell when the existence of a critical point leads us to a relative extremum? Figure 8 leads us to a test.

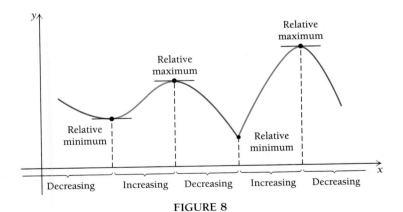

FIGURE 8

At a critical point for which there is a relative extreme value, the function is increasing on one side of the critical point and decreasing on the other side.

$f(c)$	Sign of $f'(x)$ for x in (a, c)	Sign of $f'(x)$ for x in (c, b)	Graph over the interval (a, b)
Relative minimum	−	+	
Relative maximum	+	−	
No relative maxima or minima	−	−	
No relative maxima or minima	+	+	

Derivatives tell us when a function is increasing or decreasing. This leads us to the First-Derivative Test.

THEOREM 3

The First-Derivative Test for Relative Extrema

For any continuous function f that has exactly one critical point c in an open interval (a, b):

F1. f has a relative minimum at c if $f'(x) < 0$ on (a, c) and $f'(x) > 0$ on (c, b). That is, f is decreasing to the left of c and increasing to the right of c.

F2. f has a relative maximum at c if $f'(x) > 0$ on (a, c) and $f'(x) < 0$ on (c, b). That is, f is increasing to the left of c and decreasing to the right of c.

F3. f has neither a relative maximum nor a relative minimum if $f'(x)$ has the same sign on (a, c) as on (c, b).

Now let us apply the First-Derivative Test to finding relative extrema and to graphing.

EXAMPLE 1   Find the relative extrema of the function f given by

$$f(x) = 2x^3 - 3x^2 - 12x + 12.$$

Then sketch the graph.

Solution   First, we must find the critical points. To do so we find $f'(x)$:

$$f'(x) = 6x^2 - 6x - 12.$$

We then determine where $f'(x)$ does not exist or where $f'(x) = 0$. We can replace x in $f'(x) = 6x^2 - 6x - 12$ by any real number. Thus, $f'(x)$ exists for all real numbers. So the only possibilities for critical points are where $f'(x) = 0$, at which there are horizontal tangents. To find such points, we solve $f'(x) = 0$:

$$6x^2 - 6x - 12 = 0$$

$\quad x^2 - x - 2 = 0 \qquad$ Dividing by 6 on both sides

$(x + 1)(x - 2) = 0 \qquad$ Factoring

$\quad x + 1 = 0 \quad$ *or* $\quad x - 2 = 0 \qquad$ Using the Principle of Zero Products

$\qquad x = -1 \quad$ *or* $\qquad x = 2.$

The critical points are -1 and 2. Since it is at these points that a relative maximum or minimum will exist, if there is one, we examine the intervals on each side of the critical points. We use the critical points to divide the real-number line into three intervals: $A(-\infty, -1)$, $B(-1, 2)$, and $C(2, \infty)$, as shown in Fig. 9.

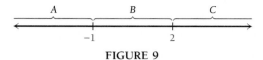

FIGURE 9

Then we analyze the sign of the derivative on each interval. If $f'(x)$ is positive for one value in the interval, then it will be positive for all numbers in the interval. Similarly, if it is negative for one value, it will be negative for all values in the interval. This is because in order for the derivative to change signs, it must become 0 or be undefined at some point. Such a point will be a critical point. Thus we merely choose a test value in each interval and make a substitution. The test values we choose are -2, 0, and 3.

A: Test -2, $\quad f'(-2) = 6(-2)^2 - 6(-2) - 12$
$\qquad\qquad\qquad = 24 + 12 - 12 = 24 > 0;$

B: Test 0, $\quad f'(0) = 6(0)^2 - 6(0) - 12 = -12 < 0;$

C: Test 3, $\quad f'(3) = 6(3)^2 - 6(3) - 12 = 54 - 18 - 12 = 24 > 0.$

Interval	$(-\infty, -1)$	$(-1, 2)$	$(2, \infty)$
Test value	$x = -2$	$x = 0$	$x = 3$
Sign of $f'(x)$	$f'(-2) > 0$	$f'(0) < 0$	$f'(3) > 0$
Result	f is increasing	f is decreasing	f is increasing

⌐—— Change —⌐   ⌐—— Change —⌐
indicates a   indicates a
relative maximum.   relative minimum.

Therefore, by the First-Derivative Test,

f has a relative maximum at $x = -1$ given by

$f(-1) = 2(-1)^3 - 3(-1)^2 - 12(-1) + 12$ Substituting into the

$\qquad = 19$ original function

and f has a relative minimum at $x = 2$ given by

$f(2) = 2(2)^3 - 3(2)^2 - 12(2) + 12 = -8.$

Thus there is a relative maximum at $(-1, 19)$ and a relative minimum at $(2, -8)$.

The information we have obtained can be very useful in sketching a graph of the function. We know that this polynomial is continuous, and we know where the function is increasing, where it is decreasing, and where it has relative extrema. We complete the graph by using a calculator to generate some additional function values. Many are listed in the table below. More can be generated by the student if there is a need. Some calculators can actually be programmed to generate function values by first entering a formula and then generating many outputs. The graph is shown in Fig. 10. It has been scaled to clearly show its curving nature.

x	$f(x)$
-3	-33
-2.5	-8
-2	8
-1.5	16.5
-1	19
-0.5	17
0	12
0.5	5.5
1	-1
1.5	-6
2	-8
2.5	-5.5
3	3
3.5	19
4	44

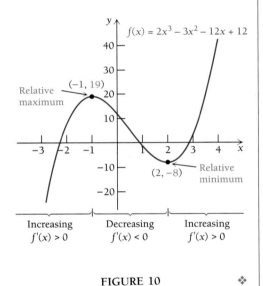

FIGURE 10   ❖

Keep the following in mind as you find relative extrema.

> The *derivative* f' is used to find the critical points of f. The test values, in the intervals defined by the critical points, are substituted into the *derivative* f', and the function values are found using the *original* function f. Use the derivative f' to find information about the shape of the graph of f.

DO EXERCISE 5.

5. Find the relative extrema of the function f given by

$$f(x) = x^3 + 3x^2 - 9x - 12.$$

List your answers in terms of ordered pairs. Sketch a graph of the function.

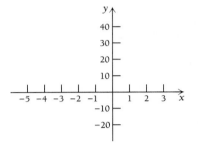

EXAMPLE 2 Find the relative extrema of the function f given by

$$f(x) = 2x^3 - x^4.$$

Then sketch the graph.

Solution   First, we must determine the critical points. To do so, we find $f'(x)$:

$$f'(x) = 6x^2 - 4x^3.$$

Next we find where $f'(x)$ does not exist or where $f'(x) = 0$. We can replace x in $f'(x) = 6x^2 - 4x^3$ by any real number. Thus, $f'(x)$ exists for all real numbers. So the only possibilities for critical points are where $f(x) = 0$, that is, where there are horizontal tangents. To find such points, we solve $f'(x) = 0$:

$$6x^2 - 4x^3 = 0$$
$$2x^2(3 - 2x) = 0 \qquad \text{Factoring}$$
$$2x^2 = 0 \quad or \quad 3 - 2x = 0$$
$$x^2 = 0 \quad or \qquad\quad 3 = 2x$$
$$x = 0 \quad or \qquad\quad x = \tfrac{3}{2}.$$

The critical points are 0 and $\tfrac{3}{2}$. We use these points to divide the real-number line into three intervals: $A(-\infty, 0)$, $B(0, \tfrac{3}{2})$, and $C(\tfrac{3}{2}, \infty)$, as shown in Fig. 11.

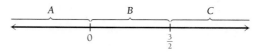

FIGURE 11

We now analyze the sign of the derivative on each interval. We begin by choosing a test value in each interval and making a substitution. We usually choose as test values numbers for which it is easy to compute outputs of the derivative—in this case, -1, 1, and 2.

A: Test -1, $f'(-1) = 6(-1)^2 - 4(-1)^3$
$\qquad\qquad\qquad = 6 + 4 = 10 > 0;$

B: Test 1, $f'(1) = 6(1)^2 - 4(1)^3$
$\qquad\qquad\qquad = 6 - 4 = 2 > 0;$

C: Test 2, $f'(2) = 6(2)^2 - 4(2)^3$
$\qquad\qquad\qquad = 24 - 32 = -8 < 0.$

Interval	$(-\infty, 0)$	$(0, \frac{3}{2})$	$(\frac{3}{2}, \infty)$
Test value	$x = -1$	$x = 1$	$x = 2$
Sign of $f'(x)$	$f'(-1) > 0$	$f'(1) > 0$	$f'(2) < 0$
Result	f is increasing	f is increasing	f is decreasing

└── No change ──┘ └── Change ──┘
indicates a
relative maximum.

Therefore, by the First-Derivative Test, f has neither a relative maximum nor a relative minimum at $x = 0$ since the function is increasing on both sides of 0, and f has a relative maximum at $x = \frac{3}{2}$ given by

$f(\frac{3}{2}) = 2(\frac{3}{2})^3 - (\frac{3}{2})^4 = \frac{27}{16}.$ Remember to substitute into the original function.

Thus there is a relative maximum at $(\frac{3}{2}, \frac{27}{16})$.

We use the information obtained to sketch the graph. Other function values are listed in the table below. (More can be generated by the student.) The graph is shown in Fig. 12.

x	$f(x)$, approximately
-1	-3
-0.75	-1.16
-0.5	-0.32
-0.25	-0.04
0	0
0.25	0.03
0.5	0.19
0.75	0.53
1	1
1.25	1.46
1.5	1.69
1.75	1.34
2	0

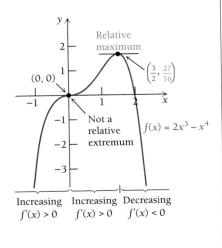

FIGURE 12

DO EXERCISE 6.

6. Find the relative extrema of the function f given by

$$f(x) = x^4 - 8x^3 + 18x^2.$$

List your answers in terms of ordered pairs. Sketch a graph of the function.

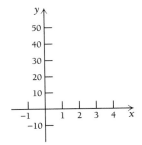

EXAMPLE 3 Find the relative extrema of the function f given by

$$f(x) = (x - 2)^{2/3} + 1.$$

Then sketch the graph.

Solution First, we determine the critical points. To do so, we find $f'(x)$:

$$f'(x) = \frac{2}{3}(x - 2)^{-1/3} = \frac{2}{3\sqrt[3]{x - 2}}.$$

Next we find where $f'(x)$ does not exist or where $f'(x) = 0$. Note that the derivative $f'(x)$ does not exist at 2, although $f(x)$ does. Thus the number 2 is a critical point. The equation $f'(x) = 0$ has no solution, so the only critical point is 2. We use 2 to divide the real-number line into two intervals: $A(-\infty, 2)$ and $B(2, \infty)$, as shown in Fig. 13.

FIGURE 13

We analyze the derivative on each interval. We begin by choosing a test value in each interval and making a substitution. We choose test points 0 and 3. It is not necessary to find an exact value of the derivative; we need only determine the sign. Sometimes we can do this by just examining the formula for the derivative:

$$A: \text{ Test } 0, \quad f'(0) = \frac{2}{3\sqrt[3]{0 - 2}} < 0;$$

$$B: \text{ Test } 3, \quad f'(3) = \frac{2}{3\sqrt[3]{3 - 2}} > 0.$$

Interval	$(-\infty, 2)$	$(2, \infty)$
Test value	$x = 0$	$x = 3$
Sign of $f'(x)$	$f'(0) < 0$	$f'(3) > 0$
Result	f is decreasing	f is increasing

Change
indicates a
relative minimum.

Since we have a change from decreasing to increasing, we conclude from the First-Derivative Test that

f has a relative minimum at $x = 2$ given by

$$f(2) = (2 - 2)^{2/3} + 1 = 1.$$

Thus there is a relative minimum at $(2, 1)$.

We use the information obtained to sketch the graph. Other function values are listed in the table below. (More can be generated by the student.) The graph is shown in Fig. 14.

x	$f(x)$, approximately
-1	3.08
-0.5	2.84
0	2.59
0.5	2.31
1	2
1.5	1.63
2	1
2.5	1.63
3	2
3.5	2.31
4	2.59

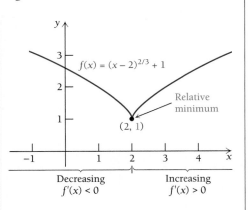

FIGURE 14

DO EXERCISE 7.

7. Find the relative extrema of the function f given by

$$f(x) = x^{2/3}.$$

List your answers in terms of ordered pairs. Sketch a graph of the function.

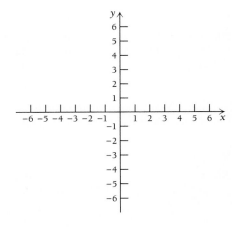

EXERCISE SET **3.1**

Find the relative extrema of the function, if they exist. List your answers in terms of ordered pairs. Then sketch a graph of the function.

1. $f(x) = x^2 - 4x + 5$
2. $f(x) = x^2 - 6x - 3$
3. $f(x) = 5 + x - x^2$
4. $f(x) = 2 - 3x - 2x^2$
5. $f(x) = 1 + 6x + 3x^2$
6. $f(x) = 0.5x^2 - 2x - 11$
7. $f(x) = x^3 - x^2 - x + 2$
8. $f(x) = x^3 + \frac{1}{2}x^2 - 2x + 5$
9. $f(x) = x^3 - 3x + 6$
10. $f(x) = x^3 - 3x^2$
11. $f(x) = 3x^2 - 2x^3$
12. $f(x) = x^3 - 3x$
13. $f(x) = 2x^3$
14. $f(x) = 1 - x^3$
15. $f(x) = x^3 - 6x^2 + 10$
16. $f(x) = 12 + 9x - 3x^2 - x^3$
17. $f(x) = x^3 - x^4$
18. $f(x) = x^4 - 2x^3$

19. $f(x) = x^4 - 8x^2 + 3$

20. $f(x) = x^4 - 2x^2 + 5$

21. $f(x) = 1 - x^{2/3}$

22. $f(x) = (x + 3)^{2/3} - 5$

23. $f(x) = \dfrac{-8}{x^2 + 1}$

24. $f(x) = \dfrac{5}{x^2 + 1}$

25. $f(x) = \dfrac{4x}{x^2 + 1}$

26. $f(x) = \dfrac{x^2}{x^2 + 1}$

27. $f(x) = \sqrt[3]{x}$

28. $f(x) = (x + 1)^{1/3}$

APPLICATIONS

❖ **Business and Economics**

29. *Advertising.* A firm estimates that it will sell N units of a product after spending a dollars on advertising, where

$$N(a) = -a^2 + 300a + 6, \quad 0 \leq a \leq 300,$$

and a is measured in thousands of dollars. Find the relative extrema and sketch a graph of the function.

❖ **Life and Physical Sciences**

30. *Temperature during an illness.* The temperature of a person during an illness is given by

$$T(t) = -0.1t^2 + 1.2t + 98.6, \quad 0 \leq t \leq 12,$$

where T is the temperature (°F) at time t, measured in days. Find the relative extrema and sketch a graph of the function.

❖ **General Interest**

The following graph shows the changes in oil prices for a recent period of days.

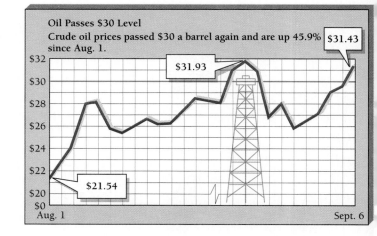

31. How many relative maxima can you find on the graph?

32. How many relative minima can you find on the graph?

COMPUTER–GRAPHING CALCULATOR EXERCISES

Sketch the graph of the function using a computer software graphing package or a graphing calculator. Then estimate any relative extrema.

33. $f(x) = x^4 + 4x^3 - 36x^2 - 160x + 400$

34. $f(x) = -x^6 - 4x^5 + 54x^4 + 160x^3 - 641x^2 - 828x + 1200$

THE CALCULUS EXPLORER:
PowerGrapher

Use the graphing program to create graphs of the functions in Exercises 33 and 34. Adjust the scaling of the axes to reveal the curving behavior, and try to guess the relative extrema. Create input–output tables to refine your estimates of relative extrema.

3.2

USING SECOND DERIVATIVES TO FIND MAXIMUM AND MINIMUM VALUES AND SKETCH GRAPHS

OBJECTIVES

a)   Find the relative extrema of a function using the Second-Derivative Test.

b)   Sketch the graph of a continuous function using the graphing strategy in this section.

Concavity: Increasing and Decreasing Derivatives

The graphs of two functions are shown in Fig. 1. The graph in Fig. 1(a) is turning up and the graph in Fig. 1(b) is turning down. Let's see if we can relate this to their derivatives.

Consider the graph of f (Fig. 1a). Take a ruler, or straightedge, and draw tangent lines as you move along the curve from left to right. What happens to the slopes of the tangent lines? Do the same for the graph of g (Fig. 1b). Look for a pattern.

1. Consider this graph.

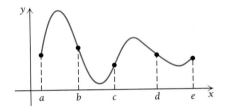

a)   Over what intervals is the derivative increasing?

b)   Over what intervals is the second derivative positive?

c)   Over what intervals is the derivative decreasing?

d)   Over what intervals is the second derivative negative?

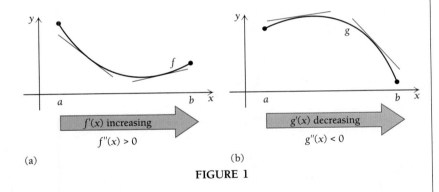

(a)   (b)

FIGURE 1

In Fig. 1(a), the slopes are increasing. That is, f' is increasing on the interval. We know that f' is increasing if we know that f'' is positive, since the relationship between f' and f'' is like the relationship between f and f'. Note also that all the tangent lines are below the graph. In Fig. 1(b), the slopes are decreasing. That is, we know that g' is decreasing if we know that g'' is negative. The tangent lines all are above the graph.

DO EXERCISE 1.

2. Consider this graph.

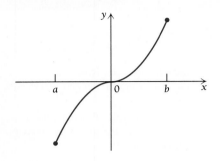

a) Over what intervals is the graph concave up?

b) Over what intervals is the graph concave down?

DEFINITION

Suppose that f is a function whose derivative f' exists at every point in an open interval I. Then:

  1. f is *concave up* on the interval I if f' is increasing on I.

  2. f is *concave down* on the interval I if f' is decreasing on I.

For example, the graph in Fig. 1(a) is concave up and the graph in Fig. 1(b) is concave down. We then have the following theorem, which allows us to use second derivatives to determine concavity.

THEOREM 4

A Test for Concavity

  1. If $f''(x) > 0$ on an interval I, then the graph of f is turning up. (f' is increasing, so f is concave up on I.)

  2. If $f''(x) < 0$ on an interval I, then the graph of f is turning down. (f' is decreasing, so f is concave down on I.)

Figure 2 shows a helpful memory device.

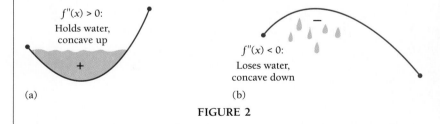

(a)   (b)

FIGURE 2

DO EXERCISE 2.

Finding Relative Extrema Using Second Derivatives

Now we see how we can use second derivatives to determine whether a function has a relative extreme value on an open interval. Refer to Fig. 3.

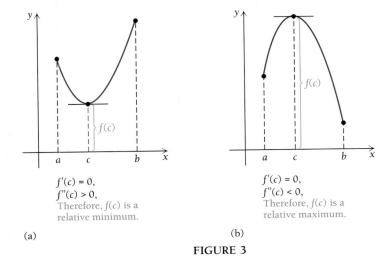

$f'(c) = 0,$
$f''(c) > 0,$
Therefore, $f(c)$ is a
relative minimum.

(a)

$f'(c) = 0,$
$f''(c) < 0,$
Therefore, $f(c)$ is a
relative maximum.

(b)

FIGURE 3

THEOREM 5

The Second-Derivative Test for Relative Extrema

Suppose that f is a function for which $f'(x)$ exists for every x in an open interval (a, b) contained in its domain, and that there is a critical point c in (a, b) for which $f'(c) = 0$. Then:

1. $f(c)$ is a relative minimum if $f''(c) > 0$.

2. $f(c)$ is a relative maximum if $f''(c) < 0$.

The test fails if $f''(c) = 0$. The First-Derivative Test would then have to be used.

Note that $f''(c) = 0$ does not tell us that there is no relative extremum. It just tells us that we do not know at this point and that we must use some other means such as the First-Derivative Test to determine whether we do.

Consider the examples in Fig. 4. In each example, f' and f'' are both 0 at $c = 2$, but the first function has an extremum and the second function has *no* extremum. Also note that if $f'(c)$ does not exist, then $f''(c)$ does not exist and the Second-Derivative Test cannot be used.

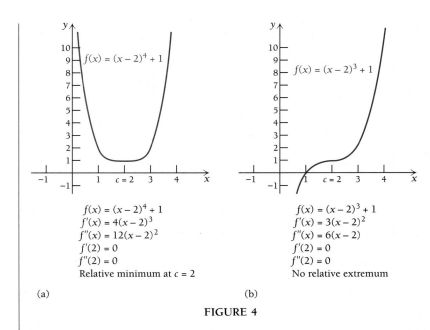

$f(x) = (x - 2)^4 + 1$
$f'(x) = 4(x - 2)^3$
$f''(x) = 12(x - 2)^2$
$f'(2) = 0$
$f''(2) = 0$
Relative minimum at $c = 2$

(a)

$f(x) = (x - 2)^3 + 1$
$f'(x) = 3(x - 2)^2$
$f''(x) = 6(x - 2)$
$f'(2) = 0$
$f''(2) = 0$
No relative extremum

(b)

FIGURE 4

EXAMPLE 1   Find the relative extrema of the function f given by

$$f(x) = x^3 + 3x^2 - 9x - 13.$$

Solution   We find both the first and second derivatives, $f'(x)$ and $f''(x)$:

$$f'(x) = 3x^2 + 6x - 9,$$
$$f''(x) = 6x + 6.$$

Then we solve $f'(x) = 0$:

$3x^2 + 6x - 9 = 0$

$x^2 + 2x - 3 = 0$ Dividing by 3 on both sides

$(x + 3)(x - 1) = 0$ Factoring

$x + 3 = 0$ *or* $x - 1 = 0$ Using the Principle of Zero Products

$x = -3$ *or* $x = 1.$

We now use the Second-Derivative Test with the numbers -3 and 1:

$$f''(-3) = 6(-3) + 6 = -12 < 0 \longrightarrow \quad \text{Relative maximum}$$
$$f''(1) = 6(1) + 6 = 12 > 0 \longrightarrow \quad \text{Relative minimum}$$

We then find second coordinates by substituting in the original function:

$$f(-3) = (-3)^3 + 3(-3)^2 - 9(-3) - 13 = 14;$$
$$f(1) = (1)^3 + 3(1)^2 - 9(1) - 13 = -18.$$

Thus there is a relative maximum at $(-3, 14)$ and a relative minimum at $(1, -18)$. The relative extrema are shown in the graph in Fig. 5.

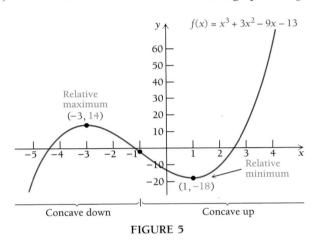

FIGURE 5

DO EXERCISE 3.

3. Find the relative extrema of the function f given by

$$f(x) = x^3 - 3x^2 - 9x - 1.$$

Use the Second-Derivative Test.

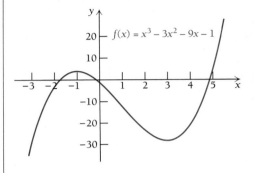

EXAMPLE 2 Find the relative extrema of the function f given by

$$f(x) = 3x^5 - 20x^3.$$

Solution We find both the first and second derivatives, $f'(x)$ and $f''(x)$:

$$f'(x) = 15x^4 - 60x^2,$$
$$f''(x) = 60x^3 - 120x.$$

Then we solve $f'(x) = 0$:

$$15x^4 - 60x^2 = 0$$
$$15x^2(x^2 - 4) = 0$$
$$15x^2(x + 2)(x - 2) = 0 \qquad \text{Factoring}$$
$$15x^2 = 0 \quad \text{or} \quad x + 2 = 0 \quad \text{or} \quad x - 2 = 0 \qquad \text{Using the Principle of Zero Products}$$
$$x = 0 \quad \text{or} \qquad x = -2 \quad \text{or} \qquad x = 2.$$

4. Find the relative extrema of the function f given by

$$f(x) = 3x^5 - 5x^3.$$

Use the Second-Derivative Test.

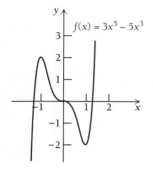

We now use the Second-Derivative Test with the numbers -2, 2, and 0:

$f''(-2) = 60(-2)^3 - 120(-2) = -240 < 0 \longrightarrow$ **Relative maximum;**

$f''(2) = 60(2)^3 - 120(2) = 240 > 0 \longrightarrow$ **Relative minimum;**

$f''(0) = 60(0)^3 - 120(0) = 0 \longrightarrow$ **Second-Derivative Test fails;**
  Use the First-Derivative Test.

We then find second coordinates by substituting in the original function:

$$f(-2) = 3(-2)^5 - 20(-2)^3 = 64;$$
$$f(2) = 3(2)^5 - 20(2)^3 = -64;$$
$$f(0) = 3(0)^5 - 20(0)^3 = 0.$$

Thus there is a relative maximum at $(-2, 64)$ and a relative minimum at $(2, -64)$. Because the function decreases to the left and to the right of $x = 0$, we know by the First-Derivative Test that it has no relative extremum at $(0, 0)$. The extrema are shown in the graph in Fig. 6.

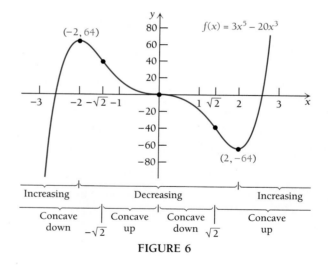

FIGURE 6

DO EXERCISE 4.

Points of Inflection

A **point of inflection**, or an **inflection point**, is a point across which the direction of concavity changes. For example, in Figs. 7–9, point P is an inflection point.

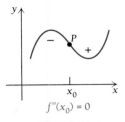

$f''(x_0) = 0$

FIGURE 7

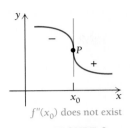

$f''(x_0)$ does not exist

FIGURE 8

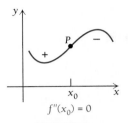

$f''(x_0) = 0$

FIGURE 9

5. What are the points of inflection?

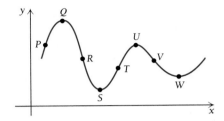

There are points of inflection in other examples that we have considered in this section. In Fig. 5, the graph has a point of inflection at $(-1, -2)$, and in Fig. 6, the graph has points of inflection at $(-\sqrt{2}, 28\sqrt{2})$, $(0, 0)$, and $(\sqrt{2}, -28\sqrt{2})$.

As we move to the right along the curve in Fig. 7, the concavity changes from concave down, $f''(x) < 0$, on the left of P to concave up, $f''(x) > 0$, on the right of P. Since, as we move through P, $f''(x)$ changes sign from $-$ to $+$, the value of $f''(x_0)$ at P must be 0, as in Fig. 7; or $f''(x_0)$ does not exist, as in Fig. 8. A similar change in concavity occurs at P in Fig. 9.

THEOREM 6

Finding Points of Inflection

If a function f has a point of inflection, it occurs at a point x_0 where

$$f''(x_0) = 0 \quad \text{or} \quad f''(x_0) \text{ does not exist.}$$

Thus we find candidates for points of inflection by looking for numbers x_0 for which $f''(x_0) = 0$ or for which $f''(x_0)$ does not exist. Then if $f''(x)$ changes sign as x moves through x_0, we have a point of inflection at $x = x_0$. If $f''(x) = k$, where $k \neq 0$, then x is not a candidate for a point of inflection.

DO EXERCISE 5.

Curve Sketching

We can use what we have learned thus far in this chapter to greatly enhance our ability to sketch curves. We use the following strategy, writing it first in abbreviated form.

> ### Strategy for Sketching Graphs
>
> a) Derivatives
> b) Critical points of f
> c) Increasing, decreasing, relative extrema
> d) Inflection points
> e) Concavity
> f) Sketch

Below we expand on the strategy.

> ### Strategy for Sketching Graphs
>
> a) *Derivatives.* Find $f'(x)$ and $f''(x)$.
>
> b) *Critical points of f.* Find the critical points of f by solving $f'(x) = 0$ and finding where $f'(x)$ does not exist. These numbers yield candidates for relative maxima or minima. Find the function values at these points.
>
> c) *Increasing, decreasing, relative extrema.* Use the critical points of f from (b) to define intervals. Determine whether f is increasing or decreasing on the intervals. Do this by selecting test values and substituting into $f'(x)$. Use this information and/or the second derivative to determine the relative maxima and minima.
>
> d) *Inflection points.* Determine candidates for inflection points by finding where $f''(x) = 0$ or where $f''(x)$ does not exist. Find the function values at these points.
>
> e) *Concavity.* Use the candidates for inflection points from (d) to define intervals. Determine the concavity by checking to see where f' is increasing and where f' is decreasing. Do this by selecting test values and substituting into $f''(x)$.
>
> f) *Sketch the graph.* Sketch the graph using the information from steps (a) through (e), plotting extra points (computing them with your calculator) if the need arises.

EXAMPLE 3   Find the relative maxima and minima for $f(x) = x^3 - 3x + 2$ and sketch the graph.

Solution

a) *Derivatives.* Find $f'(x)$ and $f''(x)$:

$$f'(x) = 3x^2 - 3,$$
$$f''(x) = 6x.$$

b) *Critical points of f.* Find the critical points of f by finding where $f'(x)$ does not exist and by solving $f'(x) = 0$. Now $f'(x) = 3x^2 - 3$ exists for all values of x, so the only critical points are where

$$3x^2 - 3 = 0$$
$$3x^2 = 3$$
$$x^2 = 1$$
$$x = \pm 1.$$

Now $f(-1) = 4$ and $f(1) = 0$. These give the points $(-1, 4)$ and $(1, 0)$ on the graph.

c) *Increasing, decreasing, relative extrema.* Find the intervals on which f is increasing and the intervals on which f is decreasing. The critical points are -1 and 1. We use these points to divide the real-number line into three intervals: $A(-\infty, -1)$, $B(-1, 1)$, and $C(1, \infty)$. We choose a test point in each interval and make a substitution. The test values we select are -2, 0, and 3:

A: Test -2, $\quad f'(-2) = 3(-2)^2 - 3 = 9 > 0$;

B: Test 0, $\quad f'(0) = 3(0)^2 - 3 = -3 < 0$;

C: Test 3, $\quad f'(3) = 3(3)^2 - 3 = 24 > 0$.

Interval	$(-\infty, -1)$	$(-1, 1)$	$(1, \infty)$
Test value	$x = -2$	$x = 0$	$x = 3$
Sign of $f'(x)$	$f'(-2) > 0$	$f'(0) < 0$	$f'(3) > 0$
Result	f is increasing	f is decreasing	f is increasing

Change indicates a relative maximum.

Change indicates a relative minimum.

Therefore, by the First-Derivative Test, there is a relative maximum at $(-1, 4)$ and a relative minimum at $(1, 0)$. That these are relative extrema can also be verified by the Second-Derivative Test: $f''(-1) < 0$ tells us that $(-1, 4)$ is a relative maximum, and $f''(1) > 0$ tells us that $(1, 0)$ is a relative minimum.

d) *Inflection points.* Find possible inflection points by finding where $f''(x)$ does not exist and by solving $f''(x) = 0$. Now $f''(x) = 6x$ exists for all values of x, so we try to solve $f''(x) = 0$:

$$6x = 0$$
$$x = 0.$$

Now $f(0) = 2$. This gives us another point, $(0, 2)$, that lies on the graph.

e) *Concavity.* Find the intervals on which f is concave up and concave down. We do this by determining where f' is increasing and decreasing using the numbers found in (d). There is only one such number, 0. We use this point to divide the real-number line into two intervals: $A(-\infty, 0)$ and $B(0, \infty)$. We choose a test point in each interval and make a substitution into f''. The test points we select are -4 and 4:

A: Test -4, $\quad f''(-4) = 6(-4) = -24 < 0$;
B: Test 4, $\quad f''(4) = 6(4) = 24 > 0$.

Interval	$(-\infty, 0)$	$(0, \infty)$
Test value	$x = -4$	$x = 4$
Sign of $f''(x)$	$f''(-4) < 0$	$f''(4) > 0$
Result	f' is decreasing	f' is increasing

Change indicates a point of inflection.

Since f' is decreasing on the interval $(-\infty, 0)$ and f' is increasing on the interval $(0, \infty)$, f is concave down on $(-\infty, 0)$ and concave up on $(0, \infty)$. The graph changes concavity across $(0, 2)$, so it is a point of inflection.

f) *Sketch the graph.* Sketch the graph using the information in the following table. Calculate some extra function values if desired. The graph is shown in Fig. 10.

x	$f(x)$
-3	-16
-2	0
-1	4
0	2
1	0
2	4
3	20

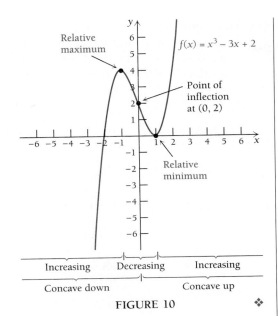

FIGURE 10

6. For the following function, find the relative maxima and minima and sketch the graph.

$$f(x) = \tfrac{1}{3}x^3 - \tfrac{1}{2}x^2 - 2x + 1$$

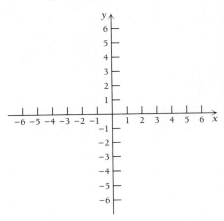

DO EXERCISE 6.

EXAMPLE 4   Find the relative maxima and minima for $f(x) = x^4 - 2x^2$ and sketch the graph.

Solution

a)   *Derivatives.* Find $f'(x)$ and $f''(x)$:

$$f'(x) = 4x^3 - 4x,$$
$$f''(x) = 12x^2 - 4.$$

b)   *Critical points of f.* Since $f'(x) = 4x^3 - 4x$ exists for all values of x, the only critical points of f are where

$$4x^3 - 4x = 0$$
$$4x(x^2 - 1) = 0$$
$$4x = 0 \quad or \quad x^2 - 1 = 0$$
$$x = 0 \quad or \quad x^2 = 1$$
$$x = \pm 1.$$

Now $f(0) = 0$, $f(-1) = -1$, and $f(1) = -1$. These give the points $(0, 0)$, $(-1, -1)$, and $(1, -1)$ on the graph.

c) *Increasing, decreasing, relative extrema.* Use the critical points of f, —namely, -1, 0, and 1—to divide the real-number line into four intervals: $A(-\infty, -1)$, $B(-1, 0)$, $C(0, 1)$, and $D(1, \infty)$. We choose a test value in each interval and make a substitution. The test values we select are -2, $-\frac{1}{2}$, $\frac{1}{2}$, and 2:

A: Test -2, $f'(-2) = 4(-2)^3 - 4(-2) = -24 < 0$;
B: Test $-\frac{1}{2}$, $f'(-\frac{1}{2}) = 4(-\frac{1}{2})^3 - 4(-\frac{1}{2}) = \frac{3}{2} > 0$;
C: Test $\frac{1}{2}$, $f'(\frac{1}{2}) = 4(\frac{1}{2})^3 - 4(\frac{1}{2}) = -\frac{3}{2} < 0$;
D: Test 2, $f'(2) = 4(2)^3 - 4(2) = 24 > 0$.

Interval	$(-\infty, -1)$	$(-1, 0)$	$(0, 1)$	$(1, \infty)$
Test value	$x = -2$	$x = -\frac{1}{2}$	$x = \frac{1}{2}$	$x = 2$
Sign of $f'(x)$	$f'(-2) < 0$	$f'(-\frac{1}{2}) > 0$	$f'(\frac{1}{2}) < 0$	$f'(2) > 0$
Result	f is decreasing	f is increasing	f is decreasing	f is increasing

Change indicates a relative minimum.   Change indicates a relative maximum.   Change indicates a relative minimum.

Thus, by the First-Derivative Test, there is a relative maximum at $(0, 0)$ and two relative minima at $(-1, -1)$ and $(1, -1)$. That these are relative extrema can also be verified by the Second-Derivative Test: $f''(0) < 0$, $f''(-1) > 0$, and $f''(1) > 0$.

d) *Inflection points.* Find where $f''(x)$ does not exist and where $f''(x) = 0$. Since $f''(x)$ exists for all real numbers, we just solve $f''(x) = 0$:

$$12x^2 - 4 = 0$$
$$4(3x^2 - 1) = 0$$
$$3x^2 - 1 = 0$$
$$3x^2 = 1$$
$$x^2 = \frac{1}{3}$$
$$x = \pm\sqrt{\frac{1}{3}} = \pm\frac{1}{\sqrt{3}}.$$

Now

$$f\left(\frac{1}{\sqrt{3}}\right) = \left(\frac{1}{\sqrt{3}}\right)^4 - 2\left(\frac{1}{\sqrt{3}}\right)^2 = \frac{1}{9} - \frac{2}{3} = -\frac{5}{9}$$

and

$$f\left(-\frac{1}{\sqrt{3}}\right) = -\frac{5}{9}.$$

This gives the points

$$\left(-\frac{1}{\sqrt{3}}, -\frac{5}{9}\right) \quad \text{and} \quad \left(\frac{1}{\sqrt{3}}, -\frac{5}{9}\right)$$

on the graph—$(-0.6, -0.6)$ and $(0.6, -0.6)$, approximately.

e) *Concavity.* Find the intervals on which f is concave up and concave down. We do this by determining where f' is increasing and decreasing using the numbers found in (d). Those numbers divide the real-number line into three intervals:

$$A\left(-\infty, -\frac{1}{\sqrt{3}}\right), B\left(-\frac{1}{\sqrt{3}}, \frac{1}{\sqrt{3}}\right), \text{ and } C\left(\frac{1}{\sqrt{3}}, \infty\right).$$

Next we choose a test value in each interval and make a substitution into f''. The test points we select are -1, 0, and 1:

A: Test -1, $f''(-1) = 12(-1)^2 - 4 = 8 > 0$;

B: Test 0, $f''(0) = 12(0)^2 - 4 = -4 < 0$;

C: Test 1, $f''(1) = 12(1)^2 - 4 = 8 > 0$.

Interval	$(-\infty, -1/\sqrt{3})$	$(-1/\sqrt{3}, 1/\sqrt{3})$	$(1/\sqrt{3}, \infty)$
Test value	$x = -1$	$x = 0$	$x = 1$
Sign of $f''(x)$	$f''(-1) > 0$	$f''(0) < 0$	$f''(1) > 0$
Result	f' is increasing	f' is decreasing	f' is increasing

└── Change ──┘ └── Change ──┘
indicates a   indicates a
point of   point of
inflection.   inflection.

7. For the following function, find the relative maxima and minima and sketch the graph.

$$f(x) = 2x^4 - 4x^2 + 2$$

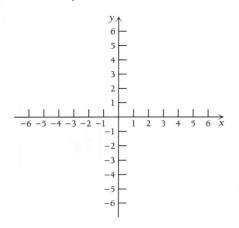

Therefore, f' is increasing on the interval $(-\infty, -1/\sqrt{3})$, decreasing on the interval $(-1/\sqrt{3}, 1/\sqrt{3})$, and increasing on the interval $(1/\sqrt{3}, \infty)$. The function f is concave up on $(-\infty, -1/\sqrt{3})$ and $(1/\sqrt{3}, \infty)$ and concave down on $(-1/\sqrt{3}, 1/\sqrt{3})$. The graph changes concavity across $(-1/\sqrt{3}, -5/9)$ and $(1/\sqrt{3}, -5/9)$, so these are points of inflection.

f) *Sketch.* Sketch the graph using the information in the following table. By solving $x^4 - 2x^2 = 0$, we can find the x-intercepts easily. They are $(-\sqrt{2}, 0)$, $(0, 0)$, and $(\sqrt{2}, 0)$. This also aids the graphing. Extra function values can be calculated if desired. The graph is shown in Fig. 11.

x	$f(x)$, approximately
-2	8
-1.5	0.56
-1	-1
-0.5	-0.44
0	0
0.5	-0.44
1	-1
1.5	0.56
2	8

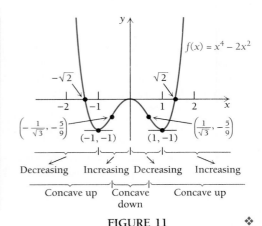

FIGURE 11

DO EXERCISE 7.

3.2

Find the relative extrema of the function. List your answers in terms of ordered pairs. Use the Second-Derivative Test, where possible. Then sketch the graph.

1. $f(x) = 2 - x^2$

2. $f(x) = 3 - x^2$

3. $f(x) = x^2 + x - 1$

4. $f(x) = x^2 - x$

5. $f(x) = -4x^2 + 3x - 1$

6. $f(x) = 7 - 8x + 5x^2$

7. $f(x) = 2x^3 - 3x^2 - 36x + 28$

8. $f(x) = 3x^3 - 36x - 3$

9. $f(x) = \frac{8}{3}x^3 - 2x + \frac{1}{3}$

10. $f(x) = 80 - 9x^2 - x^3$

11. $f(x) = -x^3 + 3x^2 - 4$

12. $f(x) = -x^3 + 3x - 2$

13. $f(x) = 3x^4 - 16x^3 + 18x^2$

14. $f(x) = 3x^4 + 4x^3 - 12x^2 + 5$

15. $f(x) = (x + 1)^{2/3}$

16. $f(x) = (x - 1)^{2/3}$

17. $f(x) = x^4 - 6x^2$

18. $f(x) = 2x^2 - x^4$

19. $f(x) = x^3 - 2x^2 - 4x + 3$

20. $f(x) = x^3 - 6x^2 + 9x + 1$

21. $f(x) = 3x^4 + 4x^3$

22. $f(x) = x^4 - 2x^3$

23. $f(x) = x^3 - 6x^2 - 135x$

24. $f(x) = x^3 - 3x^2 - 144x - 140$

25. $f(x) = \dfrac{x}{x^2 + 1}$

26. $f(x) = \dfrac{8x}{x^2 + 1}$

27. $f(x) = \dfrac{3}{x^2 + 1}$

28. $f(x) = \dfrac{-4}{x^2 + 1}$

29. $f(x) = (x - 1)^3$

30. $f(x) = (x + 2)^3$

31. $f(x) = x^2(1 - x)^2$

32. $f(x) = x^2(3 - x)^2$

33. $f(x) = 20x^3 - 3x^5$

34. $f(x) = 5x^3 - 3x^5$

35. $f(x) = x\sqrt{4 - x^2}$

36. $f(x) = -x\sqrt{1 - x^2}$

37. $f(x) = (x - 1)^{1/3} - 1$

38. $f(x) = 2 - x^{1/3}$

APPLICATIONS

❖ **Business and Economics**

Total revenue, cost, and profit. Using the same set of axes, sketch the graphs of the total-revenue, total-cost, and total-profit functions.

39. $R(x) = 50x - 0.5x^2$, $C(x) = 4x + 10$

40. $R(x) = 50x - 0.5x^2$, $C(x) = 10x + 3$

❖ **Life and Physical Sciences**

41. *Coughing velocity.* A person coughs when a foreign object is in the windpipe. The velocity of the cough depends on the size of the object. Suppose a person has a windpipe with a 20-mm radius. If a foreign object has a radius r, in millimeters, then the velocity V, in millimeters/second, needed to remove the object by a cough is given by

$$V(r) = k(20r^2 - r^3), \quad 0 \leq r \leq 20,$$

where k is some positive constant. For what size object is the maximum velocity needed to remove the object?

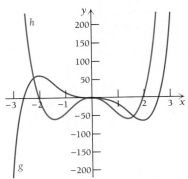

42. *Temperature in January.* Suppose that the temperature T, in degrees Fahrenheit, during a 24-hr day in January is given by

$$T(x) = 0.0027(x^3 - 34x + 240), \quad 0 \leq x \leq 24,$$

where $x =$ the number of hours since midnight. Estimate the relative minimum temperature and when it occurs.

SYNTHESIS EXERCISES

In each of Exercises 43 and 44, determine which graph is the derivative of the other.

43.

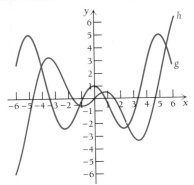

44.

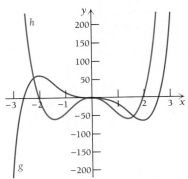

COMPUTER–GRAPHING CALCULATOR EXERCISES

Sketch the graph of the function using a computer software graphing package or a graphing calculator. Then estimate any relative extrema.

45. $f(x) = 3x^{2/3} - 2x$

46. $f(x) = 4x - 6x^{2/3}$

47. $f(x) = x^3(x - 2)^3$

48. $f(x) = x^2(1 - x)^3$

49. $f(x) = x - \sqrt{x}$

50. $f(x) = (x - 1)^{2/3} - (x + 1)^{2/3}$

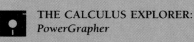

THE CALCULUS EXPLORER:
PowerGrapher

Do Exercises 45–50 using these programs.

3.3

GRAPH SKETCHING: ASYMPTOTES AND RATIONAL FUNCTIONS

OBJECTIVE

a) Sketch the graph of a rational function using the graphing strategy considered in this section.

Vertical and Horizontal Asymptotes

Thus far we have considered a strategy for graphing a continuous function using the tools of calculus. We now want to consider some discontinuous functions, most of which are rational functions. Our graphing skills will now have to take into account the discontinuities of the graph and certain lines called *asymptotes*.

Let us reconsider the definition of a rational function.

DEFINITION

A *rational function* is a function f that can be described by

$$f(x) = \frac{P(x)}{Q(x)},$$

where $P(x)$ and $Q(x)$ are polynomials having no common factor other than 1 and -1, and with $Q(x)$ not the zero polynomial. The domain of f consists of all inputs x for which $Q(x) \neq 0$.

Polynomial functions are themselves special kinds of rational functions, since $Q(x)$ can be the polynomial 1. Here we are interested in graphing rational functions in which the denominator is not a constant.

Figure 1 shows the graph of the rational function

$$f(x) = \frac{x^2 - 1}{x^2 + x - 6} = \frac{(x - 1)(x + 1)}{(x - 2)(x + 3)}.$$

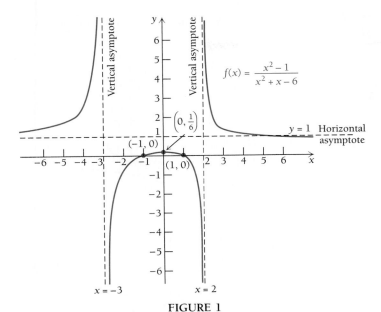

$$f(x) = \frac{x^2 - 1}{x^2 + x - 6}$$

FIGURE 1

Let us make some observations about this graph.

First, note that as x gets closer to 2 from the left, the function values get smaller and smaller negatively, approaching $-\infty$. As x gets closer to 2 from the right, the function values get larger and larger positively. Thus,

$$\lim_{x \to 2^-} f(x) = -\infty \quad \text{and} \quad \lim_{x \to 2^+} f(x) = \infty.$$

For this graph, we can think of the line $x = 2$ as a "limiting line" called a *vertical asymptote*. Similarly, the line $x = -3$ is a vertical asymptote.

The line $x = a$ is a *vertical asymptote* if any of the following limit statements is true:

$$\lim_{x \to a^-} f(x) = \infty \quad \text{or} \quad \lim_{x \to a^-} f(x) = -\infty \quad \text{or}$$

$$\lim_{x \to a^+} f(x) = \infty \quad \text{or} \quad \lim_{x \to a^+} f(x) = -\infty.$$

The graph of a rational function *never* crosses a vertical asymptote. If a is an input that makes the denominator 0, then the line $x = a$ is a vertical asymptote. (Remember that the numerator and the denominator have no common factor other than -1 or 1.)

Figure 2 shows the four ways in which a vertical asymptote can occur.

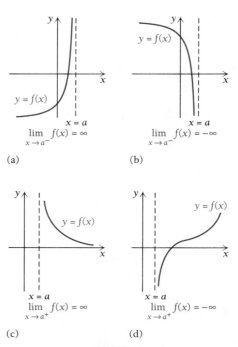

FIGURE 2

Look again at the graph in Fig. 1. Note that function values get closer and closer to 1 as x approaches $-\infty$, meaning that $f(x) \to 1$ as $x \to -\infty$. Also, function values get closer and closer to 1 as x approaches ∞, meaning that $f(x) \to 1$ as $x \to \infty$. Thus,

$$\lim_{x \to -\infty} f(x) = 1 \quad \text{and} \quad \lim_{x \to \infty} f(x) = 1.$$

The line $y = 1$ is called a *horizontal asymptote*.

The line $y = b$ is a *horizontal asymptote* if either or both of the following limit statements is true:

$$\lim_{x \to -\infty} f(x) = b \quad \text{or} \quad \lim_{x \to \infty} f(x) = b.$$

The graph of a rational function may or may not cross a horizontal asymptote. Horizontal asymptotes occur when the degree of the numerator is less than or equal to the degree of the denominator.

In Figs. 3–5, we see three ways in which horizontal asymptotes can occur.

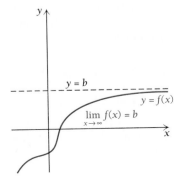

FIGURE 3

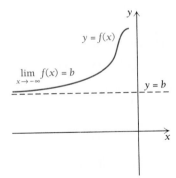

FIGURE 4

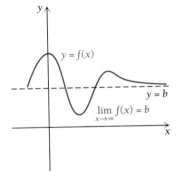

FIGURE 5

Determine the vertical asymptotes.

1. $f(x) = \dfrac{x^2 + 7x + 10}{x^2 + 3x - 28}$

2. $f(x) = \dfrac{x^2 + 5}{x^3 - x^2 - 6x}$

Occurrences of Asymptotes

It is important in graphing rational functions to determine where the asymptotes, if any, occur. Vertical asymptotes are easy to locate when a denominator can be factored. The x-inputs that make a denominator 0 give us the vertical asymptotes.

EXAMPLE 1   Determine the vertical asymptotes:

$$f(x) = \frac{3x - 2}{x(x - 5)(x + 3)}.$$

Solution   The vertical asymptotes are the lines $x = 0$, $x = 5$, and $x = -3$.   ❖

EXAMPLE 2   Determine the vertical asymptotes:

$$f(x) = \frac{x - 2}{x^3 - x}.$$

Solution   We factor the denominator:

$$x^3 - x = x(x^2 - 1) = x(x - 1)(x + 1).$$

The vertical asymptotes are $x = 0$, $x = -1$, and $x = 1$.   ❖

DO EXERCISES 1 AND 2.

EXAMPLE 3   Find the horizontal asymptotes:

$$f(x) = \frac{2x + 3}{x^3 - 2x^2 + 4}.$$

Solution   Since the degree of the numerator is less than the degree of the denominator, there is a horizontal asymptote. We then divide the numerator and the denominator by x^3 and find the limits as x approaches $-\infty$ and ∞; that is, as $|x|$ gets larger and larger. (You might want to review the part of Section 2.2 on limits involving infinity.)

$$f(x) = \frac{2x + 3}{x^3 - 2x^2 + 4} = \frac{\dfrac{2}{x^2} + \dfrac{3}{x^3}}{1 - \dfrac{2}{x} + \dfrac{4}{x^3}}$$

As x gets smaller and smaller negatively, $|x|$ gets larger and larger. Similarly, as x gets larger and larger positively, $|x|$ gets larger and larger.

Thus, as $|x|$ becomes very large, each expression with x or some power of x in the denominator takes on values ever closer to 0. Thus the numerator approaches 0 and the denominator approaches 1; hence the entire expression takes on values ever closer to 0. We have

$$f(x) \approx \frac{0 + 0}{1 - 0 + 0},$$

so

$$\lim_{x \to -\infty} f(x) = 0 \quad \text{and} \quad \lim_{x \to \infty} f(x) = 0,$$

and the x-axis, the line $y = 0$, is a horizontal asymptote. ❖

EXAMPLE 4 Find the horizontal asymptotes:

$$f(x) = \frac{3x^2 + 2x - 4}{2x^2 - x + 1}.$$

Solution The numerator and the denominator have the same degree, so there is a horizontal asymptote. We divide the numerator and the denominator by x^2 and find the limit as $|x|$ gets larger and larger:

$$f(x) = \frac{3x^2 + 2x - 4}{2x^2 - x + 1} = \frac{3 + \dfrac{2}{x} - \dfrac{4}{x^2}}{2 - \dfrac{1}{x} + \dfrac{1}{x^2}}.$$

As $|x|$ gets very large, the numerator approaches 3 and the denominator approaches 2. Therefore, the function gets very close to $\frac{3}{2}$. Thus,

$$\lim_{x \to -\infty} f(x) = \frac{3}{2} \quad \text{and} \quad \lim_{x \to \infty} f(x) = \frac{3}{2}.$$

The line $y = \frac{3}{2}$ is a horizontal asymptote. ❖

When the degree of the numerator is less than the degree of the denominator, the x-axis, or the line $y = 0$, is a horizontal asymptote. When the degree of the numerator is the same as the degree of the denominator, the line $y = a/b$ is a horizontal asymptote, where a is the leading coefficient of the numerator and b is the leading coefficient of the denominator.

DO EXERCISES 3–5.

3. For which of the following is the x-axis a horizontal asymptote?

a) $f(x) = \dfrac{3x^4 - x^2 + 4}{18x^4 - x^3 + 44}$

b) $f(x) = \dfrac{17x^2 + 14x - 5}{x^3 - 2x - 1}$

c) $f(x) = \dfrac{135x^5 - x^2}{x^7}$

Find the horizontal asymptotes.

4. $f(x) = \dfrac{3x^3 + 4x - 9}{6x^3 - 7x^2 + 3}$

5. $f(x) = \dfrac{9x^4 - 7x^2 - 9}{3x^4 + 7x^2 + 9}$

Oblique Asymptotes

There are asymptotes that are neither vertical nor horizontal. For example, in the graph of

$$f(x) = \frac{x^2 - 4}{x - 1}$$

(Fig. 6), the line $x = 1$ is a vertical asymptote. The line $y = x + 1$ is called an *oblique asymptote*. Note that as $|x|$ gets larger and larger, the curve gets closer and closer to $y = x + 1$.

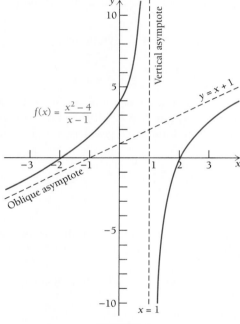

FIGURE 6

The line $y = mx + b$ is an *oblique asymptote* of the rational function $f(x) = P(x)/Q(x)$ if $f(x)$ can be expressed as

$$f(x) = (mx + b) + g(x),$$

where $g(x) \to 0$ as $|x| \to \infty$. Oblique asymptotes occur when the degree of the numerator is one more than the degree of the denominator. A graph can cross an oblique asymptote.

How can we find an oblique asymptote? One way is by division.

EXAMPLE 5   Find the oblique asymptotes:

$$f(x) = \frac{x^2 - 4}{x - 1}.$$

Solution   When we divide the numerator by the denominator, we obtain a quotient of $x + 1$ and a remainder of -3. Thus,

$$f(x) = \frac{x^2 - 4}{x - 1} = (x + 1) + \frac{-3}{x - 1}.$$

$$
\begin{array}{r}
x + 1 \\
x - 1 \overline{)x^2 - 4} \\
\underline{x^2 - x} \\
x - 4 \\
\underline{x - 1} \\
-3
\end{array}
$$

Now we can see that when $|x|$ gets very large, $-3/(x - 1)$ approaches 0. Thus, for very large $|x|$, the expression $x + 1$ is the dominant part of $(x + 1) + (-3)/(x + 1)$. Thus, $y = x + 1$ is an oblique asymptote.   ❖

DO EXERCISES 6 AND 7.

Intercepts

If they exist, the **x-intercepts** of a function occur at those values of x for which $y = f(x) = 0$ and give us points at which the graph crosses the x-axis. If it exists, the **y-intercept** of a function occurs at the value of y for which $x = 0$ and gives us the point at which the graph crosses the y-axis.

EXAMPLE 6   Find the intercepts of

$$f(x) = \frac{x^3 - x^2 - 6x}{x^2 - 3x + 2}.$$

Solution   We factor the numerator and the denominator:

$$f(x) = \frac{x(x + 2)(x - 3)}{(x - 1)(x - 2)}.$$

To find the x-intercepts, we solve the equation $f(x) = 0$. Such values occur when the numerator is 0 but the denominator is not. Thus we solve the equation

$$x(x + 2)(x - 3) = 0.$$

The x-values making the numerator 0 are 0, -2, and 3. Since none of these makes the denominator 0, they yield the x-intercepts of the function: $(0, 0)$, $(-2, 0)$, and $(3, 0)$.

Find the oblique asymptotes.

6. $f(x) = \dfrac{3x^2 - 7x + 8}{x - 2}$

7. $f(x) = \dfrac{5x^3 + 2x + 1}{x^2 - 4}$

Find the intercepts.

8. $f(x) = \dfrac{x(x-3)(x+5)}{(x+2)(x-4)}$

9. $f(x) = \dfrac{x^3 + 2x^2 - 3x}{x^2 + 5}$

To find the y-intercept, we let $x = 0$ in the equation:

$$f(0) = \frac{0^3 - 0^2 - 6(0)}{0^2 - 3(0) + 2} = 0.$$

In this case, the y-intercept is also an x-intercept, $(0, 0)$.   ❖

DO EXERCISES 8 AND 9.

Sketching Graphs

We can now refine our strategy for graphing.

Strategy for Sketching Graphs

a) *Intercepts.* Find the x-intercepts and the y-intercepts of the graph.

b) *Asymptotes.* Find the vertical, horizontal, and oblique asymptotes.

c) *Derivatives.* Find $f'(x)$ and $f''(x)$.

d) *Undefined values and critical points of f.* Find the inputs for which the function is not defined, giving denominators of 0. Find also the critical points of f.

e) *Increasing, decreasing, relative extrema.* Use the points found in (d) to determine intervals on which the function f is increasing or decreasing. Use this information and/or the second derivative to determine the relative maxima and minima. A relative extremum can occur only at a point c for which $f(c)$ exists.

f) *Inflection points.* Determine candidates for inflection points by finding points x_0 where $f''(x_0)$ does not exist or where $f''(x_0) = 0$. Find the function values at these points. If a function value $f(x)$ does not exist, then the function does not have an inflection point at x.

g) *Concavity.* Use the values c from (f) as endpoints of intervals. Determine the concavity by checking to see where f' is decreasing and where f' is increasing. Do this by selecting test points and substituting into $f''(x)$.

h) *Sketch the graph.* Use the information from steps (a) through (g) to sketch the graph, plotting extra points (computing them with your calculator) as the need arises.

EXAMPLE 7 Sketch the graph of $f(x) = \dfrac{8}{x^2 - 4}$.

Solution

a) *Intercepts.* The x-intercepts occur at the points where the numerator is 0 but the denominator is not. Since in this case the numerator is the constant 8, there are no x-intercepts. To find the y-intercept, we compute $f(0)$:

$$f(0) = \frac{8}{0^2 - 4} = \frac{8}{-4} = -2.$$

This gives us one point on the graph, $(0, -2)$.

b) *Asymptotes.*

Vertical: The denominator $x^2 - 4 = (x + 2)(x - 2)$. It is 0 for x-values of -2 and 2. Thus the graph has the lines $x = -2$ and $x = 2$ as vertical asymptotes. We plot them using dashed lines.

Horizontal: The degree of the numerator is less than the degree of the denominator, so the x-axis, or the line $y = 0$, is a horizontal asymptote. It is already plotted as an axis.

Oblique: There is no oblique asymptote since the degree of the numerator is not one more than the degree of the denominator.

c) *Derivatives.* Find $f'(x)$ and $f''(x)$. Using the Quotient Rule, we get

$$f'(x) = \frac{-16x}{(x^2 - 4)^2} \quad \text{and} \quad f''(x) = \frac{16(3x^2 + 4)}{(x^2 - 4)^3}.$$

d) *Undefined values and critical points of f.* The domain of the original function is all real numbers except -2 and 2, where the vertical asymptotes occur. We find the critical points of f by looking for values of x where $f'(x) = 0$ or where $f'(x)$ does not exist. Now $f'(x) = 0$ for values of x for which $-16x = 0$, but the denominator is not 0. The only such number is 0 itself. The derivative $f'(x)$ does not exist at -2 and 2. Thus the undefined values and the critical points are -2, 0, and 2.

e) *Increasing, decreasing, relative extrema.* Use the undefined values and the critical points to determine the intervals on which f is increasing and the intervals on which f is decreasing. The points to consider are -2, 0, and 2. These divide the real-number line into four intervals.

We choose a test point in each interval and make a substitution into the derivative f':

A: Test -3, $f'(-3) = \dfrac{-16(-3)}{[(-3)^2 - 4]^2} = \dfrac{48}{25} > 0$;

B: Test -1, $f'(-1) = \dfrac{-16(-1)}{[(-1)^2 - 4]^2} = \dfrac{16}{9} > 0$;

C: Test 1, $f'(1) = \dfrac{-16(1)}{[(1)^2 - 4]^2} = \dfrac{-16}{9} < 0$;

D: Test 3, $f'(3) = \dfrac{-16(3)}{[(3)^2 - 4]^2} = \dfrac{-48}{25} < 0$.

Interval	$(-\infty, -2)$	$(-2, 0)$	$(0, 2)$	$(2, \infty)$
Test value	$x = -3$	$x = -1$	$x = 1$	$x = 3$
Sign of $f'(x)$	$f'(-3) > 0$	$f'(-1) > 0$	$f'(1) < 0$	$f'(3) < 0$
Result	f is increasing	f is increasing	f is decreasing	f is decreasing

└ No change ┘ └ Change ┘ └ No change ┘
indicates a
relative
maximum
at 0.

Now $f(0) = -2$, so there is a relative maximum at $(0, -2)$.

f) *Inflection points.* Determine candidates for inflection points by finding where $f''(x)$ does not exist and where $f''(x) = 0$. Now $f(x)$ does not exist at -2 and 2, so $f''(x)$ does not exist at -2 and 2. We then determine where $f''(x) = 0$, or

$$16(3x^2 + 4) = 0.$$

But $16(3x^2 + 4) > 0$ for all real numbers x, so there are no points of inflection since $f(-2)$ and $f(2)$ do not exist.

g) *Concavity.* Use the values found in (f) as endpoints of intervals. Determine the concavity by checking to see where f' is increasing and decreasing. The points -2 and 2 divide the real-number line into three intervals. We choose test points in each interval and make a substitution into f'':

A: Test -3, $f''(-3) = \dfrac{16[3(-3)^2 + 4]}{[(-3)^2 - 4]^3} > 0$;

B: Test 0, $f''(0) = \dfrac{16[3(0)^2 + 4]}{[(0)^2 - 4]^3} < 0$;

C: Test 3, $f''(3) = \dfrac{16[3(3)^2 + 4]}{[(3)^2 - 4]^3} > 0$.

Interval	$(-\infty, -2)$	$(-2, 2)$	$(2, \infty)$
Test value	$x = -3$	$x = 0$	$x = 3$
Sign of $f''(x)$	$f''(-3) > 0$	$f''(0) < 0$	$f''(3) > 0$
Result	f' is increasing	f' is decreasing	f' is increasing

└─ Change does ─┘   └─ Change does ─┘
not indicate a   not indicate a
point of inflection   point of inflection
since $f(-2)$ since $f(2)$
does not exist.   does not exist.

The function is concave up on the intervals $(-\infty, -2)$ and $(2, \infty)$. The function is concave down on the interval $(-2, 2)$.

h) *Sketch.* Sketch the graph using the information in the following table, plotting extra points by computing values from your calculator as needed. The graph is shown in Fig. 7.

x	$f(x)$, approximately
-5	0.38
-4	0.67
-3	1.6
-1	-2.67
0	-2
1	-2.67
3	1.6
4	0.67
5	0.38

Sketch a graph of the function.

10. $f(x) = \dfrac{1}{x^2 - 1}$

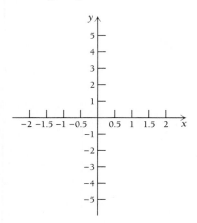

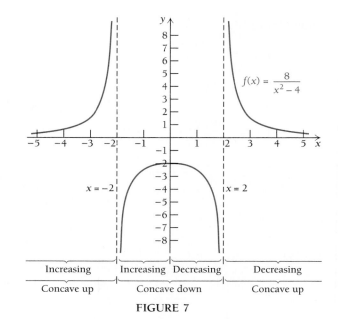

$f(x) = \dfrac{8}{x^2 - 4}$

$x = -2$ $x = 2$

| Increasing | Increasing | Decreasing | Decreasing |
| Concave up | Concave down | | Concave up |

FIGURE 7   ❖

DO EXERCISES 10 AND 11.

EXAMPLE 8   Sketch the graph of $f(x) = \dfrac{x^2 + 4}{x}$.

Solution

a) *Intercepts.* The equation $f(x) = 0$ has no real-number solution. Thus there are no x-intercepts. The number 0 is not in the domain of the function. Thus there are no y-intercepts.

b) *Asymptotes.*

   Vertical: The denominator is x. Since its replacement by 0 makes the denominator 0, the line $x = 0$ is a vertical asymptote.

   Horizontal: The degree of the numerator is greater than the degree of the denominator, so there are no horizontal asymptotes.

   Oblique: The degree of the numerator is one greater than the degree of the denominator, so there is an oblique asymptote. We do the division and express the function in the form

11. $f(x) = \dfrac{1}{x^2 + 4}$

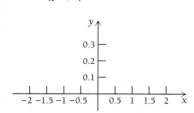

$$f(x) = x + \frac{4}{x}.$$

$$\begin{array}{r} x \\ x\overline{)x^2 + 4} \\ \underline{x^2} \\ 4 \end{array}$$

As $|x|$ gets larger, the term $4/x$ approaches 0, so the line $y = x$ is an oblique asymptote.

c)   *Derivatives.* Find $f'(x)$ and $f''(x)$:

$$f'(x) = 1 - 4x^{-2}$$

$$= 1 - \frac{4}{x^2},$$

$$f''(x) = 8x^{-3}$$

$$= \frac{8}{x^3}.$$

d)   *Undefined values and critical points of f.* The number 0 is not in the domain of f. The derivative exists for all values of x except 0. Thus, to find critical points, we solve $f'(x) = 0$, looking for solutions other than 0:

$$1 - \frac{4}{x^2} = 0$$

$$1 = \frac{4}{x^2}$$

$$x^2 = 4$$

$$x = \pm 2.$$

Thus, -2 and 2 are critical points. Now $f(0)$ does not exist, but $f(-2) = -4$ and $f(2) = 4$. These give the points $(-2, -4)$ and $(2, 4)$ on the graph.

e)   *Increasing, decreasing, relative extrema.* Use the points found in (d) to find intervals on which f is increasing and intervals on which f is decreasing. The points to consider are $-2, 0$, and 2. These divide the real-number line into four intervals. We choose test points in each interval and make a substitution into f':

A: Test -3, $f'(-3) = 1 - \dfrac{4}{(-3)^2} = \dfrac{5}{9} > 0$;

B: Test -1, $f'(-1) = 1 - \dfrac{4}{(-1)^2} = -3 < 0$;

C: Test 1, $f'(1) = 1 - \dfrac{4}{1^2} = -3 < 0$;

D: Test 3, $f'(3) = 1 - \dfrac{4}{3^2} = \dfrac{5}{9} > 0$.

Interval	$(-\infty, -2)$	$(-2, 0)$	$(0, 2)$	$(2, \infty)$
Test value	$x = -3$	$x = -1$	$x = 1$	$x = 3$
Sign of $f'(x)$	$f'(-3) > 0$	$f'(-1) < 0$	$f'(1) < 0$	$f'(3) > 0$
Result	f is increasing	f is decreasing	f is decreasing	f is increasing

Change indicates a relative maximum at -2.   No change   Change indicates a relative minimum at 2.

Now $f(-2) = -4$ and $f(2) = 4$. There is a relative maximum at $(-2, -4)$ and a relative minimum at $(2, 4)$.

f) *Inflection points.* Determine candidates for inflection points by finding where $f''(x_0)$ does not exist or where $f''(x_0) = 0$. Now $f''(0)$ does not exist, but because $f(0)$ does not exist, there cannot be an inflection point at 0. Then look for values of x for which $f''(x) = 0$:

$$\frac{8}{x^3} = 0.$$

But this equation has no solution. Thus there are no points of inflection.

g) *Concavity.* Use the values found in (f) as endpoints of intervals. Determine the concavity by checking to see where f' is increasing and decreasing. The number 0 divides the real-number line into two intervals, $(-\infty, 0)$ and $(0, \infty)$. We could choose a test value in each interval and make a substitution into f''. But note that for any $x < 0$, $x^3 < 0$, so

$$f''(x) = \frac{8}{x^3} < 0,$$

and for any $x > 0$, $x^3 > 0$, so

$$f''(x) = \frac{8}{x^3} > 0.$$

Thus, f is concave down on the interval $(-\infty, 0)$ and concave up on the interval $(0, \infty)$.

h) *Sketch.* Sketch the graph using the preceding information and any computed values of f as the need arises. The graph is shown in Fig. 8.

x	$f(x)$, approximately
-6	-6.67
-5	-5.8
-4	-5
-3	-4.3
-2	-4
-1	-5
-0.5	-8.5
0.5	8.5
1	5
2	4
3	4.3
4	5
5	5.8
6	6.67

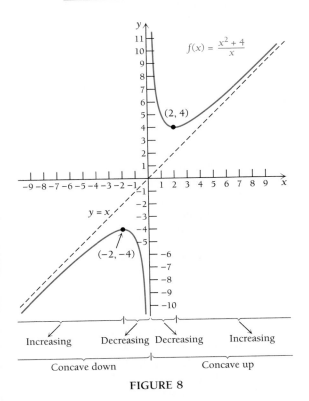

FIGURE 8

DO EXERCISES 12 AND 13.

Sketch a graph of the function.

12. $f(x) = \dfrac{x^2 + 1}{x}$

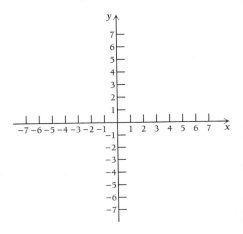

13. $f(x) = \dfrac{x^3}{x^2 - 1}$

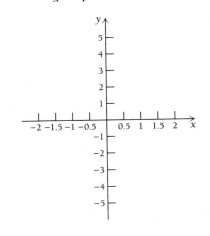

❖

EXERCISE SET 3.3

Sketch a graph of the function.

1. $f(x) = \dfrac{4}{x}$

2. $f(x) = -\dfrac{5}{x}$

3. $f(x) = \dfrac{-2}{x-5}$

4. $f(x) = \dfrac{1}{x-5}$

5. $f(x) = \dfrac{1}{x-3}$

6. $f(x) = \dfrac{1}{x+2}$

7. $f(x) = \dfrac{-2}{x+5}$

8. $f(x) = \dfrac{-3}{x-3}$

9. $f(x) = \dfrac{2x+1}{x}$

10. $f(x) = \dfrac{3x-1}{x}$

11. $f(x) = x + \dfrac{9}{x}$

12. $f(x) = x + \dfrac{2}{x}$

13. $f(x) = \dfrac{2}{x^2}$

14. $f(x) = \dfrac{-1}{x^2}$

15. $f(x) = \dfrac{x}{x-3}$

16. $f(x) = \dfrac{x}{x+2}$

17. $f(x) = \dfrac{1}{x^2+3}$

18. $f(x) = \dfrac{-1}{x^2+2}$

19. $f(x) = \dfrac{x-1}{x+2}$

20. $f(x) = \dfrac{x-2}{x+1}$

21. $f(x) = \dfrac{x^2-4}{x+3}$

22. $f(x) = \dfrac{x^2-9}{x+1}$

23. $f(x) = \dfrac{x-1}{x^2-2x-3}$

24. $f(x) = \dfrac{x+2}{x^2+2x-15}$

25. $f(x) = \dfrac{2x^2}{x^2-16}$

26. $f(x) = \dfrac{x^2+x-2}{2x^2+1}$

APPLICATIONS

❖ **Business and Economics**

27. *Depreciation.* Suppose that the value V of a certain product decreases, or depreciates, with time t, in months, where

$$V(t) = 50 - \frac{25t^2}{(t+2)^2}.$$

a) Find $V(0)$, $V(5)$, $V(10)$, and $V(70)$.

b) Find the relative maximum value of the product over the interval $[0, \infty)$.

c) Sketch a graph of V.

d) Find $\lim_{t \to \infty} V(t)$.

e) Does there seem to be a value below which V will never fall?

28. *Average cost.* The total-cost function for Acme, Inc., to produce x units of a product is given by

$$C(x) = 3x^2 + 80.$$

a) The *average cost* is given by $A(x) = C(x)/x$. Find $A(x)$.

b) Graph the average cost.

c) Find the oblique asymptote for the graph of $y = A(x)$ and interpret it.

29. *Cost of pollution control.* Cities and companies find the cost of pollution control to increase tremendously with respect to the percentage of pollutants to be removed from a situation. Suppose that the cost C of removing $p\%$ of the pollutants from a chemical dumping site is given by

$$C(p) = \frac{\$48,000}{100 - p}.$$

a) Find $C(0)$, $C(20)$, $C(80)$, and $C(90)$.

b) Find $\lim_{p \to 100^-} C(p)$.

c) Sketch a graph of C.

d) Can the company afford to remove 100% of the pollutants?

❖ **Life and Physical Sciences**

30. *Medication in the bloodstream.* After an injection, the amount of a medication A in the bloodstream decreases after time t, in hours. Suppose under certain conditions that A is given by

$$A(t) = \frac{A_0}{t^2 + 1},$$

where A_0 = the initial amount of the medication given.

Assume that an initial amount of 100 cc is injected.

a) Find $A(0)$, $A(1)$, $A(2)$, $A(7)$, and $A(10)$.

b) Find $\lim_{t \to \infty} A(t)$.

c) Find the relative maximum value of the injection over the interval $[0, \infty)$.

d) Sketch a graph of the function.

e) According to this function, does the medication ever completely leave the bloodstream?

COMPUTER–GRAPHING CALCULATOR EXERCISES

Sketch the graph of the function using a computer software graphing package or a graphing calculator.

31. $f(x) = \dfrac{x}{\sqrt{x^2 + 1}}$

32. $f(x) = x^2 + \dfrac{1}{x^2}$

33. $f(x) = \dfrac{x^3 + 2x^2 - 15x}{x^2 - 5x - 14}$

34. $f(x) = \dfrac{x^3 + 4x^2 + x - 6}{x^2 - x - 2}$

35. $f(x) = \left| \dfrac{1}{x} - 2 \right|$

36. $f(x) = \dfrac{x^3 + 2x^2 - 3x}{x^2 - 25}$

THE CALCULUS EXPLORER:
PowerGrapher

Use the program to graph the functions in Exercises 31–36.

3.4

USING DERIVATIVES TO FIND ABSOLUTE MAXIMUM AND MINIMUM VALUES

Absolute Maximum and Minimum Values

A relative minimum may or may not be an absolute minimum, meaning the smallest value of the function over its entire domain. Similarly, a relative maximum may or may not be an absolute maximum, meaning the greatest value of a function over its entire domain.

The function in Fig. 1 has relative minima at interior points c_1 and c_3 of the closed interval $[a, b]$. The relative minimum at c_1 also happens to be an absolute minimum. The function has a relative maximum at c_2 but

OBJECTIVES

a) Find absolute maximum and minimum values of functions using first derivatives and Maximum–Minimum Principle 1.

b) Find absolute maximum and minimum values of functions using second derivatives and Maximum–Minimum Principle 2.

1. In each graph, find the points at which absolute maximum and minimum values occur on $[a, b]$.

a)

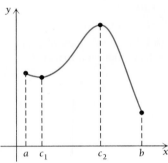

b)

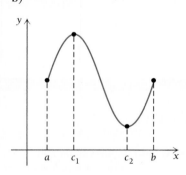

it is not an absolute maximum. The absolute maximum occurs at the endpoint b.

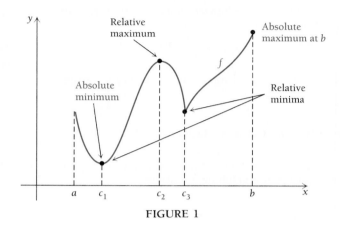

FIGURE 1

DEFINITION

Suppose that f is a function whose value $f(c)$ exists at input c in an interval I in the domain of f. Then:

$f(c)$ is an *absolute minimum* if $f(c) \leq f(x)$ for all x in I.

$f(c)$ is an *absolute maximum* if $f(x) \leq f(c)$ for all x in I.

Finding Absolute Maximum and Minimum Values on Closed Intervals

In Margin Exercise 1 you will explore absolute maxima and minima of a continuous function on a closed interval. We now have to consider the endpoints.

DO EXERCISE 1.

You may have discovered two theorems, whose proofs we will not consider. Each of the functions in Margin Exercise 1 did indeed have an absolute maximum and an absolute minimum value. This leads us to one of the theorems.

THEOREM 7

The Extreme-Value Theorem

A continuous function f defined on a closed interval $[a, b]$ must have an absolute maximum value and an absolute minimum value at points in $[a, b]$.

Look carefully at each graph in Margin Exercise 1 and consider the critical points and the endpoints. In part (a), the graph starts at $f(a)$ and falls to $f(c_1)$. Then it rises from $f(c_1)$ to $f(c_2)$. From there it falls to $f(b)$. In part (b), the graph starts at $f(a)$ and rises to $f(c_1)$. Then it falls from $f(c_1)$ to $f(c_2)$. From there it rises to $f(b)$. It seems reasonable that whatever the maximum and minimum values are, they occur among the function values $f(a), f(c_1), f(c_2)$, and $f(b)$. This leads us to another theorem regarding *absolute extrema*.

THEOREM 8

Maximum–Minimum Principle 1

Suppose that f is a continuous function over a closed interval $[a, b]$. To find the absolute maximum and minimum values of the function on $[a, b]$:

a) First find $f'(x)$.

b) Then determine the critical points of f in $[a, b]$. That is, find all points c for which

$$f'(c) = 0 \quad \text{or} \quad f'(c) \text{ does not exist.}$$

c) List the critical points of f and the endpoints of the interval:

$$a, c_1, c_2, \ldots, c_n, b.$$

d) Find the function values at the points in part (c):

$$f(a), f(c_1), f(c_2), \ldots, f(c_n), f(b).$$

The largest of these is the *absolute maximum* of f on the interval $[a, b]$. The smallest of these is the *absolute minimum* of f on the interval $[a, b]$.

EXAMPLE 1 Find the absolute maximum and minimum values of

$$f(x) = x^3 - 3x + 2$$

on the interval $[-2, \frac{3}{2}]$.

$3x^2 - 3$

Solution In each of Examples 1–6 of this section, we show the related graph (Fig. 2) so that you can see the absolute extrema. The procedures we use do not require the drawing of a graph.

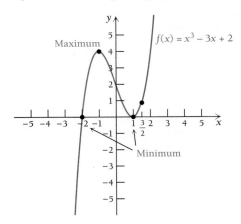

FIGURE 2

a) Find $f'(x)$:

$$f'(x) = 3x^2 - 3.$$

b) Find the critical points. The derivative exists for all real numbers. Thus we merely solve $f'(x) = 0$:

$$3x^2 - 3 = 0$$
$$3x^2 = 3$$
$$x^2 = 1$$
$$x = \pm 1.$$

c) List the critical points and the endpoints. These points are $-2, -1, 1,$ and $\frac{3}{2}$.

d) Find the function values at the points in part (c):

$$f(-2) = (-2)^3 - 3(-2) + 2 = -8 + 6 + 2 = 0; \longrightarrow \text{Minimum}$$
$$f(-1) = (-1)^3 - 3(-1) + 2 = -1 + 3 + 2 = 4; \longrightarrow \text{Maximum}$$
$$f(1) = (1)^3 - 3(1) + 2 = 1 - 3 + 2 = 0; \longrightarrow \text{Minimum}$$
$$f(\tfrac{3}{2}) = (\tfrac{3}{2})^3 - 3(\tfrac{3}{2}) + 2 = \tfrac{27}{8} - \tfrac{9}{2} + 2 = \tfrac{7}{8}.$$

The largest of these values, 4, is the maximum. It occurs at $x = -1$. The smallest of these values is 0. It occurs twice: at $x = -2$ and $x = 1$. Thus on the interval $[-2, \frac{3}{2}]$ the

$$\text{absolute maximum} = 4 \text{ at } x = -1$$

and the

$$\text{absolute minimum} = 0 \text{ at } x = -2 \text{ and } x = 1. \quad \clubsuit$$

Note that an absolute maximum or minimum value can occur at more than one point.

DO EXERCISE 2.

EXAMPLE 2   Find the absolute maximum and minimum values of

$$f(x) = x^3 - 3x + 2$$

on the interval $[-3, -\frac{3}{2}]$.

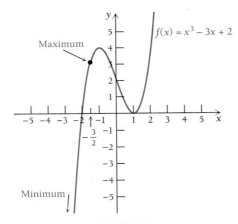

FIGURE 3

Solution   As in Example 1, the derivative is 0 at -1 and 1. But neither -1 nor 1 is in the interval $[-3, -\frac{3}{2}]$, so there are no critical points in this interval. Thus the maximum and minimum values occur at the endpoints:

$$f(-3) = (-3)^3 - 3(-3) + 2$$
$$= -27 + 9 + 2 = -16; \longrightarrow \text{Minimum}$$

$$f(-\tfrac{3}{2}) = (-\tfrac{3}{2})^3 - 3(-\tfrac{3}{2}) + 2$$
$$= -\tfrac{27}{8} + \tfrac{9}{2} + 2 = \tfrac{25}{8} = 3\tfrac{1}{8}. \longrightarrow \text{Minimum}$$

Thus, on the interval $[-3, -\frac{3}{2}]$, the

$$\text{absolute maximum} = 3\tfrac{1}{8} \text{ at } x = -\tfrac{3}{2}$$

2. Find the absolute maximum and minimum values of

$$f(x) = x^3 - x^2 - x + 2$$

on the interval $[-2, 1]$.

$3x^2 - 3$

3. Find the absolute maximum and minimum values of

$$f(x) = x^3 - x^2 - x + 2$$

on the interval $[-1, 2]$.

and the

$$\text{absolute minimum} = -16 \text{ at } x = -3.$$ ❖

DO EXERCISE 3.

Finding Absolute Maximum and Minimum Values on Other Intervals

When there is only one critical point c in I, we may not need to check endpoint values to determine whether the function has an absolute maximum or minimum at that point.

THEOREM 9

Maximum–Minimum Principle 2

Suppose that f is a function such that $f'(x)$ exists for exery x in an interval I, and that there is *exactly one* (critical) point c, interior to I, for which $f'(c) = 0$. Then:

$$f(c) \text{ is the absolute maximum on } I \text{ if } f''(c) < 0$$

or

$$f(c) \text{ is the absolute minimum value on } I \text{ if } f''(c) > 0.$$

This theorem holds no matter what the interval I is—whether open, closed, or extending to infinity. If $f''(c) = 0$, either we must use Maximum–Minimum Principle 1 or we must know more about the behavior of the function on the given interval.

EXAMPLE 3   Find the absolute maximum and minimum values of

$$f(x) = 4x - x^2.$$

Solution   When no interval is specified, we consider the entire domain of the function. In this case, the domain is the set of all real numbers.

a)   Find $f'(x)$:

$$f'(x) = 4 - 2x.$$

b)   Find the critical points. The derivative exists for all real numbers. Thus we merely solve $f'(x) = 0$:

$$4 - 2x = 0$$
$$-2x = -4$$
$$x = 2.$$

c) Since there is only one critical point, we can apply Maximum–Minimum Principle 2 using the second derivative:

$$f''(x) = -2.$$

The second derivative is constant. Thus, $f''(2) = -2$, and since this is negative, we have the

absolute maximum $= f(2) = 4 \cdot 2 - 2^2 = 8 - 4 = 4$ at $x = 2$.

The function has no minimum, as the graph in Fig. 4 indicates.

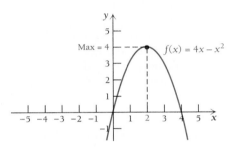

FIGURE 4

DO EXERCISE 4.

EXAMPLE 4 Find the absolute maximum and minimum values of $f(x) = 4x - x^2$ on the interval $[0, 4]$.

Solution By the reasoning in Example 3, we know that the absolute maximum of f on $(-\infty, \infty)$ is $f(2)$, or 4. Since 2 is in the interval $[0, 4]$, we know that the absolute maximum of f on $[0, 4]$ will occur at 2. To find the absolute minimum, we need to check the endpoints:

$$f(0) = 4 \cdot 0 - 0^2 = 0 \quad \text{and} \quad f(4) = 4 \cdot 4 - 4^2 = 0.$$

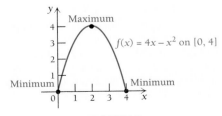

FIGURE 5

4. Find the absolute maximum and minimum values of

$$f(x) = x^2 - 4x.$$

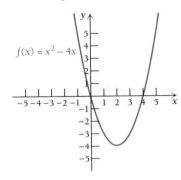

5. Find the absolute maximum and minimum values of

$$f(x) = x^2 - 4x$$

on the interval [0, 4].

We see in Fig. 5 that the minimum is 0. It occurs twice, at $x = 0$ and $x = 4$. Thus the

$$\text{absolute maximum} = 4 \text{ at } x = 2$$

and the

$$\text{absolute minimum} = 0 \text{ at } x = 0 \text{ and } x = 4. \qquad \diamond$$

DO EXERCISE 5.

We have thus far restricted the use of Maximum–Minimum Principle 2 to intervals with one critical point. Suppose a closed interval contains two critical points. Then we could break the interval up into two subintervals, consider maximums and minimums on those subintervals, and compare. But we would need to consider values at the endpoints, and since we would, in effect, be using Maximum–Minimum Principle 1, we may as well use it at the outset.

A Strategy for Finding Maximum and Minimum Values

The following general strategy can be used when finding maximum and minimum values of continuous functions.

A Strategy for Finding Absolute Maximum and Minimum Values

To find absolute maximum and minimum values of a continuous function on an interval:

a) Find $f'(x)$.

b) Find the critical points.

c) If the interval is closed and there is more than one critical point, use Maximum–Minimum Principle 1.

d) If the interval is closed and there is exactly one critical point, use either Maximum–Minimum Principle 1 or Maximum–Minimum Principle 2. If the function is easy to differentiate, use Maximum–Minimum Principle 2.

e) If the interval is not closed, does not have endpoints, or does not contain its endpoints, such as $(-\infty, \infty)$, $(0, \infty)$, or (a, b), and the function has only one critical point, use Maximum–Minimum Principle 2. In such a case, if the function has a maximum, it will have no minimum; and if it has a minimum, it will have no maximum.

The case of finding absolute maximum and minimum values when more than one critical point occurs in an interval described in part (e) above must be dealt with by a detailed graph or by techniques beyond the scope of this book.

EXAMPLE 5   Find the absolute maximum and minimum values of

$$f(x) = (x - 2)^3 + 1.$$

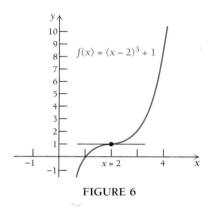

FIGURE 6

Solution

a)   Find $f'(x)$:

$$f'(x) = 3(x - 2)^2.$$

b)   Find the critical points. The derivative exists for all real numbers. Thus we solve $f'(x) = 0$:

$$3(x - 2)^2 = 0$$
$$(x - 2)^2 = 0$$
$$x - 2 = 0$$
$$x = 2.$$

c)   Since there is only one critical point and there are no endpoints, we can try to apply Maximum–Minimum Principle 2 using the second derivative:

$$f''(x) = 6(x - 2).$$

Now

$$f''(2) = 6(2 - 2)$$
$$= 0,$$

6. Find the absolute maximum and minimum values of

$$f(x) = x^3.$$

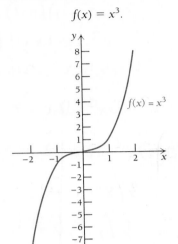

$f(x) = x^3$

7. Find the absolute maximum and minimum values of

$$f(x) = x^3$$

on the interval $[-2, 2]$. (Hint: This function must have maximum and minimum values because it is a continuous function over a closed interval.) What are the only numbers at which these can occur?

so Maximum–Minimum Principle 2 fails. We cannot use Maximum–Minimum Principle 1 because there are no endpoints. But note that $f'(x) = 3(x - 2)^2$ is never negative. Thus $f(x)$ is increasing everywhere except at $x = 2$, so there is no maximum and no minimum. For $x < 2$, $x - 2 < 0$, so $f''(x) = 6(x - 2) < 0$. Similarly, for $x > 2$, $x - 2 > 0$, so $f''(x) = 6(x - 2) > 0$. Thus, at $x = 2$, the function has a *point of inflection*.   ❖

DO EXERCISES 6 AND 7.

EXAMPLE 6   Find the absolute maximum and minimum values of $f(x) = 5x + 35/x$ on the interval $(0, \infty)$.

Solution

a) Find $f'(x)$. We first express $f(x)$ as

$$f(x) = 5x + 35x^{-1}.$$

Then

$$f'(x) = 5 - 35x^{-2} = 5 - \frac{35}{x^2}.$$

b) Find the critical points. Now $f'(x)$ exists for all values of x in $(0, \infty)$. Thus the only critical points are those for which $f'(x) = 0$:

$$5 - \frac{35}{x^2} = 0$$

$$5 = \frac{35}{x^2}$$

$$5x^2 = 35 \qquad \text{Multiplying by } x^2, \text{ since } x \neq 0$$

$$x^2 = 7$$

$$x = \pm\sqrt{7}.$$

e) The interval is not closed and is $(0, \infty)$. The only critical point is $\sqrt{7}$. Thus we can apply Maximum–Minimum Principle 2 using the second derivative,

$$f''(x) = 70x^{-3} = \frac{70}{x^3},$$

to determine whether we have a maximum or a minimum. Now $f''(x)$ is positive for all values of x in $(0, \infty)$, so $f''(\sqrt{7}) > 0$, and the

$$\text{absolute minimum} = f(\sqrt{7}) = 5 \cdot \sqrt{7} + \frac{35}{\sqrt{7}}$$

$$= 5\sqrt{7} + \frac{35}{\sqrt{7}} \cdot \frac{\sqrt{7}}{\sqrt{7}} = 5\sqrt{7} + \frac{35\sqrt{7}}{7}$$

$$= 5\sqrt{7} + 5\sqrt{7} = 10\sqrt{7}$$

at $x = \sqrt{7}$.

The function has no maximum value.

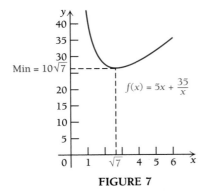

FIGURE 7

DO EXERCISE 8.

8. Find the absolute maximum and minimum values of

$$f(x) = 10x + \frac{1}{x}$$

on $(0, \infty)$.

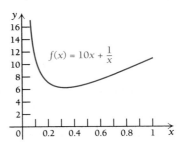

❖

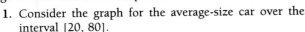

EXERCISE SET 3.4

The curves on the graph to the right show the gasoline mileage obtained when traveling at a constant speed for an average-size car and for a compact car.

1. Consider the graph for the average-size car over the interval [20, 80].

   a) Estimate the speed at which the absolute maximum gasoline mileage is obtained.

   b) Estimate the speed at which the absolute minimum gasoline mileage is obtained.

   c) What is the mileage obtained at 70 mph?

   d) What is the mileage obtained at 55 mph?

   e) What percent increase in mileage is there by traveling at 55 mph rather than at 70 mph?

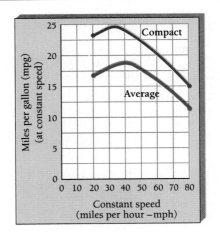

2. Answer the questions in Exercise 1 for the compact car.

Find the absolute maximum and minimum values of the function, if they exist, on the indicated interval.

3. $f(x) = 5 + x - x^2$; $[0, 2]$

4. $f(x) = 4 + x - x^2$; $[0, 2]$

5. $f(x) = x^3 - x^2 - x + 2$; $[0, 2]$

6. $f(x) = x^3 + \frac{1}{2}x^2 - 2x + 5$; $[0, 1]$

7. $f(x) = x^3 - x^2 - x + 2$; $[-1, 0]$

8. $f(x) = x^3 + \frac{1}{2}x^2 - 2x + 5$; $[-2, 0]$

9. $f(x) = 3x - 2$; $[-1, 1]$ **10.** $f(x) = 2x + 4$; $[-1, 1]$

11. $f(x) = 7 - 4x$; $[-2, 5]$

12. $f(x) = -2 + 8x$; $[-10, 10]$

13. $f(x) = -5$; $[-1, 1]$ **14.** $g(x) = 24$; $[4, 13]$

15. $f(x) = x^2 - 6x - 3$; $[-1, 5]$

16. $f(x) = x^2 - 4x + 5$; $[-1, 3]$

17. $f(x) = 3 - 2x - 5x^2$; $[-3, 3]$

18. $f(x) = 1 + 6x - 3x^2$; $[0, 4]$

19. $f(x) = x^3 - 3x^2$; $[0, 5]$

20. $f(x) = x^3 - 3x + 6$; $[-1, 3]$

21. $f(x) = x^3 - 3x$; $[-5, 1]$

22. $f(x) = 3x^2 - 2x^3$; $[-5, 1]$

23. $f(x) = 1 - x^3$; $[-8, 8]$ **24.** $f(x) = 2x^3$; $[-10, 10]$

25. $f(x) = 12 + 9x - 3x^2 - x^3$; $[-3, 1]$

26. $f(x) = x^3 - 6x^2 + 10$; $[0, 4]$

27. $f(x) = x^4 - 2x^3$; $[-2, 2]$

28. $f(x) = x^3 - x^4$; $[-1, 1]$

29. $f(x) = x^4 - 2x^2 + 5$; $[-2, 2]$

30. $f(x) = x^4 - 8x^2 + 3$; $[-3, 3]$

31. $f(x) = (x + 3)^{2/3} - 5$; $[-4, 5]$

32. $f(x) = 1 - x^{2/3}$; $[-8, 8]$

33. $f(x) = x + \dfrac{1}{x}$; $[1, 20]$ **34.** $f(x) = x + \dfrac{4}{x}$; $[-8, -1]$

35. $f(x) = \dfrac{x^2}{x^2 + 1}$; $[-2, 2]$ **36.** $f(x) = \dfrac{4x}{x^2 + 1}$; $[-3, 3]$

37. $f(x) = (x + 1)^{1/3}$; $[-2, 26]$

38. $f(x) = \sqrt[3]{x}$; $[8, 64]$

Find the absolute maximum and minimum values of the function, if they exist, on the indicated interval. When no interval is specified, use the real line $(-\infty, \infty)$.

39. $f(x) = x(70 - x)$ **40.** $f(x) = x(50 - x)$

41. $f(x) = 2x^2 - 40x + 400$

42. $f(x) = 2x^2 - 20x + 100$

43. $f(x) = x - \frac{4}{3}x^3$; $(0, \infty)$

44. $f(x) = 16x - \frac{4}{3}x^3$; $(0, \infty)$

45. $f(x) = 17x - x^2$ **46.** $f(x) = 27x - x^2$

47. $f(x) = \frac{1}{3}x^3 - 3x$; $[-2, 2]$

48. $f(x) = \frac{1}{3}x^3 - 5x$; $[-3, 3]$

49. $f(x) = -0.001x^2 + 4.8x - 60$

50. $f(x) = -0.01x^2 + 1.4x - 30$

51. $f(x) = -\frac{1}{3}x^3 + 6x^2 - 11x - 50$; $(0, 3)$

52. $f(x) = -x^3 + x^2 + 5x - 1$; $(0, \infty)$

53. $f(x) = 15x^2 - \frac{1}{2}x^3$; $[0, 30]$

54. $f(x) = 4x^2 - \frac{1}{2}x^3$; $[0, 8]$

55. $f(x) = 2x + \dfrac{72}{x}$; $(0, \infty)$

56. $f(x) = x + \dfrac{3600}{x}$; $(0, \infty)$

57. $f(x) = x^2 + \dfrac{432}{x}$; $(0, \infty)$

58. $f(x) = x^2 + \dfrac{250}{x}$; $(0, \infty)$

59. $f(x) = 2x^4 - x$; $[-1, 1]$

60. $f(x) = 2x^4 + x$; $[-1, 1]$

61. $f(x) = \sqrt[3]{x}$; $[0, 8]$ **62.** $f(x) = \sqrt{x}$; $[0, 4]$

63. $f(x) = (x + 1)^3$ **64.** $f(x) = (x - 1)^3$

65. $f(x) = 2x - 3$; $[-1, 1]$

66. $f(x) = 9 - 5x$; $[-10, 10]$

67. $f(x) = 2x - 3$ **68.** $f(x) = 9 - 5x$

69. $f(x) = x^{2/3}$; $[-1, 1]$ **70.** $g(x) = x^{2/3}$

71. $f(x) = \frac{1}{3}x^3 - x + \frac{2}{3}$

72. $f(x) = \frac{1}{3}x^3 - \frac{1}{2}x^2 - 2x + 1$

73. $f(x) = \frac{1}{3}x^3 - 2x^2 + x$; $[0, 4]$

74. $g(x) = \frac{1}{3}x^3 + 2x^2 + x$; $[-4, 0]$

75. $t(x) = x^4 - 2x^2$

76. $f(x) = 2x^4 - 4x^2 + 2$

APPLICATIONS

❖ Business and Economics

77. *Monthly productivity.* An employee's monthly productivity M, in number of units produced, is found to be a function of the number t of years of service. For a certain product, a productivity function is given by

$$M(t) = -2t^2 + 100t + 180, \quad 0 \leqslant t \leqslant 40.$$

Find the maximum productivity and the year in which it is achieved.

78. *Advertising.* A firm estimates that it will sell N units of a product after spending a dollars on advertising, where

$$N(a) = -a^2 + 300a + 6, \quad 0 \leqslant a \leqslant 300,$$

and a is measured in thousands of dollars. Find the maximum number of units that can be sold and the amount that must be spent on advertising in order to achieve that maximum.

79. *Maximizing profit.* A firm determines that its total profit in dollars from the production and sale of x units of a product is given by

$$P(x) = \frac{1500}{x^2 - 6x + 10}.$$

Find the number of units x for which the total profit is a maximum.

❖ Life and Physical Sciences

80. *Blood pressure.* For a dosage of x cubic centimeters (cc) of a certain drug, the resulting blood pressure B is approximated by

$$B(x) = 0.05x^2 - 0.3x^3, \quad 0 \leqslant x \leqslant 0.16.$$

Find the maximum blood pressure and the dosage at which it occurs.

81. *Minimizing automobile accidents.* At travel speed (constant velocity) x in miles per hour, there are y accidents at nighttime for every 100 million miles of travel, where y is given by

$$y = -6.1x^2 + 752x + 22{,}620.$$

At what travel speed does the greatest number of accidents occur?

82. *Temperature during an illness.* The temperature T of a person during an illness is given by

$$T(t) = -0.1t^2 + 1.2t + 98.6, \quad 0 \leqslant t \leqslant 12,$$

where T is the temperature (°F) at time t, in days. Find the maximum value of the temperature and when it occurs.

SYNTHESIS EXERCISES

Find the absolute maximum and minimum values of the function, if they exist, over the indicated interval.

83. $g(x) = x\sqrt{x + 3};\ [-3, 3]$

84. $h(x) = x\sqrt{1 - x};\ [0, 1]$

85. *Business: Total cost.* Several costs in a business environment can be separated into two components: those that increase with volume and those that decrease with volume. Although the quality of customer service becomes more expensive as it is increased, part of the increased cost is offset by customer goodwill. A firm has determined that its cost of service is given by the following function of "quality units,"

$$C(x) = (2x + 4) + \left(\frac{2}{x - 6}\right), \quad x > 6.$$

Find the number of "quality units" the firm should use in order to minimize its total cost of service.

86. Let

$$y = (x - a)^2 + (x - b)^2.$$

For what value of x is y a minimum?

3.5

a) Solve maximum–minimum problems using the strategy developed in this section.

MAXIMUM–MINIMUM PROBLEMS

One very important application of differential calculus is the solving of maximum–minimum problems, that is, finding the absolute maximum or minimum value of some varying quantity Q and the point at which that maximum or minimum occurs.

EXAMPLE 1   A hobby store has 20 ft of fencing to fence off a rectangular area for an electric train in one corner of its display room. The two sides up against the wall require no fence. What dimensions of the rectangle will maximize the area? What is the maximum area?

How can the area for the electric train be maximized?

Exploratory Solution   Intuitively, we might think that it does not matter what dimensions we use: They will all yield the same area. To show that this is not true, as well as to conjecture a possible solution, consider the exploratory exercises in Margin Exercise 1. Before doing so, however, let us make a drawing, as shown in Fig. 1, and express the area in terms of one variable. If we let $x =$ the length of one side and $y =$ the length of the other, then since the sum of the lengths must be 20 ft, we have

$$x + y = 20 \quad \text{and} \quad y = 20 - x.$$

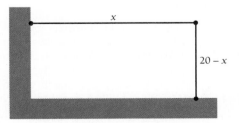

FIGURE 1

Thus the area is given by

$$A = xy$$
$$A = x(20 - x) = 20x - x^2.$$

DO EXERCISE 1.

Calculus Solution We are trying to find the maximum value of

$$A = 20x - x^2 \quad \text{on the interval} \quad (0, 20).$$

We consider the interval $(0, 20)$ because x is the length of one side and cannot be negative or 0. Since there is only 20 ft of fencing, x cannot be greater than 20. Also, x cannot be 20 because then the length of y would be 0.

a) We first find $A'(x)$, where $A(x) = 20x - x^2$:

$$A'(x) = 20 - 2x.$$

b) This derivative exists for all values of x in $(0, 20)$. Thus the only critical points are where

$$A'(x) = 20 - 2x = 0$$
$$-2x = -20$$
$$x = 10.$$

Since there is only one critical point in the interval, we can use the second derivative to determine whether we have a maximum. Note that

$$A''(x) = -2,$$

which is a constant. Thus, $A''(10)$ is negative, so $A(10)$ is a maximum. Now

$$A(10) = 10(20 - 10) = 10 \cdot 10 = 100.$$

Thus the maximum area of 100 ft^2 is obtained using 10 ft for the length of one side and $20 - 10$, or 10 ft for the other. Note that although you may have conjectured this in Margin Exercise 1, the tools of calculus allowed us to prove it. ❖

DO EXERCISE 2. (EXERCISE 2 IS ON THE FOLLOWING PAGE.)

Here is a general strategy for solving maximum–minimum problems. While it may not guarantee success, it should certainly improve your chances.

1. *Exploratory exercises*
   a) Complete this table.

x	y $20 - x$	A $x(20 - x)$
0		
4		
6.5		
8		
10		
12		
13.2		
20		

   b) Make a graph of x versus A—that is, of points (x, A) from the table—and connect them with a smooth curve.

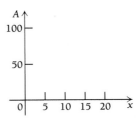

   c) Does it matter what dimensions we use?

   d) Make a conjecture about what the maximum might be and where it would occur.

2. A rancher has 50 ft of fencing to fence off a rectangular animal pen in the corner of a barn. What dimensions of the rectangle will yield the maximum area? What is the maximum area?

A Strategy For Solving Maximum–Minimum Problems

1. Read the problem carefully. If relevant, draw a picture.
2. Label the picture with appropriate variables and constants, noting what varies and what stays fixed.
3. Translate the problem to an equation involving a quantity Q to be maximized or minimized. Try to represent Q in terms of the variables of step (2).
4. Try to express Q as a function of *one* variable. Use the procedures developed in Sections 3.1, 3.2 and 3.4 to determine the maximum or minimum values and the points at which they occur.

EXAMPLE 2   From a thin piece of cardboard 8 in. by 8 in., square corners are cut out so that the sides can be folded up to make a box. What dimensions will yield a box of maximum volume? What is the maximum volume?

Exploratory Solution   We might again think at first that it does not matter what the dimensions are, but our experience with Example 1 should lead us to think otherwise. We make a drawing, as shown in Fig. 2.

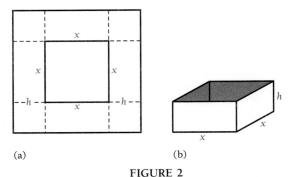

(a)   (b)

FIGURE 2

When squares of length h on a side are cut out of the corners, we are left with a square base with sides of length x. The volume of the resulting box is

$$V = lwh = x \cdot x \cdot h.$$

We want to express V in terms of one variable. Note that the overall length of a side of the cardboard is 8 in. We see from Fig. 2 that

$$h + x + h = 8,$$

or
$$x + 2h = 8.$$

Solving for h, we get

$$2h = 8 - x$$
$$h = \tfrac{1}{2}(8 - x) = \tfrac{1}{2} \cdot 8 - \tfrac{1}{2}x = 4 - \tfrac{1}{2}x.$$

Thus,

$$V = x \cdot x \cdot (4 - \tfrac{1}{2}x) = x^2(4 - \tfrac{1}{2}x) = 4x^2 - \tfrac{1}{2}x^3.$$

In Margin Exercise 3, you will compute some values of V.

DO EXERCISE 3.

Calculus Solution   You probably noted in Margin Exercise 3 that it was a bit more difficult to conjecture where the maximum occurs than it was in Margin Exercise 2. At the least it seems reasonable that the maximum occurs for some x between 5 and 6. Let us find out for certain, using calculus. We are trying to find the maximum value of

$$V(x) = 4x^2 - \tfrac{1}{2}x^3 \quad \text{on the interval} \quad (0, 8).$$

We first find $V'(x)$:

$$V'(x) = 8x - \tfrac{3}{2}x^2.$$

Now $V'(x)$ exists for all x in the interval $(0, 8)$, so we set it equal to 0 to find the critical values:

$$V'(x) = 8x - \tfrac{3}{2}x^2 = 0$$
$$x(8 - \tfrac{3}{2}x) = 0$$
$$x = 0 \quad \text{or} \quad 8 - \tfrac{3}{2}x = 0$$
$$x = 0 \quad \text{or} \quad -\tfrac{3}{2}x = -8$$
$$x = 0 \quad \text{or} \quad x = -\tfrac{2}{3}(-8) = \tfrac{16}{3}.$$

The only critical point in $(0, 8)$ is $\tfrac{16}{3}$. Thus we can use the second derivative,

$$V''(x) = 8 - 3x,$$

to determine whether we have a maximum. Since

$$V''(\tfrac{16}{3}) = 8 - 3 \cdot \tfrac{16}{3}$$
$$= -8,$$

$V''(\tfrac{16}{3})$ is negative, so $V(\tfrac{16}{3})$ is a maximum, and

$$V(\tfrac{16}{3}) = 4 \cdot (\tfrac{16}{3})^2 - \tfrac{1}{2}(\tfrac{16}{3})^3 = \tfrac{1024}{27} = 37\tfrac{25}{27}.$$

3. *Exploratory exercises*

a)   Complete this table.

x	h $4 - \tfrac{1}{2}x$	V $4x^2 - \tfrac{1}{2}x^3$
0		
1		
2		
3		
4		
4.6		
5		
6		
6.8		
7		
8		

b)   Make a graph of x versus V.

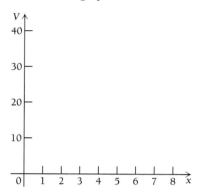

c)   Make a conjecture about what the maximum might be and where it would occur.

$$p = 1000 - x.$$

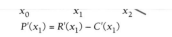

$$P'(x_1) = R'(x_1) - C'(x_1)$$

FIGURE 6

10. Repeat Margin Exercise 9 using all the data given, but change the $30 storage cost to $50.

work for all types of functions, but will work for the type we are considering here. The number of times that an order should be placed is $2500/71 \approx 35$, so there is still some estimating involved.) ❖

DO EXERCISE 10.

The lot size that minimizes total inventory costs is often referred to as the *economic ordering quantity*. There are three assumptions made in using the preceding method to determine the economic ordering quantity. The first is that the demand for the product is the same throughout the year. For television sets this may be reasonable, but for seasonal items such as clothing or skis, this assumption may not be reasonable. The second assumption is that the time between the placing of an order and its receipt will be consistent throughout the year. The third assumption is that the various costs involved, such as storage, shipping charges, and so on, do not vary. This may not be reasonable in a time of inflation, although one may account for them by anticipating what they might be and using average costs. Nevertheless, the model described above can be useful, and it allows us to analyze a seemingly difficult problem using calculus.

EXERCISE SET 3.5

1. Of all numbers whose sum is 50, find the two that have the maximum product. That is, maximize $Q = xy$, where $x + y = 50$.

2. Of all numbers whose sum is 70, find the two that have the maximum product. That is, maximize $Q = xy$, where $x + y = 70$.

3. In Exercise 1, can there be a minimum product? Explain.

4. In Exercise 2, can there be a minimum product? Explain.

5. Of all numbers whose difference is 4, find the two that have the minimum product.

6. Of all numbers whose difference is 6, find the two that have the minimum product.

7. Maximize $Q = xy^2$, where x and y are positive numbers, such that $x + y^2 = 1$.

8. Maximize $Q = xy^2$, where x and y are positive numbers, such that $x + y^2 = 4$.

9. Minimize $Q = x^2 + y^2$, where $x + y = 20$.

10. Minimize $Q = x^2 + y^2$, where $x + y = 10$.

11. Maximize $Q = xy$, where x and y are positive numbers, such that $\frac{4}{3}x^2 + y = 16$.

12. Maximize $Q = xy$, where x and y are positive numbers, such that $x + \frac{4}{3}y^2 = 1$.

13. A rancher wants to build a rectangular fence next to a river, using 120 yd of fencing. What dimensions of the rectangle will maximize the area? What is the maximum area? (Note that the rancher need not fence in the side next to the river.)

14. A rancher wants to enclose two rectangular areas near a river, one for sheep and one for cattle. There is 240 yd

of fencing available. What is the largest total area that can be enclosed?

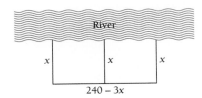

240 − 3x

15. A carpenter is building a rectangular room with a fixed perimeter of 54 ft. What are the dimensions of the largest room that can be built? What is its area?

16. Of all rectangles that have a perimeter of 34 ft, find the dimensions of the one with the largest area. What is its area?

17. From a thin piece of cardboard 30 in. by 30 in., square corners are cut out so that the sides can be folded up to make a box. What dimensions will yield a box of maximum volume? What is the maximum volume?

18. From a thin piece of cardboard 20 in. by 20 in., square corners are cut out so that the sides can be folded up to make a box. What dimensions will yield a box of maximum volume? What is the maximum volume?

19. A container company is designing an open-top, square-based, rectangular box that will have a volume of 62.5 in^3. What dimensions yield the minimum surface area? What is the minimum surface area?

20. A soup company is constructing an open-top, square-based, rectangular metal tank that will have a volume of 32 ft^3. What dimensions yield the minimum surface area? What is the minimum surface area?

APPLICATIONS

❖ **Business and Economics**

Maximizing profit. Find the maximum profit and the number of units that must be produced and sold in order to yield the maximum profit.

21. $R(x) = 50x - 0.5x^2$, $C(x) = 4x + 10$

22. $R(x) = 50x - 0.5x^2$, $C(x) = 10x + 3$

23. $R(x) = 2x$, $C(x) = 0.01x^2 + 0.6x + 30$

24. $R(x) = 5x$, $C(x) = 0.001x^2 + 1.2x + 60$

25. $R(x) = 9x - 2x^2$, $C(x) = x^3 - 3x^2 + 4x + 1$; $R(x)$ and $C(x)$ are in thousands of dollars, and x is in thousands of units.

26. $R(x) = 100x - x^2$, $C(x) = \frac{1}{3}x^3 - 6x^2 + 89x + 100$; $R(x)$ and $C(x)$ are in thousands of dollars, and x is in thousands of units.

27. *Maximizing profit.* Raggs, Ltd., a clothing firm, determines that in order to sell x suits, the price per suit must be

$$p = 150 - 0.5x.$$

It also determines that the total cost of producing x suits is given by

$$C(x) = 4000 + 0.25x^2.$$

a) Find the total revenue $R(x)$.

b) Find the total profit $P(x)$.

c) How many suits must the company produce and sell in order to maximize profit?

d) What is the maximum profit?

e) What price per suit must be charged in order to make this maximum profit?

28. *Maximizing profit.* An appliance firm is marketing a new refrigerator. It determines that in order to sell x

refrigerators, the price per refrigerator must be

$$p = 280 - 0.4x.$$

It also determines that the total cost of producing x refrigerators is given by

$$C(x) = 5000 + 0.6x^2.$$

a) Find the total revenue $R(x)$.

b) Find the total profit $P(x)$.

c) How many refrigerators must the company produce and sell in order to maximize profit?

d) What is the maximum profit?

e) What price per refrigerator must be charged in order to make this maximum profit?

29. *Maximizing revenue.* A university is trying to determine what price to charge for football tickets. At a price of $6 per ticket, it averages 70,000 people per game. For every increase of $1, it loses 10,000 people from the average number. Every person at the game spends an average of $1.50 on concessions. What price per ticket should be charged in order to maximize revenue? How many people will attend at that price?

30. *Maximizing revenue.* Suppose that you are the owner of a 30-unit motel. All units are occupied when you charge $20 a day per unit. For every increase of x dollars in the daily rate, there are x units vacant. Each occupied room costs $2 per day to service and maintain. What should you charge per unit in order to maximize profit?

31. *Maximizing yield.* An apple farm yields an average of 30 bushels of apples per tree when 20 trees are planted on an acre of ground. Each time 1 more tree is planted per acre, the yield decreases 1 bu per tree due to the extra congestion. How many trees should be planted in order to get the highest yield?

32. *Maximizing revenue.* When a theater owner charges $3 for admission, there is an average attendance of 100 people. For every $0.10 increase in admission, there is a loss of 1 customer from the average number. What admission should be charged in order to maximize revenue?

33. *Minimizing costs.* A rectangular box with a volume of 320 ft^3 is to be constructed with a square base and top. The cost per square foot for the bottom is 15¢, for the

top is 10¢, and for the sides is 2.5¢. What dimensions will minimize the cost?

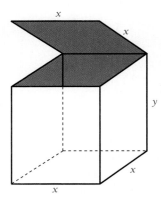

34. A merchant who was purchasing a display sign from a salesclerk said, "I want a sign 10 ft by 10 ft." The salesclerk responded, "That's just what we'll give you; only to make it more aesthetically pleasing, why don't we change it to 7 ft by 13 ft?" Comment.

35. *Maximizing profit.* The amount of money deposited in a financial institution in savings accounts is directly proportional to the interest rate that the financial institution pays on the money. Suppose that a financial institution can loan *all* the money it takes in on its savings accounts at an interest rate of 18%. What interest rate should it pay on its savings accounts in order to maximize profit?

36. ▤ *Maximizing area.* A page in this book is 73.125 in^2. On the average, there is a 0.75-in. margin at the top and at the bottom of each page and a 0.5-in. margin on each of the sides. What should the outside dimensions of each page be so that the printed area is a maximum? Measure the outside dimensions to see whether the actual dimensions maximize the printed area.

37. *Minimizing inventory costs.* A sporting goods store sells 100 pool tables per year. It costs $20 to store one pool table for one year. To reorder, there is a fixed cost of $40, plus $16 for each pool table. How many times per year should the store order pool tables, and in what lot size, in order to minimize inventory costs?

38. *Minimizing inventory costs.* A pro shop in a bowling center sells 200 bowling balls per year. It costs $4 to store one bowling ball for one year. To reorder, there is a fixed cost of $1, plus $0.50 for each bowling ball. How many times per year should the shop order bowling balls, and in what lot size, in order to minimize inventory costs?

39. *Minimizing inventory costs.* A retail outlet for Boxowitz Calculators sells 360 calculators per year. It costs $8 to store one calculator for one year. To reorder, there is a fixed cost of $10, plus $8 for each calculator. How many times per year should the store order calculators, and in what lot size, in order to minimize inventory costs?

40. *Minimizing inventory costs.* A sporting goods store in southern California sells 720 surfboards per year. It costs $2 to store one surfboard for one year. To reorder, there is a fixed cost of $5, plus $2.50 for each surfboard. How many times per year should the store order surfboards, and in what lot size, in order to minimize inventory costs?

41. *Minimizing inventory costs.* Repeat Exercise 39 using all the data given, but change the $8 storage charge to $9.

42. *Minimizing inventory costs.* Repeat Exercise 40 using all the data given, but change the $5 fixed cost to $4.

❖ **General Interest**

43. *Maximizing volume.* The postal service places a limit of 84 in. on the combined length and girth (distance around) of a package to be sent parcel post. What dimensions of a rectangular box with square cross-section will contain the largest volume that can be mailed? (*Hint:* There are two different girths.)

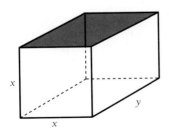

44. *Minimizing cost.* A rectangular play area is to be fenced off in a person's yard and is to contain 48 yd². The neighbor agrees to pay half the cost of the fence on the side of the play area that lines the lot. What dimensions will minimize the cost of the fence?

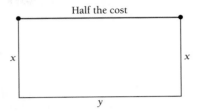

45. *Maximizing light.* A Norman window is a rectangle with a semicircle on top. Suppose that the perimeter of a particular Norman window is to be 24 ft. What should its dimensions be in order to allow the maximum amount of light to enter through the window?

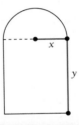

Can you find the Norman windows?

46. *Maximizing light.* Repeat Exercise 45, but assume that the semicircle is to be stained glass, which transmits only half as much light as the semicircle in Exercise 45.

47. For what positive number is the sum of its reciprocal and five times its square a minimum?

48. For what positive number is the sum of its reciprocal and four times its square a minimum?

49. *Business: Minimizing inventory costs—a general solution.* A store sells Q units of a product per year. It costs a dollars to store one unit for one year. To reorder, there is a fixed cost of b dollars, plus c dollars for each unit. How many times per year should the store reorder, and in what lot size, in order to minimize inventory costs?

50. *Business: Minimizing inventory costs.* Use the general solution found in Exercise 49 to find how many times per year a store should reorder, and in what lot size, when $Q = 2500$, $a = \$10$, $b = \$20$, and $c = \$9$.

51. A 24-in. piece of string is cut in two pieces. One piece is used to form a circle and the other to form a square. How should the string be cut so that the sum of the areas is a minimum? a maximum?

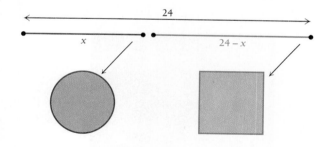

52. *Business: Minimizing costs.* A power line is to be constructed from a power station at point A to an island at point C, which is 1 mi directly out in the water from a point B on the shore. Point B is 4 mi downshore from the power station at A. It costs \$5000 per mile to lay the power line under water and \$3000 per mile to lay the line under ground. At what point S downshore from A should the line come to the shore in order to minimize cost? Note that S could very well be B or A. (*Hint:* The length of CS is $\sqrt{1 + x^2}$.) (See Fig. 1.)

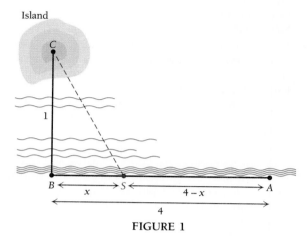

FIGURE 1

53. 🖩 *Life science: Flights of homing pigeons.* It is known that homing pigeons tend to avoid flying over water in the daytime, perhaps because the downdrafts of air over water make flying difficult. Suppose a homing pigeon is released on an island at point C, which is 3 mi directly out in the water from a point B on shore. (See Fig. 2.) Point B is 8 mi downshore from the pigeon's home loft at point A. Assume that a pigeon requires 1.28 times the rate of energy over land to fly over water. Toward what point S downshore from A should the pigeon fly in

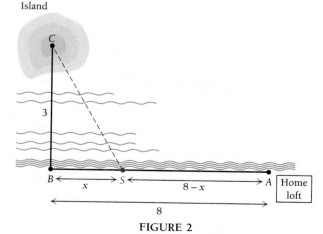

FIGURE 2

order to minimize the total energy required to get to home loft A? Assume that

(Total energy)

= (Energy rate over water) · (Distance over water)

+ (Energy rate over land) · (Distance over land).

54. *Business: Minimizing distance.* A road is to be built between two cities C_1 and C_2, which are on opposite sides of a river of uniform width r. Because of the river, a bridge must be built. C_1 is a units from the river, and C_2 is b units from the river; $a \leq b$. Where should the bridge be located in order to minimize the total distance between the cities? Give a general solution using the constants a, b, p, and r in the figure shown here.

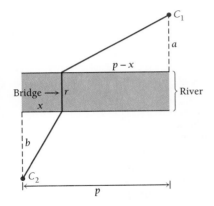

55. *Business: Minimizing cost.* The total-cost function for producing x units of a certain product is given by

$$C(x) = 8x + 20 + \frac{x^3}{100}.$$

a) Find the marginal cost $C'(x)$.

b) Find the average cost $A(x) = C(x)/x$.

c) Find the marginal average cost $A'(x)$.

d) Find the minimum of $A(x)$ and the value x_0 at which it occurs. Find the marginal cost at x_0.

e) Compare $A(x_0)$ and $C'(x_0)$.

56. *Business: Minimizing cost.* Consider $A(x) = C(x)/x$.

a) Find $A'(x)$ in terms of $C'(x)$ and $C(x)$.

b) Show that $A(x)$ has a minimum at that value of x_0 such that

$$C'(x_0) = A(x_0) = \frac{C(x_0)}{x_0}.$$

This shows that when marginal cost and average cost are the same, a product is being produced at the least average cost.

57. Minimize $Q = x^3 + 2y^3$, where x and y are positive numbers, such that $x + y = 1$.

58. Minimize $Q = 3x + y^3$, where $x^2 + y^2 = 2$.

COMPUTER—GRAPHING CALCULATOR EXERCISES

THE CALCULUS EXPLORER:
PowerGrapher

Use the program to graph the functions that arise in the maximum–minimum problems in this exercise set. Visually estimate where an extreme value occurs before finding the answer using calculus.

3.6

OBJECTIVES

a) Given a function $y = f(x)$ and a value for Δx, find Δy.

b) Given a function $y = f(x)$, find dy, and compute dy given a value of Δx or dx.

c) Use differentials to make approximations of numbers like

$$\sqrt{27} \quad \text{or} \quad \sqrt[3]{10}.$$

DIFFERENTIALS

Delta Notation

Recall the difference quotient

$$\frac{f(x + h) - f(x)}{h},$$

illustrated graphically in Fig. 1.

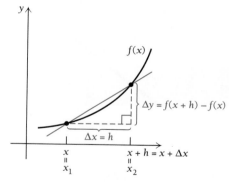

FIGURE 1

The difference quotient is used to define the derivative of a function at x. The number h is considered to be a *change* in x. Another notation for such a change is Δx, read "delta x" and called **delta notation**. The expression Δx is *not* the product of Δ and x, but is its own entity; that is, it is a new type of variable that represents the *change* in the value of x from a *first* value to a *second*. Thus,

$$\Delta x = (x + h) - x = h.$$

If subscripts are used for the first and second values of x, we have

$$\Delta x = x_2 - x_1, \quad \text{or} \quad x_2 = x_1 + \Delta x.$$

Δx can be positive or negative.

EXAMPLE 1

a) If $x_1 = 4$ and $\Delta x = 0.7$, then $x_2 = 4.7$.

b) If $x_1 = 4$ and $\Delta x = -0.7$, then $x_2 = 3.3$. ❖

We generally omit the subscripts and use x and $x + \Delta x$.

Now suppose we have a function given by $y = f(x)$. A change in x from x to $x + \Delta x$ yields a change in y from $f(x)$ to $f(x + \Delta x)$. The change in y is given by

$$\Delta y = f(x + \Delta x) - f(x).$$

EXAMPLE 2 For $y = x^2$, $x = 4$, and $\Delta x = 0.1$, find Δy.

Solution

$$\Delta y = (4 + 0.1)^2 - 4^2 = (4.1)^2 - 4^2 = 16.81 - 16 = 0.81 ❖$$

EXAMPLE 3 For $y = x^3$, $x = 2$, and $\Delta x = -0.1$, find Δy.

Solution

$$\Delta y = [2 + (-0.1)]^3 - 2^3 = (1.9)^3 - 2^3 = 6.859 - 8 = -1.141 ❖$$

DO EXERCISES 1 AND 2.

If delta notation is used, the difference quotient

$$\frac{f(x + h) - f(x)}{h}$$

becomes

$$\frac{f(x + \Delta x) - f(x)}{\Delta x} = \frac{\Delta y}{\Delta x}.$$

We can then express the derivative as

$$\frac{dy}{dx} = \lim_{\Delta x \to 0} \frac{\Delta y}{\Delta x}.$$

Note that the delta notation resembles the Leibniz notation. For values of Δx close to 0, we have the approximation

$$\frac{dy}{dx} \approx \frac{\Delta y}{\Delta x}, \quad \text{or} \quad f'(x) \approx \frac{\Delta y}{\Delta x}.$$

1. For $y = x^2$, $x = 3$, and $\Delta x = -0.1$, find Δy.

2. For $y = x^3$, $x = 2$, and $\Delta x = 1$, find Δy.

Multiplying both sides of the second expression by Δx gives us

$$\Delta y \approx f'(x)\,\Delta x.$$

We can see this in the graph in Fig. 2.

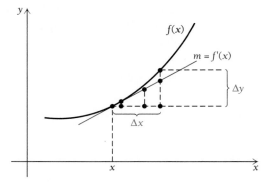

FIGURE 2

Recall that the derivative is a limit of slopes $\Delta y/\Delta x$ of secant lines. Thus as Δx gets smaller, the ratio $\Delta y/\Delta x$ gets closer to dy/dx. Note too that over small intervals, the tangent line is a good approximation to the function. Thus it is reasonable to assume that average rates of change $\Delta y/\Delta x$ of the function are approximately the same as the slope of the tangent line.

Let us use the fact that

$$\Delta y \approx f'(x)\,\Delta x$$

to make certain approximations, such as square roots.

EXAMPLE 4 Approximate $\sqrt{27}$ using $\Delta y \approx f'(x)\,\Delta x$.

Solution We first think of the number closest to 27 that is a perfect square. This is 25. What we will do is approximate how y, or $\sqrt{x}$, changes when 25 changes by $\Delta x = 2$. Let

$$y = f(x) = \sqrt{x}.$$

Then

$$\Delta y = \sqrt{x + \Delta x} - \sqrt{x} = \sqrt{x + \Delta x} - y,$$

so

$$y + \Delta y = \sqrt{x + \Delta x}.$$

Now

$$\Delta y \approx f'(x)\,\Delta x = \frac{1}{2}x^{-1/2}\,\Delta x = \frac{1}{2\sqrt{x}}\Delta x.$$

Let $x = 25$ and $\Delta x = 2$. Then

$$\Delta y \approx f'(x)\,\Delta x = \frac{1}{2\sqrt{25}}\cdot 2 = \frac{1}{\sqrt{25}} = \frac{1}{5} = 0.2.$$

Thus,

$$\sqrt{27} = \sqrt{x + \Delta x} = y + \Delta y \approx \sqrt{25} + 0.2 = 5 + 0.2 = 5.2.$$

To five decimal places, $\sqrt{27} = 5.19615$. Thus our approximation is fairly accurate. ❖

DO EXERCISE 3.

Suppose we have a total-cost function $C(x)$. When $\Delta x = 1$, we have

$$\Delta C \approx C'(x).$$

Whether this is a good approximation depends on the function and on the values of x. Let us consider an example.

EXAMPLE 5   Consider the total-cost function

$$C(x) = 2x^3 - 12x^2 + 30x + 200.$$

a)   Find ΔC and $C'(x)$ when $x = 2$ and $\Delta x = 1$.

b)   Find ΔC and $C'(x)$ when $x = 100$ and $\Delta x = 1$.

Solution

a)   We have

$$\Delta C = C(2 + 1) - C(2) = C(3) - C(2) = \$236 - \$228 = \$8.$$

Recall that $C(2)$ is the total cost of producing 2 units, and $C(3)$ is the total cost of producing 3 units, so $C(3) - C(2)$, or $\$8$, is the cost of the third unit. Now

$$C'(x) = 6x^2 - 24x + 30, \quad \text{so} \quad C'(2) = \$6.$$

b)   We have

$$\Delta C = C(100 + 1) - C(100) = C(101) - C(100) = \$58,220.$$

3.   Approximate $\sqrt{67}$ using $\Delta y \approx f'(x)\,\Delta x$. See how close the approximation is by finding $\sqrt{67}$ directly on your calculator.

4. *Business: Total cost.* Consider the total-cost function

$$C(x) = 0.01x^2 + 4x + 500.$$

a) Find ΔC and $C'(x)$ when $x = 5$ and $\Delta x = 1$.

b) Find ΔC and $C'(x)$ when $x = 100$ and $\Delta x = 1$.

Note that this is the cost of the 101st unit. Now

$$C'(100) = \$57,630.$$

Note that in part (a), the approximation of \$6 is \$2 less than the "correct" value of \$8; this is a percentage difference of \$2/\$8, or 25%, from the correct value. In part (b), the approximation of \$57,630 is \$590 less than the "correct" value of \$58,220; this is a percentage difference of \$590/\$58,220, or about 1% from the correct value, so it is a very close approximation.   ❖

We purposely used $\Delta x = 1$ in Example 5 to illustrate the following.

$$C'(x) \approx C(x + 1) - C(x)$$

Marginal cost is (approximately) the cost of the $(x + 1)$st, or next, unit.

This is the historical definition that economists have given to marginal cost.

Similarly, the following is true.

$$R'(x) \approx R(x + 1) - R(x)$$

Marginal revenue is (approximately) the revenue from the sale of the $(x + 1)$st, or next, unit.

And

$$P'(x) \approx P(x + 1) - P(x)$$

Marginal profit is (approximately) the profit from the production and sale of the $(x + 1)$st, or next, unit.

DO EXERCISE 4.

Differentials

Up to now we have not defined the symbols dy and dx as separate entities, and we have treated dy/dx as one symbol. We now define dy and dx. These symbols are called **differentials**.

DEFINITION

For $y = f(x)$, we define

$$dx, \text{ called the } differential \text{ of } x, \text{ by } dx = \Delta x$$

and

$$dy, \text{ called the } differential \text{ of } y, \text{ by } dy = f'(x)\, dx.$$

We can illustrate dx and dy as shown in Fig. 3. Note that $dx = \Delta x$, but $dy \neq \Delta y$, though $dy \approx \Delta y$, for small values of dx.

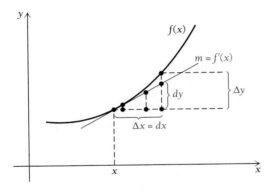

FIGURE 3

EXAMPLE 6 For $y = x(4 - x)^3$, (a) find dy and (b) find dy when $x = 5$ and $dx = 0.2$.

Solution

a) First, we find dy/dx:

$$\frac{dy}{dx} = -x[3(4 - x)^2] + (4 - x)^3 \qquad \text{Using the Product Rule and the Extended Power Rule}$$

$$= (4 - x)^2[-3x + (4 - x)]$$

$$= (4 - x)^2[-4x + 4] \qquad \text{Simplifying}$$

$$= -4(4 - x)^2(x - 1).$$

Then

$$dy = -4(4 - x)^2(x - 1)\, dx.$$

Note that the expression for dy contains *two* variables, x and dx.

5. For $y = x^3 - 5x^2 - 4x + 6$:

  a) Find dy.

  b) Find dy when $x = 3$ and $dx = 0.1$.

b) When $x = 5$ and $dx = 0.2$,

$$dy = -4(4 - 5)^2(5 - 1)0.2 = -4(-1)^2(4)(0.2) = -3.2. \quad \diamond$$

DO EXERCISE 5.

EXERCISE SET 3.6

Find Δy and $f'(x)\, \Delta x$.

1. For $y = f(x) = x^2$, $x = 2$, and $\Delta x = 0.01$.

2. For $y = f(x) = x^3$, $x = 2$, and $\Delta x = 0.01$.

3. For $y = f(x) = x + x^2$, $x = 3$, and $\Delta x = 0.04$.

4. For $y = f(x) = x - x^2$, $x = 3$, and $\Delta x = 0.02$.

5. For $y = f(x) = 1/x^2$, $x = 1$, and $\Delta x = 0.5$.

6. For $y = f(x) = 1/x$, $x = 1$, and $\Delta x = 0.2$.

7. For $y = f(x) = 3x - 1$, $x = 4$, and $\Delta x = 2$.

8. For $y = f(x) = 2x - 3$, $x = 8$, and $\Delta x = 0.5$.

9. For the total-cost function

$$C(x) = 0.01x^2 + 0.6x + 30,$$

find ΔC and $C'(x)$ when $x = 70$ and $\Delta x = 1$.

10. For the total-cost function

$$C(x) = 0.01x^2 + 1.6x + 100,$$

find ΔC and $C'(x)$ when $x = 80$ and $\Delta x = 1$.

11. For the total-revenue function

$$R(x) = 2x,$$

find ΔR and $R'(x)$ when $x = 70$ and $\Delta x = 1$.

12. For the total-revenue function

$$R(x) = 3x,$$

find ΔR and $R'(x)$ when $x = 80$ and $\Delta x = 1$.

13. a) Using $C(x)$ of Exercise 9 and $R(x)$ of Exercise 11, find the total profit $P(x)$.

   b) Find ΔP and $P'(x)$ when $x = 70$ and $\Delta x = 1$.

14. a) Using $C(x)$ of Exercise 10 and $R(x)$ of Exercise 12, find the total profit $P(x)$.

   b) Find ΔP and $P'(x)$ when $x = 80$ and $\Delta x = 1$.

Approximate using $\Delta y \approx f'(x)\, \Delta x$.

15. $\sqrt{19}$ **16.** $\sqrt{10}$ **17.** $\sqrt{102}$

18. $\sqrt{103}$ **19.** $\sqrt[3]{10}$ **20.** $\sqrt[3]{28}$

Find dy.

21. $y = (2x^3 + 1)^{3/2}$ **22.** $y = x^3(2x + 5)^2$

23. $y = \sqrt[5]{x + 27}$ **24.** $y = \dfrac{x^3 + x + 2}{x^2 + 3}$

25. $y = x^4 - 2x^3 + 5x^2 + 3x - 4$

26. $y = (7 - x)^8$

27. In Exercise 25, find dy when $x = 2$ and $dx = 0.1$.

28. In Exercise 26, find dy when $x = 1$ and $dx = 0.01$.

APPLICATIONS

❖ Business and Economics

29. *Average cost.* The average cost of a company to produce x units of a product is given by the function

$$A(x) = \frac{13x + 100}{x}.$$

By approximately how much does the average cost change as production goes from 100 units to 101 units?

30. *Supply.* A supply function for a certain product is given by

$$S(p) = 0.08p^3 + 2p^2 + 10p + 11.$$

Approximately how many more units will a seller supply when the price changes from $18.00 per unit to $18.20 per unit?

31. *Advertising.* A firm estimates that it will sell N units of a product after spending a dollars on advertising, where

$$N(a) = -a^2 + 300a + 6,$$

and a is measured in thousands of dollars. Approximately how many more products will a company sell by increasing its advertising expenditure from $100 thousand to $101 thousand?

❖ **Life and Physical Sciences**

32. *Healing wound.* The circular area of a healing wound is given by

$$A = \pi r^2,$$

where r is the radius, in centimeters. By approximately

how much does the area decrease when the radius is decreased from 2 cm to 1.9 cm? Use 3.14 for π.

33. *Tumor growth.* The spherical volume of a tumor is given by

$$V = \tfrac{4}{3}\pi r^3,$$

where r is the radius, in centimeters. By approximately how much does the volume increase when the radius is increased from 1 cm to 1.2 cm? Use 3.14 for π.

34. *Medical dosage.* The function

$$N(t) = \frac{0.8t + 1000}{5t + 4}$$

gives the bodily concentration $N(t)$, in parts per million, of a dosage of medication after time t, in hours. By approximately how much does the concentration change as time changes from 2.8 hr to 2.9 hr?

❖ **General Interest**

35. Suppose a rope surrounds the earth at the equator. The rope is lengthened by 10 ft. By about how much is the rope raised above the earth?

3.7

IMPLICIT DIFFERENTIATION AND RELATED RATES*

Implicit Differentiation

Consider the equation

$$y^3 = x.$$

This equation *implies* that y is a function of x, for if we solve for y, we get

$$y = \sqrt[3]{x}$$
$$= x^{1/3}.$$

*This section can be omitted without loss of continuity.

OBJECTIVES

a) Differentiate implicitly and find the slope of a curve at a given point.

b) Solve related-rate problems.

1. For $y^5 = 2x$, use implicit differentiation to find

$$\frac{dy}{dx}.$$

Leave the answer expressed in terms of y.

We know from our work in this chapter that

$$\frac{dy}{dx} = \frac{1}{3}x^{-2/3}. \tag{1}$$

A method known as **implicit differentiation** allows us to find dy/dx *without* solving for y. We use the Chain Rule, treating y as a function of x. We use the Extended Power Rule and differentiate both sides of

$$y^3 = x$$

with respect to x:

$$\frac{d}{dx}y^3 = \frac{d}{dx}x.$$

The derivative on the left side is found using the Extended Power Rule:

$$3y^2 \frac{dy}{dx} = 1.$$

Then

$$\frac{dy}{dx} = \frac{1}{3y^2}, \quad \text{or} \quad \frac{1}{3}y^{-2}.$$

We can show that this indeed gives us the same answer as Eq. (1) by replacing y by $x^{1/3}$:

$$\frac{dy}{dx} = \frac{1}{3}y^{-2} = \frac{1}{3}(x^{1/3})^{-2} = \frac{1}{3}x^{-2/3}.$$

DO EXERCISE 1.

Often, it is difficult or impossible to solve for y, obtaining an explicit expression in terms of x. For example, the equation

$$y^3 + x^2y^5 - x^4 = 27$$

determines y as a function of x, but it would be difficult to solve for y. We can nevertheless find a formula for the derivative of y *without* solving for y. This involves computing $\frac{d}{dx}y^n$ for various integers n, and hence involves the Extended Power Rule in the form

$$\frac{d}{dx}y^n = ny^{n-1} \cdot \frac{dy}{dx}.$$

EXAMPLE 1   For

$$y^3 + x^2y^5 - x^4 = 27:$$

a)   Find dy/dx using implicit differentiation.

b)   Find the slope of the tangent line to the curve at the point $(0, 3)$.

Solution

a)   We differentiate the term x^2y^5 using the Product Rule. Note that whenever an expression involving y is differentiated, dy/dx must be a factor of the answer. When an expression involving just x is differentiated, there is no factor dy/dx.

$$\frac{d}{dx}(y^3 + x^2y^5 - x^4) = \frac{d}{dx}(27)$$

$$\frac{d}{dx}y^3 + \frac{d}{dx}x^2y^5 - \frac{d}{dx}x^4 = 0$$

$$3y^2 \cdot \frac{dy}{dx} + x^2 \cdot 5y^4 \cdot \frac{dy}{dx} + 2x \cdot y^5 - 4x^3 = 0.$$

Then

$$3y^2 \cdot \frac{dy}{dx} + 5x^2y^4 \cdot \frac{dy}{dx} = 4x^3 - 2xy^5 \qquad \text{Only those terms involving } dy/dx \text{ should appear on one side.}$$

$$(3y^2 + 5x^2y^4)\frac{dy}{dx} = 4x^3 - 2xy^5$$

$$\frac{dy}{dx} = \frac{4x^3 - 2xy^5}{3y^2 + 5x^2y^4} \qquad \text{Solving for } dy/dx. \text{ Leave the answer in terms of } x \text{ and } y.$$

b)   To find the slope of the tangent line to the curve at $(0, 3)$, we replace x by 0 and y by 3:

$$\frac{dy}{dx} = \frac{4 \cdot 0^3 - 2 \cdot 0 \cdot 3^5}{3 \cdot 3^2 + 5 \cdot 0^2 \cdot 3^4} = 0.$$ ❖

DO EXERCISE 2.

The demand function for a product (see Sections 1.5 and 2.6) is often given implicitly.

EXAMPLE 2   For the following demand equation, differentiate implicitly to find dp/dx:

$$x = \sqrt{200 - p^3}.$$

2.   For

$$y^3 + x^2y^4 + x^3 = 8:$$

a)   Find

$$\frac{dy}{dx}$$

using implicit differentiation.

b)   Find the slope of the tangent line to the curve at the point $(0, 2)$.

3. For the following equation, differentiate implicitly to find dp/dx:

$$100\sqrt{p} = 800 - x.$$

Solution

$$\frac{d}{dx} x = \frac{d}{dx} \sqrt{200 - p^3}$$

$$1 = \frac{1}{2}(200 - p^3)^{-1/2} \cdot (-3p^2) \cdot \frac{dp}{dx}$$

$$1 = \frac{-3p^2}{2\sqrt{200 - p^3}} \cdot \frac{dp}{dx}$$

$$\frac{2\sqrt{200 - p^3}}{-3p^2} = \frac{dp}{dx}$$

❖

DO EXERCISE 3.

Related Rates

Suppose that y is a function of x, say

$$y = f(x),$$

and x varies with time t (as a function of time t). Since y depends on x and x depends on t, y also depends on t. That is, y is also a function of time t. The Chain Rule gives the following:

$$\frac{dy}{dt} = \frac{dy}{dx} \cdot \frac{dx}{dt}.$$

Thus the rate of change of y is *related* to the rate of change of x. Let us see how this comes up in problems. It helps to keep in mind that any variable can be thought of as a function of time t, even though a specific expression in terms of t may not be given.

EXAMPLE 3 *Business: Service area.* A restaurant supplier services the restaurants in a circular area in such a way that the radius r is increasing at the rate of 2 mi per year at the moment when r goes through the value $r = 5$ mi (see Fig. 1). At that moment, how fast is the area increasing?

Solution   The area A and the radius r are always related by the equation for the area of a circle:

$$A = \pi r^2.$$

We take the derivative of both sides with respect to t:

$$\frac{dA}{dt} = 2\pi r \cdot \frac{dr}{dt}.$$

FIGURE 1

At the moment in question, $dr/dt = 2$ mi/yr (miles per year) and $r = 5$ mi, so

$$\frac{dA}{dt} = 2\pi(5 \text{ mi})\left(2\,\frac{\text{mi}}{\text{yr}}\right)$$

$$= 20\pi\,\frac{\text{mi}^2}{\text{yr}}$$

$$\approx 63 \text{ square miles per year.} \quad \diamond$$

DO EXERCISE 4.

EXAMPLE 4 *Business: Rates of change of revenue, cost, and profit.* For a company making stereos, the total revenue from the sale of x stereos is given by

$$R(x) = 1000x - x^2,$$

and the total cost is given by

$$C(x) = 3000 + 20x.$$

Suppose that the company is producing and selling stereos at the rate of 10 stereos per day at the moment when the 400th stereo is produced. At that same moment, what is the rate of change of (a) total revenue? (b) total cost? (c) total profit?

Solution

a) $\dfrac{dR}{dt} = 1000 \cdot \dfrac{dx}{dt} - 2x \cdot \dfrac{dx}{dt}$ Differentiating with respect to time.

$= 1000 \cdot 10 - 2(400)10$ Substituting 10 for dx/dt and 400 for x.

$= \$2000$ per day

The radius of each ripple is a function of time.

4. *General interest: Pond ripples.* A stone is thrown into a pond. A circular ripple is spreading over the pond in such a way that the radius r is increasing at the rate of 3 ft/sec at the moment when r goes through the value $r = 4$ ft. At that moment how fast is the disturbed area increasing?

5. *Business: Total revenue, cost, and profit.* For a certain product, a company determines that the total revenue from the sale of x units is

26. A rectangular box with a square base and a cover is to contain 2500 ft^3. If the cost per square foot for the bottom is $2, for the top is $3, and for the sides is $1, what should the dimensions be in order to minimize the cost?

27. *Business: Minimizing inventory cost.* A store in California sells 360 multispeed bicycles per year. It costs $8 to store one bicycle for one year. To reorder, there is a fixed cost of $10, plus $2 for each bicycle. How many times per year should the store order bicycles, and in what lot size, in order to minimize inventory costs?

Given $y = f(x) = x^3 - x$.

28. Find Δy and $f'(x) \, \Delta x$, given that $x = 3$ and $\Delta x = -0.5$.

29. a) Find dy.

b) Find dy when $x = 2$ and $dx = 0.01$.

30. Approximate $\sqrt{69}$ using $\Delta y \approx f'(x) \, \Delta x$.

31. Differentiate the following implicitly to find dy/dx. Then find the slope of the curve at the given point.

$$2x^3 + 2y^3 = -9xy; \quad (-1, -2)$$

32. A ladder 25 ft long leans against a vertical wall. If the lower end is being moved away from the wall at the rate of 6 ft/sec, how fast is the height of the top decreasing when the lower end is 7 ft from the wall?

33. *Business: Total revenue, cost, and profit.* Find the rates of change of total revenue, cost, and profit for

$$R(x) = 120x - 0.5x^2 \quad \text{and} \quad C(x) = 15x + 6,$$

when $x = 100$ and $dx/dt = 30$ units per day.

SYNTHESIS EXERCISES

34. Find the absolute maximum and minimum values, if they exist, over the indicated interval.

$$f(x) = (x - 3)^{2/5}; \quad (-\infty, \infty)$$

35. Differentiate implicitly to find dy/dx.

$$(x - y)^4 + (x + y)^4 = x^6 + y^6$$

36. Find the relative maxima and minima of

$$y = x^4 - 8x^3 - 270x^2.$$

EXERCISES FOR THINKING AND WRITING

37. Discuss as many applications as you can of the use of differentiation considered in this chapter.

38. Using graphs and limits explain the idea of an asymptote to the graph of a function. Describe three types of asymptotes.

39. Explain how marginal revenue, cost, and profit can be used to find maximum profit.

40. Explain the usefulness of the first and second derivatives in drawing a graph of a function.

CHAPTER TEST 3

Find the relative extrema of the function. List your answers in terms of ordered pairs. Then sketch a graph of the function.

1. $f(x) = x^2 - 4x - 5$

2. $f(x) = 2x^4 - 4x^2 + 1$

3. $f(x) = (x - 2)^{2/3} - 4$

4. $f(x) = \dfrac{16}{x^2 + 4}$

5. $f(x) = x^3 + x^2 - x + 1$

6. $f(x) = 4 + 3x - x^3$

7. $f(x) = (x + 2)^3$

8. $f(x) = x\sqrt{9 - x^2}$

Sketch a graph of the function.

9. $f(x) = \dfrac{2}{x - 1}$

10. $f(x) = \dfrac{-8}{x^2 - 4}$

11. $f(x) = \dfrac{x^2 - 1}{x}$

12. $f(x) = \dfrac{x + 2}{x - 3}$

Find the absolute maximum and minimum values of the function, if they exist, over the indicated interval. Where no interval is specified, use the real line.

13. $f(x) = x(6 - x)$

14. $f(x) = x^3 + x^2 - x + 1; \quad [-2, \frac{1}{2}]$

15. $f(x) = -x^2 + 8.6x + 10$

16. $f(x) = -2x + 5; \quad [-1, 1]$

17. $f(x) = -2x + 5$

18. $f(x) = 3x^2 - x - 1$

19. $f(x) = x^2 + \dfrac{128}{x}; \quad (0, \infty)$

20. Of all numbers whose difference is 8, find the two that have the minimum product.

21. Minimize $Q = x^2 + y^2$, where $x - y = 10$.

22. *Business: Maximum profit.* Find the maximum profit and the number of units that must be produced and sold in order to yield the maximum profit.

$$R(x) = x^2 + 110x + 60,$$
$$C(x) = 1.1x^2 + 10x + 80$$

23. From a thin piece of cardboard 60 in. by 60 in., square corners are cut out so the sides can be folded up to make a box. What dimensions will yield a box of maximum volume? What is the maximum volume?

24. *Business: Minimizing inventory costs.* A sporting goods store sells 1225 tennis rackets per year. It costs $2 to store one tennis racket for one year. To reorder, there is a fixed cost of $1, plus $0.50 for each tennis racket. How many times per year should the sporting goods store order tennis rackets, and in what lot size, in order to minimize inventory costs?

25. For $y = f(x) = x^2 - 3$, $x = 5$, and $\Delta x = 0.1$, find Δy and $f'(x)\,\Delta x$.

26. Approximate $\sqrt{104}$ using $\Delta y \approx f'(x)\,\Delta x$.

27. For $y = \sqrt{x^2 + 3}$, (a) find dy and (b) find dy when $x = 2$ and $dx = 0.01$.

28. Differentiate the following implicitly to find dy/dx. Then find the slope of the curve at the given point.

$$x^3 + y^3 = 9; \quad (1, 2)$$

29. A board 13 ft long leans against a vertical wall. If the lower end is being moved away from the wall at the rate of 0.4 ft/sec, how fast is the upper end coming down when the lower end is 12 ft from the wall?

SYNTHESIS EXERCISES

30. Find the absolute maximum and minimum values of the function, if they exist, over the indicated interval.

$$f(x) = \dfrac{x^2}{1 + x^3}; \quad [0, \infty)$$

31. *Business: Minimizing average cost.* The total cost of producing x units of a product is given by

$$C(x) = 100x + 100\sqrt{x} + \dfrac{\sqrt{x^3}}{100}.$$

a) Find the average cost $A(x)$.

b) Find the minimum value of $A(x)$.

4

EXPONENTIAL AND LOGARITHMIC FUNCTIONS

AN APPLICATION

Peter Minuit of the Dutch West India Company purchased Manhattan Island from the Indians in 1626 for $24 worth of merchandise. Assuming an exponential rate of inflation of 8%, how much will Manhattan be worth in 1998?

THE MATHEMATICS

The value V of Manhattan Island is given by

$$V(t) = \$24e^{0.08t},$$

└───This is an *exponential function*.

where *t* is the number of years since 1626.

In this chapter we will consider two kinds of functions that are closely related: *exponential functions* and *logarithmic functions*. We will learn to find derivatives of such functions. Both are rich in application, with exponential functions applying particularly to problems of population growth.

4.1

EXPONENTIAL FUNCTIONS

Graphs of Exponential Functions

Have you noticed how often the price of postage increases? The graph in Fig. 1, like one that might appear in a newspaper, shows the cost of a first-class postage stamp for various years since 1900. The price became 29¢ in 1991. A curve drawn along the graph would approximate the graph of an **exponential function**. We now consider such graphs.

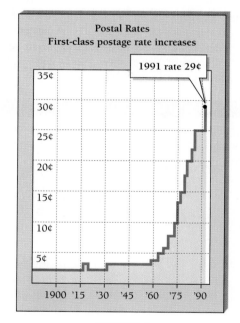

Postal Rates
First-class postage rate increases

1991 rate 29¢

FIGURE 1

In Chapter 1 we reviewed definitions of such expressions as a^x, where x is a rational number. For example,

$$a^{2.34}, \quad \text{or} \quad a^{234/100},$$

means "raise a to the 234th power and then take the 100th root."

What about expressions with irrational exponents, such as $2^{\sqrt{2}}$, 2^{π}, or $2^{-\sqrt{3}}$? An irrational number is a number named by an infinite, nonrepeating decimal. Let us consider 2^π. We know that π is irrational with infinite, nonrepeating decimal expansion:

$$3.141592654 \ldots .$$

This means that π is approached as a limit by the rational numbers

$$3,\ 3.1,\ 3.14,\ 3.141,\ 3.1415,\ \ldots ,$$

so it seems reasonable that 2^π should be approached as a limit by the rational powers

$$2^3,\ 2^{3.1},\ 2^{3.14},\ 2^{3.141},\ 2^{3.1415},\ \ldots .$$

DO EXERCISE 1.

In general, a^x is approximated by the values of a^r for rational numbers r near x; a^x is the limit of a^r as r approaches x through rational values. In summary, for $a > 0$, the definition of a^x for rational numbers x can be extended to arbitrary real numbers x in such a way that the usual laws of exponents, such as

$$a^x \cdot a^y = a^{x+y}, \qquad a^x \div a^y = a^{x-y}, \qquad (a^x)^y = a^{xy}, \quad \text{and} \quad a^{-x} = \frac{1}{a^x},$$

still hold. Moreover, the function so obtained, $f(x) = a^x$, is continuous.

DEFINITION

An *exponential function* f is given by

$$f(x) = a^x,$$

where x is any real number, $a > 0$, and $a \neq 1$. The number a is called the *base*.

The following are examples of exponential functions:

$$f(x) = 2^x, \qquad f(x) = (\tfrac{1}{2})^x, \qquad f(x) = (0.4)^x.$$

Note that in contrast to power functions like $y = x^2$ or $y = x^3$, the variable in an exponential function is in the exponent, not the base. Exponential functions have extensive application. Let us consider their graphs.

1. (📱 with a $\boxed{y^x}$ key). Complete this table. Round to six decimal places.

r	2^r
3	
3.1	
3.14	
3.141	
3.1415	
3.14159	

Considering $2^\pi = \lim_{r \to \pi} 2^r$, for rational numbers r, what seems to be the value of 2^π to two decimal places?

2. Consider $y = f(x) = 3^x$.

a) Complete this table of function values

x	0	$\frac{1}{2}$	1	2	-1	-2
3^x						

b) Graph $f(x) = 3^x$.

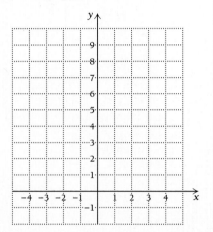

EXAMPLE 1   Graph: $y = f(x) = 2^x$.

Solution

a)   First we find some function values.

x	0	$\frac{1}{2}$	1	2	3	-1	-2
$y = f(x)$ (or 2^x)	1	1.4	2	4	8	$\frac{1}{2}$	$\frac{1}{4}$

Note: For:

$x = 0, \quad y = 2^0 = 1;$

$x = \frac{1}{2}, \quad y = 2^{1/2} = \sqrt{2} \approx 1.4;$

$x = 1, \quad y = 2^1 = 2;$

$x = 2, \quad y = 2^2 = 4;$

$x = 3, \quad y = 2^3 = 8;$

$x = -1, y = 2^{-1} = \frac{1}{2};$

$x = -2, y = 2^{-2} = \dfrac{1}{2^2} = \dfrac{1}{4}.$

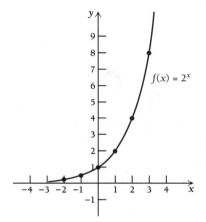

FIGURE 2

b)   Next, we plot the points and connect them with a smooth curve, as shown in Fig. 2. The graph is continuous, increasing, and concave up.   ❖

DO EXERCISE 2.

EXAMPLE 2   Graph: $y = f(x) = (\frac{1}{2})^x$.

Solution

a)   We first find some function values. Before we do this, note that

$$y = f(x) = (\tfrac{1}{2})^x = (2^{-1})^x = 2^{-x}.$$

This will ease our work.

x	0	$\frac{1}{2}$	1	2	-1	-2	-3
y	1	0.7	$\frac{1}{2}$	$\frac{1}{4}$	2	4	8

Note: For:

$x = 0, \quad y = 2^{-0} = 1;$

$x = \dfrac{1}{2}, \quad y = 2^{-1/2} = \dfrac{1}{2^{1/2}}$

$\qquad = \dfrac{1}{\sqrt{2}} \approx \dfrac{1}{1.4} \approx 0.7;$

$x = 1, \quad y = 2^{-1} = \tfrac{1}{2};$

$x = 2, \quad y = 2^{-2} = \tfrac{1}{4};$

$x = -1, y = 2^{-(-1)} = 2;$

$x = -2, y = 2^{-(-2)} = 4;$

$x = -3, y = 2^{-(-3)} = 8.$

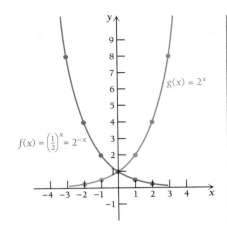

FIGURE 3

b) We plot these points and connect them with a smooth curve, as shown by the red curve in Fig. 3. The graph is continuous, decreasing, and concave up. The graph of $g(x) = 2^x$, the blue curve, is shown for comparison. ❖

DO EXERCISE 3.

The following are some properties of the exponential function for

3. Consider $y = f(x) = (\tfrac{1}{3})^x$.

a) Complete this table of function values.

x	0	$\tfrac{1}{2}$	1	2	-1	-2
y						

b) Graph $f(x) = (\tfrac{1}{3})^x$.

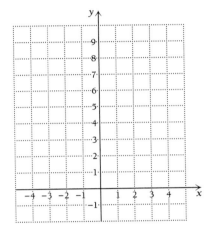

1. The function $f(x) = a^x$, where $a > 1$, is a positive, increasing, continuous function. As x gets smaller, a^x approaches 0. The graph is concave up, as shown in Fig. 4. (If you studied Section 3.3, you also know that the x-axis is an asymptote.)

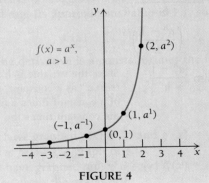

FIGURE 4

15. Graph $f(x) = 2e^{-x}$. Use your calculator. For example, for $x = 3$, $f(3) = 2e^{-3} = 2(0.049787) \approx 0.1$.

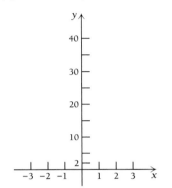

Graphs of e^x, e^{-x}, and $1 - e^{-kx}$

We use a calculator with an $\boxed{e^x}$ key to find approximate values of e^x and e^{-x}. We can also use a power key $\boxed{y^x}$ with $y = 2.7183$ or some other approximation for e. With these we can draw graphs (Figs. 9 and 10) of the functions. Note that the graph of e^{-x} (Fig. 10) is a reflection, or "mirror image," of the graph of e^x (Fig. 9) across the y-axis.

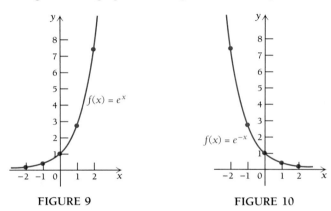

FIGURE 9   FIGURE 10

DO EXERCISE 15.

Functions of the type $f(x) = 1 - e^{-kx}$ also have important applications.

EXAMPLE 9   Graph $f(x) = 1 - e^{-2x}$ for nonnegative values of x.

Solution   We obtain these values using a calculator with an $\boxed{e^x}$ key. The graph is shown in Fig. 11.

For example,

$$f(1) = 1 - e^{-2(1)}$$
$$= 1 - e^{-2}$$
$$= 1 - 0.135335$$
$$\approx 0.86.$$

x	$f(x)$
0	0
$\frac{1}{2}$	0.63
1	0.86
2	0.98
3	0.998

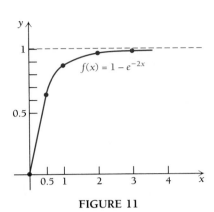

FIGURE 11

DO EXERCISE 16.

In general, the graph of $f(x) = 1 - e^{-kx}$, for $k > 0$ and $x \geqslant 0$, is increasing, which we expect since $f'(x) = ke^{-kx}$ is always positive. We also see that $f(x) \to 1$ as $x \to \infty$; that is, $\lim_{x \to \infty} (1 - e^{-kx}) = 1$.

A word of caution! Functions of the type a^x (for example, 2^x, 3^x, and e^x) are different from functions of the type x^a for example, x^2, x^3, $x^{1/2}$). For a^x, the variable is in the exponent. For x^a, the variable is in the base. The derivative of a^x is not xa^{x-1}. In particular, we have the following:

$$\frac{d}{dx}e^x \neq xe^{x-1}, \quad \text{but} \quad \frac{d}{dx}e^x = e^x.$$

16. a) Complete this table for
$$f(x) = 1 - e^{-x}.$$

x	0	$\frac{1}{2}$	1	2	3	4
$f(x)$						

b) Graph $f(x) = 1 - e^{-x}$.

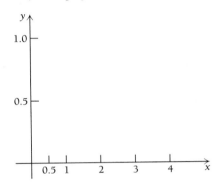

EXERCISE SET 4.1

Graph.

1. $y = 4^x$
2. $y = 5^x$
3. $y = (0.4)^x$
4. $y = (0.2)^x$
5. $x = 4^y$
6. $x = 5^y$

Differentiate.

7. $f(x) = e^{3x}$
8. $f(x) = e^{2x}$
9. $f(x) = 5e^{-2x}$
10. $f(x) = 4e^{-3x}$
11. $f(x) = 3 - e^{-x}$
12. $f(x) = 2 - e^{-x}$
13. $f(x) = -7e^x$
14. $f(x) = -4e^x$
15. $f(x) = \frac{1}{2}e^{2x}$
16. $f(x) = \frac{1}{4}e^{4x}$

17. $f(x) = x^4e^x$
18. $f(x) = x^5e^x$
19. $f(x) = \dfrac{e^x}{x^4}$
20. $f(x) = \dfrac{e^x}{x^5}$
21. $f(x) = e^{-x^2 + 7x}$
22. $f(x) = e^{-x^2 + 8x}$
23. $f(x) = e^{-x^2/2}$
24. $f(x) = e^{x^2/2}$
25. $y = e^{\sqrt{x-7}}$
26. $y = e^{\sqrt{x-4}}$
27. $y = \sqrt{e^x - 1}$
28. $y = \sqrt{e^x + 1}$
29. $y = xe^{-2x} + e^{-x} + x^3$
30. $y = e^x + x^3 - xe^x$
31. $y = 1 - e^{-x}$
32. $y = 1 - e^{-3x}$
33. $y = 1 - e^{-kx}$
34. $y = 1 - e^{-mx}$

Graph.

35. $f(x) = e^{2x}$

36. $f(x) = e^{(1/2)x}$

37. $f(x) = e^{-2x}$

38. $f(x) = e^{-(1/2)x}$

39. $f(x) = 3 - e^{-x}$, for nonnegative values of x

40. $f(x) = 2(1 - e^{-x})$, for nonnegative values of x

41. Find the tangent line to the graph of $f(x) = e^x$ at the point $(0, 1)$.

42. Find the tangent line to the graph of $f(x) = 2e^{-3x}$ at the point $(0, 2)$.

APPLICATIONS

❖ Business and Economics

43. *Marginal cost.* A company's total cost, in millions of dollars, is given by

$$C(t) = 100 - 50e^{-t},$$

where t = time. Find each of the following.

a) The marginal cost $C'(t)$
b) $C'(0)$
c) $C'(4)$

44. *Marginal cost.* A company's total cost, in millions of dollars, is given by

$$C(t) = 200 - 40e^{-t},$$

where t = time. Find each of the following.

a) The marginal cost $C'(t)$
b) $C'(0)$
c) $C'(5)$

45. *Marginal demand.* The demand function for a certain kind of VCR is given by the function

$$x = D(p) = 480e^{-0.003p}.$$

a) How many VCRs will be sold when the price is $120? $180? $340?
b) Sketch a graph of $x = D(p)$ for $0 \leq p \leq 400$.
c) Find the marginal demand $D'(p)$.

46. *Marginal supply.* The supply function for the VCR in Exercise 45 is given by the function

$$x = S(p) = 150e^{0.004p}.$$

a) How many VCRs will the seller allow to be sold when the price is $120? $180? $340?
b) Sketch a graph of $x = S(p)$ for $0 \leq p \leq 400$.
c) Find the marginal supply $S'(p)$.

❖ Life and Physical Sciences

47. *Medication concentration.* The concentration C, in parts per million, of a medication in the body t hours after ingestion is given by the function

$$C(t) = 10t^2e^{-t}.$$

a) Find the concentration after 0 hr, 1 hr, 2 hr, 3 hr, and 10 hr.
b) Sketch a graph of the function for $0 \leq t \leq 10$.
c) Find the rate of change of the concentration $C'(t)$.
d) Find the maximum value of the concentration and where it occurs.

❖ Social Sciences

48. *Ebbinghaus learning model.* Suppose you are given the task of learning 100% of a block of knowledge. Human nature would tell us that we would retain only a percentage P of the knowledge t weeks after we have learned it. The *Ebbinghaus learning model* asserts that P is given by

$$P(t) = Q + (100\% - Q)e^{-kt},$$

where Q is the percentage that we would never forget and k is a constant that depends on the knowledge learned. Suppose that $Q = 40\%$ and $k = 0.7$.

a) Find the percentage retained after 0 weeks, 1 week, 2 weeks, 6 weeks, and 10 weeks.
b) Find $\lim_{t \to \infty} P(t)$.
c) Sketch a graph of P.
d) Find the rate of change of P with respect to time t, $P'(t)$.

SYNTHESIS EXERCISES

Differentiate

49. $y = (e^{3x} + 1)^5$

50. $y = (e^{x^2} - 2)^4$

51. $y = \dfrac{e^{3t} - e^{7t}}{e^{4t}}$

52. $y = \sqrt[3]{e^{3t} + t}$

53. $y = \dfrac{e^x}{x^2 + 1}$

54. $y = \dfrac{e^x}{1 - e^x}$

55. $f(x) = e^{\sqrt{x}} + \sqrt{e^x}$

56. $f(x) = \dfrac{1}{e^x} + e^{1/x}$

57. $f(x) = e^{x/2} \cdot \sqrt{x - 1}$

58. $f(x) = \dfrac{xe^{-x}}{1 + x^2}$

59. $f(x) = \dfrac{e^x - e^{-x}}{e^x + e^{-x}}$

60. $f(x) = e^{e^x}$

(with $\boxed{y^x}$ key). Each of the following is an expression for e. Find the function values that are approximations for e. Round to five decimal places.

61. $e = \lim_{t \to 0} f(t)$; $f(t) = (1 + t)^{1/t}$. Find $f(1)$, $f(0.5)$, $f(0.2)$, $f(0.1)$, and $f(0.001)$.

62. $e = \lim_{t \to 1} g(t)$; $g(t) = t^{1/(t-1)}$. Find $g(0.5)$, $g(0.9)$, $g(0.99)$, $g(0.999)$, and $g(0.9998)$.

63. Find the maximum value of $f(x) = x^2 e^{-x}$ on $[0, 4]$.

64. Find the minimum value of $f(x) = xe^x$ on $[-2, 0]$.

COMPUTER–GRAPHING CALCULATOR EXERCISES

65. Sketch the graph of $y = x^2 e^{-x}$. Find the relative extrema of the function.

66. Sketch the graph of $y = e^{-x^2}$. Find the relative extrema of the function.

THE CALCULUS EXPLORER
Derivatives

Use the program to graph each of the following functions f and its derivative f': $f(x) = e^x$, $f(x) = e^{-x}$, $f(x) = 2e^{0.3x}$, and $f(x) = 1000e^{-0.08x}$. Then graph $f(x) = (1 + 1/x)^x$ for large values of x to see how e is approached as a limit.

4.2

LOGARITHMIC FUNCTIONS

Graphs of Logarithmic Functions

Suppose we want to solve the equation

$$10^x = 1000.$$

We are trying to find that power of 10 that will give 1000. We can see that the answer is 3. The number 3 is called the "logarithm, base 10, of 1000."

OBJECTIVES

a) Given an exponential equation, write an equivalent logarithmic equation.

b) Given a logarithmic equation, write an equivalent exponential equation.

c) Given $\log_a 3 = 1.099$ and $\log_a 5 = 1.609$, find logarithms like $\log_a 15$ and $\log_a 5a$. (continued)

d) Solve an equation like $e^t = 40$ for t.

e) Solve problems involving exponential and natural logarithmic functions.

f) Differentiate functions involving natural logarithms.

DEFINITION

A *logarithm* is as follows:

$$y = \log_a x \quad \text{means} \quad x = a^y, \quad a > 0, a \neq 1.$$

The number a is called the *logarithmic base*.

Thus for logarithms base 10, $\log_{10} x$ is that number y such that $x = 10^y$. Therefore, a logarithm can be thought of as an exponent. We can convert from a logarithmic equation to an exponential equation, and conversely, as follows.

1. Write an equivalent exponential equation.

a) $\log_b P = T$

b) $\log_9 3 = \frac{1}{2}$

c) $\log_{10} 1000 = 3$

d) $\log_{10} 0.1 = -1$

Logarithmic equation	Exponential equation
$\log_a M = N$	$a^N = M$
$\log_{10} 100 = 2$	$10^2 = 100$
$\log_{10} 0.01 = -2$	$10^{-2} = 0.01$
$\log_{49} 7 = \frac{1}{2}$	$49^{1/2} = 7$

DO EXERCISES 1 AND 2.

In order to graph a logarithmic equation, we can graph its equivalent exponential equation.

2. Write an equivalent logarithmic equation.

a) $e^k = T$

b) $16^{1/4} = 2$

c) $10^4 = 10,000$

d) $10^{-3} = 0.001$

EXAMPLE 1 Graph: $y = \log_2 x$.

Solution We first write the equivalent exponential equation:

$$x = 2^y.$$

We select values for y and find the corresponding values of 2^y. Then we plot points, remembering that x is still the first coordinate. The graph is shown in Fig. 1.

x, or 2^y	y
1	0
2	1
4	2
8	3
$\frac{1}{2}$	-1
$\frac{1}{4}$	-2

(1) — Select y.

(2) — Compute x.

$y = \log_2 x$ (or $x = 2^y$)

FIGURE 1

DO EXERCISE 3.

The graphs of $f(x) = 2^x$ and $g(x) = \log_2 x$ are shown in Fig. 2 on the same set of axes. Note that we can obtain the graph of g by reflecting the

3. Graph $y = \log_3 x$.

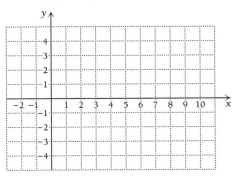

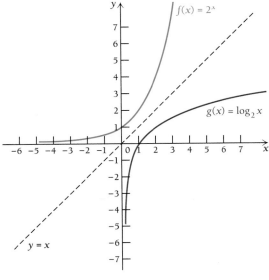

$f(x) = 2^x$

$g(x) = \log_2 x$

$y = x$

FIGURE 2

4. Consider

$$f(x) = 10^x$$

and

$$g(x) = \log_{10} x.$$

a)   Find $f(3)$.
b)   Find $g(1000)$.
c)   Use a calculator to find $g(5)$.
d)   Use a calculator to find $f(0.699)$.

graph of f across the line $y = x$. Graphs obtained in this manner are known as *inverses* of each other.

Although we cannot develop inverses in detail here, it is of interest to note that they "undo" each other. For example,

$$f(3) = 2^3 = 8 \qquad \text{The input 3 gives the output 8.}$$

and

$$g(8) = \log_2 8 = 3. \qquad \text{The input 8 gets us back to 3.}$$

DO EXERCISE 4.

Basic Properties of Logarithms

The following are some basic properties of logarithms. The proofs follow from properties of exponents.

THEOREM 4

Properties of Logarithms

For any positive numbers M, N, and a, $a \neq 1$, and any real number k:

P1. $\log_a MN = \log_a M + \log_a N$

P2. $\log_a \dfrac{M}{N} = \log_a M - \log_a N$

P3. $\log_a M^k = k \cdot \log_a M$

P4. $\log_a a = 1$

P5. $\log_a a^k = k$

P6. $\log_a 1 = 0$

Proof of P1 and P2:   Let $X = \log_a M$ and $Y = \log_a N$. Writing the equivalent exponential equations, we then have

$$M = a^X \quad \text{and} \quad N = a^Y.$$

Then by the properties of exponents (see Section 1.1), we have

$$MN = a^X \cdot a^Y = a^{X+Y},$$

so

$$\log_a MN = X + Y$$
$$= \log_a M + \log_a N.$$

Also

$$\frac{M}{N} = a^X \div a^Y = a^{X-Y},$$

so

$$\log_a \frac{M}{N} = X - Y$$

$$= \log_a M - \log_a N.$$

Proof of P3:   Let $X = \log_a M$. Then

$$M = a^X,$$

so

$$M^k = (a^X)^k, \quad \text{or} \quad M^k = a^{Xk}.$$

Thus,

$$\log_a M^k = Xk = kX = k \cdot \log_a M.$$

Proof of P4: $\log_a a = 1$ because $a^1 = a$.

Proof of P5: $\log_a (a^k) = k$ because $(a^k) = a^k$.

Proof of P6: $\log_a 1 = 0$ because $a^0 = 1$.

Let us illustrate these properties.

EXAMPLE 2   Given

$$\log_a 2 = 0.301 \quad \text{and} \quad \log_a 3 = 0.477,$$

find each of the following.

a) $\log_a 6$ $\log_a 6 = \log_a (2 \cdot 3) = \log_a 2 + \log_a 3$ By P1
$$= 0.301 + 0.477$$
$$= 0.778$$

b) $\log_a \frac{2}{3}$ $\log_a \frac{2}{3} = \log_a 2 - \log_a 3$ By P2
$$= 0.301 - 0.477$$
$$= -0.176$$

c) $\log_a 81$ $\log_a 81 = \log_a 3^4 = 4 \log_a 3$ By P3
$$= 4(0.477)$$
$$= 1.908$$

Given

$$\log_a 2 = 0.301 \quad \text{and}$$
$$\log_a 5 = 0.699,$$

find each of the following.

5. $\log_a 4$

6. $\log_a 10$

7. $\log_a \frac{2}{5}$

8. $\log_a \frac{5}{2}$

9. $\log_a \frac{1}{5}$

10. $\log_a \sqrt{a^3}$

11. $\log_a 5a$

12. $\log_a 16$

d) $\log_a \frac{1}{3}$ $\log_a \frac{1}{3} = \log_a 1 - \log_a 3$ By P2
$$= 0 - 0.477 \qquad \text{By P6}$$
$$= -0.477$$

e) $\log_a \sqrt{a}$ $\log_a \sqrt{a} = \log_a a^{1/2} = \frac{1}{2}$ By P5

f) $\log_a 2a$ $\log_a 2a = \log_a 2 + \log_a a$ By P1
$$= 0.301 + 1 \qquad \text{By P4}$$
$$= 1.301$$

g) $\log_a 5$ *No way to find using these properties.*
$(\log_a 5 \neq \log_a 2 + \log_a 3)$

h) $\dfrac{\log_a 3}{\log_a 2}$ $\dfrac{\log_a 3}{\log_a 2} = \dfrac{0.477}{0.301} \approx 1.58$

We simply divided and used none of the properties.   ❖

DO EXERCISES 5–12.

Common Logarithms

The number $\log_{10} x$ is called the **common logarithm** of x and is abbreviated log x; that is:

DEFINITION

For any positive number x,

$$\log x = \log_{10} x.$$

Thus, when we write log x with no base indicated, base 10 is understood. Note the following comparison of common logarithms and powers of 10.

$1000 = 10^3$	The common	$\log 1000 = 3$
$100 = 10^2$	logarithms	$\log 100 = 2$
$10 = 10^1$	at the right	$\log 10 = 1$
$1 = 10^0$	follow from	$\log 1 = 0$
$0.1 = 10^{-1}$	the powers at	$\log 0.1 = -1$
$0.01 = 10^{-2}$	the left.	$\log 0.01 = -2$
$0.001 = 10^{-3}$		$\log 0.001 = -3$

Since log 100 = 2 and log 1000 = 3, it seems reasonable that log 500 is somewhere between 2 and 3. Tables were generally used for such approximations, but with the advent of the calculator, that method of finding logarithms is being used more and more infrequently. Using a calculator with a $\boxed{\text{log}}$ key, we find that log 500 = 2.6990, rounded to four decimal places.

Before calculators and computers became so readily available, common logarithms were used extensively to do certain kinds of computations. In fact, computation is the reason logarithms were developed. Since the standard notation we use for numbers is based on 10, it is logical that base-10, or common, logarithms were used for computations. Today, computations with common logarithms are mainly of historical interest; the logarithmic functions, base e, are of modern importance.

DO EXERCISE 13.

Natural Logarithms

The number e, which is approximately 2.718282, was developed in Section 4.1, and has extensive application in many fields. The number $\log_e x$ is called the **natural logarithm** of x and is abbreviated ln x; that is:

DEFINITION

For any positive number x,

$$\ln x = \log_e x.$$

The following is a restatement of the basic properties of logarithms in terms of natural logarithms.

THEOREM 5

P1. $\ln MN = \ln M + \ln N$

P2. $\ln \dfrac{M}{N} = \ln M - \ln N$

P3. $\ln a^k = k \cdot \ln a$

P4. $\ln e = 1$

P5. $\ln e^k = k$

P6. $\ln 1 = 0$

---

13. ▦ Find the logarithm. Round to four decimal places.

a) log 31,456

b) log 0.9080701

c) log 78.6

d) log 7.86

e) log 0.786

f) log 0.0786

g) log 0.00786

Given

$$\ln 2 = 0.6931 \quad \text{and}$$
$$\ln 5 = 1.6094,$$

find each of the following.

14. $\ln 10$

15. $\ln \frac{2}{5}$

16. $\ln \frac{5}{2}$

17. $\ln 16$

18. $\ln 5e$

19. $\ln \sqrt{e}$

20. $\ln \frac{1}{5}$

🔲 Find the logarithm. Round to six decimal places.

21. $\ln 2$

22. $\ln 20$

23. $\ln 100$

24. $\ln 0.07432$

25. $\ln 1.08$

26. $\ln 0.9999$

Let us illustrate these properties.

EXAMPLE 3 Given

$$\ln 2 = 0.6931 \quad \text{and} \quad \ln 3 = 1.0986,$$

find each of the following.

a) $\ln 6$ 　　$\ln 6 = \ln (2 \cdot 3) = \ln 2 + \ln 3$ 　　By P1
$$= 0.6931 + 1.0986$$
$$= 1.7917$$

b) $\ln 81$ 　　$\ln 81 = \ln (3^4)$
$$= 4 \ln 3 \quad \text{By P3}$$
$$= 4(1.0986)$$
$$= 4.3944$$

c) $\ln \frac{2}{3}$ 　　$\ln \frac{2}{3} = \ln 2 - \ln 3$ 　　By P2
$$= 0.6931 - 1.0986$$
$$= -0.4055$$

d) $\ln \frac{1}{3}$ 　　$\ln \frac{1}{3} = \ln 1 - \ln 3$ 　　By P2
$$= 0 - 1.0986 \quad \text{By P6}$$
$$= -1.0986$$

e) $\ln 2e$ 　　$\ln 2e = \ln 2 + \ln e$ 　　By P1
$$= 0.6931 + 1 \quad \text{By P4}$$
$$= 1.6931$$

f) $\ln \sqrt{e^3}$ 　　$\ln \sqrt{e^3} = \ln e^{3/2}$
$$= \frac{3}{2} \quad \text{By P5}$$

DO EXERCISES 14–20.

Finding Natural Logarithms Using a Calculator

You should have a calculator with a $\boxed{\ln}$ key. You can find natural logarithms directly using this key.

EXAMPLE 4 Find each logarithm on your calculator. Round to six decimal places.

a) $\ln 5.24 = 1.656321$
b) $\ln 0.00001277 = -11.268412$

DO EXERCISES 21–26.

Exponential Equations

If an equation contains a variable in an exponent, we call the equation **exponential.** We can use logarithms to manipulate or solve exponential equations.

EXAMPLE 5 Solve $e^t = 40$ for t.

Solution

$$\ln e^t = \ln 40 \qquad \text{Taking the natural logarithm on both sides}$$
$$t = \ln 40 \qquad \text{By P5}$$
$$t = 3.688879$$
$$t \approx 3.7$$

It should be noted that this is an approximation for t even though an equals sign is often used. ❖

EXAMPLE 6 Solve $e^{-0.04t} = 0.05$ for t.

Solution

$$\ln e^{-0.04t} = \ln 0.05 \qquad \text{Taking the natural logarithm on both sides}$$
$$-0.04t = \ln 0.05 \qquad \text{By P5}$$
$$t = \frac{\ln 0.05}{-0.04}$$
$$t = \frac{-2.995732}{-0.04}$$
$$t \approx 75$$

Calculator note. For purposes of space and explanation, we have rounded the value of $\ln 0.05$ to -2.995732 in an intermediate step. When using your calculator, you should find

$$\frac{\ln 0.05}{-0.04}$$

directly, without rounding, as say,

$$\frac{-2.995732274}{-0.04}.$$

Then divide, and round at the end. Answers at the back of the book have been found in this manner. Remember, the number of places in a table or on a calculator may affect the accuracy of the answer. Usually, your answer should agree to at least three digits.

❖

DO EXERCISES 27 AND 28.

Solve for t.

27. $e^t = 80$

28. $e^{-0.06t} = 0.07$

29. a) Complete the following table using a calculator.

x	0.5	1	2	3	4
$\ln x$			0.7		

b) Graph $y = \ln x$.

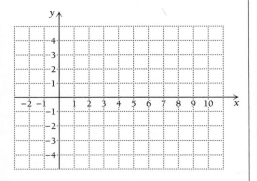

Graphs of Natural Logarithmic Functions

There are two ways in which we might obtain the graph of $y = f(x) = \ln x$. One is by writing its equivalent equation $x = e^y$. Then we select values for y and use a calculator to find the corresponding values of e^y. We then plot points, remembering that x is still the first coordinate. The graph is shown in Fig. 3.

x, or e^y	y
0.1	−2
0.4	−1
1.0	0
2.7	1
7.4	2
20.1	3

① — Select y.
② — Compute x.

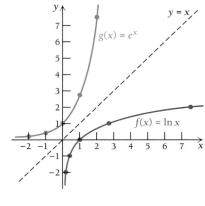

FIGURE 3

Figure 3 also shows the graph of $g(x) = e^x$ for comparison. Note again that the functions are inverses of each other. That is, the graph of $y = \ln x$, or $x = e^y$, is a reflection, or mirror image, across the line $y = x$ of the graph of $y = e^x$.

The second method of graphing $y = \ln x$ is to use a calculator to find function values. For example, when $x = 2$, then $y = \ln 2 = 0.6931 \approx 0.7$. This gives the pair $(2, 0.7)$ on the graph. (See Fig. 4.)

DO EXERCISE 29.

The following properties hold.

THEOREM 6

$\ln x$ exists only for positive numbers x. (See Figs. 3 and 4.)

$\ln x < 0$ for $0 < x < 1$.

$\ln x = 0$ when $x = 1$.

$\ln x > 0$ for $x > 1$.

Derivatives of Natural Logarithmic Functions

Consider $f(x) = \ln x$. We can show that $f'(x) = 1/x$ (the slope of the tangent line at x is just the reciprocal of x). We are trying to find the derivative of

$$f(x) = \ln x. \tag{1}$$

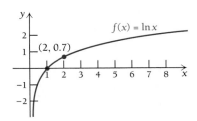

FIGURE 4

We first write its equivalent exponential equation:

$$e^{f(x)} = x. \qquad \ln x = \log_e x = f(x), \text{ so } e^{f(x)} = x, \tag{2}$$
$$\text{by the definition of logarithms}$$

Now we differentiate on both sides of this equation:

$$\frac{d}{dx} e^{f(x)} = \frac{d}{dx} x$$

$$f'(x) \cdot e^{f(x)} = 1 \qquad \text{By the Chain Rule}$$

$$f'(x) \cdot x = 1 \qquad \text{Substituting } x \text{ for } e^{f(x)} \text{ from Eq. (2)}$$

$$f'(x) = \frac{1}{x}.$$

Thus we have the following.

THEOREM 7

For any positive number x,

$$\frac{d}{dx} \ln x = \frac{1}{x}.$$

This is true only for positive values of x, since $\ln x$ is defined only for positive numbers. (For negative numbers x, this derivative formula becomes

$$\frac{d}{dx} \ln |x| = \frac{1}{x},$$

Differentiate.

30. $y = 5 \ln x$

31. $f(x) = x^3 \ln x + 4x$

32. $f(x) = \dfrac{\ln x}{x^2}$

but we will seldom consider such a case in this text.)

Let us find some derivatives.

EXAMPLE 7

$$\frac{d}{dx} \, 3 \ln x = 3 \, \frac{d}{dx} \ln x = \frac{3}{x}$$

❖

EXAMPLE 8

$$\frac{d}{dx} \, (x^2 \ln x + 5x) = x^2 \cdot \frac{1}{x} + 2x \cdot \ln x + 5 \qquad \begin{array}{l}\text{Using the Product Rule} \\ \text{on } x^2 \ln x\end{array}$$

$$= x + 2x \cdot \ln x + 5, \qquad \text{Simplifying}$$
$$\text{or } x(1 + 2 \ln x) + 5$$

❖

EXAMPLE 9

$$\frac{d}{dx} \left(\frac{\ln x}{x^3} \right) = \frac{x^3 \cdot (1/x) - (3x^2)(\ln x)}{x^6} \qquad \text{By the Quotient Rule}$$

$$= \frac{x^2 - 3x^2 \ln x}{x^6}$$

$$= \frac{x^2(1 - 3 \ln x)}{x^6} \qquad \text{Factoring}$$

$$= \frac{1 - 3 \ln x}{x^4} \qquad \text{Simplifying}$$

❖

DO EXERCISES 30–32.

Suppose we want to differentiate a more complicated function, such as

$$h(x) = \ln (x^2 - 8x).$$

This is a composition of functions. In general, we have

$$h(x) = \ln f(x) = g[f(x)], \quad \text{where} \quad g(x) = \ln x.$$

Now $g'(x) = 1/x$. Then by the Chain Rule (Section 2.8), we have

$$h'(x) = g'[f(x)] \cdot f'(x) = \frac{1}{f(x)} \cdot f'(x).$$

For the above case, $f(x) = x^2 - 8x$, so $f'(x) = 2x - 8$. Then

$$h'(x) = \frac{1}{x^2 - 8x} \cdot (2x - 8) = \frac{2x - 8}{x^2 - 8x}.$$

The following rule, which we have proven using the Chain Rule, allows us to find derivatives of functions like the one above.

THEOREM 8

$$\frac{d}{dx}\ln f(x) = f'(x) \cdot \frac{1}{f(x)} = \frac{f'(x)}{f(x)}$$

The derivative of the natural logarithm of a function is the derivative of the function divided by the function.

The following gives us a way of remembering this rule.

$$h(x) = \ln (x^2 - 8x)$$

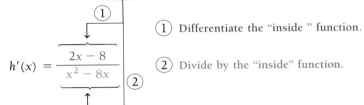

① Differentiate the "inside" function.

② Divide by the "inside" function.

EXAMPLE 10

$$\frac{d}{dx}\ln 3x = \frac{3}{3x} = \frac{1}{x}$$

Note that we could have done this another way, using Property 1:

$$\ln 3x = \ln 3 + \ln x;$$

then

$$\frac{d}{dx}\ln 3x = \frac{d}{dx}\ln 3 + \frac{d}{dx}\ln x = 0 + \frac{1}{x} = \frac{1}{x}.$$ ❖

EXAMPLE 11

$$\frac{d}{dx}\ln (x^2 - 5) = \frac{2x}{x^2 - 5}$$ ❖

EXAMPLE 12

$$\frac{d}{dx}\ln (\ln x) = \frac{1}{x} \cdot \frac{1}{\ln x} = \frac{1}{x \ln x}.$$ ❖

37. *Business: Advertising.* A model for advertising response is given by

$$N(a) = 500 + 200 \ln a, \quad a \geqslant 1,$$

where $n(a)$ = the number of units sold and a = the amount spent on advertising, in thousands of dollars.

a) How many units were sold after spending $1000? (Substitute 1, not 1000, for a.)

b) Find $N'(a)$.

c) Find the maximum and minimum values, if they exist.

DO EXERCISE 37.

EXAMPLE 15   *Business: An advertising model.* A company begins a radio advertising campaign in New York City to market a new product. The percentage of the "target market" that buys a product is normally a function of the length of the advertising campaign. The radio station estimates this percentage as $1 - e^{-0.04t}$ for this type of product, where t = the number of days of the campaign. The target market is estimated to be 1,000,000 people and the price per unit is $0.50. The costs of advertising are $1000 per day. Find the length of the advertising campaign that will result in the maximum profit.

Solution   That the percentage of the target market that buys the product can be modeled by $f(t) = 1 - e^{-0.04t}$ is justified if we look at its graph, shown in Fig. 6. The function increases from 0 (0%) toward 1 (100%). The longer the advertising campaign, the larger the percentage of the market that has bought the product.

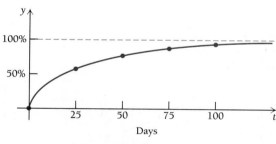

FIGURE 6

The total-profit function, here expressed in terms of time t, is given by

$$\text{Profit} = \text{Revenue} - \text{Cost}$$
$$P(t) = R(t) - C(t).$$

a) Find $R(t)$:

$$R(t) = (\text{Price per unit}) \cdot (\text{Target market}) \cdot (\text{Percentage buying})$$
$$R(t) = 0.5(1,000,000)(1 - e^{-0.04t}) = 500,000 - 500,000e^{-0.04t}.$$

b) Find $C(t)$:

$$C(t) = (\text{Advertising costs per day}) \cdot (\text{Number of days})$$
$$C(t) = 1000t.$$

c) Find $P(t)$ and take its derivative:

$$P(t) = R(t) - C(t)$$
$$P(t) = 500,000 - 500,000e^{-0.04t} - 1000t$$
$$P'(t) = (-0.04)(-500,000e^{-0.04t}) - 1000$$
$$P'(t) = 20,000e^{-0.04t} - 1000.$$

d) Set the first derivative equal to 0 and solve:

$$20,000e^{-0.04t} - 1000 = 0$$
$$20,000e^{-0.04t} = 1000$$
$$e^{-0.04t} = \frac{1000}{20,000} = 0.05$$
$$\ln e^{-0.04t} = \ln 0.05$$
$$-0.04t = \ln 0.05$$
$$t = \frac{\ln 0.05}{-0.04}$$
$$t = \frac{-2.995732}{-0.04}$$
$$t \approx 75.$$

e) We have only one critical point, so we can use the second derivative to determine whether we have a maximum:

$$P''(t) = -0.04(20,000e^{-0.04t}) = -800e^{-0.04t}.$$

Since exponential functions are positive, $e^{-0.04t} > 0$ for all numbers t. Thus, since $-800e^{-0.04t} < 0$ for all numbers t, $P''(t)$ is less than 0 for $t = 75$ and we have a maximum.

   The length of the advertising campaign must be 75 days in order to result in maximum profit. ❖

DO EXERCISE 38.

38. Business: *An advertising model.* Solve the problem in Example 15 when the price per unit is $0.80.

EXERCISE SET 4.2

Write an equivalent exponential equation.

1. $\log_2 8 = 3$

2. $\log_3 81 = 4$

3. $\log_8 2 = \frac{1}{3}$

4. $\log_{27} 3 = \frac{1}{3}$

5. $\log_a K = J$

6. $\log_a J = K$

7. $\log_b T = v$

8. $\log_c Y = t$

Write an equivalent logarithmic equation.

9. $e^M = b$

10. $e^t = p$

11. $10^2 = 100$

12. $10^3 = 1000$

13. $10^{-1} = 0.1$

14. $10^{-2} = 0.01$

15. $M^p = V$

16. $Q^n = T$

Given $\log_b 3 = 1.099$ and $\log_b 5 = 1.609$, find each of the following.

17. $\log_b 15$

18. $\log_b \frac{3}{5}$

19. $\log_b \frac{5}{3}$

20. $\log_b \frac{1}{3}$

21. $\log_b \frac{1}{5}$

22. $\log_b \sqrt{b}$

23. $\log_b \sqrt{b^3}$

24. $\log_b 3b$

25. $\log_b 5b$

26. $\log_b 9$

27. $\log_b 25$

28. $\log_b 75$

Given $\ln 4 = 1.3863$ and $\ln 5 = 1.6094$, find each of the following. Do not use a calculator.

29. $\ln 20$

30. $\ln \frac{4}{5}$

31. $\ln \frac{5}{4}$

32. $\ln \frac{1}{5}$

33. $\ln \frac{1}{4}$

34. $\ln 5e$

35. $\ln 4e$

36. $\ln \sqrt{e^6}$

37. $\ln \sqrt{e^8}$

38. $\ln 25$

39. $\ln 16$

40. $\ln 100$

📷 Find the logarithm. Round to six decimal places.

41. $\ln 5894$

42. $\ln 99{,}999$

43. $\ln 0.0182$

44. $\ln 0.00087$

45. $\ln 1.88$

46. $\ln 18.8$

47. $\ln 0.0188$

48. $\ln 0.188$

49. $\ln 906$

50. $\ln 8100$

51. $\ln 0.011$

52. $\ln 0.00056$

Solve for t.

53. $e^t = 100$

54. $e^t = 1000$

55. $e^t = 60$

56. $e^t = 90$

57. $e^{-t} = 0.1$

58. $e^{-t} = 0.01$

59. $e^{-0.02t} = 0.06$

60. $e^{0.07t} = 2$

Differentiate.

61. $y = -6 \ln x$

62. $y = -4 \ln x$

63. $y = x^4 \ln x - \frac{1}{2} x^2$

64. $y = x^5 \ln x - \frac{1}{4} x^4$

65. $y = \dfrac{\ln x}{x^4}$

66. $y = \dfrac{\ln x}{x^5}$

67. $y = \ln \dfrac{x}{4}$ $\left(Hint: \ln \dfrac{x}{4} = \ln x - \ln 4. \right)$

68. $y = \ln \dfrac{x}{2}$

69. $f(x) = \ln (5x^2 - 7)$

70. $f(x) = \ln (7x^3 + 4)$

71. $f(x) = \ln (\ln 4x)$

72. $f(x) = \ln (\ln 3x)$

73. $f(x) = \ln \left(\dfrac{x^2 - 7}{x} \right)$

74. $f(x) = \ln \left(\dfrac{x^2 + 5}{x} \right)$

75. $f(x) = e^x \ln x$

76. $f(x) = e^{2x} \ln x$

77. $f(x) = \ln (e^x + 1)$

78. $f(x) = \ln (e^x - 2)$

79. $f(x) = (\ln x)^2$ (Hint: Use the Extended Power Rule.)

80. $f(x) = (\ln x)^3$

APPLICATIONS

❖ **Business and Economics**

81. *Advertising.* A model for advertising response is given by

$$N(a) = 1000 + 200 \ln a, \quad a \geq 1,$$

where $N(a) =$ the number of units sold and $a =$ the amount spent on advertising, in thousands of dollars.

a) How many units were sold after spending $1000 ($a = 1$) on advertising?

b) Find $N'(a)$ and $N'(10)$.

c) Find the maximum and minimum values, if they exist.

82. *Advertising.* A model for advertising response is given by

$$N(a) = 2000 + 500 \ln a, \quad a \geq 1,$$

where $N(a) =$ the number of units sold and $a =$ the amount spent on advertising, in thousands of dollars.

a) How many units were sold after spending $1000 ($a = 1$) on advertising?

b) Find $N'(a)$ and $N'(10)$.

c) Find the maximum and minimum values, if they exist.

83. *An advertising model.* Solve Example 15 given that the costs of advertising are $2000 per day.

84. *An advertising model.* Solve Example 15 given that the costs of advertising are $4000 per day.

85. *Growth of a stock.* The value V of a stock is modeled by

$$V(t) = \$58(1 - e^{-1.1t}) + \$20,$$

where V is the value of the stock after time t, in months.

a) Find $V(1)$ and $V(12)$.

b) Find $V'(t)$.

c) After how many months will the value of the stock be $75?

86. *Marginal revenue.* The demand function for a certain product is given by

$$x = D(p) = 800e^{-0.125p}.$$

Recall that total revenue is given by $R(p) = p\, D(p)$.

a) Find $R(p)$.

b) Find the marginal revenue, $R'(p)$.

c) At what price per unit p will the revenue be maximum?

❖ Life and Physical Sciences

87. *Acceptance of a new medicine.* The percentage P of doctors who accept a new medicine is given by

$$P(t) = 1 - e^{-0.2t},$$

where t = the time, in months.

a) Find $P(1)$ and $P(6)$.

b) Find $P'(t)$.

c) How many months will it take for 90% of the doctors to accept the new medicine?

88. *The Reynolds number.* For many kinds of animals, the Reynolds number R is given by

$$R = A \ln r - Br,$$

where A and B are positive constants and r is the radius of the aorta. Find the maximum value of R.

❖ Social Sciences

89. *Forgetting.* Students in college botany took a final exam. They took equivalent forms of the exam in monthly intervals thereafter. The average score $S(t)$, in percent, after t months was found to be given by

$$S(t) = 68 - 20 \ln (t + 1), \quad t \geq 0.$$

a) What was the average score when they initially took the test, $t = 0$?

b) What was the average score after 4 months?

c) What was the average score after 24 months?

d) What percentage of the initial score did they retain after 2 years (24 months)?

e) Find $S'(t)$.

f) Find the maximum and minimum values, if they exist.

90. *Forgetting.* Students in college zoology took a final exam. They took equivalent forms of the exam in monthly intervals thereafter. The average score $S(t)$, in percent, after t months was found to be given by

$$S(t) = 78 - 15 \ln (t + 1), \quad t \geq 0.$$

a) What was the average score when they initially took the test, $t = 0$?

b) What was the average score after 4 months?

c) What was the average score after 24 months?

d) What percentage of the initial score did they retain after 2 years (24 months)?

e) Find $S'(t)$.

f) Find the maximum and minimum values, if they exist.

91. *Walking speed.* Bornstein and Bornstein found in a study that the average walking speed v of a person living in a city of population p, in thousands, is given by

$$v(p) = 0.37 \ln p + 0.05,$$

where v is in feet per second.

a) The population of Seattle is 531,000. What is the average walking speed of a person living in Seattle? [*Hint:* Find $v(531)$.]

b) The population of New York is 7,900,000. What is the average walking speed of a person living in New York?

c) Find $v'(p)$. Interpret $v'(p)$.

92. *The Hullian learning model.* A typist learns to type W words per minute after t weeks of practice, where W is

given by

$$W(t) = 100(1 - e^{-0.3t}).$$

a) Find $W(1)$ and $W(8)$.

b) Find $W'(t)$.

c) After how many weeks will the typist's speed be 95 words per minute?

SYNTHESIS EXERCISES

Differentiate.

93. $y = (\ln x)^{-4}$

94. $y = (\ln x)^n$

95. $f(t) = \ln (t^3 + 1)^5$

96. $f(t) = \ln (t^2 + t)^3$

97. $f(x) = [\ln (x + 5)]^4$

98. $f(x) = \ln [\ln (\ln 3x)]$

99. $f(t) = \ln [(t^3 + 3)(t^2 - 1)]$

100. $f(t) = \ln \dfrac{1 - t}{1 + t}$

101. $y = \ln \dfrac{x^5}{(8x + 5)^2}$

102. $y = \ln \sqrt{5 + x^2}$

103. $f(t) = \dfrac{\ln t^2}{t^2}$

104. $f(x) = \dfrac{1}{5}x^5 \left(\ln x - \dfrac{1}{5} \right)$

105. $y = \dfrac{x^{n+1}}{n + 1} \left(\ln x - \dfrac{1}{n + 1} \right)$

106. $y = \dfrac{x \ln x - x}{x^2 + 1}$

107. $y = \ln (t + \sqrt{1 + t^2})$

108. $f(x) = \ln \dfrac{1 + \sqrt{x}}{1 - \sqrt{x}}$

109. $f(x) = \ln [\ln x]^3$

110. $f(x) = \dfrac{\ln x}{1 + (\ln x)^2}$

111. Find $\lim\limits_{h \to 0} \dfrac{\ln (1 + h)}{h}$.

112. ▦ Which is larger, e^π or π^e?

113. ▦ Find $\sqrt[e]{e}$. Compare it to other expressions of the type $\sqrt[x]{x}$, $x > 0$. What can you conclude?

Solve for t.

114. $P = P_0 e^{-kt}$

115. $P = P_0 e^{kt}$

▦ Use input–output tables to find the limit.

116. $\lim\limits_{x \to 1} \ln x$

117. $\lim\limits_{x \to \infty} \ln x$

Verify each of the following.

118. $\log x = \dfrac{\ln x}{\ln 10} \approx 0.4343 \ln x$

119. $\ln x = \dfrac{\log x}{\log e} \approx 2.3026 \log x$

COMPUTER-GRAPHING CALCULATOR EXERCISES

120. Find the minimum value of $f(x) = x \ln x$.

121. Find the minimum value of $f(x) = x^2 \ln x$.

122. Sketch the graph of $y = \dfrac{\ln x}{x^2}$.

THE CALCULUS EXPLORER:
Derivatives

Use the program to graph each of the following functions f and its derivative f': $f(x) = \ln x$, $f(x) = x \ln x$, $f(x) = x^2 \ln x$, and $f(x) = (\ln x)/x^2$.

4.3

APPLICATIONS: THE UNINHIBITED GROWTH MODEL, $dP/dt = kP$

Exponential Growth

Consider the function

$$f(x) = 2e^{3x}.$$

Differentiating, we get

$$f'(x) = 3 \cdot 2e^{3x} = 3 \cdot f(x).$$

Graphically, this says that the derivative, or slope of the tangent line, is simply the constant 3 times the function value (Fig. 1).

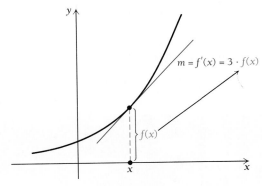

FIGURE 1

DO EXERCISE 1.

In general, we have the following.

THEOREM 9

A function $y = f(x)$ satisfies the equation

$$\frac{dy}{dx} = ky \qquad [f'(x) = k \cdot f(x)]$$

if and only if

$$y = ce^{kx} \qquad [f(x) = ce^{kx}]$$

for some constant c.

OBJECTIVES

a) Find the function that satisfies the equation

$$\frac{dP}{dt} = kP.$$

b) Given a growth rate, find the doubling time.

c) Given a doubling time, find the growth rate.

d) Solve applied problems involving exponential growth.

1. a) Differentiate $y = 5e^{4x}$.

   b) Express dy/dx in terms of y.

2. Find the function that satisfies each equation.

a) $\dfrac{dN}{dt} = kN$

b) $f'(t) = k \cdot f(t)$

No matter what the variables, you should be able to write the function that satisfies what is called a *differential equation*.

EXAMPLE 1   Find the function that satisfies the equation

$$\frac{dA}{dt} = kA.$$

Solution   The function is $A = ce^{kt}$, or $A(t) = ce^{kt}$.   ❖

EXAMPLE 2   Find the function that satisfies the equation

$$\frac{dP}{dt} = kP.$$

Solution   The function is $P = ce^{kt}$, or $P(t) = ce^{kt}$.   ❖

EXAMPLE 3   Find the function that satisfies the equation

$$f'(Q) = k \cdot f(Q).$$

Solution   The function is $f(Q) = ce^{kQ}$.   ❖

DO EXERCISE 2.

What will the world population be in 2000?

The equation

$$\frac{dP}{dt} = kP, \quad k > 0 \qquad [P'(t) = k \cdot P(t), \quad k > 0]$$

is the basic model of uninhibited population growth, whether it be a population of humans, a bacteria culture, or money invested at interest compounded continuously. Neglecting special inhibiting and stimulating factors, we know that a population normally reproduces itself at a rate proportional to its size, and this is exactly what the equation $dP/dt = kP$ says. The solution of the equation is

$$P(t) = ce^{kt}, \tag{1}$$

where t = the time. At $t = 0$, we have some "initial" population $P(0)$ that we will represent by P_0. We can rewrite Eq. (1) in terms of P_0 as

$$P_0 = P(0) = ce^{k \cdot 0} = ce^0 = c \cdot 1 = c.$$

Thus, $P_0 = c$, so we can express $P(t)$ as

$$P(t) = P_0 e^{kt}.$$

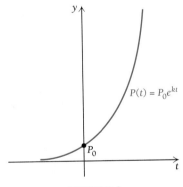

FIGURE 2

Its graph is the curve in Fig. 2, which shows how uninhibited growth produces a "population explosion."

DO EXERCISE 3.

The constant k is called the **rate of exponential growth**, or simply the **growth rate**. This is not the rate of change of the population size, which is

$$\frac{dP}{dt} = kP,$$

3. *Exploratory exercises: Growth.* Use a sheet of $8\frac{1}{2}'' \times 11''$ paper. Cut it into two equal pieces. Then cut these into four equal pieces. Then cut these into eight equal pieces, and so on, performing five cutting steps.

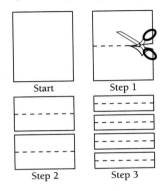

a) Place all the pieces in a stack and measure the thickness.

b) A piece of paper is typically 0.004 in. thick. Check the calculation in part (a) by completing this table.

	t	$0.004 \cdot 2^t$
Start	0	$0.004 \cdot 2^0$, or 0.004
Step 1	1	$0.004 \cdot 2^1$, or 0.008
Step 2	2	$0.004 \cdot 2^2$, or 0.016
Step 3	3	
Step 4	4	
Step 5	5	

c) Compute the thickness of the paper (in miles) after 25 steps.

4. *Business: Interest compounded continuously.* Suppose an amount P_0 is invested in a savings account in which interest is compounded continuously at 13% per year. That is, the balance P grows at the rate given by

$$\frac{dP}{dt} = 0.13P.$$

a) Find the function that satisfies the equation. List it in terms of P_0 and 0.13.

b) Suppose $1000 is invested. What is the balance after 1 year?

c) After what period of time will an investment of $1000 double itself?

but the constant by which P must be multiplied in order to get its rate of change. It is thus a different use of the word *rate*. It is like the *interest rate* paid by a bank. If the interest rate is 12%, or 0.12, we do not mean that your bank balance P is growing at the rate of 0.12 dollars per year, but at the rate of $0.12P$ dollars per year. We therefore express the rate as 12% per year, rather than 0.12 dollars per year. We could say that the rate is 0.12 dollars *per dollar* per year. When interest is compounded continuously, the interest rate is a true exponential growth rate.

EXAMPLE 4   *Business: Interest compounded continuously.* Suppose an amount P_0 is invested in a savings account where interest is compounded continuously at 12% per year. That is, the balance P grows at the rate given by

$$\frac{dP}{dt} = 0.12P.$$

a) Find the function that satisfies the equation. List it in terms of P_0 and 0.12.

b) Suppose $100 is invested. What is the balance after 1 year?

c) After what period of time will an investment of $100 double itself?

Solution

a) $P(t) = P_0 e^{0.12t}$

b) $P(1) = \quad e^{0.12(1)} = 100e^{0.12} = 100(1.127497)$
$$\approx \$112.75$$

c) We are looking for that time T for which $P(T) = \$200$. The number T is called the **doubling time**. To find T, we solve the equation

$$200 = 100e^{0.12 \cdot T}$$
$$2 = e^{0.12T}.$$

We use natural logarithms to solve this equation:

$$\ln 2 = \ln e^{0.12T}$$
$$\ln 2 = 0.12T \qquad \text{By P5: } \ln e^k = k$$
$$\frac{\ln 2}{0.12} = T$$
$$\frac{0.693147}{0.12} = T$$
$$5.8 \approx T.$$

Thus, $100 will double itself in 5.8 years.

DO EXERCISE 4.

Let us consider another way of developing the formula

$$P(t) = P_0 e^{kt}$$

by starting with

$$A = P\left(1 + \frac{i}{n}\right)^{nt}$$

and compounding interest continuously. Let $P = P_0$ and $i = k$ to obtain

$$A = P_0\left(1 + \frac{k}{n}\right)^{nt}.$$

We are interested in what happens as n gets very large, that is, as n approaches ∞. To determine this limit, we first let

$$\frac{k}{n} = \frac{1}{h}.$$

Then

$$hk = n \quad \text{and} \quad h = \frac{n}{k},$$

which shows that since k is a positive constant, as n gets large, so does h. To find a formula for continuously compounded interest, we evaluate the following limit:

Under ideal conditions, the growth rate of this rabbit population might be 11.7% per day. When will this population of rabbits double?

$$P(t) = \lim_{n \to \infty} \left[P_0\left(1 + \frac{k}{n}\right)^{nt} \right]$$ Letting the number of compounding periods become infinite

$$= P_0 \lim_{h \to \infty} \left[\left(1 + \frac{1}{h}\right)^{hkt} \right]$$ The limit of a constant times a function is the constant times the limit. We also substitute $1/h$ for k/n and hk for n. Also, $h \to \infty$ because $n \to \infty$.

$$= P_0 \left[\lim_{h \to \infty} \left(1 + \frac{1}{h}\right)^{h} \right]^{kt}$$ The limit of a power is the power of the limit: a form of L2 in Section 2.2.

$$= P_0 [e]^{kt}. \quad \text{Theorem 1}$$

We can find a general expression relating the growth rate k and the doubling time T by solving the following equation:

$$2P_0 = P_0 e^{kT}$$
$$2 = e^{kT} \quad \text{Dividing by } P_0$$
$$\ln 2 = \ln e^{kT}$$
$$\ln 2 = kT.$$

5. Complete this table relating growth rate k and doubling time T.

Growth rate k (% per year)	Doubling time T (in years)
2%	
	10
14%	
	15
1%	

THEOREM 10

The *growth rate k* and the *doubling time T* are related by

$$kT = \ln 2 = 0.693147,$$

or

$$k = \frac{\ln 2}{T} = \frac{0.693147}{T},$$

and

$$T = \frac{\ln 2}{k} = \frac{0.693147}{k}.$$

Note that this relationship between k and T does not depend on P_0.

EXAMPLE 5   *Business: Doubling time.* At one time in Canada, a bank advertised that it would double your money in 6.6 years. What was the interest rate on such an account, assuming interest were compounded continuously?

Solution

$$k = \frac{\ln 2}{T} = \frac{0.693147}{6.6} \approx 0.105 = 10.5\% \qquad ❖$$

DO EXERCISE 5.

EXAMPLE 6   *Life science: World population growth.* The population of the world was 5.2 billion in 1990. On the basis of data available at that time, it was estimated that the population P was growing exponentially at the rate of 1.6% per year. That is, $dP/dt = 0.016P$, where t = the time, in years, from 1990. (To facilitate computations, we assume that the population was 5.2 billion at the beginning of 1990.)

a) Find the function that satisfies the equation. Assume that $P_0 = 5.2$ and $k = 0.016$.

b) Estimate the world population in 2000 ($t = 10$).

c) After what period of time will the population be double that in 1990?

Solution

a) $P(t) = 5.2e^{0.016t}$

b) $P(10) = 5.2e^{0.016(10)} = 5.2e^{0.16} = 5.2(1.173511)$

    ≈ 6.1 billion

c) $T = \dfrac{\ln 2}{k} = \dfrac{0.693147}{0.016} \approx 43.3 \text{ yr}$

Thus, according to this model, the 1990 population will double in the year 2033. (No wonder ecologists are alarmed!) ❖

DO EXERCISE 6.

The Rule of 70

The relationship between doubling time T and interest rate k is the basis of a rule often used in the investment world, called the *Rule of 70*. To estimate how long it will take to double your money at varying rates of return, divide 70 by the rate of return. To see how this works, let the interest rate $k = r\%$. Then

$$T = \frac{\ln 2}{k} = \frac{0.693147}{r\%} = \frac{0.693147}{r \times 0.01}$$

$$= \frac{0.693147}{r \times 0.01} \cdot \frac{100}{100} = \frac{69.3147}{r} \approx \frac{70}{r}.$$

Modeling Other Phenomena

EXAMPLE 7 *Life science: Alcohol absorption and the risk of having an accident.* Extensive research has provided data relating the risk R (in percent) of having an automobile accident to the blood alcohol level b (in percent). Note in Fig. 3 that these data are not a perfect fit (see the part of the graph between $b = 0$ and $b = 0.05$), but we can approximate the data with an exponential function. The modeling assumption is that the rate of change of the risk R with respect to the blood alcohol level b is given by

$$\frac{dR}{db} = kR.$$

a) Find the function that satisfies the equation. Assume that $R_0 = 1\%$.

b) Find k using the data point $R(0.14) = 20$. (This is how one might fit the data to an exponential equation.)

6. *Life science: Population growth.* The population of the United States was 241 million in 1986. It was estimated that the population P was growing exponentially at the rate of 0.9% per year. That is,

$$\frac{dP}{dt} = 0.009P,$$

where t = the time in years.

a) Find the function that satisfies the equation. Assume that $P_0 = 241$ and $k = 0.009$.

b) Estimate the population of the United States in 1998 ($t = 12$).

c) After what period of time will the population be double that in 1986?

Some myths about alcohol (Indianapolis Alcohol Safety Action Project). It's a fact—the blood alcohol concentration (BAC) in the human body is measurable. And there's no cure for its effect on the central nervous system except time. It takes time for the body's metabolism to recover. That means a cup of coffee, a cold shower, and fresh air can't erase the effect of several drinks.

There are variables, of course: a person's body weight, how many drinks have been consumed in a given time, how much has been eaten, and so on. These account for different BAC levels. But the myth that some people can "handle their liquor" better than others is a gross rationalization—especially when it comes to driving. Some people can act more sober than others. But an automobile doesn't act; it reacts.

c) Rewrite $R(b)$ in terms of k.

d) At what blood alcohol level will the risk of having an accident be 100%? Round to the nearest hundredth.

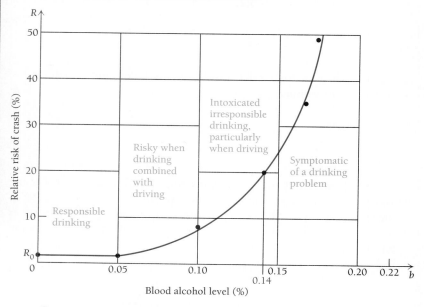

Number of 1-oz drinks of 86-proof whiskey for 160-lb man within 2 hrs of eating

FIGURE 3

Solution

a) Since both R and b are percents, we omit the % symbol for ease of computation. The solution is

$$R(b) = e^{kb}, \quad \text{since } R_0 = 1.$$

b) We solve this equation for k, using natural logarithms:

$$20 = e^{k(0.14)} = e^{0.14k}$$

$$\ln 20 = \ln e^{0.14k}$$

$$\ln 20 = 0.14k$$

$$\frac{\ln 20}{0.14} = k$$

$$\frac{2.995732}{0.14} = k$$

$$21.4 \approx k. \qquad \text{Rounded to the nearest tenth}$$

c) $R(b) = e^{21.4b}$

d) We substitute 100 for $R(b)$ and solve the equation for b:

$$100 = e^{21.4b}$$
$$\ln 100 = \ln e^{21.4b}$$
$$\ln 100 = 21.4b$$
$$\frac{\ln 100}{21.4} = b$$
$$\frac{4.605170}{21.4} = b$$
$$0.22 \approx b. \qquad \text{Rounded to the nearest hundredth}$$

Calculator note: The calculations done in this problem can be performed more conveniently on your calculator if you do not stop to round. For example, in part (b) we find ln 20 and divide by 0.14, obtaining 21.39808767. . . . We then use that value for k in part (d). Answers will be found that way in the exercises. You may note some variance in the last one or two decimal places if you round as you go.

Thus when the blood alcohol level is 0.22%, according to this model, the risk of an accident is 100%. From the graph, we see that this would occur after 12 1-oz drinks of 86-proof whiskey. "Theoretically," the model tells us that after 12 drinks of whiskey, one is "sure" to have an accident. This might be questioned in reality, since a person who has had 12 drinks might not be able to drive at all. ❖

DO EXERCISE 7.

Models of Limited Growth

The growth model $P(t) = P_0e^{kt}$ has many applications, as we have seen thus far in this section. It seems reasonable that there can be factors that prevent a population from exceeding some limiting value L—perhaps a limitation on food, living space, or other natural resources. One model of such growth is

$$P(t) = \frac{a}{1 + be^{-kt}},$$

which is called the *logistic equation*.

7. *Business: Cost of a 60-second commercial during the Super Bowl.* Past data on the cost of a 60-sec commercial during the Super Bowl are given below.

Year	Cost of a 60-sec TV commercial during the Super Bowl
1967	$80,000
1983	$800,000
1991	$1,600,000

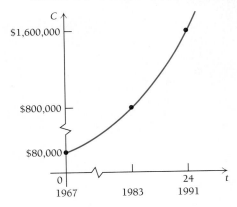

It appears that we can fit an exponential function to the data. We accept the modeling assumption that the rate of change of the cost of a Super Bowl commercial C with respect to time t is given by $dC/dt = kC$.

a) Find the solution to the equation, assuming $C_0 = \$80,000$ [at $t = 0$ (1967), $C = \$80$ (in thousands)].

b) Find k using the data point $C(24) = \$1600$ thousand. That is, in 1985, $1600 thousand, or $1,600,000, was the cost of a 60-sec commercial.

c) Rewrite $C(t)$ in terms of k.

d) What will be the cost of a 60-sec commercial in 1995?

EXAMPLE 8 *Life science: Limited population growth.* A ship carrying 1000 passengers has the misfortune to be shipwrecked on a small island from which the passengers are never rescued. The natural resources of the island limit the growth of the population to a *limiting value* of 5780, to which the population gets closer and closer but which it never reaches. The population of the island after time t, in years, is given by the logistic equation

$$P(t) = \frac{5780}{1 + 4.78e^{-0.4t}}.$$

a) Find the population after 0 years, 1 year, 2 years, 5 years, 10 years, and 20 years.

b) Find the rate of change $P'(t)$.

c) Sketch a graph of the function.

Solution

a) We use a calculator to find the function values, listing them in a table, as follows.

t	$P(t)$, approximately
0	1000
1	1375
2	1836
5	3510
10	5315
20	5771

b) We find the rate of change $P'(t)$ as follows, using the Quotient Rule:

$$P'(t) = \frac{(1 + 4.78e^{-0.4t}) \cdot 0 - [(-0.4)(4.78)e^{-0.4t}](5780)}{(1 + 4.78e^{-0.4t})^2}$$

$$= \frac{11{,}051.36e^{-0.4t}}{(1 + 4.78e^{-0.4t})^2}.$$

c) We consider the curve-sketching procedure discussed in Section 3.2. The derivative $P'(t)$ exists for all real numbers t; thus there are no

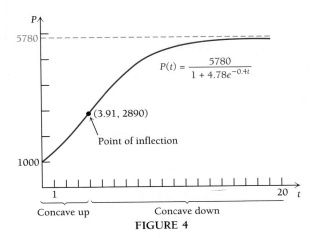

$$P(t) = \frac{5780}{1 + 4.78e^{-0.4t}}$$

(3.91, 2890)

Point of inflection

1000

1

20 *t*

Concave up   Concave down

FIGURE 4

critical points for which the derivative does not exist. The equation $P'(t) = 0$ holds only when the numerator is 0. But the numerator is positive for all real numbers t. Thus the function has no critical points, and hence no relative extrema.

Since $e^{-0.4t}$ is positive for all real numbers t, $P'(t)$ is positive for all real numbers t. Thus, P is increasing over the entire real line. Though we will not develop it here, the second derivative can be used to show that the graph has a point of inflection at (3.91, 2890). This function is concave up on the interval (0, 3.91) and concave down on the interval (3.91, ∞). The graph is the S-shaped curve shown in Fig. 4.   ❖

DO EXERCISE 8.

Another model of limited growth is provided by the function

$$P(t) = L(1 - e^{-kt}),$$

which is shown graphed in Fig. 5. This function also increases over the entire interval [0, ∞).

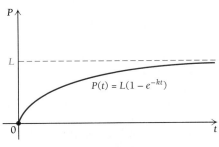

$$P(t) = L(1 - e^{-kt})$$

FIGURE 5

8. *Life science: Spread of an epidemic.* In a town whose population is 2000, a disease *Baditch* creates an epidemic. The number of people N infected t days after the disease begins is given by the function

$$N(t) = \frac{2000}{1 + 19.9e^{-0.6t}}.$$

a) How many are initially infected with the disease ($t = 0$)?

b) Find the number infected after 2 days, 5 days, 8 days, 12 days, and 16 days.

c) Find the rate of change $N'(t)$.

d) Sketch a graph of the function.

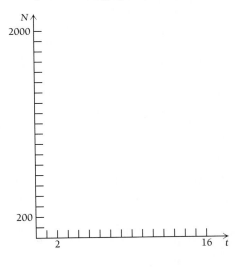

1. Find the function that satisfies the equation $dQ/dt = kQ$.

2. Find the function that satisfies the equation $dR/dt = kR$.

APPLICATIONS

❖ **Business and Economics**

3. *Compound interest.* Suppose P_0 is invested in a savings account in which interest is compounded continuously at 9% per year. That is, the balance P grows at the rate given by

$$\frac{dP}{dt} = 0.09P.$$

a) Find the function that satisfies the equation. List it in terms of P_0 and 0.09.

b) Suppose $1000 is invested. What is the balance after 1 year? after 2 years?

c) When will an investment of $1000 double itself?

4. *Compound interest.* Suppose P_0 is invested in a savings account in which interest is compounded continuously at 10% per year. That is, the balance P grows at the rate given by

$$\frac{dP}{dt} = 0.10P.$$

a) Find the function that satisfies the equation. List it in terms of P_0 and 0.10.

b) Suppose $20,000 is invested. What is the balance after 1 year? after 2 years?

c) When will an investment of $20,000 double itself?

5. *Annual interest rate.* A bank advertises that it compounds interest continuously and that it will double your money in 10 years. What is its annual interest rate?

6. *Annual interest rate.* A bank advertises that it compounds interest continuously and that it will double your money in 12 years. What is its annual interest rate?

7. *Franchise expansion.* A national hamburger firm is selling franchises throughout the country. The president estimates that the number of franchises N will increase

at the rate of 10% per year, that is,

$$\frac{dN}{dt} = 0.10N.$$

a) Find the function that satisfies the equation. Assume that the number of franchises at $t = 0$ is 50.

b) How many franchises will there be in 20 years?

c) After what period of time will the initial number of 50 franchises double?

8. *Franchise expansion.* Pizza, Unltd., a national pizza firm, is selling franchises throughout the country. The president estimates that the number of franchises N will increase at the rate of 15% per year, that is,

$$\frac{dN}{dt} = 0.15N.$$

a) Find the function that satisfies the equation. Assume that the number of franchises at $t = 0$ is 40.

b) How many franchises will there be in 20 years?

c) After what period of time will the initial number of 40 franchises double?

9. *Oil demand.* The growth rate of the demand for oil in the United States is 10% per year. When will the demand be double that of 1990?

10. *Coal demand.* The growth rate of the demand for coal in the world is 4% per year. When will the demand be double that of 1992?

11. *Value of a Van Gogh painting.* The Van Gogh painting "Irises," shown below, sold for $84,000 in 1947, but sold again for $53,900,000 in 1987. Assuming the exponential model:

a) Find the value k ($V_0 = \$84,000$), and write the function.

b) Estimate the value of the painting in the year 1997.

c)   What is the doubling time for the value of the painting?

d)   How long after 1947 will the value of the painting be $1 billion?

Van Gogh's "Irises," a 28-by-32-inch oil on canvas.

12. *Cost of a double-dip ice cream cone.* In 1970 the cost of a double-dip ice cream cone was 52¢. In 1978 it was 66¢. Assuming the exponential model:

a)   Find the value k ($P_0 = 52$), and write the function.

b)   Estimate the cost of a cone in 1994.

c)   After what period of time will the cost of a cone be twice that of 1970?

13. *Consumer price index.* The *consumer price index* compares the costs of goods and services over various years, where 1967 is used as a base (P_0). The same goods and services that cost $100 in 1967 cost $184.50 in 1977. Assuming the exponential model:

a)   Find the value k ($P_0 = \$100$), and write the function.

b)   Estimate what the same goods and services will cost in 1997.

c)   After what period of time did the same goods and services cost double that of 1967?

14. *Job opportunities.* It is estimated that there were 714,000 accountants employed in 1972 and 935,000 in 1985. Assuming the exponential model:

a)   Find the value k ($P_0 = 714,000$), and write the function.

b)   Estimate the number of accountants needed in 1996.

c)   After what period of time will the need for accountants be double that of 1972?

15. *Value of Manhattan Island.* Peter Minuit of the Dutch West India Company purchased Manhattan Island from the Indians in 1626 for $24 worth of merchandise. Assuming an exponential rate of inflation of 8%, how much will Manhattan be worth in 1998?

16. *Cost of a first-class postage stamp.* The cost of a first-class postage stamp in 1962 was 4¢. In 1991, it was 29¢. This was exponential growth. What was the growth rate? What will be the cost of a first-class postage stamp in 1996? in 2000?

17. *Average salary of major-league baseball players.* The average salary of major-league baseball players in 1970 was $29,303. In 1985 the average salary was $363,000. This was exponential growth. What was the growth rate? What will the average salary be in 1997? in 2000?

18. *Cost of a Hershey bar.* The cost of a Hershey bar in 1962 was $0.05, and was increasing at an exponential growth rate of 9.7%. What will the cost of a Hershey bar be in 1996? in 1998?

19. *Effect of advertising.* A company introduces a new product on a trial run in a city. They advertised the

product on television and found that the percentage P of people who bought the product after t ads were run satisfied the function

$$P(t) = \frac{100\%}{1 + 49e^{-0.13t}}.$$

a) What percentage buy the product without seeing the ad $(t = 0)$?

b) What percentage buy the product after the ad is run 5 times, 10 times, 20 times, 30 times, 50 times, 60 times?

c) Find the rate of change $P'(t)$.

d) Sketch a graph of the function.

❖ **Life and Physical Sciences**

20. *Population growth.* The growth rate of the population of Alaska is 2.8% per year (one of the highest of the fifty states). What is the doubling time?

21. *Population growth.* The growth rate of the population of Central America is 3.5% per year (one of the highest in the world). What is the doubling time?

22. *Population growth.* The population of Europe west of the USSR was 430 million in 1961. It was estimated that the population was growing exponentially at the rate of 1% per year, that is,

$$\frac{dP}{dt} = 0.01P.$$

a) Find the function that satisfies the equation. Assume that $P_0 = 430$ and $k = 0.01$.

b) Estimate the population of Europe in 1996.

c) After what period of time will the population be double that of 1961?

23. *Population growth.* The population of the USSR was 209 million in 1959. It was estimated that the population P was growing exponentially at the rate of 1% per year, that is,

$$\frac{dP}{dt} = 0.01P.$$

a) Find the function that satisfies the equation. Assume that $P_0 = 209$ and $k = 0.01$.

b) Estimate the population of the USSR in 1999.

c) After what period of time will the population be double that of 1959?

24. *Blood alcohol level.* In Example 7 (on alcohol absorption), at what blood alcohol level will the risk of an accident be 90%?

25. *Blood alcohol level.* In Example 7, at what blood alcohol level will the risk of an accident be 80%?

26. *Population growth.* The population of Austin, Texas, was 253 thousand in 1970. In 1980, it was 345 thousand. Assuming the exponential model:

a) Find the value of k ($P_0 = 253$), and write the function.

b) Estimate the population of Austin in 2000.

27. *Population growth.* The population of Tempe, Arizona, was 107 thousand in 1980. In 1984, it was 118 thousand. Assuming the exponential model:

a) Find the value k ($P_0 = 107$), and write the function.

b) Estimate the population of Tempe in 1996.

28. *Population growth in the Virgin Islands.* The U.S. Virgin Islands have one of the highest growth rates in the world, 9.6%. In 1970, the population was 75,150. The land area of the Virgin Islands is 3,097,600 square yards. Assuming this growth rate continues and is exponential, after what period of time will the population of the Virgin Islands be such that there is one person for every square yard of land?

29. *Limited population growth.* A lake is stocked with 400 fish of a new variety. The size of the lake, the availability of food, and the number of other fish restrict growth in the lake to a *limiting value* of 2500. The population of fish in the lake after time t, in months, is given by

$$P(t) = \frac{2500}{1 + 5.25e^{-0.32t}}.$$

a) Find the population after 0 months, 1 month, 5 months, 10 months, 15 months, and 20 months.

b) Find the rate of change $P'(t)$.

c) Sketch a graph of the function.

30. *Bicentennial growth of the United States.* The population of the United States in 1776 was about 2,508,000. In its bicentennial year the population was about 216,000,000. Assuming the exponential model, what

was the growth rate of the United States through its bicentennial year?

❖ Social Sciences

31. *Diffusion of information.* Pharmaceutical firms spend an immense amount of money in order to test a new medication. After the drug is approved by the Federal Drug Administration, it still takes time for physicians to fully accept and make use of the medication. The use approaches a *limiting value* of 100%, or 1, after time t, in months. Suppose that for a new cancer medication the percentage P of physicians using the product after t months is given by

$$P(t) = 100\%(1 - e^{-0.4t}).$$

a) What percentage of doctors have accepted the medication after 0 months? 1 month? 2 months? 3 months? 5 months? 12 months? 16 months?

b) Find the rate of change $P'(t)$.

c) Sketch a graph of the function.

32. *Hullian learning model.* The Hullian learning model asserts that the probability p of mastering a task after t learning trials is given by

$$p(t) = 1 - e^{-kt},$$

where k is a constant that depends on the task to be learned. Suppose a new task is introduced in a factory on an assembly line. For that particular task, the constant $k = 0.28$.

a) What is the probability of learning the task after 1 trial? 2 trials? 5 trials? 11 trials? 16 trials? 20 trials?

b) Find the rate of change $p'(t)$.

c) Sketch a graph of the function.

SYNTHESIS EXERCISES

Business: Effective annual yield. Suppose $100 is invested at 12% compounded continuously for 1 year. We know from Example 4 that the balance will be $112.75. This is the same as if $100 were invested at 12.75% and compounded once a year (simple interest). The 12.75% is called the *effective annual yield*. In general, if P_0 is invested at $k\%$ compounded continuously, then the effective annual yield is that number i satisfying $P_0(1 + i) = P_0 e^k$. Then $1 + i = e^k$, or

Effective annual yield $= i = e^k - 1$.

33. An amount is invested at 14% per year compounded continuously. What is the effective annual yield?

34. An amount is invested at 8% per year compounded continuously. What is the effective annual yield?

35. The effective annual yield on an investment is 9.42% compounded continuously. At what rate was it invested?

36. The effective annual yield on an investment is 10.52% compounded continuously. At what rate was it invested?

37. Find an expression relating the growth rate k and the *tripling time* T_3.

38. Find an expression relating the growth rate k and the *quadrupling time* T_4.

39. Gather data concerning population growth in your city. Estimate the population in 1996; in 2010.

40. A quantity Q_1 grows exponentially with a doubling time of 1 year. A quantity Q_2 grows exponentially with a doubling time of 2 years. If the initial amounts of Q_1 and Q_2 are the same, when will Q_1 be twice the size of Q_2?

41. A growth rate of 100% per day corresponds to what exponential growth rate per hour?

42. Show that any two measurements of an exponentially growing population will determine k. That is, show that if y has the values y_1 at t_1 and y_2 at t_2, then

$$k = \frac{\ln(y_2/y_1)}{t_2 - t_1}.$$

OBJECTIVES

a) Find the function that satisfies the equation

$$\frac{dP}{dt} = -kP.$$

b) Given a decay rate, find the half-life.

c) Given a half-life, find the decay rate.

d) Solve applied problems involving decay.

e) Solve applied problems involving Newton's law of cooling.

1. Using the same set of axes, graph $y = e^{2x}$ and $y = e^{-2x}$.

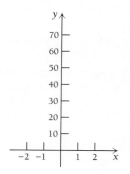

4.4

APPLICATIONS: DECAY

DO EXERCISE 1.

In the equation of population growth $dP/dt = kP$, the constant k is actually given by

$$k = (\text{Birth rate}) - (\text{Death rate}).$$

Thus a population "grows" only when the *birth rate* is greater than the *death rate*. When the birth rate is less than the death rate, k will be negative so the population will be decreasing, or "decaying," at a rate proportional to its size. For convenience in our computations, we will express such a negative value as $-k$, where $k > 0$. The equation

$$\frac{dP}{dt} = -kP, \quad \text{where } k > 0,$$

shows P to be *decreasing* as a function of time, and the solution

$$P(t) = P_0 e^{-kt}$$

shows it to be decreasing exponentially (Fig. 2). This is called **exponential decay**. The amount present initially at $t = 0$ is again P_0.

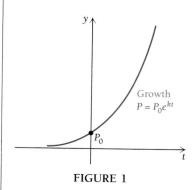

FIGURE 1

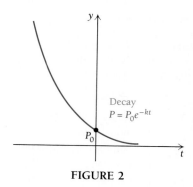

FIGURE 2

Radioactive Decay

Radioactive elements decay exponentially; that is, they disintegrate at a rate that is proportional to the amount present.

EXAMPLE 1   *Life science: Decay.* Strontium-90 has a decay rate of 2.8% per year. The rate of change of an amount N is given by

$$\frac{dN}{dt} = -0.028N.$$

a)   Find the function that satisfies the equation in terms of N_0 (the amount present at $t = 0$).

b)   Suppose 1000 grams of strontium-90 is present at $t = 0$. How much will remain after 100 years?

c)   After how long will half of the 1000 grams remain?

Solution

a) $N(t) = N_0 e^{-0.028t}$

b) $N(100) = 1000 e^{-0.028(100)} = 1000 e^{-2.8}$

$$= 1000(0.060810) \qquad \boxed{}$$

$$\approx 60.8 \text{ grams}$$

c)   We are asking, "At what time T will $N(T) = 500$?" The number T is called the **half-life**. To find T, we solve the equation

$$500 = 1000 e^{-0.028T}$$

$$\frac{1}{2} = e^{-0.028T}$$

$$\ln \frac{1}{2} = \ln e^{-0.028T}$$

$$\ln 1 - \ln 2 = -0.028T$$

$$0 - \ln 2 = -0.028T$$

$$\frac{-\ln 2}{-0.028} = T$$

$$\frac{\ln 2}{0.028} = T$$

$$\frac{0.693147}{0.028} = T$$

$$25 \approx T.$$

Thus the half-life of strontium-90 is 25 years.   ❖

DO EXERCISE 2.

2.   *Life science: Decay rate.* Xenon-133 has a decay rate of 14% per day. The rate of change of an amount N is given by

$$\frac{dN}{dt} = -0.14N.$$

a)   Find the function that satisfies the equation. List it in terms of N_0.

b)   Suppose 1000 grams of xenon-133 is present at $t = 0$. How much will remain after 10 days?

c)   After how long will half of the 1000 grams remain?

We can find a general expression relating the decay rate k and the half-life T by solving the equation

$$\tfrac{1}{2}P_0 = P_0 e^{-kT}$$
$$\tfrac{1}{2} = e^{-kT}$$
$$\ln \tfrac{1}{2} = \ln e^{-kT}$$
$$\ln 1 - \ln 2 = -kT$$
$$0 - \ln 2 = -kT$$
$$-\ln 2 = -kT$$
$$\ln 2 = kT.$$

Again, we have the following.

THEOREM 11

The *decay rate* k and the *half-life* T are related by

$$kT = \ln 2 = 0.693147,$$

or

$$k = \frac{\ln 2}{T} \quad \text{and} \quad T = \frac{\ln 2}{k}.$$

Thus the half-life T depends only on the decay rate k. In particular, it is independent of the initial population size.

The effect of half-life is shown in the radioactive decay curve in Fig. 3.

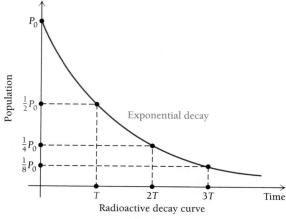

Radioactive decay curve

FIGURE 3

Note that the exponential function gets close to, but never reaches, 0 as t gets larger. Thus, in theory, a radioactive substance never completely decays.

EXAMPLE 2 *Life science: Half-life.* Plutonium, a common product and ingredient of nuclear reactors, is of great concern to those who are against the building of nuclear reactors. Its decay rate is 0.003% per year. What is its half-life?

Solution

$$T = \frac{\ln 2}{k} = \frac{0.693147}{0.00003} \approx 23,100 \text{ years} \qquad \diamondsuit$$

DO EXERCISES 3 AND 4.

EXAMPLE 3 *Life science: Carbon dating.* The radioactive element carbon-14 has a half-life of 5750 years. The percentage of carbon-14 present in the remains of plants and animals can be used to determine age. Archaeologists found that the linen wrapping from one of the Dead Sea Scrolls had lost 22.3% of its carbon-14. How old was the linen wrapping?

Solution

a) Find the decay rate k:

$$k = \frac{\ln 2}{T} = \frac{0.693147}{5750} \approx 0.0001205, \quad \text{or} \quad 0.01205\% \text{ per year.}$$

b) Find the exponential equation for the amount $N(t)$ that remains from an initial amount N_0 after t years:

$$N(t) = N_0 e^{-0.0001205t}.$$

(*Note:* This equation can be used for any subsequent carbon-dating problem.)

c) If the Dead Sea Scrolls lost 22.3% of their carbon-14 from an initial amount P_0, then 77.7% (P_0) is the amount present. To find the age t of the scrolls, we solve the following equation for t:

$$77.7\% \ P_0 = P_0 e^{-0.0001205t}$$
$$0.777 = e^{-0.0001205t}$$
$$\ln 0.777 = \ln e^{-0.0001205t}$$
$$\ln 0.777 = -0.0001205t$$
$$-0.2523149 = -0.0001205t$$
$$\frac{0.2523149}{0.0001205} = t$$
$$2094 \approx t.$$

3. *Life science: Half-life.* The decay rate of cesium-137 is 2.3% per year. What is its half-life?

4. *Life science: Decay rate.* The half-life of barium-140 is 13 days. What is its decay rate?

How can scientists determine that an animal bone has lost 30% of its carbon-14? The assumption is that the percentage of carbon-14 in the atmosphere and in living plants and animals is the same. When a plant or animal dies, the amount of carbon-14 decays exponentially. The scientist burns the animal bone and uses a Geiger counter to determine the percentage of the smoke that is carbon-14. The amount by which this varies from the percentage in the atmosphere indicates how much carbon-14 has been lost.

The process of carbon-14 dating was developed by the American chemist Willard F. Libby in 1952. It is known that the radioactivity in a living plant is 16 disintegrations per gram per minute. Since the half-life of carbon-14 is 5750 years, a dead plant with an activity of 8 disintegrations per gram per minute is 5750 years old, one with an activity of 4 disintegrations per gram per minute is 11,500 years old, and so on. Carbon-14 dating can be used to measure the age of objects from 30,000 to 40,000 years old. Beyond such an age it is too difficult to measure the radioactivity, and some other method would have to be used.

5. *Life science: Carbon dating.* How old is a skeleton that has lost 80% of its carbon-14?

In 1947, a Bedouin youth looking for a stray goat climbed into a cave at Kirbet Qumran on the shores of the Dead Sea near Jericho and came upon earthenware jars containing an incalculable treasure of ancient manuscripts. Shown here are fragments of those so-called Dead Sea Scrolls, a portion of some 600 or so texts found so far and which concern the Jewish books of the Bible. Officials date them before 70 A.D., making them the oldest Biblical manuscripts by 1000 years.

Thus the linen wrapping of the Dead Sea Scrolls is about 2094 years old. ❖

DO EXERCISE 5.

Present Value

A representative of a financial institution is often asked to solve a problem like the following.

EXAMPLE 4 *Business: Present value.* Following the birth of a child, a parent wants to make an initial investment of P_0 that will grow to $10,000 by the child's 20th birthday. Interest is compounded continuously at 8%. What should the initial investment be?

Solution Using the equation $P = P_0 e^{kt}$, we find P_0 such that

$$10,000 = P_0 e^{0.08 \cdot 20}, \quad \text{or} \quad 10,000 = P_0 e^{1.6}.$$

Now

$$\frac{10,000}{e^{1.6}} = P_0, \quad \text{or} \quad 10,000 e^{-1.6} = P_0,$$

and, using a calculator, we have

$$P_0 = 10,000 e^{-1.6} = 10,000(0.201897) = \$2018.97.$$

Thus the parent must deposit $2018.97, which will grow to $10,000 by the child's 20th birthday.   ❖

Economists call $2018.97 the *present value* of $10,000 due 20 years from now at 8% compounded continuously. The process of computing present value is called *discounting*.

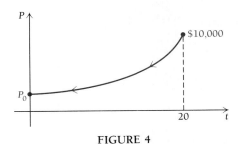

FIGURE 4

Computing present value can be interpreted as exponential decay from the future back to the present, as shown in Fig. 4.

DO EXERCISE 6.

In general, the present value P_0 of an amount P due t years later is found by solving the following equation for P_0:

$$P_0 e^{kt} = P$$

$$P_0 = \frac{P}{e^{kt}} = Pe^{-kt}.$$

THEOREM 12

The *present value* P_0 of an amount P due t years later at interest rate k, compounded continuously, is given by

$$P_0 = Pe^{-kt}.$$

DO EXERCISE 7.

Newton's Law of Cooling

Before you study the following, do the exploratory exercises in the margin.

6. *Business: Present value.* Following the birth of a child, a parent wants to make an initial investment P_0 that will grow to $10,000 by the child's 20th birthday. Interest is compounded continuously at 7.5%. What should the initial investment be?

7. *Business: Present value.* Find the present value of $40,000 due 5 years later at 10%, compounded continuously.

Exploratory exercises: Cooling. Draw a glass of hot tap water. Place a thermometer in the glass and check the temperature. Check the temperature every 30 minutes thereafter. Plot your data on this graph, and connect the points with a smooth curve.

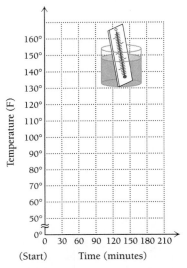

(Start)   Time (minutes)

a)   What was the temperature when you began?

b)   At what temperature does there seem to be a leveling off of the graph?

c)   What is the difference between your answers to parts (a) and (b)?

d)   How does the temperature in part (b) compare with the room temperature?

Newton's Law of Cooling

The temperature T of a cooling object drops at a rate that is proportional to the difference $T - C$, where C is the constant temperature of the surrounding medium. Thus,

$$\frac{dT}{dt} = -k(T - C). \tag{1}$$

The function that satisfies Eq. (1) is

$$T = T(t) = ae^{-kt} + C. \tag{2}$$

We can check this by differentiating:

$$\frac{dT}{dt} = -kae^{-kt} = -k(ae^{-kt}) = -k(T - C).$$

EXAMPLE 5   *Physical science: Cooling.* The temperature of a cup of freshly brewed coffee is 200° and the room temperature is 70°. The temperature of the coffee cools to 190° in 5 min.

a)   What is the temperature after 10 min?

b)   How long does it take the cup of coffee to cool to 90°?

Solution

a)   We first find the value of a in Eq. (2). At $t = 0$, $T = 200°$. We solve the following equation for a:

$$200 = ae^{-k \cdot 0} + 70$$
$$200 = a \cdot 1 + 70$$
$$130 = a.$$

Now we find k using the fact that at $t = 5$, $T = 190°$. We solve the following equation for k:

$$190 = 130e^{-k \cdot 5} + 70$$
$$120 = 130e^{-5k}$$
$$\frac{12}{13} = e^{-5k}$$
$$\ln \frac{12}{13} = \ln e^{-5k}$$
$$\ln \frac{12}{13} = -5k$$
$$-5k = -0.080043 \qquad \text{▦   Divide 12 by 13 and take the}$$
$$\text{natural logarithm.}$$
$$k \approx 0.016.$$

Now we find the temperature at $t = 10$:

$$T(10) = 130e^{-0.016(10)} + 70 = 130e^{-0.16} + 70$$
$$= 130(0.852144) + 70$$
$$\approx 181°.$$

b) To find how long it will take for the cup of coffee to cool to $90°$, we solve for t:

$$90 = 130e^{-0.016t} + 70$$
$$20 = 130e^{-0.016t}$$
$$\frac{2}{13} = e^{-0.016t}.$$

Thus,

$$\ln \frac{2}{13} = \ln e^{-0.016t}$$
$$-1.871802 = -0.016t$$
$$t \approx 117 \text{ min.} \qquad \diamondsuit$$

DO EXERCISES 8 AND 9.

The graph of $T(t) = ae^{-kt} + C$ is shown in Fig. 5. Note that $\lim_{t \to \infty} T(t) = C$. The temperature of the object decreases toward the temperature of the surrounding medium.

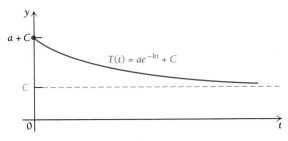

FIGURE 5

Mathematically, this model tells us that the temperature never reaches C, but in practice this happens eventually. At least, the temperature of the cooling object gets so close to that of the surrounding medium that no device could detect a difference. Let us now see how Newton's Law of Cooling can be used in solving a crime.

8. *Physical science: Cooling.* The temperature of a hot cup of soup is $200°$. The room temperature is $70°$. The temperature of the soup cools to $190°$ in 8 min.

   a) Find the value of the constant a in Newton's Law of Cooling.

   b) Find the value of the constant k.

   c) What is the temperature of the soup after 10 min?

   d) How long does it take for the soup to cool to $80°$?

9. Return to the data you found in the exploratory exercises. Find an equation that "fits" the data. Use this equation to check values of other data points. How do they compare? Is it ever "theoretically" possible for the temperature of the water to be the same as the room temperature?

EXAMPLE 6   *Life science: When was the murder committed?* The police discover the body of a calculus professor. Critical to solving the crime is determining when the murder was committed. The police call the coroner, who arrives at 12 noon. He immediately takes the temperature of the body and finds it to be 94.6°. He waits 1 hour, takes the temperature again, and finds it to be 93.4°. He also notes that the temperature of the room is 70°. When was the murder committed?

Solution   We first find a in the equation $T(t) = ae^{-kt} + C$. Assuming that the temperature of the body was normal when the murder occurred, we have $T = 98.6°$ at $t = 0$. Thus,

$$98.6° = ae^{-k \cdot 0} + 70°,$$

so

$$a = 28.6°.$$

Thus, T is given by $T(t) = 28.6e^{-kt} + 70$.

We want to find the number of hours N since the murder was committed. To find N, we must first determine k. From the two temperature readings the coroner made, we have

$$94.6 = 28.6e^{-kN} + 70, \qquad \text{or} \quad 24.6 = 28.6e^{-kN}; \qquad (1)$$
$$93.4 = 28.6e^{-k(N + 1)} + 70, \qquad \text{or} \quad 23.4 = 28.6e^{-k(N + 1)}. \qquad (2)$$

Dividing Eq. (1) by Eq. (2), we get

$$\frac{24.6}{23.4} = \frac{28.6e^{-kN}}{28.6e^{-k(N + 1)}}$$
$$= e^{-kN + k(N + 1)} = e^{-kN + kN + k} = e^{k}.$$

We solve this equation for k:

$$\ln \frac{24.6}{23.4} = \ln e^{k} \qquad \begin{array}{l}\text{Taking the natural logarithm} \\ \text{on both sides}\end{array}$$

$$0.05 \approx k.$$

Now we substitute back into Eq. (1) and solve for N:

$$24.6 = 28.6e^{-0.05N}$$
$$\frac{24.6}{28.6} = e^{-0.05N}$$
$$\ln \frac{24.6}{28.6} = \ln e^{-0.05N}$$
$$-0.150660 = -0.05N$$
$$3 \approx N.$$

The coroner arrived at 12 noon, so the murder was committed at about 9:00 A.M.

DO EXERCISE 10.

10. *Life science: Cooling body.* A butcher is murdered and the body is placed in a cooler, where the temperature is 40°. The coroner, arriving at 3:00 P.M., takes the temperature of the body, finds it to be 73.8°, waits 1 hour, takes the temperature again, and finds it to be 70.3°. When was the murder committed?

EXERCISE SET 4.4

APPLICATIONS

Business and Economics

1. *Present value.* Following the birth of a child, a parent wants to make an initial investment P_0 that will grow to $5000 by the child's 20th birthday. Interest is compounded continuously at 9%. What should the initial investment be?

2. *Present value.* Following the birth of a child, a parent wants to make an initial investment P_0 that will grow to $5000 by the child's 20th birthday. Interest is compounded continuously at 10%. What should the initial investment be?

3. *Present value.* Find the present value of $60,000 due 8 years later at 12%, compounded continuously.

4. *Present value.* Find the present value of $50,000 due 16 years later at 14%, compounded continuously.

5. *Salvage value.* A business estimates that the salvage value V of a piece of machinery after t years is given by

$$V(t) = \$40,000e^{-t}.$$

   a) What did the machinery cost initially?
   b) What is the salvage value after 2 years?

6. *Supply and demand.* The supply and demand for the sale of stereos by a sound company are given by

$$x = S(p) = \ln p \quad \text{and} \quad x = D(p) = \ln \frac{163,000}{p},$$

   where $S(p)$ = the number of stereos that the company is willing to sell at price p and $D(p)$ = the quantity that

the public is willing to buy at price p. Find the equilibrium point. (For reference, see Section 1.5.)

7. *Present value.* A person knows that a trust fund will yield $80,000 in 13 years. A CPA is preparing a financial statement for this client and wants to take into account the present value of the trust fund in computing the client's net worth. Interest is compounded continuously at 8.3%. What is the present value of the trust fund?

8. *Consumer price index.* The consumer price index compares the costs of goods and services over various years, where 1967 is used as a base. The same goods and services that cost $100 in 1967 cost $42 in 1940. Assuming the exponential-decay model:

   a) Find the value k, and write the equation. Round to the nearest hundredth.

   b) Estimate what the same goods and services cost in 1900.

❖ Life and Physical Sciences

9. *Decay rate.* The decay rate of iodine-131 is 9.6% per day. What is its half-life?

10. *Decay rate.* The decay rate of krypton-85 is 6.3% per year. What is its half-life?

11. *Half-life.* The half-life of polonium is 3 min. What is its decay rate?

12. *Half-life.* The half-life of lead is 22 yr. What is its decay rate?

13. *Half-life.* Of an initial amount of 1000 g of polonium, how much will remain after 20 min? See Exercise 11 for the value of k.

14. *Half-life.* Of an initial amount of 1000 g of lead, how much will remain after 100 yr? See Exercise 12 for the value of k.

15. *Carbon dating.* How old is a piece of wood that has lost 90% of its carbon-14?

16. *Carbon dating.* How old is an ivory tusk that has lost 40% of its carbon-14?

17. *Carbon dating.* How old is a Chinese artifact that has lost 60% of its carbon-14?

18. *Carbon dating.* How old is a skeleton that has lost 50% of its carbon-14?

19. In a *chemical reaction,* substance A decomposes at a rate proportional to the amount of A present.

    a) Write an equation relating A to the amount left of an initial amount A_0 after time t.

    b) It is found that 8 g of A will reduce to 4 g in 3 hr. After how long will there be only 1 g left?

20. In a *chemical reaction,* substance A decomposes at a rate proportional to the amount of A present.

    a) Write an equation relating A to the amount left of an initial amount A_0 after time t.

    b) It is found that 10 lb of A will reduce to 5 lb in 3.3 hr. After how long will there be only 1 lb left?

21. *Weight loss.* The initial weight of a starving animal is W_0. Its weight W after t days is given by

$$W = W_0 e^{-0.008t}.$$

    a) What percentage of its weight does it lose each day?

    b) What percentage of its initial weight remains after 30 days?

22. *Weight loss.* The initial weight of a starving animal is W_0. Its weight W after t days is given by

$$W = W_0 e^{-0.009t}.$$

    a) What percentage of its weight does it lose each day?

    b) What percentage of its initial weight remains after 30 days?

BEER—LAMBERT LAW

A beam of light enters a medium such as water or smoky air with initial intensity I_0. Its intensity is decreased depending on the thickness (or concentration) of the medium. The intensity I at a depth (or concentration) of x units is given by

$$I = I_0 e^{-\mu x}.$$

The constant μ ("mu"), called the *coefficient of absorption,* varies with the medium.

---

23. *Light through sea water* has $\mu = 1.4$ when x is measured in meters (m).

    a) What percentage of I_0 remains at a depth of sea water that is 1 m? 2 m? 3 m?

    b) Plant life cannot exist below 10 m. What percentage of I_0 remains at 10 m?

24. *Light through smog.* Particulate concentrations of pollution reduce sunlight. In a smoggy area, $\mu = 0.01$ and $x =$ the concentration of particulates measured in micrograms per cubic meter. What percentage of an initial amount I_0 of sunlight passes through smog that has a concentration of 100 micrograms per cubic meter?

25. *Cooling.* The temperature of a hot liquid is 100° and the room temperature is 75°. The liquid cools to 90° in 10 min.

    a) Find the value of the constant a in Newton's Law of Cooling.

    b) Find the value of the constant k. Round to the nearest hundredth.

    c) What is the temperature after 20 min?

    d) How long does it take the liquid to cool to 80°?

26. *Cooling.* The temperature of a hot liquid is 100°. The liquid is placed in a refrigerator where the temperature is 40°, and it cools to 90° in 5 min.

    a) Find the value of the constant a in Newton's Law of Cooling.

    b) Find the value of the constant k. Round to the nearest hundredth.

    c) What is the temperature after 10 min?

    d) How long does it take the liquid to cool to 41°?

27. *Cooling body.* The coroner arrives at the scene of a murder at 11 P.M. She takes the temperature of the body and finds it to be 85.9°. She waits 1 hour, takes the temperature again, and finds it to be 83.4°. She notes that the room temperature is 60°. When was the murder committed?

28. *Cooling body.* The coroner arrives at the scene of a murder at 2 A.M. He takes the temperature of the body and finds it to be 61.6°. He waits 1 hour, takes the temperature again, and finds it to be 57.2°. The body is in a meat freezer, where the temperature is 10°. When was the murder committed?

29. *Population decrease of Cincinnati.* The population of Cincinnati was 453,000 in 1970 and 385,000 in 1980. Assuming the population is decreasing according to the exponential-decay model:
    a) Find the value k, and write the equation.
    b) Estimate the population of Cincinnati in 2000.
    c) After how long (theoretically) will Cincinnati have just 1 person?

30. *Population decrease of Panama.* The population of Panama was 1,464,000 in 1970 and 1,260,000 in 1980.

Assuming the population is decreasing according to the exponential-decay model:
a) Find the value k, and write the equation.
b) Estimate the population of Panama in 2000.
c) After how long will the population of Panama be 100,000?

31. *Population of the USSR in an earlier year.* The population of the USSR was 258 million in 1980 and was growing at the rate of 1% per year. What was the population in 1970? in 1940?

32. *Atmospheric pressure.* Atmospheric pressure P at altitude a is given by
$$P = P_0 e^{-0.00005a},$$
where P_0 = the pressure at sea level. Assume that P_0 = 14.7 lb/in² (pounds per square inch).
a) Find the pressure at an altitude of 1000 ft.
b) Find the pressure at an altitude of 20,000 ft.
c) At what altitude is the pressure 1.47 lb/in²?

33. *Satellite power.* The power supply of a satellite is a radioisotope. The power output P, in watts, decreases at a rate proportional to the amount present; P is given by
$$P = 50e^{-0.004t},$$
where t = the time, in days.
a) How much power will be available after 375 days?
b) What is the half-life of the power supply?
c) The satellite's equipment cannot operate on fewer than 10 watts of power. How long can the satellite stay in operation?
d) How much power did the satellite have to begin with?

34. *Carbon dating.* Recently, while digging in Chaco Canyon, New Mexico, archeologists found corn pollen that was 4000 years old. This was evidence that Indians had begun cultivating crops in the Southwest centuries earlier than scientists had thought. What percent of the carbon-14 had been lost from the pollen?

SYNTHESIS EXERCISE

35. *Economics: Supply and demand.* The demand and supply functions for a certain type of VCR are given by

$$x = D(p) = 480e^{-0.003p} \quad \text{and} \quad x = S(p) = 150e^{0.004p}.$$

Find the equilibrium point.

4.5

THE DERIVATIVES OF a^x AND $\log_a x$

a) Express a power like 2^3 as a power of e.

b) Differentiate functions involving a^x.

c) Differentiate functions involving $\log_a x$.

The Derivative of a^x

To find the derivative of a^x, for any base a, we first express a^x as a power of e. In order to do this, we first prove the following additional property of a logarithm.

P7. $b^{\log_b x} = x$

To prove this, let

$$y = \log_b x.$$

Then, by the definition of a logarithm,

$$b^y = x.$$

Substituting $\log_b x$ for y, we have

$$b^{\log_b x} = x.$$

We can now express a^x as a power of e. Using Property 7, where $b = e$ and $x = a$, we have

$$a = e^{\ln a}. \qquad \text{Remember: } \ln a = \log_e a.$$

Raising both sides to the power x, we get

$$a^x = (e^{\ln a})^x$$
$$= e^{x \cdot \ln a}. \qquad \text{Multiplying exponents}$$

Thus we have the following.

THEOREM 13

$$a^x = e^{x \cdot \ln a}$$

EXAMPLE 1   Express as a power of e.

a) 3^2 $3^2 = e^{2 \cdot \ln 3}$

$\qquad\qquad = e^{2(1.098612)}$

$\qquad\qquad \approx e^{2.1972}$

b) 10^x $10^x = e^{x \cdot \ln 10}$

$\qquad\qquad = e^{x(2.302585)}$

$\qquad\qquad \approx e^{2.3026x}$

DO EXERCISES 1 AND 2.

Now we can differentiate.

EXAMPLE 2

$$\frac{d}{dx} 2^x = \frac{d}{dx} e^{x \cdot \ln 2} \qquad \text{Theorem 13}$$

$$= \left[\frac{d}{dx} (x \cdot \ln 2) \right] \cdot e^{x \cdot \ln 2}$$

$$= (\ln 2)(e^{\ln 2})^x$$

$$= (\ln 2)2^x$$

We completed this by taking the derivative of $x \ln 2$ and replacing $e^{x \ln 2}$ by 2^x. Note that $\ln 2 \approx 0.7$, so the above verifies our earlier approximation of the derivative of 2^x as $(0.7)2^x$ (see Section 4.1).

DO EXERCISE 3.

In general,

$$\frac{d}{dx} a^x = \frac{d}{dx} e^{x \cdot \ln a} \qquad \text{Theorem 13}$$

$$= \left[\frac{d}{dx} (x \cdot \ln a) \right] \cdot e^{x \cdot \ln a}$$

$$= (\ln a) \, a^x.$$

Thus we have the following.

THEOREM 14

$$\frac{d}{dx} a^x = (\ln a) \, a^x$$

Express as a power of e.

1. 4^5

2. 2^x

3. Differentiate $y = 5^x$.

Differentiate.

4. $f(x) = 4^x$

5. $f(x) = (4.3)^x$

EXAMPLE 3

$$\frac{d}{dx}\, 3^x = (\ln 3)\, 3^x \qquad \diamondsuit$$

EXAMPLE 4

$$\frac{d}{dx}\, (1.4)^x = (\ln 1.4)(1.4)^x \qquad \diamondsuit$$

Compare these formulas:

$$\frac{d}{dx}\, a^x = (\ln a)\, a^x \quad \text{and} \quad \frac{d}{dx}\, e^x = e^x.$$

It is the simplicity of the latter formula that is a reason for the use of base e in calculus. The many applications of e in natural phenomena provide other reasons.

One other result also follows from what we have done. If

$$f(x) = a^x,$$

we know that

$$f'(x) = a^x\, (\ln a).$$

In Section 4.1, we also showed that

$$f'(x) = a^x \left(\lim_{h \to 0} \frac{a^h - 1}{h} \right).$$

Since $a^x > 0$, we have the following.

THEOREM 15

$$\ln a = \lim_{h \to 0} \frac{a^h - 1}{h}$$

DO EXERCISES 4 AND 5.

The Derivative of $\log_a x$

Just as the derivative of a^x is expressed in terms of $\ln a$, so too is the derivative of $\log_a x$. To find this derivative, we first express $\log_a x$ in terms of $\ln a$ using Property 7:

$$a^{\log_a x} = x.$$

Then

$$\ln (a^{\log_a x}) = \ln x$$

$$(\log_a x) \cdot \ln a = \ln x \qquad \text{By P3, treating } \log_a x \text{ as an exponent}$$

and

$$\log_a x = \boxed{\frac{1}{\ln a}} \cdot \ln x.$$

$$\text{constant}$$

The derivative of $\log_a x$ follows.

THEOREM 16

$$\frac{d}{dx} \log_a x = \frac{1}{\ln a} \cdot \frac{1}{x}$$

Comparing this with

$$\frac{d}{dx} \ln x = \frac{1}{x},$$

we again see a reason for the use of base e in calculus.

EXAMPLE 5

$$\frac{d}{dx} \log_3 x = \frac{1}{\ln 3} \cdot \frac{1}{x} \qquad \qquad ❖$$

EXAMPLE 6

$$\frac{d}{dx} \log x = \frac{1}{\ln 10} \cdot \frac{1}{x} \qquad \log x = \log_{10} x \qquad ❖$$

EXAMPLE 7

$$\frac{d}{dx} x^2 \log x = x^2 \frac{1}{\ln 10} \cdot \frac{1}{x} + 2x \log x \qquad \text{By the Product Rule}$$

$$= \frac{x}{\ln 10} + 2x \log x, \quad \text{or} \quad x \left(\frac{1}{\ln 10} + 2 \log x \right) \qquad ❖$$

DO EXERCISES 6–8.

Differentiate.

6. $y = \log_2 x$

7. $y = -7 \log x$

8. $f(x) = x^6 \log x$

EXERCISE SET 4.5

Express as a power of e.

1. 5^4
2. 2^3
3. $(3.4)^{10}$
4. $(5.3)^{20}$
5. 4^k
6. 5^R
7. 8^{kT}
8. 10^{kR}

Differentiate.

9. $y = 6^x$
10. $y = 7^x$
11. $f(x) = 10^x$
12. $f(x) = 100^x$

13. $f(x) = x(6.2)^x$
14. $f(x) = x(5.4)^x$
15. $y = x^3 10^x$
16. $y = x^4 5^x$
17. $y = \log_4 x$
18. $y = \log_5 x$
19. $f(x) = 2 \log x$
20. $f(x) = 5 \log x$
21. $f(x) = \log \dfrac{x}{3}$
22. $f(x) = \log \dfrac{x}{5}$
23. $y = x^3 \log_8 x$
24. $y = x \log_6 x$

APPLICATIONS

❖ **Business and Economics**

25. *Recycling aluminum cans.* It is known that one fourth of all aluminum cans distributed will be recycled each year. A beverage company distributes 250,000 cans. The number still in use after time t, in years, is given by

$$N(t) = 250,000(\tfrac{1}{4})^t.$$

Find $N'(t)$.

26. *Double declining-balance depreciation.* An office machine is purchased for $5200. Under certain assumptions, its salvage value V depreciates according to a method called double declining balance, basically 80% each year, and is given by

$$V(t) = \$5200(0.80)^t,$$

where t is the time, in years. Find $V'(t)$.

❖ **Life and Physical Sciences**

Earthquake magnitude. The magnitude R (measured on the Richter scale) of an earthquake of intensity I is defined as

$$R = \log \frac{I}{I_0},$$

where I_0 is a minimum intensity used for comparison. When one earthquake is 10 times as intense as another, its magnitude on the Richter scale is 1 higher. If one earthquake is 100 times as intense as another, its magnitude on the Richter scale is 2 higher, and so on. Thus an earthquake whose magnitude is 7 on the Richter scale is 10 times as intense as an earthquake whose magnitude is 6. Earthquakes can be interpreted as multiples of the minimum intensity I_0.

27. In 1986, there was an earthquake near Cleveland, Ohio. It had an intensity of $10^5 \cdot I_0$. What was its magnitude on the Richter scale?

28. The Anchorage, Alaska, earthquake on March 27, 1964, had an intensity of $10^{8.4} \cdot I_0$. What was its magnitude on the Richter scale?

This photograph shows part of the damage of the earthquake in Anchorage, Alaska, in 1964.

29. *Earthquake intensity.* The intensity I of an earthquake is given by

$$I = I_0 10^R,$$

where R = the magnitude on the Richter scale and I_0 = the minimum intensity, where $R = 0$, used for comparison.

a) Find I, in terms of I_0, for an earthquake of magnitude 7 on the Richter scale.

b) Find I, in terms of I_0, for an earthquake of magnitude 8 on the Richter scale.

c) Compare your answers to parts (a) and (b).

d) Find the rate of change dI/dR.

30. *Intensity of sound.* The intensity of a sound is given by

$$I = I_0 10^{0.1L},$$

where L = the loudness of the sound as measured in decibels and I_0 = the minimum intensity detectable by the human ear.

a) Find I, in terms of I_0, for the loudness of a power mower, which is 100 decibels.

b) Find I, in terms of I_0, for the loudness of just audible sound, which is 10 decibels.

c) Compare your answers to parts (a) and (b).

d) Find the rate of change dI/dL.

31. *Earthquake magnitude.* The magnitude R (measured on the Richter scale) of an earthquake of intensity I is defined as

$$R = \log \frac{I}{I_0},$$

where I_0 = the minimum intensity (used for comparison). (The exponential form of this definition is given in Exercise 29.) Find the rate of change dR/dI.

32. *Loudness of sound.* The loudness L of a sound of intensity I is defined as

$$L = 10 \log \frac{I}{I_0},$$

where I_0 = the minimum intensity detectable by the human ear and L = the loudness measured in decibels. (The exponential form of this definition is given in Exercise 30.) Find the rate of change dL/dI.

33. *Response to drug dosage.* The response y to a dosage x of a drug is given by

$$y = m \log x + b.$$

The response may be hard to measure with a number. The patient might perspire more, have an increase in temperature, or faint. Find the rate of change dy/dx.

SYNTHESIS EXERCISES

34. Find $\lim\limits_{h \to 0} \dfrac{3^h - 1}{h}$.

Use the Chain Rule and other formulas given in this section to differentiate each of the following. In some cases, it may help to take the logarithm on both sides of the equation before differentiating.

35. $f(x) = 3^{2x}$

36. $y = 2^{x^4}$

37. $y = x^x, x > 0$

38. $y = \log_3 (x^2 + 1)$

39. $f(x) = x^{e^x}, x > 0$

40. $y = a^{f(x)}$

41. $y = \log_a f(x)$, $f(x)$ positive

42. $y = [f(x)]^{g(x)}$, $f(x)$ positive

4.6

AN ECONOMIC APPLICATION: ELASTICITY OF DEMAND

Suppose that x represents a quantity of goods sold and p is the price per unit of the goods. Recall that x and p are related by the demand

OBJECTIVE

a) Given a demand function, find the elasticity and the value(s) of p for which total revenue is maximized.

(continued)

b) Characterize a demand as elastic, inelastic, or having unit elasticity. Then interpret the effect of elasticity on total revenue.

function

$$x = D(p).$$

Suppose that there is a change Δp in the price per unit of a product. The percent change in price is

$$\frac{\Delta p}{p}.$$

A change in the price of a product produces a change Δx in the quantity sold. The percent change in quantity is

$$\frac{\Delta x}{x}.$$

The ratio of the percent change in quantity to the percent change in price is

$$\frac{(\Delta x/x)}{(\Delta p/p)},$$

which can be expressed as

$$\frac{p}{x} \cdot \frac{\Delta x}{\Delta p}. \tag{1}$$

For continuous functions,

$$\lim_{\Delta p \to 0} \frac{\Delta x}{\Delta p} = \frac{dx}{dp},$$

and the limit as Δp approaches 0 of the expression in Eq. (1) becomes

$$\frac{p}{x} \cdot \frac{dx}{dp} = \frac{p}{D(p)} \cdot D'(p) = \frac{p\, D'(p)}{D(p)}.$$

DEFINITION

The *elasticity of demand* E is given as a function of price p by

$$E(p) = -\frac{p\, D'(p)}{D(p)}.$$

The numbers x, or $D(p)$, and p are always nonnegative. The slope of the demand curve $dx/dp = D'(p)$ is always negative, since the demand curve is decreasing, and $D(p)$ is always nonnegative and must be positive for the elasticity to exist. Economists have used the $-$ sign in the definition of elasticity to make $E(p)$ nonnegative and easier to work with.

EXAMPLE 1   *Economics: Demand for videotape rentals.* A videotape store works out a demand function for its videotape rentals and finds it to be

$$x = D(p) = 120 - 20p,$$

where x = the number of videotapes rented per day when p is the price per rental. Find each of the following.

a)   The quantity demanded when the price is $2 per rental

b)   The elasticity as a function of p

c)   The elasticity at $p = \$2$ and at $p = \$5$. Interpret the meaning of these values of the elasticity.

d)   The value of p for which $E(p) = 1$. Interpret the meaning of this price.

e)   The total-revenue function $R(p) = p\,D(p)$

f)   The price p at which total revenue is a maximum

Solution

a)   At $p = \$2$, $x = D(2) = 120 - 20(2) = 80$. Thus, 80 videotapes will be rented per day when the price per rental is $2.

b)   To find the elasticity, we first find the derivative $D'(p)$:

$$D'(p) = -20.$$

Then we substitute -20 for $D'(p)$ and $120 - 20p$ for $D(p)$ in the expression for elasticity:

$$E(p) = -\frac{p\,D'(p)}{D(p)} = -\frac{p \cdot (-20)}{120 - 20p} = \frac{20p}{120 - 20p} = \frac{p}{6 - p}.$$

c) $E(\$2) = \dfrac{2}{6 - 2} = \dfrac{1}{2}$

At $p = \$2$, the elasticity is $\frac{1}{2}$, which is less than 1. Thus the ratio of the percent change in quantity to the percent change in price is less than 1. A small increase in price will cause a percentage decrease in the quantity that is smaller than the percentage change in price.

$$E(\$5) = \frac{5}{6 - 5} = 5$$

At $p = \$5$, the elasticity is 5, which is greater than 1. Thus the ratio of the percent change in quantity to the percent change in price is greater than 1. A small increase in price will cause a percentage decrease in the quantity that is greater than the percentage change in price.

1. *Economics: Demand for hand radios.* A company determines that the demand function for hand radios is

$$x = D(p) = 300 - p,$$

where x = the number sold per day when the price is p dollars per radio. Find each of the following.

a) The quantity demanded when the price is $55 per unit

b) The elasticity as a function of p

c) The elasticity at $p = \$100$ and $p = \$200$

d) The value of p for which $E(p) = 1$

e) The total-revenue function

f) The value of p for which the total revenue is a maximum

d) We set $E(p) = 1$ and solve for p:

$$\frac{p}{6 - p} = 1$$

$$p = 6 - p \qquad \text{We multiply by } 6 - p, \text{ assuming } p \ne 6.$$

$$2p = 6$$

$$p = 3.$$

Thus when the price is $3 per rental, the ratio of the percent change in quantity to the percent change in price is 1.

e) Recall that the total revenue $R(p)$ is given by $p\,D(p)$. Then

$$R(p) = p\,D(p)$$
$$= p(120 - 20p)$$
$$= 120p - 20p^2.$$

f) To find the price p that maximizes total revenue, we find $R'(p)$:

$$R'(p) = 120 - 40p.$$

Now $R'(p)$ exists for all values of p in the interval $[0, \infty)$. Thus we solve:

$$R'(p) = 120 - 40p = 0$$
$$-40p = -120$$
$$p = 3.$$

Since there is only one critical point, we can try to use the second derivative to see if we have a maximum:

$$R''(p) = -40, \quad \text{a constant.}$$

Thus, $R''(3)$ is negative, so $R(3)$ is a maximum. That is, total revenue is a maximum at $p = \$3$ per rental. ❖

DO EXERCISE 1.

Note in both Example 1 and Margin Exercise 1 that the value of p for which $E(p) = 1$ is the same as the value of p for which the total revenue is a maximum. This is always true.

THEOREM 17

Total revenue is a maximum at the value(s) of p for which $E(p) = 1$.

We can prove this as follows. We know that

$$R(p) = p\,D(p),$$

so

$$R'(p) = p \cdot D'(p) + 1 \cdot D(p)$$

$$= D(p)\left[\frac{p\,D'(p)}{D(p)} + 1\right] \quad \text{Check this by multiplying.}$$

$$= D(p)[-E(p) + 1]$$

$$= D(p)[1 - E(p)] \tag{2}$$

$$R'(p) = 0 \quad \text{when} \quad 1 - E(p) = 0, \quad \text{or} \quad E(p) = 1.$$

We have shown that total revenue is a maximum when $E = 1$. What do we know when $E < 1$ and $E > 1$? We answer this by looking at Eq. (2) in the preceding proof.

For prices p for which $D(p) > 0$ and for which $E(p) < 1$, it follows that $1 - E(p) > 0$, so by Eq. (2), $R'(p) > 0$. This tells us that the total-revenue function is increasing.

For values of p for which $D(p) > 0$ and for which $E(p) > 1$, it follows that $1 - E(p) < 0$, so by Eq. (2), $R'(p) < 0$. This tells us that the total-revenue function is decreasing.

The following statement and the graphs shown in Figs. 1 and 2 summarize our results.

For a particular value of the price p:

1. The demand is *inelastic* if $E(p) < 1$. An increase in price will bring an increase in revenue. If demand is inelastic, then revenue is increasing.
2. The demand has *unit elasticity* if $E(p) = 1$. The demand has unit elasticity when revenue is at a maximum.
3. The demand is *elastic* if $E(p) > 1$. An increase in price will bring a decrease in revenue. If demand is elastic, then revenue is decreasing.

Price Elasticities in the U.S. Economy

Industry	Elasticity
Elastic Demands	
Metals	1.52
Electrical engineering products	1.39
Mechanical engineering products	1.30
Furniture	1.26
Motor vehicles	1.14
Instrument engineering products	1.10
Professional services	1.09
Transportation services	1.03
Inelastic Demands	
Gas, electricity, and water	0.92
Oil	0.91
Chemicals	0.89
Beverages (all types)	0.78
Tobacco	0.61
Food	0.58
Banking and insurance services	0.56
Housing services	0.55
Clothing	0.49
Agricultural and fish products	0.42
Books, magazines, and newspapers	0.34
Coal	0.32

Source: Ahsan Mansur and John Whalley, "Numerical specification of applied general equilibrium models: Estimation, calibration, and data." In H. E. Scarf and J. B. Shoven (eds.), *Applied General Equilibrium Analysis.* (New York: Cambridge University Press, 1984), p. 109.

2. *Economics: Demand for hand radios.*
For Margin Exercise 1, determine
each of the following.

a) The elasticity of demand at a
price of $55 per radio. Interpret
your answer.

b) The elasticity of demand at a
price of $200 per radio.
Interpret your answer.

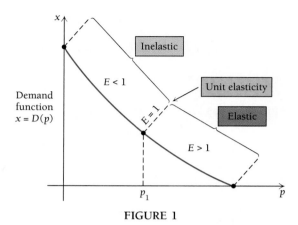

FIGURE 1

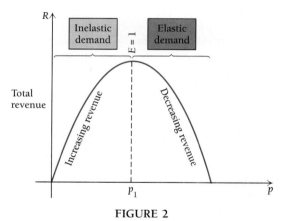

FIGURE 2

In summary, suppose that the videotape store in Example 1 raises the
price per rental and that the total revenue increases. Then we say the
demand is *inelastic*. If the total revenue decreases, we say the demand is
elastic. Some price elasticities in the U.S. economy are listed in the table
in the margin.

DO EXERCISE 2.

For each demand function, find:

a)   the elasticity;

b)   the elasticity at the given price, stating whether the demand is elastic or inelastic; and

c)   the value(s) of p for which total revenue is a maximum.

1. $x = D(p) = 400 - p; \quad p = \$125$

2. $x = D(p) = 500 - p; \quad p = \$38$

3. $x = D(p) = 200 - 4p; \quad p = \$46$

4. $x = D(p) = 500 - 2p; \quad p = \$57$

5. $x = D(p) = \dfrac{400}{p}; \quad p = \$50$

6. $x = D(p) = \dfrac{3000}{p}; \quad p = \$60$

7. $x = D(p) = \sqrt{500 - p}; \quad p = \$400$

8. $x = D(p) = \sqrt{300 - p}; \quad p = \$250$

9. $x = D(p) = 100e^{-0.25p}; \quad p = \$10$

10. $x = D(p) = 200e^{-0.05p}; \quad p = \$80$

11. $x = D(p) = \dfrac{100}{(p + 3)^2}; \quad p = \$1$

12. $x = D(p) = \dfrac{300}{(p + 8)^2}; \quad p = \$4.50$

APPLICATIONS

❖ **Business and Economics**

13. *Demand for chocolate chip cookies.* A bakery works out a demand function for its sale of chocolate chip cookies and finds it to be

$$x = D(p) = 967 - 25p,$$

where $x =$ the quantity of cookies sold when the price per cookie, in cents, is p.

a)   Find the elasticity.

b)   At what price is the elasticity of demand equal to 1?

c)   At what prices is the elasticity of demand elastic?

d)   At what prices is the elasticity of demand inelastic?

e)   At what price is the revenue a maximum?

f)   At a price of 20¢ per cookie, will a small increase in price cause the total revenue to increase or decrease?

14. *Demand for oil.* Suppose you have been hired by OPEC as an economic consultant for the world demand for oil. The demand function is

$$x = D(p) = 63{,}000 + 50p - 25p^2, \quad 0 \leqslant p \leqslant 50,$$

where x is measured in millions of barrels of oil per day at a price of p dollars per barrel.

a)   Find the elasticity.

b)   Find the elasticity at a price of $10 per barrel, stating whether the demand is elastic or inelastic.

c)   Find the elasticity at a price of $20 per barrel, stating whether the demand is elastic or inelastic.

d)   Find the elasticity at a price of $30 per barrel, stating whether the demand is elastic or inelastic.

e)   At what price is the revenue a maximum?

f)   What quantity of oil will be sold at the price that maximizes revenue? Check current world prices to see how this compares with your answer.

g)   At a price of $30 per barrel, will a small increase in price cause the total revenue to increase or decrease?

15. *Demand for videogames.* A computer software store determines the following demand function for a new videogame:

$$x = D(p) = \sqrt{200 - p^3},$$

where $x =$ the number of videogames sold per day

when the price is p dollars per game.

a) Find the elasticity.

b) Find the elasticity when $p = \$3$ per game.

c) At $p = \$3$, will a small increase in price cause the total revenue to increase or decrease?

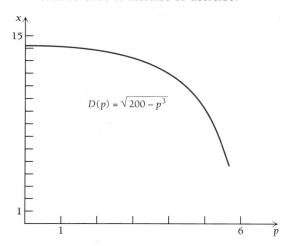

16. *Demand for tomato plants.* A garden store determines the following demand function during early summer for a certain size of tomato plant:

$$x = D(p) = \frac{2p + 300}{10p + 11},$$

where $x = $ the number of plants sold per day when the price is p dollars per plant.

a) Find the elasticity.

b) Find the elasticity when $p = \$3$ per plant.

c) At $p = \$3$, will a small increase in price cause the total revenue to increase or decrease?

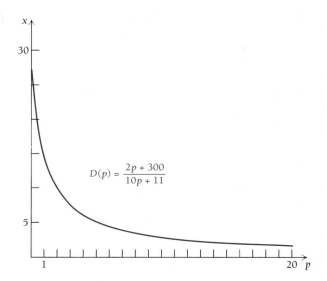

SYNTHESIS EXERCISES

17. *Economics: Constant elasticity curve*

a) Find the elasticity of the demand function

$$x = D(p) = \frac{k}{p^n},$$

where k is a positive constant and n is an integer greater than 0.

b) Is the value of the elasticity dependent on the price per unit?

c) Does the total revenue have a maximum? When?

18. *Economics: Exponential demand curve*

a) Find the elasticity of the demand function

$$x = D(p) = Ae^{-kp},$$

where A and k are positive constants.

b) Is the value of the elasticity dependent on the price per unit?

c) Does the total revenue have a maximum? At what value of p?

19. Let

$$L(p) = \ln D(p).$$

Describe the elasticity in terms of $L'(p)$.

THE CALCULUS EXPLORER
PowerGrapher

Graph each of the demand functions D in Exercises 1, 3, 5, 7, and 11. Then find the related total-revenue function R and the elasticity E, and graph all three using the same set of axes.

CHAPTER SUMMARY AND REVIEW

4

TERMS TO KNOW

Exponential function, p. 277
Base, p. 277
The number e, p. 282, 283
Logarithm, p. 290
Logarithmic base, p. 290
Common logarithm, p. 294
Natural logarithm, p. 295
Exponential equation, p. 297

Exponential growth, p. 309
Exponential growth rate, p. 311
Interest compounded continuously,
   p. 312
Doubling time, p. 314
Models of limited growth, p. 317
Exponential decay, p. 324
Half-life, p. 326

Decay rate, p. 326
Carbon dating, p. 327
Present value, p. 329
Newton's Law of Cooling, p. 330
Elasticity of demand, p. 342
Elastic demand, p. 346
Inelastic demand, p. 346
Unit elasticity, p. 346

REVIEW EXERCISES

These review exercises are for test preparation. They can also be used as a lengthened practice test. Answers are at the back of the book. The answers also contain bracketed section references, which tell you where to restudy if your answer is incorrect.

Differentiate.

1. $y = \ln x$

2. $y = e^x$

3. $y = \ln (x^4 + 5)$

4. $y = e^{2\sqrt{x}}$

5. $f(x) = \ln x^6$

6. $f(x) = e^{4x} + x^4$

7. $f(x) = \dfrac{\ln x}{x^3}$

8. $f(x) = e^{x^2} \cdot \ln 4x$

9. $f(x) = e^{4x} - \ln \dfrac{x}{4}$

10. $g(x) = x^8 - 8 \ln x$

11. $y = \dfrac{\ln e^x}{e^x}$

Given $\log_a 2 = 1.8301$ and $\log_a 7 = 5.0999$, find each of the following.

12. $\log_a 14$

13. $\log_a \dfrac{2}{7}$

14. $\log_a 28$

15. $\log_a 3.5$

16. $\log_a \sqrt{7}$

17. $\log_a \frac{1}{4}$

18. Find the function that satisfies $dQ/dt = kQ$. List the answer in terms of Q_0.

19. *Life science: Population growth.* The population of a certain city doubled in 16 years. What was the growth rate of the city? Round to the nearest tenth of a percent.

20. *Business: Interest compounded continuously.* Suppose $8300 is deposited in a savings and loan association in which the interest rate is 8%, compounded continuously. How long will it take for the $8300 to double itself? Round to the nearest tenth of a year.

21. *Business: Cost of a prime-rib dinner.* The average cost C of a prime-rib dinner was $4.65 in 1962. In 1986, it was $15.81. Assuming that the exponential-growth model applies:

a) Find the exponential-growth rate, and write the equation.

b) What will the cost of such a dinner be in 1997? in 2007?

22. *Business: Franchise growth.* A clothing firm is selling franchises throughout the United States and Canada. It is estimated that the number of franchises N will increase at the rate of 12% per year; that is,

$$\frac{dN}{dt} = 0.12N,$$

where t is the time, in years.

a) Find the function that satisfies the equation, assuming that the number of franchises in 1988 ($t = 0$) is 60.

b) How many franchises will there be in 1998?

c) When will the number of franchises be 120? Round to the nearest tenth of a year.

23. *Life science: Decay rate.* The decay rate of a certain radioactive isotope is 13% per year. What is its half-life? Round to the nearest tenth of a year.

24. *Life science: Half-life.* The half-life of radon-222 is 3.8 days. What is its decay rate? Round to the nearest tenth of a percent.

25. *Life science: Decay rate.* A certain radioactive element has a decay rate of 7% per day; that is,

$$\frac{dA}{dt} = -0.07A,$$

where $A =$ the amount of the element present at time t, in days.

a) Find a function that satisfies the equation if the amount of the element present at $t = 0$ is 800 grams.

b) After 20 days, how much of the 800 grams will remain? Round to the nearest gram.

c) After how long does half the original amount remain?

26. *Social science: The Hullian learning model.* The probability p of mastering a certain assembly line task after t learning trials is given by

$$p(t) = 1 - e^{-0.7t}.$$

a) What is the probability of learning the task after 1 trial? 2 trials? 5 trials? 10 trials? 14 trials?

b) Find the rate of change $p'(t)$.

c) Sketch a graph of the function.

27. *Business: Present value.* Find the present value of $1,000,000 due 40 years later at 8.4%, compounded continuously.

Differentiate.

28. $y = 3^x$

29. $f(x) = \log_{15} x$

30. *Economics: Elasticity of demand.* Consider the demand function

$$x = D(p) = \frac{600}{(p + 4)^2}.$$

a) Find the elasticity.

b) Find the elasticity at $p = 1, stating whether the demand is elastic or inelastic.

c) Find the elasticity at $p = 12, stating whether the demand is elastic or inelastic.

d) At a price of $12, will a small increase in price cause the total revenue to increase or decrease?

e) Find the value of p for which the total revenue is a maximum.

SYNTHESIS EXERCISES

31. Differentiate: $y = \dfrac{e^{2x} + e^{-2x}}{e^{2x} - e^{-2x}}$.

32. Find the minimum value of $f(x) = x^4 \ln 4x$.

EXERCISES FOR THINKING AND WRITING

33. Describe as many applications as you can to convince another student of the usefulness of exponential functions.

34. Explain why e is important as a base of exponential and logarithmic functions.

35. A student made the following error on a test. Explain the error and how to correct it.

$$\frac{d}{dx} e^x = xe^{x-1}.$$

36. Describe the concept of elasticity in your own words and explain its usefulness in economics. Do some library research or consult an economist in order to determine when and how this concept was first developed.

CHAPTER TEST   4

Differentiate.

1. $y = e^x$

2. $y = \ln x$

3. $f(x) = e^{-x^2}$

4. $f(x) = \ln \dfrac{x}{7}$

5. $f(x) = e^x - 5x^3$

6. $f(x) = 3e^x \ln x$

7. $y = \ln (e^x - x^3)$

8. $y = \dfrac{\ln x}{e^x}$

Given $\log_b 2 = 0.2560$ and $\log_b 9 = 0.8114$, find each of the following.

9. $\log_b 18$ **10.** $\log_b 4.5$ **11.** $\log_b 3$

12. Find the function that satisfies $dM/dt = kM$. List the answer in terms of M_0.

13. The doubling time of a certain bacteria culture is 4 hours. What is the growth rate? Round to the nearest tenth of a percent.

14. *Business: Interest compounded continuously.* An investment is made at 6.931% per year, compounded continuously. What is the doubling time? Round to the nearest year.

15. *Business: Cost of milk.* The cost C of a gallon of milk was $0.54 in 1941. In 1985, the cost of a gallon of milk was $2.31. Assuming the exponential-growth model applies:

a) Find the exponential-growth rate, and write the equation.

b) Find the cost of a gallon of milk in 2000; in 2010.

16. *Life science: Drug dosage.* A dose of a drug is injected into the body of a patient. The drug amount in the body decreases at the rate of 10% per hour; that is,

$$\frac{dA}{dt} = -0.1A,$$

where A = the amount in the body and t is the time, in hours.

a) A dose of 3 cubic centimeters (cc) is administered. Assuming $A_0 = 3$ and $k = 0.1$, find the function that satisfies the equation.

b) How much of the initial dose of 3 cc will remain after 10 hr?

c) After how long does half the original dose remain?

17. *Life science: Decay rate.* The decay rate of zirconium is 1.1% per day. What is its half-life?

18. *Life science: Half-life.* The half-life of tellurium is 1,000,000 years. What is its decay rate? As a percent, round to six decimal places.

19. *Business: Effect of advertising.* A company introduces a new product on a trial run in a city. They advertised the product on television and found that the percentage P of people who bought the product after t ads had been run satisfied the function

$$P(t) = \frac{100\%}{1 + 24e^{-0.28t}}.$$

a) What percentage buys the product without having seen the ad ($t = 0$)?

b) What percentage buys the product after the ad is run 1 time? 5 times? 10 times? 15 times? 20 times? 30 times? 35 times?

c) Find the rate of change $P'(t)$.

d) Sketch a graph of the function.

20. *Business: Present value.* Find the present value of $80,000 due 12 years later at 8.6%, compounded continuously.

Differentiate.

21. $f(x) = 20^x$

22. $y = \log_{20} x$

23. *Economics: Elasticity of demand.* Consider the demand function

$$p = D(p) = 400e^{-0.2p}.$$

a) Find the elasticity.

b) Find the elasticity at $p = \$3$, stating whether the demand is elastic or inelastic.

c) Find the elasticity at $p = \$18$, stating whether the demand is elastic or inelastic.

d) At a price of $3, will a small increase in price cause the total revenue to increase or decrease?

e) Find the value of p for which the total revenue is a maximum.

SYNTHESIS EXERCISES

24. Differentiate: $y = x (\ln x)^2 - 2x \ln x + 2x$.

25. Find the maximum and minimum values of $f(x) = x^4 e^{-x}$ on $[0, 10]$.

CHAPTER

5

INTEGRATION

AN APPLICATION

Suppose that in a certain memory experiment the rate of memorizing is given by

$$M'(t) = -0.003t^2 + 0.2t,$$

where $M'(t)$ = the memory rate, in words per minute. How many words are memorized in the first 10 min (from $t = 0$ to $t = 10$)?

THE MATHEMATICS

The number of words memorized in the first 10 min is given by

$$\int_0^{10} (-0.003t^2 + 0.2t)\ dt.$$

└──── This is an *integral*.

This expression also gives the area under the graph of $M'(t) = -0.003t^2 + 0.2t$ over the interval [0, 10].

Suppose we do the reverse of differentiating, that is, suppose we try to find a function whose derivative is a given function. This process is called *antidifferentiation,* or *integration*; it is the main topic of this chapter. We will see that we can use integration to find the area under a curve over a closed interval as well as to find the accumulation of a certain quantity over an interval.

We first consider the meaning of integration and then we learn several techniques for integrating.

5.1

a) Find the indefinite integral (antiderivative) of a given function.

b) Find a function f with a given derivative and function value (boundary or initial value).

c) Solve applied problems involving antiderivatives.

THE ANTIDERIVATIVE

In Chapters 2, 3, and 4, we have considered several interpretations of the derivative, some of which are listed below.

Function	Derivative
Distance	Velocity
Revenue	Marginal revenue
Cost	Marginal cost
Population	Rate of growth of population

For population we actually considered the derivative first and then the function. Many problems can be solved by doing the reverse of differentiation, called *antidifferentiation*.

The Antiderivative

Suppose that y is a function of x and that the derivative is the constant 8. Can we find y? It is easy to see that one such function is $8x$. That is, $8x$ is a function whose derivative is 8. Is there another function whose derivative is 8? In fact, there are many. Here are some examples:

$$8x + 3, \qquad 8x - 10, \qquad 8x + 7.4, \qquad 8x + \sqrt{2}.$$

All these functions are $8x$ plus some constant. There are no other functions having a derivative of 8 other than those of the form $8x + C$. Another way of saying this is that any two functions having a derivative of 8 must differ by a constant. In general, any two functions having the same derivative differ by a constant.

THEOREM 1

If two functions F and G have the same derivative on an interval, then

$$F(x) = G(x) + C, \quad \text{where } C \text{ is a constant.}$$

The reverse of differentiating is *antidifferentiating,* and the result of antidifferentiating is called an **antiderivative**. Above, we found antiderivatives of the function 8. There are several of them, but they are all $8x$ plus some constant.

EXAMPLE 1   Antidifferentiate (find the antiderivatives of) x^2.

Solution   One antiderivative is $x^3/3$. All other antiderivatives differ from this by a constant, so we can denote them as follows:

$$\frac{x^3}{3} + C.$$

❖

The solution to Example 1 is the *general form* of the antiderivative of x^2.

DO EXERCISES 1–6.

Integrals and Integration

To indicate that the antiderivative of x^2 is $x^3/3 + C$, we write

$$\int x^2 \, dx = \frac{x^3}{3} + C, \tag{1}$$

and we note that

$$\int \ldots \, dx$$

is the symbolism we will use from now on to call for the antiderivative of a function $f(x)$. More generally, we can write

$$\int f(x) \, dx = F(x) + C, \tag{2}$$

where $F(x) + C$ is the general form of the antiderivative of $f(x)$. Equation (2) is read "the *indefinite integral* of $f(x)$ is $F(x) + C$." The constant C is called the *constant of integration* and can have any fixed value; hence the use of the word "indefinite."

The symbolism $\int f(x) \, dx$, from Leibniz, is called an *integral,* or more precisely, an **indefinite integral.** The symbol $\int$ is called an *integral sign.* The left side of Eq. (2) is often read "the integral of $f(x)$, dx" or more briefly, "the integral of $f(x)$." In this context, $f(x)$ is called the *integrand.* We illustrate this notation using the preceding example.

EXAMPLE 2   Evaluate $\int x^9 \, dx$.

Solution

$$\int x^9 \, dx = \frac{x^{10}}{10} + C$$

Find three antiderivatives.

1. $\dfrac{dy}{dx} = 7$

2. $\dfrac{dy}{dx} = -2$

Find the general form of the antiderivative.

3. x

4. x^3

5. e^x

6. $\dfrac{1}{x}$

Evaluate. Don't forget the constant of integration!

7. $\int x^3 \, dx$

8. $\int e^x \, dx$

9. $\int 8e^{3x} \, dx$

10. $\int \dfrac{1}{x} \, dx, \; x > 0$

The symbol on the left is read "the integral of x, dx." (The "dx" is often omitted in the reading.) In this case, the integrand is x^9. The constant C is called the constant of integration. ❖

EXAMPLE 3 Evaluate $\int 5e^{4x} \, dx$.

Solution

$$\int 5e^{4x} \, dx = \frac{5}{4}e^{4x} + C$$ ❖

DO EXERCISES 7–10.

To integrate (or antidifferentiate), we make use of differentiation formulas, in effect, reading them in reverse. Below are some of these, stated in reverse, as integration formulas. These can be checked by differentiating the right-hand side and noting that the result is, in each case, the integrand.

THEOREM 2

Basic Integration Formulas

1. $\int k \, dx = kx + C$ (k a constant)

2. $\int x^r \, dx = \dfrac{x^{r+1}}{r+1} + C, \quad$ provided $r \neq -1$

   (To integrate a power of x other than -1, increase the power by 1 and divide by the increased power.)

3. $\int x^{-1} \, dx = \int \dfrac{1}{x} \, dx = \int \dfrac{dx}{x} = \ln x + C, \, x > 0$

   $\int x^{-1} \, dx = \ln |x| + C, \quad x < 0$

   (We will generally assume that $x > 0$.)

4. $\int be^{ax} \, dx = \dfrac{b}{a}e^{ax} + C$

The following rules combined with the preceding formulas allow us to find many integrals. They can be derived by reversing two familiar differentiation rules.

THEOREM 3

Rule A. $\displaystyle\int kf(x)\ dx = k\int f(x)\ dx$

(The integral of a constant times a function is the constant times the integral of the function.)

Rule B. $\displaystyle\int [f(x) + g(x)]\ dx = \int f(x)\ dx + \int g(x)\ dx$

(The integral of a sum is the sum of the integrals.)

EXAMPLE 4

$$\int (5x + 4x^3)\ dx = \int 5x\ dx + \int 4x^3\ dx \qquad \text{Rule B}$$

$$= 5\int x\ dx + 4\int x^3\ dx \qquad \text{Rule A}$$

Then

$$\int (5x + 4x^3)\ dx = 5\cdot\frac{x^2}{2} + 4\cdot\frac{x^4}{4} + C = \frac{5}{2}x^2 + x^4 + C.$$

Don't forget the constant of integration. It is not necessary to write two constants of integration. If we did, we could add them and consider C as the sum. ❖

Note:

We can always check an integration by differentiating.

Thus, in Example 4,

$$\frac{d}{dx}\left(\frac{5}{2}x^2 + x^4 + C\right) = 2\cdot\frac{5}{2}\cdot x + 4x^3 = 5x + 4x^3.$$

DO EXERCISE 11.

11. Evaluate. Don't forget the constant of integration!

$$\int (7x^4 + 2x)\ dx$$

Evaluate. Don't forget the constant of integration.

12. $\displaystyle\int (6e^{2x} - x^{2/5})\, dx$

13. $\displaystyle\int \left(\frac{5}{x} - 7 + \frac{1}{x^6}\right) dx$

14. Using the same set of axes, graph $y = \dfrac{x^3}{3}$, $y = \dfrac{x^3}{3} + 1$, and $y = \dfrac{x^3}{3} - 1$.

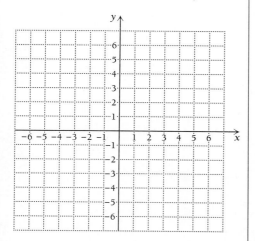

EXAMPLE 5

$$\int (7e^{6x} - \sqrt{x})\, dx = \int 7e^{6x}\, dx - \int \sqrt{x}\, dx$$

$$= \int 7e^{6x}\, dx - \int x^{1/2}\, dx$$

$$= \frac{7}{6}e^{6x} - \frac{x^{(1/2)+1}}{\frac{1}{2}+1} + C = \frac{7}{6}e^{6x} - \frac{x^{3/2}}{\frac{3}{2}} + C$$

$$= \frac{7}{6}e^{6x} - \frac{2}{3}x^{3/2} + C \qquad ❖$$

EXAMPLE 6

$$\int \left(1 - \frac{3}{x} + \frac{1}{x^4}\right) dx = \int 1\, dx - 3\int \frac{dx}{x} + \int x^{-4}\, dx$$

$$= x - 3\ln x + \frac{x^{-4+1}}{-4+1} + C$$

$$= x - 3\ln x - \frac{x^{-3}}{3} + C \qquad ❖$$

DO EXERCISES 12 AND 13.

Another Look at Antiderivatives

The graphs of the antiderivatives of x^2 are the graphs of the functions

$$y = \int x^2\, dx = \frac{x^3}{3} + C$$

for the various values of the constant C.

DO EXERCISE 14.

As we can see in Figs. 1 and 2, x^2 is the derivative of each function. That is, the tangent line at the point

$$\left(a, \frac{a^3}{3} + C\right)$$

has slope a^2. The curves $(x^3/3) + C$ fill up the plane, with exactly one curve going through any given point (x_0, y_0).

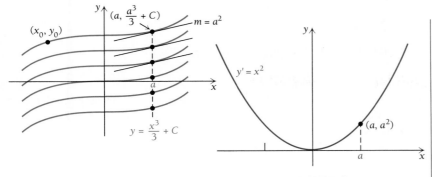

FIGURE 1   **FIGURE 2**

15. Find f such that
$$f'(x) = x^2 \quad \text{and} \quad f(-2) = 5.$$

Suppose we look for an antiderivative of x^2 with a specified value at a certain point, say $f(-1) = 2$. We find that there is only one such function.

EXAMPLE 7   Find the function f such that
$$f'(x) = x^2 \quad \text{and} \quad f(-1) = 2.$$

Solution

a) We find $f(x)$ by integrating:

16. Find g such that
$$g'(x) = 2x - 4 \quad \text{and} \quad g(2) = 9.$$

$$f(x) = \int x^2 \, dx = \frac{x^3}{3} + C.$$

b) The condition $f(-1) = 2$ allows us to find C,
$$f(-1) = \frac{(-1)^3}{3} + C = 2,$$

and solving for C, we get
$$-\tfrac{1}{3} + C = 2$$
$$C = 2 + \tfrac{1}{3}, \quad \text{or} \quad \tfrac{7}{3}.$$

Thus, $f(x) = \dfrac{x^3}{3} + \dfrac{7}{3}.$ ❖

We are often given a solution with a function, such as $f(-1) = 2$ or $(-1, 2)$ in Example 7. The fact that $f(-1) = 2$ is called a *boundary condition*, or an *initial condition*. Sometimes the terms *boundary value*, or *initial value*, are used.

DO EXERCISES 15 AND 16.

17. *Business: Determining total cost from marginal cost.* A company determines that the marginal cost C' of producing the xth unit of a certain product is given by

$$C'(x) = x^2 + 5x.$$

Find the total-cost function C, assuming fixed costs to be $35.

Applications

Determining Total Cost from Marginal Cost

EXAMPLE 8 *Business: Total cost from marginal cost.* A company determines that the marginal cost C' of producing the xth unit of a certain product is given by

$$C'(x) = x^3 + 2x.$$

Find the total-cost function C, assuming that fixed costs (costs when 0 units are produced) are $45. The boundary condition, or initial condition, is that $C = 45 when $x = 0$; that is, $C(0) = 45.

Solution

a) We integrate to find $C(x)$, using K for the integration constant to avoid confusion with the cost function C:

$$C(x) = \int C'(x)\, dx = \int (x^3 + 2x)\, dx = \frac{x^4}{4} + x^2 + K.$$

b) Fixed costs are $45; that is, $C(0) = 45$. This allows us to determine the value of K:

$$C(0) = \frac{0^4}{4} + 0^2 + K = 45$$

$$K = 45.$$

Thus, $C(x) = \dfrac{x^4}{4} + x^2 + 45.$ ❖

DO EXERCISE 17.

Finding Velocity and Distance from Acceleration

Recall that the position coordinate at time t of an object moving along a number line is $s(t)$. Then

$$s'(t) = v(t) = \text{the } velocity \text{ at time } t,$$
$$v'(t) = a(t) = \text{the } acceleration \text{ at time } t.$$

EXAMPLE 9 *Physical science: Distance.* Suppose that $v(t) = 5t^4$ and $s(0) = 9$. Find $s(t)$.

Solution

a)   We find $s(t)$ by integrating:

$$s(t) = \int v(t)\ dt$$

$$= \int 5t^4\ dt = t^5 + C.$$

b)   We determine C by using the boundary condition $s(0) = 9$, which is the initial value, or position, of s at time $t = 0$:

$$s(0) = 0^5 + C = 9$$
$$C = 9.$$

Thus, $s(t) = t^5 + 9$.   ❖

DO EXERCISE 18.

EXAMPLE 10   *Physical science: Distance.* Suppose that $a(t) = 12t^2 - 6$, $v(0) =$ the initial velocity $= 5$, and $s(0) =$ the initial position $= 10$. Find $s(t)$.

Solution

a)   We find $v(t)$ by integrating $a(t)$:

$$v(t) = \int a(t)\ dt = \int (12t^2 - 6)\ dt = 4t^3 - 6t + C_1.$$

b)   The condition $v(0) = 5$ allows us to find C_1:

$$v(0) = 4 \cdot 0^3 - 6 \cdot 0 + C_1 = 5$$
$$C_1 = 5.$$

Thus, $v(t) = 4t^3 - 6t + 5$.

c)   We find $s(t)$ by integrating $v(t)$:

$$s(t) = \int v(t)\ dt = \int (4t^3 - 6t + 5)\ dt = t^4 - 3t^2 + 5t + C_2.$$

d)   The condition $s(0) = 10$ allows us to find C_2:

$$s(0) = 0^4 - 3 \cdot 0^2 + 5 \cdot 0 + C_2 = 10$$
$$C_2 = 10.$$

Thus, $s(t) = t^4 - 3t^2 + 5t + 10$.   ❖

DO EXERCISE 19.

18. *Physical science: Distance.* Suppose that $v(t) = 4t^3$ and $s(0) = 13$. Find $s(t)$.

19. Suppose that $a(t) = 24t^2 - 12$, $v(0) = 7$, and $s(0) = 8$. Find $s(t)$.

EXERCISE SET 5.1

Evaluate.

1. $\int x^6 \, dx$

2. $\int x^7 \, dx$

3. $\int 2 \, dx$

4. $\int 4 \, dx$

5. $\int x^{1/4} \, dx$

6. $\int x^{1/3} \, dx$

7. $\int (x^2 + x - 1) \, dx$

8. $\int (x^2 - x + 2) \, dx$

9. $\int (t^2 - 2t + 3) \, dt$

10. $\int (3t^2 - 4t + 7) \, dt$

11. $\int 5e^{8x} \, dx$

12. $\int 3e^{5x} \, dx$

13. $\int (x^3 - x^{8/7}) \, dx$

14. $\int (x^4 - x^{6/5}) \, dx$

15. $\int \frac{1000}{x} \, dx$

16. $\int \frac{500}{x} \, dx$

17. $\int \frac{dx}{x^2} \left(\text{or} \int \frac{1}{x^2} \, dx \right)$

18. $\int \frac{dx}{x^3}$

19. $\int \sqrt{x} \, dx$

20. $\int \sqrt[3]{x^2} \, dx$

21. $\int \frac{-6}{\sqrt[3]{x^2}} \, dx$

22. $\int \frac{20}{\sqrt[5]{x^4}} \, dx$

23. $\int 8e^{-2x} \, dx$

24. $\int 7e^{-0.25x} \, dx$

25. $\int \left(x^2 - \frac{3}{2}\sqrt{x} + x^{-4/3} \right) dx$

26. $\int \left(x^4 + \frac{1}{8\sqrt{x}} - \frac{4}{5}x^{-2/5} \right) dx$

Find f such that:

27. $f'(x) = x - 3, f(2) = 9$
28. $f'(x) = x - 5, f(1) = 6$
29. $f'(x) = x^2 - 4, f(0) = 7$
30. $f'(x) = x^2 + 1, f(0) = 8$

APPLICATIONS

❖ **Business and Economics**

31. *Total cost from marginal cost.* A company determines that the marginal cost C' of producing the xth unit of a certain product is given by

$$C'(x) = x^3 - 2x.$$

Find the total-cost function C, assuming fixed costs to be $100.

32. *Total cost from marginal cost.* A company determines that the marginal cost C' of producing the xth unit of a certain product is given by

$$C'(x) = x^3 - x.$$

Find the total-cost function C, assuming fixed costs to be $200.

33. *Total revenue from marginal revenue.* A company determines that the marginal revenue R' from selling the xth unit of a certain product is given by

$$R'(x) = x^2 - 3.$$

a) Find the total-revenue function R, assuming that $R(0) = 0$.

b) Why is $R(0) = 0$ a reasonable assumption?

34. *Total revenue from marginal revenue.* A company determines that the marginal revenue R' from selling the xth unit of a certain product is given by

$$R'(x) = x^2 - 1.$$

a) Find the total-revenue function R, assuming that $R(0) = 0$.

b) Why is $R(0) = 0$ a reasonable assumption?

35. *Demand from marginal demand.* A company finds that the rate at which consumer demand quantity changes with respect to price is given by the marginal demand function

$$D'(p) = -\frac{4000}{p^2}.$$

Find the demand function if it is known that 1003 units of the product are demanded by consumers when the price is $4 per unit.

36. *Supply from marginal supply.* A company finds that the rate at which a seller's quantity changes with respect to price is given by the marginal supply function

$$S'(p) = 0.24p^2 + 4p + 10.$$

Find the supply function if it is known that the seller will sell 121 units of the product when the price is $5 per unit.

37. *Efficiency of a machine operator.* The rate at which a machine operator's efficiency E (expressed as a percentage) changes with respect to time t is given by

$$\frac{dE}{dt} = 30 - 10t,$$

where t = the number of hours that the operator has been at work.

a) Find $E(t)$, given that the operator's efficiency after working 2 hr is 72%; that is, $E(2) = 72$.

b) Use the answer to part (a) to find the operator's efficiency after 3 hr; after 5 hr.

A machine operator's efficiency changes with respect to time.

38. *Efficiency of a machine operator.* The rate at which a machine operator's efficiency E (expressed as a percentage) changes with respect to time t is given by

$$\frac{dE}{dt} = 40 - 10t,$$

where t = the number of hours that the operator has been at work.

a) Find $E(t)$, given that the operator's efficiency after working 2 hr is 72%; that is, $E(2) = 72$.

b) Use the answer to part (a) to find the operator's efficiency after 4 hr; after 8 hr.

❖ **Life and Physical Sciences**

Find $s(t)$.

39. $v(t) = 3t^2$, $s(0) = 4$ **40.** $v(t) = 2t$, $s(0) = 10$

Find $v(t)$.

41. $a(t) = 4t$, $v(0) = 20$ **42.** $a(t) = 6t$, $v(0) = 30$

Find $s(t)$.

43. $a(t) = -2t + 6$, $v(0) = 6$, and $s(0) = 10$

44. $a(t) = -6t + 7$, $v(0) = 10$, and $s(0) = 20$

45. *Distance.* For a freely falling object, $a(t) = -32$ ft/sec^2, $v(0) = $ initial velocity $= v_0$, and $s(0) = $ initial height $= s_0$. Find a general expression for $s(t)$ in terms of v_0 and s_0.

46. *Time.* A ball is thrown from a height of 10 ft, where $s(0) = 10$, at an initial velocity of 80 ft/sec, where $v(0) = 80$. How long will it take before the ball hits the ground? (See Exercise 45.)

47. *Distance.* A car with constant acceleration goes from 0 to 60 mph in $\frac{1}{2}$ min. How far does the car travel during that time?

48. *Area of a healing wound.* The area A of a healing wound is decreasing at a rate given by

$$A'(t) = -43.4t^{-2}, \quad 1 \leqslant t \leqslant 7,$$

where t is the time in days and A is in square centimeters.

a) Find $A(t)$ if $A(1) = 39.7$.

b) Find the area of the wound after 7 days.

❖ **Social Sciences**

49. *Memory.* In a certain memory experiment, the rate of memorizing is given by

$$M'(t) = 0.2t - 0.003t^2,$$

where $M(t)$ is the number of Spanish words memorized in t minutes.

a) Find $M(t)$ if it is known that $M(0) = 0$.

b) How many words are memorized in 8 min?

SYNTHESIS EXERCISES

Find f.

50. $f'(t) = \sqrt{t} + \dfrac{1}{\sqrt{t}}, \quad f(4) = 0$

51. $f'(t) = t^{\sqrt{3}}, \quad f(0) = 8$

Evaluate. Each of the following can be integrated using the rules developed in this section, but some algebra may be required beforehand.

52. $\displaystyle\int (5t + 4)^2 \, dt$

53. $\displaystyle\int (x - 1)^2 x^3 \, dx$

54. $\displaystyle\int (1 - t)\sqrt{t} \, dt$

55. $\displaystyle\int \frac{(t + 3)^2}{\sqrt{t}} \, dt$

56. $\displaystyle\int \frac{x^4 - 6x^2 - 7}{x^3} \, dx$

57. $\displaystyle\int (t + 1)^3 \, dt$

58. $\displaystyle\int \frac{1}{\ln 10} \frac{dx}{x}$

59. $\displaystyle\int be^{ax} \, dx$

60. $\displaystyle\int (3x - 5)(2x + 1) \, dx$

61. $\displaystyle\int \sqrt[3]{64x^4} \, dx$

62. $\displaystyle\int \frac{x^2 - 1}{x + 1} \, dx$

63. $\displaystyle\int \frac{t^3 + 8}{t + 2} \, dt$

5.2

OBJECTIVES

a) Find the area under a curve on a given closed interval.

b) Interpret the area under a curve in two other ways.

FINDING AREA USING ANTIDERIVATIVES

We now consider the application of integration to finding areas of certain regions. Consider a function whose outputs are positive in an interval (the function might be 0 at one of the endpoints). We wish to find the area of the region between the graph of the function and the x-axis on that interval, as illustrated in Figs. 1 and 2.

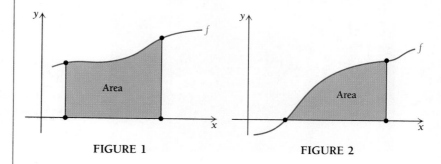

FIGURE 1   FIGURE 2

Let us first consider a constant function $f(x) = m$ on the interval from 0 to x, $[0, x]$.

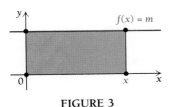

FIGURE 3

Refer to Fig. 3 which shows a rectangle whose area is mx. Suppose that we allow x to vary, giving us rectangles of different areas. The area of each rectangle is still mx. We now have an area function:

$$A(x) = mx.$$

Its graph is shown in Fig. 4.

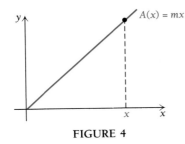

FIGURE 4

DO EXERCISES 1–3. (Exer. 3 is on the following page.)

Let us consider next the linear function $f(x) = mx$ on the interval from 0 to x, $[0, x]$, as shown in Fig. 5.

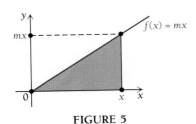

FIGURE 5

The figure formed this time is a triangle, and its area is one-half the base times the height, $\frac{1}{2} \cdot x \cdot (mx)$, or $\frac{1}{2}mx^2$. If we allow x to vary, we again

1. Consider the constant function $f(x) = 3$.

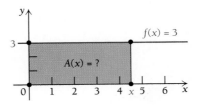

a) Find $A(x)$.
b) Find $A(1)$, $A(2)$, and $A(5)$.
c) Graph $A(x)$.
d) How are $f(x)$ and $A(x)$ related?

2. *Business: Total cost from marginal cost.* A clothing firm, Raggs, Ltd., determines that the marginal cost of each suit that it produces is $50.

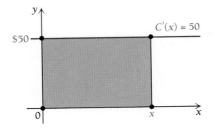

a) Find the total cost $C(x)$ of producing x suits, assuming that fixed costs are $0 (that is, ignore fixed costs).
b) Find the area of the shaded rectangle in the figure. Compare your answer with (a).
c) Graph $C(x)$. Why is this an increasing function?

3. *Business: Total cost from marginal cost.* With better management, Raggs, Ltd., of Margin Exercise 2, is able to decrease its production costs by $10 per suit for every 100 suits it produces. This is shown below.

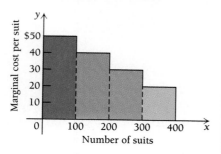

a) Find the total cost of producing 400 suits.

b) Find the total area of the rectangles. Compare your answer with that of part (a).

4. Consider the function $f(x) = 3x$.

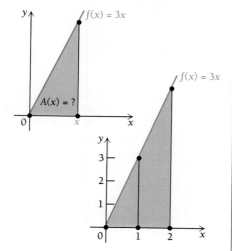

a) Find $A(x)$.

b) Find $A(1)$.　　　　(continued)

get an area function:

$$A(x) = \tfrac{1}{2}mx^2.$$

Its graph is shown in Fig. 6.

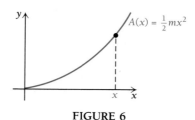

FIGURE 6

DO EXERCISE 4.

Now consider the linear function $f(x) = mx + b$ on the interval from 0 to x, $[0, x]$, as shown in Fig. 7.

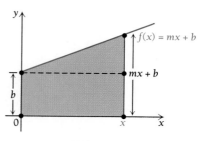

FIGURE 7

The figure formed this time is a trapezoid, and its area is one-half the height times the sum of the lengths of its parallel sides (or, noting the dashed line, the area of the triangle plus the area of the rectangle):

$$\tfrac{1}{2} \cdot x \cdot [b + (mx + b)],$$

or

$$\tfrac{1}{2} \cdot x \cdot (mx + 2b),$$

or

$$\tfrac{1}{2}mx^2 + bx.$$

If we allow x to vary, we again get an area function:

$$A(x) = \tfrac{1}{2}mx^2 + bx.$$

Its graph is as shown in Fig. 8.

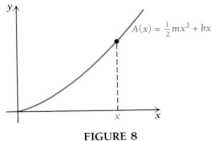

FIGURE 8

DO EXERCISE 5.

Now we consider the function $f(x) = x^2 + 1$ on the interval from 0 to x, $[0, x]$. The graph of the region in question is shown in Fig. 9, but it is not so easy this time to find the area function because the graph of $f(x)$ is not a straight line. Let us tabulate our previous results and look for a pattern.

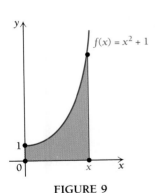

FIGURE 9

$f(x)$	$A(x)$
$f(x) = 3$	$A(x) = 3x$
$f(x) = m$	$A(x) = mx$
$f(x) = 3x$	$A(x) = \tfrac{3}{2}x^2$
$f(x) = mx$	$A(x) = \tfrac{1}{2}mx^2$
$f(x) = mx + b$	$A(x) = \tfrac{1}{2}mx^2 + bx$

You may have conjectured that the area function $A(x)$ is an antiderivative of $f(x)$. In the following exploratory exercises, you will investigate further.

4. c)   Find $A(2)$.

d)   Find $A(3.5)$.

e)   Graph $A(x)$.

f)   How are $f(x)$ and $A(x)$ related?

5. *Business: Total cost from marginal cost.* Raggs, Ltd., of Margin Exercise 3, installs new sewing machines. This allows the marginal cost per suit to decrease continually in such a way that

$$C'(x) = -0.1x + 50.$$

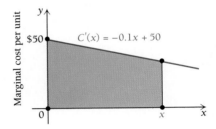

a)   Find the total cost of producing x suits, ignoring fixed costs.

b)   Find the area of the shaded trapezoid in the figure above.

c)   Find the total cost of producing 400 suits. Compare your answer with that of Margin Exercise 3.

EXPLORATORY EXERCISES: FINDING AREAS

1. The region under the graph of $f(x) = x^2 + 1$, on the interval $[0, 2]$, is shown in Fig. 10.

   a) Make a copy of the shaded region on thin paper.

   b) Cut up the shaded region in any way you wish in order to fill up squares in the grid shown in Fig. 11. Make an estimate of the total area.

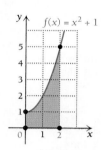

FIGURE 10

FIGURE 11

   c) Using the antiderivative

$$F(x) = \frac{x^3}{3} + x,$$

   find $F(2)$.

   d) Compare your answers in parts (b) and (c).

2. Repeat Exercises 1(a) and (b) for the shaded region of the graph shown in Fig. 12, using the grid shown in Fig. 13.

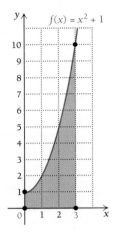

FIGURE 12

FIGURE 13

   c) Using the antiderivative

$$F(x) = \frac{x^3}{3} + x,$$

   find $F(3)$.

   d) Compare your answers in parts (b) and (c).

   The conjecture concerning areas and antiderivatives (or integrals) is true. It is expressed as follows.

THEOREM 4

Let f be a positive, continuous function on an interval $[a, b]$, and let $A(x)$ be the area of the region between the graph of f and the x-axis on the interval $[a, x]$. Then $A(x)$ is a differentiable function of x and

$$A'(x) = f(x).$$

Proof.   The situation described in the theorem is shown in Fig. 14.

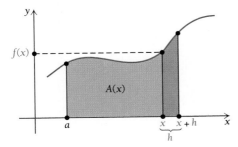

FIGURE 14

The derivative of $A(x)$ is, by the definition of a derivative,

$$A'(x) = \lim_{h \to 0} \frac{A(x + h) - A(x)}{h}.$$

Note from the figure that $A(x + h) - A(x)$ is the area of the small, orange, vertical strip. The area of this small strip is approximately that of a rectangle of base h and height $f(x)$, especially for small values of h. Thus we have

$$A(x + h) - A(x) \approx f(x) \cdot h.$$

Now

$$A'(x) = \lim_{h \to 0} \frac{A(x + h) - A(x)}{h} = \lim_{h \to 0} \frac{f(x) \cdot h}{h} = \lim_{h \to 0} f(x) = f(x),$$

since $f(x)$ does not involve h.

The theorem above also holds if $f(x) = 0$ at one or both endpoints of the interval $[a, b]$.

Since the area function A is an antiderivative of f, and since any two antiderivatives differ by a constant, we easily conclude that the area function and any antiderivative differ by a constant.

We can think of the function A as given by

$$A(x) = \text{the area on the interval } [a, x],$$

where a is some fixed point and x varies as shown in Fig. 15.

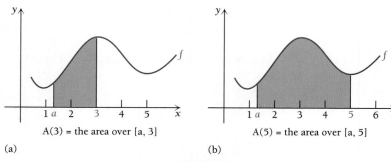

A(3) = the area over [a, 3]   A(5) = the area over [a, 5]

(a)   (b)

FIGURE 15

Now let us find some areas.

EXAMPLE 1   Find the area under the graph of $y = x^2 + 1$ on the interval $[-1, 2]$.

Solution

a) We first make a drawing. This includes a graph of the function as well as the region in question, as seen in Fig. 16.

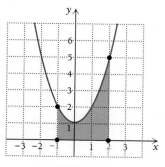

FIGURE 16

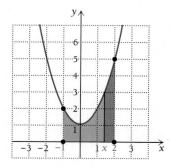

FIGURE 17

b) Next we make a drawing (Fig. 17) showing a portion of the region from -1 to x. Now $A(x)$ is the area of this portion, that is, on the interval $[-1, x]$.

c) Now

$$A(x) = \int (x^2 + 1)\, dx = \frac{x^3}{3} + x + C,$$

where C must be determined. Since we know that $A(-1) = 0$ (there is no area above the number -1), we can substitute -1 for x in $A(x)$, as follows:

$$A(-1) = \frac{(-1)^3}{3} + (-1) + C = 0$$

$$-\frac{1}{3} - 1 + C = 0$$

$$C = \frac{4}{3}.$$

This determines that $C = \frac{4}{3}$, so we have

$$A(x) = \frac{x^3}{3} + x + \frac{4}{3}.$$

Then the area on the interval $[-1, 2]$ is $A(2)$. We compute $A(2)$ as follows:

$$A(2) = \frac{2^3}{3} + 2 + \frac{4}{3}$$

$$= \frac{8}{3} + 2 + \frac{4}{3}$$

$$= \frac{12}{3} + 2$$

$$= 6. \qquad \qquad \text{❖}$$

DO EXERCISE 6.

EXAMPLE 2 Find the area under the graph of $y = x^3$ on the interval $[0, 5]$.

6. Find the area under the graph of $y = x^2 + 3$ on the interval $[1, 2]$.

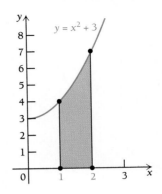

7. Find the area under the graph of $y = x^2 + x$ on the interval $[0, 3]$.

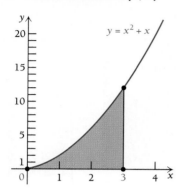

Solution

a) We first make a drawing (Fig. 18) that includes a graph of the function and the region in question.

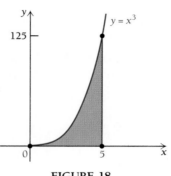

FIGURE 18

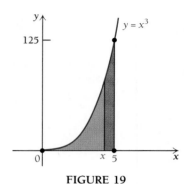

FIGURE 19

b) Next we make a drawing showing a portion of that region from 0 to x (Fig. 19). Now $A(x)$ is the area of this portion, that is, on the interval $[0, x]$.

c) Now

$$A(x) = \int x^3 \, dx = \frac{x^4}{4} + C,$$

where C must be determined. Since we know that $A(0) = 0$, we can substitute 0 for x in $A(x)$, as follows:

$$A(0) = \frac{0^4}{4} + C = 0$$

$$C = 0.$$

This determines C. Thus,

$$A(x) = \frac{x^4}{4}.$$

Then the area on the interval $[0, 5]$ is $A(5)$. We can compute $A(5)$ as follows:

$$A(5) = \frac{5^4}{4} = \frac{625}{4} = 156\tfrac{1}{4}.$$ ❖

DO EXERCISE 7.

We have seen that the antiderivative of $f(x)$, where f is positive, on $[a, x]$ can be interpreted as an accumulating area, which (as x moves from a to b) finally gives the total area on $[a, b]$.

In a similar way, we can interpret the antiderivative of a velocity function $v(t)$ over $[0, b]$ as the area under the curve $y = v(t)$ on $[0, b]$; and the area represents the total distance, assuming that $v(t) > 0$. For example, if we have a velocity function over an interval $[0, b]$, then the area under the curve in that interval is the total distance. Suppose the velocity function is

$$v(t) = t^3.$$

In 5 hr, the total distance covered is $156\frac{1}{4}$. We can see this in Example 2, simply by changing the variable from x to t.

For a marginal cost function $C'(x)$ over the interval $[0, x]$, the area under the curve is the total cost, or the accumulated cost, of producing x units.

EXAMPLE 3 *Business: Total cost from marginal cost.* Raggs, Ltd., goes even further to reduce production costs. The president of the company takes a calculus course and then manages the business in such a way that the marginal cost per suit becomes

$$C'(x) = 0.0003x^2 - 0.2x + 50.$$

Find the total cost of producing 400 suits (ignore fixed costs).

Solution

a) First we make a drawing (Fig. 20) that includes a graph of the function and the region in question.

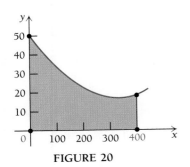

FIGURE 20

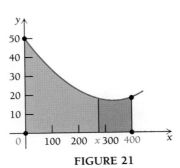

FIGURE 21

b) Next we make a drawing (Fig. 21) showing a portion of that region from 0 to x. Now $A(x)$ is the area of that portion on the interval $[0, x]$.

8. *Business: Total cost from marginal cost.* Refer to Example 3.

a) Compare $10,400 to your answer for Margin Exercise 5. Has the company reduced total costs?

b) Find the total cost of producing 100 suits.

c) Now

$$C(x) = A(x) = \int (0.0003x^2 - 0.2x + 50)\, dx$$

$$= 0.0001x^3 - 0.1x^2 + 50x + K,$$

where K must be determined. Since we are ignoring fixed costs (that is, $K = 0$), we have

$$C(x) = 0.0001x^3 - 0.1x^2 + 50x.$$

Then the area in the interval $[0, 400]$ is $A(400)$, or $C(400)$. We can compute $C(400)$ as follows:

$$C(400) = 0.0001 \cdot 400^3 - 0.1 \cdot 400^2 + 50 \cdot 400, \quad \text{or} \quad \$10,\!400. \quad \clubsuit$$

DO EXERCISE 8.

EXERCISE SET 5.2

Find the area under the given curve on the interval indicated.

1. $y = 4$; $[1, 3]$

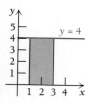

2. $y = 5$; $[1, 3]$

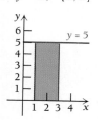

3. $y = 2x$; $[1, 3]$

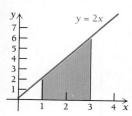

4. $y = x^2$; $[0, 3]$

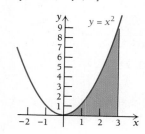

5. $y = x^2$; $[0, 5]$

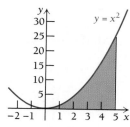

6. $y = x^3$; $[0, 2]$

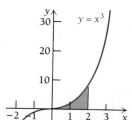

7. $y = x^3$; $[0, 1]$

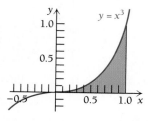

8. $y = 1 - x^2$; $[-1, 1]$

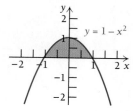

9. $y = 4 - x^2$; $[-2, 2]$

10. $y = e^x$; $[0, 2]$

11. $y = e^x$; $[0, 3]$

12. $y = \dfrac{1}{x}$; $[1, 2]$

13. $y = \dfrac{1}{x}$; $[1, 3]$

14. $y = x^2 - 4x$; $[-4, -2]$

15. $y = x^2 - 4x$; $[-4, -1]$

In each case, give two interpretations of the shaded region other than as area.

16.

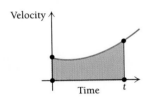

17.

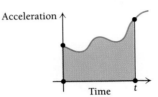

18.

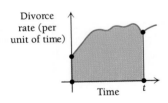

19.

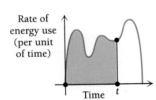

20.

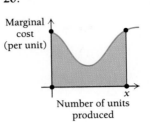

21.

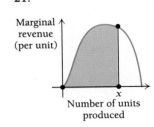

22.

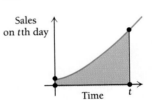

23.

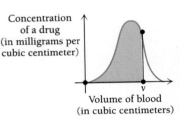

24.

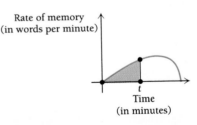

APPLICATIONS

❖ **Business and Economics**

25. *Total cost, revenue, and profit.* A sound company determines that the marginal cost of producing the xth stereo is given by

$$C'(x) = 100 - 0.2x, \quad C(0) = 0.$$

It also determines that its marginal revenue from the sale of the xth stereo is given by

$$R'(x) = 100 + 0.2x, \quad R(0) = 0.$$

a) Find the total cost of producing x stereos.

b) Find the total revenue from selling x stereos.

c) Find the total profit from the production and sale of x stereos.

d) Find the total profit from the production and sale of 1000 stereos.

26. *Total cost, revenue, and profit.* A refrigeration company determines that the marginal cost of producing the xth refrigerator is given by

$$C'(x) = 50 - 0.4x, \quad C(0) = 0.$$

It also determines that its marginal revenue from the sale of the xth refrigerator is given by

$$R'(x) = 50 + 0.4x, \quad R(0) = 0.$$

a) Find the total cost of producing x refrigerators.

b) Find the total revenue from selling x refrigerators.

c) Find the total profit from the production and sale of x refrigerators.

d) Find the total profit from the production and sale of 1000 refrigerators.

❖ Life and Physical Sciences

27. *Distance from velocity.* A particle starts out from the origin. Its velocity at time t is given by

$$v(t) = 3t^2 + 2t.$$

a) Find the distance that the particle has traveled after t hr.

b) Find the distance that the particle has traveled after 5 hr.

28. *Distance from velocity.* A particle starts out from the origin. Its velocity at time t is given by

$$v(t) = 4t^3 + 2t.$$

a) Find the distance that the particle has traveled after t hr.

b) Find the distance that the particle has traveled after 2 hr.

SYNTHESIS EXERCISES

Find the area under the curve on the interval indicated.

29. $y = \dfrac{x^2 - 1}{x - 1}$; [2, 3]

30. $y = \dfrac{x^5 - x^{-1}}{x^2}$; [1, 5]

31. $y = (x - 1)\sqrt{x}$; [4, 16]

32. $y = (x + 2)^3$; [0, 1]

33. $y = \dfrac{\sqrt[3]{x^2} - 1}{\sqrt[3]{x}}$; [1, 8]

34. $y = \dfrac{x^3 + 8}{x + 2}$; [0, 1]

5.3

OBJECTIVES

a) Evaluate a definite integral.

b) Find the area under a graph on the interval [a, b].

c) Solve applied problems involving definite integrals.

INTEGRATION ON AN INTERVAL: THE DEFINITE INTEGRAL

Let f be a positive, continuous function on an interval $[a, b]$. Let F and G be any two antiderivatives of f. Then

$$F(b) - F(a) = G(b) - G(a).$$

To understand this, recall that F and G differ by a constant; that is, $F(x) = G(x) + C$. Then

$$F(b) - F(a) = [G(b) + C] - [G(a) + C] = G(b) - G(a).$$

Thus the difference $F(b) - F(a)$ has the same value for all antiderivatives of f. It is called the **definite integral** of f from a to b.

Definite integrals are generally symbolized as follows:

$$\int_a^b f(x)\, dx, \quad \text{where } a < b.$$

This is read "the integral from a to b of $f(x)\, dx$." (The "dx" is often omitted from the reading.) From the preceding development, we see that to find a definite integral $\int_a^b f(x)\, dx$, we first find an antiderivative $F(x)$. The simplest one is the one for which the constant of integration is 0. From the value of F at b, we subtract the value of F at a. Then we have $F(b) - F(a)$.

DEFINITION

Let f be any positive, continuous function on the interval $[a, b]$, and let F be any antiderivative of f. Then

$$\int_a^b f(x)\, dx = F(b) - F(a).$$

Evaluating definite integrals is called *integrating*. The numbers a and b are known as the **limits of integration**.

EXAMPLE 1 Evaluate $\int_a^b x^2\, dx$.

Solution Using the antiderivative $F(x) = x^3/3$, we have

$$\int_a^b x^2\, dx = \frac{b^3}{3} - \frac{a^3}{3}.$$

❖

DO EXERCISES 1 AND 2.

It is convenient to use an intermediate notation:

$$\int_a^b f(x)\, dx = [F(x)]_a^b = F(b) - F(a).$$

We now evaluate several definite integrals.

EXAMPLE 2

$$\int_{-1}^{2} x^2\, dx = \left[\frac{x^3}{3}\right]_{-1}^{2} = \frac{2^3}{3} - \frac{(-1)^3}{3}$$

$$= \frac{8}{3} - \left(-\frac{1}{3}\right) = \frac{8}{3} + \frac{1}{3} = 3$$

❖

Evaluate.

1. $\displaystyle\int_a^b 2x\, dx$

2. $\displaystyle\int_a^b e^x\, dx$

Evaluate.

3. $\displaystyle\int_1^3 2x\ dx$

4. $\displaystyle\int_{-2}^0 e^x\ dx$

5. $\displaystyle\int_0^1 (2x - x^2\ dx$

6. $\displaystyle\int_1^e \left(1 + 3x^2 - \frac{1}{x}\right) dx$

EXAMPLE 3

$$\int_0^3 e^x\ dx = [e^x]_0^3 = e^3 - e^0 = e^3 - 1 \qquad \diamond$$

EXAMPLE 4

$$\int_1^4 (x^2 - x)\ dx = \left[\frac{x^3}{3} - \frac{x^2}{2}\right]_1^4 = \left(\frac{4^3}{3} - \frac{4^2}{2}\right) - \left(\frac{1^3}{3} - \frac{1^2}{2}\right)$$

$$= \left(\frac{64}{3} - \frac{16}{2}\right) - \left(\frac{1}{3} - \frac{1}{2}\right)$$

$$= \frac{64}{3} - 8 - \frac{1}{3} + \frac{1}{2} = 13\frac{1}{2} \qquad \diamond$$

EXAMPLE 5

$$\int_1^e \left(1 + 2x - \frac{1}{x}\right) dx = [x + x^2 - \ln x]_1^e$$

$$= (e + e^2 - \ln e) - (1 + 1^2 - \ln 1)$$

$$= (e + e^2 - 1) - (1 + 1 - 0)$$

$$= e + e^2 - 1 - 1 - 1$$

$$= e + e^2 - 3 \qquad \diamond$$

For some applications that we will soon consider, it is important to keep in mind that in $\int_a^b f(x)\ dx$, we assumed that $a < b$. That is, the larger number is on the top!

DO EXERCISES 3–6.

The area under a curve can be expressed by a definite integral.

THEOREM 5

Let f be a positive, continuous function over the closed interval $[a, b]$. The area under the graph of f on the interval $[a, b]$ is

$$\int_a^b f(x)\ dx.$$

Proof. Let

$$A(x) = \text{the area of the region over } [0, x].$$

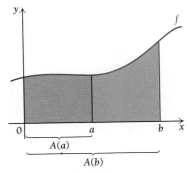

FIGURE 1

Then

$$A'(x) = f(x),$$

so $A(x)$ is an antiderivative of $f(x)$. Then

$$\int_a^b f(x) \ dx = A(b) - A(a).$$

But $A(b) - A(a)$ is the area over $[0, b]$ minus the area over $[0, a]$, which is the area over $[a, b]$.

Let us now find some areas.

EXAMPLE 6   Find the area under $y = x^2 + 1$ on $[-1, 2]$.

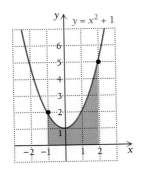

FIGURE 2

7. Find the area under $y = x^2 + 3$ on $[1, 2]$. Don't forget that it helps to draw the graph. (Compare the result with Margin Exercise 6 of Section 5.2.)

Solution

$$\int_{-1}^{2} (x^2 + 1)\, dx = \left[\frac{x^3}{3} + x\right]_{-1}^{2}$$

$$= \left(\frac{2^3}{3} + 2\right) - \left(\frac{(-1)^3}{3} + (-1)\right)$$

$$= \left(\frac{8}{3} + 2\right) - \left(-\frac{1}{3} - 1\right)$$

$$= \frac{8}{3} + 2 + \frac{1}{3} + 1 = 6$$

Compare this example with Example 1 of Section 5.2.   ❖

DO EXERCISE 7.

EXAMPLE 7   Find the area under $y = x^3$ on $[0, 5]$.

Solution

$$\int_{0}^{5} x^3\, dx = \left[\frac{x^4}{4}\right]_{0}^{5}$$

$$= \frac{5^4}{4} - \frac{0^4}{4} = \frac{625}{4}$$

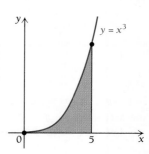

FIGURE 3

8. Find the area under $y = x^2 + x$ on $[0, 3]$. (Compare the result with Margin Exercise 7 of Section 5.2.)

Compare this example with Example 2 of Section 5.2.   ❖

DO EXERCISE 8.

EXAMPLE 8   Find the area under $y = 1/x$ on $[1, 4]$.

Solution

$$\int_{1}^{4} \frac{dx}{x} = [\ln x]_{1}^{4} = \ln 4 - \ln 1$$

$$= \ln 4 \approx 1.3863$$

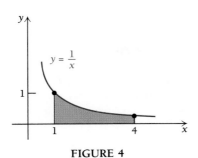

FIGURE 4

9. Find the area under $y = 1/x$ on $[1, 7]$.

❖

EXAMPLE 9   Find the area under $y = 1/x^2$ on $[1, b]$.

Solution

$$\int_1^b \frac{dx}{x^2} = \int_1^b x^{-2} \, dx = \left[\frac{x^{-2+1}}{-2+1} \right]_1^b$$

$$= \left[\frac{x^{-1}}{-1} \right]_1^b = \left[-\frac{1}{x} \right]_1^b$$

$$= \left(-\frac{1}{b} \right) - \left(-\frac{1}{1} \right)$$

$$= 1 - \frac{1}{b}$$

10. Find the area under $y = 1/x^4$ on $[1, b]$.

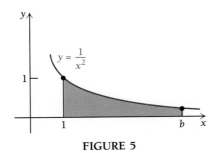

FIGURE 5

❖

DO EXERCISES 9 AND 10.

The following properties of definite integrals can be derived rather easily from the definition of a definite integral and from the properties of the indefinite integral.

11. Evaluate $\int_1^2 20x^3 \ dx$.

PROPERTY 1

$$\int_a^b k \cdot f(x) \ dx = k \cdot \int_a^b f(x) \ dx$$

The integral of a constant times a function is the constant times the integral of the function. That is, we can "factor out" a constant from the integrand.

EXAMPLE 10

$$\int_0^5 100e^x \ dx = 100 \int_0^5 e^x \ dx$$
$$= 100[e^x]_0^5$$
$$= 100(e^5 - e^0)$$
$$= 100(e^5 - 1)$$
$$\approx 14{,}741.32$$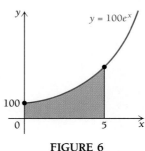

FIGURE 6

DO EXERCISE 11.

PROPERTY 2

$$\int_a^b [f(x) + g(x)] \ dx = \int_a^b f(x) \ dx + \int_a^b g(x) \ dx$$

The integral of a sum is the sum of the integrals.

PROPERTY 3

For $a < c < b$,

$$\int_a^b f(x) \ dx = \int_a^c f(x) \ dx + \int_c^b f(x) \ dx.$$

For any number c between a and b, the integral from a to b is the integral from a to c plus the integral from c to b (see Fig. 7).

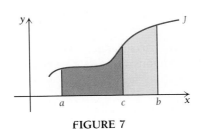

FIGURE 7

Property 3 has particular application when a function is defined piecewise in different ways over subintervals.

EXAMPLE 11 Find the area under the graph of $y = f(x)$ from -4 to 5, where

$$f(x) = \begin{cases} 9, & \text{if } x < 3, \\ x^2, & \text{if } x \geqslant 3. \end{cases}$$

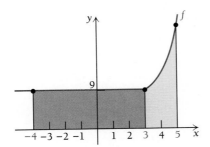

FIGURE 8

Solution

$$\int_{-4}^{5} f(x)\ dx = \int_{-4}^{3} f(x)\ dx + \int_{3}^{5} f(x)\ dx$$

$$= \int_{-4}^{3} 9\ dx + \int_{3}^{5} x^2\ dx$$

$$= 9\int_{-4}^{3} dx + \int_{3}^{5} x^2\ dx$$

$$= 9[x]_{-4}^{3} + \left[\frac{x^3}{3}\right]_{3}^{5}$$

$$= 9[3 - (-4)] + \left(\frac{5^3}{3} - \frac{3^3}{3}\right)$$

$$= 95\tfrac{2}{3}$$

❖

DO EXERCISE 12.

12. Find the area under the graph of $y = f(x)$ from -3 to 2, where

$$f(x) = \begin{cases} 4, & \text{if } x < 0, \\ 4 - x^2, & \text{if } x \geqslant 0. \end{cases}$$

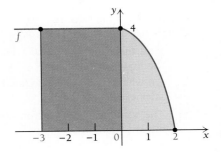

13. *Business: Accumulated sales.* The sales of a company are expected to grow continuously at a rate given by the function

$$S'(t) = 200e^t,$$

where $S'(t)$ = the sales rate, in dollars per day, at time t, in days.

a) Find the accumulated sales for the first 8 days.

b) On what day will accumulated sales exceed $300,000?

c) The accumulated sales from the 8th through the 10th day is given by the integral from 7 to 10,

$$\int_7^{10} S'(t)\ dt.$$

Find this.

An Application

EXAMPLE 12 *Business: Accumulated sales.* The sales of a company are expected to grow continuously at a rate given by the function

$$S'(t) = 100e^t,$$

where $S'(t)$ = the sales rate, in dollars per day, at time t, in days.

a) Find the accumulated sales for the first 7 days.

b) On what day will accumulated sales exceed $810,000?

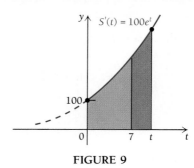

FIGURE 9

Solution

a) Accumulated sales through day 7 are

$$\int_0^7 S'(t)\ dt = \int_0^7 100e^t\ dt$$

$$= 100 \int_0^7 e^t\ dt$$

$$= 100[e^t]_0^7$$

$$= 100(e^7 - e^0)$$

$$= 100(1096.633158 - 1)$$

$$= 100(1095.633158)$$

$$\approx \$109,563.32.$$

b) Accumulated sales through day k are

$$\int_0^k S'(t)\ dt = \int_0^k 100e^t\ dt$$

$$= 100 \int_0^k e^t\ dt$$

$$= 100[e^t]_0^k = 100(e^k - e^0)$$

$$= 100(e^k - 1).$$

We set this equal to $810,000 and solve for k:

$$100(e^k - 1) = 810{,}000$$
$$e^k - 1 = 8100$$
$$e^k = 8101.$$

We solve this equation for k using natural logarithms:

$$\ln e^k = \ln 8101$$
$$k = 8.999743$$
$$k \approx 9.$$

Accumulated sales will exceed $810,000 approximately on day 9.

❖

DO EXERCISE 13 ON PAGE 384.

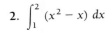

EXERCISE SET 5.3

Evaluate.

1. $\displaystyle\int_0^1 (x - x^2)\, dx$

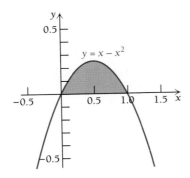

$y = x - x^2$

2. $\displaystyle\int_1^2 (x^2 - x)\, dx$

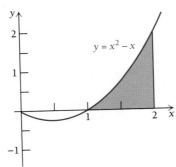

$y = x^2 - x$

3. $\displaystyle\int_{-1}^1 (x^2 - x^4)\, dx$

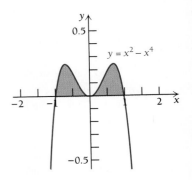

$y = x^2 - x^4$

4. $\displaystyle\int_0^b 2e^{3x}\, dx$

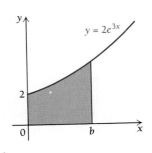

$y = 2e^{3x}$

5. $\displaystyle\int_a^b e^t \, dt$ 6. $\displaystyle\int_0^a (ax - x^2) \, dx$

7. $\displaystyle\int_a^b 3t^2 \, dt$ 8. $\displaystyle\int_a^b 4t^3 \, dt$

9. $\displaystyle\int_1^e \left(x + \frac{1}{x}\right) dx$ 10. $\displaystyle\int_1^e \left(x - \frac{1}{x}\right) dx$

11. $\displaystyle\int_0^1 \sqrt{x} \, dx$ 12. $\displaystyle\int_0^1 3\sqrt{x} \, dx$

13. $\displaystyle\int_0^1 \frac{10}{17} t^3 \, dt$ 14. $\displaystyle\int_0^1 \frac{12}{13} t^2 \, dt$

Find the area under the graph on the interval indicated.

15. $y = x^3$;   [0, 2]   16. $y = x^4$;   [0, 1]

17. $y = x^2 + x + 1$;   [2, 3]

18. $y = 2 - x - x^2$;   [−2, 1]

19. $y = 5 - x^2$;   [−1, 2]   20. $y = e^x$;   [−2, 3]

21. $y = e^x$;   [−1, 5]   22. $y = 2x + \dfrac{1}{x^2}$;   [1, 4]

23. $y = 2x - \dfrac{1}{x^2}$;   [1, 3]

Find the area under the graph on [−2, 3], where:

24. $f(x) = \begin{cases} x^2, & \text{if } x < 1, \\ 1, & \text{if } x \geq 1. \end{cases}$

25. $f(x) = \begin{cases} 4 - x^2, & \text{if } x < 0, \\ 4, & \text{if } x \geq 0. \end{cases}$

APPLICATIONS

❖ **Business and Economics**

26. *Accumulated sales.* Raggs, Ltd., estimates that its sales will grow continuously at a rate given by the function

$$S'(t) = 10e^t,$$

where $S'(t)$ = the sales rate, in dollars per day, at time t, in days.

  a) Find the accumulated sales for the first 5 days.

  b) Find the sales from the 2nd day through the 5th day. (This is the integral from 1 to 5.)

  c) On what day will accumulated sales exceed $40,000?

27. *Accumulated sales.* A company estimates that its sales will grow continuously at a rate given by the function

$$S'(t) = 20e^t,$$

where $S'(t)$ = the sales rate, in dollars per day, at time t, in days.

  a) Find the accumulated sales for the first 5 days.

  b) Find the sales from the 2nd day through the 5th day. (This is the integral from 1 to 5.)

  c) On what day will accumulated sales exceed $20,000?

28. *Total cost from marginal cost.* Raggs, Ltd., determines that the marginal cost per suit is given by

$$C'(x) = 0.0003x^2 - 0.2x + 50.$$

Ignoring fixed costs, find the total cost of producing the 101st suit through the 400th suit (that is, integrate from $x = 100$ to $x = 400$).

29. *Total cost from marginal cost.* Using the information in Exercise 28, find the cost of producing the 201st suit through the 400th suit (that is, integrate from $x = 200$ to $x = 400$).

30. *Total profit from marginal profit.* A company has the marginal-profit function given by

$$P'(x) = -2x + 980.$$

This means that the rate of change of total profit with respect to the number of units x produced is $P'(x)$. Find the total profit from the production and sale of the 101st unit through the 800th unit of the product.

31. *Total revenue from marginal revenue.* A fight promoter sells x tickets and has a marginal-revenue function given by

$$R'(x) = 200x - 1080.$$

This means that the rate of change of total revenue with respect to the number of tickets sold x is $R'(x)$. Find the total revenue from the sale of the 1001st ticket through the 1300th ticket.

❖ **Life and Physical Sciences**

32. A particle starts out from the origin. Its velocity at time t is given by

$$v(t) = 3t^2 + 2t.$$

How far does it travel from the 2nd hour through the 5th hour (from $t = 1$ to $t = 5$)?

33. A particle starts out from the origin. Its velocity at time t is given by

$$v(t) = 4t^3 + 2t.$$

How far does it travel from the start through the 3rd hour (from $t = 0$ to $t = 3$)?

34. *Total pollution.* A factory is polluting a lake in such a way that the rate of pollutants entering the lake at time t, in months, is given by

$$N'(t) = 280t^{3/2},$$

where N = the total number of pounds of pollutants in the lake at time t.

a) How many pounds of pollutants enter the lake in 16 months?

b) An environmental expert tells the factory that it will have to begin cleanup procedures after 50,000 lb of pollutants have entered the lake. After what amount of time will this occur?

35. *Bacteria growth.* A population of bacteria in a biological experiment grows at the rate of

$$P'(t) = 1200e^{0.32t},$$

where $P(t)$ = the total population at time t, in days.

a) How many bacteria will there be after 20 days?

b) The experiment will cease when the population reaches 4 million. After what amount of time will this occur?

❖ **Social Sciences**

Memorizing. In the psychological process of memorizing, the rate of memorizing (say, in words per minute) initially increases with respect to time. Eventually, however, a maximum rate of memorizing is reached, after which the memory rate begins to decrease.

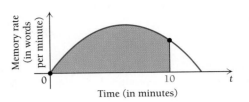

36. Suppose in a certain memory experiment that the rate of memorizing is given by

$$M'(t) = -0.009t^2 + 0.2t,$$

where $M'(t)$ = the memory rate, in words per minute. How many words are memorized in the first 10 min (from $t = 0$ to $t = 10$)?

37. Suppose in a certain memory experiment that the rate of memorizing is given by

$$M'(t) = -0.003t^2 + 0.2t,$$

where $M'(t)$ = the memory rate, in words per minute. How many words are memorized in the first 10 min (from $t = 0$ to $t = 10$)?

38. *Industrial learning curve.* A company is producing a new product. However, due to the nature of the product, it is felt that the time required to produce each unit will decrease as the workers become more familiar with production procedure. It is determined that the function for the learning process is

$$T(x) = ax^b,$$

where $T(x)$ = the average time to produce x units, x = the number of units produced, a = the number of hours required to produce the first unit, and b = the slope of the learning curve.

a) Find an expression for the total time required to produce 100 units.

b) Suppose that $a = 100$ hr and $b = -0.322$. Find the total time required to produce 100 units.

SYNTHESIS EXERCISES

Evaluate.

39. $\int_1^2 (4x + 3)(5x - 2)\, dx$

40. $\int_2^5 (t + \sqrt{3})(t - \sqrt{3})\, dt$

41. $\int_0^1 (t + 1)^3\, dt$

42. $\int_1^3 \left(x - \dfrac{1}{x}\right)^2 dx$

43. $\int_1^3 \dfrac{t^5 - t}{t^3}\, dt$

44. $\int_4^9 \dfrac{t + 1}{\sqrt{t}}\, dt$

45. $\int_3^5 \dfrac{x^2 - 4}{x - 2}\, dx$

46. $\int_0^1 \dfrac{t^3 + 1}{t + 1}\, dt$

COMPUTER-GRAPHING CALCULATOR EXERCISES

Use a computer software package or a graphing calculator to graph each of the following functions over the indicated interval. Then find the area under the graph.

47. $f(x) = x^3 - 4x$; $[-2, 0]$

48. $f(x) = x - x^3$; $[0, 1]$

49. $f(x) = x^3 - 9x^2 + 27x + 50$; $[-1.2, 6.3]$

50. $f(x) = x^4 + 4x^3 - 36x^2 - 160x + 300$; $[-8, 1.4]$

THE CALCULUS EXPLORER
Riemann Sums

Use this program to evaluate various integrals in this exercise set. Approximation methods that we will develop in Section 5.8 are used.

5.4

THE DEFINITE INTEGRAL: THE AREA BETWEEN CURVES

OBJECTIVES

a) Evaluate the definite integral of a continuous function.

b) Find the area of a region bounded by two graphs.

c) Solve applied problems involving the area between two graphs.

We have considered the definite integral for functions that are positive on an interval $[a, b]$ (the function might be 0 at one or both endpoints). Now we will consider functions that have negative values. First, let us evaluate the integral of a function without negative values (see Fig. 1):

$$\int_0^2 x^2\, dx = \left[\dfrac{x^3}{3}\right]_0^2 = \dfrac{2^3}{3} - \dfrac{0^3}{3} = \dfrac{8}{3}.$$

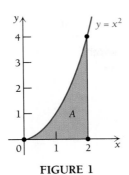

FIGURE 1

Thus the area of the shaded region is $\frac{8}{3}$.

Now let us consider the function $y = -x^2$ on the interval $[0, 2]$ (see Fig. 2). Even though we have not defined the definite integral for functions with negative values, let us apply the evaluation procedures and see what we get:

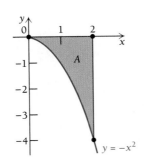

FIGURE 2

The graphs of the two functions shown in Figs. 1 and 2 are reflections of each other across the x-axis. Thus the areas of the shaded regions are the same—that is, $\frac{8}{3}$. The evaluation procedure in the second case gave us $-\frac{8}{3}$. This illustrates that for negative-valued functions, the definite integral gives us the additive inverse of the area between the curve and the x-axis.

DO EXERCISES 1 AND 2.

1. To find the area of the shaded region, evaluate

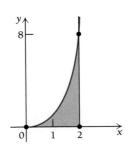

2. To find the area of the shaded region, evaluate

$$\int_0^2 -x^3 \, dx$$

and take the additive inverse, or the opposite, of the answer.

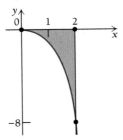

Now let us consider the function $x^2 - 1$ on $[-1, 2]$ (see Fig. 3). It has both positive and negative values. We will apply the preceding evaluation procedures, even though function values are not all nonnegative. We will do this in two ways.

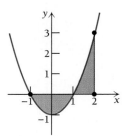

FIGURE 3

First,

$$\int_{-1}^{2} (x^2 - 1) \, dx = \int_{-1}^{1} (x^2 - 1) \, dx + \int_{1}^{2} (x^2 - 1) \, dx$$

$$= \left[\frac{x^3}{3} - x \right]_{-1}^{1} + \left[\frac{x^3}{3} - x \right]_{1}^{2}$$

$$= \left[\left(\frac{1^3}{3} - 1 \right) - \left(\frac{(-1)^3}{3} - (-1) \right) \right]$$

$$+ \left[\left(\frac{2^3}{3} - 2 \right) - \left(\frac{1^3}{3} - 1 \right) \right]$$

$$= \left[-\frac{4}{3} \right] + \left[\frac{4}{3} \right] = 0.$$

This shows that the area of the region under the x-axis is the same as the area of the region over the x-axis.

Now let us evaluate in another way:

$$\int_{-1}^{2} (x^2 - 1) \, dx = \left[\frac{x^3}{3} - x \right]_{-1}^{2}$$

$$= \left(\frac{2^3}{3} - 2 \right) - \left[\frac{(-1)^3}{3} - (-1) \right]$$

$$= \left(\frac{8}{3} - 2 \right) - \left(-\frac{1}{3} + 1 \right)$$

$$= \frac{8}{3} - 2 + \frac{1}{3} - 1 = 0.$$

This result is consistent with the first. Thus we are motivated to extend our definition of definite integral to include any continuous function having positive, negative, or zero values.

DEFINITION

For any function with antiderivative F on an interval $[a, b]$,

$$\int_a^b f(x)\, dx = F(b) - F(a).$$

DO EXERCISES 3 AND 4.

The definite integral turns out to be the area above the x-axis minus the area below.

EXAMPLE 1   Decide whether $\int_a^b f(x)\, dx$ is positive, negative, or zero.

a)

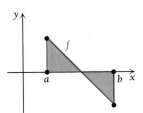

b)

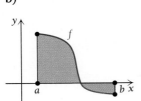

c)

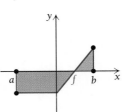

Solution

a) In this figure, there is the same area above the x-axis as below. Thus,

$$\int_a^b f(x)\, dx = 0.$$

b) In this figure, there is more area above the x-axis than below. Thus,

$$\int_a^b f(x)\, dx > 0.$$

3. Evaluate.

a) $\displaystyle\int_0^1 (x^2 - x)\, dx$

b) $\displaystyle\int_1^2 (x^2 - x)\, dx$

c) $\displaystyle\int_0^2 (x^2 - x)\, dx$

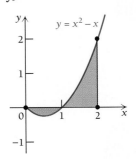

4. Evaluate.

a) $\displaystyle\int_{-2}^0 x^3\, dx$

b) $\displaystyle\int_0^2 x^3\, dx$

c) $\displaystyle\int_{-2}^2 x^3\, dx$

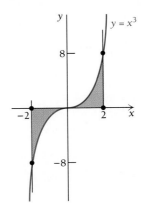

In each exercise:

a) Decide whether $\int_a^b f(x)\, dx$ is positive, negative, or zero.

b) Express $\int_a^b f(x)\, dx$ in terms of A.

5.
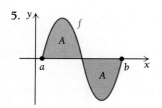

6.
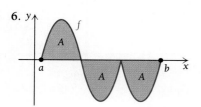

7.
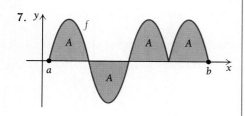

8.
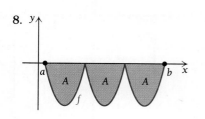

c) In this figure, there is more area below the x-axis than above. Thus,

$$\int_a^b f(x)\, dx < 0.$$ ❖

DO EXERCISES 5–8.

The Area of a Region Bounded by Two Graphs

Suppose we want to find the area of a region bounded by the graphs of two functions, $y = f(x)$ and $y = g(x)$, as shown in Fig. 4.

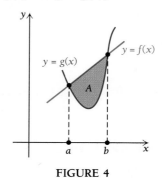

FIGURE 4

Note that the area of the desired region A in Fig. 5(a) is that of A_2 in Fig. 5(b) minus that of A_1 in Fig. 5(c).

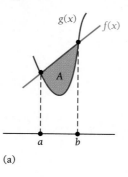
(a)

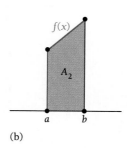
(b)

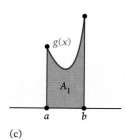
(c)

FIGURE 5

Thus,

$$A = \int_a^b f(x)\, dx - \int_a^b g(x)\, dx, \quad \text{or} \quad A = \int_a^b [f(x) - g(x)]\, dx.$$

In general, we have the following.

THEOREM 6

Let f and g be continuous functions and suppose that $f(x) \geqslant g(x)$ over the interval $[a, b]$. Then the area of the region between the two curves, from $x = a$ to $x = b$, is

$$\int_a^b [f(x) - g(x)] \, dx.$$

EXAMPLE 2   Find the area of the region bounded by the graphs of $y = 2x + 1$ and $y = x^2 + 1$.

Solution

a)   First, make a reasonably accurate sketch, as in Fig. 6, to ensure that you have the right configuration. Note which is the *upper* graph. In this case, $2x + 1 \geqslant x^2 + 1$ over the interval $[0, 2]$.

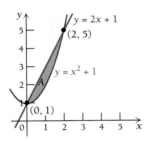

FIGURE 6

b)   Second, if boundaries are not stated, determine the first coordinates of possible points of intersection. Occasionally you can do this just by looking at the graph. If not, you can solve the system of equations as follows. At the points of intersection, $y = x^2 + 1$ and $y = 2x + 1$, so

$$x^2 + 1 = 2x + 1$$
$$x^2 - 2x = 0$$
$$x(x - 2) = 0$$
$$x = 0 \quad \text{or} \quad x = 2.$$

Thus the interval with which we are concerned is $[0, 2]$.

9. Find the area of the region bounded by the graphs of

$$y = x \quad \text{and} \quad y = x^2.$$

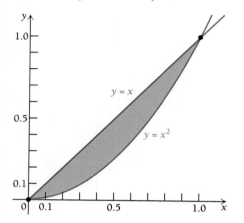

c) Compute the area as follows:

$$\int_0^2 [(2x + 1) - (x^2 + 1)] \, dx = \int_0^2 (2x - x^2) \, dx$$

$$= \left[x^2 - \frac{x^3}{3} \right]_0^2$$

$$= \left(2^2 - \frac{2^3}{3} \right) - \left(0^2 - \frac{0^3}{3} \right)$$

$$= 4 - \frac{8}{3} = \frac{4}{3}.$$

DO EXERCISE 9.

An Application

EXAMPLE 3 *Life science: Emission control.* A clever college student develops an engine that is believed to meet federal standards for emission control. The engine's rate of emission is given by

$$E(t) = 2t^2,$$

where $E(t) =$ the emissions, in billions of pollution particulates per year, at time t, in years. The emission rate of a conventional engine is given by

$$C(t) = 9 + t^2.$$

The curves for both are shown in Fig. 7.

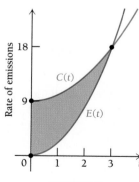

FIGURE 7

a) At what point in time will the emission rates be the same?

b) What is the reduction in emissions resulting from using the student's engine?

Solution

a)   The rates of emission will be the same when $E(t) = C(t)$, or

$$2t^2 = 9 + t^2$$

$$t^2 - 9 = 0$$

$$(t - 3)(t + 3) = 0$$

$$t = 3 \quad \text{or} \quad t = -3.$$

Since negative time has no meaning in this problem, the emission rates will be the same when $t = 3$ yr.

b)   The reduction in emissions is represented by the area of the shaded region in Fig. 7. It is the area between $y = 9 + t^2$ and $y = 2t^2$, from $t = 0$ to $t = 3$, and is computed as follows:

$$\int_0^3 [(9 + t^2) - 2t^2] \, dt = \int_0^3 (9 - t^2) \, dt$$

$$= \left[9t - \frac{t^3}{3} \right]_0^3$$

$$= \left(9 \cdot 3 - \frac{3^3}{3} \right) - \left(9 \cdot 0 - \frac{0^3}{3} \right)$$

$$= 27 - 9$$

$$= 18 \text{ billion pollution particulates.} \quad \diamondsuit$$

DO EXERCISE 10.

10. *Business: Integrating to find profit.* A company determines that its marginal revenue per unit is given by

$$R'(x) = 14x, \quad R(0) = 0.$$

Its marginal cost per unit is given by

$$C'(x) = 4x - 10, \quad C(0) = 0.$$

Total profit from the production and sale of k units is given by

$$P(k) = R(k) - C(k)$$

$$= \int_0^k [R'(x) - C'(x)] \, dx.$$

a)   Find $P(k)$.
b)   Find $P(5)$.

EXERCISE SET   5.4

Find the area of the region bounded by the given graphs.

1. $y = x$, $y = x^3$, $x = 0$, $x = 1$
2. $y = x$, $y = x^4$
3. $y = x + 2$, $y = x^2$
4. $y = x^2 - 2x$, $y = x$
5. $y = 6x - x^2$, $y = x$
6. $y = x^2 - 6x$, $y = -x$
7. $y = 2x - x^2$, $y = -x$

8. $y = x^2$, $y = \sqrt{x}$
9. $y = x$, $y = \sqrt{x}$
10. $y = 3$, $y = x$, $x = 0$
11. $y = 5$, $y = \sqrt{x}$, $x = 0$
12. $y = x^2$, $y = x^3$
13. $y = 4 - x^2$, $y = 4 - 4x$
14. $y = x^2 + 1$, $y = x^2$, $x = 1$, $x = 3$
15. $y = x^2 + 3$, $y = x^2$, $x = 1$, $x = 2$

Find the area of the shaded region.

16. $f(x) = 2x + x^2 - x^3$, $g(x) = 0$

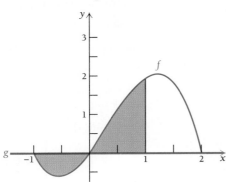

17. $f(x) = x^3 + 3x^2 - 9x - 12$, $g(x) = 4x + 3$

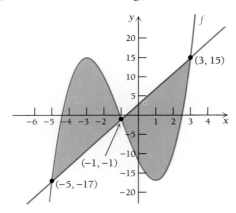

18. $f(x) = x^4 - 8x^3 + 18x^2$, $g(x) = x + 28$

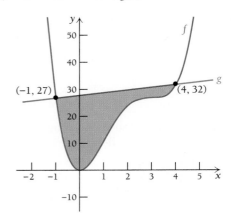

19. $f(x) = 4x - x^2$, $g(x) = x^2 - 6x + 8$

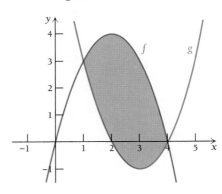

APPLICATIONS

❖ **Business and Economics**

20. *Total profit.* A company determines that its marginal revenue per day is given by

$$R'(t) = 100e^t, \quad R(0) = 0,$$

where $R(t)$ = the revenue, in dollars, on the tth day. The company's marginal cost per day is given by

$$C'(t) = 100 - 0.2t, \quad C(0) = 0,$$

where $C(t)$ = the cost, in dollars, on the tth day. Find the total profit from $t = 0$ to $t = 10$ (the first 10 days).

Note:

$$P(T) = R(T) - C(T) = \int_0^T [R'(t) - C'(t)] \, dt.$$

❖ **Social Sciences**

21. *Memorizing.* In a certain memory experiment, subject A is able to memorize words at the rate given by

$$m'(t) = -0.009t^2 + 0.2t \quad \text{(words per minute)}.$$

In the same memory experiment, subject B is able to

memorize at the rate given by

$$M'(t) = -0.003t^2 + 0.2t \quad \text{(words per minute)}.$$

a) Which subject has the higher rate of memorization?

b) How many more words does that subject memorize from $t = 0$ to $t = 10$ (during the first 10 min)?

SYNTHESIS EXERCISES

Find the area of the region bounded by the given graphs.

22. $y = x^2$, $y = x^{-2}$, $x = 1$, $x = 5$

23. $y = e^x$, $y = e^{-x}$, $x = 0$, $x = 1$

24. $y = x^2$, $y = \sqrt[3]{x^2}$, $x = 1$, $x = 8$

25. $y = x^2$, $y = x^3$, $x = -1$, $x = 1$

26. $x + 2y = 2$, $y - x = 1$, $2x + y = 7$

27. $y = x + 6$, $y = -2x$, $y = x^3$

28. Find the area of the region bounded by $y = x^3 - 3x + 2$, the x-axis, and the first coordinates of the relative maximum and minimum values of the function.

29. Find the area of the region bounded by $y = 3x^5 - 20x^3$, the x-axis, and the first coordinates of the relative maximum and minimum values of the function.

30. *Life science: Poiseuille's Law.* The flow of blood in a blood vessel is faster toward the center of the vessel and slower toward the outside. The speed of the blood is given by

$$V = \frac{p}{4Lv}(R^2 - r^2),$$

where R = the radius of the blood vessel, r = the distance of the blood from the center of the vessel, and p, v, and L are physical constants related to the pressure and viscosity of the blood and the length of the blood vessel. If R is constant, we can think of V as a function of r:

$$V(r) = \frac{p}{4Lv}(R^2 - r^2).$$

The *total blood flow* Q is given by

$$Q = \int_0^R 2\pi \cdot V(r) \cdot r \cdot dr.$$

Find Q.

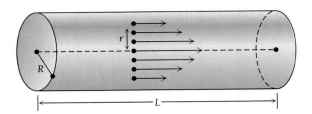

COMPUTER-GRAPHING CALCULATOR EXERCISES

31. Consider the following functions:

$$f(x) = 3.8x^5 - 18.6x^3,$$
$$g(x) = 19x^4 - 55.8x^2.$$

a) Use a computer software package or a graphing calculator to graph each of these functions over the interval $[-3, 3]$.

b) Estimate the first coordinates a, b, and c of the three points of intersection of the two graphs.

c) Estimate the area between the curves over the interval $[a, b]$.

d) Estimate the area between the curves over the interval $[b, c]$.

5.5

a) Evaluate integrals using substitution.

b) Solve applied problems involving integration by substitution.

INTEGRATION TECHNIQUES: SUBSTITUTION

The following formulas provide a basis for an integration technique called *substitution*.

A. $\displaystyle\int u^r \, du = \frac{u^{r+1}}{r+1} + C, \quad$ provided $r \neq -1$

B. $\displaystyle\int e^u \, du = e^u + C$

C. $\displaystyle\int \frac{1}{u} \, du = \ln u + C, \quad u > 0; \quad$ or $\quad \displaystyle\int \frac{1}{u} \, du = \ln |u| + C, \quad u < 0$
(We will generally consider $u > 0$.)

Recall the Leibniz notation, dy/dx, for a derivative. We gave specific definitions of the differentials dy and dx in Section 3.6. Recall that

$$\frac{dy}{dx} = f'(x)$$

and

$$dy = f'(x) \, dx.$$

We will make extensive use of this notation in this section.

EXAMPLE 1 For $y = f(x) = x^3$, find dy.

Solution We have

$$\frac{dy}{dx} = f'(x) = 3x^2,$$

so

$$dy = f'(x) \, dx = 3x^2 \, dx. \qquad\qquad ❖$$

EXAMPLE 2 For $u = g(x) = \ln x$, find du.

Solution We have

$$\frac{du}{dx} = g'(x) = \frac{1}{x},$$

so

$$du = g'(x) \ dx = \frac{1}{x} \ dx, \quad \text{or} \quad \frac{dx}{x}.$$

❖

EXAMPLE 3 For $y = f(x) = e^{x^2}$, find dy.

Solution Using the Chain Rule, we have

$$\frac{dy}{dx} = f'(x) = 2xe^{x^2},$$

so

$$dy = f'(x) \ dx = 2xe^{x^2} \ dx.$$

❖

DO EXERCISES 1–3.

So far the dx in

$$\int f(x) \ dx$$

has played no role in integrating other than to indicate the variable of integration. Now it will be convenient to make use of dx. Consider the integral

$$\int 2xe^{x^2} \ dx.$$

If we set

$$u = x^2,$$

then

$$du = 2x \ dx.$$

If we substitute u for x^2 and du for $2x \ dx$, the integral takes on the form

$$\int e^u \ du.$$

Since

$$\int e^u \ du = e^u + C,$$

1. For $y = f(x) = 6x^2 + x$, find dy.

2. For $u = g(x) = x + 3$, find du.

3. For $y = f(x) = e^{x^3}$, find dy.

4. Evaluate $\int 3x^2 \cdot e^{x^3}\, dx$.

it follows that

$$\int 2xe^{x^2}\, dx = \int e^u\, du$$

$$= e^u + C$$

$$= e^{x^2} + C.$$

In effect, we have used the Chain Rule in reverse. We can check the result by differentiating. The procedure is referred to as *substitution*, or *change of variable*. It is a *trial-and-error* procedure that you will become more proficient with after much practice. If you try a substitution that doesn't result in an integrand that can be easily integrated, try another. There are many integrations that cannot be carried out using substitution. We do know that any integrations that fit rules A, B, or C can be done with substitution.

DO EXERCISE 4.

Let us consider some additional examples.

EXAMPLE 4 Evaluate $\int \dfrac{2x\, dx}{1 + x^2}$.

Solution

5. Evaluate $\int \dfrac{2x\, dx}{5 + x^2}$.

$$\int \frac{2x\, dx}{1 + x^2} = \int \frac{du}{u} \qquad \underline{\text{Substitution}} \qquad \boxed{\text{Let } u = 1 + x^2;\\ \text{then } du = 2x\, dx.}$$

$$= \ln u + C$$

$$= \ln (1 + x^2) + C \qquad \qquad \diamondsuit$$

DO EXERCISE 5.

EXAMPLE 5 Evaluate $\int \dfrac{2x\, dx}{(1 + x^2)^2}$.

Solution

$$\int \frac{2x\, dx}{(1 + x^2)^2} = \int \frac{du}{u^2} \qquad \underline{\text{Substitution}} \qquad \boxed{u = 1 + x^2,\\ du = 2x\, dx}$$

$$= \int u^{-2}\, du$$

$$= -u^{-1} + C$$

$$= -\frac{1}{u} + C$$

$$= \frac{-1}{1 + x^2} + C \qquad \qquad \diamondsuit$$

DO EXERCISE 6.

EXAMPLE 6   Evaluate $\displaystyle\int \frac{\ln 3x \, dx}{x}$.

Solution

$$\int \frac{\ln 3x \, dx}{x} = \int u \, du \quad \underline{\text{Substitution}} \quad \boxed{\begin{array}{l} u = \ln 3x, \\ du = \dfrac{1}{x} \, dx \end{array}}$$

$$= \frac{u^2}{2} + C$$

$$= \frac{(\ln 3x)^2}{2} + C \qquad \qquad ❖$$

DO EXERCISE 7.

EXAMPLE 7   Evaluate $\displaystyle\int x e^{x^2} dx$.

Solution   Suppose we try

$$u = x^2;$$

then we have

$$du = 2x \, dx.$$

We don't have $2x \, dx$ in $\int x e^{x^2} dx$. We have $x \, dx$ and need to supply a 2. We do this by multiplying by 1, using $\frac{1}{2} \cdot 2$:

$$\frac{1}{2} \cdot 2 \cdot \int x e^{x^2} dx = \frac{1}{2} \int 2x e^{x^2} dx$$

$$= \frac{1}{2} \int e^{x^2} (2x \, dx)$$

$$= \frac{1}{2} \int e^u \, du$$

$$= \frac{1}{2} e^u + C$$

$$= \frac{1}{2} e^{x^2} + C. \qquad ❖$$

DO EXERCISES 8–11.

EXAMPLE 8   Evaluate $\displaystyle\int \frac{dx}{x + 3}$.

6. Evaluate $\displaystyle\int \frac{2x \, dx}{(3 + x^2)^2}$.

7. Evaluate $\displaystyle\int \frac{\ln x \, dx}{x}$.

Evaluate.

8. $\displaystyle\int x^2 \cdot e^{x^3} \, dx$

9. $\displaystyle\int e^{5x} \, dx$

10. $\displaystyle\int e^{0.02x} \, dx$

11. $\displaystyle\int e^{-x} \, dx$

12. Evaluate $\int x^3(x^4 + 5)^{19} \, dx$.

13. Evaluate $\int_1^e \dfrac{\ln x \, dx}{x}$.

(See Margin Exercise 7.)

Solution

$$\int \frac{dx}{x + 3} = \int \frac{du}{u} \qquad \underline{\text{Substitution}} \quad \boxed{u = x + 3, \\ du = dx}$$

$$= \ln u + C$$

$$= \ln (x + 3) + C \qquad \qquad ❖$$

With practice, you will be able to make certain substitutions mentally and just write down the answer. Example 8 is a good illustration of this.

EXAMPLE 9 Evaluate $\int x^2(x^3 + 1)^{10} \, dx$.

Solution

$$\int x^2(x^3 + 1)^{10} \, dx$$

$$= \frac{1}{3} \int (x^3 + 1)^{10}(3x^2 \, dx) \qquad \underline{\text{Substitution}} \quad \boxed{u = x^3 + 1, \\ du = 3x^2 \, dx}$$

$$= \frac{1}{3} \int u^{10} \, du$$

$$= \frac{1}{3} \cdot \frac{u^{11}}{11} + C$$

$$= \frac{1}{33}(x^3 + 1)^{11} + C \qquad \qquad ❖$$

DO EXERCISE 12.

EXAMPLE 10 Evaluate $\int_0^1 x^2(x^3 + 1)^{10} \, dx$.

Solution

a) First we find the indefinite integral (shown in Example 9).

b) Then we evaluate the definite integral on $[0, 1]$:

$$\int_0^1 x^2(x^3 + 1)^{10} \, dx = \left[\frac{1}{33}(x^3 + 1)^{11} \right]_0^1$$

$$= \frac{1}{33}[(1^3 + 1)^{11} - (0^3 + 1)^{11}]$$

$$= \frac{1}{33}(2^{11} - 1^{11})$$

$$= \frac{2^{11} - 1}{33}. \qquad \qquad ❖$$

DO EXERCISE 13.

Evaluate. (Be sure to check by differentiating!)

1. $\int \dfrac{3x^2\,dx}{7+x^3}$

2. $\int \dfrac{3x^2\,dx}{1+x^3}$

3. $\int e^{4x}\,dx$

4. $\int e^{3x}\,dx$

5. $\int e^{x/2}\,dx$

6. $\int e^{x/3}\,dx$

7. $\int x^3 e^{x^4}\,dx$

8. $\int x^4 e^{x^5}\,dx$

9. $\int t^2 e^{-t^3}\,dx$

10. $\int t e^{-t^2}\,dt$

11. $\int \dfrac{\ln 4x\,dx}{x}$

12. $\int \dfrac{\ln 5x\,dx}{x}$

13. $\int \dfrac{dx}{1+x}$

14. $\int \dfrac{dx}{5+x}$

15. $\int \dfrac{dx}{4-x}$

16. $\int \dfrac{dx}{1-x}$

17. $\int t^2(t^3-1)^7\,dt$

18. $\int t(t^2-1)^5\,dt$

19. $\int (x^4+x^3+x^2)^7(4x^3+3x^2+2x)\,dx$

20. $\int (x^3-x^2-x)^9(3x^2-2x-1)\,dx$

21. $\int \dfrac{e^x\,dx}{4+e^x}$

22. $\int \dfrac{e^t\,dt}{3+e^t}$

23. $\int \dfrac{\ln x^2}{x}\,dx$

24. $\int \dfrac{(\ln x)^2}{x}\,dx$

25. $\int \dfrac{dx}{x\ln x}$

26. $\int \dfrac{dx}{x\ln x^2}$

27. $\int \sqrt{ax+b}\,dx$

28. $\int x\sqrt{ax^2+b}\,dx$

29. $\int b e^{ax}\,dx$

30. $\int P_0 e^{kt}\,dt$

31. $\int \dfrac{3x^2\,dx}{(1+x^3)^5}$

32. $\int \dfrac{x^3\,dx}{(2-x^4)^7}$

33. $\int 7x\sqrt[3]{4-x^2}\,dx$

34. $\int 12x\sqrt[5]{1+6x^2}\,dx$

Evaluate.

35. $\int_0^1 2xe^{x^2}\,dx$

36. $\int_0^1 3x^2 e^{x^3}\,dx$

37. $\int_0^1 x(x^2+1)^5\,dx$

38. $\int_1^2 x(x^2-1)^7\,dx$

39. $\int_1^3 \dfrac{dt}{1+t}$

40. $\int_1^3 e^{2x}\,dx$

41. $\int_1^4 \dfrac{2x+1}{x^2+x-1}\,dx$

42. $\int_1^3 \dfrac{2x+3}{x^2+3x}\,dx$

43. $\int_0^b e^{-x}\,dx$

44. $\int_0^b 2e^{-2x}\,dx$

45. $\int_0^b me^{-mx}\,dx$

46. $\int_0^b ke^{-kx}\,dx$

47. $\int_0^4 (x-6)^2\,dx$

48. $\int_0^3 (x-5)^2\,dx$

49. $\int_0^2 \dfrac{3x^2\,dx}{(1+x^3)^5}$

50. $\int_{-1}^0 \dfrac{x^3\,dx}{(2-x^4)^7}$

51. $\int_0^{\sqrt{7}} 7x\sqrt[3]{1+x^2}\,dx$

52. $\int_0^1 12x\sqrt[5]{1-x^2}\,dx$

APPLICATIONS

❖ **Business and Economics**

53. *Demand from marginal demand.* A firm has the marginal-demand function

$$D'(p) = \dfrac{-2000p}{\sqrt{25-p^2}}.$$

Find the demand function given that $D = 13{,}000$ when $p = \$3$ per unit.

54. *Profit from marginal profit.* A firm has the marginal-profit function

$$\frac{dP}{dx} = \frac{9000 - 3000x}{(x^2 - 6x + 10)^2}.$$

Find the total-profit function given that $P = \$1500$ at $x = 3$.

55. *Value of an investment.* A company buys a new machine for $250,000. The marginal revenue from the sale of products produced by the machine is projected to be

$$R'(t) = 4000t.$$

The salvage value of the machine decreases at the rate of

$$V'(t) = 25,000e^{-0.1t}.$$

The total profit from the machine after T years is given by

$$P(T) = \begin{pmatrix} \text{Revenue} \\ \text{from} \\ \text{sale of} \\ \text{product} \end{pmatrix} + \begin{pmatrix} \text{Revenue} \\ \text{from} \\ \text{sale of} \\ \text{machine} \end{pmatrix} - \begin{pmatrix} \text{Cost} \\ \text{of} \\ \text{machine} \end{pmatrix}$$

$$= \int_0^T R'(t)\, dt + \int_0^T V'(t)\, dt - \$250,000.$$

a)   Find $P(T)$.
b)   Find $P(10)$.

❖ **Social Sciences**

56. *Divorce.* The U.S. divorce rate is approximated by

$$D(t) = 100,000e^{0.025t},$$

where $D(t) =$ the number of divorces occurring at time t and $t =$ the number of years measured from 1900. That is, $t = 0$ corresponds to 1900, $t = 91\frac{9}{365}$ corresponds to January 9, 1991, and so on.

a)   Find the total number of divorces from 1900 to 1992. Note that this is

$$\int_0^{92} D(t)\, dt.$$

b)   Find the total number of divorces from 1980 to 1992. Note that this is

$$\int_{80}^{92} D(t)\, dt.$$

SYNTHESIS EXERCISES

Find the area of the shaded region.

57.   **58.**

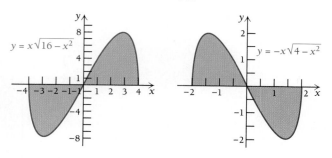

$y = x\sqrt{16 - x^2}$

$y = -x\sqrt{4 - x^2}$

Evaluate.

59. $\displaystyle\int 5x\sqrt{1 - 4x^2}\, dx$

60. $\displaystyle\int \frac{dx}{ax + b}$

61. $\displaystyle\int \frac{x^2}{e^{x^3}}\, dx$

62. $\displaystyle\int \frac{e^{\sqrt{t}}}{\sqrt{t}}\, dt$

63. $\displaystyle\int \frac{e^{1/t}}{t^2}\, dt$

64. $\displaystyle\int \frac{(\ln x)^{99}}{x}\, dx$

65. $\displaystyle\int \frac{dx}{x(\ln x)^4}$

66. $\displaystyle\int (e^t + 2)e^t\, dt$

67. $\displaystyle\int x^2\sqrt{x^3 + 1}\, dx$

68. $\displaystyle\int \frac{t^2}{\sqrt[4]{2 + t^3}}\, dt$

69. $\displaystyle\int \frac{x - 3}{(x^2 - 6x)^{1/3}}\, dx$

70. $\displaystyle\int \frac{[(\ln x)^2 + 3(\ln x) + 4]}{x}\, dx$

71. $\displaystyle\int \frac{t^3 \ln\,(t^4 + 8)}{t^4 + 8}\, dt$

72. $\int \dfrac{t^2 + 2t}{(t+1)^2}\, dt$

$\left(\text{Hint: } \dfrac{t^2 + 2t}{(t+1)^2} = \dfrac{t^2 + 2t + 1 - 1}{t^2 + 2t + 1} = 1 - \dfrac{1}{(t+1)^2}. \right)$

73. $\int \dfrac{x^2 + 6x}{(x+3)^2}\, dx$

74. $\int \dfrac{x+3}{x+1}\, dx$

$\left(\text{Hint: Divide } \dfrac{x+3}{x+1} = 1 + \dfrac{2}{x+1}. \right)$

75. $\int \dfrac{t-5}{t-4}\, dt$

76. $\int \dfrac{dx}{x(\ln x)^n}$

77. $\int \dfrac{dx}{e^x + 1}$

$\left(\text{Hint: } \dfrac{1}{e^x + 1} = \dfrac{e^{-x}}{1 + e^{-x}}. \right)$

78. $\int \dfrac{e^x - e^{-x}}{e^x + e^{-x}}\, dx$

79. $\int \dfrac{(\ln x)^n}{x}\, dx$

80. $\int \dfrac{e^{-mx}}{1 + ae^{-mx}}\, dx$

81. $\int \dfrac{dx}{x \ln x\, [\ln\,(\ln x)]}$

82. $\int 5x^2(2x^3 - 7)^n\, dx$

83. $\int 9x(7x^2 + 9)^n\, dx$

5.6

INTEGRATION TECHNIQUES: INTEGRATION BY PARTS

Recall the Product Rule for derivatives:

$$\frac{d}{dx}\, uv = u\,\frac{dv}{dx} + v\,\frac{du}{dx}.$$

Integrating both sides with respect to x, we get

$$uv = \int u\,\frac{dv}{dx}\, dx + \int v\,\frac{du}{dx}\, dx$$

$$= \int u\, dv + \int v\, du.$$

Solving for $\int u\, dv$, we get the following.

THEOREM 7

The Integration-by-Parts Formula

$$\int u\, dv = uv - \int v\, du$$

OBJECTIVES

a)   Evaluate integrals using the formula for integration by parts.

b)   Solve applied problems involving integration by parts.

1. Evaluate $\int 3xe^{3x}\,dx$.

This equation can be used as a formula for integrating in certain situations—that is, situations in which an integrand is a product of two functions, and one of the functions can be integrated using the techniques we have already developed. For example,

$$\int xe^x\,dx$$

can be considered as

$$\int x(e^x\,dx) = \int u\,dv,$$

where we let

$$u = x \quad \text{and} \quad dv = e^x\,dx.$$

If so, we have

$$du = dx, \quad \text{by differentiating}$$

and

$$v = e^x, \quad \text{by integrating and using the simplest antiderivative.}$$

Then the integration-by-parts formula gives us

$$\int \overset{u}{(x)}\overset{dv}{(e^x\,dx)} = \overset{u}{(x)}\overset{v}{(e^x)} - \int \overset{v}{(e^x)}\overset{du}{(dx)}$$

$$= xe^x - e^x + C.$$

This method of integrating is called **integration by parts**.

Note that integration by parts, like substitution, is a trial-and-error process. In the preceding example, suppose we had reversed the roles of x and e^x. We would have obtained

$$u = e^x, \qquad dv = x\,dx,$$

$$du = e^x\,dx, \qquad v = \frac{x^2}{2},$$

and

$$\int (e^x)(x\,dx) = (e^x)\left(\frac{x^2}{2}\right) - \int \left(\frac{x^2}{2}\right)(e^x\,dx).$$

Now the integrand on the right is more difficult to integrate than the one with which we began. When we can integrate *both* factors of an integrand, and thus have a choice as to how to apply the integration-by-parts formula, it can happen that only one (and maybe none) of the possibilities will work.

DO EXERCISE 1.

Tips on Using Integration by Parts

1. If you have tried substitution, and have had no success, then try integration by parts.

2. Use integration by parts when an integral is of the form

$$\int f(x)\, g(x)\, dx.$$

Then match it with an integral of the form

$$\int u\, dv$$

by choosing a function to be $u = f(x)$, where $f(x)$ can be differentiated, and the remaining factor to be $dv = g(x)\, dx$, where $g(x)$ can be integrated.

3. Find du by differentiating and v by integrating.

4. If the resulting integral is more complicated than the original, make some other choice for $u = f(x)$ and $dv = g(x)\, dx$.

Let us consider some additional examples.

EXAMPLE 1   Evaluate $\int \ln x\, dx$.

Solution   Note that $\int (dx/x) = \ln x + C$, but we do not yet know how to find $\int \ln x\, dx$ since we do not know how to find an antiderivative of $\ln x$. Since we can differentiate $\ln x$, we let

$$u = \ln x \quad \text{and } dv = dx.$$

Then

$$du = \frac{1}{x}\, dx \quad \text{and} \quad v = x.$$

Using the integration-by-parts formula gives

$$\int \overset{u}{(\ln x)}\,\overset{dv}{(dx)} = \overset{u}{(\ln x)}\overset{v}{x} - \int \overset{v}{x}\left(\overset{du}{\frac{1}{x}\, dx}\right)$$

$$= x \ln x - \int dx$$

$$= x \ln x - x + C.$$

DO EXERCISE 2.

2. Evaluate $\int \ln 5x\, dx$.

3. Evaluate $\int x \ln 4x \, dx$.

EXAMPLE 2   Evaluate $\int x \ln x \, dx$.

Solution   Let us examine several choices, as follows.

   Choice 1   We let

$$u = 1 \quad \text{and} \quad dv = x \ln x \, dx.$$

This will not work because we are back to our original integral, in which we do not know how to integrate $dv = x \ln x \, dx$.

   Choice 2   We let

$$u = x \ln x \qquad\qquad \text{and} \quad dv = dx.$$

Then

$$du = \left[x \left(\frac{1}{x} \right) + 1(\ln x) \right] dx \quad \text{and} \quad v = x.$$
$$= (1 + \ln x) \, dx.$$

Using the integration-by-parts formula, we have

$$\int u \, dv = uv - \int v \, du = (x \ln x)x - \int x(1 + \ln x) \, dx$$
$$= x^2 \ln x - \int (x + x \ln x) \, dx.$$

This integral is worse than the original.

   Choice 3   We let

$$u = \ln x \quad \text{and} \quad dv = x \, dx.$$

Then

$$du = \frac{1}{x} \, dx \quad \text{and} \quad v = \frac{x^2}{2}.$$

Using the integration-by-parts formula, we have

$$\int u \, dv = uv - \int v \, du = (\ln x) \frac{x^2}{2} - \int \frac{x^2}{2} \cdot \frac{1}{x} \, dx$$
$$= \frac{x^2}{2} \ln x - \int \frac{x}{2} \, dx$$
$$= \frac{x^2}{2} \ln x - \frac{x^2}{4} + C.$$

This choice allows us to evaluate the integral.   ❖

DO EXERCISE 3.

EXAMPLE 3 Evaluate $\int x\sqrt{x+1}\ dx$.

Solution We let

$$u = x \quad \text{and} \quad dv = (x+1)^{1/2}\ dx.$$

Then

$$du = dx \quad \text{and} \quad v = \tfrac{2}{3}(x+1)^{3/2}.$$

Note that we had to use substitution in order to integrate dv. We see this as follows:

$$\int (x+1)^{1/2}\ dx = \int w^{1/2}\ dw \quad \xrightarrow{\text{Substitution}} \boxed{\begin{array}{l} w = x + 1, \\ dw = dx \end{array}}$$

$$= \frac{w^{1/2+1}}{\tfrac{1}{2}+1} = \tfrac{2}{3}w^{3/2} = \tfrac{2}{3}(x+1)^{3/2}.$$

Using the integration-by-parts formula gives us

$$\int x\sqrt{x+1}\ dx = x \cdot \tfrac{2}{3}(x+1)^{3/2} - \int \tfrac{2}{3}(x+1)^{3/2}\ dx$$

$$= \tfrac{2}{3}x(x+1)^{3/2} - \tfrac{2}{3} \cdot \tfrac{2}{5}(x+1)^{5/2} + C$$

$$= \tfrac{2}{3}x(x+1)^{3/2} - \tfrac{4}{15}(x+1)^{5/2} + C. \qquad ❖$$

DO EXERCISE 4.

EXAMPLE 4 Evaluate $\int_1^2 \ln x\ dx$.

Solution

a) First find the indefinite integral (Example 1).

b) Then evaluate the definite integral:

$$\int_1^2 \ln x\ dx = [x \ln x - x]_1^2$$

$$= (2 \ln 2 - 2) - (1 \cdot \ln 1 - 1)$$

$$= 2 \ln 2 - 2 + 1$$

$$= 2 \ln 2 - 1. \qquad ❖$$

DO EXERCISE 5.

Repeated Integration by Parts

Sometimes we may need to apply the integration-by-parts formula more than once.

DO EXERCISE 7.

4. Evaluate $\int x\sqrt{x+3}\ dx$.

5. Evaluate $\displaystyle\int_1^2 x \ln 4x\ dx$.

(See Margin Exercise 3.)

Evaluate using integration by parts. Check by differentiating.

1. $\int 5xe^{5x} \, dx$

2. $\int 2xe^{2x} \, dx$

3. $\int x^3(3x^2 \, dx)$

4. $\int x^2(2x \, dx)$

5. $\int xe^{2x} \, dx$

6. $\int xe^{3x} \, dx$

7. $\int xe^{-2x} \, dx$

8. $\int xe^{-x} \, dx$

9. $\int x^2 \ln x \, dx$

10. $\int x^3 \ln x \, dx$

11. $\int x \ln x^2 \, dx$

12. $\int x^2 \ln x^3 \, dx$

13. $\int \ln (x + 3) \, dx$

14. $\int \ln (x + 1) \, dx$

15. $\int (x + 2) \ln x \, dx$

16. $\int (x + 1) \ln x \, dx$

17. $\int (x - 1) \ln x \, dx$

18. $\int (x - 2) \ln x \, dx$

19. $\int x\sqrt{x + 2} \, dx$

20. $\int x\sqrt{x + 4} \, dx$

21. $\int x^3 \ln 2x \, dx$

22. $\int x^2 \ln 5x \, dx$

23. $\int x^2 e^x \, dx$

24. $\int (\ln x)^2 \, dx$

25. $\int x^2 e^{2x} \, dx$

26. $\int x^{-5} \ln x \, dx$

27. $\int x^3 e^{-2x} \, dx$

28. $\int x^5 e^{4x} \, dx$

29. $\int (x^4 + 1)e^{3x} \, dx$

30. $\int (x^3 - x + 1)e^{-x} \, dx$

Evaluate using integration by parts.

31. $\int_1^2 x^2 \ln x \, dx$

32. $\int_1^2 x^3 \ln x \, dx$

33. $\int_2^6 \ln (x + 3) \, dx$

34. $\int_0^5 \ln (x + 1) \, dx$

35. $\int_0^1 xe^x \, dx$

36. $\int_0^1 xe^{-x} \, dx$

APPLICATIONS

❖ **Business and Economics**

37. *Cost from marginal cost.* A company determines that its marginal-cost function is given by

$$C'(x) = 4x\sqrt{x + 3}.$$

Find the total cost given that $C(13) = \$1126.40$.

38. *Profit from marginal profit.* A firm determines that its marginal-profit function is given by

$$P'(x) = 1000x^2 e^{-0.2x}.$$

Find the total profit given that $P = -\$2000$ when $x = 0$.

❖ **Life and Physical Sciences**

39. *Electrical energy use.* The rate of electrical energy used by a family, in kilowatt hours per day, is given by

$$K(t) = 10te^{-t},$$

where t is the time, in hours. That is, t is in the interval $[0, 24]$.

a) How many kilowatt hours does the family use in the first T hours of a day ($t = 0$ to $t = T$)?

b) How many kilowatt hours does the family use in the first 4 hours of the day?

40. *Drug dosage.* Suppose that an oral dose of a drug is taken. From that time, the drug is assimilated in the body and excreted through the urine. The total amount of the drug that has passed through the body in time T is given by

$$\int_0^T E(t)\ dt,$$

where E is the rate of excretion of the drug. A typical rate-of-excretion function is

$$E(t) = te^{-kt},$$

where $k > 0$ and t is the time, in hours.

a) Use integration by parts to find a formula for

$$\int_0^T E(t)\ dt.$$

b) ▦ Find

$$\int_0^{10} E(t)\ dt, \quad \text{when } k = 0.2 \text{ mg/hr.}$$

SYNTHESIS EXERCISES

Evaluate using integration by parts.

41. $\int \sqrt{x}\ \ln x\ dx$

42. $\int x^n \ln x\ dx$

43. $\int \dfrac{te^t}{(t+1)^2}\ dt$

44. $\int x^2\ (\ln x)^2\ dx$

45. $\int \dfrac{\ln x}{\sqrt{x}}\ dx$

46. $\int x^n\ (\ln x)^2\ dx$

47. $\int \dfrac{13t^2 - 48}{\sqrt[5]{4t+7}}\ dt$

48. $\int (27x^3 + 83x - 2)\ \sqrt[6]{3x+8}\ dx$

49. Verify that for any positive integer n,

$$\int x^n e^x\ dx = x^n e^x - n \int x^{n-1} e^x\ dx.$$

50. Verify that for any positive integer n,

$$\int (\ln x)^n\ dx = x(\ln x)^n - n \int (\ln x)^{n-1}\ dx.$$

COMPUTER-GRAPHING CALCULATOR EXERCISES

51. Sketch a graph of the function

$$f(x) = x^5 \ln x$$

over the interval $[1, 10]$ and approximate the area.

5.7

INTEGRATION TECHNIQUES: USING TABLES

Tables of Integration Formulas

You have probably noticed that, generally speaking, integration is more difficult and "tricky" than differentiation. Because of this, integral formulas that are reasonable and/or important have been gathered into

OBJECTIVE

a)   Evaluate integrals using a table of integration formulas.

1. Using Table 1, evaluate

$$\int \frac{1}{x(7 + 2x)} \, dx.$$

tables. Table 1 at the back of the book, though quite brief, is such an example. Entire books of integration formulas are available in libraries, and lengthy tables are also available in mathematics handbooks. Such tables are usually classified by the form of the integrand. The idea is to properly match the integral in question with a formula in the table. Sometimes some algebra or a technique such as integration by substitution or parts may be needed as well as a table.

EXAMPLE 1 Evaluate

$$\int \frac{dx}{x(3 - x)}.$$

Solution This integral fits *Formula 20* in Table 1:

$$\int \frac{1}{x(ax + b)} \, dx = \frac{1}{b} \ln \left(\frac{x}{ax + b} \right) + C.$$

In our integral, $a = -1$ and $b = 3$, so we have, by the formula,

$$\int \frac{1}{x(3 - x)} \, dx = \int \frac{dx}{x(-1 \cdot x + 3)}$$

$$= \frac{1}{3} \ln \left(\frac{x}{-1 \cdot x + 3} \right) + C$$

$$= \frac{1}{3} \ln \left(\frac{x}{3 - x} \right) + C. \qquad \diamondsuit$$

DO EXERCISE 1.

EXAMPLE 2 Evaluate

$$\int \frac{5x}{7x - 8} \, dx.$$

Solution If we first factor 5 out of the integral, then the integral fits *Formula 18* in Table 1:

$$\int \frac{x}{ax + b} \, dx = \frac{b}{a^2} + \frac{x}{a} - \frac{b}{a^2} \ln (ax + b) + C.$$

In our integral, $a = 7$ and $b = -8$, so we have, by the formula,

$$\int \frac{5x}{7x - 8} \, dx = 5 \int \frac{x}{7x - 8} \, dx$$

$$= 5 \left[\frac{-8}{7^2} + \frac{x}{7} - \frac{-8}{7^2} \ln (7x - 8) \right] + C$$

$$= 5\left[\frac{-8}{49} + \frac{x}{7} + \frac{8}{49} \ln (7x - 8)\right] + C$$

$$= -\frac{40}{49} + \frac{5x}{7} + \frac{40}{49} \ln (7x - 8) + C.$$

❖

DO EXERCISES 2 AND 3.

EXAMPLE 3 Evaluate

$$\int \sqrt{16x^2 + 3} \, dx.$$

Solution This integral almost fits *Formula 22* in Table 1:

$$\int \sqrt{x^2 \pm a^2} \, dx = \tfrac{1}{2}\left[x\sqrt{x^2 \pm a^2} \pm a^2 \ln (x + \sqrt{x^2 \pm a^2})\right] + C.$$

But the x^2-coefficient needs to be 1. To achieve this, we first factor out 16, as follows. Then we apply *Formula 22* in Table 1:

$$\int \sqrt{16x^2 + 3} \, dx = \int \sqrt{16(x^2 + \tfrac{3}{16})} \, dx$$

$$= \int 4\sqrt{x^2 + \tfrac{3}{16}} \, dx$$

$$= 4 \int \sqrt{x^2 + \tfrac{3}{16}} \, dx$$

$$= 4 \cdot \tfrac{1}{2}\left[x\sqrt{x^2 + \tfrac{3}{16}} + \tfrac{3}{16} \ln (x + \sqrt{x^2 + \tfrac{3}{16}})\right] + C$$

$$= 2\left[x\sqrt{x^2 + \tfrac{3}{16}} + \tfrac{3}{16} \ln (x + \sqrt{x^2 + \tfrac{3}{16}})\right] + C.$$

In our integral, $a^2 = 3/16$ and $a = \sqrt{3}/4$, though we did not need to use a in this form when applying the formula. ❖

DO EXERCISE 4.

EXAMPLE 4 Evaluate

$$\int \frac{dx}{x^2 - 25}.$$

Solution This integral fits *Formula 14* in Table 1:

$$\int \frac{1}{x^2 - a^2} \, dx = \frac{1}{2a} \ln \left(\frac{x - a}{x + a}\right) + C.$$

Using Table 1, evaluate.

2. $\displaystyle\int \frac{3x}{5 - 2x} \, dx$

3. $\displaystyle\int \frac{x}{(5 - 2x)^2} \, dx$

4. Using Table 1, evaluate

$$\int \sqrt{25y^2 - 4} \, dy.$$

5. Using Table 1, evaluate

$$\int \frac{2}{49 - x^2}\, dx.$$

In our integral, $a = 5$, so we have, by the formula,

$$\int \frac{dx}{x^2 - 25} = \frac{1}{10} \ln \left(\frac{x - 5}{x + 5} \right) + C.$$

❖

DO EXERCISE 5.

EXAMPLE 5   Evaluate $\int (\ln x)^3\, dx$.

Solution   This integral fits *Formula 9* in Table 1:

$$\int (\ln x)^n\, dx = x(\ln x)^n - n \int (\ln x)^{n-1}\, dx + C, \quad n \ne -1.$$

We must apply the formula three times:

$$\int (\ln x)^3\, dx$$

$$= x(\ln x)^3 - 3 \int (\ln x)^2\, dx + C \qquad \text{Formula 9}$$

$$= x(\ln x)^3 - 3\left[x(\ln x)^2 - 2 \int \ln x\, dx \right] + C \qquad \begin{array}{l} \text{Applying Formula 9} \\ \text{again} \end{array}$$

$$= x(\ln x)^3 - 3\left[x(\ln x)^2 - 2(x \ln x - \int dx) \right] + C \qquad \begin{array}{l} \text{Applying} \\ \text{Formula 9 for} \\ \text{the third time} \end{array}$$

$$= x(\ln x)^3 - 3x(\ln x)^2 + 6x \ln x - 6x + C.$$

❖

DO EXERCISE 6.

6. Using Table 1, evaluate

$$\int (\ln x)^2\, dx.$$

EXERCISE SET **5.7**

Evaluate using Table 1.

1. $\int xe^{-3x}\, dx$

2. $\int xe^{4x}\, dx$

3. $\int 5^x\, dx$

4. $\int \frac{1}{\sqrt{x^2 - 9}}\, dx$

5. $\int \frac{1}{16 - x^2}\, dx$

6. $\int \frac{1}{x\sqrt{4 + x^2}}\, dx$

7. $\int \frac{x}{5 - x}\, dx$

8. $\int \frac{x}{(1 - x)^2}\, dx$

9. $\int \frac{1}{x(5 - x)^2}\, dx$

10. $\int \sqrt{x^2 + 9}\, dx$

11. $\int \ln 3x\, dx$

12. $\int \ln \frac{4}{5}x\, dx$

13. $\int x^4 e^{5x}\, dx$

14. $\int x^3 e^{-2x}\, dx$

15. $\int x^3 \ln x\, dx$

16. $\int 5x^4 \ln x\, dx$

17. $\int \dfrac{dx}{\sqrt{x^2 + 7}}$

18. $\int \dfrac{3\ dx}{x\sqrt{1 - x^2}}$

19. $\int \dfrac{10\ dx}{x(5 - 7x)^2}$

20. $\int \dfrac{2}{5x(7x + 2)}\ dx$

21. $\int \dfrac{-5}{4x^2 - 1}\ dx$

22. $\int \sqrt{9t^2 - 1}\ dt$

23. $\int \sqrt{4m^2 + 16}\ dm$

24. $\int \dfrac{3 \ln x}{x^2}\ dx$

25. $\int \dfrac{-5 \ln x}{x^3}\ dx$

26. $\int (\ln x)^4\ dx$

27. $\int \dfrac{e^x}{x^{-3}}\ dx$

28. $\int \dfrac{3}{\sqrt{4x^2 + 100}}\ dx$

APPLICATIONS

❖ **Business and Economics**

29. *Supply from marginal supply.* A lawn machinery company introduces a new kind of lawn seeder. It finds that its marginal supply for the seeder satisfies the function

$$S'(p) = \frac{100p}{(20 - p)^2}, \quad 0 \le p \le 19,$$

where S = the quantity purchased when the price is p thousand dollars per seeder. Find the supply function $S(p)$ given that the company will sell 2000 seeders when the price is $19 thousand.

❖ **Social Sciences**

30. *Learning rate.* The rate of change of the probability that an employee learns a task on a new assembly line is given by

$$p'(t) = \frac{1}{t(2 + t)^2},$$

where p = the probability of learning the task after time t, in months. Find the function $p(t)$ given that $p = 0.8267$ when $t = 2$.

SYNTHESIS EXERCISES

Evaluate using Table 1.

31. $\int \dfrac{8}{3x^2 - 2x}\ dx$

32. $\int \dfrac{x\ dx}{4x^2 - 12x + 9}$

33. $\int \dfrac{dx}{x^3 - 4x^2 + 4x}$

34. $\int e^x \sqrt{e^{2x} + 1}\ dx$

35. $\int \dfrac{-e^{-2x}\ dx}{9 - 6e^{-x} + e^{-2x}}$

36. $\int \dfrac{\sqrt{(\ln x)^2 + 49}}{2x}\ dx$

5.8

THE DEFINITE INTEGRAL AS A LIMIT OF SUMS

We now consider approximating the area of a region by dividing that region into subregions that are almost rectangles. In Fig. 1, $[a, b]$ has been divided into 4 subintervals, each having width Δx, or $(b - a)/4$.

OBJECTIVES

a) Approximate

$$\int_a^b f(x)\ dx$$

by adding areas of rectangles.

b) Find the average value of a function over a given interval.

The area under a curve can be approximated by a sum of rectangular areas.

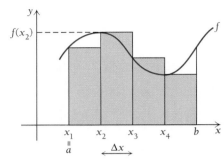

FIGURE 1

The heights of the rectangles shown are

$$f(x_1), \qquad f(x_2), \qquad f(x_3), \quad \text{and} \quad f(x_4).$$

The area of the region under the curve is approximately the sum of the areas of the four rectangles:

$$f(x_1)\,\Delta x + f(x_2)\,\Delta x + f(x_3)\,\Delta x + f(x_4)\,\Delta x.$$

We can name this sum using **summation notation**, which utilizes the Greek capital letter sigma, Σ:

$$\sum_{i=1}^{4} f(x_i)\,\Delta x, \quad \text{or} \quad \sum_{i=1}^{4} f(x_i)\,\Delta x.$$

This is read "the sum of the numbers $f(x_i)\,\Delta x$ from $i = 1$ to $i = 4$." To recover the original expression, we substitute the numbers 1 through 4 successively into $f(x_i)\,\Delta x$ and write plus signs between the results.

EXAMPLE 1   Write summation notation for $2 + 4 + 6 + 8 + 10$.

Solution

$$2 + 4 + 6 + 8 + 10 = \sum_{i=1}^{5} 2i \qquad \text{❖}$$

EXAMPLE 2   Write summation notation for

$$g(x_1)\,\Delta x + g(x_2)\,\Delta x + \cdots + g(x_{19})\,\Delta x.$$

Solution

$$g(x_1)\,\Delta x + g(x_2)\,\Delta x + \cdots + g(x_{19})\,\Delta x = \sum_{i=1}^{19} g(x_i)\,\Delta x \qquad \text{❖}$$

DO EXERCISES 1–3.

Write the summation notation.

1. $1 + 4 + 9 + 16 + 25 + 36$

2. $e + e^2 + e^3 + e^4$

3. $P(x_1)\,\Delta x + P(x_2)\,\Delta x + \cdots + P(x_{38})\,\Delta x$

EXAMPLE 3   Express $\sum_{i=1}^{4} 3^i$ without using summation notation.

Solution

$$\sum_{i=1}^{4} 3^i = 3^1 + 3^2 + 3^3 + 3^4, \quad \text{or} \quad 120 \qquad ❖$$

EXAMPLE 4   Express $\sum_{i=1}^{30} h(x_i)\, \Delta x$ without using summation notation.

Solution

$$\sum_{i=1}^{30} h(x_i)\, \Delta x = h(x_1)\, \Delta x + h(x_2)\, \Delta x + \cdots + h(x_{30})\, \Delta x \qquad ❖$$

DO EXERCISES 4–6.

Approximation of area by rectangles becomes more accurate as we use more rectangles and smaller subintervals, as shown in Fig. 2.

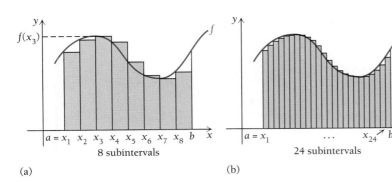

8 subintervals

(a)

24 subintervals

(b)

(c)

FIGURE 2

Express without using summation notation.

4. $\displaystyle\sum_{i=1}^{3} 4^i$

5. $\displaystyle\sum_{i=1}^{5} i e^i$

6. $\displaystyle\sum_{i=1}^{20} t(x_i)\, \Delta x$

In general, the interval $[a, b]$ is divided into n equal subintervals, each of width $\Delta x = (b - a)/n$. The heights of the rectangles are

$$f(x_1), f(x_2), \ldots, f(x_n).$$

The width of each rectangle is Δx, so the first rectangle has area

$$f(x_1) \, \Delta x,$$

the second rectangle has area

$$f(x_2) \, \Delta x,$$

and so on. The area of the region under the curve is approximated by the sum of the areas of the rectangles:

$$\sum_{i=1}^{n} f(x_i) \, \Delta x.$$

We now obtain the actual area by letting the number of intervals increase indefinitely and then by taking the limit. The exact area is thus given by

$$A = \lim_{n \to \infty} \sum_{i=1}^{n} f(x_i) \, \Delta x.$$

The area is also given by a definite integral:

$$\int_{a}^{b} f(x) \, dx = \lim_{n \to \infty} \sum_{i=1}^{n} f(x_i) \, \Delta x.$$

The fact that we can express the integral of a function (positive or otherwise) as a limit of a sum or in terms of an antiderivative is so important that it has a name: *The Fundamental Theorem of Integral Calculus.*

The Fundamental Theorem of Integral Calculus

If a function f has an antiderivative F on $[a, b]$, then

$$\lim_{n \to \infty} \sum_{i=1}^{n} f(x_i) \, \Delta x = \int_{a}^{b} f(x) \, dx = F(b) - F(a).$$

It is interesting to envision that, as we take the limit, the summation sign stretches into something reminiscent of an S (the integral sign) and Δx is defined to be dx. This is why dx appears in the integral notation.

This theorem allows us to approximate the value of a definite integral by a sum, making it as good as we please by taking n sufficiently large.

EXAMPLE 5   Consider the function

$$f(x) = 600x - x^2$$

over the interval $[0, 600]$ (Fig. 3).

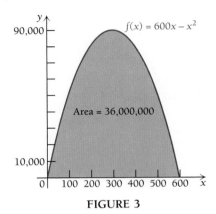

FIGURE 3

a)   Approximate the area by dividing the interval into 6 subintervals.
b)   Approximate the area by dividing the interval into 12 subintervals.
c)   Find the exact area using integration.

Solution

a)   The interval $[0, 600]$ is divided into 6 subintervals, each of length $\Delta x = (600 - 0)/6 = 100$ (Fig. 4). Now x_i is varying from $x_1 = 0$ to $x_6 = 500$ and $b = 600$. Thus we have

$$\sum_{i=1}^{6} f(x_i)\,\Delta x = f(0) \cdot 100 + f(100) \cdot 100 + f(200) \cdot 100$$
$$+ f(300) \cdot 100 + f(400) \cdot 100 + f(500) \cdot 100$$
$$= 0 \cdot 100 + 50{,}000 \cdot 100 + 80{,}000 \cdot 100$$
$$+ 90{,}000 \cdot 100 + 80{,}000 \cdot 100 + 50{,}000 \cdot 100$$
$$= 35{,}000{,}000.$$

b)   The interval $[0, 600]$ is divided into 12 subintervals, each of length $\Delta x = (600 - 0)/12 = 50$ (Fig. 5). Now x_i is varying from $x_1 = 0$ to

7. Approximate the area in Example 5 by dividing the interval into 24 subintervals.

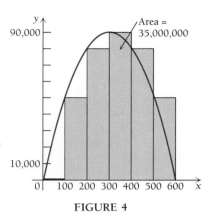

FIGURE 4

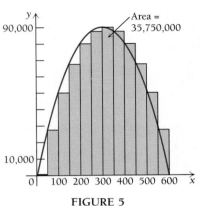

FIGURE 5

8. Referring to Example 6, find

$$\sum_{i=1}^{8} C'(x_i)\, \Delta x,$$

where the interval [0, 400] is divided into 8 equal subintervals of length

$$\Delta x = \frac{400 - 0}{8} = 50.$$

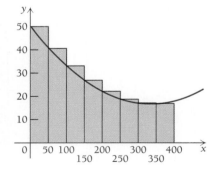

$x_{12} = 550$ and $b = 600$. Thus we have

$$
\begin{aligned}
\sum_{i=1}^{12} f(x_i)\, \Delta x &= f(0) \cdot 50 + f(50) \cdot 50 + f(100) \cdot 50 + f(150) \cdot 50 \\
&\quad + f(200) \cdot 50 + f(250) \cdot 50 + f(300) \cdot 50 \\
&\quad + f(350) \cdot 50 + f(400) \cdot 50 + f(450) \cdot 50 + f(500) \cdot 50 \\
&\quad + f(550) \cdot 50 \\
&= 0 \cdot 50 + 27{,}500 \cdot 50 + 50{,}000 \cdot 50 + 67{,}500 \cdot 50 \\
&\quad + 80{,}000 \cdot 50 + 87{,}500 \cdot 50 + 90{,}000 \cdot 50 \\
&\quad + 87{,}500 \cdot 50 + 80{,}000 \cdot 50 + 67{,}500 \cdot 50 \\
&\quad + 50{,}000 \cdot 50 + 27{,}500 \cdot 50 \\
&= 35{,}750{,}000.
\end{aligned}
$$

c) We find the exact area by integrating as follows:

$$
\int_0^{600} (600x - x^2)\, dx = \left[300x^2 - \frac{x^3}{3} \right]_0^{600}
$$

$$
= \left(300 \cdot 600^2 - \frac{600^3}{3} \right) - \left(300 \cdot 0^2 - \frac{0^3}{3} \right)
$$

$$
= 36{,}000{,}000. \qquad \diamondsuit
$$

DO EXERCISE 7.

EXAMPLE 6 *Business: Cost from marginal cost.* Raggs, Ltd., determines that the marginal cost per suit is given by

$$C'(x) = 0.0003x^2 - 0.2x + 50.$$

a)   Approximate the total cost of producing 400 suits by computing the sum $\sum_{i=1}^{4} C'(x_i) \, \Delta x$.

b)   Find the exact total cost of producing 400 suits by integrating.

Solution

a)   The interval [0, 400] is divided into 4 subintervals, each of length $\Delta x = (400 - 0)/4 = 100$ (Fig. 7). Now x_i is varying from $x_1 = 0$ to $x_4 = 300$ and $b = 400$.

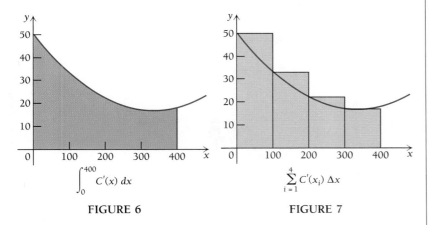

$$\int_0^{400} C'(x) \, dx$$

FIGURE 6

$$\sum_{i=1}^{4} C'(x_i) \, \Delta x$$

FIGURE 7

Thus we have

$$\sum_{x=1}^{4} C'(x_i) \, \Delta x = C'(0) \cdot 100 + C'(100) \cdot 100 + C'(200) \cdot 100$$
$$+ \, C'(300) \cdot 100$$
$$= 50 \cdot 100 + 33 \cdot 100 + 22 \cdot 100 + 17 \cdot 100$$
$$= \$12{,}200.$$

b)   The exact total cost is

$$\int_0^{400} C'(x) \, dx = \$10{,}400. \qquad \text{See Example 3 of Section 5.2.}$$

Thus the approximation is not too far off, even though the number of subintervals is small. In Margin Exercise 8, you will obtain a better approximation using 8 subintervals.   ❖

DO EXERCISES 8 AND 9.

9. In graphs (a) and (b), compute the areas of each rectangle to four decimal places. Then add them to approximate the area under the curve $y = 1/x$ over [1, 7].

a)

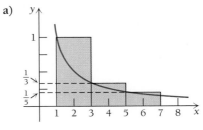

b)

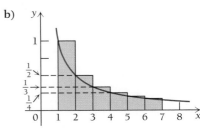

c)   Evaluate

$$\int_1^7 \frac{1}{x} \, dx.$$

Find this answer using a calculator and compare it to your answers to parts (a) and (b).

The fact that an integral can be approximated by a sum is useful when the antiderivative of a function does not have an elementary formula. For example, for the function $e^{-x^2/2}$, important in probability, there is no formula for the antiderivative. Thus tables of approximate values of its integral have been computed using summation methods.

The Average Value of a Function

Suppose that

$$T = f(t)$$

is the temperature at time t recorded at a weather station on a certain day. The station uses a 24-hr clock, so the domain of the temperature function is the interval [0, 24]. The function is continuous (Fig. 8).

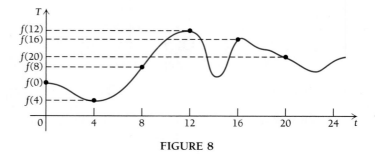

FIGURE 8

To find the average temperature for the given day, we might take six temperature readings at 4-hr intervals, starting at midnight:

$$T_0 = f(0), \qquad T_1 = f(4),$$
$$T_2 = f(8), \qquad T_3 = f(12),$$
$$T_4 = f(16), \qquad T_5 = f(20).$$

The average reading would then be the sum of these six readings divided by 6:

$$T_{av} = \frac{T_0 + T_1 + T_2 + T_3 + T_4 + T_5}{6}.$$

This computation of the average temperature may not give the most useful answer. For example, suppose it is a hot summer day, and at 2:00 in the afternoon (hour 14 on the 24-hr clock), there is a short thunderstorm that cools the air for an hour between our readings. This temporary dip would not show up in the average computed above.

What can we do? We could take 48 readings at half-hour intervals. This should give a better result. In fact, the shorter the time between readings, the better the result should be. It seems reasonable that we might define the **average value** of T over the interval $[0, 24]$ to be the limit, as n approaches ∞, of the average of n values:

$$\text{Average value of } T = \lim_{n \to \infty} \frac{1}{n} \sum_{i=1}^{n} T_i$$

$$= \lim_{n \to \infty} \frac{1}{n} \sum_{i=1}^{n} f(t_i).$$

Note that this is not too far from our definition of an integral. All we would need is to get Δt, which is $(24 - 0)/n$, or $24/n$, into the summation. We do this by expressing $1/n$ as $(1/24) \cdot (24/n)$. Then

$$\text{Average value of } T = \lim_{n \to \infty} \frac{1}{n} \sum_{i=1}^{n} f(t_i)$$

$$= \lim_{n \to \infty} \frac{1}{\Delta t} \cdot \frac{1}{n} \sum_{i=1}^{n} f(t_i) \, \Delta t$$

$$= \lim_{n \to \infty} \frac{n}{24} \cdot \frac{1}{n} \sum_{i=1}^{n} f(t_i) \, \Delta t \qquad \Delta t = \frac{24}{n}, \text{ or } \frac{1}{\Delta t} = \frac{n}{24}$$

$$= \frac{1}{24} \lim_{n \to \infty} \sum_{i=1}^{n} f(t_i) \, \Delta t$$

$$= \frac{1}{24} \int_{0}^{24} f(t) \, dt.$$

DEFINITION

Let f be a continuous function over a closed interval $[a, b]$. Its *average value*, y_{av}, is given by

$$y_{av} = \frac{1}{b-a} \int_{a}^{b} f(x) \, dx.$$

Let us consider average value in another way. If we multiply on both sides of

$$y_{av} = \frac{1}{b-a} \int_{a}^{b} f(x) \, dx$$

by $b - a$, we get

$$(b - a)y_{av} = \int_a^b f(x)\ dx.$$

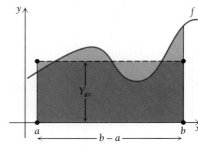

FIGURE 9

Now the expression on the left will give the area of a rectangle of length $b - a$ and height y_{av}. The area of such a rectangle is the same as the area bounded by $y = f(x)$ on the interval $[a, b]$ (Fig. 9).

EXAMPLE 7   Find the average value of $f(x) = x^2$ over the interval $[0, 2]$.

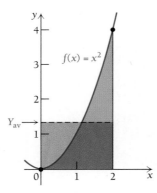

FIGURE 10

Solution   The average value is

$$\frac{1}{2 - 0} \int_0^2 x^2\ dx = \frac{1}{2}\left[\frac{x^3}{3}\right]_0^2$$

$$= \frac{1}{2}\left(\frac{2^3}{3} - \frac{0^3}{3}\right)$$

$$= \frac{1}{2} \cdot \frac{8}{3} = \frac{4}{3}, \quad \text{or} \quad 1\frac{1}{3}.$$

Note that although the values of $f(x)$ increase from 0 to 4, we would not expect the average value to be 2, because we see from the graph in Fig. 10 that $f(x)$ is less than 2 over more than half the interval. ❖

DO EXERCISES 10 AND 11.

EXAMPLE 8 *Life science: Engine emissions.* The emissions of an engine are given by

$$E(t) = 2t^2,$$

where $E(t)$ = the engine's rate of emission, in billions of pollution particulates, at time t, in years. Find the average emissions from $t = 1$ to $t = 5$.

Solution The average emissions are

$$\frac{1}{5-1} \int_1^5 2t^2 \, dt = \frac{1}{4}\left[\frac{2}{3}t^3\right]_1^5$$

$$= \frac{1}{4} \cdot \frac{2}{3}(5^3 - 1^3)$$

$$= \frac{1}{6}(125 - 1)$$

$$= 20\frac{2}{3} \text{ billion pollution particulates.} ❖$$

DO EXERCISE 12.

10. Find the average value of $f(x) = x^3$ over the interval $[0, 2]$.

11. The temperature over a 10-hr period is given by

$$f(t) = -t^2 + 5t + 40, \quad 0 \le t \le 10.$$

   a)   Find the average temperature.
   b)   Find the minimum temperature.
   c)   Find the maximum temperature.

12. *Business: Total sales.* The sales of a company are expected to grow according to the function

$$S(t) = 100t + t^2,$$

where $S(t)$ = the sales, in dollars, on the tth day. Find the average sales from $t = 1$ to $t = 4$ (from the 1st to the 4th day).

EXERCISE SET **5.8**

1. a)   Approximate

$$\int_1^7 \frac{dx}{x^2}$$

   by computing the area of each rectangle to four decimal places and then adding.

   b)   Evaluate

$$\int_1^7 \frac{dx}{x^2}.$$

   Compare the answer to part (a).

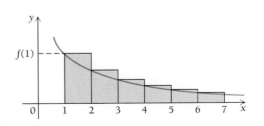

2. a) Approximate

$$\int_0^5 (x^2 + 1) \, dx$$

by computing the area of each rectangle and then adding.

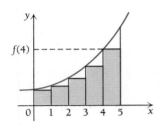

b) Evaluate

$$\int_0^5 (x^2 + 1) \, dx.$$

Compare the answer to part (a).

Find the average value over the given interval.

3. $y = 2x^3$; $[-1, 1]$ **4.** $y = 4 - x^2$; $[-2, 2]$

5. $y = e^x$; $[0, 1]$ **6.** $y = e^{-x}$; $[0, 1]$

7. $y = x^2 - x + 1$; $[0, 2]$

8. $y = x^2 + x - 2$; $[0, 4]$

9. $y = 3x + 1$; $[2, 6]$ **10.** $y = 4x + 1$; $[3, 7]$

11. $y = x^n$; $[0, 1]$ **12.** $y = x^n$; $[1, 2]$

APPLICATIONS

❖ **Business and Economics**

13. *Results of practice.* A typist's speed over a 4-min interval is given by

$$W(t) = -6t^2 + 12t + 90, \quad t \text{ in } [0, 4],$$

where $W(t)$ = the speed, in words per minute, at time t.

a) Find the speed at the beginning of the interval.

b) Find the maximum speed and when it occurs.

c) Find the average speed over the 4-min interval.

❖ **Life and Physical Sciences**

14. *Average drug dosage.* The amount of a drug in the body at time t is given by

$$A(t) = 3e^{-0.1t},$$

where A is in cubic centimeters and t is the time, in hours.

a) What is the initial dosage of the drug?

b) What is the average amount in the body over the first 2 hr?

❖ **Social Sciences**

15. *Average population.* The population of the United States is given by

$$P(t) = 241e^{0.009t},$$

where P is in millions and t is the number of years since 1986. Find the average value of the population from 1986 to 1996.

16. *Average population of a city.* The population of a city increased and then decreased over an 8-yr period according to the function

$$P(t) = -0.1t^2 + t + 3, \quad 0 \leq t \leq 8,$$

where P is in millions and t is the time.

a) Find the average population.

b) Find the minimum population.

c) Find the maximum population.

17. *Results of studying.* A student's score on a test is given by the function

$$S(t) = t^2, \quad t \text{ in } [0, 10],$$

where $S(t)$ = the score after t hours of study.

a) Find the maximum score that the student can achieve and the number of hours of study required to attain it.

b) Find the average score over the 10-hr interval.

A student's test score is a function of the time spent studying.

SYNTHESIS EXERCISES

The Trapezoidal Rule. Another way to approximate an integral is to replace each rectangle in the sum (see Fig. 1) by a trapezoid (see Fig. 2). The area of a trapezoid is $h(c_1 + c_2)/2$, where c_1 and c_2 are the lengths of the parallel sides. Thus, in Fig. 2,

$$\int_a^b f(x) \, dx = \int_a^m f(x) \, dx + \int_m^b f(x) \, dx$$

$$\approx \Delta x \frac{f(a) + f(m)}{2} + \Delta x \frac{f(m) + f(b)}{2}$$

$$\approx \Delta x \left[\frac{f(a)}{2} + f(m) + \frac{f(b)}{2} \right].$$

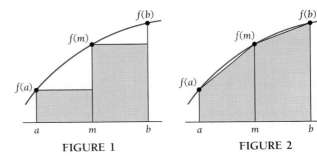

FIGURE 1   FIGURE 2

For an interval $[a, b]$ subdivided into n equal subintervals of length $\Delta x = (b - a)/n$, we get the approximation

$$\int_a^b f(x) \, dx \approx$$

$$\Delta x \left[\frac{f(a)}{2} + f(x_2) + f(x_3) + \cdots + f(x_n) + \frac{f(b)}{2} \right],$$

where $x_1 = a$ and $x_n = x_{n-1} + \Delta x$ or $x_n = a + (n - 1)\Delta x$. This is called the **Trapezoidal Rule**.

18. Use the Trapezoidal Rule and the interval subdivision of Exercise 1 to approximate

$$\int_1^7 \frac{dx}{x^2}.$$

19. Use the Trapezoidal Rule and the interval subdivision of Exercise 2 to approximate

$$\int_0^5 (x^2 + 1) \, dx.$$

COMPUTER EXERCISES

THE CALCULUS EXPLORER
Integral Evaluation

The program uses the Trapezoidal Rule to approximate definite integrals. Use it to approximate the following:

$$\int_0^4 \sqrt{x}\, dx, \quad \int_0^2 (x^4 + 1)\, dx,$$

and $\int_0^4 \ln x\, dx.$

CHAPTER SUMMARY AND REVIEW

TERMS TO KNOW

Antiderivative, p. 354
Indefinite integral, p. 355
Area, p. 364
Definite integral, p. 376
Limits of integration, p. 377

Integration by substitution, p. 398
Integration by parts, p. 406
Tabular integration by parts, p. 411
Integration using tables, p. 413
Summation notation, p. 418

Fundamental Theorem of Integral
  Calculus, p. 420
Average value, p. 425
Trapezoidal Rule, p. 429

REVIEW EXERCISES

These review exercises are for test preparation. They can also be used as a lengthened practice test. Answers are at the back of the book. The answers also contain bracketed section references, which tell you where to restudy if your answer is incorrect.

Evaluate.

1. $\int 8x^4\, dx$

2. $\int (3e^x + 2)\, dx$

3. $\int \left(3t^2 + 7t + \frac{1}{t}\right) dt$

Find the area under the curve on the interval indicated.

4. $y = 4 - x^2;\quad [-2, 1]$

5. $y = x^2 + 3x + 6;\quad [0, 2]$

In each case, give two interpretations of the shaded region.

6.

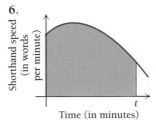

7.

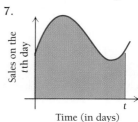

Evaluate.

8. $\int_a^b x^5 \, dx$

9. $\int_{-1}^1 (x^3 - x^4) \, dx$

10. $\int_0^1 (e^x + x) \, dx$

11. $\int_1^3 \frac{3}{x} \, dx$

Decide whether $\int_a^b f(x) \, dx$ is positive, negative, or zero.

12.

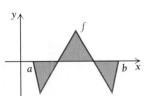

13.

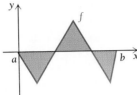

14.

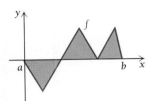

15. Find the area of the region bounded by $y = 3x^2$ and $y = 9x$.

Evaluate using substitution. Do not use Table 1.

16. $\int x^3 e^{x^4} \, dx$

17. $\int \frac{24t^5}{4t^6 + 3} \, dt$

18. $\int \frac{\ln 4x}{2x} \, dx$

19. $\int 2e^{-3x} \, dx$

Evaluate using integration by parts. Do not use Table 1.

20. $\int 3x \, e^{3x} \, dx$

21. $\int \ln x^7 \, dx$

22. $\int 3x^2 \ln x \, dx$

Evaluate using Table 1.

23. $\int \frac{1}{49 - x^2} \, dx$

24. $\int x^2 e^{5x} \, dx$

25. $\int \frac{x}{7x + 1} \, dx$

26. $\int \frac{dx}{\sqrt{x^2 - 36}}$

27. $\int x^6 \ln x \, dx$

28. $\int xe^{8x} \, dx$

29. Approximate $\int_1^4 (2/x) \, dx$ by computing the area of each rectangle to three decimal places and adding.

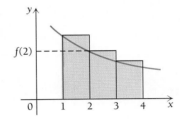

30. Find the average value of $y = e^{-x} + 5$ over $[0, 2]$.

31. A particle starts out from the origin. Its velocity at any time t, $t \geq 0$, is given by $v(t) = 3t^2 + 2t$. Find the distance that the particle travels during the first 4 hr (from $t = 0$ to $t = 4$).

32. *Business: Accumulated sales.* A company estimates that its sales will grow continuously according to the function $S'(t) = 3e^{3t}$, where $S'(t) =$ the sales, in dollars, on the tth day. Find the accumulated sales for the first 4 days.

Integrate using any method.

33. $\int x^3 e^{0.1x} \, dx$

34. $\int \frac{12t^2}{4t^3 + 7} \, dt$

35. $\int \frac{x \, dx}{\sqrt{4 + 5x}}$

36. $\int 5x^4 e^{x^5} \, dx$

37. $\int \frac{dx}{x + 9}$

38. $\int t^7 (t^8 + 3)^{11} \, dt$

39. $\int \ln 7x \, dx$

40. $\int x \ln 8x \, dx$

SYNTHESIS EXERCISES

Evaluate.

41. $\displaystyle\int \frac{t^4 \ln (t^5 + 3)}{t^5 + 3}\, dt$

42. $\displaystyle\int \frac{dx}{e^x + 2}$

45. $\displaystyle\int \ln \left(\frac{x - 3}{x - 4} \right) dx$

46. $\displaystyle\int \frac{dx}{x\,(\ln x)^4}$

43. $\displaystyle\int \frac{\ln \sqrt{x}}{x}\, dx$

44. $\displaystyle\int x^{91} \ln x\, dx$

EXERCISES FOR THINKING AND WRITING

47. Discuss as many interpretations as you can of the integral of a positive function over a closed interval.

48. On a test, a student makes the statement "The function $f(x) = x^2$ has a unique integral." Discuss.

49. Compare the procedures of differentiation and integration. Which seems to be the most complicated or difficult and why?

50. Determine whether the following is a theorem:

$$\int f(x)\, g(x)\, dx = \int f(x)\, dx \cdot \int g(x)\, dx.$$

CHAPTER TEST 5

Evaluate.

1. $\displaystyle\int dx$

2. $\displaystyle\int 1000x^4\, dx$

3. $\displaystyle\int \left(e^x + \frac{1}{x} + x^{3/8} \right) dx$

Find the area under the curve on the interval indicated.

4. $y = x - x^2$; [0, 1]

5. $y = \dfrac{4}{x}$; [1, 3]

6. Give two interpretations of the shaded area.

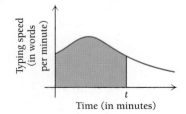

Typing speed (in words per minute)

Time (in minutes)

Evaluate.

7. $\displaystyle\int_{-1}^{2} (2x + 3x^2)\, dx$

8. $\displaystyle\int_{0}^{1} e^{-2x}\, dx$

9. $\displaystyle\int_{a}^{b} \frac{dx}{x}$

10. Decide whether $\int_{a}^{b} f(x)\, dx$ is positive, negative, or zero.

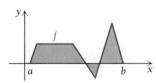

Evaluate using substitution. Do not use Table 1.

11. $\displaystyle\int \frac{dx}{x + 8}$

12. $\displaystyle\int e^{-0.5x}\, dx$

13. $\displaystyle\int t^3(t^4 + 1)^9\, dt$

Evaluate using integration by parts. Do not use Table 1.

14. $\int xe^{5x}\,dx$

15. $\int x^3 \ln x^4\,dx$

Evaluate using Table 1.

16. $\int 2^x\,dx$

17. $\int \dfrac{dx}{x(7-x)}$

18. Find the average value of $y = 4t^3 + 2t$ over $[-1, 2]$.

19. Find the area of the region bounded by $y = x$, $y = x^5$, $x = 0$, and $x = 1$.

20. Approximate

$$\int_0^5 (25 - x^2)\,dx$$

by computing the area of each rectangle and adding.

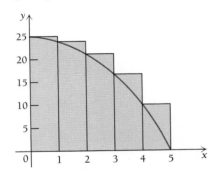

21. *Business: Cost from marginal cost.* An air conditioning company determines that the marginal cost of the xth air conditioner is given by

$$C'(x) = -0.2x + 500, \quad C(0) = 0.$$

Find the total cost of producing 100 air conditioners.

22. *Social science: Learning curve.* A typist's speed over a 4-min interval is given by

$$W(t) = -6t^2 + 12t + 90, \quad t \text{ in } [0, 4],$$

where $W(t) =$ the speed, in words per minute, at time t. How many words are typed during the second minute (from $t = 1$ to $t = 2$)?

Integrate using any method.

23. $\int \dfrac{dx}{x(10-x)}$

24. $\int x^5 e^x\,dx$

25. $\int x^5 e^{x^6}\,dx$

26. $\int \sqrt{x}\,\ln x\,dx$

27. $\int x^3 \sqrt{x^2 + 4}\,dx$

28. $\int \dfrac{dx}{64 - x^2}$

29. $\int x^4 e^{0.1x}\,dx$

30. $\int x \ln 13x\,dx$

Evaluate using any method.

31. $\int \dfrac{[(\ln x)^3 - 4(\ln x)^2 + 5]}{x}\,dx$

32. $\int \ln\left(\dfrac{x+3}{x+5}\right) dx$

33. $\int \dfrac{8x^3 + 10}{\sqrt[3]{5x - 4}}\,dx$

6

APPLICATIONS

OF INTEGRATION

AN APPLICATION

The depletion of oil. The world reserves of oil are 920,700 million barrels. In 1990 ($t = 0$), the world use of oil was 6570 million barrels, and the growth rate for the use of oil was 10%. Assuming that this growth rate continues and that no new reserves are discovered, when will the world reserves of oil be exhausted?

THE MATHEMATICS

To find the time T at which the world reserves of oil will be exhausted, we solve the equation

$$920,700 = \int_0^T 6570e^{0.1t}\, dt.$$

In this chapter we study a wide variety of applications of integration. We first consider an application in economics to finding consumer's and producer's surplus. Then we study the integration of functions involved in exponential growth and decay. Here we find applications not only to business, but also to such environmental concerns as the depletion of resources and the buildup of radioactivity in the atmosphere. Integration has extensive application to probability and statistics, to finding volume, and to many situations involving the solution of differential equations.

Topics in this chapter can be chosen to fit the needs of the student and the course.

6.1

a) Given a demand function $D(x)$ and a supply function $S(x)$, find the equilibrium point and the consumer's surplus and the producer's surplus at the equilibrium point.

ECONOMIC APPLICATION: CONSUMER'S SURPLUS AND PRODUCER'S SURPLUS

In our preceding functions, it was convenient to think of demand and supply as functions of price. For purposes of this section, it will be convenient to think of them as functions of quantity: $p = D(x)$ and $p = S(x)$. Indeed, such interpretation is common in a study of economics.

The consumer's demand curve, $p = D(x)$, gives the demand price per unit that the consumer is willing to pay for x units. It is a decreasing function. The producer's supply curve, $p = S(x)$, gives the price per unit at which the seller is willing to supply x units. It is an increasing function. The equilibrium point (x_E, p_E) is the intersection of the two curves. See Fig. 1.

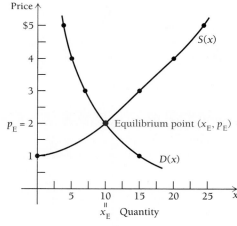

FIGURE 1

Suppose the graphs in Figs. 2 and 3 show the demand curve of college students for movies. Suppose we consider this curve for just one such student, Samantha.

We want to examine the *utility*, or consumer satisfaction, that Samantha receives from going to the movies. Samantha goes to 0 movies per week if the price is $6. She will go to 1 movie per week if the price is $4.80. Suppose she goes to 1 movie. Then the total expenditure to her is $4.80 · 1, or $4.80, as shown in Fig. 2. The area of the blue region represents her total expenditure, $4.80. Look at the area of the orange

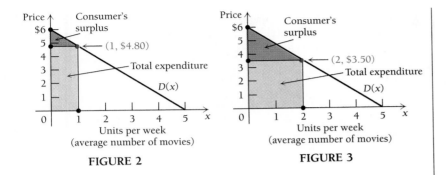

FIGURE 2

FIGURE 3

region. It is $\frac{1}{2}(1)(\$1.20)$, or $0.60. The total area under the curve is a measure of the *total utility* of going to 1 movie, and is $5.40. The area of the orange region, $0.60, is a measure of the satisfaction Samantha gets but for which she does not have to pay. Economists define this amount as **consumer's surplus.** It is the utility that consumers derive from living in a society in which the price consumers are willing to pay decreases when more units are purchased. It is the extra utility that is gained from this purchase. The consumer surplus is the source of revenue that becomes available to a firm that discriminates in its prices, say, by offering discounts to college students, senior citizens, and children, or by selling discount coupons.

Suppose Samantha goes to 2 movies per week if the price is $3.50 (Fig. 3). The total amount Samantha actually spends is $3.50(2), or $7, and is the area of the blue region. It is her total expenditure. The total area under the curve is $9.50 and represents the total satisfaction to Samantha of going to 2 movies. Look at the area of the orange region. It (the consumer's surplus) is $\frac{1}{2}(2)(\$2.50)$, or $2.50. It is a measure of the utility Samantha received but for which she did not have to pay.

Suppose the graph of the demand function is a curve, as in Fig. 4.

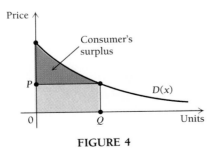

FIGURE 4

1. Find the consumer's surplus for the demand function

$$D(x) = (x - 6)^2$$

when $x = 4$.

If Samantha goes to Q movies when the price is P, then Samantha's total expenditure is QP. The total area under the curve is the total utility, or the total satisfaction received, and is

$$\int_0^Q D(x)\, dx.$$

The *consumer's surplus* is the total area under the curve minus the total expenditure, quantity times price, or QP, and it is the total utility minus the total cost, which is given by

$$\int_0^Q D(x)\, dx - QP.$$

DEFINITION

Suppose that $p = D(x)$ describes the demand function for a commodity. Then the *consumer's surplus* is defined for point (Q, P) as

$$\int_0^Q D(x)\, dx - QP.$$

EXAMPLE 1   Find the consumer's surplus for the demand function $D(x) = (x - 5)^2$ when $x = 3$.

Solution   When $x = 3$, $D(3) = (3 - 5)^2 = (-2)^2 = 4$. Then

$$\text{Consumer's surplus} = \int_0^3 (x - 5)^2\, dx - 3 \cdot 4$$

$$= \int_0^3 (x^2 - 10x + 25)\, dx - 12$$

$$= \left[\frac{x^3}{3} - 5x^2 + 25x \right]_0^3 - 12$$

$$= \left(\frac{3^3}{3} - 5(3)^2 + 25(3) \right)$$

$$- \left(\frac{0^3}{3} - 5(0)^2 + 25(0) \right) - 12$$

$$= (9 - 45 + 75) - 0 - 12$$

$$= \$27. \qquad \qquad ❖$$

DO EXERCISE 1.

Suppose we now look at the supply curve for the movies, as shown in Figs. 5 and 6. At a price of $0 per movie, the producer is willing to supply 0 movies. At a price of $1.75 per movie, the producer is willing to supply 1 movie and makes total receipts of 1($1.75), or $1.75. The area of the beige triangle in Fig. 5 represents the total cost to the producer of producing 1 movie, and is $\frac{1}{2}(1)(\$1.75)$, or $0.875. The area of the green triangle is also $0.875, and represents the surplus over cost. It is a contribution to profit. Economists call this number the **producer's surplus**. It is the utility, or satisfaction, to the producer of living in a society in which a greater amount of a commodity will be supplied when the price increases. It is the extra revenue that the producer receives when the consumer cannot purchase the individual units along the producer's supply curve.

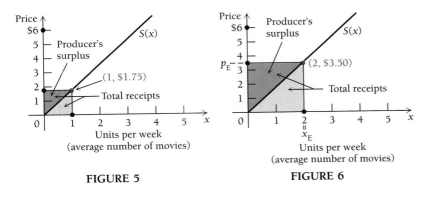

FIGURE 5   FIGURE 6

At a price of $3.50, the producer is willing to supply 2 movies and makes total receipts of 2($3.50), or $7. The area of the beige triangle in Fig. 6 represents the total cost to the producer of actually making the 2 movies, and is $\frac{1}{2}(2)(\$3.50)$, or $3.50. The area of the green triangle is also $3.50 and is the producer's surplus. It is a contribution to the profit of the producer.

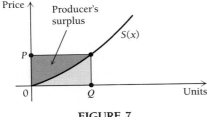

FIGURE 7

2. Find the producer's surplus for the supply function

$$S(x) = x^2 + x + 10$$

when $x = 1$.

Suppose the graph of the supply function is a curve, as shown in Fig. 7. The producer will supply Q movies if the price is P. The total receipts are QP. The *producer's surplus* is the total receipts minus the area under the curve and is given by

$$QP - \int_0^Q S(x)\ dx.$$

DEFINITION

Suppose that $p = S(x)$ is the supply function for a commodity. Then the *producer's surplus* is defined for the point (Q, P) as

$$QP - \int_0^Q S(x)\ dx.$$

EXAMPLE 2 Find the producer's surplus for $S(x) = x^2 + x + 3$ when $x = 3$.

Solution When $x = 3$, $S(3) = 3^2 + 3 + 3 = 15$. Then

$$\text{Producer's surplus} = 3 \cdot 15 - \int_0^3 (x^2 + x + 3)\ dx$$

$$= 45 - \left[\frac{x^3}{3} + \frac{x^2}{2} + 3x \right]_0^3$$

$$= 45 - \left(\left[\frac{3^3}{3} + \frac{3^2}{2} + 3(3) \right] - \left[\frac{0^3}{3} + \frac{0^2}{2} + 3(0) \right] \right)$$

$$= 45 - \left(9 + \frac{9}{2} + 9 \right) + 0$$

$$= \$22.50. \qquad \qquad \qquad ❖$$

DO EXERCISE 2.

The **equilibrium point** (x_E, p_E) in Fig. 8 is the point at which the supply and the demand curves intersect. It is that point at which the sellers and buyers come together and purchases and sales actually occur. In Fig. 9, we see the equilibrium point and the consumer's and producer's surpluses for the movie curves.

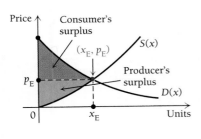

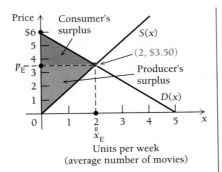

FIGURE 8   **FIGURE 9**

EXAMPLE 3   Given

$$D(x) = (x - 5)^2$$

and

$$S(x) = x^2 + x + 3,$$

find each of the following (Fig. 10).

a) The equilibrium point

b) The consumer's surplus at the equilibrium point

c) The producer's surplus at the equilibrium point

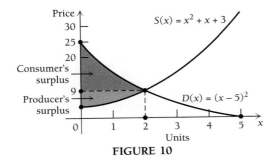

FIGURE 10

Solution

a) To find the equilibrium point, we set $D(x) = S(x)$ and solve:

$$(x - 5)^2 = x^2 + x + 3$$
$$x^2 - 10x + 25 = x^2 + x + 3$$
$$-10x + 25 = x + 3$$
$$22 = 11x$$
$$2 = x.$$

Thus, $x_E = 2$. To find p_E, we substitute x_E into either $D(x)$ or $S(x)$. If we choose $D(x)$, then

$$p_E = D(x_E) = D(2) = (2 - 5)^2 = (-3)^2 = \$9 \text{ per unit.}$$

Thus the equilibrium point is $(2, \$9)$.

b) The consumer's surplus at the equilibrium point is

$$\int_0^{x_E} D(x) \, dx - x_E p_E,$$

or

$$\int_0^2 (x - 5)^2 \, dx - 9 \cdot 2 = \left[\frac{(x - 5)^3}{3} \right]_0^2 - 18$$

$$= \left[\frac{(2 - 5)^3}{3} - \frac{(0 - 5)^3}{3} \right] - 18$$

$$= \frac{(-3)^3}{3} - \frac{(-5)^3}{3} - 18$$

$$= -\frac{27}{3} + \frac{125}{3} - \frac{54}{3}$$

$$= \frac{44}{3} \approx \$14.67.$$

c) The producer's surplus at the equilibrium point is

$$x_E p_E - \int_0^{x_E} S(x) \, dx,$$

or

$$2 \cdot 9 - \int_0^2 (x^2 + x + 3)\, dx$$

$$= 2 \cdot 9 - \left[\frac{x^3}{3} + \frac{x^2}{2} + 3x \right]_0^2$$

$$= 18 - \left[\left(\frac{2^3}{3} + \frac{2^2}{2} + 3 \cdot 2 \right) - \left(\frac{0^3}{3} + \frac{0^2}{2} + 3 \cdot 0 \right) \right]$$

$$= 18 - \left(\frac{8}{3} + 2 + 6 \right)$$

$$= \frac{22}{3} \approx \$7.33.$$

DO EXERCISE 3.

3. Given

$$D(x) = (x - 6)^2 \quad \text{and}$$
$$S(x) = x^2 + x + 10,$$

find each of the following.
a) The equilibrium point
b) The consumer's surplus at the equilibrium point
c) The producer's surplus at the equilibrium point

❖

EXERCISE SET 6.1

In each exercise, find (a) the equilibrium point, (b) the consumer's surplus at the equilibrium point, and (c) the producer's surplus at the equilibrium point.

1. $D(x) = -\frac{5}{6}x + 10, \quad S(x) = \frac{1}{2}x + 2$
2. $D(x) = -2x + 8, \quad S(x) = x + 2$
3. $D(x) = (x - 4)^2, \quad S(x) = x^2 + 2x + 6$
4. $D(x) = (x - 3)^2, \quad S(x) = x^2 + 2x + 1$

5. $D(x) = (x - 6)^2, \quad S(x) = x^2$
6. $D(x) = (x - 8)^2, \quad S(x) = x^2$
7. $D(x) = 1000 - 10x, \quad S(x) = 250 + 5x$
8. $D(x) = 8800 - 30x, \quad S(x) = 7000 + 15x$
9. $D(x) = 5 - x,\ 0 \leqslant x \leqslant 5; \quad S(x) = \sqrt{x + 7}$
10. $D(x) = 7 - x,\ 0 \leqslant x \leqslant 7; \quad S(x) = 2\sqrt{x + 1}$

SYNTHESIS EXERCISES

11. $D(x) = e^{-x + 4.5}, \quad S(x) = e^{x - 5.5}$

12. $D(x) = \sqrt{56 - x}, \quad S(x) = x$

COMPUTER-GRAPHING CALCULATOR EXERCISES

Sketch each of the following demand and supply functions. Then determine the appropriate horizontal line, sketch it, and shade the regions of both the consumer's and the producer's surplus.

13. $D(x) = (x - 4)^2,$
    $S(x) = x^2 + 2x + 6$
14. $D(x) = 5 - x,$
    $S(x) = \sqrt{x + 7}$

6.2

APPLICATIONS OF THE MODELS $\displaystyle\int_0^T P_0 e^{kt}\, dt$

AND $\displaystyle\int_0^T P_0 e^{-kt}\, dt$

Recall the basic model of exponential growth (Section 4.3):

$$P'(t) = k \cdot P(t), \quad \text{or} \quad \frac{dP}{dt} = kP, \quad \text{where } P = P_0 \text{ when } t = 0.$$

The function that satisfies the equation is

$$P(t) = P_0 e^{kt}. \tag{1}$$

One application of Eq. (1) is to compute the balance of a savings account after t years from an initial investment of P_0 at interest rate k, compounded continuously.

EXAMPLE 1　*Business: Growth in an investment.* Find the balance in a savings account after 3 years from an initial investment of $1000 at interest rate 8%, compounded continuously.

Solution　Using Eq. (1) with $k = 0.08$, $t = 3$, and $P_0 = \$1000$, we get

$$\begin{aligned}
P(3) &= 1000 e^{0.08\,(3)} \\
&= 1000 e^{0.24} \\
&= 1000(1.271249) \quad \blacksquare \\
&\approx \$1271.25.
\end{aligned}$$

DO EXERCISE 1.

1. *Business: Growth in an investment.* Find the balance in a savings account after 2 years from an initial investment of $1000 at interest rate 7.5%, compounded continuously.

The Integral $\displaystyle\int_0^T P_0 e^{kt}\, dt$

Suppose we consider the integral of $P_0 e^{kt}$ over the interval $[0, T]$:

$$\int_0^T P_0 e^{kt}\, dt = \left[\frac{P_0}{k} \cdot e^{kt} \right]_0^T = \frac{P_0}{k}(e^{kT} - e^{k \cdot 0}) = \frac{P_0}{k}(e^{kT} - 1).$$

We have a formula for this integral, given as

$$\int_0^T P(t)\, dt = \int_0^T P_0 e^{kt}\, dt = \frac{P_0}{k}(e^{kT} - 1). \tag{2}$$

We know that this formula represents the area under the graph of $P(t) = P_0 e^{kt}$ over the interval $[0, T]$ (Fig. 1).

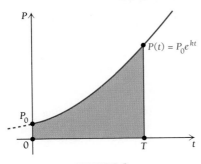

FIGURE 1

In the remainder of this section, we consider two other applications or interpretations of this integral.

Continuous Money Flow

Suppose that money is flowing continuously into a savings account at an annual rate of $1000 per year at interest rate 8%, compounded continuously. This means that over a small amount of time dt, the bank pays interest on all money accumulated in the account up to this time (since $t = 0$). The amount that is paid in over time dt is

$$\$1000 e^{0.08t}\, dt.$$

Suppose we want to find the accumulation of all these amounts over a 5-year period. That accumulation is given by the integral

$$\int_0^5 \$1000 e^{0.08t}\, dt = \left[\frac{1000}{0.08} e^{0.08t} \right]_0^5$$

$$= 12{,}500(e^{0.08(5)} - e^{0.08(0)})$$

$$= 12{,}500(e^{0.4} - 1)$$

$$= 12{,}500(1.491825 - 1) \qquad \text{▦}$$

$$\approx \$6147.81.$$

Economists call $6147.81 the *amount of a continuous money flow*. In this case the money is flowing according to a constant function $R(t) = \$1000$. Money could also be flowing according to some variable function, say, $R(t) = 2t - 7$ or $R(t) = t^2$.

2. *Business: Amount of a continuous money flow.* Find the amount of a continuous money flow where $2000 per year is being invested at 7.5%, compounded continuously, for 10 years.

THEOREM 1

If the rate of flow of money into an investment is given by some constant function $R(t)$, then the *amount of a continuous money flow* at interest rate k, compounded continuously, over time T is given by

$$\int_0^T R(t)e^{kt}\, dt.$$

EXAMPLE 2 *Business: Amount of a continuous money flow.* Find the amount of a continuous money flow where $1000 per year is being invested at 8%, compounded continuously, for 15 years.

Solution

$$\int_0^{15} \$1000e^{0.08t}\, dt = \frac{1000}{0.08}(e^{0.08(15)} - 1) \qquad \text{Using Eq. (2)}$$

$$= 12{,}500(e^{1.2} - 1)$$

$$= 12{,}500(3.320117 - 1) \qquad \text{▦}$$

$$\approx \$29{,}001.46 \qquad \qquad \qquad ❖$$

DO EXERCISE 2.

Sometimes we might want to know how money should be flowing into an investment so that we end up with a specified amount.

3. *Business: Continuous money flow.* Consider a continuous flow of money into an investment at the constant rate of P_0 dollars per year. What should P_0 be so that the amount of a continuous money flow over 20 years at interest rate 7.5%, compounded continuously, will be $10,000?

EXAMPLE 3 *Business: Continuous money flow.* Consider a continuous flow of money into an investment at the constant rate of P_0 dollars per year. What should P_0 be so that the amount of a continuous money flow over 20 years at interest rate 8%, compounded continuously, will be $10,000?

Solution We find P_0 such that

$$10{,}000 = \int_0^{20} P_0 e^{0.08t}\, dt.$$

We solve the following equation:

$$10{,}000 = \frac{P_0}{0.08}(e^{0.08(20)} - 1) \qquad \text{Using Eq. (2)}$$

$$800 = P_0(e^{1.6} - 1)$$

$$800 = P_0(4.953032 - 1) \qquad \text{▦}$$

$$800 = P_0(3.953032)$$

$$\$202.38 \approx P_0. \qquad \qquad ❖$$

DO EXERCISE 3.

Life and Physical Sciences: Depletion of Natural Resources

Another application of the integral of exponential growth concerns

$$P(t) = P_0 e^{kt}$$

as a model of the demand for natural resources. Suppose that P_0 represents the amount of a natural resource (such as coal or oil) used at time $t = 0$ and that the growth rate for the use of this resource is k. Then, assuming exponential growth (which is the case for the use of many resources), the amount to be used at time t is $P(t)$, given by

$$P(t) = P_0 e^{kt}.$$

The total amount used during an interval $[0, T]$ is given by

$$\int_0^T P(t)\, dt = \int_0^T P_0 e^{kt}\, dt = \frac{P_0}{k}(e^{kT} - 1). \tag{2}$$

EXAMPLE 4 *Physical science: Demand for copper.* In 1990 ($t = 0$), the world use of copper was 8,830,000 metric tons, and the demand for it was growing exponentially at the rate of 15% per year. If the growth continues at this rate, how many tons of copper will the world use from 1990 to 2000?

Solution   Using Eq. (2), we have

$$\int_0^{10} 8{,}830{,}000 e^{0.15t}\, dt = \frac{8{,}830{,}000}{0.15}(e^{0.15(10)} - 1) \qquad \text{Using Eq. (2)}$$

$$\approx 58{,}866{,}667(e^{1.5} - 1)$$

$$= 58{,}866{,}667(4.481689 - 1) \qquad \text{▥}$$

$$= 58{,}866{,}667(3.481689)$$

$$\approx 204{,}955{,}400. \qquad \text{Rounded to the nearest hundred}$$

Thus from 1990 to 2000, the world will use 204,955,400 metric tons of copper.   ❖

DO EXERCISE 4.

EXAMPLE 5 *Physical science: Depletion of copper ore.* The world reserves of copper ore are estimated to be 2,300,000,000 metric tons. Assuming that the growth rate in Example 4 continues and that no new reserves are discovered, when will the world reserves of copper ore be exhausted?

4. *Physical science: Demand for oil.* In 1990 ($t = 0$), the world use of oil was 6570 million barrels, and the demand for it was growing exponentially at the rate of 10% per year. If the demand continues at this rate, how many barrels of oil will the world use from 1990 to 2000?

Bingham Canyon mine in Utah has produced more copper than any other mine in history. The grade of ore, however, has dropped from 1.93 percent copper in 1906 to 0.6 percent today. At present, it is planned that mining will cease when the percentage of copper reaches 0.4.

Solution Using Eq. (2), we want to find T such that

$$2{,}300{,}000{,}000 = \frac{8{,}830{,}000}{0.15}(e^{0.15T} - 1).$$

We solve for T as follows:

$$2{,}300{,}000{,}000 \approx 58{,}866{,}667(e^{0.15T} - 1)$$

$$\frac{2{,}300{,}000{,}000}{58{,}866{,}667} = e^{0.15T} - 1$$

$$39.071347 \approx e^{0.15T} - 1$$

$$40.071347 = e^{0.15T}$$

$$\ln 40.071347 = \ln e^{0.15T} \qquad \text{Taking the natural logarithm on both sides}$$

$$\ln 40.071347 = 0.15T \qquad \text{Recall: } \ln e^k = k.$$

$$\frac{\ln 40.071347}{0.15} = T$$

$$\frac{3.690661}{0.15} = T \qquad \text{\small▣}$$

$$25 \approx T. \qquad \text{Rounding to the nearest one}$$

Thus 25 years from 1990 (or by 2015), the world reserves of copper ore will be exhausted. ❖

DO EXERCISE 5.

Applications of the Model $\int_0^T P e^{-kt}\, dt$

Recall our study of present value in Section 4.4. If you did not study that topic there, you should do so now. The **present value** P_0 of an amount P due t years later is found by solving the following equation for P_0:

$$P_0 e^{kt} = P$$

$$P_0 = \frac{P}{e^{kt}} = P e^{-kt}.$$

THEOREM 2

The *present value* P_0 of an amount P due t years later at interest rate k, compounded continuously, is given by

$$P_0 = P e^{-kt}.$$

Note that this can be interpreted as exponential decay from the future back to the present.

EXAMPLE 6 *Business: Present value.* Find the present value of $200,000 due 25 years from now at 8.7%, compounded continuously.

Solution We substitute $200,000 for P, 0.087 for k, and 25 for t in the equation for present value:

$$P_0 = 200{,}000 e^{-0.087(25)} \approx \$22{,}721.63.$$

Thus the present value is $22,721.63. ❖

DO EXERCISES 6 AND 7.

Suppose we know that a continuous flow of money will go into an investment at the constant rate of P dollars per year, from now until some time T in the future. If an infinitesimal amount of time dt passes,

$$P \cdot dt$$

5. *Physical science: Depletion of oil.* The world reserves of oil are 920,700 million barrels. In 1990 ($t = 0$), the world use of oil was 6570 million barrels, and the growth rate for the use of oil was 10%. Assuming that this growth rate continues and that no new reserves are discovered, when will the world reserves of oil be exhausted?

6. *Business: Present value.* Following the birth of a child, a parent wants to make an initial investment P_0 that will grow to $40,000 by the child's 18th birthday. Interest is compounded continuously at 7.5%. What should this initial investment be?

7. *Business: Present value.* Find the present value of $100,000 due 5 years later at 8%, compounded continuously.

8. *Business: Accumulated present value.* Find the accumulated present value of an investment over a 20-year period if there is a continuous money flow of $1800 per year and the current interest rate is 10%, compounded continuously.

dollars will have accumulated. The present value of that amount is

$$(P \cdot dt)e^{-kt},$$

where k is the current interest rate, compounded continuously. The accumulation of all the present values is given by the integral

$$\int_0^T Pe^{-kt}\, dt$$

and is called the **accumulated present value**. Evaluating this integral, we get

$$\int_0^T Pe^{-kt}\, dt = \frac{P}{-k}(e^{-kT} - e^{-k \cdot 0}) = \frac{P}{k}(1 - e^{-kT}). \qquad (3)$$

In the preceding case, money is flowing according to a constant function $R(t) = P$. Money could also be flowing according to some variable function, such as $R(t) = 2t + 8$ or $R(t) = t^3$.

THEOREM 3

The *accumulated present value* of a continuous money flow into an investment at a constant rate of $R(t)$ dollars per year from now until some time T in the future is given by

$$\int_0^T R(t)e^{-kt}\, dt,$$

where k is the current interest rate, compounded continuously.

EXAMPLE 7 *Business: Accumulated present value.* Find the accumulated present value of an investment over a 5-year period if there is a continuous money flow of $2400 per year and the current interest rate is 14%, compounded continuously.

Solution   The accumulated present value is

$$\int_0^5 \$2400e^{-0.14t}\, dt = \frac{2400}{0.14}(1 - e^{-0.14 \cdot 5}) \qquad \text{Using Eq. (3)}$$

$$\approx 17{,}142.86(1 - e^{-0.7})$$

$$= 17{,}142.86(1 - 0.496585) \qquad \text{▨}$$

$$\approx \$8629.97. \qquad \qquad \text{❖}$$

DO EXERCISE 8.

The preceding example is an application of the model

$$\int_0^T Pe^{-kt} \, dt = \frac{P}{k}(1 - e^{-kT}).$$

This model can be applied to a calculation of the buildup of a specific amount of radioactive material released into the atmosphere annually. Some of the material decays, but more continues to be released. The amount present at time T is given by the integral above.

EXERCISE SET 6.2

APPLICATIONS

❖ Business and Economics

1. Find the amount in a savings account after 3 years from an initial investment of $100 at interest rate 9%, compounded continuously.

2. Find the amount in a savings account after 4 years from an initial investment of $100 at interest rate 10%, compounded continuously.

3. Find the amount of a continuous money flow in which $100 per year is being invested at 9%, compounded continuously, for 20 years.

4. Find the amount of a continuous money flow in which $100 per year is being invested at 10%, compounded continuously, for 20 years.

5. Find the amount of a continuous money flow in which $1000 per year is being invested at 8.5%, compounded continuously, for 40 years.

6. Find the amount of a continuous money flow in which $1000 per year is being invested at 7.5%, compounded continuously, for 40 years.

7. What should P_0 be so that the amount of a continuous money flow over 20 years at interest rate 8.5%, compounded continuously, will be $50,000?

8. What should P_0 be so that the amount of a continuous money flow over 20 years at interest rate 7.5%, compounded continuously, will be $50,000?

9. What should P_0 be so that the amount of a continuous money flow over 30 years at interest rate 9%, compounded continuously, will be $40,000?

10. What should P_0 be so that the amount of a continuous money flow over 30 years at interest rate 10%, compounded continuously, will be $40,000?

11. Following the birth of a child, a parent wants to make an initial investment P_0 that will grow to $50,000 by the child's 20th birthday. Interest is compounded continuously at 9%. What should the initial investment be?

12. Following the birth of a child, a parent wants to make an initial investment P_0 that will grow to $60,000 by the child's 20th birthday. Interest is compounded continuously at 10%. What should the initial investment be?

13. Find the present value of $60,000 due 8 years later at 8.8%, compounded continuously.

14. Find the present value of $50,000 due 16 years later at 7.4%, compounded continuously.

15. Find the accumulated present value of an investment over a 10-year period if there is a continuous money flow of $2700 per year and the current interest rate is 9%, compounded continuously.

16. Find the accumulated present value of an investment over a 10-year period if there is a continuous money flow of $2700 per year and the current interest rate is 10%, compounded continuously.

17. A woman with an MBA accepts a position as president of a company at the age of 35. Assuming retirement at age 65 and an annual salary of $45,000 that is paid in a continuous money flow, what is the president's accu-

mulated present value? The current interest rate is 8%, compounded continuously.

18. A college dropout takes a job as a truck driver at the age of 25. Assuming retirement at age 65 and an annual salary of $14,000 that is paid in a continuous money flow, what is the truck driver's accumulated present value? The current interest rate is 7%, compounded continuously.

❖ **Life and Physical Sciences**

19. *The demand for aluminum ore (bauxite).* In 1990 ($t = 0$), the world use of aluminum ore was 101,000,000 tons, and the demand for it was growing exponentially at the rate of 12% per year. If the demand continues to grow at this rate, how many tons of aluminum ore will the world use from 1990 to 2004?

20. *The demand for natural gas.* In 1990 ($t = 0$), the world use of natural gas was 72,137 billion cubic feet, and the demand for it was growing exponentially at the rate of 4% per year. If the demand continues to grow at this rate, how many cubic feet of natural gas will the world use from 1990 to 2020?

21. *The depletion of aluminum ore.* The world reserves of aluminum ore are 75,000,000,000 tons. Assuming that the growth rate of Exercise 19 continues and that no new reserves are discovered, when will the world reserves of aluminum ore be exhausted?

22. *The depletion of natural gas.* The world reserves of natural gas are 3,926,000 billion cubic feet. Assuming that the growth rate of Exercise 20 continues and that no new reserves are discovered, when will the world reserves of natural gas be exhausted?

23. *Radioactive buildup.* Plutonium has a decay rate of 0.003% per year. Suppose plutonium is released into the atmosphere each year for 20 years at the rate of 1 pound per year. What is the total amount of radioactive buildup?

24. *Radioactive buildup.* Cesium-137 has a decay rate of 2.3% per year. Suppose cesium-137 is released into the atmosphere each year for 20 years at the rate of 1 pound per year. What is the total amount of radioactive buildup?

SYNTHESIS EXERCISES

For a nonconstant function $R(t)$, the amount of a continuous money flow is

$$\int_0^T R(t)e^{k(T-t)}\,dt.$$

For a nonconstant function $R(t)$, the accumulated present value is

$$\int_0^T R(t)e^{-k(T-t)}\,dt.$$

Find the amount of a continuous money flow for each of the following.

25. $R(t) = 2000t + 7$, $k = 8\%$, and $T = 30$ years
26. $R(t) = t^2$, $k = 7\%$, and $T = 40$ years

Find the accumulated present value for each of the following.

27. $R(t) = t$, $k = 8\%$, and $T = 20$ years
28. $R(t) = e^t$, $k = 7\%$, and $T = 10$ years

COMPUTER-GRAPHING CALCULATOR EXERCISES

29. Sketch the graph and the area under the curve represented by the continuous money flow in Exercise 3.

30. Sketch the graph and the area under the curve represented by the continuous money flow in Exercise 19.

6.3

IMPROPER INTEGRALS

Let us try to find the area of the region under the graph of $y = 1/x^2$ on the interval $[1, \infty)$.

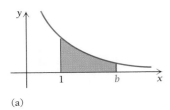

(a)

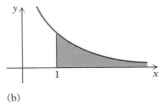

(b)

FIGURE 1

Note in Fig. 1 that this region is of infinite extent. We have not yet considered how to find the area of such a region. Let us find the area under the curve on the interval from 1 to b, and then see what happens as b gets very large. The area under the graph on $[1, b]$ is

$$\int_1^b \frac{dx}{x^2} = \left[-\frac{1}{x} \right]_1^b$$

$$= \left(-\frac{1}{b} \right) - \left(-\frac{1}{1} \right)$$

$$= -\frac{1}{b} + 1 = 1 - \frac{1}{b}.$$

Then

$$\lim_{b \to \infty} [\text{area from 1 to } b] = \lim_{b \to \infty} \left(1 - \frac{1}{b} \right).$$

Let us investigate this limit in Margin Exercise 1 by computing the area when $b = 2, 3, 10, 100,$ and 200.

DO EXERCISE 1.

Note that as $b \to \infty$, $1/b \to 0$, so $(1 - 1/b) \to 1$. Thus,

$$\lim_{b \to \infty} [\text{area from 1 to } b] = \lim_{b \to \infty} \left(1 - \frac{1}{b} \right) = 1.$$

OBJECTIVES

a)   Determine whether an improper integral is convergent or divergent, and calculate its value if it is convergent.

b)   Solve applied problems involving improper integrals.

1. Complete.

b	$1 - \dfrac{1}{b}$
2	$1 - \frac{1}{2}$, or $\frac{1}{2}$
3	
10	
100	
200	

2. Find the area of the region under the graph of

$$y = \frac{1}{x^2}$$

on the interval $[2, \infty)$.

We *define* the area from 1 to infinity to be this limit. Here we have an example of an infinitely long region with a finite area.

DO EXERCISE 2.

Such areas may not always be finite. Let us try to find the area of the region under the graph of $y = 1/x$ on the interval $[1, \infty)$ (Fig. 2).

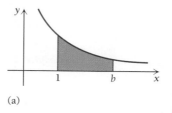 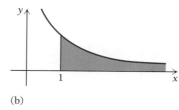

(a)   (b)

FIGURE 2

By definition, the area A from 1 to infinity is the limit as $b \to \infty$ of the area from 1 to b, so

$$A = \lim_{b \to \infty} \int_1^b \frac{dx}{x} = \lim_{b \to \infty} [\ln x]_1^b = \lim_{b \to \infty} (\ln b - \ln 1) = \lim_{b \to \infty} \ln b.$$

In Section 4.2, we learned that $\ln b$ increases indefinitely as b increases. Therefore, the limit does not exist.

Thus we have an infinitely long region with an infinite area. Note that the graphs of $y = 1/x^2$ and $y = 1/x$ have similar shapes, but the region under one of them has a finite area and the other does not.

3. Find the area under the graph of

$$y = \frac{1}{x}$$

from $x = 2$ to $x = \infty$.

DO EXERCISE 3.

An integral such as

$$\int_a^\infty f(x) \, dx,$$

with an upper limit of infinity, is called an **improper integral**. Its value is defined to be the following limit.

DEFINITION

$$\int_a^\infty f(x) \, dx = \lim_{b \to \infty} \int_a^b f(x) \, dx$$

If the limit exists, then we say that the improper integral **converges**, or is **convergent**. If the limit does not exist, then we say that the improper integral **diverges**, or is **divergent**. Thus,

$$\int_1^\infty \frac{dx}{x^2} = 1 \ converges; \quad \text{and} \quad \int_1^\infty \frac{dx}{x} \ diverges.$$

EXAMPLE 1   Determine whether the following integral is convergent or divergent, and calculate its value if it is convergent:

$$\int_0^\infty 2e^{-2x} \, dx.$$

Solution   We have

$$\int_0^\infty 2e^{-2x} \, dx = \lim_{b \to \infty} \int_0^b 2e^{-2x} \, dx$$

$$= \lim_{b \to \infty} \left[\frac{2}{-2} e^{-2x} \right]_0^b$$

$$= \lim_{b \to \infty} \left[-e^{-2x} \right]_0^b$$

$$= \lim_{b \to \infty} \left[-e^{-2b} - (-e^{-2 \cdot 0}) \right]$$

$$= \lim_{b \to \infty} \left(-e^{-2b} + 1 \right)$$

$$= \lim_{b \to \infty} \left(1 - \frac{1}{e^{2b}} \right).$$

Now as $b \to \infty$, we know that $e^{2b} \to \infty$ (from Chapter 4), so

$$\frac{1}{e^{2b}} \to 0 \quad \text{and} \quad \left(1 - \frac{1}{e^{2b}} \right) \to 1.$$

Thus,

$$\int_0^\infty 2e^{-2x} \, dx = \lim_{b \to \infty} \left(1 - \frac{1}{e^{2b}} \right) = 1.$$

The integral is convergent.   ❖

DO EXERCISES 4 AND 5.

Following are definitions of two other types of improper integrals.

Determine whether the improper integral is convergent or divergent, and calculate its value if it is convergent.

4. $\int_0^\infty 5e^{-5x} \, dx$

5. $\int_0^\infty 2x \, dx$

DEFINITIONS

1. $\displaystyle\int_{-\infty}^{b} f(x)\ dx = \lim_{a \to -\infty} \int_{a}^{b} f(x)\ dx$

2. $\displaystyle\int_{-\infty}^{\infty} f(x)\ dx = \int_{-\infty}^{c} f(x)\ dx + \int_{c}^{\infty} f(x)\ dx$

In order for $\int_{-\infty}^{\infty} f(x)\ dx$ to converge, both integrals on the right above must converge.

Applications

In Section 6.2, we learned that the accumulated present value of a continuous money flow of P dollars per year from now until time T in the future is given by

$$\int_{0}^{T} Pe^{-kt}\ dt = \frac{P}{k}(1 - e^{-kT}),$$

where k is the current interest rate. Suppose the money flow is to continue perpetually. Under this assumption, the accumulated present value over this infinite time period would be

$$\int_{0}^{\infty} Pe^{-kt}\ dt = \lim_{T \to \infty} \int_{0}^{T} Pe^{-kt}\ dt$$

$$= \lim_{T \to \infty} \frac{P}{k}(1 - e^{-kT})$$

$$= \lim_{T \to \infty} \frac{P}{k}\left(1 - \frac{1}{e^{kT}}\right)$$

$$= \frac{P}{k}.$$

THEOREM 4

The *accumulated present value* of a continuous money flow into an investment at the rate of P dollars per year perpetually is given by

$$\frac{P}{k},$$

where k is the current interest rate, compounded continuously.

EXAMPLE 2   *Business: Accumulated present value.* Find the accumulated present value of an investment for which there is a perpetual continuous money flow of $2000 per year. The current interest rate is 8%, compounded continuously.

Solution   The accumulated present value is 2000/0.08, or $25,000. ❖

DO EXERCISE 6.

When an amount P of radioactive material is being released into the atmosphere annually, the amount present at time T is given by

$$\int_0^T Pe^{-kt}\,dt = \frac{P}{k}(1 - e^{-kT}).$$

As $T \to \infty$ (the radioactive material is to be released forever), the buildup of radioactive material approaches a limiting value P/k, as shown in Fig. 3. It is no wonder that scientists and environmentalists are so concerned about continued nuclear detonations. Eventually the buildup is "here to stay."

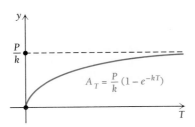

FIGURE 3

6. *Business: Accumulated present value.* Find the accumulated present value of an investment for which there is a perpetual continuous money flow of $3600 per year. The current interest rate is 10%, compounded continuously.

EXERCISE SET 6.3

Determine whether the improper integral is convergent or divergent, and calculate its value if it is convergent.

1. $\int_3^\infty \frac{dx}{x^2}$

2. $\int_4^\infty \frac{dx}{x^2}$

3. $\int_3^\infty \frac{dx}{x}$

4. $\int_4^\infty \frac{dx}{x}$

5. $\int_0^\infty 3e^{-3x}\,dx$

6. $\int_0^\infty 4e^{-4x}\,dx$

7. $\int_1^\infty \frac{dx}{x^3}$

8. $\int_1^\infty \frac{dx}{x^4}$

9. $\int_0^\infty \frac{dx}{1+x}$

10. $\int_0^\infty \frac{4\,dx}{1+x}$

11. $\int_1^\infty 5x^{-2}\,dx$

12. $\int_1^\infty 7x^{-2}\,dx$

13. $\int_0^\infty e^x\,dx$

14. $\int_0^\infty e^{2x}\,dx$

15. $\int_3^\infty x^2 \, dx$

16. $\int_5^\infty x^4 \, dx$

17. $\int_0^\infty xe^x \, dx$

18. $\int_1^\infty \ln x \, dx$

19. $\int_0^\infty me^{-mx} \, dx, \ m > 0$

20. $\int_0^\infty Qe^{-kt} \, dt, \ k > 0$

21. Find the area, if it exists, of the region bounded by $y = 2xe^{-x^2}$ and the lines $x = 0$ and $y = 0$.

22. Find the area, if it exists, of the region bounded by $y = 1/\sqrt{(3x - 2)^3}$ and the lines $x = 6$ and $y = 0$.

APPLICATIONS

❖ Business and Economics

23. *Accumulated present value.* Find the accumulated present value of an investment for which there is a perpetual continuous money flow of $3600 per year. The current interest rate is 8%.

24. *Accumulated present value.* Find the accumulated present value of an investment for which there is a perpetual continuous money flow of $4500 per year. The current interest rate is 9%.

25. *Total profit from marginal profit.* A firm is able to determine that its marginal profit is given by

$$P'(x) = 200e^{-0.032x}.$$

Suppose it were possible for the firm to make infinitely many units of this product. What would its total profit be?

26. *Total profit from marginal profit.* In Exercise 25, find the total profit if

$$P'(x) = 200x^{-0.032}, \quad \text{where } x \geqslant 1.$$

27. *Total cost from marginal cost.* A company determines that its marginal cost for the production of x units of a

product is given by

$$C'(x) = 3600x^{-1.8}, \quad \text{where } x \geqslant 1.$$

Suppose it were possible for the firm to make infinitely many units of this product. What would the total cost be?

28. *Total production.* A firm determines that it can produce tires at a rate of

$$r(t) = 2000e^{-0.42t}.$$

where t is the time, in years. Assuming the firm endures forever, how many tires can it make?

❖ Life and Physical Sciences

29. *Radioactive buildup.* Plutonium has a decay rate of 0.003% per year. Suppose plutonium is released into the atmosphere each year perpetually at the rate of 1 pound per year. What is the limiting value of the radioactive buildup?

30. *Radioactive buildup.* Cesium-137 has a decay rate of 2.3% per year. Suppose cesium-137 is released into the atmosphere each year perpetually at the rate of 1 pound per year. What is the limiting value of the radioactive buildup?

SYNTHESIS EXERCISES

Determine whether the improper integral is convergent or divergent, and calculate its value if it is convergent.

31. $\int_0^\infty \dfrac{dx}{x^{2/3}}$

32. $\int_1^\infty \dfrac{dx}{\sqrt{x}}$

33. $\int_0^\infty \dfrac{dx}{(x + 1)^{3/2}}$

34. $\int_{-\infty}^0 e^{2x} \, dx$

35. $\int_0^\infty xe^{-x^2} \, dx$

36. $\int_{-\infty}^\infty xe^{-x^2} \, dx$

Life science: Drug dosage. Suppose an oral dose of a drug is taken. From that time, the drug is assimilated in the body and excreted through the urine. The total amount of the drug that has passed through the body in time T is given by

$$\int_0^T E(t) \, dt,$$

where E is the rate of excretion of the drug. A typical rate-of-excretion function is $E(t) = te^{-kt}$, where $k > 0$ and t is the time, in hours. (Now do Exercises 37 and 38.)

37. Find $\int_0^\infty E(t)\,dt$ and interpret the answer. That is, what does the integral represent?

38. A physician prescribes a dosage of 100 mg. Find k.

39. Sketch a graph of the function E and shade the area under the curve for the situation in Exercises 37 and 38.

6.4

PROBABILITY

The definite integral plays a role in the theory of probability. Briefly, the **probability** of an event is a number from 0 to 1 that represents the chances of the event occurring. It is the "relative frequency" of occurrence—that is, the proportion or percentage of times that we might expect an event to occur in a large number of trials.

EXAMPLE 1   What is the probability of drawing an ace from a well-shuffled deck of cards?

A desire to calculate odds in games of chance gave rise to the theory of probability

OBJECTIVES

a)   Verify that a given function satisfies the property

$$\int_a^b f(x)\,dx = 1$$

for being a probability density function.

b)   Find k such that a function like

$$f(x) = kx^2$$

is a probability density function over an interval $[a, b]$.

c)   Solve applied problems involving probability density functions.

1. In Example 2, what is the probability that the ball you are holding is:

   a) Black?

   b) Yellow?

   c) Green?

   d) Purple?

2. Consider this dartboard.

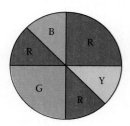

   You throw a dart at the board without aiming at any particular region. What is the probability that you will land on:

   a) Red?

   b) Blue?

   c) Green?

   d) Yellow?

Solution Since there are 52 possible outcomes and each card has the same chance of being drawn, and since there are 4 aces, the probability of drawing an ace is $\frac{4}{52}$ or $\frac{1}{13}$, or about 7.7%.

In practice, we may not draw an ace 7.7% of the time, but in a large number of trials, after shuffling the cards and drawing one, replacing the card, and shuffling the cards and drawing one, we would expect to draw an ace about 7.7% of the time. That is, the more draws we make, the closer we expect to get to 7.7%. ❖

EXAMPLE 2 A jar contains 7 black balls, 6 yellow balls, 4 green balls, and 3 red balls (Fig. 1). The jar is shaken well and you remove 1 ball without looking. What is the probability that the ball is red? that it is white?

FIGURE 1

Solution There are 20 balls altogether and of these 3 are red, so the probability of drawing a red ball is $\frac{3}{20}$. There are no white balls, so the probability of drawing a white one is $\frac{0}{20}$, or 0. ❖

DO EXERCISES 1 AND 2.

Let us consider a table of probabilities from Example 2.

Color	Probability
Black (B)	$\frac{7}{20}$
Yellow (Y)	$\frac{6}{20}$
Green (G)	$\frac{4}{20}$
Red (R)	$\frac{3}{20}$

Note that the sum of these probabilities is 1. We are certain that we will draw either a black, yellow, green, or red ball. The probability of that event is 1. Let us arrange these data from the table into what is called a *relative frequency graph* (Fig. 2), or *histogram,* which shows the proportion of times each event occurs (the probability of each event).

If we assign a width of 1 to each rectangle, then the sum of the areas of the rectangles is 1.

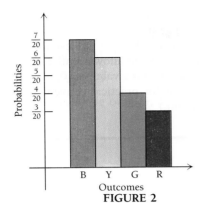

FIGURE 2

Continuous Random Variables

Suppose we throw a dart at a number line in such a way that it always lands in the interval [1, 3] (Fig. 3). Let x be the number that the dart hits. Note that x is a quantity that can be observed (or measured) repeatedly and whose possible values consist of an entire interval of real numbers. Such a variable is called a **continuous random variable**.

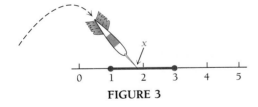

FIGURE 3

Suppose we throw the dart a large number of times and it lands 43% of the time in the subinterval [1.6, 2.8] of the main interval [1, 3]. The probability, then, that the dart lands in the interval [1.6, 2.8] is 0.43.

Let us consider some other examples of continuous random variables.

EXAMPLE 3 Suppose that x is the arrival time of buses at a bus stop in a 3-hr period from 2:00 P.M. to 5:00 P.M. The interval is [2, 5]. Then x is a continuous random variable distributed over the interval [2, 5] (Fig. 4).

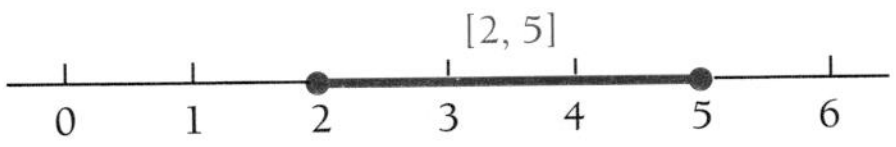

[2, 5]

0 1 2 3 4 5 6

FIGURE 4

Thus,

$$P([4, 5]) = \int_4^5 f(x)\, dx$$

$$= \int_4^5 \frac{3}{117} x^2\, dx$$

$$= \frac{3}{117} \left[\frac{x^3}{3} \right]_4^5$$

$$= \frac{1}{117} \left[x^3 \right]_4^5$$

$$= \frac{1}{117} (5^3 - 4^3)$$

$$= \frac{61}{117} \approx 0.52.$$

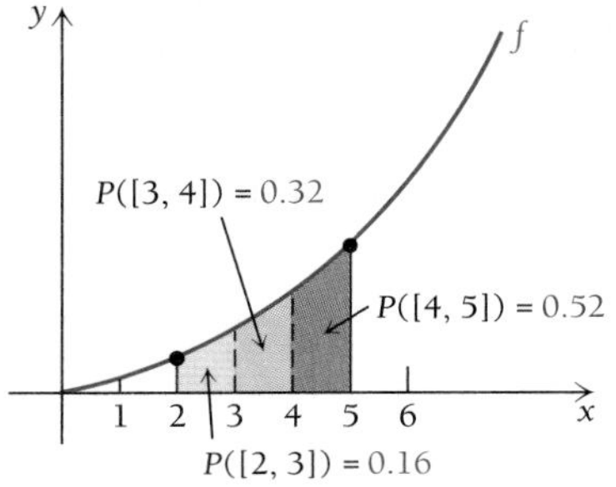

FIGURE 9

Thus there is a probability of 0.52 that at least one bus will arrive between 4:00 P.M. and 5:00 P.M. That is, 52% of the buses will arrive between 4:00 P.M. and 5:00 P.M. The function f is called a **probability density function**. Its integral over *any* subinterval gives the probability that x "lands" in that subinterval.

Similar calculations are shown in the following table.

Time interval	Probability a bus arrives during interval
2:00 P.M. and 3:00 P.M.	$\int_2^3 \frac{3}{117} x^2\, dx = 0.16 = P([2, 3])$
3:00 P.M. and 4:00 P.M.	$\int_3^4 \frac{3}{117} x^2\, dx = 0.32 = P([3, 4])$
4:00 P.M. and 5:00 P.M.	$\int_4^5 \frac{3}{117} x^2\, dx = 0.52 = P([4, 5])$
2:00 P.M. and 5:00 P.M.	$\int_2^5 \frac{3}{117} x^2\, dx = 1.00 = P([2, 5])$

The results in the table lead us to the following definition of a probability density function.

DEFINITION

Let x be a continuous random variable distributed over some interval $[a, b]$. A function f is said to be a *probability density function* for x if:

1. f is nonnegative over $[a, b]$; that is, $f(x) \geq 0$ for all x in $[a, b]$.

2. The probability that x lands in $[a, b]$ is 1:

$$\int_a^b f(x)\, dx = 1;$$

that is, we are "certain" that x is in the interval $[a, b]$. (See Fig. 10.)

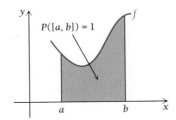

P([a, b]) = 1

FIGURE 10

3. For any subinterval $[c, d]$ of $[a, b]$, the probability $P([c, d])$, or $P(c \leq x \leq d)$, that x lands in that subinterval is given by

$$P([c, d]) = \int_c^d f(x)\, dx.$$

(See Fig. 11.)

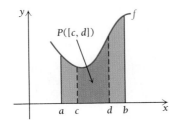

P([c, d])

FIGURE 11

6. Verify Property 2 of the definition of a probability density function for

$$f(x) = \tfrac{2}{3}x \quad \text{over } [1, 2].$$

EXAMPLE 5   Verify Property 2 of the definition above for

$$f(x) = \frac{3}{117}x^2, \quad \text{for } 2 \leqslant x \leqslant 5.$$

Solution

$$\int_2^5 \frac{3}{117}x^2 \, dx = \frac{3}{117}\left[\frac{1}{3}x^3\right]_2^5$$

$$= \frac{1}{117}\left[x^3\right]_2^5$$

$$= \frac{1}{117}(5^3 - 2^3)$$

$$= \frac{117}{117}$$

$$= 1 \qquad\qquad ❖$$

DO EXERCISE 6.

EXAMPLE 6   *Business: Life of a product.* A company that produces transistors determines that the life t of a transistor is from 3 to 6 years and that the probability density function for t is given by

$$f(t) = \frac{24}{t^3}, \quad \text{for } 3 \leqslant t \leqslant 6.$$

a)   Verify Property 2.

b)   Find the probability that a transistor will last no more than 4 years.

c)   Find the probability that a transistor will last at least 4 years and at most 5 years.

Solution

a)   We want to show that $\displaystyle\int_3^6 f(t) \, dt = 1$.

Now

$$\int_3^6 \frac{24}{t^3}\, dt = 24 \int_3^6 t^{-3}\, dt$$

$$= 24 \left[\frac{t^{-2}}{-2}\right]_3^6$$

$$= -12 \left[\frac{1}{t^2}\right]_3^6$$

$$= -12 \left(\frac{1}{6^2} - \frac{1}{3^2}\right)$$

$$= -12 \left(\frac{1}{36} - \frac{1}{9}\right)$$

$$= -12 \left(-\frac{3}{36}\right)$$

$$= 1.$$

b)   The probability that a transistor will last no more than 4 years is

$$P(3 \le t \le 4) = \int_3^4 \frac{24}{t^3}\, dt$$

$$= 24 \int_3^4 t^{-3}\, dt$$

$$= 24 \left[\frac{t^{-2}}{-2}\right]_3^4$$

$$= -12 \left[\frac{1}{t^2}\right]_3^4$$

$$= -12 \left(\frac{1}{4^2} - \frac{1}{3^2}\right)$$

$$= -12 \left(\frac{1}{16} - \frac{1}{9}\right)$$

$$= -12 \left(-\frac{7}{144}\right)$$

$$= \frac{7}{12} \approx 0.58.$$

7. *Business: Life of a product.* A different transistor company determines that the life t of its transistor is from 3 to 8 years and that the probability density function for t is given by

$$f(t) = \frac{24}{5t^2}, \quad \text{for } 3 \leq t \leq 8.$$

a) Verify Property 2 of the definition of a probability density function.

b) Find the probability that a transistor will last no more than 5 years.

c) Find the probability that a transistor will last at least 4 years and at most 6 years.

c) The probability that a transistor will last at least 4 years and at most 5 years is

$$P(4 \leq t \leq 5) = \int_4^5 \frac{24}{t^3}\, dt$$

$$= 24 \int_4^5 t^{-3}\, dt$$

$$= 24 \left[\frac{t^{-2}}{-2} \right]_4^5$$

$$= -12 \left[\frac{1}{t^2} \right]_4^5$$

$$= -12 \left(\frac{1}{5^2} - \frac{1}{4^2} \right)$$

$$= -12 \left(\frac{1}{25} - \frac{1}{16} \right)$$

$$= -12 \left(-\frac{9}{400} \right)$$

$$= \frac{27}{100} = 0.27. \qquad \diamond$$

DO EXERCISE 7.

Constructing Probability Density Functions

Suppose you have an arbitrary nonnegative function $f(x)$ whose definite integral over some interval $[a, b]$ is K. Then

$$\int_a^b f(x)\, dx = K.$$

Multiplying on both sides by $1/K$ gives us

$$\frac{1}{K} \int_a^b f(x)\, dx = \frac{1}{K} \cdot K = 1, \quad \text{or} \quad \int_a^b \frac{1}{K} \cdot f(x)\, dx = 1.$$

Thus when we multiply the function $f(x)$ by $1/K$, we have a function whose area over the given interval is 1.

EXAMPLE 7 Find k such that

$$f(x) = kx^2$$

is a probability density function over the interval [2, 5]. Then write the probability density function.

Solution We have

$$\int_2^5 x^2 \, dx = \left[\frac{x^3}{3}\right]_2^5 = \frac{5^3}{3} - \frac{2^3}{3} = \frac{125}{3} - \frac{8}{3} = \frac{117}{3}.$$

Thus,

$$k = \frac{1}{\frac{117}{3}} = \frac{3}{117}$$

and the probability density function is

$$f(x) = \frac{3}{117}x^2. \qquad \qquad \text{❖}$$

DO EXERCISES 8 AND 9.

Uniform Distributions

Suppose that the probability density function of a continuous random variable is constant. How is it described? Consider the graph in Fig. 12. The length of the shaded rectangle is the length of the interval [2, 5], which is 3. In order for the shaded area to be 1, the height of the rectangle must be $\frac{1}{3}$. Thus, $f(x) = \frac{1}{3}$.

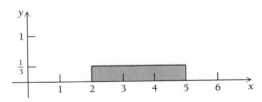

FIGURE 12

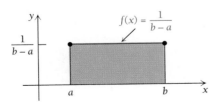

FIGURE 13

8. Find k such that

$$f(x) = kx^2$$

is a probability density function over the interval [1, 3]. Then write the probability density function.

9. Find k such that

$$f(x) = kx^3$$

is a probability density function over the interval [0, 1]. Then write the probability density function.

10. A number x is selected at random from the interval $[7, 15]$. The probability density function for x is given by

$$f(x) = \tfrac{1}{8}, \quad \text{for } 7 \leqslant x \leqslant 15.$$

Find the probability that a number selected is in the subinterval $[11, 13]$.

11. *Business: Waiting time.* A person arrives at a bus stop. The waiting time t for a bus is 0 to 20 min. The probability density function for t is

$$f(t) = \tfrac{1}{20}, \quad \text{for } 0 \leqslant t \leqslant 20.$$

What is the probability that the person will have to wait no more than 5 min for a bus?

The length of the shaded rectangle is the length of the interval $[a, b]$, which is $b - a$. In order for the shaded area to be 1, the height of the rectangle must be $1/(b - a)$. Thus, $f(x) = 1/(b - a)$.

DEFINITION

A continuous random variable x is said to be *uniformly distributed* over an interval $[a, b]$ if it has a probability density function f given by

$$f(x) = \frac{1}{b - a}, \quad \text{for } a \leqslant x \leqslant b.$$

EXAMPLE 8 A number x is selected at random from the interval $[40, 50]$. The probability density function for x is given by

$$f(x) = \frac{1}{10}, \quad \text{for } 40 \leqslant x \leqslant 50.$$

Find the probability that a number selected is in the subinterval $[42, 48]$.

Solution The probability is

$$P(42 \leqslant x \leqslant 48) = \int_{42}^{48} \tfrac{1}{10}\, dx = \tfrac{1}{10}[x]_{42}^{48}$$

$$= \tfrac{1}{10}(48 - 42) = \tfrac{6}{10} = 0.6. \quad ❖$$

DO EXERCISE 10.

EXAMPLE 9 *Business: Quality control.* A company produces sirens for tornado warnings. The maximum loudness L of the sirens ranges from 70 to 100 decibels. The probability density function for L is

$$f(L) = \tfrac{1}{30}, \quad \text{for } 70 \leqslant L \leqslant 100.$$

A siren is selected at random off the assembly line. Find the probability that its maximum loudness is from 70 to 92 decibels.

Solution The probability is

$$P(70 \leqslant L \leqslant 92) = \int_{70}^{92} \tfrac{1}{30}\, dL$$

$$= \tfrac{1}{30}[L]_{70}^{92}$$

$$= \tfrac{1}{30}(92 - 70)$$

$$= \tfrac{22}{30} = \tfrac{11}{15} \approx 0.73. \quad ❖$$

DO EXERCISE 11.

Exponential Distributions

The duration of a phone call, the distance between successive cars on a highway, and the amount of time required to learn a task are all examples of exponentially distributed random variables. That is, their probability density functions are exponential.

DEFINITION

A continuous random variable is *exponentially distributed* if it has a probability density function given by

$$f(x) = ke^{-kx}, \quad \text{over the interval } [0, \infty).$$

The function $f(x) = 2e^{-2x}$ is such a probability density function. That

$$\int_0^\infty 2e^{-2x}\,dx = 1$$

is shown in Section 6.3. The general case

$$\int_0^\infty ke^{-kx}\,dx = 1$$

can be verified in a similar way. (Refer to Fig. 14.)

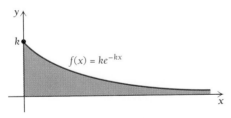

FIGURE 14

DO EXERCISE 12.

Why is it reasonable to assume that the distance between cars is exponentially distributed? Part of the reason is that there are many more cases in which distances are small, though we can find other distributions that are "skewed" in this manner. The same argument holds for the duration of a phone call. That is, there are more short calls than long ones. The rest of the reason might lie in an analysis of the data involving such distances or phone calls.

12. Verify that

$$\int_0^\infty ke^{-kx}\,dx = 1.$$

A transportation planner can determine probabilities that cars are certain distances apart.

13. *Business: Transportation planning.* A transportation planner determines that the average distance between cars on a certain stretch of highway is 125 ft. What is the probability that if we chose two successive cars at random, the distance between those cars is 50 ft or less?

EXAMPLE 10 *Business: Transportation planning.* The distance x, in feet, between successive cars on a certain stretch of highway has a probability density function

$$f(x) = ke^{-kx}, \quad \text{for } 0 \leqslant x < \infty,$$

where $k = 1/a$ and $a = $ the average distance between successive cars over some period of time.

A transportation planner determines that the average distance between cars on a certain stretch of highway is 166 ft. What is the probability that if we choose two successive cars at random, the distance between those cars is 50 ft or less?

Solution We first determine k:

$$k = \tfrac{1}{166}$$

$$\approx 0.006024.$$

The probability density function for x is

$$f(x) = 0.006024e^{-0.006024x}, \quad \text{for } 0 \leqslant x < \infty.$$

The probability that the distance between the cars is 50 ft or less is

$$P(0 \leqslant x \leqslant 50) = \int_0^{50} 0.006024e^{-0.006024x}\, dx$$

$$= \left[\frac{0.006024}{-0.006024}e^{-0.006024x} \right]_0^{50}$$

$$= \left[-e^{-0.006024x} \right]_0^{50}$$

$$= (-e^{-0.006024 \cdot 50}) - (-e^{-0.006024 \cdot 0})$$

$$= -e^{-0.301200} + 1$$

$$= 1 - e^{-0.301200}$$

$$= 1 - 0.739930$$

$$\approx 0.260.$$

DO EXERCISE 13.

EXERCISE SET 6.4

Verify Property 2 of the definition of a probability density function over the given interval.

1. $f(x) = 2x$, $[0, 1]$

2. $f(x) = \frac{1}{4}x$, $[1, 3]$

3. $f(x) = \frac{1}{3}$, $[4, 7]$

4. $f(x) = \frac{1}{4}$, $[9, 13]$

5. $f(x) = \frac{3}{26}x^2$, $[1, 3]$

6. $f(x) = \frac{3}{64}x^2$, $[0, 4]$

7. $f(x) = \dfrac{1}{x}$, $[1, e]$

8. $f(x) = \dfrac{1}{e-1}e^x$, $[0, 1]$

9. $f(x) = \frac{3}{2}x^2$, $[-1, 1]$ **10.** $f(x) = \frac{1}{3}x^2$, $[-2, 1]$

11. $f(x) = 3e^{-3x}$, $[0, \infty)$ **12.** $f(x) = 4e^{-4x}$, $[0, \infty)$

Find k such that the function is a probability density function over the given interval. Then write the probability density function.

13. $f(x) = kx$, $[1, 3]$ **14.** $f(x) = kx$, $[1, 4]$

15. $f(x) = kx^2$, $[-1, 1]$ **16.** $f(x) = kx^2$, $[-2, 2]$

17. $f(x) = k$, $[2, 7]$ **18.** $f(x) = k$, $[3, 9]$

19. $f(x) = k(2 - x)$, $[0, 2]$ **20.** $f(x) = k(4 - x)$, $[0, 4]$

21. $f(x) = \dfrac{k}{x}$, $[1, 3]$ **22.** $f(x) = \dfrac{k}{x}$, $[1, 2]$

23. $f(x) = ke^x$, $[0, 3]$ **24.** $f(x) = ke^x$, $[0, 2]$

25. A dart is thrown at a number line in such a way that it always lands in the interval $[0, 10]$. Let $x =$ the number that the dart hits. Suppose the probability density function for x is given by

$$f(x) = \tfrac{1}{50}x, \quad \text{for } 0 \leq x \leq 10.$$

Find $P(2 \leq x \leq 6)$, the probability that it lands in $[2, 6]$.

26. In Exercise 25, suppose that the dart always lands in the interval $[0, 5]$, and that the probability density function for x is given by

$$f(x) = \tfrac{3}{125}x^2, \quad \text{for } 0 \leq x \leq 5.$$

Find $P(1 \leq x \leq 4)$, the probability that it lands in $[1, 4]$.

27. A number x is selected at random from the interval $[4, 20]$. The probability density function for x is given by

$$f(x) = \tfrac{1}{16}, \quad \text{for } 4 \leq x \leq 20.$$

Find the probability that a number selected is in the subinterval $[9, 17]$.

28. A number x is selected at random from the interval $[5, 29]$. The probability density function for x is given by

$$f(x) = \tfrac{1}{24}, \quad \text{for } 5 \leq x \leq 29.$$

Find the probability that a number selected is in the subinterval $[13, 29]$.

APPLICATIONS

❖ **Business and Economics**

29. *Transportation planning.* A transportation planner determines that the average distance between cars on a certain highway is 100 ft. What is the probability that if we chose two successive cars at random, the distance between those cars is 40 ft or less?

30. *Transportation planning.* A transportation planner determines that the average distance between cars on a certain highway is 200 ft. What is the probability that if we chose two successive cars at random, the distance between those cars is 10 ft or less?

31. *Duration of a phone call.* A telephone company determines that the duration t of a phone call is an exponentially distributed random variable with probability density function

$$f(t) = 2e^{-2t}, \quad 0 \leq t < \infty.$$

Find the probability that a phone call will last no more than 5 min.

32. *Duration of a phone call.* Referring to Exercise 31, find the probability that a phone call will last no more than 2 min.

33. *Time to failure.* The *time to failure t*, in hours, of a certain machine can often be assumed to be exponentially distributed with probability density function

$$f(t) = ke^{-kt}, \quad 0 \leq t < \infty,$$

where $k = 1/a$ and $a =$ the average time that will pass before a failure occurs. Suppose the average time that will pass before a failure occurs is 100 hr. What is the probability that a failure will occur in 50 hr or less?

34. *Reliability of a machine.* The *reliability* of the machine (the probability that it will work) in Exercise 33 is defined as

$$R(T) = 1 - \int_0^T 0.01e^{-0.01t} \, dt,$$

where $R(T)$ is the reliability at time T. Find $R(T)$.

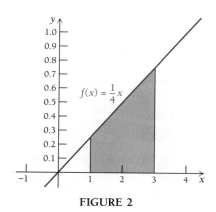

FIGURE 2

Suppose we continue to throw the dart and compute averages. The more times we throw the dart, the closer we expect the averages to come to 2.17.

Let x be a continuous random variable over the interval $[a, b]$ with probability density function f.

DEFINITION

The *expected value* of x is defined by

$$E(x) = \int_a^b x \cdot f(x) \, dx,$$

where f is a probability density function for x.

The concept of expected value of a random variable generalizes to other functions of a random variable. Suppose $y = g(x)$ is a function of the random variable x. Then we have the following.

DEFINITION

The *expected value* of $g(x)$ is defined by

$$E(g(x)) = \int_a^b g(x) \cdot f(x) \, dx,$$

where f is a probability density function for x.

For example,

$$E(x) = \int_a^b xf(x)\ dx,$$

$$E(x^2) = \int_a^b x^2 f(x)\ dx,$$

$$E(e^x) = \int_a^b e^x f(x)\ dx,$$

and

$$E(2x + 3) = \int_a^b (2x + 3)f(x)\ dx.$$

EXAMPLE 1   Given the probability density function

$$f(x) = \tfrac{1}{2}x \quad \text{over } [0, 2],$$

find $E(x)$ and $E(x^2)$.

Solution

$$E(x) = \int_0^2 x \cdot \frac{1}{2}x\ dx = \int_0^2 \frac{1}{2}x^2\ dx$$

$$= \frac{1}{2}\left[\frac{x^3}{3}\right]_0^2 = \frac{1}{6}\left[x^3\right]_0^2$$

$$= \frac{1}{6}(2^3 - 0^3) = \frac{1}{6} \cdot 8 = \frac{4}{3};$$

$$E(x^2) = \int_0^2 x^2 \cdot \frac{1}{2}x\ dx = \int_0^2 \frac{1}{2}x^3\ dx = \frac{1}{2}\left[\frac{x^4}{4}\right]_0^2$$

$$= \frac{1}{8}[x^4]_0^2 = \frac{1}{8}(2^4 - 0^4) = \frac{1}{8} \cdot 16 = 2 \qquad \text{❖}$$

DO EXERCISE 1.

DEFINITION

The *mean* μ of a continuous random variable is defined to be $E(x)$. That is,

$$\mu = E(x) = \int_a^b xf(x)\ dx,$$

where f is a probability density function for x. (The symbol μ is the lower-case Greek letter "mu.")

1. Given the probability density function

$$f(x) = 2x \quad \text{over } [0, 1],$$

find $E(x)$ and $E(x^2)$.

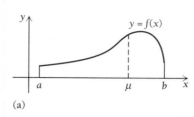 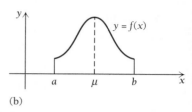

(a)   (b)

FIGURE 3

We can get a physical idea of a mean by pasting the graph on cardboard and cutting out the area under the curve over the interval $[a, b]$ (see Fig. 3). Then we try to find a fulcrum, or balance point, on the x-axis. That balance point is the mean.

DEFINITION

The *variance* σ^2 of a continuous random variable is defined as

$$\sigma^2 = E(x^2) - \mu^2 = E(x^2) - [E(x)]^2$$

$$= \int_a^b x^2 f(x) \, dx - \left[\int_a^b x f(x) \, dx \right]^2.$$

The *standard deviation* σ of a continuous random variable is defined as

$$\sigma = \sqrt{\text{variance}}.$$

(The symbol σ is the lower-case Greek letter "sigma.")

EXAMPLE 2   Given the probability density function

$$f(x) = \tfrac{1}{2} x \quad \text{over } [0, 2],$$

find the mean, the variance, and the standard deviation.

Solution   From Example 1, we have

$$E(x) = \tfrac{4}{3} \quad \text{and} \quad E(x^2) = 2.$$

Then

$$\text{the mean} = \mu = E(x) = \tfrac{4}{3};$$
$$\text{the variance} = \sigma^2 = E(x^2) - [E(x)]^2$$
$$= 2 - \left(\tfrac{4}{3}\right)^2 = 2 - \tfrac{16}{9}$$
$$= \tfrac{18}{9} - \tfrac{16}{9} = \tfrac{2}{9};$$
$$\text{the standard deviation} = \sigma = \sqrt{\tfrac{2}{9}} = \tfrac{1}{3}\sqrt{2} \approx 0.47.$$

Loosely speaking, we say that the standard deviation is a measure of how close the graph of f is to the mean. Note the examples shown in Fig. 4.

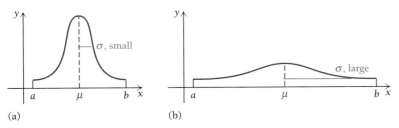

(a)   (b)

FIGURE 4

DO EXERCISE 2.

The Normal Distribution

Suppose the average on a test is 70. Usually there are about as many scores above the average as there are below the average; and the further away from the average a particular score is, the fewer people there are who get that score. In this example, it is probable that more people would score in the 80s than in the 90s, and more people would score in the 60s than in the 50s. Test scores, heights of human beings, and weights of human beings are all examples of random variables that are often *normally* distributed.

Consider the function

$$g(x) = e^{-x^2/2} \quad \text{over the interval } (-\infty, \infty).$$

This function has the entire set of real numbers as its domain. Its graph is the bell-shaped curve shown in Fig. 5. We can find function values by using a calculator:

$$y = e^{-x^2/2}.$$

x	0	1	2	3	-1	-2	-3
y	1	0.6	0.1	0.01	0.6	0.1	0.01

2. Given the probability density function

$$f(x) = 2x \quad \text{over } [0, 1],$$

find the mean, the variance, and the standard deviation.

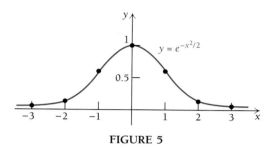

FIGURE 5

This function has an antiderivative, but that antiderivative has no elementary formula. Nevertheless, it has been shown that its improper integral converges over the interval $(-\infty, \infty)$ to a number given by

$$\int_{-\infty}^{\infty} e^{-x^2/2} \, dx = \sqrt{2\pi}.$$

That is, although an elementary expression for the antiderivative cannot be found, there is a numerical value for the improper integral evaluated over the set of real numbers. Note that since the area is not 1, the function g is not a probability density function, but the following is:

$$\frac{1}{\sqrt{2\pi}} e^{-x^2/2}.$$

DEFINITION

A continuous random variable x has a *standard normal distribution* if its probability density function is

$$f(x) = \frac{1}{\sqrt{2\pi}} e^{-x^2/2} \quad \text{over } (-\infty, \infty).$$

This distribution has a mean of 0 and a standard deviation of 1. Its graph is shown in Fig. 6.

The general case is defined as follows.

DEFINITION

A continuous random variable x is *normally distributed* with mean μ and standard deviation σ if its probability density function is given by

$$f(x) = \frac{1}{\sigma\sqrt{2\pi}} e^{-(1/2)[(x - \mu)/\sigma]^2} \quad \text{over } (-\infty, \infty).$$

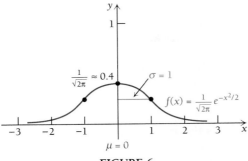

FIGURE 6

The graph is a transformation of the graph of the standard density function. This can be shown by translating the graph along the x-axis and changing the way in which the graph is clustered about the mean. Some examples are shown in Fig. 7.

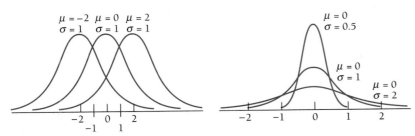

Normal distributions with same
standard deviations but different
means

Normal distributions with same
means but different standard
deviations

FIGURE 7

The normal distribution is extremely important in statistics; it underlies much of the research in the behavioral and social sciences. Because of this, tables of approximate values of the definite integral of the standard density functions have been prepared using numerical approximation methods like the Trapezoidal Rule given in Exercise Set 5.8. Table 2 at the back of the book is such a table. In contains values of

$$P(0 \leq x \leq t) = \int_0^t \frac{1}{\sqrt{2\pi}} e^{-x^2/2} \, dx.$$

The symmetry of the graph about the mean allows many types of probabilities to be computed from the table.

EXAMPLE 3 Let x be a continuous random variable with standard normal density. Using Table 2, find each of the following.

a) $P(0 \leqslant x \leqslant 1.68)$ b) $P(-0.97 \leqslant x \leqslant 0)$

c) $P(-2.43 \leqslant x \leqslant 1.01)$ d) $P(1.90 \leqslant x \leqslant 2.74)$

e) $P(-2.98 \leqslant x \leqslant -0.42)$ f) $P(x \geqslant 0.61)$

Solution

a) $P(0 \leqslant x \leqslant 1.68)$ is the area bounded by the standard normal curve and the lines $x = 0$ and $x = 1.68$. We look this up in Table 2 by going down the left column to 1.6, then moving to the right to the column headed 0.08. There we read 0.4535. Thus,

$$P(0 \leqslant x \leqslant 1.68) = 0.4535.$$

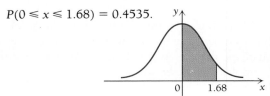

b) Because of the symmetry of the graph,

$P(-0.97 \leqslant x \leqslant 0)$
$= P(0 \leqslant x \leqslant 0.97)$
$= 0.3340.$

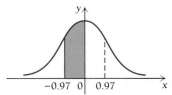

c) $P(-2.43 \leqslant x \leqslant 1.01)$
$= P(-2.43 \leqslant x \leqslant 0) + P(0 \leqslant x \leqslant 1.01)$
$= P(0 \leqslant x \leqslant 2.43) + P(0 \leqslant x \leqslant 1.01)$
$= 0.4925 + 0.3438$
$= 0.8363$

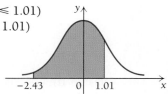

d) $P(1.90 \leqslant x \leqslant 2.74)$
$= P(0 \leqslant x \leqslant 2.74) - P(0 \leqslant x \leqslant 1.90)$
$= 0.4969 - 0.4713$
$= 0.0256$

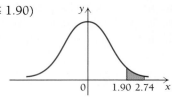

e) $P(-2.98 \leqslant x \leqslant -0.42)$
$$= P(0.42 \leqslant x \leqslant 2.98)$$
$$= P(0 \leqslant x \leqslant 2.98) - P(0 \leqslant x \leqslant 0.42)$$
$$= 0.4986 - 0.1628$$
$$= 0.3358$$

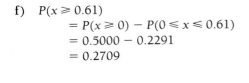

f) $P(x \geqslant 0.61)$
$$= P(x \geqslant 0) - P(0 \leqslant x \leqslant 0.61)$$
$$= 0.5000 - 0.2291$$
$$= 0.2709$$

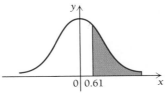

(Because of the symmetry about the line $x = 0$, half the area is on each side of the line, and since the entire area is 1, $P(x \geqslant 0) = 0.5000$.)   ❖

DO EXERCISE 3.

In many applications, a normal distribution is not standard. It would be a hopeless task to make tables for all values of the mean μ and the standard deviation σ. In such cases, the transformation

$$X = \frac{x - \mu}{\sigma}$$

standardizes the distribution, permitting the use of Table 2 at the back of the book. That is,

$$P(a \leqslant x \leqslant b) = P\left(\frac{a - \mu}{\sigma} \leqslant X \leqslant \frac{b - \mu}{\sigma}\right),$$

and the probability on the right can be found using Table 2. To see this, consider

$$P(a \leqslant x \leqslant b) = \int_a^b \frac{1}{\sigma\sqrt{2\pi}} e^{-(1/2)[(x-\mu)/\sigma]^2} \, dx,$$

and make the substitution

$$X = \frac{x - \mu}{\sigma} = \frac{x}{\sigma} - \frac{\mu}{\sigma}.$$

3. Let x be a continuous random variable with standard normal density. Using Table 2, find each of the following.

a) $P(0 \leqslant x \leqslant 2.17)$

b) $P(-1.76 \leqslant x \leqslant 0)$

c) $P(-1.77 \leqslant x \leqslant 2.53)$

d) $P(0.49 \leqslant x \leqslant 1.75)$

e) $P(-1.66 \leqslant x \leqslant -1.00)$

f) $P(x \geqslant 1.87)$

Then

$$dX = \frac{1}{\sigma}\, dx.$$

When $x = a$, $X = (a - \mu)/\sigma$; and when $x = b$, $X = (b - \mu)/\sigma$. Then

$$P(a \leqslant x \leqslant b) = \int_a^b \frac{1}{\sigma\sqrt{2\pi}}\, e^{-(1/2)[(x-\mu)/\sigma]^2}\, dx$$

and

$$P(a \leqslant x \leqslant b) = \int_{(a-\mu)/\sigma}^{(b-\mu)/\sigma} \frac{1}{\sqrt{2\pi}}\, e^{-(1/2)X^2}\, dX$$

The integrand is now in the form of the standard density.

$$= P\left(\frac{a - \mu}{\sigma} \leqslant X \leqslant \frac{b - \mu}{\sigma}\right).$$

We can look this up in Table 2.

EXAMPLE 4   The weights w of the students in a calculus class are normally distributed with mean 150 lb and standard deviation 25 lb. Find the probability that a student's weight is from 160 lb to 180 lb.

Solution   We first standardize the weights (see Fig. 8):

$$180 \text{ is standardized to } \frac{b - \mu}{\sigma} = \frac{180 - 150}{25} = 1.2;$$

$$160 \text{ is standardized to } \frac{a - \mu}{\sigma} = \frac{160 - 150}{25} = 0.4.$$

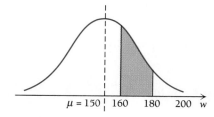

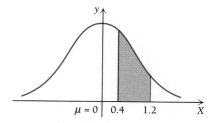

FIGURE 8

Then

$$P(160 \leqslant w \leqslant 180) = P(0.4 \leqslant X \leqslant 1.2) \qquad \text{Now we can use Table 2.}$$
$$= P(0 \leqslant X \leqslant 1.2) - P(0 \leqslant X \leqslant 0.4)$$
$$= 0.3849 - 0.1554$$
$$= 0.2295.$$

Thus the probability that a student's weight is from 160 lb to 180 lb is 0.2295. That is, about 23% of the students have weights from 160 lb to 180 lb. ❖

DO EXERCISE 4.

4. *Business: Daily profits.* The daily profits *p* of a small business firm are normally distributed with mean $200 and standard deviation $40. Find the probability that the daily profit will be from $230 and $250.

EXERCISE SET # 6.5

For each probability density function, over the given interval, find $E(x)$, $E(x^2)$, the mean, the variance, and the standard deviation.

1. $f(x) = \frac{1}{3}$, $[2, 5]$

2. $f(x) = \frac{1}{4}$, $[3, 7]$

3. $f(x) = \frac{2}{9}x$, $[0, 3]$

4. $f(x) = \frac{1}{8}x$, $[0, 4]$

5. $f(x) = \frac{2}{3}x$, $[1, 2]$

6. $f(x) = \frac{1}{4}x$, $[1, 3]$

7. $f(x) = \frac{1}{3}x^2$, $[-2, 1]$

8. $f(x) = \frac{3}{2}x^2$, $[-1, 1]$

9. $f(x) = \frac{1}{\ln 3} \cdot \frac{1}{x}$, $[1, 3]$

10. $f(x) = \frac{1}{\ln 2} \cdot \frac{1}{x}$, $[1, 2]$

Let *x* be a continuous random variable with standard normal density. Using Table 2, find each of the following.

11. $P(0 \leqslant x \leqslant 2.69)$

12. $P(0 \leqslant x \leqslant 0.04)$

13. $P(-1.11 \leqslant x \leqslant 0)$

14. $P(-2.61 \leqslant x \leqslant 0)$

15. $P(-1.89 \leqslant x \leqslant 0.45)$

16. $P(-2.94 \leqslant x \leqslant 2.00)$

17. $P(1.76 \leqslant x \leqslant 1.86)$

18. $P(0.76 \leqslant x \leqslant 1.45)$

19. $P(-1.45 \leqslant x \leqslant -0.69)$

20. $P(-2.45 \leqslant x \leqslant -1.69)$

21. $P(x \geqslant 3.01)$

22. $P(x \geqslant 1.01)$

23. a) $P(-1 \leqslant x \leqslant 1)$

  b) What percentage of the area is from -1 to 1?

24. a) $P(-2 \leqslant x \leqslant 2)$

  b) What percentage of the area is from -2 to 2?

Let *x* be a continuous random variable that is normally distributed with mean $\mu = 22$ and standard deviation $\sigma = 5$. Using Table 2, find each of the following.

25. $P(24 \leqslant x \leqslant 30)$

26. $P(22 \leqslant x \leqslant 27)$

27. $P(19 \leqslant x \leqslant 25)$

28. $P(18 \leqslant x \leqslant 26)$

APPLICATIONS

❖ **Business and Economics**

29. *Mail orders.* The number of daily orders *N* received by a mail-order firm is normally distributed with mean 250 and standard deviation 20. The company has to hire extra help or pay overtime on those days when the number of orders received is 300 or higher. What percentage of the days will the company have to hire extra help or pay overtime?

30. *Stereo production.* The daily production *N* of stereos by a recording company is normally distributed with mean 1000 and standard deviation 50. The company promises to pay bonuses to its employees on those days when the production of stereos is 1100 or more. What percentage of the days will the company have to pay a bonus?

❖ Social Sciences

31. *Test score distribution.* The scores S on a psychology test are normally distributed with mean 65 and standard deviation 20. A score of 80 to 89 is a B. What is the probability of getting a B?

❖ General Interest

32. *Bowling scores.* At the time of this writing, the bowling scores S of the author of this text were normally distributed with mean 203 and standard deviation 23.

a) Find the probability that a score is from 190 to 213.

b) Find the probability that a score is from 160 to 175.

c) Find the probability that a score is greater than 200.

SYNTHESIS EXERCISES

For each probability density function, over the given interval, find $E(x)$, $E(x^2)$, the mean, the variance, and the standard deviation.

33. The uniform probability density

$$f(x) = \frac{1}{b-a} \quad \text{over } [a, b]$$

34. The probability density

$$f(x) = \frac{3a^3}{x^4} \quad \text{over } [a, \infty)$$

Median. Let x be a continuous random variable over $[a, b]$ with probability density function f. Then the *median* of x is that number m for which

$$\int_a^m f(x) \, dx = \tfrac{1}{2}.$$

Find the median.

35. $f(x) = \tfrac{1}{2}x$, $[0, 2]$

36. $f(x) = \tfrac{3}{2}x^2$, $[-1, 1]$

37. $f(x) = ke^{-kx}$, $[0, \infty)$

38. *Business: Sugar production.* Suppose the amount of sugar going in a sack has a mean μ, which can be adjusted on the filling machine. Suppose the amount dispensed is normally distributed with $\sigma = 0.1$ oz. What should be the setting of μ to ensure that only one bag in 20 will have less than 60 oz?

39. *Business: Does thy cup overflow?* Suppose the mean amount of coffee μ dispensed by a coffee machine can be set. If a cup holds 6.5 oz and the amount of coffee dispensed is normally distributed with $\sigma = 0.3$ oz, what should the setting of μ be to ensure that the cup will overflow only one time in a hundred?

COMPUTER EXERCISES

CALCULUS EXPLORER:
Riemann Sums

Use the program to devise a way to
approximate the integral

$$\int_{-\infty}^{\infty} \frac{1}{\sqrt{2\pi}}\, e^{-x^2/2}\, dx.$$

6.6

VOLUME

Consider the graph of $y = f(x)$. If the upper half-plane is rotated about
the x-axis, then each point on the graph has a circular path, and the
whole graph sweeps out a certain surface, called a *surface of revolution*
(Fig. 1).

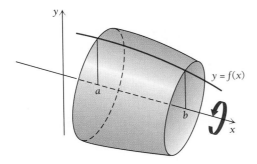

FIGURE 1

The plane region bounded by the graph, the x-axis, and the interval
$[a, b]$ sweeps out a *solid of revolution*. To calculate the volume of this
solid, we first approximate it by a finite sum of thin right circular cylinders (Fig. 2). We divide the interval $[a, b]$ into equal subintervals, each of
length Δx. Thus the height of each cylinder is Δx (Fig. 3). The radius of
each cylinder is $f(x_i)$, where x_i is the righthand endpoint of the subinterval that determines that cylinder.

OBJECTIVE

a)   Use integration to find the volume of
a solid of revolution.

Find a curve that can be rotated to form the solid of revolution.

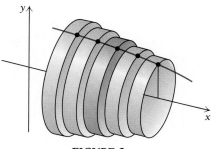

FIGURE 2

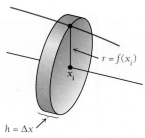

FIGURE 3

Since the volume of a right circular cylinder is given by

$$V = \pi r^2 h,$$

each of the approximating cylinders has volume

$$\pi [f(x_i)]^2 \, \Delta x.$$

The volume of the solid of revolution is approximated by the sum of the volumes of all the cylinders:

$$V \approx \sum_{i=1}^{n} \pi [f(x_i)]^2 \, \Delta x.$$

The actual volume is the limit as the thickness of the cylinders approaches zero or the number of them approaches infinity:

$$V = \lim_{n \to \infty} \sum_{i=1}^{n} \pi [f(x_i)]^2 \, \Delta x.$$

This is just the definite integral of the function $y = \pi [f(x)]^2$.

THEOREM 5

$$V = \int_a^b \pi [f(x)]^2 \, dx \qquad \begin{array}{l} \textit{Volume of a solid of revolution} \\ \textit{rotated about the x-axis} \end{array}$$

EXAMPLE 1   Find the volume of the solid of revolution generated by rotating the region under the graph of

$$y = \sqrt{x}$$

from $x = 0$ to $x = 1$ about the x-axis (Fig. 4).

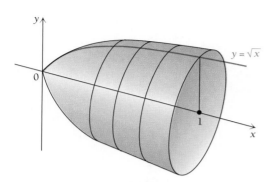

FIGURE 4

1. Find the volume of the solid of revolution generated by rotating the region under the graph of

$$y = x$$

from $x = 0$ to $x = 1$ about the x-axis.

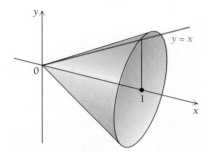

Solution

$$V = \int_0^1 \pi[f(x)]^2 \, dx$$

$$= \int_0^1 \pi[\sqrt{x}]^2 \, dx$$

$$= \int_0^1 \pi x \, dx$$

$$= \pi\left[\frac{x^2}{2}\right]_0^1$$

$$= \frac{\pi}{2}\left[x^2\right]_0^1$$

$$= \frac{\pi}{2}(1^2 - 0^2) = \frac{\pi}{2}$$

DO EXERCISE 1.

EXAMPLE 2 Find the volume of the solid of revolution generated by rotating the region under the graph of

$$y = e^x$$

from $x = -1$ to $x = 2$ about the x-axis (Fig. 5).

2. Find the volume of the solid of revolution generated by rotating the region under the graph of

$$y = e^x$$

from $x = -2$ to $x = 1$ about the x-axis.

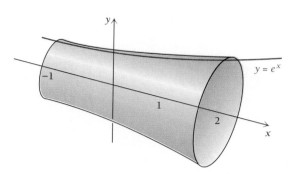

FIGURE 5

Explain how this could be interpreted as a solid of revolution.

Solution

$$V = \int_{-1}^{2} \pi [f(x)]^2 \, dx$$

$$= \int_{-1}^{2} \pi [e^x]^2 \, dx$$

$$= \int_{-1}^{2} \pi e^{2x} \, dx$$

$$= \left[\frac{\pi}{2} e^{2x} \right]_{-1}^{2}$$

$$= \frac{\pi}{2} \left[e^{2x} \right]_{-1}^{2}$$

$$= \frac{\pi}{2} (e^{2 \cdot 2} - e^{2(-1)})$$

$$= \frac{\pi}{2} (e^4 - e^{-2})$$

❖

DO EXERCISE 2.

EXERCISE SET **6.6**

Find the volume generated by revolving about the x-axis the regions bounded by the following graphs.

1. $y = \sqrt{x}$, $x = 0$, $x = 3$
2. $y = \sqrt{x}$, $x = 0$, $x = 2$
3. $y = x$, $x = 1$, $x = 2$
4. $y = x$, $x = 1$, $x = 3$
5. $y = e^x$, $x = -2$, $x = 5$
6. $y = e^x$, $x = -3$, $x = 2$
7. $y = \frac{1}{x}$, $x = 1$, $x = 3$
8. $y = \frac{1}{x}$, $x = 1$, $x = 4$

9. $y = \dfrac{1}{\sqrt{x}}$, $x = 1$, $x = 3$ 10. $y = \dfrac{1}{\sqrt{x}}$, $x = 1$, $x = 4$

11. $y = 4$, $x = 1$, $x = 3$ 12. $y = 5$, $x = 1$, $x = 3$

13. $y = x^2$, $x = 0$, $x = 2$

14. $y = x + 1$, $x = -1$, $x = 2$

15. $y = \sqrt{1 + x}$, $x = 2$, $x = 10$

16. $y = 2\sqrt{x}$, $x = 1$, $x = 2$

17. $y = \sqrt{4 - x^2}$, $x = -2$, $x = 2$

18. $y = \sqrt{r^2 - x^2}$, $x = -r$, $x = r$

Make a drawing for Exercise 18. Here you will derive a general formula for the volume of a sphere.

SYNTHESIS EXERCISES

Find the volume generated by revolving about the x-axis the regions bounded by the following graphs.

19. $y = \sqrt{\ln x}$, $x = e$, $x = e^3$

20. $y = \sqrt{xe^{-x}}$, $x = 1$, $x = 2$

21. Consider $y = \dfrac{1}{x}$ (Now do **a**) and **b**)).

a) Find $\displaystyle\int_1^\infty \dfrac{1}{x}\,dx$.

b) Find the volume generated by revolving about the x-axis the region under the graph of $y = 1/x$ for $x \geqslant 1$. Note that the area under the graph does not exist, although the volume does. This is like a can of paint that has a finite volume, but not enough paint to cover a cross section of the can.

6.7

DIFFERENTIAL EQUATIONS

A **differential equation** is an equation that involves derivatives or differentials. In Chapter 4, we studied one very important differential equation,

$$\frac{dP}{dt} = kP,$$

where P, or $P(t)$, is the population at time t. This equation is a model of uninhibited population growth. Its solution is the function

$$P = P_0 e^{kt},$$

where the constant P_0 is the size of the initial population, that is, at $t = 0$. As this one equation illustrates, differential equations are rich in application.

OBJECTIVES

a) Solve certain differential equations, giving both general and particular solutions.

b) Solve certain differential equations, given a condition $f(a) = b$.

c) Verify that a given function is a solution of a given differential equation.

d) Solve certain differential equations using separation of variables.

1. Solve $y' = 3x^2$.

Solving Certain Differential Equations

In this chapter, we will frequently use the notation y' for a derivative—mainly because it is simple. Thus, if $y = f(x)$, then

$$y' = \frac{dy}{dx} = f'(x).$$

We have already found solutions of certain differential equations when we found antiderivatives or indefinite integrals. The differential equation

$$\frac{dy}{dx} = g(x), \quad \text{or} \quad y' = g(x),$$

has the solution

$$y = \int g(x) \, dx.$$

EXAMPLE 1 Solve $y' = 2x$.

Solution

$$y = \int 2x \, dx = x^2 + C \qquad \qquad ❖$$

DO EXERCISE 1.

Look again at the solution to Example 1. Note the constant of integration. This solution is called a *general solution* because taking all values of C gives *all* the solutions. Taking specific values of C gives *particular solutions*. For example, the following are particular solutions to $y' = 2x$:

$$y = x^2 + 3,$$
$$y = x^2,$$
$$y = x^2 - 3.$$

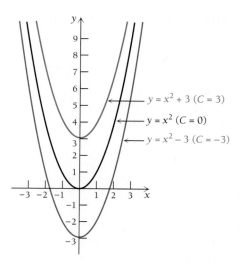

$y = x^2 + 3 \ (C = 3)$

$y = x^2 \ (C = 0)$

$y = x^2 - 3 \ (C = -3)$

FIGURE 1

Waves can be represented by differential equations.

The graph in Fig. 1 shows the curves of a few particular solutions. The general solution can be regarded as the set of all particular solutions, a *family* of curves.

DO EXERCISE 2.

Knowing the value of a function at a particular point may allow us to select a particular solution from the general solution.

EXAMPLE 2 Solve
$$f'(x) = e^x + 5x - x^{1/2},$$
given that $f(0) = 8$.

Solution

a) We first find the general solution:
$$f(x) = \int f'(x)\, dx$$
$$= e^x + \tfrac{5}{2}x^2 - \tfrac{2}{3}x^{3/2} + C.$$

b) Since $f(0) = 8$, we substitute to find C:
$$8 = e^0 + \tfrac{5}{2}\cdot 0^2 - \tfrac{2}{3}\cdot 0^{3/2} + C$$
$$8 = 1 + C$$
$$7 = C.$$

Thus the particular solution is
$$f(x) = e^x + \tfrac{5}{2}x^2 - \tfrac{2}{3}x^{3/2} + 7.$$ ❖

DO EXERCISES 3 AND 4.

Verifying Solutions

To verify that a function is a solution to a differential equation, we find the necessary derivatives and substitute.

EXAMPLE 3 Show that $y = 4e^x + 5e^{3x}$ is a solution to
$$y'' - 4y' + 3y = 0.$$

2. For
$$y' = 3x^2,$$
a) write the general solution;
b) write three particular solutions.

3. Solve
$$y' = 2x,$$
given that $y = 7$ when $x = 1$.

4. Solve
$$f'(x) = \frac{1}{x} - 2x + x^{1/2},$$
given that $f(1) = 4$.

At the 1968 Olympic Games in Mexico City, Bob Beamon made a miraculous long jump of 29 ft, $2\frac{1}{2}$ in. Many believed this was due to the altitude, which was 7400 ft. Using differential equations for analysis, M. N. Bearley refuted the altitude theory in "The Long Jump Miracle of Mexico City," (*Mathematics Magazine,* **45**, November 1972, pp. 241–246). Bearley argues that the world record jump was a result of Beamon's exceptional speed (9.5 sec in the 100-yd dash) and the fact that he hit the take-off board in perfect position.

5. Show that
$$y = 2e^x - 7e^{3x}$$
is a solution to
$$y'' - 4y' + 3y = 0.$$

Solution

a) We first find y' and y'':
$$y' = 4e^x + 15e^{3x};$$
$$y'' = 4e^x + 45e^{3x}.$$

b) Then we substitute in the differential equation, as follows:

$$y'' - 4y' + 3y = 0$$

$(4e^x + 45e^{3x}) - 4(4e^x + 15e^{3x}) + 3(4e^x + 5e^{3x})$	0
$4e^x + 45e^{3x} - 16e^x - 60e^{3x} + 12e^x + 15e^{3x}$	
	0

❖

DO EXERCISES 5 AND 6.

Separation of Variables

Consider the differential equation

$$\frac{dy}{dx} = 2xy. \tag{1}$$

We treat dy/dx as a quotient, as we did in Chapter 5. We multiply Eq. (1) by dx and then by $1/y$ to get

$$\frac{dy}{y} = 2x \, dx, \quad y \neq 0. \tag{2}$$

We say that we have **separated the variables**, meaning that all the expressions involving y are on one side and all those involving x are on the other. We then integrate both sides of Eq. (2):

$$\int \frac{dy}{y} = \int 2x \, dx$$
$$\ln y = x^2 + C, \quad y > 0.$$

We use only one constant because any two antiderivatives differ by a constant. Recall that the definition of logarithms says that if $\log_a b = t$, then $b = a^t$. Now, $\ln y = \log_e y = x^2 + C$, so by the definition of logarithms, we have

$$y = e^{x^2 + C} = e^{x^2} \cdot e^C.$$

Thus the solution to differential equation (1) is

$$y = C_1 e^{x^2}, \quad \text{where } C_1 = e^C.$$

In fact, C_1 is still an arbitrary constant.

DO EXERCISE 7.

EXAMPLE 4 Solve

$$3y^2 \frac{dy}{dx} + x = 0, \quad \text{where } y = 5 \text{ when } x = 0.$$

Solution

a) We first separate the variables as follows:

$$3y^2 \frac{dy}{dx} = -x$$

$$3y^2 \, dy = -x \, dx.$$

We then integrate both sides:

$$\int 3y^2 \, dy = \int -x \, dx$$

$$y^3 = -\frac{x^2}{2} + C = C - \frac{x^2}{2}$$

$$y = \sqrt[3]{C - \frac{x^2}{2}}. \qquad \text{Taking the cube root}$$

b) Since $y = 5$ when $x = 0$, we substitute to find C:

$$5 = \sqrt[3]{C - \frac{0^2}{2}} \qquad \text{Substituting 5 for } y \text{ and 0 for } x$$

$$5 = \sqrt[3]{C}$$

$$125 = C. \qquad \text{Cubing both sides}$$

The particular solution is

$$y = \sqrt[3]{125 - \frac{x^2}{2}}.$$

DO EXERCISE 8.

EXAMPLE 5 Solve

$$\frac{dy}{dx} = \frac{x}{y}.$$

6. Show that
$$y = xe^{2x}$$
is a solution to
$$\frac{dy}{dx} - 2y = e^{2x}.$$

7. Use separation of variables to solve
$$\frac{dy}{dx} = 3x^2 y.$$

8. Solve
$$3y^2 \frac{dy}{dx} - 2x = 0,$$
where $y = 7$ when $x = 2$.

9. Solve $\dfrac{dy}{dx} = \dfrac{5}{y}$.

Solution   We first separate the variables:

$$y\,\frac{dy}{dx} = x$$

$$y\,dy = x\,dx.$$

We then integrate both sides:

$$\int y\,dy = \int x\,dx$$

$$\frac{y^2}{2} = \frac{x^2}{2} + C$$

$$y^2 = x^2 + 2C$$

$$y^2 = x^2 + C_1,$$

where $C_1 = 2C$. We make this substitution in order to simplify the equation. We then obtain the solutions

$$y = \sqrt{x^2 + C_1}$$

and

$$y = -\sqrt{x^2 + C_1}$$ ❖

DO EXERCISE 9.

EXAMPLE 6   Solve $y' = x - xy$.

Solution   Before we separate the variables, we replace y' by dy/dx:

$$\frac{dy}{dx} = x - xy.$$

Then we separate the variables:

$$dy = (x - xy)\,dx$$

$$dy = x(1 - y)\,dx$$

$$\frac{dy}{1 - y} = x\,dx.$$

Next we integrate both sides:

$$\int \frac{dy}{1 - y} = \int x\,dx$$

$$-\ln(1 - y) = \frac{x^2}{2} + C \qquad 1 - y > 0$$

$$\ln (1 - y) = -\frac{x^2}{2} - C$$

$$1 - y = e^{-x^2/2 - C}$$

$$-y = e^{-x^2/2 - C} - 1$$

$$y = -e^{-x^2/2 - C} + 1$$

$$y = -e^{-x^2/2} \cdot e^{-C} + 1.$$

Thus,

$$y = 1 + C_1 e^{-x^2/2}, \quad \text{where } C_1 = -e^{-C}.$$

DO EXERCISE 10.

An Economic Application: Elasticity

EXAMPLE 7   Suppose for a certain product that the elasticity of demand is 1 for all $p > 0$. That is, $E(p) = 1$ for all $p > 0$. Find the demand function $x = D(p)$. (See Section 4.6.)

Solution   Since $E(p) = 1$ for all $p > 0$,

$$1 = E(p) = -\frac{p \, D'(p)}{D(p)} = -\frac{p}{x} \cdot \frac{dx}{dp}.$$

Then

$$\frac{dx}{dp} = -\frac{x}{p}. \tag{3}$$

Separating the variables, we get

$$\frac{dp}{p} = -\frac{dx}{x}.$$

Now we integrate both sides:

$$\int \frac{dp}{p} = -\int \frac{dx}{x}$$

$$\ln p = -\ln x + C. \qquad \begin{array}{l} p > 0 \text{ and } dx/dp < 0, \\ \text{since demand functions} \\ \text{are decreasing. So } x > 0. \\ \text{See Eq. (3).} \end{array}$$

10. Solve $y' = 2x + xy$.

11. *Economics: Elasticity.* Find the demand function, $x = D(p)$, given the elasticity condition

$$E(p) = 3 \quad \text{for all } p > 0.$$

We can express $C = \ln C_1$, since any real number C is the natural logarithm of some number C_1. Then

$$\ln p = \ln C_1 - \ln x$$
$$= \ln \frac{C_1}{x},$$

so

$$p = \frac{C_1}{x} \quad \text{and} \quad x = \frac{C_1}{p}.$$

This characterizes those demand functions for which the elasticity is always 1. ❖

DO EXERCISE 11.

A Psychological Application: Reaction to a Stimulus

The Weber–Fechner Law

In psychology, one model of stimulus–response asserts that the rate of change dR/dS of the reaction R with respect to a stimulus S is inversely proportional to the stimulus; that is,

$$\frac{dR}{dS} = \frac{k}{S},$$

where k is some positive constant.

To solve this equation, we first separate the variables:

$$dR = k \cdot \frac{dS}{S}.$$

We then integrate both sides:

$$\int dR = \int k \cdot \frac{dS}{S}$$
$$R = k \ln S + C. \tag{4}$$

Now suppose we let S_0 be the lowest level of the stimulus that can be detected consistently. This the *threshold value,* or the *detection threshold.* For example, the lowest level of sound that can be consistently detected is the tick of a watch at 20 feet, under very quiet conditions. If S_0 is the

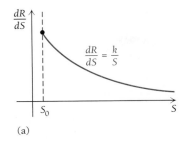

(a)

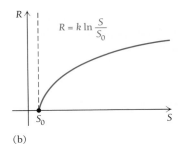

(b)

FIGURE 2

lowest level of sound that can be detected, it seems reasonable that the reaction to it would be 0. That is, $R(S_0) = 0$. Substituting this condition into Eq. (4), we get

$$0 = k \ln S_0 + C,$$

or

$$-k \ln S_0 = C.$$

Replacing C in Eq. (4) by $-k \ln S_0$ gives us

$$R = k \cdot \ln S - k \cdot \ln S_0$$
$$R = k(\ln S - \ln S_0).$$

Using a property of logarithms, we have

$$R = k \cdot \ln \frac{S}{S_0}.$$

Look at the graphs of dR/dS and R shown in Fig. 2. Note that as the stimulus gets larger, the rate of change decreases; that is, the reaction becomes smaller as the stimulation received becomes stronger. For example, suppose you are in a room with one lamp and that lamp has a 50-watt bulb in it. If the bulb were suddenly changed to 100 watts, you would

12. *Social science: The Brentano–Stevens Law.* The Weber–Fechner Law has been the subject of great debate among psychologists as to its validity. The model

$$\frac{dR}{dS} = k \cdot \frac{R}{S},$$

where k is a positive constant, has also been conjectured and experimented with. Find the general solution to this equation. (This has also been referred to as the *Power Law of Stimulus–Response.*)

probably be very aware of the difference. That is, your reaction would be strong. If the bulb were then changed to 150 watts, your reaction would not be as great as it was to the change from 50 to 100 watts. A change from a 150- to a 200-watt bulb would cause even less reaction, and so on.

For your interest, here are some other detection thresholds.

Stimulus	Detection threshold
Light	The flame of a candle 30 miles away on a dark night
Taste	Water diluted with sugar in the ratio of one teaspoon to two gallons
Smell	One drop of perfume diffused into the volume of three average-size rooms
Touch	The wing of a bee dropped on your cheek at a distance of 1 centimeter (about $\frac{3}{8}$ of an inch)

DO EXERCISE 12.

EXERCISE SET 6.7

Find the general solution and three particular solutions.

1. $y' = 4x^3$
2. $y' = 6x^5$
3. $y' = e^{2x} + x$
4. $y' = e^{3x} - x$
5. $y' = \dfrac{3}{x} - x^2 + x^5$
6. $y' = \dfrac{5}{x} + x^2 - x^4$

Find the particular solution determined by the given condition.

7. $y' = x^2 + 2x - 3$; $y = 4$ when $x = 0$
8. $y' = 3x^2 - x + 5$; $y = 6$ when $x = 0$
9. $f'(x) = x^{2/3} - x$; $f(1) = -6$
10. $f'(x) = x^{2/5} + x$; $f(1) = -7$
11. Show that $y = x \ln x + 3x - 2$ is a solution to

$$y'' - \frac{1}{x} = 0.$$

12. Show that $y = x \ln x - 5x + 7$ is a solution to

$$y'' - \frac{1}{x} = 0.$$

13. Show that $y = e^x + 3xe^x$ is a solution to

$$y'' - 2y' + y = 0.$$

14. Show that $y = -2e^x + xe^x$ is a solution to

$$y'' - 2y' + y = 0.$$

Solve.

15. $\dfrac{dy}{dx} = 4x^3 y$
16. $\dfrac{dy}{dx} = 5x^4 y$
17. $3y^2 \dfrac{dy}{dx} = 5x$
18. $3y^2 \dfrac{dy}{dx} = 7x$

19. $\dfrac{dy}{dx} = \dfrac{2x}{y}$ **20.** $\dfrac{dy}{dx} = \dfrac{x}{2y}$

21. $\dfrac{dy}{dx} = \dfrac{3}{y}$ **22.** $\dfrac{dy}{dx} = \dfrac{4}{y}$

23. $y' = 3x + xy;$ $y = 5$ when $x = 0$

24. $y' = 2x - xy;$ $y = 9$ when $x = 0$

25. $y' = 5y^{-2};$ $y = 3$ when $x = 2$

26. $y' = 7y^{-2};$ $y = 3$ when $x = 1$

27. $\dfrac{dy}{dx} = 3y$ **28.** $\dfrac{dy}{dx} = 4y$

29. $\dfrac{dP}{dt} = 2P$ **30.** $\dfrac{dP}{dt} = 4P$

31. Solve

$$f'(x) = \dfrac{1}{x} - 4x + \sqrt{x},$$

given that $f(1) = \tfrac{23}{3}$.

APPLICATIONS

❖ **Business and Economics**

32. *Total revenue from marginal revenue.* The marginal revenue for a certain product is given by $R'(x) = 300 - 2x$. Find the total-revenue function $R(x)$, assuming that $R(0) = 0$.

33. *Total cost from marginal cost.* The marginal cost for a certain product is given by $C'(x) = 2.6 - 0.02x$. Find the total-cost function $C(x)$ and the average cost $A(x)$, assuming that fixed costs are \$120; that is, $C(0) = \$120$.

34. *Domar's capital expansion model* is

$$\dfrac{dI}{dt} = hkI,$$

where $I =$ the investment, $h =$ the investment productivity (constant), $k =$ the marginal productivity to the consumer (constant), and $t =$ the time.

a) Use separation of variables to solve the differential equation.

b) Rewrite the solution in terms of the condition $I_0 = I(0)$.

35. *Total profit from marginal profit.* A firm's marginal profit P as a function of its total cost C is given by

$$\dfrac{dP}{dC} = \dfrac{-200}{(C + 3)^{3/2}}.$$

a) Find the profit function $P(C)$ if $P = \$10$ when $C = \$61$.

b) At what cost will the firm break even $(P = 0)$?

36. *Stock growth.* The *growth rate of a certain stock* is modeled by

$$\dfrac{dV}{dt} = k(L - V), \; V = \$20 \text{ when } t = 0,$$

where $V =$ the value of the stock, per share, after time t (in months), $L = \$24.81$, the *limiting value* of the stock, and $k =$ a constant. Find the solution to the differential equation in terms of t and k.

37. *Utility.* The reaction R in pleasure units by a consumer receiving S units of a product can be modeled by the differential equation

$$\dfrac{dR}{dS} = \dfrac{k}{S + 1},$$

where k is a positive constant.

a) Use separation of variables to solve the differential equation.

b) Rewrite the solution in terms of the initial condition $R(0) = 0$.

c) Explain why the condition $R(0) = 0$ is reasonable.

Elasticity. Find the demand function $p = D(x)$ given the following elasticity conditions.

38. $E(p) = \dfrac{4}{p};$ $x = e$ when $p = 4$

39. $E(p) = \dfrac{p}{200 - p};$ $x = 190$ when $p = 10$

40. $E(p) = 2$ for all $p > 0$

41. $E(p) = n$ for some constant n and all $p > 0$

❖ **Life and Physical Sciences**

42. a) Use separation of variables to solve the differential-

equation model of uninhibited growth,

$$\frac{dP}{dt} = kP.$$

b) Rewrite the solution in terms of the condition $P_0 = P(0)$.

CHAPTER SUMMARY AND REVIEW

TERMS TO KNOW

Consumer's surplus, p. 438
Producer's surplus, p. 440
Equilibrium point, p. 440
Continuous money flow, p. 445
Amount of a continuous money flow, p. 446
Present value, p. 449
Accumulated present value, p. 450
Improper integral, p. 454

Convergent integral, p. 455
Divergent integral, p. 455
Probability, p. 459
Continuous random variables, p. 461
Probability density function, p. 465
Uniform distribution, p. 470
Exponential distribution, p. 471
Expected value, p. 476

Mean, p. 477
Variance, p. 478
Standard deviation, p. 478
Standard normal distribution, p. 480
Normal distribution p. 480
Volume, p. 488
Differential equation, p. 491
Separation of variables, p. 494

REVIEW EXERCISES

These review exercises are for test preparation. They can also be used as a lengthened practice test. Answers are at the back of the book. The answers also contain bracketed section references, which tell you where to restudy if your answer is incorrect.

Given $D(p) = (p - 6)^2$ and $S(p) = p^2 + 12$, find each of the following.

1. The equilibrium point

2. The consumer's surplus at the equilibrium point

3. The producer's surplus at the equilibrium point

4. *Business: Amount of a continuous money flow.* Find the amount of a continuous money flow in which $2400 per year is being invested at 10%, compounded continuously, for 15 years.

5. *Business: Continuous money flow.* Consider a continuous money flow into an investment at the constant rate of P_0 dollars per year. What should P_0 be so that the amount of a continuous money flow over 25 years at 12%, compounded continuously, will be $40,000?

6. *Physical science: Demand for potash.* In 1990 ($t = 0$), the world use of potash was 31,270 thousand tons, and the demand for it was growing at the rate of 3% per year. If the demand continues to grow at this rate, how many tons of potash will the world use from 1990 to 1998?

7. *Physical science: Depletion of potash.* The world reserves of potash is 250,000,000 thousand tons. Assuming that the growth rate in Exercise 6 continues and that no new reserves are discovered, when will the world reserves of potash be exhausted?

8. *Business: Present value.* Find the present value of $100,000 due 50 years later at 12%, compounded continuously.

9. *Business: Accumulated present value.* Find the accumulated present value of an investment over a 20-year period if there is a continuous money flow of $4800 per year and the current interest rate is 9%.

10. *Business: Accumulated present value.* Find the accumulated present value of an investment for which the continuous money flow in Exercise 9 is perpetual.

Determine whether the improper integral is convergent or divergent, and calculate its value if it is convergent.

11. $\int_1^\infty \frac{1}{x^2} \, dx$

12. $\int_1^\infty e^{4x} \, dx$

13. $\int_0^\infty e^{-2x} \, dx$

14. Find k such that $f(x) = k/x^3$ is a probability density function over the interval $[1, 2]$. Then write the probability density function.

15. *Business: Waiting time.* A person randomly arrives at a bus stop where the waiting time t for a bus is no more than 25 min. The probability density function for t is $f(t) = \frac{1}{25}$, for $0 \leqslant t \leqslant 25$. Find the probability that a person will have to wait no more than 15 min for a bus.

Given the probability density function $f(x) = 3x^2$ over $[0, 1]$, find each of the following.

16. $E(x^2)$

17. $E(x)$

18. The mean

19. The variance

20. The standard deviation

Let x be a continuous random variable with standard normal density. Using Table 2, find each of the following.

21. $P(0 \leqslant x \leqslant 1.85)$

22. $P(-1.74 \leqslant x \leqslant 1.43)$

23. $P(-2.08 \leqslant x \leqslant -1.18)$

24. $P(x \geqslant 0)$

25. *Business: Pizza sales.* The number of pizzas sold daily at Benito's Pizzeria is normally distributed with mean 400 and standard deviation 60. What is the probability that the number sold during one day is 480 or more?

Find the volume generated by revolving about the x-axis the region bounded by each of the following graphs.

26. $y = x^3$, $x = 1$, $x = 2$

27. $y = \dfrac{1}{x + 2}$, $x = 0$, $x = 1$

Solve the differential equation.

28. $\dfrac{dy}{dx} = 11x^{10}y$

29. $\dfrac{dy}{dx} = \dfrac{2}{y}$

30. $\dfrac{dy}{dx} = 4y$; $y = 5$ when $x = 0$

31. $\dfrac{dv}{dt} = 5v^{-2}$; $v = 4$ when $t = 3$

32. $y' = \dfrac{3x}{y}$

33. $y' = 8x - xy$

34. *Economics: Elasticity.* Find the demand function given the elasticity condition

$$E(p) = \frac{p}{p - 100}; \quad x = 70 \text{ when } p = 30.$$

35. *Business: Stock growth.* The growth rate of a stock is modeled by

$$\frac{dV}{dt} = k(L - V), \quad V = \$30 \text{ when } t = 0,$$

where $V = $ the value of the stock, per share, after time t (in months), $L = \$36.37$, the limiting value of the stock, and $k = $ a constant. Find the solution to the differential equation in terms of t and k.

SYNTHESIS EXERCISES

36. The function $f(x) = x^8$ is a probability density function on the interval $[-c, c]$. Find c.

Determine whether the improper integral is convergent or divergent, and calculate its value if it is convergent.

37. $\int_{-\infty}^0 x^4 e^{-x^5} \, dx$

38. $\int_0^\infty \dfrac{dx}{(x + 1)^{4/3}}$

EXERCISES FOR THINKING AND WRITING

39. Discuss as many applications as you can of the use of integration considered in this chapter.

40. Explain the uses of integration in the study of probability.

41. The statement is made in the text that "The antiderivative of the function $f(x) = e^{-x^2/2}$ has no elementary formula." Make some guesses of functions that might seem reasonable to you as antiderivatives and show why they are not.

42. Consider a soda pop bottle. Explain what you might do in order to use integration to find its volume.

CHAPTER TEST

Given the demand and supply functions $D(p) = (p - 7)^2$ and $S(p) = p^2 + p + 4$, find each of the following.

1. The equilibrium point

2. The consumer's surplus at the equilibrium point

3. The producer's surplus at the equilibrium point

4. *Business: Amount of a continuous money flow.* Find the amount of a continuous money flow if $1200 per year is being invested at 6%, compounded continuously, for 15 years.

5. *Business: Continuous money flow.* Consider a continuous money flow into an investment at the rate of P_0 dollars per year. What should P_0 be so that the amount of a continuous money flow over 25 years at 6%, compounded continuously, will be $20,000?

6. *Physical science: Demand for iron ore.* In 1990 ($t = 0$), the world use of iron ore was 960,000 thousand tons, and the demand for it was growing exponentially at the rate of 6% per year. If the demand continues to grow at this rate, how many tons of iron ore will the world use from 1990 to 2010?

7. *Physical science: Depletion of iron ore.* The world reserves of iron ore are 147,400,000 thousand tons. Assuming that the growth rate of 6% per year continues and that no new reserves are discovered, when will the world reserves of iron ore be exhausted?

8. *Business: Present value.* Following the birth of a child, a parent wants to make an initial investment P_0 that will grow to $10,000 by the child's 20th birthday. Interest is compounded continuously at 7%. What should the initial investment be?

9. *Business: Accumulated present value.* Find the accumulated present value of an investment over a 20-year period if there is a continuous money flow of $3800 per year and the current interest rate is 11%.

10. *Business: Accumulated present value.* Find the accumulated present value of an investment for which the continuous money flow in Question 9 is perpetual.

Determine whether the improper integral is convergent or divergent, and calculate its value if it is convergent.

11. $\int_1^\infty \frac{dx}{x^5}$

12. $\int_0^\infty \frac{3}{1 + x} \, dx$

13. Find k such that $f(x) = kx^3$ is a probability density function over the interval $[0, 2]$. Then find the probability density function.

14. *Business: Times of telephone calls.* A telephone company determines that the length of time t of a phone call is an exponentially distributed random variable with probability density function

$$f(t) = 2e^{-2t}, \quad 0 \leq t < \infty.$$

Find the probability that a phone call will last no more than 1 minute.

Given the probability density function $f(x) = \frac{1}{4}x$ over $[1, 3]$, find each of the following.

15. $E(x)$

16. $E(x^2)$

17. The mean

18. The variance

19. The standard deviation

Let x be a continuous random variable with standard normal density. Using Table 2, find each of the following.

20. $P(0 \leqslant x \leqslant 1.5)$

21. $P(0.12 \leqslant x \leqslant 2.32)$

22. $P(-1.61 \leqslant x \leqslant 1.76)$

23. The price per pound p of a T-bone steak at various stores in a certain city is normally distributed with mean \$4.75 and standard deviation \$0.25. What is the probability that the price per pound is \$4.80 or more?

Find the volume generated by revolving about the x-axis the regions bounded by the following graphs.

24. $y = \dfrac{1}{\sqrt{x}}, \quad x = 1, \quad x = 5$

25. $y = \sqrt{2 + x}, \quad x = 0, \quad x = 1$

Solve the differential equation.

26. $\dfrac{dy}{dx} = 8x^7 y$

27. $\dfrac{dy}{dx} = \dfrac{9}{y}$

28. $\dfrac{dy}{dt} = 6y; \quad y = 11$ when $t = 0$

29. $y' = 5x^2 - x^2 y$

30. $\dfrac{dv}{dt} = 2v^{-3}$

31. $y' = 4y + xy$

32. *Economics: Elasticity.* Find the demand function given the elasticity condition

$$E(p) = 4 \quad \text{for all } p > 0.$$

33. *Business: Stock growth.* The growth rate of stock for Glamour Industries is modeled by

$$\frac{dV}{dt} = k(L - V),$$

where $V =$ the value of the stock per share, after time t (in months), $L = \$36$, the *limiting value* of the stock, $k = $ a constant, and $V(0) = \$0$.

a) Write the solution $V(t)$ in terms of L and k.

b) If $V(6) = \$18$, determine k to the nearest hundredth.

c) Rewrite $V(t)$ in terms of t and k using the value of k found in part (b).

d) Use the equation in part (c) to find $V(12)$, the value of the stock after 12 months.

e) In how many months will the value be \$30?

SYNTHESIS EXERCISES

36. The function $f(x) = x^8$ is a probability density function on the interval $[-c, c]$. Find c.

Determine whether the improper integral is convergent or divergent, and calculate its value if it is convergent.

37. $\displaystyle\int_{-\infty}^{0} x^4 e^{-x^5}\, dx$

38. $\displaystyle\int_{0}^{\infty} \dfrac{dx}{(x + 1)^{4/3}}$

7

FUNCTIONS OF
SEVERAL VARIABLES

AN APPLICATION

The wind speed of a tornado. Under certain conditions, the *wind speed* of a tornado at a distance *d* from its center can be approximated by the function

$$S = \frac{aV}{0.51d^2},$$

where *a* is an atmospheric constant that depends on certain atmospheric conditions and V is the approximate volume of the tornado, in cubic feet.

Approximate the wind speed 100 ft from the center of a tornado when its volume is 1,600,000 ft^3 and $a = 0.78$.

THE MATHEMATICS

To solve the problem, we substitute 100 for *d*, 1,600,000 for V, and 0.78 for *a* in the equation

$$S = \frac{aV}{0.51d^2}.$$

└── This is a *function of several variables*.

Functions that have more than one input are called *functions of several variables*. We introduce these functions in this chapter and learn to differentiate them to find what are called *partial derivatives*. Then we use these functions and their partial derivatives to solve maximum–minimum problems. Finally, we consider the integration of such functions.

7.1

a) Find a function value for a function
    of several variables.

FUNCTIONS OF SEVERAL VARIABLES

Suppose that a one-product firm produces x items of its product at a profit of $4 per item. Then its total profit $P(x)$ is given by

$$P(x) = 4x.$$

This is a function of one variable.

Suppose that a two-product firm produces x items of one product at a profit of $4 per item and y items of a second at a profit of $6 per item. Then its total profit P is a function of the *two* variables x and y, and is given by

$$P(x, y) = 4x + 6y.$$

This function assigns to the input pair (x, y) a unique output number $4x + 6y$.

DEFINITION

A *function of two variables* assigns to each input pair (x, y) **exactly one output number** $f(x, y)$.

We can also think of a function of two variables as a machine that has two inputs (Fig. 1). The domain of a function of two variables is a set of pairs (x, y) in the plane. Unless otherwise restricted, when such a function is given by a formula, the domain consists of all ordered pairs (x, y) that are meaningful replacements in the formula.

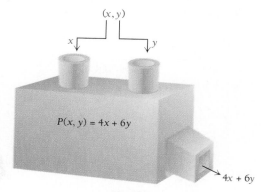

FIGURE 1

EXAMPLE 1 For $P(x, y) = 4x + 6y$, find $P(25, 10)$.

Solution $P(25, 10)$ is defined to be the value of the function found by substituting 25 for x and 10 for y:

$$P(25, 10) = 4 \cdot 25 + 6 \cdot 10$$
$$= 100 + 60$$
$$= \$160.$$

This means that by selling 25 items of the first product and 10 of the second, the two-product firm will make a profit of \$160. ❖

DO EXERCISES 1 AND 2.

The following are examples of **functions of several variables**, that is, functions of two or more variables. If there are n variables, then there are n inputs for such a function.

EXAMPLE 2 *Business: Total cost.* The total cost of a company, in thousands of dollars, is given by

$$C(x, y, z, w) = 4x^2 + 5y + z - \ln{(w + 1)},$$

where x dollars is spent for labor, y dollars for raw materials, z dollars for advertising, and w dollars for machinery. This is a function of four variables. Find $C(3, 2, 0, 10)$.

Solution We substitute 3 for x, 2 for y, 0 for z, and 10 for w:

$$C(3, 2, 0, 10) = 4 \cdot 3^2 + 5 \cdot 2 + 0 - \ln{(10 + 1)}$$
$$= 4 \cdot 9 + 10 + 0 - 2.397895$$
$$\approx \$43.6 \text{ thousand.}$$ ❖

DO EXERCISES 3 AND 4.

EXAMPLE 3 *Business: Cost of storage equipment.* A business purchases a piece of storage equipment that costs C_1 dollars and has capacity V_1. Later it wishes to replace the original with a new piece of equipment that costs C_2 dollars and has capacity V_2. The ratio of the new capacity to the original is

$$\frac{V_2}{V_1} = k,$$

so

$$V_2 = kV_1.$$

1. For $P(x, y) = 4x + 6y$:

   a) Find $P(14, 12)$ and interpret its meaning.

   b) Find $P(0, 8)$ and interpret its meaning.

2. *General interest: Earned-run average.* A pitcher's *earned-run average* is given by

   $$A(n, i) = 9 \cdot \left(\frac{n}{i}\right),$$

   where n is the total number of earned runs given up in i innings of pitching. Find each of the following.

   a) $A(4, 5)$

   b) $A(7, \frac{2}{3})$

   c) $A(1, 15)$

3. For $V(x, y, z) = xyz$, find $V(5, 10, 40)$.

4. *Business: Total cost.* For the function C of Example 2, find each of the following.

   a) $C(9, 5, 3, 0)$

   b) $C(1, 2, 3, 4)$

5. *Business: Cost of storage equipment.* Repeat Example 3 given that the original tank cost $65,000 and the new tank is triple the capacity of the original.

It has been found in industrial economics that in this case, the cost of the new piece of equipment can be estimated by the function of three variables:

$$C_2 = \left(\frac{V_2}{V_1}\right)^{0.6} C_1 = k^{0.6} C_1. \tag{1}$$

For $45,000, a beverage company buys a manufacturing tank that has a capacity of 10,000 gallons. Later it decides to buy a tank with double the capacity of the original. Estimate the cost of the new tank.

Solution   We substitute 20,000 for V_2, 10,000 for V_1, and 45,000 for C_1 in Eq. (1):

$$C_2 = \left(\frac{20,000}{10,000}\right)^{0.6} (45,000)$$

$$= 2^{0.6}(45,000)$$

$$= (1.515717)(45,000)$$

$$\approx \$68,207.25.$$

Note that a 100% increase in capacity was achieved by about a 52% increase in cost. This is independent of any increase in the costs of labor, management, or other equipment resulting from the purchase of the tank.   ❖

DO EXERCISE 5.

EXAMPLE 4   *Social science: The gravity model.* The number of telephone calls between two cities is given by

$$N(d, P_1, P_2) = \frac{2.8 P_1 P_2}{d^{2.4}},$$

where d is the distance between the cities and P_1 and P_2 are their populations.   ❖

A constant can also be thought of as a function of several variables.

EXAMPLE 5   The constant function f is given by

$$f(x, y) = -3 \quad \text{for all inputs } x \text{ and } y.$$

Find $f(5, 7)$ and $f(-2, 0)$.

Solution   Since this is a constant function, it has the value -3 for any x and y. Thus,

Sociologists say that as two cities merge, the communication between them increases.

$$f(5, 7) = -3 \quad \text{and} \quad f(-2, 0) = -3.$$ ❖

DO EXERCISE 6.

Geometric Interpretations

Consider a function of two variables

$$z = f(x, y).$$

Recall the mapping interpretation of function that we considered in Chapter 1. As a mapping, a function of two variables can be thought of as mapping a point (x_1, y_1) in an xy-plane onto a point z_1 on a number line (Fig. 2).

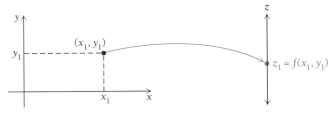

FIGURE 2

6. The constant function g is given by

$$g(x, y) = 4$$

for all inputs x and y. Find each of the following.

a) $g(-9, 10)$

b) $g(560, 43)$

To graph a function of two variables, we need a three-dimensional coordinate system. The axes are generally placed as shown in Fig. 3. The line z, called the z-axis, is placed perpendicular to the xy-plane at the origin.

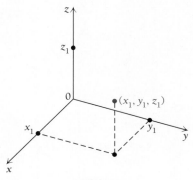

FIGURE 3

To help visualize this, think of looking into the corner of a room, where the floor is the xy-plane and the z-axis is the intersection of the two walls. To plot a point (x_1, y_1, z_1), we locate the point (x_1, y_1) in the xy-plane and move up or down in space according to the value of z_1.

EXAMPLE 6   Plot these points:

$$P_1(2, 3, 5), \quad P_2(2, -2, -4), \quad P_3(0, 5, 2), \quad \text{and} \quad P_4(2, 3, 0).$$

Solution   The solution is shown in Fig. 4.

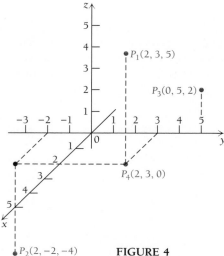

FIGURE 4

DO EXERCISE 7.

The *graph* of a function of two variables

$$z = f(x, y)$$

consists of ordered triples (x_1, y_1, z_1), where $z_1 = f(x_1, y_1)$. The domain of f is a region D in the xy-plane, and the graph of f is a surface S, as illustrated in Fig. 5.

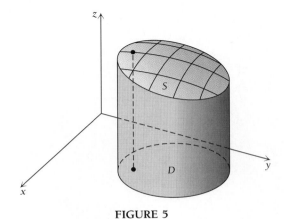

FIGURE 5

The equations and their graphs shown in Figs. 6–14 have been generated by a computer graphics program called *Mathematica*. There are many excellent graphics programs for generating graphs of functions of two variables. The program *3D Grapher* in *The Calculus Explorer* is another. The graphs can be generated with very few keystrokes. Seeing one of these graphs often reveals more insight to the behavior of the function than does its formula.

7. Using the axes shown below, graph $P_1(3, 2, 5)$, $P_2(2, 3, 1)$, and $P_3(-3, 2, 0)$.

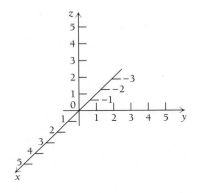

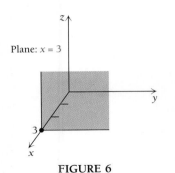

Plane: $x = 3$

FIGURE 6

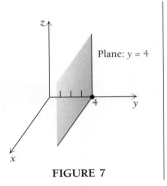

Plane: $y = 4$

FIGURE 7

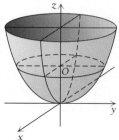

FIGURE 8

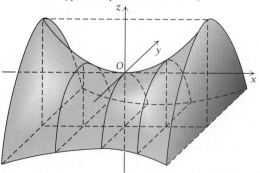

FIGURE 9

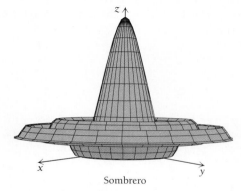

Sombrero
FIGURE 10

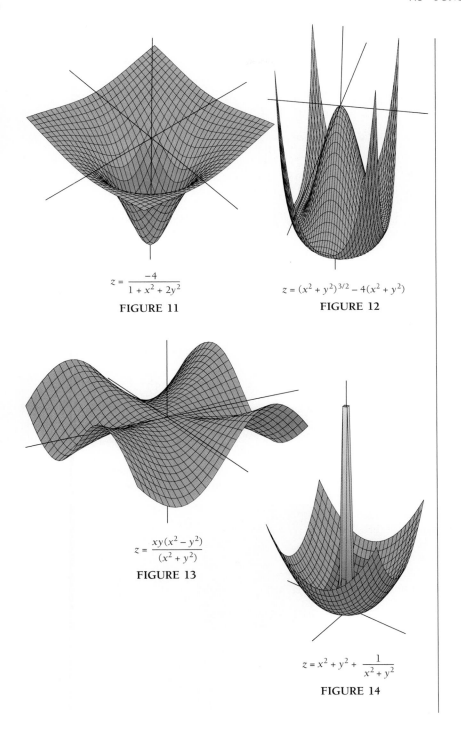

$$z = \frac{-4}{1 + x^2 + 2y^2}$$
FIGURE 11

$$z = (x^2 + y^2)^{3/2} - 4(x^2 + y^2)$$
FIGURE 12

$$z = \frac{xy(x^2 - y^2)}{(x^2 + y^2)}$$
FIGURE 13

$$z = x^2 + y^2 + \frac{1}{x^2 + y^2}$$
FIGURE 14

1. For $f(x, y) = x^2 - 2xy$, find $f(0, -2)$, $f(2, 3)$, and $f(10, -5)$.

2. For $f(x, y) = (y^2 + 3xy)^3$, find $f(-2, 0)$, $f(3, 2)$, and $f(-5, 10)$.

3. For $f(x, y) = 3^x + 7xy$, find $f(0, -2)$, $f(-2, 1)$, and $f(2, 1)$.

4. For $f(x, y) = \log_{10} x - 5y^2$, find $f(10, 2)$, $f(1, -3)$, and $f(100, 4)$.

5. For $f(x, y) = \ln x + y^3$, find $f(e, 2)$, $f(e^2, 4)$, and $f(e^3, 5)$.

6. For $f(x, y) = 2^x - 3^y$, find $f(0, 0)$, $f(1, 1)$, and $f(2, 2)$.

7. For $f(x, y, z) = x^2 - y^2 + z^2$, find $f(-1, 2, 3)$ and $f(2, -1, 3)$.

8. For $f(x, y, z) = 2^x + 5zy - x$, find $f(0, 1, -3)$ and $f(1, 0, -3)$.

APPLICATIONS

❖ **Business and Economics**

9. *Cost of storage equipment.* Consider the storage model in Example 3. For $100,000, a company buys a storage tank that has a capacity of 80,000 gallons. Later it replaces the tank with a new tank with double the capacity of the original. Estimate the cost of the new tank.

10. *Price-earnings ratio.* The *price-earnings ratio* of a stock is given by

$$R(P, E) = \frac{P}{E},$$

where P is the price per share of the stock and E is the earnings per share. Recently, the price per share of IBM stock was $287\frac{3}{8}$ and the earnings per share were $23.30. Find the price-earnings ratio. Give decimal notation to the nearest tenth.

11. *Yield of a stock.* The *yield* of a stock is given by

$$Y(D, P) = \frac{D}{P},$$

where D is the dividends per share of a stock and P is the price per share. Recently, the price per share of Goodyear stock was $16\frac{7}{8}$ and the dividends per share were $1.30. Find the yield. Give percent notation to the nearest tenth of a percent.

❖ **Life and Physical Sciences**

12. *Poiseuille's Law.* The speed of blood in a vessel is given by

$$V(L, p, R, r, v) = \frac{p}{4Lv}(R^2 - r^2),$$

where R is the radius of the vessel, r is the distance of the blood from the center of the vessel, L is the length of the blood vessel, p is pressure, and v is viscosity. Find $V(1, 100, 0.0075, 0.0025, 0.05)$.

13. *Wind speed of a tornado.* Under certain conditions, the *wind speed* of a tornado at a distance d from its center

can be approximated by the function

$$S = \frac{aV}{0.51d^2},$$

where a is an atmospheric constant that depends on certain atmospheric conditions and V is the approximate volume of the tornado, in cubic feet. Approximate the wind speed 100 ft from the center of a tornado when its volume is 1,600,000 ft^3 and $a = 0.78$.

❖ **Social Sciences**

14. *Intelligence quotient.* The *intelligence quotient* in psychology is given by

$$Q(m, c) = 100 \cdot \frac{m}{c},$$

where m is a person's mental age and c is his or her chronological, or actual, age. Find $Q(21, 20)$ and $Q(19, 20)$.

SYNTHESIS EXERCISES

General interest: Wind chill temperature. Wind speed affects the actual temperature, making a person colder due to extra heat loss from the skin. The *wind chill temperature* is what the temperature would have to be with no wind in order to give the same chilling effect. The wind chill temperature W is given by

$$W(v, T) = 91.4 - \frac{(10.45 + 6.68\sqrt{v} - 0.447v)(457 - 5T)}{110},$$

where T is the actual temperature as given by a thermometer, in degrees Fahrenheit, and v is the speed of the wind, in miles per hour. Find the wind chill temperature in each case. Round to the nearest one degree.

15. $T = 30°F, \quad v = 25$ mph
16. $T = 20°F, \quad v = 20$ mph
17. $T = 20°F, \quad v = 40$ mph
18. $T = -10°F, \quad v = 30$ mph

COMPUTER EXERCISES

Use a 3D graphing software program to generate the graph of the function.

19. $f(x, y) = y^2$
20. $f(x, y) = x^2 + y^2$
21. $f(x, y) = (x^4 - 16x^2)e^{-y^2}$
22. $f(x, y) = 4(x^2 + y^2) - (x^2 + y^2)^2$
23. $f(x, y) = x^3 - 3xy^2$
24. $f(x, y) = \dfrac{1}{x^2 + 4y^2}$

THE CALCULUS EXPLORER: *3D Grapher*

Use the program to graph the surfaces given by the functions in Exercises 19–24. Also, try graphing other functions in this exercise set.

7.2

OBJECTIVES

a)   Find the partial derivatives of a given function.

b)   Evaluate the partial derivatives of a function at a given point.

1. Consider

$$f(x, y) = 1 - x^2 - y^2.$$

a)   Fix y at 4 and find $f(x, 4)$.

b)   The answer to part (a) could be interpreted as a function of one variable, x. Find the first derivative.

2. For $f(x, y) = 1 - x^2 - y^2$, find $\partial f/\partial x$.

PARTIAL DERIVATIVES

Finding Partial Derivatives

Consider the function f given by

$$z = f(x, y) = x^2 y^3 + xy + 4y^2.$$

Suppose for the moment that we fix y at 3. Then

$$f(x, 3) = x^2(3^3) + x(3) + 4(3^2) = 27x^2 + 3x + 36.$$

Note that we now have a function of only one variable. Taking the first derivative with respect to x, we have

$$54x + 3.$$

DO EXERCISE 1.

Now, without replacing y by a specific number, let us consider y fixed. Then f becomes a function of x alone and we can calculate its derivative with respect to x. This derivative is called the *partial derivative of f with respect to x*. Notation for this partial derivative is

$$\frac{\partial f}{\partial x} \quad \text{or} \quad \frac{\partial z}{\partial x}.$$

Thus let us again consider the function

$$z = f(x, y) = x^2 y^3 + xy + 4y^2.$$

The color blue indicates the variable x when we fix y and treat it as a constant. The expressions y^3, y, and y^2 are then constants. We have

$$\frac{\partial f}{\partial x} = \frac{\partial z}{\partial x} = 2xy^3 + y.$$

DO EXERCISE 2.

Similarly, we find $\partial f/\partial y$ or $\partial z/\partial y$ by fixing x (treating it as a constant) and calculating the derivative with respect to y. From

$$z = f(x, y) = x^2 y^3 + xy + 4y^2, \qquad \text{The color blue indicates the variable.}$$

we get

$$\frac{\partial f}{\partial y} = \frac{\partial z}{\partial y} = 3x^2y^2 + x + 8y.$$

DO EXERCISE 3.

A definition of partial derivatives is as follows.

<hr>

DEFINITION

For $z = f(x, y)$,

$$\frac{\partial z}{\partial x} = \lim_{h \to 0} \frac{f(x + h, y) - f(x, y)}{h},$$

$$\frac{\partial z}{\partial y} = \lim_{k \to 0} \frac{f(x, y + k) - f(x, y)}{k}.$$

<hr>

We can find partial derivatives of functions of any number of variables.

EXAMPLE 1   For $w = x^2 - xy + y^2 + 2yz + 2z^2 + z$, find

$$\frac{\partial w}{\partial x}, \quad \frac{\partial w}{\partial y}, \quad \text{and} \quad \frac{\partial w}{\partial z}.$$

Solution   In order to find $\partial w/\partial x$, we consider x the variable and the other letters the constants. From

$$w = x^2 - xy + y^2 + 2yz + 2z^2 + z,$$

we get

$$\frac{\partial w}{\partial x} = 2x - y.$$

From

$$w = x^2 - xy + y^2 + 2yz + 2z^2 + z,$$

we get

$$\frac{\partial w}{\partial y} = -x + 2y + 2z;$$

and from

$$w = x^2 - xy + y^2 + 2yz + 2z^2 + z,$$

3. For $z = 3x^2y + 5x^3$, find each of the following.

a) $\dfrac{\partial z}{\partial x}$

b) $\dfrac{\partial z}{\partial y}$

4. For $t = xy + xz + x^2 + y^3$, find each of the following.

a) $\dfrac{\partial t}{\partial x}$

b) $\dfrac{\partial t}{\partial y}$

c) $\dfrac{\partial t}{\partial z}$

5. For $f(x, y) = 3x^3y + 2xy$, find each of the following.

a) f_x
b) $f_x(-4, 1)$
c) f_y
d) $f_y(2, 6)$

6. For $f(x, y) = \ln(xy) + ye^x$, find f_x and f_y.

we get

$$\frac{\partial w}{\partial z} = 2y + 4z + 1. \qquad \diamond$$

DO EXERCISE 4.

We will often make use of a simpler notation f_x for the partial derivative of f with respect to x and f_y for the partial derivative of f with respect to y.

EXAMPLE 2   For $f(x, y) = 3x^2y + xy$, find f_x and f_y.

Solution   We have

$$f_x = 6xy + y,$$
$$f_y = 3x^2 + x. \qquad \diamond$$

For the function in the preceding example, let us evaluate f_x at $(2, -3)$:

$$f_x(2, -3) = 6 \cdot 2 \cdot (-3) + (-3) = -39.$$

If we use the notation $\partial f/\partial x = 6xy + y$, where $f = 3x^2y + xy$, the value of the partial derivative at $(2, -3)$ is given by

$$\left.\frac{\partial f}{\partial x}\right|_{(2, -3)} = 6 \cdot 2 \cdot (-3) + (-3) = -39,$$

but this notation is not as convenient as $f_x(2, -3)$.

DO EXERCISE 5.

EXAMPLE 3   For $f(x, y) = e^{xy} + y \ln x$, find f_x and f_y.

Solution

$$f_x = y \cdot e^{xy} + y \cdot \frac{1}{x} = ye^{xy} + \frac{y}{x},$$
$$f_y = x \cdot e^{xy} + 1 \cdot \ln x = xe^{xy} + \ln x \qquad \diamond$$

DO EXERCISE 6.

The Geometric Interpretation of Partial Derivatives

The *graph* of a function of two variables $z = f(x, y)$ is a surface S, as shown in Fig. 1.

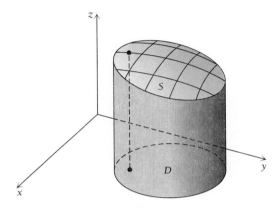

FIGURE 1

Now suppose we hold x fixed, say, at the value a. The set of all points for which $x = a$ is a plane parallel to the yz-plane, so when x is fixed at a, y and z vary along the plane, as shown in Fig. 2. The plane shown cuts the surface in some curve C_1. The partial derivative f_y gives the slope of tangent lines to this curve.

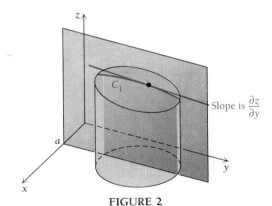

Slope is $\dfrac{\partial z}{\partial y}$

FIGURE 2

Similarly, if we hold y fixed, say, at the value b, we obtain a curve C_2, as shown in Fig. 3. The partial derivative f_x gives the slope of tangent lines to this curve.

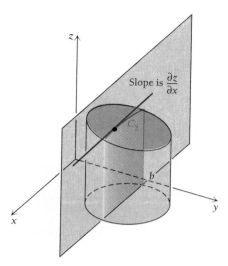

FIGURE 3

An Economic Application: The Cobb–Douglas Production Function

One model of production that is frequently considered in business and economics is the *Cobb–Douglas production function*:

$$p(x, y) = Ax^a y^{1-a}, \qquad A > 0 \quad \text{and} \quad 0 < a < 1,$$

where p is the number of units produced with x units of labor and y units of capital. (Capital is the cost of machinery, buildings, tools, and other supplies.) The partial derivatives

$$\frac{\partial p}{\partial x} \quad \text{and} \quad \frac{\partial p}{\partial y}$$

are called, respectively, the *marginal productivity of labor* and the *marginal productivity of capital*.

EXAMPLE 4 A company has the following production function for a certain product:

$$p(x, y) = 50x^{2/3} y^{1/3}.$$

a) Find the production from 125 units of labor and 64 units of capital.

b) Find the marginal productivities.

c) Evaluate the marginal productivities at $x = 125$ and $y = 64$.

Solution

a) $p(125, 64) = 50(125)^{2/3}(64)^{1/3} = 50(25)(4) = 5000$ units

b) $\dfrac{\partial p}{\partial x} = 50\left(\dfrac{2}{3}\right)x^{-1/3}y^{1/3} = \dfrac{100y^{1/3}}{3x^{1/3}}, \quad$ or $\quad \dfrac{100}{3}\left(\dfrac{y}{x}\right)^{1/3};$

$\dfrac{\partial p}{\partial y} = 50\left(\dfrac{1}{3}\right)x^{2/3}y^{-2/3} = \dfrac{50x^{2/3}}{3y^{2/3}}, \quad$ or $\quad \dfrac{50}{3}\left(\dfrac{x}{y}\right)^{2/3}$

c) $\left.\dfrac{\partial p}{\partial x}\right|_{(125,\,64)} = \dfrac{100(64)^{1/3}}{3(125)^{1/3}} = \dfrac{100(4)}{3(5)} = 26\dfrac{2}{3};$

$\left.\dfrac{\partial p}{\partial y}\right|_{(125,\,64)} = \dfrac{50(125)^{2/3}}{3(64)^{2/3}} = \dfrac{50(25)}{3(16)} = 26\dfrac{1}{24}$ ❖

How can we interpret marginal productivities? Suppose the amount spent on capital is fixed at, say, $y = 64$. Then if the amount of labor changes by one unit, production will change by about 27 units. Suppose the amount of labor is held fixed at, say, $x = 125$. Then if the amount of capital spent changes by one unit, production will change by about 26 units.

A Cobb–Douglas production function is consistent with the law of diminishing returns. That is, if one input (of either labor or capital) is held fixed while the other increases infinitely, then production will eventually increase at a decreasing rate. With such functions, it also turns out that if a certain maximum production is possible, then the expense of more labor, for example, will not prevent that maximum output from still being attainable.

DO EXERCISE 7.

7. *Economics: The Cobb–Douglas model.* A company has the following production function for a certain product:

$$p(x, y) = 800x^{3/4}y^{1/4}.$$

a) Find the production from 81 units of labor and 625 units of capital.

b) Find the marginal productivities.

c) Evaluate the marginal productivities at $x = 81$ and $y = 625$.

<hr>

EXERCISE SET 7.2

Find $\dfrac{\partial z}{\partial x}, \dfrac{\partial z}{\partial y}, \left.\dfrac{\partial z}{\partial x}\right|_{(-2,\,-3)}$ and $\left.\dfrac{\partial z}{\partial y}\right|_{(0,\,-5)}$

1. $z = 2x - 3xy$

2. $z = (x - y)^3$

3. $z = 3x^2 - 2xy + y$

4. $z = 2x^3 + 3xy - x$

Find $f_x, f_y, f_x(-2, 4)$, and $f_y(4, -3)$.

5. $f(x, y) = 2x - 3y$

6. $f(x, y) = 5x + 7y$

Find $f_x, f_y, f_x(-2, 1)$, and $f_y(-3, -2)$.

7. $f(x, y) = \sqrt{x^2 + y^2}$

8. $f(x, y) = \sqrt{x^2 - y^2}$

Find f_x and f_y.

9. $f(x, y) = e^{2x + 3y}$

10. $f(x, y) = e^{3x - 2y}$

11. $f(x, y) = e^{xy}$

12. $f(x, y) = e^{2xy}$

13. $f(x, y) = y \ln (x + y)$

14. $f(x, y) = x \ln (x + y)$

15. $f(x, y) = x \ln (xy)$

16. $f(x, y) = y \ln (xy)$

17. $f(x, y) = \dfrac{x}{y} - \dfrac{y}{x}$

18. $f(x, y) = \dfrac{x}{y} + \dfrac{y}{x}$

19. $f(x, y) = 3(2x + y - 5)^2$

20. $f(x, y) = 4(3x + y - 8)^2$

Find $\dfrac{\partial f}{\partial b}$ and $\dfrac{\partial f}{\partial m}$.

21. $f(b, m) = (m + b - 4)^2 + (2m + b - 5)^2 +$
   $(3m + b - 6)^2$

22. $f(b, m) = (m + b - 6)^2 + (2m + b - 8)^2 +$
   $(3m + b - 9)^2$

Find f_x, f_y, and f_λ.

23. $f(x, y, \lambda) = 3xy - \lambda(2x + y - 8)$

24. $f(x, y, \lambda) = 4xy - \lambda(3x - y + 7)$

25. $f(x, y, \lambda) = x^2 + y^2 - \lambda(10x + 2y - 4)$

26. $f(x, y, \lambda) = x^2 - y^2 - \lambda(4x - 7y - 10)$

APPLICATIONS

❖ Business and Economics

27. *The Cobb–Douglas model.* A company has the following production function for a certain product:

$$p(x, y) = 1800x^{0.621}y^{0.379},$$

where p is the number of units produced with x units of labor and y units of capital.

a) Find the production from 2500 units of labor and 1700 units of capital.

b) Find the marginal productivities.

c) Evaluate the marginal productivities at $x = 2500$ and $y = 1700$.

28. *The Cobb–Douglas model.* A company has the following production function for a certain product:

$$p(x, y) = 2400x^{2/5}y^{3/5},$$

where p is the number of units produced with x units of labor and y units of capital.

a) Find the production from 32 units of labor and 1024 units of capital.

b) Find the marginal productivities.

c) Evaluate the marginal productivities at $x = 32$ and $y = 1024$.

❖ Life and Physical Sciences

29. *Temperature–humidity heat index.* In the summer, humidity affects the actual temperature, making a person feel hotter due to a reduced heat loss from the skin caused by higher humidity. The *temperature–humidity index*, T_h, is what the temperature would have to be with no humidity in order to give the same heat effect. One index often used is given by

$$T_h = 1.98T - 1.09(1 - H)(T - 58) - 56.9,$$

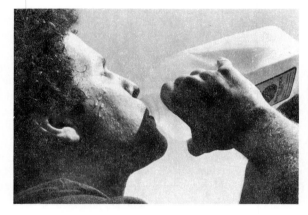

where $T =$ the air temperature, in degrees Fahrenheit, and $H =$ the relative humidity, which is the ratio of the amount of water vapor in the air to the maximum amount of water vapor possible in the air at that temperature. H is usually expressed as a percentage. Find the temperature–humidity index in each case. Round to the nearest tenth of a degree.

a) $T = 85°$ and $H = 60\%$

b) $T = 90°$ and $H = 90\%$

c) $T = 90°$ and $H = 100\%$

d) $T = 78°$ and $H = 100\%$

e) Find $\dfrac{\partial T_h}{\partial H}$.

f) Find $\dfrac{\partial T_h}{\partial T}$.

❖ Social Sciences

30. *Reading ease.* The following formula is used by psychologists and educators to predict the *reading ease E* of a passage of words:

$$E = 206.835 - 0.846w - 1.015s,$$

where w = the number of syllables in a 100-word section and s = the average number of words per sentence. Find the reading ease in each case.

a) $w = 146$ and $s = 5$

b) $w = 180$ and $s = 6$

c) Find $\dfrac{\partial E}{\partial w}$.

d) Find $\dfrac{\partial E}{\partial s}$.

SYNTHESIS EXERCISES

Find f_x and f_t.

31. $f(x, t) = \dfrac{x^2 + t^2}{x^2 - t^2}$

32. $f(x, t) = \dfrac{x^2 - t}{x^3 + t}$

33. $f(x, t) = \dfrac{2\sqrt{x} - 2\sqrt{t}}{1 + 2\sqrt{t}}$

34. $f(x, t) = \sqrt[4]{x^3 t^5}$

35. $f(x, t) = 6x^{2/3} - 8x^{1/4}t^{1/2} - 12x^{-1/2}t^{3/2}$

36. $f(x, t) = \left(\dfrac{x^2 + t^2}{x^2 - t^2}\right)^5$

7.3

HIGHER-ORDER PARTIAL DERIVATIVES

Consider

$$z = f(x, y) = 3xy^2 + 2xy + x^2.$$

Then

$$\frac{\partial z}{\partial x} = \frac{\partial f}{\partial x} = 3y^2 + 2y + 2x.$$

Suppose we continue and find the first partial derivative of $\partial z/\partial x$ with respect to y. This will be a **second-order partial derivative** of the original function z. Its notation is as follows:

$$\frac{\partial}{\partial y}\left(\frac{\partial z}{\partial x}\right) = \frac{\partial}{\partial y}\left(\frac{\partial f}{\partial x}\right)$$

$$= \frac{\partial^2 z}{\partial y\, \partial x}$$

$$= \frac{\partial^2 f}{\partial y\, \partial x} = 6y + 2.$$

DO EXERCISE 1.

OBJECTIVE

a) Find the four second-order partial derivatives of a function.

1. Consider

$$z = 3xy^2 + 2xy + x^2.$$

a) Find $\partial z/\partial y$.

b) For the function in part (a), find the first partial derivative with respect to x.

c) For the function in part (a), find the first partial derivative with respect to y; that is, differentiate "twice" with respect to y.

18. Consider $f(x, y) = x^3 - 5xy^2$. Show that f is a solution to the partial differential equation

$$xf_{xy} - f_y = 0.$$

19. Consider the function f defined as follows:

$$f(x, y) = \begin{cases} \dfrac{xy(x^2 - y^2)}{x^2 + y^2}, & \text{for } (x, y) \neq (0, 0), \\ 0, & \text{for } (x, y) = (0, 0). \end{cases}$$

a) Find $f_x(0, y)$ by evaluating the limit

$$\lim_{h \to 0} \frac{f(h, y) - f(0, y)}{h}.$$

b) Find $f_y(x, 0)$ by evaluating the limit

$$\lim_{h \to 0} \frac{f(x, h) - f(x, 0)}{h}.$$

c) Now find and compare $f_{yx}(0, 0)$ and $f_{xy}(0, 0)$.

20. For $f = [\ln (x^3 + e^y)]^5$, find $f_{xx}, f_{xy}, f_{yx},$ and f_{yy}.

7.4

a) Find relative maximum and minimum values of a function of two variables.

MAXIMUM–MINIMUM PROBLEMS

We will now find maximum and minimum values of functions of two variables.

DEFINITION

A function f of two variables:

1. has a *relative maximum* at (a, b) if

$$f(x, y) \leqslant f(a, b)$$

for all points in a rectangular region containing (a, b);

2. has a *relative minimum* at (a, b) if

$$f(x, y) \geqslant f(a, b)$$

for all points in a rectangular region containing (a, b).

This definition is illustrated in Figs. 1 and 2. A relative maximum (minimum) may not be an "absolute" maximum (minimum), as illustrated in Fig. 3.

Determining Maximum and Minimum Values

Suppose a function f assumes a relative maximum (or minimum) value at some point (a, b) inside its domain. We assume that f and its partial derivatives exist and are "continuous" inside its domain, though

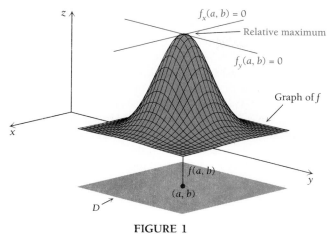

$f_x(a, b) = 0$

Relative maximum

$f_y(a, b) = 0$

Graph of f

$f(a, b)$

(a, b)

D

FIGURE 1

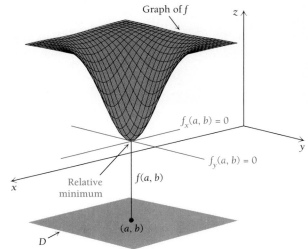

Graph of f

$f_x(a, b) = 0$

$f_y(a, b) = 0$

Relative minimum

$f(a, b)$

(a, b)

D

FIGURE 2

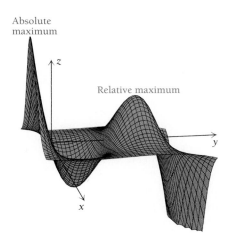

Absolute maximum

Relative maximum

FIGURE 3

we will not take the space to define continuity. If we hold y fixed at the value b, then $f(x, b)$ is a function of one variable x having its relative maximum value at $x = a$. Thus its derivative must be 0 there—that is, $f_x = 0$ at the point (a, b). Similarly, $f_y = 0$ at (a, b). The equations

$$f_x = 0 \quad \text{and} \quad f_y = 0 \tag{1}$$

Where is the saddle point?

are thus satisfied by the point (a, b) at which the relative maximum occurs. We call a point (a, b) at which both partial derivatives are 0 a *critical point*. This is comparable to the earlier definition for functions of one variable. Thus one strategy for finding relative maximum or minimum values is to solve the system of equations (1) above to find critical points. Just as for functions of one variable, this strategy does *not* guarantee that we will have a relative maximum or minimum value. We have argued only that *if* f has a maximum or minimum value at (a, b), *then* both its partial derivatives must be 0 at that point. Look at Figs. 1 and 2. Then note Fig. 4, which illustrates a case in which the partial derivatives are 0 but the function does not have a relative maximum or minimum value at (a, b).

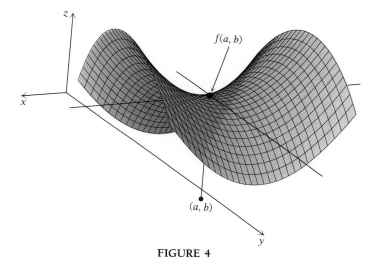

$f(a, b)$

(a, b)

FIGURE 4

In Fig. 4, suppose we fix y at a value b. Then $f(x, b)$, considered as a function of one variable, has a minimum at a, but f does not. Similarly, if we fix x at a, then $f(a, y)$, considered as a function of one variable, has a maximum at b, but f does not. The point $f(a, b)$ is called a **saddle point**. In other words, $f_x(a, b) = 0$ and $f_y(a, b) = 0$ [the point (a, b) is a critical point], but f does not attain a relative maximum or minimum value at (a, b). A saddle point for a function of two variables is comparable to a point of inflection (which is simultaneously a critical point) for a function of one variable.

A test for finding relative maximum and minimum values that involves the use of first- and second-order partial derivatives is stated below. We will not prove this theorem.

THEOREM 1

The D-test

To find the relative maximum and minimum values of f:

1. Find f_x, f_y, f_{xx}, f_{yy}, and f_{xy}.
2. Solve the system of equations $f_x = 0, f_y = 0$. Let (a, b) represent a solution.
3. Evaluate D, where $D = f_{xx}(a, b) \cdot f_{yy}(a, b) - [f_{xy}(a, b)]^2$.
4. Then:
   a) f has a maximum at (a, b) if $D > 0$ and $f_{xx}(a, b) < 0$.
   b) f has a minimum at (a, b) if $D > 0$ and $f_{xx}(a, b) > 0$.
   c) f has neither a maximum nor a minimum at (a, b) if $D < 0$. The function has a *saddle point* at (a, b). See Fig. 4.
   d) This test is not applicable if $D = 0$.

The shape of a perfect tent. To give a tent roof the maximum strength possible, designers draw the fabric into a series of three-dimensional shapes that, viewed in profile, resemble a horse's saddle and that mathematicians call an anticlastic curve. Two people with a stretchy piece of fabric such as Lycra Spandex can duplicate the shape, as shown above. One person pulls up and out on two diagonal corners; the other person pulls down and out on the other two corners. The opposing tensions draw each point of the fabric's surface into rigid equilibrium. The more pronounced the curve, the stiffer the surface.

A relative maximum or minimum *may not be an absolute maximum or minimum value.* Tests for absolute maximum or minimum values are rather complicated. We will restrict our attention to finding *relative* maximum or minimum values. Fortunately, in most of our applications, relative maximum and minimum values turn out to be absolute as well.

EXAMPLE 1   Find the relative maximum and minimum values of

$$f(x, y) = x^2 + xy + y^2 - 3x.$$

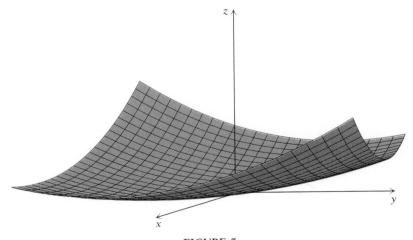

FIGURE 5

1. Find the relative maximum and minimum values of

$$f(x, y) = x^2 + 2xy + 2y^2 - 6y.$$

Solution

1. Find f_x, f_y, f_{xx}, f_{yy}, and f_{xy}.

$$f_x = 2x + y - 3, \qquad f_y = x + 2y,$$
$$f_{xx} = 2; \qquad f_{yy} = 2;$$
$$f_{xy} = 1.$$

2. Solve the system of equations $f_x = 0$, $f_y = 0$:

$$2x + y - 3 = 0, \qquad (1)$$
$$x + 2y = 0. \qquad (2)$$

Solving Eq. (2) for x, we get $x = -2y$. Substituting $-2y$ for x in Eq. (1) and solving, we get

$$2(-2y) + y - 3 = 0$$
$$-4y + y - 3 = 0$$
$$-3y = 3$$
$$y = -1.$$

To find x when $y = -1$, we substitute -1 for y in either Eq. (1) or Eq. (2). We choose Eq. (2):

$$x + 2(-1) = 0$$
$$x = 2.$$

Thus, $f(2, -1)$ is our candidate for a maximum or minimum value.

3. We must check to see whether $f(2, -1)$ is a maximum or minimum value:

$$D = f_{xx}(2, -1) \cdot f_{yy}(2, -1) - [f_{xy}(2, -1)]^2$$
$$= 2 \cdot 2 - [1]^2$$
$$= 3.$$

4. Thus, $D = 3$ and $f_{xx}(2, -1) = 2$. Since $D > 0$ and $f_{xx}(2, -1) > 0$, it follows that f has a relative minimum at $(2, -1)$ and that the minimum value is found as follows:

$$f(2, -1) = 2^2 + 2(-1) + (-1)^2 - 3 \cdot 2$$
$$= 4 - 2 + 1 - 6 = -3. \qquad ❖$$

DO EXERCISE 1.

EXAMPLE 2 Find the relative maximum and minimum values of

$$f(x, y) = xy - x^3 - y^2.$$

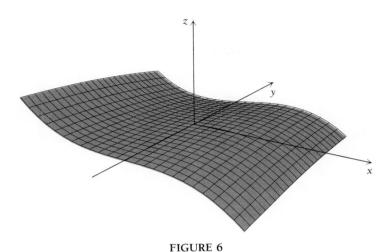

FIGURE 6

Solution

1. Find f_x, f_y, f_{xx}, f_{yy}, and f_{xy}:

$$f_x = y - 3x^2, \qquad f_y = x - 2y,$$
$$f_{xx} = -6x; \qquad f_{yy} = -2;$$
$$f_{xy} = 1.$$

2. Solve the system of equations $f_x = 0$, $f_y = 0$:

$$y - 3x^2 = 0, \tag{1}$$
$$x - 2y = 0. \tag{2}$$

Solving Eq. (1) for y, we get $y = 3x^2$. Substituting $3x^2$ for y in Eq. (2) and solving, we get

$$x - 2(3x^2) = 0$$
$$x - 6x^2 = 0$$
$$x(1 - 6x) = 0. \qquad \text{Factoring}$$

Setting each factor equal to 0 and solving, we have

$$x = 0 \quad \text{or} \quad 1 - 6x = 0$$
$$x = 0 \quad \text{or} \qquad x = \tfrac{1}{6}.$$

To find y when $x = 0$, we substitute 0 for x in either Eq. (1) or Eq. (2). We choose Eq. (2):

$$0 - 2y = 0$$
$$-2y = 0$$
$$y = 0.$$

2. Find the relative maximum and minimum values of

$$f(x, y) = 2xy - 4x^3 - y^2.$$

Thus, $f(0, 0)$ is one candidate for a maximum or minimum value. To find the other, we substitute $\frac{1}{6}$ for x in either Eq. (1) or Eq. (2). We choose Eq. (2):

$$\frac{1}{6} - 2y = 0$$
$$-2y = -\frac{1}{6}$$
$$y = \frac{1}{12}.$$

Thus, $f(\frac{1}{6}, \frac{1}{12})$ is another candidate for a maximum or minimum value.

3. We must check both $(0, 0)$ and $(\frac{1}{6}, \frac{1}{12})$ to see whether they yield maximum or minimum values:

For $(0, 0)$: $\quad D = f_{xx}(0, 0) \cdot f_{yy}(0, 0) - [f_{xy}(0, 0)]^2$
$$= (-6 \cdot 0) \cdot (-2) - [1]^2$$
$$= -1.$$

Since $D < 0$, it follows that $f(0, 0)$ is neither a maximum nor a minimum value, but a saddle point.

For $(\frac{1}{6}, \frac{1}{12})$: $\quad D = f_{xx}(\frac{1}{6}, \frac{1}{12}) \cdot f_{yy}(\frac{1}{6}, \frac{1}{12}) - [f_{xy}(\frac{1}{6}, \frac{1}{12})]^2$
$$= (-6 \cdot \frac{1}{6}) \cdot (-2) - [1]^2$$
$$= -1(-2) - 1$$
$$= 1.$$

4. Thus, $D = 1$ and $f_{xx}(\frac{1}{6}, \frac{1}{12}) = -1$. Since $D > 0$ and $f_{xx}(\frac{1}{6}, \frac{1}{12}) < 0$, it follows that f has a relative maximum at $(\frac{1}{6}, \frac{1}{12})$ and that maximum value is found as follows:

$$f(\tfrac{1}{6}, \tfrac{1}{12}) = \tfrac{1}{6} \cdot \tfrac{1}{12} - (\tfrac{1}{6})^3 - (\tfrac{1}{12})^2$$
$$= \tfrac{1}{72} - \tfrac{1}{216} - \tfrac{1}{144} = \tfrac{1}{432}. \qquad \diamondsuit$$

DO EXERCISE 2.

EXAMPLE 3 *Business: Maximizing profit.* A firm produces two kinds of golf ball, one that sells for $3 each and the other for $2 each. The total revenue, in thousands of dollars, from the sale of x thousand balls at $3 each and y thousand at $2 each is given by

$$R(x, y) = 3x + 2y.$$

The company determines that the total cost, in thousands of dollars, of producing x thousand of the $3 ball and y thousand of the $2 ball is given by

$$C(x, y) = 2x^2 - 2xy + y^2 - 9x + 6y + 7.$$

Find the amount of each type of ball that must be produced and sold in order to maximize profit.

Solution   The total profit $P(x, y)$ is given by

$$P(x, y) = R(x, y) - C(x, y)$$
$$= 3x + 2y - (2x^2 - 2xy + y^2 - 9x + 6y + 7)$$
$$P(x, y) = -2x^2 + 2xy - y^2 + 12x - 4y - 7.$$

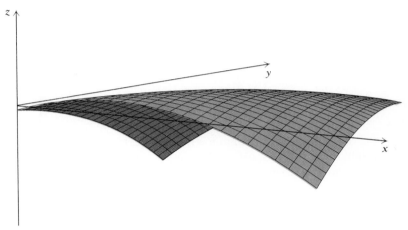

FIGURE 7

1. Find P_x, P_y, P_{xx}, P_{yy}, and P_{xy}:

$$P_x = -4x + 2y + 12, \qquad P_y = 2x - 2y - 4,$$
$$P_{xx} = -4; \qquad\qquad P_{yy} = -2;$$
$$P_{xy} = 2.$$

2. Solve the system of equations $P_x = 0$, $P_y = 0$:

$$-4x + 2y + 12 = 0, \tag{1}$$
$$2x - 2y - 4 = 0. \tag{2}$$

Adding these equations, we get

$$-2x + 8 = 0.$$

Then

$$-2x = -8$$
$$x = 4.$$

3. *Business: Maximizing profit.* A firm produces two kinds of calculator, one that sells for $15 each and the other for $20 each. The total revenue from the sale of x thousand calculators at $15 each and y thousand at $20 each is given by

$$R(x, y) = 15x + 20y.$$

The company determines that the total cost, in thousands of dollars, of producing x thousand of the $15 calculator and y thousand of the $20 calculator is given by

$$C(x, y) =$$
$$3x^2 - 3xy + \tfrac{3}{2}y^2 + 6x + 14y - 50.$$

Find the amount of each type of calculator that must be produced and sold in order to maximize profit.

To find y when $x = 4$, we substitute 4 for x in either Eq. (1) or Eq. (2). We choose Eq. (2):

$$2 \cdot 4 - 2y - 4 = 0$$
$$-2y + 4 = 0$$
$$-2y = -4$$
$$y = 2.$$

Thus, $P(4, 2)$ is a candidate for a maximum or minimum value.

3. We must check to see whether $P(4, 2)$ is a maximum or minimum value:

$$D = P_{xx}(4, 2) \cdot P_{yy}(4, 2) - [P_{xy}(4, 2)]^2$$
$$= (-4)(-2) - 2^2$$
$$= 4.$$

4. Thus, $D = 4$ and $P_{xx} = -4$. Since $D > 0$ and $P_{xx}(4, 2) < 0$, it follows that P has a relative maximum at $(4, 2)$. Thus in order to maximize profit, the company must produce and sell 4 thousand of the $3 golf balls and 2 thousand of the $2 golf balls.   ❖

DO EXERCISE 3.

EXERCISE SET **7.4**

Find the relative maximum and minimum values.

1. $f(x, y) = x^2 + xy + y^2 - y$

2. $f(x, y) = x^2 + xy + y^2 - 5y$

3. $f(x, y) = 2xy - x^3 - y^2$

4. $f(x, y) = 4xy - x^3 - y^2$

5. $f(x, y) = x^3 + y^3 - 3xy$

6. $f(x, y) = x^3 + y^3 - 6xy$

7. $f(x, y) = x^2 + y^2 - 2x + 4y - 2$

8. $f(x, y) = x^2 + 2xy + 2y^2 - 6y + 2$

9. $f(x, y) = x^2 + y^2 + 2x - 4y$

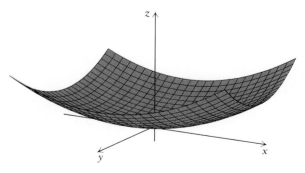

10. $f(x, y) = 4y + 6x - x^2 - y^2$

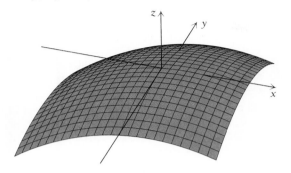

12. $f(x, y) = x^2 - y^2$

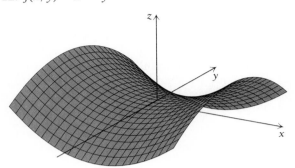

11. $f(x, y) = 4x^2 - y^2$

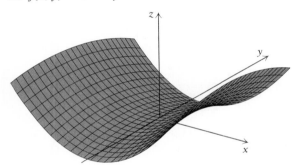

APPLICATIONS

❖ **Business and Economics**

In these problems, assume that relative maximum and minimum values are absolute maximum and minimum values.

13. *Maximizing profit.* A firm produces two kinds of radio, one that sells for $17 each and the other for $21 each. The total revenue from the sale of x thousand radios at $17 each and y thousand at $21 each is given by

$$R(x, y) = 17x + 21y.$$

The company determines that the total cost, in thousands of dollars, of producing x thousand of the $17 radio and y thousand of the $21 radio is given by

$$C(x, y) = 4x^2 - 4xy + 2y^2 - 11x + 25y - 3.$$

Find the amount of each type of radio that must be produced and sold in order to maximize profit.

14. *Maximizing profit.* A firm produces two kinds of baseball glove, one that sells for $18 each and the other for $25 each. The total revenue from the sale of x thousand gloves at $18 each and y thousand at $25 each is given by

$$R(x, y) = 18x + 25y.$$

The company determines that the total cost, in thousands of dollars, of producing x thousand of the $18 glove and y thousand of the $25 glove is given by

$$C(x, y) = 4x^2 - 6xy + 3y^2 + 20x + 19y - 12.$$

Find the amount of each type of glove that must be produced and sold in order to maximize profit.

15. *Maximizing profit.* A one-product company finds that its profit, in millions of dollars, is a function P given by

$$P(a, p) = 2ap + 80p - 15p^2 - \tfrac{1}{10}a^2p - 100,$$

where a = the amount spent on advertising, in millions of dollars, and p = the price charged per item of the product, in dollars. Find the maximum value of P and the values of a and p at which it is attained.

16. *Maximizing profit.* A one-product company finds that its profit, in millions of dollars, is a function P given by

$$P(a, n) = -5a^2 - 3n^2 + 48a - 4n + 2an + 300,$$

where a = the amount spent on advertising, in millions of dollars, and n = the number of items sold, in thousands. Find the maximum value of P and the values of a and n at which it is attained.

17. *Minimizing the cost of a container.* A trash company is designing an open-top, rectangular container that will have a volume of 320 ft^3. The cost of making the bottom of the container is $5 per square foot, and the cost of the sides is $4 per square foot. Find the dimensions of the container that will minimize total cost. (*Hint:* Make a substitution using the formula for volume.)

18. *Two-variable revenue maximization.* Boxowitz, Inc., a computer firm, markets two kinds of electronic calculator that compete with one another. Their demand functions are expressed by the following relation-

ships:

$$q_1 = 78 - 6p_1 - 3p_2, \tag{1}$$
$$q_2 = 66 - 3p_1 - 6p_2, \tag{2}$$

where p_1 and p_2 = the price of each calculator, in multiples of $10, and q_1 and q_2 = the quantity of each calculator demanded, in hundreds of units.

a) Find a formula for the total-revenue function R in terms of the variables p_1 and p_2. [*Hint:* $R = p_1 q_1 + p_2 q_2$; then substitute expressions from Eqs. (1) and (2) to find $R(p_1, p_2)$.]

b) What prices p_1 and p_2 should be charged for each product in order to maximize total revenue?

c) How many units will be demanded?

d) What is the maximum total revenue?

19. *Two-variable revenue maximization.* Repeat Exercise 18, where

$$q_1 = 64 - 4p_1 - 2p_2 \quad \text{and}$$
$$q_2 = 56 - 2p_1 - 4p_2.$$

20. *Temperature.* A flat metal plate is located on a coordinate plane. The temperature of the plate, in degrees Fahrenheit, at point (x, y) is given by

$$T(x, y) = x^2 + 2y^2 - 8x + 4y.$$

Find the minimum temperature and where it occurs. Is there a maximum temperature?

SYNTHESIS EXERCISES

Find the relative maximum and minimum values and the saddle points.

21. $f(x, y) = e^x + e^y - e^{x+y}$

22. $f(x, y) = xy + \dfrac{2}{x} + \dfrac{4}{y}$

23. $f(x, y) = 2y^2 + x^2 - x^2y$

24. $S(b, m) = (m + b - 72)^2 + (2m + b - 73)^2 + (3m + b - 75)^2$

7.5

TOTAL DIFFERENTIAL

If $y = f(x)$, then the *differential of y*, *dy*, is defined by

$$dy = f'(x)\ dx,$$

where *dx* is the *differential of x*. Note that *dy* is a function of two variables, *x* and *dx*.

In Section 3.6, we saw that linear approximation is a useful tool for certain problems involving functions of one variable, and so we'd like to extend the techniques developed there to functions of two or more variables. To do so, we'll use differentials, beginning with the following definition.

a) Compute the total differential of a function of two or more variables.

b) Use the differential *dw* of a function *w* to approximate the actual change Δw in that function.

1. Find *dz* if $z = xe^y$.

DEFINITION

Let $z = f(x, y)$. Then the *total differential dz* is given by

$$dz = f_x\ dx + f_y\ dy.$$

2. Find *dz* if $z = \ln(xy)$.

EXAMPLE 1 Find *dz* if $z = x^2y^3$.

Solution Here, $f(x, y) = x^2y^3$, so that $f_x = 2xy^3$ and $f_y = 3x^2y^2$. According to the definition,

$$dz = f_x\ dx + f_y\ dy$$
$$= 2xy^3\ dx + 3x^2y^2\ dy. \qquad \diamond$$

DO EXERCISES 1 AND 2.

EXAMPLE 2 Evaluate *dz* for $x = 2$, $y = -1$ if $z = xy - 2x^2$.

Solution Since $f(x, y) = xy - 2x^2$, it follows that $f_x = y - 4x$ and $f_y = x$, so

$$dz = f_x\ dx + f_y\ dy = (y - 4x)\ dx + x\ dy$$

is valid for all values of *x* and *y*. If we substitute in $x = 2$ and $y = -1$, we find that

$$dz = (-1 - 4 \cdot 2)\ dx + 2\ dy = (-1 - 8)\ dx + 2\ dy$$
$$= -9\ dx + 2\ dy. \qquad \diamond$$

3. Evaluate *dz* for $x = 2$, $y = 3$ if $z = xy^2$.

DO EXERCISE 3.

4. Evaluate dz for $z = x^3y + 5y^2$ if $x = 1$, $y = -1$, $dx = 0.25$, and $dy = -0.01$.

Note in Example 2 that after specific values have been substituted for x and y, dz is still a function of the variables dx and dy. In general, if $z = f(x, y)$, then dz is a function of four variables: x, y, dx, and dy.

EXAMPLE 3   Evaluate dz for $z = x^2 - y^2$ if $x = 2$, $y = 1$, $dx = 0.1$, and $dy = 0.15$.

Solution   Since $f(x, y) = x^2 - y^2$, we have $f_x = 2x$ and $f_y = -2y$. From the definition,

$$dz = 2x\,dx + -2y\,dy.$$

If we let $x = 2$, $y = 1$, $dx = 0.1$, and $dy = 0.15$ in the last equation, we get

$$dz = 2 \cdot 2 \cdot (0.1) - 2 \cdot 1 \cdot (0.15)$$
$$= 4(0.1) - 2(0.15)$$
$$= 0.4 - 0.3 = 0.1. \qquad \diamondsuit$$

5. Find dw if $w = \dfrac{1}{x} - xy + 2z$.

DO EXERCISE 4.

Our definition can be extended easily to functions of three or more variables.

DEFINITION

Let $w = f(x, y, z)$. Then the *total differential dw* is given by

$$dw = f_x\,dx + f_y\,dy + f_z\,dz.$$

If w is a function of n variables, say $w = f(x_1, x_2, \ldots, x_n)$, then we define the *total differential dw* as

$$dw = f_{x_1}\,dx_1 + f_{x_2}\,dx_2 + \cdots + f_{x_n}\,dx_n$$
$$= \sum_{i=1}^{n} f_{x_i}\,dx_i.$$

6. Find dw if $w = (x^2 + y^2 + z^2)^{1/2}$.

EXAMPLE 4   Find dw given that $w = x^2 + y^3 + z^4$.

Solution   Here, $f(x, y, z) = x^2 + y^3 + z^4$. Thus, $f_x = 2x$, $f_y = 3y^2$, and $f_z = 4z^3$. Substituting these quantities into the definition gives

$$dw = f_x\,dx + f_y\,dy + f_z\,dz$$
$$= 2x\,dx + 3y^2\,dy + 4z^3\,dz. \qquad \diamondsuit$$

DO EXERCISES 5 AND 6.

EXAMPLE 5 Find dw if $w = x_1^2 + 2x_2x_3 - x_4$.

Solution Here, $f(x_1, x_2, x_3, x_4) = x_1^2 + 2x_2x_3 - x_4$, so that $f_{x_1} = 2x_1$, $f_{x_2} = 2x_3$, $f_{x_3} = 2x_2$, and $f_{x_4} = -1$. Substitution yields

$$dw = f_{x_1}\, dx_1 + f_{x_2}\, dx_2 + f_{x_3}\, dx_3 + f_{x_4}\, dx_4$$
$$= 2x_1\, dx_1 + 2x_3\, dx_2 + 2x_2\, dx_3 - 1\, dx_4$$
$$= 2x_1\, dx_1 + 2x_3\, dx_2 + 2x_2\, dx_3 - dx_4. \qquad \diamond$$

DO EXERCISE 7.

EXAMPLE 6 Evaluate dw for $x = 1$, $y = 2$, $z = 3$, if $w = xy - xz + 2yz$.

Solution Since $f(x, y, z) = xy - xz + 2yz$ and $f_x = y - z$, $f_y = x + 2z$, and $f_z = -x + 2y$, we get

$$dw = (y - z)\, dx + (x + 2z)\, dy + (2y - x)\, dz.$$

So, letting $x = 1$, $y = 2$, and $z = 3$, we have

$$dw = (2 - 3)\, dx + (1 + 2 \cdot 3)\, dy + (2 \cdot 2 - 1)\, dz$$

or

$$dw = -dx + 7\, dy + 3\, dz. \qquad \diamond$$

DO EXERCISE 8.

Note that after the three variables x, y, and z have been assigned specific values in Example 6 above, w is still a function of the three variables dx, dy, and dz. If x, y, and z are allowed to vary, then we see that dw is actually a function of the six independent variables x, y, z, dx, dy, and dz. If w is a function of n variables, say $w = f(x_1, x_2, \ldots, x_n)$, then dw is a function of the $2n$ variables $x_1, x_2, \ldots, x_n, dx_1, dx_2, \ldots, dx_n$.

EXAMPLE 7 Evaluate dw for $x = 2$, $y = 5$, $z = -3$, $dx = -0.01$, $dy = 0.02$, and $dz = -0.03$ if $w = 4x^2 - 2y^2 + 3z^2$.

Solution Since $w = f(x, y, z) = 4x^2 - 2y^2 + 3z^2$, we have $f_x = 8x$, $f_y = -4y$, and $f_z = 6z$, so that $dw = 8x\, dx - 4y\, dy + 6z\, dz$. Substituting gives us

$$dw = 8 \cdot 2 \cdot (-0.01) - 4 \cdot 5 \cdot (0.02) + 6 \cdot (-3) \cdot (-0.03)$$
$$= 16(-0.01) - 20(0.02) - 18(-0.03)$$
$$= -0.16 - 0.40 + 0.54 = -0.02. \qquad \diamond$$

7. Find dw if $w = x_1^2 + x_2^2 + x_3^2 + x_4^2$.

8. Evaluate dw for $x = -1$, $y = 1$, and $z = 1$ if $w = x^2y^3 - z^2$.

9. Evaluate dw for $x = 1$, $y = -1$, $z = 3$, $dx = 0.1$, $dy = -0.2$, and $dz = 0$ if $w = \ln(xyz)$.

DO EXERCISE 9.

If $z = f(x, y)$, then the change Δz in z corresponding to changes Δx and Δy in x and y, respectively, is defined by

$$\Delta z = f(x + \Delta x, y + \Delta y) - f(x, y).$$

EXAMPLE 8   Calculate Δz for $f(x, y) = x^2 - y^2$ if $x = 1$, $y = 2$, $\Delta x = 0.01$, and $\Delta y = -0.03$.

Solution   Since $x = 1$, $y = 2$, $\Delta x = 0.01$, and $\Delta y = -0.03$, we have $x + \Delta x = 1 + 0.01 = 1.01$ and $y + \Delta y = 2 + -0.03 = 1.97$. So,

$$
\begin{aligned}
\Delta z &= f(x + \Delta x, y + \Delta y) - f(x, y) \\
&= f(1.01, 1.97) - f(1, 2) \\
&= ((1.01)^2 - (1.97)^2) - (1^2 - 2^2) \\
&= (1.0201 - 3.8809) - (1 - 4) \\
&= (-2.8608) - (-3) = 0.1392.
\end{aligned}
$$

10. Calculate Δz for

$$f(x, y) = 3x^2 + 4y$$

if $x = 2$, $y = 1$, $\Delta x = 0.07$, and $\Delta y = -0.015$.

DO EXERCISE 10.

In Example 8 above, we calculated the exact change in $z = f(x, y)$ for certain values of x, y, Δx, and Δy. Just as we used the differential dy to approximate the exact change Δy when y is a function of x, we now want to use the differential dz to approximate the exact change Δz in $z = f(x, y)$.

EXAMPLE 9   Use the differential dz to approximate Δz in Example 8 above.

11. Use the differential dz to approximate the exact change Δz in Margin Exercise 10.

Solution   We had $z = f(x, y) = x^2 - y^2$ in Example 8, so that $f_x = 2x$, $f_y = -2y$, and $dz = 2x\,dx - 2y\,dy$. If we use the values that were given in Example 8, setting $\Delta x = dx = 0.01$ and $\Delta y = dy = -0.03$, then we get

$$
\begin{aligned}
dz &= 2 \cdot 1 \cdot (0.01) - 2 \cdot (2) \cdot (-0.03) \\
&= 2(0.01) - 4(-0.03) \\
&= 0.02 + 0.12 \\
&= 0.14.
\end{aligned}
$$

Comparing this with the exact value $\Delta z = 0.1392$ from Example 8, we see that $dz \approx \Delta z$.

DO EXERCISE 11.

These ideas extend immediately to functions of three or more variables.

EXAMPLE 10 If $w = f(x, y, z) = 2x - y^2 + xz$, calculate both the exact change Δw and the approximate change dw for $x = 1$, $y = 1$, $z = 2$, $\Delta x = 0.01$, $\Delta y = -0.02$, and $\Delta z = 0.04$.

Solution Note that $x + \Delta x = 1.01$, $y + \Delta y = 0.98$, and $z + \Delta z = 2.04$, so that the exact change Δw is given by

$$\Delta w = f(x + \Delta x, y + \Delta y, z + \Delta z) - f(x, y, z)$$
$$= f(1.01, 0.98, 2.04) - f(1, 1, 2)$$
$$= [2(1.01) - (0.98)^2 + (1.01)(2.04)] - [2 \cdot 1 - 1^2 + 1 \cdot 2]$$
$$= [2.02 - 0.9604 + 2.0604] - [2 - 1 + 2]$$

or

$$\Delta w = 3.12 - 3 = 0.12.$$

To compute dw, note that $f_x = 2 + z$, $f_y = -2y$, and $f_z = x$, so that

$$dw = (2 + z)\, dx - 2y\, dy + x\, dz.$$

Setting $dx = \Delta x = 0.01$, $dy = \Delta y = -0.02$, and $dz = \Delta z = 0.04$, we get

$$dw = (2 + 2) \cdot (0.01) - 2 \cdot 1 \cdot (-0.02) + 1 \cdot (0.04)$$
$$= 4(0.01) - 2(-0.02) + (0.04)$$
$$= 0.04 + 0.04 + 0.04$$
$$= 0.12.$$

In this case, coincidentally, $dw = \Delta w$. ❖

DO EXERCISE 12.

If $z = f(x, y)$, then we had defined $\Delta z = f(x + \Delta x, y + \Delta y) - f(x, y)$. Rewriting, we get

$$f(x + \Delta x, y + \Delta y) = f(x, y) + \Delta z.$$

Now, if we use the approximation $dz \approx \Delta z$, we obtain

$$f(x + \Delta x, y + \Delta y) \approx f(x, y) + dz.$$

In a similar way, if $w = f(x, y, z)$, then we can show that

$$f(x + \Delta x, y + \Delta y, z + \Delta z) \approx f(x, y, z) + dw.$$

12. If $w = x^2 + y^2 + z^2$, calculate both the exact change Δw and the approximate change dw for $x = 4$, $y = -1$, $z = 5$, $\Delta x = 0.03$, $\Delta y = -0.04$, and $\Delta z = -0.01$.

13. If $z = \ln(xy)$, use differentials to calculate an approximate value for $\ln((1.01)(0.97))$.

EXAMPLE 11 If $z = f(x, y) = x^2y^3$, use differentials to calculate an approximate value for $(1.01)^2(0.98)^3$.

Solution We are asked to calculate $f(1.01, 0.98)$. If we set $x = 1, y = 1$, $\Delta x = 0.01$, and $\Delta y = -0.02$, then we must calculate

$$f(x + \Delta x, y + \Delta y).$$

From above, we already know that

$$f(x + \Delta x, y + \Delta y) \approx f(x, y) + dz.$$

But

$$f(x, y) = f(1, 1)$$
$$= 1^2(1)^3 = 1,$$

while

$$dz = f_x\, dx + f_y\, dy$$
$$= 2xy^3\, dx + 3x^2y^2\, dy$$
$$= 2 \cdot 1 \cdot (1)^3\, dx + 3(1)^2(1)^2\, dy$$
$$= 2\, dx + 3\, dy.$$

Setting $dx = \Delta x = 0.01$ and $dy = \Delta y = -0.02$, we find that

$$dz = 2(0.01) + 3(-0.02)$$
$$= 0.02 - 0.06$$
$$= -0.04.$$

So $f(1.01, 0.98) \approx f(1, 1) + dz \approx 1 - 0.04 = 0.96$. In this case, we can calculate the actual value of $f(1.01, 0.98)$ with a calculator. We find that

$$f(1.01, 0.98) = (1.01)^2(0.98)^3$$
$$\approx 0.9601.$$

In this case, our approximation was rather good. ❖

14. If $z = \ln(x/y)$, use differentials to calculate an approximate value for $\ln(1.01/0.97)$.

DO EXERCISES 13 AND 14.

EXAMPLE 12 The temperature T at the point (x, y, z) of a certain region of space is given by $T(x, y, z) = (x^2 + y^2 + z^2)^{3/2}$. Use differentials to calculate the approximate temperature at $(2.014, 2.15, 1.06)$.

Solution We use

$$T(x + \Delta x, y + \Delta y, z + \Delta z) \approx T(x, y, z) + dT.$$

Since $T(2, 2, 1)$ is easy to evaluate, we let $x = 2, y = 2$, and $z = 1$; this forces us to choose $\Delta x = 0.014, \Delta y = 0.15$, and $\Delta z = 0.06$. Proceeding, we get

$$T(2, 2, 1) = (2^2 + 2^2 + 1^2)^{3/2}$$
$$= (4 + 4 + 1)^{3/2} = 9^{3/2} = 27.$$

Now,

$$dT = T_x \, dx + T_y \, dy + T_z \, dz$$
$$= \tfrac{3}{2}(x^2 + y^2 + z^2)^{1/2} \, 2x \, dx + \tfrac{3}{2}(x^2 + y^2 + z^2)^{1/2} \, 2y \, dy$$
$$+ \tfrac{3}{2}(x^2 + y^2 + z^2)^{1/2} \, 2z \, dz$$
$$= (x^2 + y^2 + z^2)^{1/2} \, (3x \, dx + 3y \, dy + 3z \, dz).$$

If we let $dx = \Delta x = 0.014$, $dy = \Delta y = 0.15$, $dz = \Delta z = 0.06$, then we get

$$dT = (2^2 + 2^2 + 1^2)^{1/2}[3 \cdot 2 \cdot (0.014) + 3 \cdot 2 \cdot (0.15) + 3 \cdot 1 \cdot (0.06)]$$
$$= (9)^{1/2}[6(0.014) + 6(0.15) + 3(0.06)]$$
$$= 3[1.164] = 3.492.$$

Thus,

$$T(2.014, 2.15, 1.06) \approx T(2, 2, 1) + dT$$
$$= 27 + 3.492 = 30.492.$$

DO EXERCISE 15.

EXERCISE SET 7.5

Find dz.

1. $z = x^2 + 4y^2$

2. $z = 3x - 4y + xy$

3. $z = xe^{4y}$

4. $z = e^{xy}$

5. $z = x^4 + \dfrac{x}{y} + y^3$

6. $z = x \ln y$

Find dw.

7. $w = x^2 - y^2 + 3z^2$

8. $w = xy + 2xz + yz$

9. $w = xy^2z^4$

10. $w = \ln (xyz)$

11. $w = x_1^2 + x_2x_3 - x_4^2$

12. $w = x_1x_2x_3x_4$

Evaluate both Δz and dz.

13. $z = x^2 + y^2$, $x = 5$, $y = 12$; $dx = \Delta x = 0.01$, $dy = \Delta y = 0.02$

14. $z = 4x + 3y$, $x = 1$, $y = -1$; $dx = \Delta x = 0.03$, $dy = \Delta y = -0.02$

15. $z = e^{x + y}$, $x = y = 0$; $dx = \Delta x = 0.01$, $dy = \Delta y = 0.0005$

16. $z = \ln (xy)$, $x = 2$, $y = 0.5$; $dx = \Delta x = 0.02$, $dy = \Delta y = 0.01$

Evaluate both Δw and dw.

17. $w = x^2 + y^2 - z^2$, $x = 2$, $y = 2$, $z = 1$; $dx = \Delta x = 0.01$, $dy = \Delta y = 0.02$, $dz = \Delta z = 0.02$

18. $w = 4x - 2y + 3z$, $x = 1$, $y = 1$, $z = 2$; $dx = \Delta x = 0.06$, $dy = \Delta y = 0.13$, $dz = \Delta z = -0.04$

19. $w = xyz$, $x = 1$, $y = 0$, $z = 3$; $dx = \Delta x = 0.04$, $dy = \Delta y = -0.02$, $dz = \Delta z = 0.01$

15. The pressure P at the point (x, y, z) in a certain region of space is given (in pounds per square inch) by

$$P(x, y, z) = x^2 + 3xy + z^2.$$

Use differentials to calculate an approximate value for the pressure at $(1.99, 1.52, 3.98)$.

20. $w = \ln(xz/y)$, $x = 2$, $y = 6$, $z = 3$; $dx = \Delta x = 0.05$, $dy = \Delta y = 0.08$, $dz = \Delta z = 0.01$

Use differentials to calculate an approximate value.

21. $(5.01)^2 + (11.94)^2$

22. $(1.06)(0.99)^3 + (2.03)^2$

23. $(1.04)^2 + (0.97)^2 + (1.02)^2 + (1.015)^2$

24. $(1.05)^3 - (0.93)^2$

25. Use differentials to find an approximate value for the area of a right triangle whose base and altitude are 2.013 ft and 4.501 ft, respectively.

26. A rectangular box has dimensions 1.54 ft by 1.01 ft by 7.05 ft. Use differentials to approximate its volume.

27. Use differentials to approximate the surface area of the box in Exercise 26.

28. Calculate dz if $z = x^y$, $x > 0$.

APPLICATIONS

❖ Business and Economics

29. *Business: Estimated profit.* A firm estimates that its profit from selling gold medallions at a price of g dollars each and silver medallions at a price of s dollars each is given by

$$P(g, s) = -50{,}000 + 300s + 1000g - 3s^2 - g^2 + sg.$$

a) Show that if the firm sells the silver medallions at $100 each and the gold ones at $500 each, its estimated profit is $250,000.

b) Use the total differential to approximate the effect on estimated profit of a $1 increase in the price of both medallions.

30. *Business: Price-earnings ratio.* Referring to Exercise 10 in Exercise Set 7.1, use differentials to estimate the change in price-earnings ratio $R(P, E)$ of IBM common stock if the price per share increases from $287\frac{3}{8}$ to $288\frac{5}{8}$ while the earnings per share decreases from $23.30 to $22.80.

31. *Business: Increase in production.* Referring to Exercise 28 in Exercise Set 7.2, use differentials to calculate the approximate increase in production for the company if the number of units of labor is increased from 32 to 35 and the number of units of capital is increased from 1024 to 1040. What is the exact increase in production?

32. *Investments: Dividends.* If the price P of a certain stock is given by

$$P = D\left(1 + \frac{1}{i}\right),$$

where D is the annual dividend per share and i is the annual interest rate, use differentials to estimate what happens to the price of the stock on the day its dividend is cut from $5 to $4 per share if the interest rate i increases from 6% to 6.5% on the same day.

33. *Business: Ticket prices.* Tridelta Airlines flies from Atlanta to Miami. They discover that for round-trip ticket prices T (in dollars) in a certain range, the number N of riders is given by

$$N = \frac{1000}{T - 100}.$$

a) Find a formula for the total revenue R (in dollars) in terms of both N and T.

b) Show that

$$\frac{dR}{R} = \frac{dN}{N} + \frac{dT}{T}.$$

c) Estimate the change in R if ticket prices increase from $200 to $250.

❖ Life and Physical Sciences

34. *Biomedical: Poiseuille's law.* Referring to Exercise 12 in Exercise Set 7.1, use differentials to find the approximate change in the speed of blood if p increases from 100 to 103, v decreases from 0.05 to 0.049, and R, L, and r are held constant.

35. If $w = x_1 x_2 \ldots x_n$, show that

$$dw = w\left[\frac{1}{x_1}\,dx_1 + \frac{1}{x_2}\,dx_2 + \cdots + \frac{1}{x_n}\,dx_n\right].$$

7.6

AN APPLICATION: THE LEAST-SQUARES TECHNIQUE

The problem of fitting an equation to a set of data occurs frequently. We considered one procedure for doing this in Section 1.6. Such an equation provides a model of the phenomena from which predictions can be made. For example, in business, one might want to predict future sales on the basis of past data. In ecology, one might want to predict future demands for natural gas on the basis of past need. Suppose we are trying to find a linear equation

$$y = mx + b$$

to fit the data. To determine this equation is to determine the values of m and b. But how? Let us consider some factual data.

The graph shown in Fig. 1 appeared in a newspaper advertisement for the Indianapolis Life Insurance Company. It pertains to the total amount of life insurance in force in various years. The same data are compiled in the table at the top of the next page.

OBJECTIVE

a)   Find the regression line for a given set of data points and use the regression line to make predictions regarding further data.

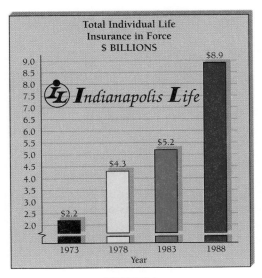

FIGURE 1

1. a) Use Fig. 2 to predict the total amount of life insurance in force in 1993.

b) Use Fig. 3 to predict the total amount of life insurance in force in 1993.

c) Compare your answers.

Year, x	1. 1973	2. 1978	3. 1983	4. 1988	5. 1993
Total amount of individual life insurance in force (in billions), y	$2.2	$4.3	$5.2	$8.9	?

Suppose we plot these points and try to draw a line through them that fits. Note that there are several ways in which this might be done (see Figs. 2 and 3). Each would give a different estimate of the total amount of insurance in force in 1993.

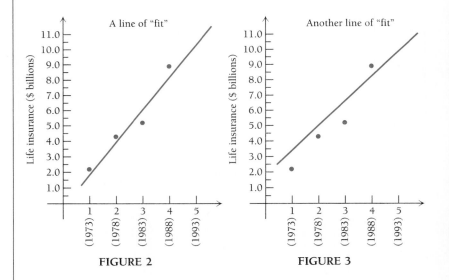

FIGURE 2 · FIGURE 3

DO EXERCISE 1.

Note that the time is incremented in groups of five years, making computations easier. Consider the data points $(1, 2.2)$, $(2, 4.3)$, $(3, 5.2)$, and $(4, 8.9)$, as plotted in Fig. 4.

We will try to fit these data with a line

$$y = mx + b$$

by determining the values of m and b. Note the y-errors, or y-deviations, $y_1 - 2.2$, $y_2 - 4.3$, $y_3 - 5.2$, and $y_4 - 8.9$ between the observed points

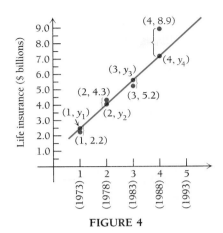

FIGURE 4

$(1, 2.2)$, $(2, 4.3)$, $(3, 5.2)$, and $(4, 8.9)$ and the points $(1, y_1)$, $(2, y_2)$, $(3, y_3)$, and $(4, y_4)$ on the line. We would like, somehow, to minimize these deviations in order to have a good fit. One way of minimizing the deviations is based on the *least-squares assumption*.

The Least-Squares Assumption

The line of best fit is the line for which the sum of the squares of the y-deviations is a minimum. This is called the *regression line*.

Using the least-squares assumption for the life insurance data, we would minimize

$$(y_1 - 2.2)^2 + (y_2 - 4.3)^2 + (y_3 - 5.2)^2 + (y_4 - 8.9)^2, \qquad (1)$$

and since the points $(1, y_1)$, $(2, y_2)$, $(3, y_3)$, and $(4, y_4)$ must be solutions of $y = mx + b$, it follows that

$$y_1 = m(1) + b = m + b,$$
$$y_2 = m(2) + b = 2m + b,$$
$$y_3 = m(3) + b = 3m + b,$$
$$y_4 = m(4) + b = 4m + b.$$

Substituting $m + b$ for y_1, $2m + b$ for y_2, $3m + b$ for y_3, and $4m + b$ for y_4 in Eq. (1), we have

$$(m + b - 2.2)^2 + (2m + b - 4.3)^2 + (3m + b - 5.2)^2$$
$$+ (4m + b - 8.9)^2. \qquad (2)$$

2. *Business: Predicting the number of 1-800 telephone listings.* In recent years, there have been more and more 1-800 toll-free telephone listings. The following data relate the number of 1-800 telephone listings in certain years.

Year, x	Number of 1-800 telephone listings (in thousands), y
1. 1983	260
2. 1984	329
3. 1985	398
4. 1986	466
5. 1987	535

a) Find the regression line.

b) Use the regression line to predict the number of 1-800 listings in 1998; in 2004.

Thus, to find the regression line for the given set of data, we must find the values of m and b that minimize the function S given by the sum in Eq. (2).

To apply the D-test, we first find the partial derivatives $\partial S/\partial b$ and $\partial S/\partial m$:

$$\frac{\partial S}{\partial b} = 2(m + b - 2.2) + 2(2m + b - 4.3) + 2(3m + b - 5.2)$$
$$+ 2(4m + b - 8.9)$$
$$= 20m + 8b - 41.2;$$

$$\frac{\partial S}{\partial m} = 2(m + b - 2.2) + 2(2m + b - 4.3)2 + 2(3m + b - 5.2)3$$
$$+ 2(4m + b - 8.9)4$$
$$= 60m + 20b - 124.$$

We set these derivatives equal to 0 and solve the resulting system:

$$20m + 8b - 41.2 = 0, \qquad 5m + 2b = 10.3,$$
$$60m + 20b - 124 = 0; \qquad \text{or} \qquad 15m + 5b = 31.$$

The solution of this system is

$$b = -0.1, \qquad m = 2.1.$$

We leave it to the reader to complete the D-test to verify that $(-0.1, 2.1)$ does, in fact, yield the minimum of S. We need not bother to compute $S(-0.1, 2.1)$.

The values of m and b are all we need to determine $y = mx + b$. The regression line is

$$y = 2.1x - 0.1.$$

We can extrapolate from the data to predict the total amount of life insurance in force in 1993:

$$y = 2.1(5) - 0.1 = \$10.4.$$

Thus the total amount of life insurance in force in 1993 will be about $10.4 billion.

The method of least squares is a statistical process illustrated here with only four data points in order to simplify the explanation. Most statistical researchers would warn that many more than four data points should be used to get a "good" regression line. Furthermore, making predictions too far in the future from any linear model may not be valid. It can be done, but the further into the future the prediction is made, the more dubious one should be about the prediction.

DO EXERCISE 2.

*The Regression Line for an Arbitrary Collection of Data
Points (c_1, d_1), (c_2, d_2), . . . , (c_n, d_n)

Look again at the regression line

$$y = 2.1x - 0.1$$

for the data points $(1, 2.2)$, $(2, 4.3)$, $(3, 5.2)$, and $(4, 8.9)$. Let us consider
the arithmetic averages, or means, of the x-coordinates, denoted $\bar{x}$, and
the y-coordinates, denoted $\bar{y}$:

$$\bar{x} = \frac{1 + 2 + 3 + 4}{4} = 2.5,$$

$$\bar{y} = \frac{2.2 + 4.3 + 5.2 + 8.9}{4} = 5.15.$$

It turns out that the point $(\bar{x}, \bar{y})$, or $(2.5, 5.15)$, is on the regression line
since

$$5.15 = 2.1(2.5) - 0.1.$$

Thus the regression line is

$$y - \bar{y} = m(x - \bar{x}), \quad \text{or} \quad y - 5.15 = m(x - 2.5).$$

All that remains, in general, is to determine m.

Suppose we want to find the regression line for an arbitrary number
of points (c_1, d_1), (c_2, d_2), . . . , (c_n, d_n). To do so, we find the values m and
b that minimize the function S given by

$$S(b, m) = (y_1 - d_1)^2 + (y_2 - d_2)^2 + \cdots + (y_n - d_n)^2$$

$$= \sum_{i=1}^{n} (y_i - d_i)^2,$$

where $y_i = mc_i + b$.

Using a procedure like the one we used earlier to minimize S, we can
show that $y = mx + b$ takes the form

$$y - \bar{y} = m(x - \bar{x}),$$

where

$$\bar{x} = \frac{\sum_{i=1}^{n} c_i}{n}, \quad \bar{y} = \frac{\sum_{i=1}^{n} d_i}{n}, \quad \text{and} \quad m = \frac{\sum_{i=1}^{n} (c_i - \bar{x})(d_i - \bar{y})}{\sum_{i=1}^{n} (c_i - \bar{x})^2}.$$

Let us see how this works out for the individual life-insurance example
used previously.

---

*This part is considered optional and can be omitted without loss of continuity.

3. Repeat Margin Exercise 2(a) using the procedure just outlined in the optional part of this section.

c_i	d_i	$c_i - \bar{x}$	$(c_i - \bar{x})^2$	$(d_i - \bar{y})$	$(c_i - \bar{x})(d_i - \bar{y})$
1	2.2	-1.5	2.25	-2.95	4.425
2	4.3	-0.5	0.25	-0.85	0.425
3	5.2	0.5	0.25	0.05	0.025
4	8.9	1.5	2.25	3.75	5.625

$$\sum_{i=1}^{4} c_i = 10 \qquad \sum_{i=1}^{4} d_i = 20.6 \qquad \sum_{i=1}^{4} (c_i - \bar{x})^2 = 5 \qquad \sum_{i=1}^{4} (c_i - \bar{x})(d_i - \bar{y}) = 10.5$$

$$\bar{x} = 2.5 \qquad \bar{y} = 5.15 \qquad\qquad m = \frac{10.5}{5} = 2.1$$

Thus the regression line is

$$y - 5.15 = 2.1(x - 2.5),$$

which simplifies to

$$y = 2.1x - 0.1.$$

DO EXERCISE 3.

*Nonlinear Regression

It can happen that data do not seem to fit a linear equation, but when logarithms of either the x-values or the y-values (or both) are taken, a linear relationship will exist. Indeed, when considering the graph in Fig. 1, it is not unreasonable to expect these data to fit an exponential function.

EXAMPLE 1   Use logarithms and regression to find an exponential function

$$y = Be^{kx}$$

that fits the data. Then estimate the total amount of life insurance in force in 1993.

———————

*This part is considered optional and can be omitted without loss of continuity.

Year, x	1. 1973	2. 1978	3. 1983	4. 1988
Total individual life insurance in force (in billions), y	$2.2	$4.3	$5.2	$8.9

If we take the natural logarithm of both sides of

$$y = Be^{kx},$$

we get

$$\ln y = \ln B + kx, \quad \text{or} \quad kx + \ln B.$$

Note that $\ln B$ and k are constants. Thus, if we replace $\ln y$ by a new variable Y, the equation takes the form of a linear function

$$Y = mx + B,$$

where $m = k$ and $b = \ln B$.

We are going to find this regression line, but before beginning, we need to find the logarithms of the y-values.

x	1	2	3	4
$Y = \ln y$	0.7885	1.4586	1.6487	2.1861

To find the regression line, we use the abbreviated procedure described in the preceding part of this section.

c_i	d_i	$c_i - \bar{x}$	$(c_i - \bar{x})^2$	$(d_i - \bar{Y})$	$(c_i - \bar{x})(d_i - \bar{Y})$
1	0.7885	-1.5	2.25	-0.731975	1.097963
2	1.4586	-0.5	0.25	-0.061875	0.030938
3	1.6487	0.5	0.25	0.128225	0.064113
4	2.1861	1.5	2.25	0.665625	0.998438

$$\sum_{i=1}^{4} c_i = 10 \qquad \sum_{i=1}^{4} d_i \qquad \sum_{i=1}^{4} (c_i - \bar{x})^2 \qquad \sum_{i=1}^{4} (c_i - \bar{x})(d_i - \bar{Y})$$

$$\bar{x} = \frac{10}{4} = 2.5 \qquad = 6.0819 \qquad = 5 \qquad = 2.191452$$

$$\bar{Y} = \frac{6.0819}{4} \qquad\qquad m = \frac{2.191452}{5}$$

$$= 1.520475 \qquad\qquad = 0.4382904$$

took the same course with the same instructor the previous semester (see the following table).

Midterm score (%), x	Final exam score (%), y
70	75
60	62
85	89

4. *Business: Predicting the number of 1-800 telephone listings.*

a)   Use natural logarithms and regression to find an equation

$$y = Be^{kx}$$

Thus the regression line is

$$Y - 1.520475 = 0.4382904(x - 2.5),$$

which simplifies to

$$Y = 0.4382904x + 0.424749.$$

Recall that we were to find k and B. From this equation, we know

a)   Find the regression line $y = mx + b$. (*Hint:* The y-deviations are $70m + b - 75$, $60m + b - 62$, and so on.)

b)   The midterm score of a student was 81. Use the regression line to predict the student's final exam score.

6. *Predicting the world record in the high jump.* On July 29, 1989, Javier Sotomayer of Cuba set an astounding

world record of 8 ft in the high jump. It has been established that most world records in track and field can be modeled by a linear function. The following table shows world records for various years.

Year, x (Use the actual year for x.)	World record in high jump (in inches), y
1912	78.0
1956	84.5
1973	90.5
1989	96.0

a)   Find the regression line $y = mx + b$.

b)   Use the regression line to predict the world record in the high jump in 2000; in 2050.

SYNTHESIS EXERCISES

7. *General interest: Predicting the world record in the one-mile run.*

a)   🖩 Find the regression line $y = mx + b$ that fits the set of data in the following table. (*Hint:* Convert each time to decimal notation; for example, $4{:}24.5 = 4\frac{24.5}{60} = 4.4083$.)

Year, x (Use the actual year for x.)	World record in mile (min:sec), y
1875 (Walter Slade)	4:24.5
1894 (Fred Bacon)	4:18.2
1923 (Paavo Nurmi)	4:10.4
1937 (Sidney Wooderson)	4:06.4
1942 (Gunder Haegg)	4:06.2
1945 (Gunder Haegg)	4:01.4
1954 (Roger Bannister)	3:59.4
1964 (Peter Snell)	3:54.4
1967 (Jim Ryun)	3:51.1
1975 (John Walker)	3:49.4
1979 (Sebastian Coe)	3:49.0
1980 (Steve Ovett)	3:48.8

b)   Use the regression line to predict the world record in the mile in 1996; in 2000.

c)   In August 1985, Steve Cram set a new world record of 3:46.31 for the mile. How does this compare with what can be predicted by the regression?

8. *Life science: Predicting the population of the United States.*

a)   Use logarithms and regression to find an exponential function

$$y = Be^{kx}$$

that fits the following set of data.

Year (from 1986), x	Population of U.S. (in millions), y
0	241
1	243
2	245
3	248

b) Use the regression equation to find the population of the United States in 2000.

9. *Business: Predicting the cost of a ticket to a Broadway musical.*

   a) 🖩 Even though the data in Exercise 1 do not support the assumption that an exponential function can be used as a model, use logarithms and regression to find an exponential function

$$y = Be^{kx}$$

   that fits the set of data in Exercise 1.

   b) Use the regression equation to predict the average ticket price in 1994; in 2000.

   c) Compare your answers with those of Exercise 1.

10. *Social science: Predicting the death rate from number of hours of sleep.*

   a) Use regression (but not logarithms) to find a quadratic function

$$y = ax^2 + bx + c$$

   that fits the following set of data.*

(*Hint:* Find the partial derivatives, $\frac{\partial f}{\partial a}, \frac{\partial f}{\partial b},$ and $\frac{\partial f}{\partial c},$ set each equal to zero, solve the resulting system, and assume there is a minimum.)

Average number of hours of sleep, x	Death rate in one year (per 100,000 males), y
5	1121
6	805
7	626
8	813
9	967

   b) Find the death rate of those who average 4 hr of sleep, 10 hr of sleep, and 7.5 hr of sleep.

---

*The set of data in Exercise 10 comes from a study by Dr. Harold J. Morowitz.

7.7

CONSTRAINED MAXIMUM AND MINIMUM VALUES: LAGRANGE MULTIPLIERS

Before we proceed in detail, let us return to a problem we considered in Chapter 3.

PROBLEM   A hobby store has 20 ft of fencing to fence off a rectangular electric-train area in one corner of its display room. The two sides up against the wall require no fence. What dimensions of the rectangle will maximize the area? (See Fig. 1.)

OBJECTIVES

a) Find a maximum or minimum value of a given function subject to a given constraint, using the method of Lagrange multipliers.

b) Solve applied problems involving Lagrange multipliers.

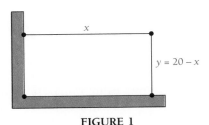

FIGURE 1

We maximize the function

$$A = xy$$

subject to the condition, or *constraint*, $x + y = 20$. Note that A is a function of two variables.

When we solved this earlier, we first solved the constraint for y:

$$y = 20 - x.$$

We then substituted $20 - x$ for y to obtain

$$A(x, y) = x(20 - x) = 20x - x^2,$$

which is a function of one variable. Next, we found a maximum value using Maximum–Minimum Principle 1 (see Section 3.5). By itself, the function of two variables

$$A(x, y) = xy$$

has no maximum value. This can be checked using the D-test. With the constraint $x + y = 20$, however, the function does have a maximum. We see this in Fig. 2.

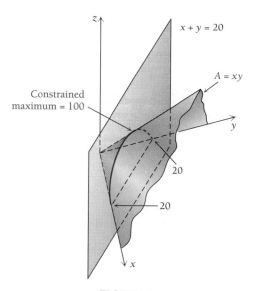

FIGURE 2

It may be quite difficult to solve a constraint for one variable. The procedure outlined below allows us to proceed without doing so.

THEOREM 2

The Method of Lagrange Multipliers

To find a maximum or minimum value of a function $f(x, y)$ subject to the constraint $g(x, y) = 0$:

1. Form a new function:

$$F(x, y, \lambda) = f(x, y) - \lambda g(x, y).$$

2. Find the first partial derivatives F_x, F_y, and F_λ.

3. Solve the system

$$F_x = 0, \qquad F_y = 0, \quad \text{and} \quad F_\lambda = 0.$$

Let (a, b, λ) represent a solution of this system. We still must determine whether (a, b, λ) yields a maximum or minimum of the function f, but we will assume such a maximum or minimum for each of the problems considered here.

The variable λ (lambda) is called a **Lagrange multiplier**. We first illustrate the method of Lagrange multipliers by resolving the hobby-store problem.

EXAMPLE 1   Find the maximum value of

$$A(x, y) = xy$$

subject to the constraint $x + y = 20$.

Solution

1. We form the new function F given by

$$F(x, y, \lambda) = xy - \lambda \cdot (x + y - 20).$$

Note that we first had to express $x + y = 20$ as $x + y - 20 = 0$.

2. We find the first partial derivatives:

$$F_x = y - \lambda,$$
$$F_y = x - \lambda,$$
$$F_\lambda = -(x + y - 20).$$

3. We set these derivatives equal to 0 and solve the resulting system:

$$y - \lambda = 0, \qquad\qquad (1)$$
$$x - \lambda = 0, \qquad\qquad (2)$$
$$-(x + y - 20) = 0, \quad \text{or} \quad x + y - 20 = 0. \qquad (3)$$

1. A rancher has 50 ft of fencing to fence off a rectangular animal pen in the corner of a barn. What dimensions of the rectangle will yield the maximum area?

a) Express the area as a function of two variables with a constraint.

b) Find the maximum value of the function in part (a) using the method of Lagrange multipliers.

From Eqs. (1) and (2), it follows that

$$x = y = \lambda.$$

Substituting λ for x and y in Eq. (3), we get

$$\lambda + \lambda - 20 = 0$$
$$2\lambda = 20$$
$$\lambda = 10.$$

Thus, $x = \lambda = 10$ and $y = \lambda = 10$. The maximum value of A subject to the constraint occurs at $(10, 10)$ and is

$$A(10, 10) = 10 \cdot 10 = 100. \qquad \qquad ❖$$

DO EXERCISE 1.

EXAMPLE 2   Find the maximum value of

$$f(x, y) = 3xy$$

subject to the constraint

$$2x + y = 8.$$

(*Note:* f can be interpreted as a production function with budget constraint $2x + y = 8$.)

Solution

1. We form the new function F given by

$$F(x, y, \lambda) = 3xy - \lambda(2x + y - 8).$$

Note that we first had to express $2x + y = 8$ as $2x + y - 8 = 0$.

2. We find the first partial derivatives:

$$F_x = 3y - 2\lambda,$$
$$F_y = 3x - \lambda,$$
$$F_\lambda = -(2x + y - 8).$$

3. We set these derivatives equal to 0 and solve the resulting system:

$$3y - 2\lambda = 0, \qquad \qquad (4)$$
$$3x - \lambda = 0, \qquad \qquad (5)$$
$$-(2x + y - 8) = 0, \quad \text{or} \quad 2x + y - 8 = 0. \qquad (6)$$

Solving Eq. (4) for y, we get

$$y = \frac{2}{3}\lambda.$$

Solving Eq. (5) for x, we get

$$x = \frac{\lambda}{3}.$$

Substituting $(2/3)\lambda$ for y and $(\lambda/3)$ for x in Eq. (6), we get

$$2\left(\frac{\lambda}{3}\right) + \left(\frac{2}{3}\lambda\right) - 8 = 0$$

$$\frac{4}{3}\lambda = 8$$

$$\lambda = \frac{3}{4} \cdot 8 = 6.$$

Then

$$x = \frac{\lambda}{3} = \frac{6}{3} = 2 \quad \text{and} \quad y = \frac{2}{3}\lambda = \frac{2}{3} \cdot 6 = 4.$$

The maximum value of f subject to the constraint occurs at $(2, 4)$ and is

$$f(2, 4) = 3 \cdot 2 \cdot 4 = 24. \qquad \diamondsuit$$

DO EXERCISE 2.

EXAMPLE 3 *Business: The beverage-can problem (12 oz).* The standard beverage can has a volume of 12 oz, or 21.66 in³. What dimensions yield the minimum surface area? Find the minimum surface area. (See Fig. 3.)

FIGURE 3

2. Find the maximum value of

$$f(x, y) = 5xy$$

subject to the constraint

$$4x + y = 20.$$

Solution We want to minimize the function s given by

$$s(h, r) = 2\pi rh + 2\pi r^2$$

subject to the volume constraint

$$\pi r^2 h = 21.66, \quad \text{or} \quad \pi r^2 h - 21.66 = 0.$$

Note that s does not have a minimum without the constraint.

1. We form the new function S given by

$$S(h, r, \lambda) = 2\pi rh + 2\pi r^2 - \lambda(\pi r^2 h - 21.66).$$

2. We find the first partial derivatives:

$$\frac{\partial S}{\partial h} = 2\pi r - \lambda \pi r^2,$$

$$\frac{\partial S}{\partial r} = 2\pi h + 4\pi r - 2\lambda \pi rh,$$

$$\frac{\partial S}{\partial \lambda} = -(\pi r^2 h - 21.66).$$

3. We set these derivatives equal to 0 and solve the resulting system:

$$2\pi r - \lambda \pi r^2 = 0, \tag{7}$$

$$2\pi h + 4\pi r - 2\lambda \pi rh = 0, \tag{8}$$

$$-(\pi r^2 h - 21.66) = 0, \quad \text{or} \quad \pi r^2 h - 21.66 = 0. \tag{9}$$

Note that we can solve Eq. (7) for r:

$$\pi r(2 - \lambda r) = 0$$

$$\pi r = 0 \quad \text{or} \quad 2 - \lambda r = 0$$

$$r = 0 \quad \text{or} \quad r = \frac{2}{\lambda}.$$

Since $r = 0$ cannot be a solution to the original problem, we continue by substituting $2/\lambda$ for r in Eq. (8):

$$2\pi h + 4\pi \cdot \frac{2}{\lambda} - 2\lambda \pi \cdot \frac{2}{\lambda} \cdot h = 0$$

$$2\pi h + \frac{8\pi}{\lambda} - 4\pi h = 0$$

$$\frac{8\pi}{\lambda} - 2\pi h = 0$$

$$-2\pi h = -\frac{8\pi}{\lambda},$$

so

$$h = \frac{4}{\lambda}.$$

Since $h = 4/\lambda$ and $r = 2/\lambda$, it follows that $h = 2r$. Substituting $2r$ for h in Eq. (9) yields

$$\pi r^2(2r) - 21.66 = 0$$
$$2\pi r^3 - 21.66 = 0$$
$$2\pi r^3 = 21.66$$
$$\pi r^3 = 10.83$$
$$r^3 = \frac{10.83}{\pi}$$
$$r = \sqrt[3]{\frac{10.83}{\pi}} \approx 1.51 \text{ in.} \quad \blacksquare$$

Thus when $r = 1.51$ in., $h = 3.02$ in. The surface area is a minimum and is approximately

$$2\pi(1.51)(3.02) + 2\pi(1.51)^2, \quad \text{or about} \quad 42.98 \text{ in}^2. \qquad \clubsuit$$

DO EXERCISE 3.

The actual dimensions of a standard-sized 12-oz beverage can are $r = 1.25$ in. and $h = 4.875$ in. A natural question after studying Example 3 is, "Why don't beverage companies make cans using the dimensions found in that example?" To do this at this time would mean an enormous cost in retooling. New can-making machines would have to be purchased at a cost of millions. New beverage-filling machines would have to be purchased. Vending machines would no longer be the correct size. A partial response to the desire to save aluminum has been found in recycling and in manufacturing cans with a rippled effect at the top. These cans require less aluminum. As a result of many engineering ideas, the amount of aluminum required to make 1000 cans has been reduced from 36.5 lb to 28.1 lb. The consumer is actually a very important factor in the shape of the can. Market research has shown that a can with the dimensions found in Example 3 is not as comfortable to hold and might not be accepted by consumers.*

_____

*Many thanks to Don Hauser, formerly of the Pepsi-Cola Co., and Bobby Ryals of the Continental Can Co. for the ideas in this paragraph.

3. *Business: The beverage-can problem (16 oz).* Repeat Example 4 for a can of 16 oz, or 28.88 in^3.

Find the maximum value of f subject to the given constraint.

1. $f(x, y) = xy;$ $2x + y = 8$
2. $f(x, y) = 2xy;$ $4x + y = 16$
3. $f(x, y) = 4 - x^2 - y^2;$ $x + 2y = 10$
4. $f(x, y) = 3 - x^2 - y^2;$ $x + 6y = 37$

Find the minimum value of f subject to the given constraint.

5. $f(x, y) = x^2 + y^2;$ $2x + y = 10$
6. $f(x, y) = x^2 + y^2;$ $x + 4y = 17$
7. $f(x, y) = 2y^2 - 6x^2;$ $2x + y = 4$
8. $f(x, y) = 2x^2 + y^2 - xy;$ $x + y = 8$

9. $f(x, y, z) = x^2 + y^2 + z^2;$ $y + 2x - z = 3$
10. $f(x, y, z) = x^2 + y^2 + z^2;$ $x + y + z = 1$

Use the method of Lagrange multipliers to solve each of the following.

11. Of all numbers whose sum is 70, find the two that have the maximum product.
12. Of all numbers whose sum is 50, find the two that have the maximum product.
13. Of all numbers whose difference is 6, find the two that have the minimum product.
14. Of all numbers whose difference is 4, find the two that have the minimum product.

APPLICATIONS

❖ **Business and Economics**

15. *Maximizing typing area.* A standard piece of typing paper has a perimeter of 39 in. Find the dimensions of the paper that will give the most typing area, subject to the perimeter constraint of 39 in. What is its area? Does the standard $8\frac{1}{2}$-in. $\times$ 11-in. paper have maximum area?

16. *Maximizing room area.* A carpenter is building a rectangular room with a fixed perimeter of 80 ft. What are the dimensions of the largest room that can be built? What is its area?

17. *Minimizing surface area.* An oil drum of standard size has a volume of 200 gal, or 27 ft^3. What dimensions yield the minimum surface area? Find the minimum surface area.

18. *Juice-can problem.* A standard-sized juice can has a volume of 99 in^3. What dimensions yield the minimum surface area? Find the minimum surface area.

Do these drums appear to be made in such a way as to minimize surface area?

19. *Maximizing total sales.* The total sales S of a one-product firm are given by

$$S(L, M) = ML - L^2,$$

where M = the cost of materials and L = the cost of labor. Find the maximum value of this function subject to the budget constraint

$$M + L = 80.$$

20. *Maximizing total sales.* The total sales S of a one-product firm are given by

$$S(L, M) = 2ML - L^2,$$

where M = the cost of materials and L = the cost of labor. Find the maximum value of this function subject to the budget constraint

$$M + L = 60.$$

21. *Minimizing construction costs.* A company is planning to construct a warehouse whose cubic footage is to be $252{,}000$ ft^3. Construction costs per square foot are estimated to be as follows:

Walls: $3.00
Floor: $4.00
Ceiling: $3.00

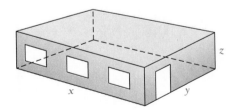

a) The total cost of the building is a function $C(x, y, z)$, where x is the length, y is the width, and z is the height. Find a formula for $C(x, y, z)$.

b) What dimensions of the building will minimize the total cost? What is the minimum cost?

22. *Minimizing the costs of container construction.* A container company is going to construct a shipping container of volume 12 ft^3 with a square bottom and top. The cost of the top and the sides is $2 per square foot and for the bottom is $3 per square foot. What dimensions will minimize the cost of the container?

23. *Minimizing total cost.* A product can be made entirely on either machine A or machine B, or both. The nature of the machines make their cost functions differ:

Machine A: $C(x) = 10 + \dfrac{x^2}{6}$,

Machine B: $C(y) = 200 + \dfrac{y^3}{9}$.

Total cost is given by $C(x, y) = C(x) + C(y)$. How many units should be made on each machine in order to minimize total costs if $x + y = 10{,}100$ units are required?

SYNTHESIS EXERCISES

Find the indicated maximum or minimum values of f subject to the given constraint.

24. Minimum: $f(x, y) = xy$; $x^2 + y^2 = 4$

25. Minimum: $f(x, y) = 2x^2 + y^2 + 2xy + 3x + 2y$; $y^2 = x + 1$

26. Maximum: $f(x, y, z) = x + y + z$; $x^2 + y^2 + z^2 = 1$

27. Maximum: $f(x, y, z) = x^2 y^2 z^2$; $x^2 + y^2 + z^2 = 1$

28. Maximum: $f(x, y, z) = x + 2y - 2z$; $x^2 + y^2 + z^2 = 4$

29. Maximum: $f(x, y, z, t) = x + y + z + t$; $x^2 + y^2 + z^2 + t^2 = 1$

30. Minimum: $f(x, y, z) = x^2 + y^2 + z^2$; $x - 2y + 5z = 1$

31. *Economics: The law of equimarginal productivity.* Suppose $p(x, y)$ represents the production of a two-product firm. We give no formula for p. The company produces x items of the first product at a cost of c_1 each and y items of the second product at a cost of c_2 each. The budget constraint B is a constant given by

$$B = c_1 x + c_2 y.$$

Find the value of λ using the Lagrange-multiplier method in terms of p_x, p_y, c_1, and c_2. The resulting equation is called the *Law of Equimarginal Productivity.*

32. *Business: Maximizing production.* A company has the following Cobb–Douglas production function for a certain product:

$$p(x, y) = 800x^{3/4}y^{1/4},$$

where $x =$ the labor, measured in dollars, and $y =$ the capital, measured in dollars. Suppose a company can make a total investment in labor and capital of $1,000,000. How should it allocate the investment between labor and capital in order to maximize production?

7.8

OBJECTIVE

a) Evaluate the double integral of $f(x, y)$ over a rectangle R whose sides are parallel to the axes by evaluating two appropriately chosen iterated integrals.

DOUBLE INTEGRATION: RECTANGULAR REGIONS

Consider a rectangle R in the plane whose sides are parallel to the x- and y-axes. Such a rectangle is shown in Fig. 1.

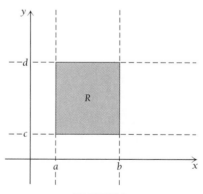

FIGURE 1

Now, let $z = f(x, y)$ be a continuous function* whose domain contains R. For the moment, assume that $f(x, y)$ is a positive function on R and consider the problem of calculating the volume of the solid figure (see Fig. 2) whose base is R and whose "top" is the portion of $z = f(x, y)$

---

*Intuitively speaking, a function of two variables is continuous if the surface that is formed by its graph has no "breaks." Formally, a function f is continuous at (a, b) if (1) $f(a, b)$ exists; (2) $\lim_{(x, y) \to (a, b)} f(x, y)$ exists; and (3) $\lim_{(x, y) \to (a, b)} f(x, y) = f(a, b)$.

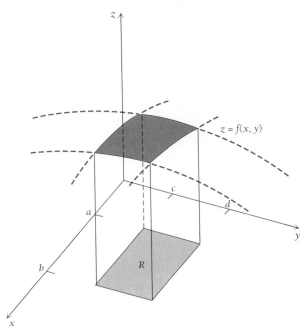

FIGURE 2

that lies directly above R. Since the top is a curved surface, the volume cannot be computed using the standard formula $V = A \cdot h$, where V is the volume, A is the area of the base, and h is the height.

To calculate the volume, we proceed as follows. First, we divide the interval $[a, b]$ along the x-axis into n subintervals, all with the same length Δx. This means that $\Delta x = (b - a)/n$. It also means that the endpoints of the subintervals, $x_1 = a, x_2, x_3, \ldots, x_{n+1} = b$, are given by $x_{k+1} = a + k \cdot \Delta x$ for $k = 0, 1, 2, \ldots, n$.

Next, we divide the interval $[c, d]$ along the y-axis into m subintervals, all having the same length Δy. Then $\Delta y = (d - c)/m$ and $y_{k+1} = c + k \Delta y$ for $k = 0, 1, 2, \ldots, m$.

Note in Fig. 3 that if we draw the lines passing through $x_2, x_3, \ldots, x_n$ that are parallel to the y-axis and the lines through $y_2, y_3, \ldots, y_m$ that are parallel to the x-axis, then these lines will divide R into small rectangles. There are exactly $m \cdot n$ of these smaller rectangles, each having area $\Delta x \cdot \Delta y$. To keep track of the rectangles, we label the rectangle that has x_i and y_j for its "smallest" x- and y-endpoints R_{ij}. So, the amber rectangle whose x-endpoints are x_1 and x_2 and whose y-endpoints are y_1 and y_2 is called R_{11}, since x_1 and y_1 are the smaller x- and y-endpoints. Similarly, the orange rectangle is R_{22}.

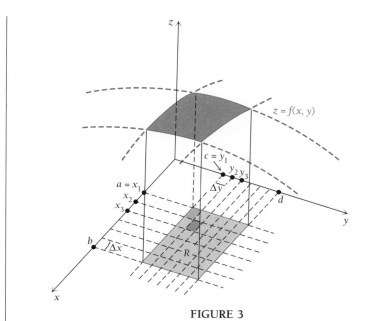

FIGURE 3

Continuing our calculation, look at Fig. 4.

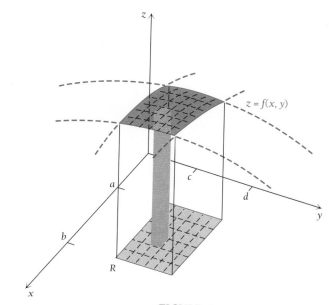

FIGURE 4

The portion of the volume that is taupe has the rectangle R_{43} for its base. The pictured volume looks like a tall, thin box, but it is not a (rectangular) box, since its top is slightly curved. Nevertheless, there is so little variation in its altitude that we can approximate its volume using the volume of the box whose base is R_{43} and whose altitude or height is $f(x_4, y_3)$, the value of $f(x, y)$ at the corner point of R_{43} nearest the origin in the figure. Thus the volume of the colored portion in the figure is approximately $f(x_4, y_3) \cdot \Delta x \cdot \Delta y$ since the area of the base rectangle R_{43} is $\Delta x \cdot \Delta y$. We can obtain a good approximation to the volume V of the original solid by constructing rectangular boxes for each base rectangle R_{ij} and then adding the volumes of these boxes. The volume of the box whose base is R_{ij} will be approximately equal to $f(x_i, y_j) \cdot \Delta x \cdot \Delta y$ for all $i = 1, \ldots, n$ and all $j = 1, \ldots, m$. Adding these numbers gives

$$V \approx \sum_{i=1}^{n} \sum_{j=1}^{m} f(x_i, y_j) \, \Delta x \, \Delta y.$$

We can increase the accuracy of the approximation by increasing m or n (or both). We expect that if we take the limit of these sums as m and n approach infinity, then we would obtain the exact volume. That is true.

DEFINITION

Let $z = f(x, y)$ be a continuous function on the rectangle R, where the sides of R are parallel to the x- and y-axes. Then the *double integral of $f(x, y)$ over R*, denoted

$$\iint_R f(x, y) \, dx \, dy,$$

is the limit

$$\lim_{m, n \to \infty} \sum_{i=1}^{n} \sum_{j=1}^{m} f(x_i, y_j) \, \Delta x \, \Delta y.$$

If $f(x, y)$ is continuous, we can show that the limit above always exists. If $f(x, y) \geq 0$ on R, then we define the volume of the solid whose base is R and whose altitude at each point (x, y) of R is $f(x, y)$ to be the double integral defined above.

Some double integrals can be calculated directly from the definition, but the calculations are tedious. We'd like to have a method of evaluating double integrals that is fairly easy to apply. Recall that although the integral

$$\int_a^b f(x) \, dx$$

1. Evaluate $\int_1^3 xy\, dx$ by factoring out the y-term and evaluating the remaining integral in the usual way.

2. Evaluate $\int_0^1 xy^2\, dx$.

3. Evaluate $\int_0^1 xy^2\, dy$.

4. Evaluate $\int_{-1}^2 (xy + x^2)\, dy$.

5. Evaluate $\int_0^2 (rx^2 + ys)\, ds$.

can be defined as a limit of sums (see Section 5.8), it is usually evaluated using antiderivatives. There is a corresponding method for handling double integrals. The next few examples lay the groundwork for this method.

EXAMPLE 1 Evaluate $\int_1^3 xy\, dx$.

Solution First, we consider the key to the evaluation. The dx in the integral tells us that we are to carry out the integration with respect to x. This means that any other independent variables (like y above) that occur in the integral are to be treated as constants (much as you do when you calculate partial derivatives). Here, if we think of y as being a fixed number, then the antiderivative of xy is $y \cdot (x^2/2)$. Proceeding as usual, we get

$$\int_1^3 xy\, dx = \left[\frac{y \cdot x^2}{2} \right]_1^3 = \frac{y \cdot (3)^2}{2} - \frac{y \cdot (1)^2}{2}$$

$$= \frac{y \cdot 9}{2} - \frac{y \cdot 1}{2} = \frac{8y}{2} = 4y.$$

Note that the answer is a function of y alone. ❖

DO EXERCISES 1 AND 2.

EXAMPLE 2 Evaluate $\int_0^1 (xy + t)\, dy$.

Solution Here, the dy tells us that we are to integrate with respect to y, so that x and t are to be regarded as constants. In other words, we integrate from $y = 0$ to $y = 1$. Just as the antiderivative of $3y + 5$ is $3 \cdot (y^2/2) + 5y$, the antiderivative of $xy + t$ is $x \cdot (y^2/2) + ty$. Using the limits of integration, we get

$$\left[x \cdot \frac{y^2}{2} + ty \right]_0^1 = \left(x \cdot \frac{(1)^2}{2} + t \cdot 1 \right) - \left(x \cdot \frac{0^2}{2} + t \cdot 0 \right)$$

$$= \left(\frac{x}{2} + t \right) - (0 + 0)$$

$$= \frac{x}{2} + t.$$

Note again that there is no y in the answer. The expression is a function of x and t only. ❖

DO EXERCISES 3–5.

EXAMPLE 3   Evaluate $\int_0^2 \left\{ \int_1^3 xy\, dx \right\} dy.$

Solution   Note that there are two integral signs here. This is our first example of a *repeated*, or *iterated, integral*. In order to evaluate it, we work from the inside out. We first attack the integral in brackets,

$$\int_1^3 xy\, dx,$$

which we recognize as the integral from Example 1 above. Using the answer from that example, we get

$$\int_0^2 \left\{ \int_1^3 xy\, dx \right\} dy = \int_0^2 4y\, dy.$$

Since the antiderivative of $4y$ (we are integrating with respect to y here) is $4 \cdot (y^2/2) = 2y^2$, we get

$$\int_0^2 \left\{ \int_1^3 xy\, dx \right\} dy = \int_0^2 4y\, dy$$
$$= [2y^2]_0^2 = 2(2)^2 - 2(0)^2$$
$$= 2 \cdot 4 - 2 \cdot 0 = 8. \qquad ❖$$

DO EXERCISES 6 AND 7.

EXAMPLE 4   Evaluate

$$\int_1^3 \left\{ \int_0^2 xy\, dy \right\} dx.$$

Solution   As before, we work from the inside out. So, we look first at

$$\int_0^2 xy\, dy.$$

The dy tells us to integrate with respect to y. The antiderivative will then be $x \cdot (y^2/2)$. So,

$$\int_0^2 xy\, dy = \left[x \cdot \frac{y^2}{2} \right]_0^2 = x \cdot \frac{(2)^2}{2} - x \cdot \frac{(0)^2}{2} = 2x.$$

Then,

$$\int_1^3 \left\{ \int_0^2 xy\, dy \right\} dx = \int_1^3 2x\, dx$$
$$= [x^2]_1^3 = (3)^2 - (1)^2$$
$$= 8.$$

6. Evaluate

$$\int_0^1 \left\{ \int_0^1 xy^2\, dx \right\} dy.$$

(See Margin Exercise 2.)

7. Evaluate

$$\int_0^5 \left\{ \int_0^1 xy^2\, dy \right\} dx.$$

8. Evaluate

$$\int_0^1 \left\{ \int_0^1 xy^2\, dy \right\} dx$$

and compare your answer with that of Margin Exercise 6.

That is,

$$\int_1^3 \left\{ \int_0^2 xy\, dy \right\} dx = 8.$$

Note now that the integrals in Examples 3 and 4 are the same except that the order of integration has been reversed.   ❖

DO EXERCISES 8 AND 9.

The relationship between the double integral

$$\iint_R f(x,\, y)\, dx\, dy$$

and the *iterated* (or repeated) *integral*, such as

$$\int_a^b \left\{ \int_c^d f(x,\, y)\, dy \right\} dx,$$

is given in the following theorem.

9. Evaluate

$$\int_0^1 \left\{ \int_0^5 xy^2\, dx \right\} dy$$

and compare your answer with that of Margin Exercise 7.

THEOREM 3

Let $f(x,\, y)$ be continuous on the rectangle R (R as in the definition of double integral above). Then

$$\iint_R f(x,\, y)\, dx\, dy = \int_c^d \left[\int_a^b f(x,\, y)\, dx \right] dy$$

$$= \int_a^b \left[\int_c^d f(x,\, y)\, dy \right] dx.$$

That is, the double integral of $f(x,\, y)$ over R can be evaluated as an *iterated integral* (in two different ways).

EXAMPLE 5   Evaluate the double integral of $f(x,\, y) = x^2 y$ over the rectangle shown in Fig. 5.

Solution   We apply the theorem ($x^2 y$ is certainly continuous) to obtain

$$I = \iint_R x^2 y\, dx\, dy$$

$$= \int_1^3 \left\{ \int_2^4 x^2 y\, dx \right\} dy.$$

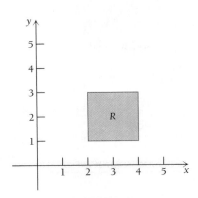

FIGURE 5

10. a)   Evaluate $\displaystyle\int_1^3 x^2 y \, dy$.

  b)   Evaluate the double integral I from Example 5 as

$$\int_2^4 \left\{ \int_1^3 x^2 y \, dy \right\} dx.$$

We evaluate the inside integral first, getting

$$I = \int_1^3 \left\{ \left[\frac{x^3}{3} \cdot y \right]_2^4 \right\} dy.$$

As before, we treat y as a constant in finding the antiderivative above, since we are integrating with respect to x. So,

$$I = \int_1^3 \left\{ \frac{4^3}{3} y - \frac{2^3}{3} y \right\} dy$$

$$= \int_1^3 \left\{ \frac{64}{3} y - \frac{8}{3} y \right\} dy$$

$$= \int_1^3 \frac{56}{3} y \, dy = \frac{56}{3} \int_1^3 y \, dy$$

and

$$I = \frac{56}{3} \left[\frac{y^2}{2} \right]_1^3 = \frac{56}{3} \left[\frac{3^2}{2} - \frac{1^2}{2} \right]$$

$$= \frac{56}{3} \left[\frac{9}{2} - \frac{1}{2} \right] = \frac{56}{3} \cdot 4 = \frac{224}{3}.$$ ❖

11.   Evaluate $\iint_R 3xy \, dx \, dy$, where R is the rectangular region in Example 5.

DO EXERCISES 10 AND 11.

EXAMPLE 6   Evaluate

$$I = \iint_R \frac{x}{y} \, dx \, dy$$

(over the rectangle $R = \{(x, y) | 0 \leqslant x \leqslant 1, 1 \leqslant y \leqslant 2\}$ shown in Fig. 6) in

12. Evaluate $I = \iint_R ye^x \, dx \, dy$ in two ways, where R is the rectangle in Example 6.

two ways. This notation means: R is the set of all points (x, y) such that $0 \leqslant x \leqslant 1$ and $1 \leqslant y \leqslant 2$.

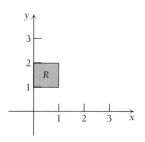

FIGURE 6

Solution If we integrate first with respect to x and then with respect to y, we get

$$I = \int_1^2 \left\{ \int_0^1 \frac{x}{y} \, dx \right\} dy = \int_1^2 \frac{1}{y} \left\{ \int_0^1 x \, dx \right\} dy$$

$$= \int_1^2 \frac{1}{y} \cdot \left\{ \left[\frac{x^2}{2} \right] \right\}_0^1 dy = \int_1^2 \frac{1}{y} \left\{ \frac{1^2}{2} - \frac{0^2}{2} \right\} dy$$

$$= \int_1^2 \frac{1}{y} \cdot \frac{1}{2} \, dy = \left[\frac{1}{2} \ln |y| \right]_1^2$$

$$= \frac{1}{2} \ln |2| - \frac{1}{2} \ln |1|$$

$$= \frac{1}{2} \ln 2 - 0 = \frac{1}{2} \ln 2.$$

13. Evaluate $I = \iint_R xy^3 \, dx \, dy$ in two ways, where R is the rectangle in Example 6.

If we integrate first with respect to y and then with respect to x, we get

$$I = \int_0^1 \left\{ \int_1^2 \frac{x}{y} \, dy \right\} dx = \int_0^1 x \left\{ \int_1^2 \frac{1}{y} \, dy \right\} dx$$

$$= \int_0^1 x \{ [\ln |y|]_1^2 \} \, dx = \int_0^1 x \{ \ln 2 \} \, dx$$

$$= \ln 2 \cdot \left[\frac{x^2}{2} \right]_0^1$$

$$= \frac{1}{2} \ln 2.$$

❖

DO EXERCISES 12 AND 13.

EXAMPLE 7   Find the volume of the solid (shown in Fig. 7) whose base is the rectangle $R = \{(x, y) \mid 0 \leq x \leq 1,\ 0 \leq y \leq 1\}$ and whose height at (x, y) is given by $f(x, y) = x + y$.

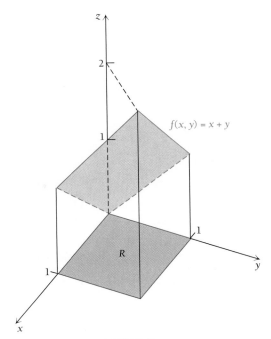

FIGURE 7

Solution   According to the definition, the volume V is given by the double integral

$$V = \iint_R (x + y)\ dx\ dy.$$

Writing the double integral as two iterated integrals, we get

$$V = \int_0^1 \left\{ \int_0^1 (x + y)\ dx \right\} dy = \int_0^1 \left\{ \left[\frac{x^2}{2} + xy \right]_0^1 \right\} dy$$

$$= \int_0^1 \left\{ \left(\frac{1^2}{2} + 1 \cdot y \right) - \left(\frac{(0)^2}{2} + 0 \cdot y \right) \right\} dy = \int_0^1 \left(\frac{1}{2} + y \right) dy$$

$$= \left[\frac{1}{2} y + \frac{y^2}{2} \right]_0^1 = \left(\frac{1}{2} \cdot 1 + \frac{1^2}{2} \right) - \left(\frac{1}{2} \cdot 0 - \frac{(0)^2}{2} \right)$$

$$= \left(\frac{1}{2} + \frac{1}{2} \right) - (0 - 0) = 1.$$

14. Find the volume of the solid whose base is the rectangle R in Example 7 and whose height at (x, y) is $f(x, y) = x^2 + y$.

Geometrically, we can see that this is correct by noting that our solid comprises one half of the rectangular box whose base is R and whose altitude is 2. This box has volume $1 \cdot 1 \cdot 2 = 2$. Thus, $1 = \frac{1}{2} \cdot 2$ is correct.

❖

DO EXERCISE 14.

EXERCISE SET 7.8

Evaluate each of the following integrals.

1. $\int_0^1 x^2 y \, dx$

2. $\int_{-1}^1 (x^2 + xy + 1) \, dy$

3. $\int_{-1}^1 (x^2 + xy + 1) \, dx$

4. $\int_0^3 (ax^2 + bx + y) \, dx$

Evaluate each of the following iterated integrals.

5. $\int_0^1 \left\{ \int_0^2 x^2 y^3 \, dy \right\} dx$

6. $\int_{-1}^0 \left\{ \int_0^1 (3y^2 - 1 + 2x) \, dx \right\} dy$

7. $\int_0^1 \left\{ \int_0^{\ln 2} 2xe^y \, dy \right\} dx$

8. $\int_0^1 \left\{ \int_0^1 (2axy + x^2) \, dy \right\} dx$

Evaluate the double integral $\iint_R f(x, y) \, dx \, dy$ over the indicated rectangle by reducing to an appropriate iterated integral.

9. $f(x, y) = x + y$, $R = \{(x, y)|0 \leqslant x \leqslant 2, 1 \leqslant y \leqslant 3\}$

10. $f(x, y) = x$, $R = \{(x, y)| -1 \leqslant x \leqslant 1, 0 \leqslant y \leqslant 1\}$

11. $f(x, y) = \dfrac{x^2}{y}$, $R = \{(x, y)|0 \leqslant x \leqslant 1, 2 \leqslant y \leqslant 4\}$

12. $f(x, y) = 1$, $R = \{(x, y)|a \leqslant x \leqslant b, c \leqslant y \leqslant d\}$

13. Find the volume of the solid whose base is the rectan-

gle

$$R = \{(x, y)|0 \leqslant x \leqslant 1, 0 \leqslant y \leqslant 1\}$$

and whose height at (x, y) is $f(x, y) = 3x^2 + 3y^2$.

14. Find the volume of the solid whose base is the rectangle

$$R = \{(x, y)|0 \leqslant x \leqslant 1, 0 \leqslant y \leqslant 1\}$$

and whose height at (x, y) is $f(x, y) = e^x$.

15. Find the volume of the solid whose base is the rectangle

$$\{(x, y)|1 \leqslant x \leqslant 2, 0 \leqslant y \leqslant 4\}$$

and whose height at (x, y) is $f(x, y) = \dfrac{y}{x}$.

DEFINITION

If $f(x, y)$ is a continuous function on the rectangle R, then f_{av}, *the average value of* $f(x, y)$ *on* R, is

$$\frac{1}{\text{area } R} \iint_R f(x, y) \, dx \, dy.$$

Calculate f_{av} for each of the following choices of $f(x, y)$ and R.

16. $f(x, y) = x^2 + y^2$, $R = \{(x, y)|0 \leqslant x \leqslant 1, 0 \leqslant y \leqslant 1\}$

17. $f(x, y) = xy$, $R = \{(x, y)|0 \leqslant x \leqslant 2, 0 \leqslant y \leqslant 2\}$

18. $f(x, y) = xy$, $R = \{(x, y)| -1 \leqslant x \leqslant 1, 0 \leqslant y \leqslant 1\}$

19. Let $R = \{(x, y)|a \le x \le b, \ c \le y \le d\}$, and let $g(x)$ be continuous on $[a, b]$ and $h(y)$ be continuous on $[c, d]$. Let $f(x, y) = g(x)h(y)$ on R. Show that

$$f_{av} = g_{av}h_{av},$$

where f_{av} is the average value of $f(x, y)$ on R and g_{av} and h_{av} are the average values of $g(x)$ on $[a, b]$ and $h(y)$ on $[c, d]$, respectively.

7.9

DOUBLE INTEGRATION: MORE GENERAL REGIONS

In Section 7.8, we saw how to define and evaluate the double integral of $f(x, y)$ over R, where R is a rectangle in the plane whose sides are parallel to the x- and y-axes. It is useful, however, to be able to define and evaluate the double integral of $f(x, y)$ over more general regions. To avoid technical problems, we will restrict ourselves for the time being to integration over regions G that look like those shown in Fig. 1. In these cases, the region G lies between the graphs of $y = g(x)$, $y = h(x)$, $x = a$, and $x = b$, where $g(x)$ and $h(x)$ are smooth functions with $g(x) < h(x)$ for $a < x < b$.

OBJECTIVE

a) Evaluate the double integral $f(x, y)$ over a suitable region G by evaluating two appropriately chosen iterated integrals.

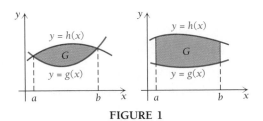

FIGURE 1

For such a region G, we can cover G with rectangles as shown in Fig. 2, taking only those rectangles that are entirely contained in G, form the sum $\sum_{i=1}^{n} \sum_{j=1}^{m} f(x_i, y_j) \, \Delta x \, \Delta y$ as before, and take the limit as m and n both tend to infinity. If $f(x, y)$ is continuous on G and if both $g(x)$ and $h(x)$ have a continuous derivative on $[a, b]$ and are as in Fig. 1 above, then this limit always exists. We denote it by

$$\int\int_G f(x, y) \, dx \, dy.$$

1. Evaluate $\displaystyle\int_x^1 3y^2 \, dy$.

We will soon see that we can evaluate the double integral by an appropriate iterated integral.

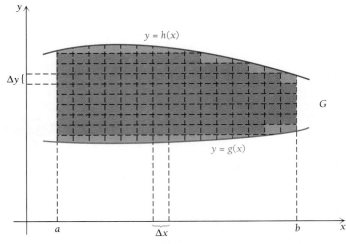

FIGURE 2

2. Evaluate $\displaystyle\int_{-x}^x (2y + 4) \, dy$.

EXAMPLE 1 Evaluate

$$\int_{x^2}^x xy \, dy.$$

Solution We proceed just as we did in Section 7.8. The dy indicates that we are to integrate with respect to y. The antiderivative of $f(x, y) = xy$ (with x regarded as a constant) is $x \cdot (y^2/2)$. So,

$$\int_{x^2}^x xy \, dy = \left[x \cdot \frac{y^2}{2} \right]_{x^2}^x$$

$$= x \cdot \frac{(x)^2}{2} - \frac{x \cdot (x^2)^2}{2}$$

$$= \frac{x^3}{2} - \frac{x^5}{2}$$

$$= \frac{x^3 - x^5}{2}.$$

Note that we use the upper and lower limits of integration (to evaluate the integral) just as we would if they were constants. ❖

DO EXERCISES 1 AND 2.

EXAMPLE 2 Evaluate the iterated integral

$$\int_0^1 \left\{ \int_{x^2}^x xy \, dy \right\} dx.$$

Solution We proceed as usual, evaluating the inside integral first. Using the result of Example 1, we get

$$\int_0^1 \left\{ \int_{x^2}^x xy \, dy \right\} dx = \int_0^1 \frac{x^3 - x^5}{2} \, dx$$

$$= \int_0^1 \left(\frac{1}{2} x^3 - \frac{1}{2} x^5 \right) dx$$

$$= \left[\frac{1}{8} x^4 - \frac{1}{12} x^6 \right]_0^1$$

$$= \left[\frac{1}{8} (1)^4 - \frac{1}{12} (1)^6 \right] - \left[\frac{1}{8} (0)^4 - \frac{1}{12} (0)^6 \right]$$

$$= \left(\frac{1}{8} - \frac{1}{12} \right) - (0) = \frac{3}{24} - \frac{2}{24}$$

$$= \frac{1}{24}. \qquad \diamond$$

DO EXERCISES 3 AND 4.

To see why we can evaluate the double integral by iterated integrals, we consider the volume of the solid based on *G* and capped above by the piece of the surface $z = f(x, y)$ lying over *G*, where *f* is a positive (continuous) function of two variables, as in Fig. 3.

3. Evaluate the iterated integral

$$\int_0^1 \left\{ \int_x^1 3y^2 \, dy \right\} dx.$$

4. Evaluate the iterated integral

$$\int_0^1 \left\{ \int_{-x}^x (2y + 4) \, dy \right\} dx.$$

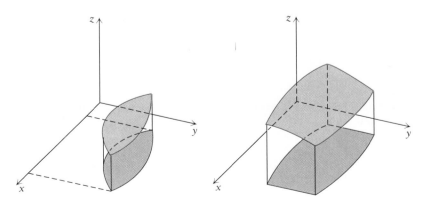

FIGURE 3

The cross section of this solid in the plane $x = x_0$ is the plane region under the graph of $z = f(x_0, y)$ from $y_1 = g(x_0)$ to $y_2 = h(x_0)$ (see Fig. 4). Its area is

$$A(x_0) = \int_{y_1}^{y_2} f(x_0, y)\ dy = \int_{g(x_0)}^{h(x_0)} f(x_0, y)\ dy.$$

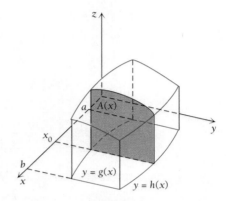

FIGURE 4

As we vary the slicing plane, the cross section also changes. It follows that the volume in question is given by

$$V = \int_a^b A(x)\ dx = \int_a^b \left\{ \int_{g(x)}^{h(x)} f(x, y)\ dy \right\} dx.$$

In the following theorem, we give the relationship between the double integral $\iint_G f(x, y)\ dx\ dy$ and an iterated integral.

THEOREM 4

Let $g(x)$ and $h(x)$ have continuous derivatives and satisfy $g(x) < h(x)$ for $a \leqslant x \leqslant b$. Let G be either of the regions shown in Fig. 1, so that

$$G = \{(x, y) | a \leqslant x \leqslant b \quad \text{and} \quad g(x) \leqslant y \leqslant h(x)\}.$$

Let $f(x, y)$ be continuous at every point of G. Then the double integral

$$\iint_G f(x, y)\ dx\ dy$$

exists and is equal to the iterated integral

$$\int_a^b \left\{ \int_{g(x)}^{h(x)} f(x, y)\ dy \right\} dx.$$

Note that the iterated integral is taken from the lower boundary of G, $y = g(x)$, to the upper boundary of G, $y = h(x)$. In particular, the first (inside) integration takes place with respect to y. The second (outside) integration is taken from the leftmost value of x in G, $x = a$, to the rightmost value of x in G, $x = b$. The second integration, then, is taken with respect to the variable x.

EXAMPLE 3   Evaluate $\iint_G f(x, y) \, dx \, dy$, where

$$G = \{(x, y) | 0 \leqslant x \leqslant 1 \text{ and } x^2 \leqslant y \leqslant x\} \quad \text{and} \quad f(x, y) = xy.$$

Solution   We apply Theorem 4 to reduce the double integral to an iterated integral. The region G is shown in Fig. 5. As in the last section, the notation in the statement of the example is read: G is the set of all points (x, y) such that $0 \leqslant x \leqslant 1$ and $x^2 \leqslant y \leqslant x$. The first, or inside, integral will extend from the lower boundary $y = x^2$ to the upper boundary $y = x$. So it will be

$$\int_{x^2}^{x} f(x, y) \, dy = \int_{x^2}^{x} xy \, dy.$$

The second, or outside, integral will be taken from $x = 0$ to $x = 1$. So

$$\iint_G xy \, dx \, dy = \int_0^1 \left\{ \int_{x^2}^{x} xy \, dy \right\} dx.$$

For the record, note that the order of integration is important in this case.

To complete the problem, note that the iterated integral above is the integral from Example 2. Using the answer from that example, we get

$$\iint_G xy \, dy \, dx = \frac{1}{24}.$$

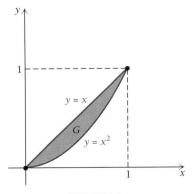

FIGURE 5

582   FUNCTIONS OF SEVERAL VARIABLES

5. a) Sketch the region
$G = \{(x, y)|0 \leqslant x \leqslant 1$ and $x \leqslant y \leqslant 1\}$.

b) Then evaluate

$$\int\int_G 3y^2\, dx\, dy.$$

6. a) Sketch the region
$G = \{(x, y)|0 \leqslant x \leqslant 1$ and $-x \leqslant y \leqslant x\}$.

b) Evaluate

$$\int\int_G (2y + 4)\, dx\, dy.$$

7. a) Sketch the region
$G = \{(x, y)|0 \leqslant x \leqslant 2$ and $-x \leqslant y \leqslant 1\}$.

b) Evaluate

$$\int\int_G 6x\, dx\, dy.$$

DO EXERCISES 5–7.

EXAMPLE 4 Find the volume of the solid whose base is the triangular region $G = \{(x, y)|0 \leqslant x \leqslant 1, 0 \leqslant y \leqslant x\}$ and whose height at each point of G is $z = f(x, y) = x - y$.

Solution   The solid is shown in Fig. 6. As in Section 7.8, we define the volume V of the solid to be the double integral of $z = f(x, y) = x - y$ over G. Note that $x - y \geqslant 0$ on G. So

$$V = \int\int_G f(x, y)\, dx\, dy.$$

We can evaluate the double integral by reducing it to an iterated integral. The first, or inside, integral will extend from the x-axis ($y = 0$) to the upper boundary $y = x$. The first integral will be

$$\int_0^x (x - y)\, dy.$$

The outer integral is taken from $x = 0$ to $x = 1$. So

$$V = \int_0^1 \left\{\int_0^x (x - y)\, dy\right\} dx.$$

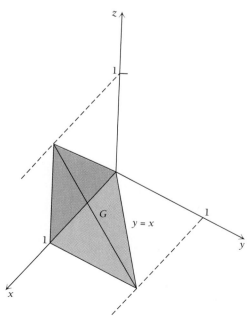

FIGURE 6

Evaluating, we get

$$V = \int_0^1 \left\{ \left[xy - \frac{y^2}{2} \right]_0^x \right\} dx$$

$$= \int_0^1 \left\{ \left(x \cdot x - \frac{(x)^2}{2} \right) - \left(x \cdot 0 - \frac{(0)^2}{2} \right) \right\} dx$$

$$= \int_0^1 \left\{ \left(x^2 - \frac{x^2}{2} \right) - 0 \right\} dx = \int_0^1 \frac{1}{2} x^2 \, dx$$

$$= \left[\frac{1}{2} \cdot \frac{x^3}{3} \right]_0^1 = \left(\frac{1}{2} \cdot \frac{1^3}{3} \right) - \left(\frac{1}{2} \cdot \frac{0^3}{3} \right) = \frac{1}{2} \cdot \frac{1}{3} - \frac{1}{2} \cdot 0 = \frac{1}{6}. \quad \diamondsuit$$

DO EXERCISES 8 AND 9.

DEFINITION

Let $f(x, y)$ be continuous on G, where G satisfies the condition of the previous theorem. Then *the average value of $f(x, y)$ on G,* denoted f_{av}, is defined by

$$f_{av} = \frac{1}{\text{area } (G)} \int \int_G f(x, y) \, dx \, dy.$$

EXAMPLE 5 Calculate the average value of $f(x, y) = x - y$ over the region $G = \{(x, y) \mid 0 \leqslant x \leqslant 1, \, 0 \leqslant y \leqslant x\}$.

Solution The region G is triangular. The triangle has base and height 1 so that its area is $\frac{1}{2}$. So

$$f_{av} = \frac{1}{\text{area } (G)} \int \int_G f(x, y) \, dx \, dy$$

$$= \frac{1}{\frac{1}{2}} \int \int_G (x - y) \, dx \, dy.$$

The double integral that appears here was evaluated in Example 4. Using that result, we have

$$f_{av} = 2 \cdot \frac{1}{6} = \frac{1}{3}.$$

Note that in most cases, we need to use calculus to compute the area of G. $\diamondsuit$

DO EXERCISES 10 AND 11.

8. Find the volume of the solid whose base is the triangular region
$$G = \{(x, y) \mid 0 \leqslant x \leqslant 1, \, x \leqslant y \leqslant 1\}$$
and whose height at (x, y) is $z = f(x, y) = 3y^2$.

9. Find the volume of the solid whose base is the region
$$G = \{(x, y) \mid 0 \leqslant x \leqslant 1, \, 0 \leqslant y \leqslant e^x\}$$
and whose height is given by $z = 2y$.

10. Calculate the average value of $f(x, y) = 3y^2$ over
$$G = \{(x, y) \mid 0 \leqslant x \leqslant 1, \, x \leqslant y \leqslant 1\}.$$

11. Calculate the average value of $f(x, y) = x$ over
$$G = \{(x, y) \mid 0 \leqslant x \leqslant 1, \, x^2 \leqslant y \leqslant 1\}.$$

12. Calculate $\iint_G 2y\, dx\, dy$, where G is the portion of the plane above the parabola $y = x^2 + x$ and below the parabola $y = 1 - x^2$.

EXAMPLE 6   Evaluate $\iint_G x\, dx\, dy$, where G is the portion of the plane above $y = x^2$ and below $y = x + 2$.

Solution   This problem differs from previous ones only in that the x-limits of integration are not given. The region G is shown in Fig. 7.

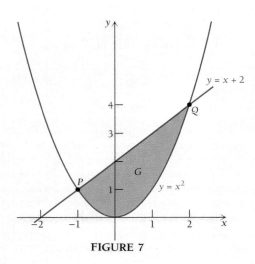

FIGURE 7

We must find the points P and Q of intersection of the two curves. Setting $x^2 = x + 2$ and solving gives

$$x^2 - x - 2 = 0 \quad \text{or} \quad (x - 2)(x + 1) = 0.$$

We see that P is $(-1, 1)$ and Q is $(2, 4)$. So we obtain

$$\iint_G x\, dx\, dy = \int_{-1}^{2} \left\{ \int_{x^2}^{x+2} x\, dy \right\} dx = \int_{-1}^{2} \left\{ [xy]_{x^2}^{x+2} \right\} dx$$

$$= \int_{-1}^{2} (x^2 + 2x - x^3)\, dx = \left[\frac{x^3}{3} + x^2 - \frac{x^4}{4} \right]_{-1}^{2}$$

$$= \left(\frac{8}{3} + 4 - \frac{16}{4} \right) - \left(\frac{-1}{3} + 1 - \frac{1}{4} \right)$$

$$= \frac{9}{3} + 3 - \frac{15}{4} = \frac{9}{4}. \qquad \qquad ❖$$

DO EXERCISE 12.

In the iterated integrals that have appeared so far in this section, the first, or inside, integral has always been taken with respect to y. If the region G looks like those in Fig. 8, however, it is more convenient to integrate first with respect to x.

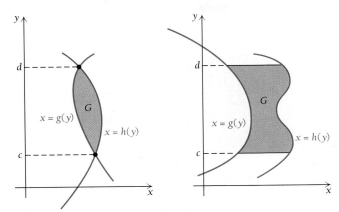

FIGURE 8

In the case shown, we integrate from the left-hand boundary $x = g(y)$ to the right-hand boundary $x = h(y)$. So

$$\int_{g(y)}^{h(y)} f(x, y) \, dx$$

is the inside, or first, integral. Then we integrate with respect to y, integrating from the smallest y-value in G, $y = c$, to the largest, $y = d$. The result is the equality

$$\iint_G f(x, y) \, dx \, dy = \int_c^d \left\{ \int_{g(y)}^{h(y)} f(x, y) \, dx \right\} dy.$$

EXAMPLE 7 Evaluate $\iint_G 6xy \, dx \, dy$, where G is the region shown between $x = y^2 - 2$ and $y = x$.

Solution Solving as in Example 6, we find that the region G looks like Fig. 9.

In this case, we integrate from the left-hand boundary, $x = y^2 - 2$, to the right-hand boundary, $x = y$, getting

$$\int_{y^2 - 2}^{y} 6xy \, dx.$$

Then we integrate from the smallest y-value in G, -1, to the largest, 2, obtaining

$$\iint_G 6xy \, dx \, dy = \int_{-1}^{2} \left\{ \int_{y^2-2}^{y} 6xy \, dx \right\} dy.$$

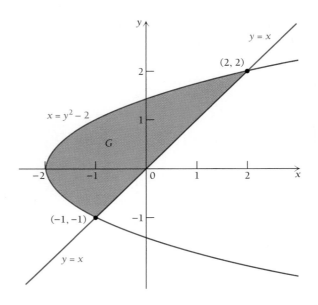

FIGURE 9

Evaluating, we get

$$\int_{-1}^{2} \{[3x^2y]_{y^2-2}^{y}\} \, dy = \int_{-1}^{2} \{3(y)^2 y - 3(y^2 - 2)^2 y\} \, dy$$

$$= \int_{-1}^{2} \{3y^3 - 3(y^4 - 4y^2 + 4)y\} \, dy$$

$$= \int_{-1}^{2} \{3y^3 - 3y^5 + 12y^3 - 12y\} \, dy$$

$$= \int_{-1}^{2} \{-3y^5 + 15y^3 - 12y\} \, dy$$

$$= \left[-\frac{1}{2} y^6 + \frac{15}{4} y^4 - 6y^2 \right]_{-1}^{2}$$

$$= \left\{ -\frac{1}{2} (2)^6 + \frac{15}{4} (2)^4 - 6(2)^2 \right\}$$

$$\quad - \left\{ -\frac{1}{2} (-1)^6 + \frac{15}{4} (-1)^4 - 6(-1)^2 \right\}$$

$$= \{-32 + 60 - 24\} - \left\{ -\frac{1}{2} + \frac{15}{4} - 6 \right\}$$

$$= 4 + \frac{1}{2} - \frac{15}{4} + 6 = 10 + \frac{2}{4} - \frac{15}{4} = \frac{27}{4}. \quad ❖$$

DO EXERCISE 13.

In Example 7 and Margin Exercise 13, it is much more convenient to integrate first with respect to x. In some cases, we can apply both methods. In these cases, we obtain the same answer, as the following example shows.

EXAMPLE 8 Rework Example 3, taking the inside integral with respect to x.

Solution We integrate from the left-hand boundary, $x = y$, to the right-hand boundary, $x = \sqrt{y}$, getting

$$\int_y^{\sqrt{y}} xy \, dx.$$

Integrating next from $y = 0$ to $y = 1$, we find that

$$\iint_G xy \, dx \, dy = \int_0^1 \left\{ \int_y^{\sqrt{y}} xy \, dx \right\} dy = \int_0^1 \left\{ \left[\frac{1}{2} x^2 y \right]_y^{\sqrt{y}} \right\} dy$$

$$= \int_0^1 \left\{ \frac{1}{2} (\sqrt{y})^2 y - \frac{1}{2} (y)^2 y \right\} dy$$

$$= \int_0^1 \left\{ \frac{1}{2} y^2 - \frac{1}{2} y^3 \right\} dy$$

$$= \left[\frac{1}{6} y^3 - \frac{1}{8} y^4 \right]_0^1$$

$$= \frac{1}{6} - \frac{1}{8} = \frac{1}{24}.$$

DO EXERCISES 14 AND 15.

Application to Probability

Suppose we throw a dart at a region G in a plane. It lands on a point (x, y). We can think of (x, y) as a continuous random variable that assumes all values in some region G. A function f is said to be a *joint probability density* if

$$f(x, y) \geq 0 \quad \text{for all } (x, y) \text{ in } G$$

and

$$\iint_G f(x, y) \, dx \, dy = 1.$$

13. Evaluate $\iint_G 2y \, dx \, dy$, where G is the region in the first quadrant to the left of $x = 2y - y^2$.

14. Rework Example 4, integrating first with respect to x.

15. Rework Example 6 by first breaking up the given region into two smaller regions, integrating first with respect to x in both cases.

Now if G_0 is a subregion of G and is the sort of region we discussed earlier, then the probability that the dart "lands in G_0" is given by

$$\iint_{G_0} f(x, y) \, dx \, dy.$$

For appropriate regions, we can evaluate this double integral as an iterated integral.

EXERCISE SET 7.9

Evaluate each of the following.

1. $\displaystyle\int_0^1 \int_0^1 2y \, dx \, dy$ 2. $\displaystyle\int_0^1 \int_0^1 2x \, dx \, dy$

3. $\displaystyle\int_{-1}^1 \int_x^1 xy \, dy \, dx$ 4. $\displaystyle\int_{-1}^1 \int_x^2 (x + y) \, dy \, dx$

5. $\displaystyle\int_0^1 \int_{-1}^3 (x + y) \, dy \, dx$ 6. $\displaystyle\int_0^1 \int_{-1}^1 (x + y) \, dy \, dx$

7. $\displaystyle\int_0^1 \int_{x^2}^x (x + y) \, dy \, dx$

8. $\displaystyle\int_0^1 \int_{-1}^x (x^2 + y^2) \, dy \, dx$

9. $\displaystyle\int_0^2 \int_0^x (x + y^2) \, dy \, dx$

10. $\displaystyle\int_1^3 \int_0^x 2e^{x^2} \, dy \, dx$

Evaluate $\iint_G f(x, y) \, dx \, dy$ for the given function $f(x, y)$ and region G. Sketch G.

11. $f(x, y) = 1$, $G = \{(x, y) \mid 0 \leqslant x \leqslant 2, 0 \leqslant y \leqslant x\}$
12. $f(x, y) = x$, $G = \{(x, y) \mid 0 \leqslant y \leqslant 1 - x^2\}$
13. $f(x, y) = 1 + 2x$, $G = \{(x, y) \mid \frac{1}{2}y \leqslant x \leqslant y, y \leqslant 1\}$
14. $f(x, y) = 4x$, $G = \{(x, y) \mid y \geqslant 0, x^3 \leqslant y \leqslant x\}$
15. Find the volume of the solid capped by the surface $z = 1 - y - x^2$ over the region bounded above and below by $y = 0$ and $y = 1 - x^2$, and left and right by $x = 0$ and $x = 1$, by evaluating the integral

$$\int_0^1 \int_0^{1 - x^2} (1 - y - x^2) \, dy \, dx.$$

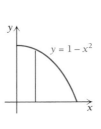

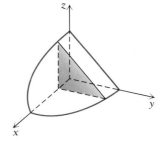

16. Find the volume of the solid capped by the surface $z = x + y$ over the region bounded above and below by $y = 0$ and $y = 1 - x$, and left and right by $x = 0$ and $x = 1$, by evaluating the integral

$$\int_0^1 \int_0^{1 - x} (x + y) \, dy \, dx.$$

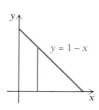

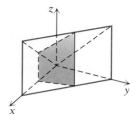

17. Find the volume of the solid whose base is the region in the xy-plane that is bounded by the parabola $y = 4 - x^2$ and the line $y = 3x$, while the top of the solid is capped by $z = x + 6$.

18. Find the volume of the solid whose base is the region in the xy-plane above the x-axis and below $y = 2 - x^2$ and which is capped by $z = 3y^2$.

Find the average value f_{av} of the given function over the region R.

19. $f(x, y) = x^2 + y^2$, $G = \{(x, y)|0 \leqslant x \leqslant 2, 0 \leqslant y \leqslant 2x\}$

20. $f(x, y) = e^{x+y}$, $G = \{(x, y)|0 \leqslant y \leqslant x, 0 \leqslant x \leqslant 1\}$

In each of the following, write an equivalent iterated integral with the order of integration reversed. As a check, evaluate *both* iterated integrals.

21. $\int_0^2 \int_1^{e^x} dy\, dx$

22. $\int_0^2 \int_0^y dx\, dy$

In Exercises 23–26, suppose that a continuous random variable has a joint probability function given by

$$f(x, y) = x^2 + \tfrac{1}{3}xy, \quad 0 \leqslant x \leqslant 1, \quad 0 \leqslant y \leqslant 2.$$

23. Find $\int_0^2 \int_0^1 f(x, y)\, dx\, dy$.

24. Find the probability that a point (x, y) is in the region bounded by $0 \leqslant x \leqslant \tfrac{1}{2}$, $1 \leqslant y \leqslant 2$, by evaluating the integral

$$\int_1^2 \int_0^{1/2} f(x, y)\, dx\, dy.$$

25. Find the probability that a point (x, y) is in the region $\{(x, y)|0 \leqslant x \leqslant 1, 0 \leqslant y \leqslant 1\}$, that is, the lower half of the rectangle.

26. Find the probability that a point (x, y) is in the part of the rectangle above the line $y = 2 - 2x$.

SYNTHESIS EXERCISES

A *triple integral* of a continuous function $f(x, y, z)$ over an appropriate region G in three dimensions, say

$$\iiint_G f(x, y, z)\, dx\, dy\, dz,$$

can be defined in a manner similar to that in which the double integral was defined. Triple integrals can frequently be evaluated by the iterated integral

$$\int_a^b \left\{ \int_{G(x)}^{H(x)} \left\{ \int_{g(x, y)}^{h(x, y)} f(x, y, z)\, dz \right\} dy \right\} dz,$$

where the region G is bounded above and below by the surfaces $z = g(x, y)$ and $z = h(x, y)$ and where the y- and x-limits of integration are obtained from the boundaries of the region bounded by $g(x, y) = h(x, y)$ in the xy-plane. Evaluate each of the following triple integrals.

27. $\int_0^1 \int_1^3 \int_{-1}^2 (2x + 3y - z)\, dx\, dy\, dz$

28. $\int_0^2 \int_1^4 \int_{-1}^6 (8x - 2y + z)\, dx\, dy\, dz$

29. $\int_0^1 \int_0^{1-x} \int_0^{2-x} xyz\, dz\, dy\, dx$

30. $\int_0^2 \int_{2-y}^{6-2y} \int_0^{\sqrt{4-y^2}} z\, dz\, dx\, dy$

If $f(x, y, z)$ is continuous on a region G in three dimensions, we define *the average value f_{av} of $f(x, y, z)$ on G* by

$$f_{av} = \frac{1}{\text{vol }(G)} \iiint_G f(x, y, z)\, dx\, dy\, dz,$$

where vol (G) is the volume of G.

31. Find the average value of $f(x, y, z) = xyz$ on the unit cube

$$\{(x, y, z)|0 \leqslant x \leqslant 1, 0 \leqslant y \leqslant 1 \text{ and } 0 \leqslant z \leqslant 1\}.$$

COMPUTER EXERCISES

THE CALCULUS EXPLORER:
Double Integral Evaluation

Use this program to evaluate the double integrals in Exercises 1–8 above.

CHAPTER SUMMARY AND REVIEW 7

TERMS TO KNOW

Function of two variables, p. 508
Function of several variables, p. 509
Partial derivative, p. 519
Second-order partial derivative,
  p. 525

Relative maximum, p. 528
Relative minimum, p. 528
Saddle point, p. 530
D-test, p. 531
Total differential, p. 539

Least-squares technique, p. 549
Method of LaGrange multipliers,
  p. 559
Multiple integration, p. 569
Double iterated integral, p. 571

REVIEW EXERCISES

These exercises are for test preparation. They can also be used as a lengthened practice test. Answers are at the back of the book. The answers also contain bracketed section references, which tell you where to restudy if your answer is incorrect.

Given $f(x, y) = e^y + 3xy^3 + 2y$, find each of the following.

1. f_x
2. f_y
3. f_{xy}
4. f_{yx}
5. f_{xx}
6. f_{yy}

Given $z = x^2 \ln y + y^4$, find each of the following.

7. $\dfrac{\partial z}{\partial x}$
8. $\dfrac{\partial z}{\partial y}$
9. $\dfrac{\partial^2 z}{\partial y\, \partial x}$
10. $\dfrac{\partial^2 z}{\partial x\, \partial y}$
11. $\dfrac{\partial^2 z}{\partial x^2}$
12. $\dfrac{\partial^2 z}{\partial y^2}$

Find the relative maximum and minimum values.

13. $f(x, y) = x^3 - 6xy + y^2 + 6x + 3y - \frac{1}{5}$
14. $f(x, y) = x^2 - xy + y^2 - 2x + 4y$
15. $f(x, y) = 3x - 6y - x^2 - y^2$
16. $f(x, y) = x^4 + y^4 + 4x - 32y + 80$

17. Calculate the total differential dz given that
$$z = 5x^2 - e^{2y}.$$

18. Calculate the total differential dz given that
$$z = 2x^2 y.$$

19. The legs of a right triangle are measured to be 5 ft and 12 ft. Find the maximum possible error in measuring the area of the triangle if errors of measurement of 0.1 ft and 0.4 ft, respectively, are possible in the measurement of the legs.

20. A rectangular box is found to measure 2 ft by 4 ft by 5 ft. If errors of 0.01 ft, 0.02 ft, and 0.04 ft are possible in the three measurements, respectively, use differentials to calculate the maximum possible error in measuring the volume of the box.

21. Consider these data regarding the average cost of a ticket to a Broadway show over a recent five-year period.

Year, x	Average cost of a ticket to a Broadway show, y
1. 1980	$22.50
2. 1981	$27.50
3. 1982	$30.00
4. 1983	$32.50
5. 1984	$35.00
6. 1985	$35.00

a) Find the regression line $y = mx + b$.

b) Use the regression line to predict the average cost of a ticket in 1989; in 2000.

22. Consider these data regarding enrollment in colleges and universities during a recent three-year period.

Year, x	Enrollment (in millions), y
1	7.2
2	8.0
3	8.4

   a) Find the regression line $y = mx + b$.
   b) Use the regression line to predict enrollment in the fourth year.

23. Find the minimum value of

$$f(x, y) = x^2 - 2xy + 2y^2 + 20$$

subject to the constraint $2x - 6y = 15$.

24. Find the maximum value of

$$f(x, y) = 6xy$$

subject to the constraint $2x + y = 20$.

Evaluate.

25. $\int_0^1 \int_{x^2}^{3x} (x^3 + 2y)\, dy\, dx$ 26. $\int_0^1 \int_{x^2}^{x} (x - y)\, dy\, dx$

27. Evaluate $\int_0^1 \int_{x^2}^{x} (x - y)\, dx\, dy$.

28. Find the average value f_{av} of $f(x, y) = 4xy$ on the rectangle in the plane whose vertices are the four points $(0, 0)$, $(1, 0)$, $(0, 2)$, and $(1, 2)$.

29. Evaluate $\iint_G 12x^2\, dx\, dy$, where G is the region in the plane bounded by the positive x-axis, the positive y-axis, and the line $x + y = 2$.

30. Find the volume of the solid whose base is the semicircular region $\{(x, y)|\, y \geqslant 0,\ x^2 + y^2 \leqslant 1\}$ and which is capped by the plane $z = 4y$.

SYNTHESIS EXERCISES

31. Evaluate $\int_0^2 \int_{1 - 2x}^{1 - x} \int_0^{\sqrt{2 - x^2}} z\, dz\, dy\, dx$.

32. Suppose beverages could be packaged in either a cylin-drical container or a rectangular container with a square top and bottom. If we assume a volume of 26 in³, which container would have the minimum surface area?

EXERCISES FOR THINKING AND WRITING

33. Find some applied examples of functions of several variables not considered in the text, even some that may not have formulas.

34. Do some library research on the Cobb–Douglas production function. Can you find how it was developed?

35. Explain the meaning of the first partial derivatives of a function of two variables in terms of slopes of tangent lines.

36. Describe the geometric meaning of the multiple integral of a function of two variables.

CHAPTER TEST 7

Given $f(x, y) = e^x + 2x^3y + y$, find each of the following.

1. $f(-1, 2)$

2. $\dfrac{\partial f}{\partial x}$

3. $\dfrac{\partial f}{\partial y}$

4. $\dfrac{\partial^2 f}{\partial x^2}$

5. $\dfrac{\partial^2 f}{\partial x\, \partial y}$

6. $\dfrac{\partial^2 f}{\partial y\, \partial x}$

7. $\dfrac{\partial^2 f}{\partial y^2}$

8. Find the relative maximum and minimum values of
$$f(x, y) = x^2 - xy + y^3 - x.$$

9. Find the relative maximum and minimum values of
$$f(x, y) = y^2 - x^2.$$

10. *Business: Predicting total sales.* Consider the data in the following table regarding the total sales of a company during the first three years of operation.

Year, x	Sales (in millions), y
1	$10
2	15
3	19

a) Find the regression line $y = mx + b$.

b) Use the regression line to predict sales in the fourth year.

11. Find the maximum value of
$$f(x, y) = 6xy - 4x^2 - 3y^2$$
subject to the constraint $x + 3y = 19$.

12. Evaluate
$$\int_0^2 \int_1^x (x^2 - y)\, dy\, dx.$$

13. Calculate the total differential dz given that $z = x^3 + e^y$.

14. Calculate the total differential dz given that $z = 3xy^2$.

15. A rectangular box is found to measure 2 ft by 3 ft by 5 ft. If errors of 0.01 ft, 0.02 ft, and 0.04 ft are possible in the three measurements, respectively, use differentials to calculate the maximum possible error in measuring the volume of the box.

16. Find the volume of the solid whose base is the semicircular region
$$\{(x, y) | 0 \leqslant y,\ x^2 + y^2 \leqslant 1\}$$
and which is capped by the plane $z = 2y$.

17. Find the average value f_{av} of $f(x, y) = xy$ on the region of the plane bounded by the positive x- and y-axes and the curve $y = 1 - x^3$.

SYNTHESIS EXERCISES

18. *Business: Cobb–Douglas Model.* A company has the following Cobb–Douglas production function for a certain product,
$$p(x, y) = 50x^{2/3}y^{1/3},$$
where x = labor, measured in dollars, and y = capital, measured in dollars. Suppose a company can make a total investment in labor and capital of $600,000. How should it allocate the investment between labor and capital in order to maximize production?

19. Find f_x and f_t:
$$f(x, t) = \frac{x^2 - 2t}{x^3 + 2t}.$$

8

TRIGONOMETRIC FUNCTIONS

AN APPLICATION

A company in a northern climate has sales of skis as given by

$$S(t) = 7\left(1 - \cos \frac{\pi}{6} t\right),$$

where S is sales, in thousands of dollars, during the tth month. Find the total sales for the first year.

THE MATHEMATICS

Total sales for the first year are given by

$$\int_0^{12} S(t) \, dt = \int_0^{12} 7\underbrace{\left(1 - \cos \frac{\pi}{6} t\right)}_{\uparrow} \, dt.$$

This is a *trigonometric function*.

Functions for which the graphs repeat themselves are called *periodic*. Of special importance in mathematics are the periodic *trigonometric functions*. These functions originated as a means of indirect measurement. For example, a surveyor might use trigonometry to measure the distance across a river without actually crossing the river. We will learn to differentiate and integrate these functions and to solve related problems.

8.1

INTRODUCTION TO TRIGONOMETRY

Angles and Rotations

We now introduce the trigonometric functions together with their derivatives and integrals.

We first consider a rotating ray, with its endpoint at the origin in an xy-plane. The ray starts in position along the positive half of the x-axis. A counterclockwise rotation is called *positive*, and a clockwise rotation is called *negative* (Fig. 1).

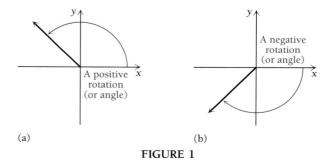

(a)   (b)

FIGURE 1

Note that the rotating ray and the positive half of the x-axis form an **angle**. Thus we often speak of "rotations" and "angles" interchangeably. The rotating ray is often called the *terminal side* of the angle, and the positive half of the x-axis is called the *initial side*.

Measures of Rotations or Angles

The size, or *measure*, of an angle, or rotation, may be given in **degrees**. A complete revolution has a measure of 360°, half a revolution has a measure of 180°, and so on. We can also speak of an *angle* of 90° or 720° or −240°.

An angle between 0° and 90° has its terminal side in the first quadrant. An angle between 90° and 180° has its terminal side in the second quadrant. An angle between 180° and 270° has its terminal side in the third quadrant. An angle between 0° and −90° has its terminal side in the fourth quadrant.

Note that angles with measures 0°, 360°, and 720° have the same terminal side, as do 270° and −90°.

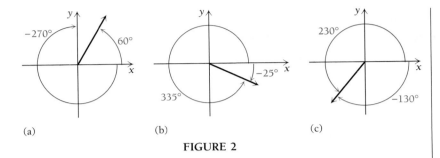

(a)   (b)   (c)

FIGURE 2

1. In which quadrant does the terminal side of the angle lie?

   a)   47°

   b)   212°

   c)   −43°

   d)   −135°

   e)   365°

   f)   −365°

   g)   740°

DO EXERCISE 1.

Measurements in Radians

A unit of angle or rotation measure other than the degree is very useful for many purposes. This unit is called the **radian**. Consider a circle with radius of length 1, centered at the origin. The measure of an angle, or rotation, in radians is the distance around this circle from the initial side of the angle to the terminal side (Fig. 3).

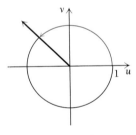

FIGURE 3

Since the circumference of the circle is $2\pi \cdot 1$, or 2π, a complete revolution (360°) has a measure of 2π radians. Half of this (180°) is π radians, and a fourth of this (90°) is $\pi/2$ radians. In general, we can convert from one measure to the other using this proportion.

THEOREM 1

$$\frac{\text{Radian measure}}{\pi} = \frac{\text{Degree measure}}{180}$$

EXAMPLE 1   Convert 270° to radians.

2. Convert to radian measure. Leave answers in terms of π.

   a) $135°$

   b) $315°$

   c) $-90°$

   d) $720°$

   e) $-225°$

   f) $-315°$

   g) $405°$

   h) $480°$

3. Convert to degree measure.

   a) $\dfrac{\pi}{3}$

   b) $\dfrac{3}{4}\pi$

   c) $\dfrac{5}{2}\pi$

   d) 10π

   e) $-\dfrac{7\pi}{6}$

   f) 300π

   g) -270π

   h) $\dfrac{25\pi}{4}$

Solution   We use the proportion in Theorem 1, multiplying by π on both sides:

$$\frac{\text{Radian measure}}{\pi} = \frac{270}{180}$$

$$\text{Radian measure} = \frac{270}{180} \cdot \pi, \quad \text{or} \quad \frac{3}{2}\pi. \qquad \text{❖}$$

When no unit is specified for an angle measure, it is understood to be given in radians.

EXAMPLE 2   Convert $\pi/4$ radians to degrees.

Solution

$$\frac{\pi/4}{\pi} = \frac{\text{Degree measure}}{180}$$

$$\text{Degree measure} = 180 \cdot \frac{\pi/4}{\pi}, \quad \text{or} \quad 45°. \qquad \text{❖}$$

DO EXERCISES 2 AND 3.

Trigonometric Functions

The concept of rotation or angle is important to functions called **trigonometric**, or **circular**, functions.

Consider an angle t, measured in radians, shown in Fig. 4 on a circle with radius 1. The unit circle has equation $u^2 + v^2 = 1$.

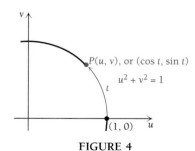

FIGURE 4

The terminal side of the angle intersects this circle at point $P(u, v)$. The arc length *around the circle* from $(1, 0)$ to P is t. We define $\cos t$ (cosine t) and $\sin t$ (sine t) as the first and second coordinates of P, respectively.

DEFINITION

$$\cos t = \text{the first coordinate of } P = u,$$
$$\sin t = \text{the second coordinate of } P = v$$

Certain values of these functions are easy to determine. When $t = \pi$, for example, the terminal side of the angle is on the horizontal axis and P is one unit to the left; hence the first coordinate is -1 and the second coordinate is 0 (Fig. 5). Thus,

$$\cos \pi = u = -1 \quad \text{and} \quad \sin \pi = v = 0,$$

which checks because $(-1)^2 + 0^2 = 1$.

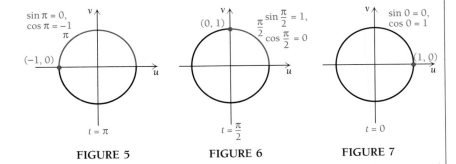

| FIGURE 5 | FIGURE 6 | FIGURE 7 |

Similarly, when $t = \pi/2$, the point P is one unit up on the vertical axis; hence the first coordinate is 0 and the second coordinate is 1 (Fig. 6). Thus,

$$\cos \frac{\pi}{2} = u = 0 \quad \text{and} \quad \sin \frac{\pi}{2} = v = 1,$$

which checks because $0^2 + 1^2 = 1$.

When $t = 0$, the terminal side is on the horizontal axis and P is one unit to the right; hence the first coordinate is 1 and the second coordinate is 0 (Fig. 7). Thus,

$$\cos 0 = u = 1 \quad \text{and} \quad \sin 0 = v = 0.$$

We can also define the trigonometric functions by using right triangles (Fig. 8). Suppose t is an angle of a right triangle, measured in degrees. Then we have the following.

DEFINITION

$$\sin t = \frac{\text{length of side opposite } t}{\text{length of hypotenuse}},$$

$$\cos t = \frac{\text{length of side adjacent to } t}{\text{length of hypotenuse}}$$

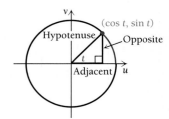

FIGURE 8

Function Values for Special Angles

Using properties of right triangles, we can develop values of the sine and cosine functions for the special angles 45°, 30°, and 60°. First, recall the Pythagorean theorem. It says that in any right triangle, $a^2 + b^2 = c^2$, where c is the length of the hypotenuse and a and b are the lengths of the legs (Fig. 9).

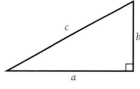

FIGURE 9

Consider $t = \pi/4 = 45°$ (Fig. 10). A right triangle with a 45° acute angle actually has two 45° angles. Thus the triangle is isosceles, and the legs are the same length.

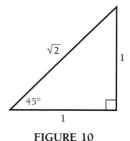

FIGURE 10

Then using the Pythagorean theorem or the equation of the unit circle, we have

$$u^2 + v^2 = 1$$
$$u^2 + u^2 = 1$$
$$2u^2 = 1$$
$$u^2 = \frac{1}{2}, \quad \text{so} \quad u = \sqrt{\frac{1}{2}} = \frac{\sqrt{2}}{2} = v.$$

Thus,

$$\cos \frac{\pi}{4} = \cos 45° = \frac{\sqrt{2}}{2} \approx 0.707$$

and

$$\sin \frac{\pi}{4} = \sin 45° = \frac{\sqrt{2}}{2} \approx 0.707.$$

In a similar way, we can determine function values for 30° and 60°. A right triangle with 30° and 60° acute angles is half of an equilateral triangle, as shown in Fig. 11. Thus if we choose an equilateral triangle whose sides have length 2 and take half of it, we obtain a right triangle that has a hypotenuse of length 2 and a leg of length 1. The other leg has length a, which can be found using the Pythagorean theorem, as follows:

$$a^2 + 1^2 = 2^2$$
$$a^2 = 3$$
$$a = \sqrt{3}.$$

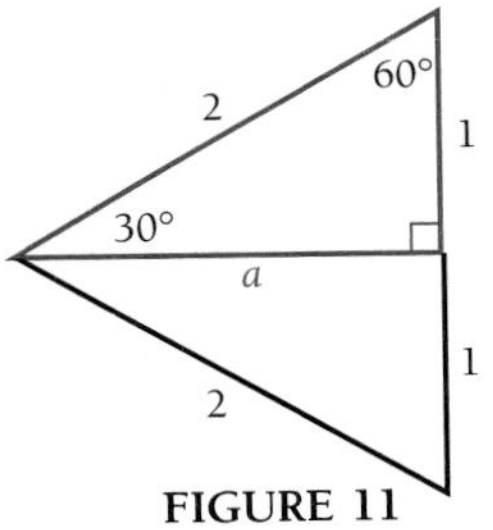

FIGURE 11

We can now determine function values for 60° and 30° (Fig. 12).
For $t = \pi/3 = 60°$,

$$\cos \frac{\pi}{3} = \cos 60° = \frac{1}{2} = 0.5$$

and

$$\sin \frac{\pi}{3} = \sin 60° = \frac{\sqrt{3}}{2} \approx 0.866.$$

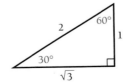

FIGURE 12

For $t = \pi/6 = 30°$,

$$\cos \frac{\pi}{6} = \cos 30° = \frac{\sqrt{3}}{2} \approx 0.866$$

and

$$\sin \frac{\pi}{6} = \sin 30° = \frac{1}{2} = 0.5.$$

The following table summarizes these important values of the sine and cosine functions. It should be memorized.

t	t	$\sin t$	$\cos t$
0°	0	0	1
30°	$\dfrac{\pi}{6}$	$\dfrac{1}{2}$	$\dfrac{\sqrt{3}}{2}$
45°	$\dfrac{\pi}{4}$	$\dfrac{\sqrt{2}}{2}$	$\dfrac{\sqrt{2}}{2}$
60°	$\dfrac{\pi}{3}$	$\dfrac{\sqrt{3}}{2}$	$\dfrac{1}{2}$

Other function values follow from certain symmetries on the unit circle. Some are shown in Figs. 13 and 14.

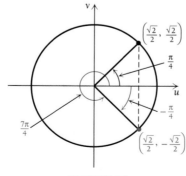

FIGURE 13

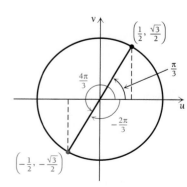

FIGURE 14

From Fig. 13, we get

$$\cos(-45°) = \cos\left(-\frac{\pi}{4}\right) = \frac{\sqrt{2}}{2} \approx 0.707,$$

$$\sin(-45°) = \sin\left(-\frac{\pi}{4}\right) = -\frac{\sqrt{2}}{2} \approx -0.707,$$

and

$$\cos 315° = \cos\frac{7\pi}{4} = \frac{\sqrt{2}}{2} \approx 0.707,$$

$$\sin 315° = \sin\frac{7\pi}{4} = -\frac{\sqrt{2}}{2} \approx -0.707.$$

From Fig. 14, we get

$$\cos(-120°) = \cos\left(-\frac{2\pi}{3}\right) = -\frac{1}{2} = -0.5,$$

$$\sin(-120°) = \sin\left(-\frac{2\pi}{3}\right) = -\frac{\sqrt{3}}{2} \approx -0.866,$$

and

$$\cos 240° = \cos\frac{4\pi}{3} = -\frac{1}{2} = -0.5,$$

$$\sin 240° = \sin\frac{4\pi}{3} = -\frac{\sqrt{3}}{2} \approx -0.866.$$

Approximations of the trigonometric values can be found on your calculator. Be sure to check whether it uses degrees or radians. Usually you can adapt to consider either.

DO EXERCISE 4.

Graphs of cos t and sin t

Note that $\cos t$ and $\sin t$ are functions of t defined for all real numbers t. For very large $|t|$, we may "wrap around" the circle several times before coming to the terminal point P. Nevertheless, P still has one first coordinate, $\cos t$, and one second coordinate, $\sin t$. For example,

$$\cos 3\pi = \cos \pi = -1 \quad \text{and} \quad \sin 3\pi = \sin \pi = 0.$$

Also,

$$\cos\left(\frac{15\pi}{4}\right) = \cos\left(\frac{7\pi}{4}\right) = \frac{\sqrt{2}}{2} \quad \text{and} \quad \sin\left(\frac{15\pi}{4}\right) = \sin\left(\frac{7\pi}{4}\right) = -\frac{\sqrt{2}}{2}.$$

4. Using this figure, find each of the following.

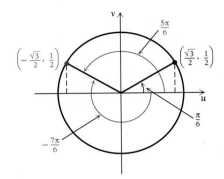

a) $\cos\left(\dfrac{5\pi}{6}\right)$

b) $\sin\left(\dfrac{5\pi}{6}\right)$

c) $\cos\left(-\dfrac{7\pi}{6}\right)$

d) $\sin\left(-\dfrac{7\pi}{6}\right)$

Plotting points previously obtained, and finding others with a calculator, we graph the cosine and sine functions, as shown in Figs. 15 and 16.

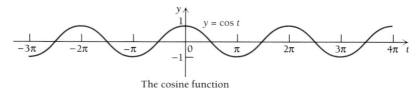

The cosine function

FIGURE 15

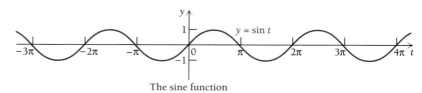

The sine function

FIGURE 16

At the origin, t is, of course, 0. Thus the point P is on the horizontal axis, so cos $0 = 1$ and sin $0 = 0$. Moving to the right on the graphs corresponds to rotating the terminal side of the angle counterclockwise. Moving to the left corresponds to rotating the terminal side clockwise. Note in particular that sin $t = 0$ has solutions $t = 0$, $\pm\pi$, $\pm2\pi$, . . . , and cos $t = 0$ has solutions $t = \pm(\pi/2)$, $\pm(3\pi/2)$, $\pm(5\pi/2)$, ... Note that each curve repeats itself as the terminal side makes successive revolutions. From 0 to 2π is one complete revolution, or *cycle*. The cycle repeats itself from there on. We say that the *period* of each function is 2π. Algebraically, this means that for any t,

$$\cos (t + 2\pi) = \cos t \quad \text{and} \quad \sin (t + 2\pi) = \sin t.$$

DEFINITION

A function f is *periodic* if there exists a positive number p such that

$$f(t + p) = f(t)$$

for every t in the domain of f. This means that adding p to an input does not change the output. The smallest such number p is called the *period*.

A printout from an electrocardiogram may form a periodic function.

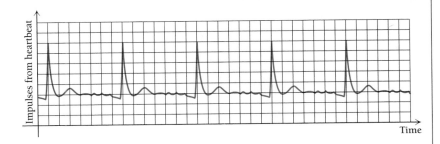

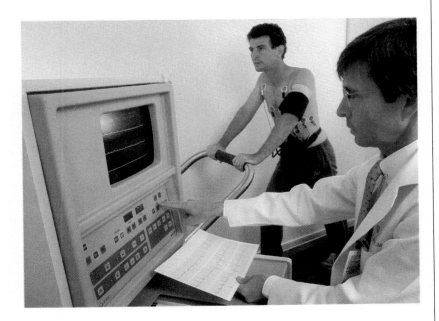

Other Trigonometric Functions

The functions sin *t* and cos *t* are the basic trigonometric functions, but there are four others—the tangent, cotangent, secant, and cosecant functions—defined as follows.

5. Find each of the following.

   a) $\tan \dfrac{\pi}{4}$

   b) $\tan \dfrac{\pi}{3}$

   c) $\tan 0$

   d) $\sec \dfrac{\pi}{4}$

   e) $\sec \dfrac{\pi}{3}$

   f) $\sec 0$

6. Find each of the following.

   a) $\cot \dfrac{\pi}{6}$

   b) $\cot 0$

   c) $\cot \dfrac{\pi}{3}$

   d) $\csc \dfrac{\pi}{6}$

   e) $\csc 0$

   f) $\csc \dfrac{\pi}{4}$

DEFINITION

$$\tan t = \frac{\sin t}{\cos t},$$

$$\cot t = \frac{\cos t}{\sin t} = \frac{1}{\tan t},$$

$$\sec t = \frac{1}{\cos t},$$

$$\csc t = \frac{1}{\sin t},$$

provided the denominators are not equal to 0.

Let us find some values of the tangent function.

EXAMPLE 3 Find $\tan \pi/6$ and $\tan \pi/2$.

Solution

$$\tan \frac{\pi}{6} = \frac{\sin (\pi/6)}{\cos (\pi/6)} = \frac{\frac{1}{2}}{\sqrt{3}/2} = \frac{1}{\sqrt{3}} = \frac{1}{\sqrt{3}} \cdot \frac{\sqrt{3}}{\sqrt{3}} = \frac{\sqrt{3}}{3} \approx 0.577;$$

$$\tan \frac{\pi}{2} = \frac{\sin (\pi/2)}{\cos (\pi/2)} = \frac{1}{0} = \text{undefined} \qquad ❖$$

DO EXERCISES 5 AND 6.

The graph of $\tan t$ is shown in Fig. 17. The vertical dashed lines are asymptotes. They occur at values t for which $\cos t = 0$. These are $t = \pi/2 + k\pi$, where k is any integer.

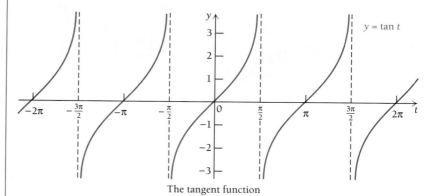

The tangent function

FIGURE 17

Identities

An **identity** is an equation that holds for all meaningful replacements of its variables by real numbers. An example is $t + 3 = 3 + t$, which is true for all real numbers. Another example is $(t^2 - 4)/(t - 2) = t + 2$. It is true for all real numbers except 2, which is *not* a meaningful replacement in the expression on the left.

The properties

$$\cos (t + 2\pi) = \cos t \quad \text{and} \quad \sin (t + 2\pi) = \sin t$$

hold for all real numbers t. They are examples of **trigonometric identities**. Another identity that holds for all real numbers is the following:

THEOREM 2

$$\sin^2 t + \cos^2 t = 1, \tag{1}$$

---

where $\sin^2 t$ means $(\sin t)^2$ and $\cos^2 t$ means $(\cos t)^2$. To see why this holds, note the right triangle inside the unit circle shown in Fig. 18.

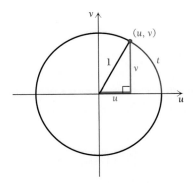

FIGURE 18

Since the length of the radius is 1, we know from the Pythagorean theorem that

$$u^2 + v^2 = 1^2, \quad \text{or} \quad u^2 + v^2 = 1;$$

and, since for any point (u, v) on the unit circle, $\cos t = u$ and $\sin t = v$, the identity follows.

If we multiply Identity (1) by $1/(\cos^2 t)$, we get another identity:

$$\frac{\sin^2 t}{\cos^2 t} + \frac{\cos^2 t}{\cos^2 t} = \frac{1}{\cos^2 t},$$

7. Multiply Identity (1) by

$$\frac{1}{\sin^2 t}$$

to develop another identity.

or

$$\tan^2 t + 1 = \sec^2 t,$$

where, as before, $\tan^2 t = (\tan t)^2$ and $\sec^2 t = (\sec t)^2$. Another identity can be proved in a similar way and you are asked to do so in Margin Exercise 7.

DO EXERCISE 7.

The new identities we have proven are as follows.

THEOREM 3

$$\tan^2 t + 1 = \sec^2 t, \tag{2}$$
$$\cot^2 t + 1 = \csc^2 t \tag{3}$$

---

8. Use Identity (7) to find sin 15°.

The following are called **sum–difference identities**.

THEOREM 4

$$\cos (u + v) = \cos u \cos v - \sin u \sin v, \tag{4}$$
$$\cos (u - v) = \cos u \cos v + \sin u \sin v, \tag{5}$$
$$\sin (u + v) = \sin u \cos v + \cos u \sin v, \tag{6}$$
$$\sin (u - v) = \sin u \cos v - \cos u \sin v \tag{7}$$

---

9. Use Identity (6) to find

$$\sin \left(\frac{\pi}{4} + \frac{\pi}{3} \right).$$

EXAMPLE 4   Use Identity (5) to find cos 15°.

Solution   If we think of 15° as 45° − 30°, then

$$\cos 15° = \cos (45° - 30°)$$
$$= \cos 45° \cos 30° + \sin 45° \sin 30°$$
$$= \frac{\sqrt{2}}{2} \cdot \frac{\sqrt{3}}{2} + \frac{\sqrt{2}}{2} \cdot \frac{1}{2}$$
$$= \frac{\sqrt{6} + \sqrt{2}}{4}.$$

❖

DO EXERCISES 8 AND 9.

If we let $u = v$ in Identity (4), we obtain a **double-angle identity**:

$$\cos 2u = \cos (u + u)$$
$$= \cos u \cos u - \sin u \sin u$$
$$= \cos^2 u - \sin^2 u.$$

You will prove a similar identity for $\sin 2u$ in Margin Exercise 10. We have proven the following.

THEOREM 5

$$\cos 2u = \cos^2 u - \sin^2 u,$$
$$\sin 2u = 2 \sin u \cos u$$

DO EXERCISE 10.

10. Let $u = v$ in Identity (6) to find an identity for $\sin 2u$.

EXERCISE SET 8.1

In what quadrant does the terminal side of each angle lie?

1. $34°$ 2. $320°$

3. $-120°$ 4. $-205°$

Convert to radian measure. Leave answers in terms of π.

5. $30°$ 6. $15°$ 7. $60°$

8. $200°$ 9. $75°$ 10. $300°$

Convert to degree measure.

11. $\dfrac{3}{2}\pi$ 12. $\dfrac{5}{4}\pi$ 13. $-\dfrac{\pi}{4}$

14. $-\dfrac{\pi}{6}$ 15. 8π 16. -12π

17. 1 radian   18. 2 radians

Find each of the following.

19. $\sin \dfrac{\pi}{3}$ 20. $\cos \dfrac{\pi}{4}$ 21. $\cos \dfrac{3\pi}{2}$

22. $\sin \dfrac{5\pi}{4}$ 23. $\sin \dfrac{\pi}{6}$ 24. $\cos \pi$

25. $\tan \dfrac{\pi}{2}$ 26. $\tan \dfrac{\pi}{6}$ 27. $\cot \dfrac{\pi}{3}$

28. $\cot \dfrac{\pi}{6}$ 29. $\sec \pi$ 30. $\csc \dfrac{\pi}{4}$

Verify each of the following identities.

31. $\tan (u + v) = \dfrac{\tan u + \tan v}{1 - \tan u \tan v}$

32. $\tan 2u = \dfrac{2 \tan u}{1 - \tan^2 u}$

🖩 If your calculator has trigonometric keys, find answers to the following exercises to four decimal places. Check to see if your calculator is using degree or radian measure for angles.

33. $\sin 31.4°$ 34. $\cos 1.07°$

35. $\tan 139.2°$ 36. $\cot 153.5°$

37. $\cos (-1.91)$ 38. $\sin (-11.2\pi)$

39. $\cot 49\pi$ 40. $\tan (-17.4)$

APPLICATIONS

❖ **Business and Economics**

41. 🖩 *Sales.* Sales of certain products fluctuate in cycles, as shown in this graph. A company in a northern climate has total sales of skis as given by

$$S(t) = 7\left(1 - \cos\frac{\pi}{6}t\right),$$

where S is sales, in thousands of dollars, during the tth month. Find $S(0)$, $S(1)$, $S(2)$, $S(3)$, $S(6)$, $S(12)$, and $S(15)$. Round to the nearest tenth.

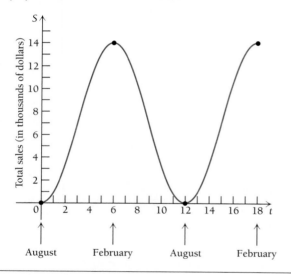

❖ **Life and Physical Sciences**

42. 🖩 *Temperature during an illness.* The temperature of a patient during a 12-day illness is given by

$$T(t) = 101.6° + 3\sin\left(\frac{\pi}{8}t\right).$$

The graph is shown here. Find $T(0)$, $T(1)$, $T(2)$, $T(4)$, and $T(12)$. Round to the nearest tenth of a degree.

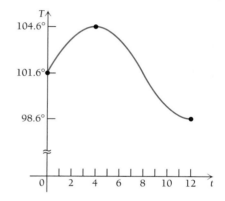

8.2

OBJECTIVES

a) Differentiate trigonometric functions.

b) Solve problems involving derivatives of trigonometric functions.

DERIVATIVES OF THE TRIGONOMETRIC FUNCTIONS

The development of the derivatives of $\sin x$ and $\cos x$ is comparable to the development of the derivative of e^x. We first find the value of the derivative at 0, and then extend this to the general formula. The derivatives at 0 are given by the following limits:

$$\sin'(0) = \lim_{h \to 0} \frac{\sin(0 + h) - \sin 0}{h} = \lim_{h \to 0} \frac{\sin h}{h};$$

$$\cos'(0) = \lim_{h \to 0} \frac{\cos(0 + h) - \cos 0}{h} = \lim_{h \to 0} \frac{\cos h - 1}{h}.$$

We find the limits using the following input–output table as well as the graphs of each function shown in Figs. 1 and 2.

h (in radians) with decimal approximation	$\sin h$	$\cos h$	$\dfrac{\sin h}{h}$	$\dfrac{\cos h - 1}{h}$
$\dfrac{\pi}{2}(\approx 1.5708)$	1	0	0.6366	-0.6366
$\dfrac{\pi}{3}(\approx 1.0472)$	$\dfrac{\sqrt{3}}{2}(\approx 0.8660)$	$\dfrac{1}{2}(\approx 0.5000)$	0.8270	-0.4775
$\dfrac{\pi}{4}(\approx 0.7854)$	$\dfrac{\sqrt{2}}{2}(\approx 0.7071)$	$\dfrac{\sqrt{2}}{2}(\approx 0.7071)$	0.9003	-0.3729
$\dfrac{\pi}{6}(\approx 0.5236)$	$\dfrac{1}{2}(\approx 0.5000)$	$\dfrac{\sqrt{3}}{2}(\approx 0.8660)$	0.9549	-0.2559
$\dfrac{\pi}{20}(\approx 0.1571)$	0.1564	0.9877	0.9959	-0.0784
$\dfrac{\pi}{50}(\approx 0.0628)$	0.0628	0.9980	0.9993	-0.0314
$\dfrac{\pi}{4000}(\approx 0.0008)$	0.0008	0.9999997	1.0000	-0.00039

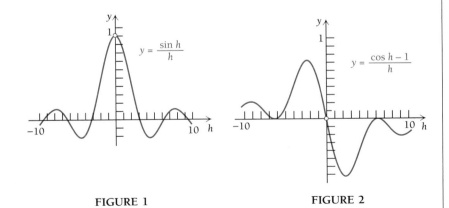

FIGURE 1   FIGURE 2

   Values approaching 0 from the left can also be checked. The following values of the limits can be conjectured from the table and the graphs:

$$\sin'(0) = \lim_{h \to 0} \frac{\sin h}{h} = 1 \quad \text{and} \quad \cos'(0) = \lim_{h \to 0} \frac{\cos h - 1}{h} = 0.$$

The results can be proved more formally, but we will not do so here. Now let us consider the general derivatives:

$$\frac{d}{dx} \sin x = \lim_{h \to 0} \frac{\sin (x + h) - \sin x}{h}.$$

Using Identity (6), we get

$$\frac{\sin (x + h) - \sin x}{h} = \frac{\sin x \cos h + \cos x \sin h - \sin x}{h}$$

$$= \frac{\sin x \cos h - \sin x}{h} + \frac{\cos x \sin h}{h}$$

$$= \sin x \left(\frac{\cos h - 1}{h} \right) + \cos x \left(\frac{\sin h}{h} \right).$$

Then using the limits just developed, we have

$$\frac{d}{dx} \sin x = \lim_{h \to 0} \frac{\sin (x + h) - \sin x}{h}$$

$$= \lim_{h \to 0} \left[\sin x \left(\frac{\cos h - 1}{h} \right) + \cos x \left(\frac{\sin h}{h} \right) \right]$$

$$= (\sin x) \cdot 0 + (\cos x) \cdot 1$$

$$= \cos x.$$

A development for the derivative of $\cos x$ is similar but uses Identity (4). The result is $-\sin x$.

In summary, we have the following.

THEOREM 6

$$\frac{d}{dx} \sin x = \cos x \quad \text{and} \quad \frac{d}{dx} \cos x = -\sin x$$

The derivatives of the remaining trigonometric functions are computed from their definitions in terms of $\sin x$ and $\cos x$, together with the Quotient Rule and/or the Extended Power Rule. The remaining formulas are as follows.

THEOREM 7

$$\frac{d}{dx}\tan x = \sec^2 x, \qquad \frac{d}{dx}\cot x = -\csc^2 x,$$

$$\frac{d}{dx}\sec x = \tan x \sec x, \qquad \frac{d}{dx}\csc x = -\cot x \csc x$$

1. Prove that $\dfrac{d}{dx}\cot x = -\csc^2 x$.

EXAMPLE 1   Prove that

$$\frac{d}{dx}\tan x = \sec^2 x.$$

Solution   Recall that $(\sec x)^2 = \sec^2 x$, $(\sin x)^2 = \sin^2 x$, and $(\cos x)^2 = \cos^2 x$. By definition,

$$\tan x = \frac{\sin x}{\cos x}.$$

Thus we can find its derivative using the Quotient Rule:

$$\begin{aligned}
\frac{d}{dx}\tan x &= \frac{d}{dx}\left[\frac{\sin x}{\cos x}\right] \\
&= \frac{\cos x\,(\cos x) - (-\sin x)(\sin x)}{\cos^2 x} \\
&= \frac{\cos^2 x + \sin^2 x}{\cos^2 x} \\
&= \frac{1}{\cos^2 x} \\
&= \sec^2 x.
\end{aligned}$$

❖

DO EXERCISE 1.

EXAMPLE 2   Find the derivative of $y = \sec^3 x$.

Solution   We use the Chain Rule:

$$\begin{aligned}
\frac{dy}{dx} &= 3\sec^2 x \cdot \left(\frac{d}{dx}\sec x\right) \\
&= 3\sec^2 x \cdot \tan x \cdot \sec x. \qquad \text{By Theorem 7}
\end{aligned}$$

2. Differentiate $y = \tan^3 x$.

Replacing the factors by the definitions in terms of $\sin x$ and $\cos x$, we can simplify this as follows:

$$\frac{dy}{dx} = 3 \sec^2 x \cdot \tan x \cdot \sec x$$

$$= 3 \cdot \frac{1}{\cos^2 x} \cdot \frac{\sin x}{\cos x} \cdot \frac{1}{\cos x}$$

$$= \frac{3 \sin x}{\cos^4 x}.$$ ❖

DO EXERCISE 2.

Using the Chain Rule, we can find other derivatives.

Differentiate.

3. $f(x) = \sin (x^4 + 3x^2)$

EXAMPLE 3   Differentiate $f(x) = \sin (x^3 - 5x)$.

Solution

$$f'(x) = [\cos (x^3 - 5x)] \cdot (3x^2 - 5)$$
$$= (3x^2 - 5) \cos (x^3 - 5x)$$ ❖

EXAMPLE 4   Differentiate $y = \cos (e^{4x}) \sin x^2$.

Solution   We use the Product Rule and the Chain Rule:

$$\frac{dy}{dx} = \cos (e^{4x}) \cdot 2x \cdot \cos x^2 - \sin (e^{4x}) \cdot (4e^{4x}) \cdot \sin x^2$$

$$= 2x \cos (e^{4x}) \cos x^2 - 4e^{4x} \sin (e^{4x}) \sin x^2.$$ ❖

4. $y = \cos (e^{x^2}) \sin x^3$

DO EXERCISES 3 AND 4.

Applications

Equations of the type

$$y = A \sin (Bx - C) + D$$

have many applications. We can also express this equation as

$$y = A \sin \left[B \left(x - \frac{C}{B} \right) \right] + D.$$

The numbers A, B, C, and D play an important role in graphing such an equation. The number A corresponds to a vertical stretching or shrinking of the graph of $y = \sin x$, B corresponds to a horizontal stretching or shrinking, C/B corresponds to a shift of the entire graph to the right or left, and D corresponds to a vertical translation. These particular expressions are given the following names:

$$Amplitude = |A|,$$

$$Period = \frac{2\pi}{B},$$

$$Phase\ shift = \frac{C}{B}.$$

EXAMPLE 5 *Physical science: Spring oscillation.* A weight is attached to the end of a spring. When the weight is disturbed, it bobs up and down with a definite frequency. If the motion were to occur in a perfect vacuum and if the spring were perfectly elastic, then the oscillatory motion would continue undiminished forever (Fig. 3). Suppose a spring oscillates in such a way that its vertical position at time t, from its position at rest, is given by

$$y = 3 \sin\left(2t + \frac{\pi}{2}\right) = 3 \sin\left[2\left(t - \left(-\frac{\pi}{4}\right)\right)\right].$$

a) Find the amplitude, the period, and the phase shift.

b) Graph the equation.

c) Find dy/dt.

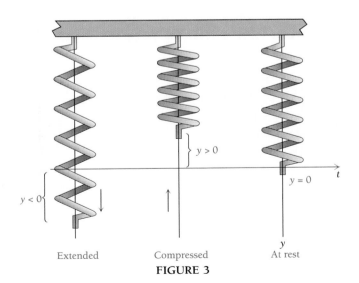

Extended   Compressed   At rest

FIGURE 3

Solution

a) The amplitude = $|3| = 3$. The period = $2\pi/2 = \pi$. The phase shift = $-(\pi/4)$.

b) We plot the various equations needed to get the final graph. The graph of $y = 3 \sin t$ is a vertical stretching, by a factor of 3, of the graph of $y = \sin t$ (Fig. 4). The period of $y = 3 \sin t$ is still 2π.

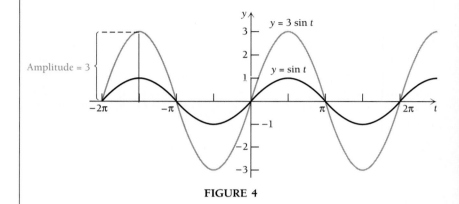

FIGURE 4

The graph of $y = 3 \sin 2t$ is a horizontal shrinking, by a factor of $\frac{1}{2}$, of the graph of $y = 3 \sin t$. The period of $y = 3 \sin 2t$ is π.

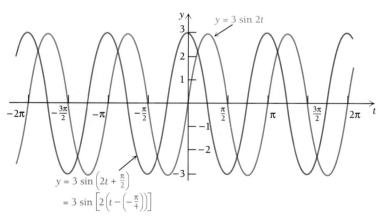

FIGURE 5

The graph of

$$y = 3 \sin \left[2\left(t - \left(-\frac{\pi}{4} \right) \right) \right]$$

is shifted $\pi/4$ units to the left of the graph of $y = 3 \sin 2t$. If the phase shift were positive, the shift would be to the right (Fig. 5).

c) $\frac{dy}{dt} = \left[3 \cos \left(2t + \frac{\pi}{2} \right) \right] 2 = 6 \cos \left(2t + \frac{\pi}{2} \right)$ ❖

DO EXERCISE 5.

The motion of the spring in the preceding example is called *simple harmonic motion.* Other types of simple harmonic motion are sound waves and light waves. For sound waves, the amplitude is the loudness. For light waves, the amplitude is the brightness. The *frequency,* which is the reciprocal of the period, is the pitch of a sound wave and the color of a light wave.

DO EXERCISES 6 AND 7.

A Biological Application: Biorhythms

Some people conjecture that a person's life has cycles of good and bad days that begin at birth. There are supposedly three such cycles, or *biorhythms.*

1. *Physical.* A cycle represented by a sine function with a period of 23 days:

$$y = \sin \frac{2\pi}{23} t.$$

This cycle is related to physical qualities such as vitality, strength, and energy.

2. *Emotional (sensitivity).* A cycle represented by a sine function with a period of 28 days:

$$y = \sin \frac{2\pi}{28} t.$$

This cycle is related to creativity, moodiness, intuition, and cheerfulness.

5. *Physical science: Spring oscillation.* Suppose a spring oscillates in such a way that its vertical position at time t, from its position at rest, is given by

$$y = 4 \sin \left(2t - \frac{\pi}{2} \right).$$

a) Find the amplitude, the period, and the phase shift.

b) Graph the equation.

c) Find dy/dt.

6. *Physical science: Sound waves.* A sound wave is given by

$$y = 0.03 \sin 1.198\pi x.$$

Find dy/dx.

7. *Physical science: Light waves.* A light wave is given by

$$L = A \sin \left(2\pi f T - \frac{d}{\omega} \right).$$

Find dL/dT.

3. *Mental (intellectual).* A cycle represented by a sine function with a period of 33 days:

$$y = \sin \frac{2\pi}{33}t.$$

This cycle is related to the ability to study, think, react, and remember. (The positive part of *this* cycle would be a good time to study calculus.)

Because these cycles have periods of different lengths, there are varying combinations of highs and lows throughout one's life. This is illustrated in Fig. 6.*

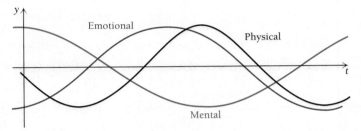

FIGURE 6

EXAMPLE 6   *Biology: Minimizing the surface area of a bee's cell.* Did you know that a honey bee constructs the cells in its comb in such a way that the minimum amount of wax is used?[†] One cell of a honeycomb is shown in Fig. 7. It is a solid whose upper base is a regular hexagon. The bottom comes together at point *A*. The surface area is given by

$$S(\theta) = 6ab + \frac{3}{2}a^2 \left(\frac{\sqrt{3} - \cos\theta}{\sin\theta} \right),$$

where θ is the measure of what is called the angle of inclination. To maintain the prism shape of the cell, we can vary the values of θ between $0°$ and $90°$. Thus we want to minimize S on the open interval $(0, 90)$.

---

*For more on these biorhythms, write to Biorhythm Computers, Inc., 298 Fifth Ave, New York, NY 10001.

[†]This discovery was published in a study by Sir D'Arcy Wentworth Thompson, *On Growth and Form* (Cambridge University Press, 1917).

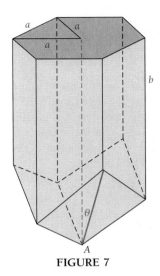

FIGURE 7

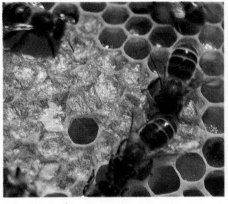

A honeycomb.

Solution   We first find $dS/d\theta$:

$$\frac{dS}{d\theta} = \frac{3}{2}a^2 \left[\frac{\sin\theta \, (\sin\theta) - \cos\theta \, (\sqrt{3} - \cos\theta)}{\sin^2\theta} \right]$$

$$= \frac{3}{2}a^2 \left[\frac{\sin^2\theta - \sqrt{3}\cos\theta + \cos^2\theta}{\sin^2\theta} \right]$$

$$= \frac{3}{2}a^2 \left[\frac{1 - \sqrt{3}\cos\theta}{\sin^2\theta} \right].$$

Critical points occur where $\sin\theta$ is 0. But if we check such values, we see that they are multiples of 90°, and that at such values the bottom of the cell would be flat, which is not the way it is to be constructed. Thus we set $dS/d\theta = 0$ and solve for θ with the assumption that $\sin\theta \neq 0$:

$$\frac{dS}{d\theta} = \frac{3}{2}a^2 \left[\frac{1 - \sqrt{3}\cos\theta}{\sin^2\theta} \right] = 0$$

$$\frac{1 - \sqrt{3}\cos\theta}{\sin^2\theta} = 0$$

$$1 - \sqrt{3}\cos\theta = 0 \qquad \text{Multiplying by } \sin^2\theta$$

$$\cos\theta = \frac{1}{\sqrt{3}} = \frac{\sqrt{3}}{3} \approx 0.5774.$$

8. *Physical science: Corridor width.* A corridor of width *a* meets a corridor of width *b* at right angles. Workers wish to push a heavy beam of length *L* on dollies around the corner. Before starting, however, they want to be sure they can make the turn. How long a beam will go around the corner? (Disregard the width of the beam.)

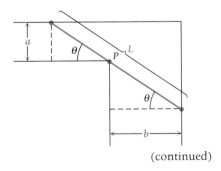

(continued)

The length L of the beam is given by

$$L = \frac{a}{\sin \theta} + \frac{b}{\cos \theta},$$

where θ is the angle shown above.

a) Find L in terms of a and b such that L is of maximum length.

b) Use the formula in part (a) to find how long a beam will go around the corner when $a = 8$ and $b = 5\sqrt{5}$.

We need to find θ such that $\cos \theta \approx 0.5774$. We can work backward on a calculator using a $\boxed{\cos^{-1}}$ key. It follows that

$$\theta \approx 54.7°.$$

The study showed that bees tend to use this angle. That they do may be explained by a genetic theory that those who survive to pass on their genetic traits are those who are more "successful" at living. Perhaps bees that waste less time and wax by making cells with the minimum amount of wax were, at one time, such successful genetic survivors. ❖

DO EXERCISE 8. (STARTS ON PAGE 617)

EXERCISE SET 8.2

Prove the following derivative formulas.

1. $\dfrac{d}{dx} \sec x = \tan x \sec x$

2. $\dfrac{d}{dx} \csc x = -\cot x \csc x$

Differentiate.

3. $y = x \sin x$
4. $y = x \cos x$
5. $f(x) = e^x \sin x$
6. $f(x) = e^x \cos x$
7. $y = \dfrac{\sin x}{x}$
8. $y = \dfrac{\cos x}{x}$
9. $f(x) = \sin^2 x$
10. $f(x) = \cos^2 x$
11. $y = \sin x \cos x$
12. $y = \cos^2 x + \sin^2 x$
13. $f(x) = \dfrac{\sin x}{1 + \cos x}$
14. $f(x) = \dfrac{1 - \cos x}{\sin x}$
15. $y = \tan^2 x$
16. $y = \sec^2 x$
17. $f(x) = \sqrt{1 + \cos x}$
18. $f(x) = \sqrt{1 - \sin x}$
19. $y = x^2 \cos x - 2x \sin x - 2 \cos x$

20. $y = x^2 \sin x - 2x \cos x + 2 \sin x$
21. $y = e^{\sin x}$
22. $y = e^{\cos x}$
23. Find d^2y/dx^2 if $y = \sin x$.
24. Find d^2y/dx^2 if $y = \cos x$.
25. $y = \sin (x^2 + x^3)$
26. $y = \sin (x^5 - x^4)$
27. $f(x) = \cos (x^5 - x^4)$
28. $f(x) = \cos (x^2 + x^3)$
29. $f(x) = \cos \sqrt{x}$
30. $f(x) = \sin \sqrt{x}$
31. $y = \sin (\cos x)$
32. $y = \cos (\sin x)$
33. $y = \sqrt{\cos 4x}$
34. $y = \sin^2 5x$
35. $f(x) = \cot \sqrt[3]{5 - 2x}$
36. $f(x) = \sec (\tan 7x)$
37. $y = \tan^4 3x - \sec^4 3x$
38. $y = \sqrt[5]{\cot 5x - \cos 5x}$

Differentiate. Use the formula $\dfrac{d}{du} \ln |u| = \dfrac{1}{u} du$.

39. $y = \ln |\sin x|$
40. $f(x) = \ln |x - \cos x|$

APPLICATIONS

❖ **Business and Economics**

41. *Total sales.* A company determines that sales during the tth month are given by

$$S(t) = 40{,}000(\sin t + \cos t).$$

The sales are seasonal and fluctuate. Find $S'(t)$.

42. *Total sales.* A company in a northern climate has total sales of skis as given by

$$S(t) = 7\left(1 - \cos\frac{\pi}{6}t\right).$$

Find $S'(t)$.

❖ **Life and Physical Sciences**

43. *Temperature during an illness.* The temperature of a patient during a 12-day illness is given by

$$T(t) = 101.6° + 3\sin\frac{\pi}{8}t.$$

Find $T'(t)$.

44. *Rollercoaster layout.* A rollercoaster is constructed in such a way that it is y meters above ground x meters from the starting point, where

$$y = 15 + 15\sin\frac{\pi}{50}x.$$

Find dy/dx.

45. *Satellite location.* A satellite circles the earth in such a manner that it is y miles from the equator (north or south, height not considered) t minutes after its launch, where

$$y = 5000\left[\cos\frac{\pi}{45}(t - 10)\right].$$

Find dy/dt.

46. *Electric current.* The current i at time t of a wire passing through a magnetic field is given by

$$i = I\sin(\omega t + a).$$

a)   Find the amplitude, the period, and the phase shift.

b)   Find di/dt.

47. *Spring oscillation.* A spring oscillates in such a way that its vertical position at time t, from its position at rest, is given by

$$y = 5\sin(4t + \pi).$$

a)   Find the amplitude, the period, and the phase shift.

b)   Find dy/dt.

48. *Piston movement.* A piston connected to a crankshaft (see the figure) moves up and down in such a way that its second coordinate after time t is given by

$$y(t) = \sin t + \sqrt{25 - \cos^2 t}.$$

Find the rate of change $y'(t)$.

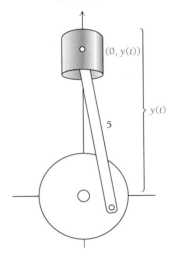

49. *Corridor width.* Referring to Margin Exercise 8, determine how long a beam will go around the corner when $a = 8$ and $b = 8$.

50. *Corridor width.* Referring to Margin Exercise 8, determine how long a beam will go around the corner when $a = 3\sqrt{3}$ and $b = 5\sqrt{5}$.

❖ **General Interest**

51. A V-shaped water trough is to be made with sides that are 10 in. wide and 500 in. long. At what angle θ will the trough be able to carry the greatest volume of water?

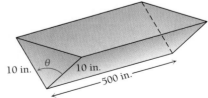

52. The function f given by $f(x) = x + \sin x$ achieves an absolute maximum on the interval $[0, \pi]$. Find the maximum value and where it occurs.

SYNTHESIS EXERCISES

Find the partial derivatives f_x, f_y, f_{xx}, f_{yx}, f_{xy}, and f_{yy}.

53. $f(x, y) = \sin 2y - ye^x$ **54.** $f(x, y) = e^{-x} \cos y$

55. $f(x, y) = \cos (2x + 3y)$

56. $f(x, y) = y \ln (\sin x)$

57. $f(x, y) = x^3 \tan (5xy)$ **58.** $f(x, y) = \sin x \cos y$

Differentiate implicitly to find y'.

59. $y = x \cos y$ **60.** $xy = e^x \sin y$

COMPUTER–GRAPHING CALCULATOR EXERCISES

Use a computer software graphing package or a graphing calculator to graph each of these functions.

61. $f(x) = x \cos x$

62. $f(x) = x \sin x$

63. $f(x) = 2 \sin x + \sin 2x$

64. $f(x) = e^{-x/2} \sin x$

65. $f(x) = \ln |\sec x + \tan x|$

66. $f(x) = \sin x + \ln x$

Use a 3D graphing software program to graph each of these surfaces.

67. $z = 1 + \cos (x^2 + y^2)$

68. $z = e^{-y} \cos x$

 THE CALCULUS EXPLORER:
PowerGrapher, 3D Grapher

Use these programs to graph the functions in Exercises 61–68.

8.3

a) Integrate trigonometric functions.

INTEGRATION OF THE TRIGONOMETRIC FUNCTIONS

Each of the previously developed differentiation formulas yields an integration formula. For example, we have the following.

THEOREM 8

$$\int \sin x \, dx = -\cos x + C,$$

$$\int \cos x \, dx = \sin x + C,$$

$$\int \sec^2 x \, dx = \tan x + C,$$

$$\int \csc^2 x \, dx = -\cot x + C$$

EXAMPLE 1   Find the area under the graph of $y = \cos x$ on the interval $[0, \pi/2]$ (Fig. 1).

Solution

$$\int_0^{\pi/2} \cos x \, dx = [\sin x]_0^{\pi/2} = \sin \frac{\pi}{2} - \sin 0 = 1 - 0 = 1$$

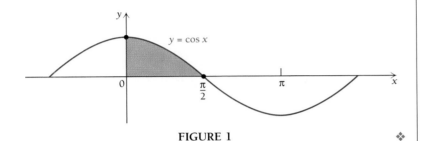

FIGURE 1

DO EXERCISE 1.

We use substitution in the following examples.

EXAMPLE 2   Evaluate $\int \sin^2 x \cos x \, dx$.

Solution

$$\int \sin^2 x \cos x \, dx = \int u^2 \, du \qquad \boxed{\text{Substitution}} \quad \boxed{\begin{array}{l} u = \sin x, \\ du = \cos x \, dx \end{array}}$$

$$= \frac{u^3}{3} + C$$

$$= \frac{\sin^3 x}{3} + C$$

DO EXERCISE 2.

EXAMPLE 3   Evaluate $\int \sin 3x \, dx$.

Solution

$$\int \sin 3x \, dx = \frac{1}{3} \int \sin u \, du \qquad \boxed{\text{Substitution}} \quad \boxed{\begin{array}{l} u = 3x, \\ du = 3 \, dx \end{array}}$$

$$= -\frac{1}{3} \cos u + C$$

$$= -\frac{1}{3} \cos 3x + C$$

DO EXERCISE 3.

1. Find the area under the graph of $y = \sin x$ on the interval $[0, \pi]$.

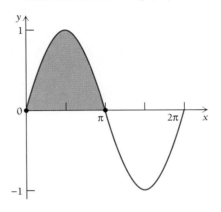

2. Integrate $\int \sin^3 x \cos x \, dx$.

3. Integrate $\int \cos 4x \, dx$.

4. Integrate $\displaystyle\int \frac{2x - \sin x}{x^2 + \cos x}\, dx.$

In an expression where natural logarithms of trigonometric functions are involved, absolute-value signs are usually necessary because the trigonometric functions alternate periodically between positive and negative values. Accordingly, we use the integration formula

$$\int \frac{du}{u} = \ln |u| + C.$$

EXAMPLE 4   Evaluate $\displaystyle\int \frac{1 + \cos x}{x + \sin x}\, dx.$

Solution

$$\int \frac{1 + \cos x}{x + \sin x}\, dx = \int \frac{du}{u} \qquad \underline{\text{Substitution}} \qquad \boxed{\begin{array}{l} u = x + \sin x, \\ du = (1 + \cos x)\, dx \end{array}}$$

$$= \ln |u| + C$$

$$= \ln |x + \sin x| + C \qquad ❖$$

DO EXERCISE 4.

We use integration by parts in the following example.

5. Integrate $\displaystyle\int x \sin x\, dx.$

EXAMPLE 5   Integrate $\int x \cos x\, dx.$

Solution   Let

$$u = x \qquad \text{and} \qquad dv = \cos x\, dx.$$

Then

$$du = dx \qquad \text{and} \qquad v = \sin x.$$

Using the integration-by-parts formula, we get

$$\int \overset{u}{(x)}\overset{dv}{(\cos x\, dx)} = \overset{u}{(x)}\overset{v}{(\sin x)} - \int \overset{v}{(\sin x)}\overset{du}{(dx)}$$

$$= x \sin x - (-\cos x) + C$$

$$= x \sin x + \cos x + C. \qquad ❖$$

DO EXERCISE 5.

To find an integral such as $\int \sec x\, dx$, we must first use some algebra to rename the expression. Then we use substitution.

EXAMPLE 6   Integrate $\int \sec x \, dx$.

Solution

$$\int \sec x \, dx = \int (\sec x) \cdot 1 \, dx \qquad \text{Multiplying by 1}$$

$$= \int \sec x \cdot \frac{\sec x + \tan x}{\sec x + \tan x} \, dx \qquad \text{Substituting } \frac{\sec x + \tan x}{\sec x + \tan x} \text{ for 1}$$

$$= \int \frac{\sec x \, (\sec x + \tan x)}{\sec x + \tan x} \, dx$$

$$= \int \frac{\sec x \tan x + \sec^2 x}{\sec x + \tan x} \, dx$$

$$= \int \frac{du}{u} \qquad \underline{\text{Substitution}} \qquad \boxed{\begin{array}{l} u = \sec x + \tan x, \\ du = (\sec x \tan x + \sec^2 x) \, dx \end{array}}$$

$$= \ln |u| + C$$

$$= \ln |\sec x + \tan x| + C$$

DO EXERCISE 6.

6. Integrate $\int \csc x \, dx$.

$$\left(\text{\emph{Hint:} Multiply by 1 using} \right.$$
$$\left. \frac{\csc x + \cot x}{\csc x + \cot x}. \right)$$

EXERCISE SET   8.3

1. Find the area under the graph of $y = \sin x$ on the interval $[0, \pi/3]$.

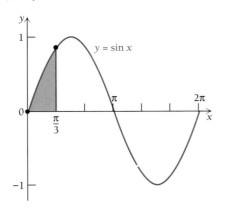

2. Find the area under the graph of $y = \cos x$ on the interval $[-\pi/2, \pi/2]$.

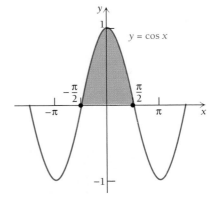

Evaluate.

3. $\displaystyle\int_{-\pi}^{\pi} \cos x \, dx$

4. $\displaystyle\int_{-\pi/2}^{\pi/2} \sin x \, dx$

5. $\displaystyle\int_{-\pi/4}^{2\pi} \sec^2 x \, dx$

6. $\displaystyle\int_{\pi/4}^{\pi/3} \csc^2 x \, dx$

Integrate using substitution.

7. $\displaystyle\int \sin^4 x \cos x \, dx$

8. $\displaystyle\int \sin^5 x \cos x \, dx$

9. $\displaystyle\int -\cos^2 x \sin x \, dx$

10. $\displaystyle\int (\cos^3 x)(-\sin x) \, dx$

11. $\displaystyle\int \cos (x + 3) \, dx$

12. $\displaystyle\int \sin (x + 4) \, dx$

13. $\displaystyle\int \sin 2x \, dx$

14. $\displaystyle\int \cos 3x \, dx$

15. $\displaystyle\int x \cos x^2 \, dx$

16. $\displaystyle\int x \sin x^2 \, dx$

17. $\displaystyle\int e^x \sin (e^x) \, dx$

18. $\displaystyle\int e^x \cos (e^x) \, dx$

19. $\displaystyle\int \tan x \, dx \left(Hint:\ \tan x = \frac{\sin x}{\cos x}. \right)$

20. $\displaystyle\int \cot x \, dx \left(Hint:\ \cot x = \frac{\cos x}{\sin x}. \right)$

Integrate by parts.

21. $\displaystyle\int x \cos 4x \, dx$

22. $\displaystyle\int x \sin 3x \, dx$

23. $\displaystyle\int 3x \cos x \, dx$

24. $\displaystyle\int 2x \sin x \, dx$

25. $\displaystyle\int x^2 \sin x \, dx$ (*Hint:* Let $u = x$ and $dv = x \sin x \, dx$ and use the result of Margin Exercise 5, or use tabular integration.)

26. $\displaystyle\int x^2 \cos x \, dx$ (*Hint:* Let $u = x$ and $dv = x \cos x \, dx$ and use the result of Example 5, or use tabular integration.)

27. $\displaystyle\int \tan^2 x \, dx$

28. $\displaystyle\int \cot^2 x \, dx$

APPLICATIONS

❖ **Business and Economics**

29. *Total sales.* A company in a northern climate has sales of skis as given by

$$S(t) = 7\left(1 - \cos \frac{\pi}{6} t \right),$$

where S is sales, in thousands of dollars, during the tth month. The total sales for the first year are given by

$$\int_0^{12} S(t) \, dt.$$

Find the total sales.

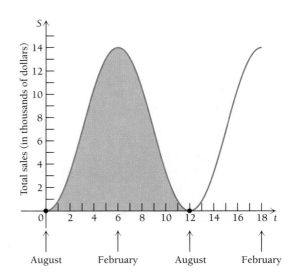

❖ **General Interest**

30. *Total area under a rollercoaster.* A rollercoaster is made in such a way that it is y meters above the ground x meters from the starting point, where

$$y = 15 + 15 \sin \frac{\pi}{50} x.$$

The area under the rollercoaster, from the starting point to a point 100 meters away, is given by

$$\int_0^{100} y \, dx.$$

Find this area.

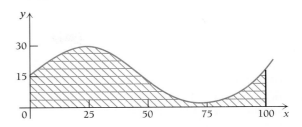

SYNTHESIS EXERCISES

Integrate by parts.

31. $\int \sin (\ln x) \, dx$

32. $\int \cos (\ln x) \, dx$

33. $\int e^x \cos x \, dx$

34. $\int e^x \sin x \, dx$

35. $\int x^3 \sin x \, dx$

36. $\int x^4 \cos x \, dx$

Integrate.

37. $\int \sec^2 7x \, dx$

38. $\int e^x \csc^2 (e^x) \, dx$

39. $\int \sec x \, (\sec x + \tan x) \, dx$

40. $\int \sec u \tan u \, du$

41. $\int \csc u \cot u \, du$

42. $\int e^{2x} \sec (e^{2x}) \, dx$

43. $\int \frac{\cos^2 x}{\sin x} \, dx$

44. $\int \cot 7x \sin 7x \, dx$

45. $\int (1 + \sec x)^2 \, dx$

46. $\int (1 - \csc x)^2 \, dx$

47. $\int_{\pi/4}^{\pi/2} \int_0^{\pi/2} \sin x \cos y \, dy \, dx$

48. $\int_0^\pi \int_0^x \frac{\sin x}{x} \, dy \, dx$

49. $\int_0^\pi \int_0^x x \cos y \, dy \, dx$

50. $\int_\pi^{2\pi} \int_0^\pi (\sin y + \cos x) \, dx \, dy$

COMPUTER EXERCISES

THE CALCULUS EXPLORER:
Riemann Sums, Double Integral Evaluation

Use the programs to evaluate the integrals in Exercises 3–6 and 47–50.

8.4

a) Find certain values of inverse trigonometric functions.

b) Integrate certain functions whose antiderivatives involve inverse trigonometric functions.

Find each of the following.

1. $\sin^{-1}\left(\dfrac{1}{2}\right)$

2. $\sin^{-1}(0)$

3. $\sin^{-1}\left(-\dfrac{\sqrt{3}}{2}\right)$

INVERSE TRIGONOMETRIC FUNCTIONS

Look back at the graph of $y = \sin x$. Note that outputs ranged from -1 to 1. Suppose we wanted to work backward from an output to an input. More specifically, suppose x is some number such that $-1 \le x \le 1$ and that we wanted to find a number y such that

$$-\frac{\pi}{2} \le y \le \frac{\pi}{2} \quad \text{and} \quad \sin y = x.$$

This determines a function, called the *inverse sine function,* given by

$$y = \sin^{-1} x \quad \text{or} \quad \arcsin x.$$

Caution! The -1 is *not* an exponent in this context and $\sin^{-1} x$ is *not* the reciprocal of $\sin x$. The function $\sin^{-1} x$ is an **inverse trigonometric function.**

Let us find a function value.

EXAMPLE 1   Find $\sin^{-1}(\sqrt{3}/2)$.

Solution   *Think:* What number between $-(\pi/2)$ and $\pi/2$ is such that its sine is $\sqrt{3}/2$? That number is $\pi/3$. Thus,

$$\sin^{-1}\left(\frac{\sqrt{3}}{2}\right) = \frac{\pi}{3}, \quad \text{which means that} \quad \sin\frac{\pi}{3} = \frac{\sqrt{3}}{2}. \qquad ❖$$

DO EXERCISES 1–3.

The *inverse cosine function* is given by

$$y = \cos^{-1} x \quad \text{or} \quad y = \arccos x,$$

where

$$x = \cos y, \quad 0 \le y \le \pi, \quad \text{and} \quad -1 \le x \le 1.$$

The *inverse tangent function* is given by

$$y = \tan^{-1} x \quad \text{or} \quad y = \arctan x,$$

where

$$x = \tan y, \quad -\frac{\pi}{2} < y < \frac{\pi}{2}, \quad \text{and} \quad x \text{ is any real number.}$$

DO EXERCISES 4–6.

Graphs of these functions are shown in Figs. 1–3.

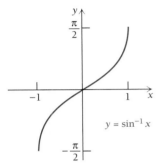

FIGURE 1

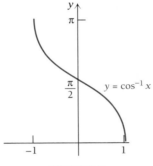

FIGURE 2

Find each of the following.

4. $\tan^{-1}(\sqrt{3})$

5. $\cos^{-1}\left(\frac{\sqrt{3}}{2}\right)$

6. $\cos^{-1}\left(-\frac{1}{2}\right)$

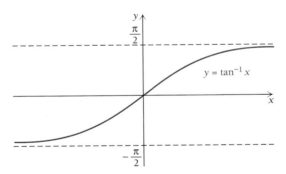

FIGURE 3

Inverse trigonometric functions are integrals of certain functions. You will often see them in tables of integrals.

EXAMPLE 2 Prove that

$$\int \frac{1}{1 + x^2} \, dx = \tan^{-1} x + C.$$

7. Find

$$\int \frac{1}{1 + 25x^2} \, dx$$

using the substitution $u = 5x$ in the integral formula

$$\int \frac{1}{1 + u^2} \, du = \tan^{-1} u + C.$$

8. Find

$$\int - \frac{1}{\sqrt{1 - 4x^2}} \, dx$$

using the substitution $u = 2x$ in the integral formula

$$\int - \frac{1}{\sqrt{1 - u^2}} \, du = \cos^{-1} u + C.$$

Solution   We use a method of integration known as *trigonometric substitution*. We simplify the integral using the identity $1 + \tan^2 u = \sec^2 u$:

$$\int \frac{1}{1 + x^2} \, dx$$

$$= \int \frac{1}{1 + \tan^2 u} \sec^2 u \, du$$

Substitution	$u = \tan^{-1} x,$
	$x = \tan u,$
	$dx = \sec^2 u \, du$

$$= \int \frac{1}{\sec^2 u} \sec^2 u \, du \qquad \text{By Identity (2)}$$

$$= \int du = u + C = \tan^{-1} x + C. \qquad \quad ❖$$

DO EXERCISE 7.

EXAMPLE 3   Prove that

$$\int - \frac{1}{\sqrt{1 - x^2}} \, dx = \cos^{-1} x + C.$$

Solution   Again we use substitution:

$$\int - \frac{1}{\sqrt{1 - x^2}} \, dx$$

$$= \int \frac{1}{\sqrt{1 - \cos^2 u}} \sin u \, du.$$

Substitution	$u = \cos^{-1} x,$
	$x = \cos u,$
	$dx = -\sin u \, du$

By Identity (1), $\sin^2 u + \cos^2 u = 1$, so

$$1 - \cos^2 u = \sin^2 u.$$

Then, since u is such that $0 \le u \le \pi$, $\sin u \ge 0$, so $\sqrt{\sin^2 u} = \sin u$ and the integral becomes

$$\int \frac{1}{\sqrt{\sin^2 u}} \cdot \sin u \, du \qquad \text{By Identity (1)}$$

$$= \int \frac{1}{\sin u} \cdot \sin u \, du = \int du = u + C = \cos^{-1} x + C. \qquad ❖$$

DO EXERCISE 8.

EXERCISE SET 8.4

Find each of the following.

1. $\sin^{-1}\left(\dfrac{\sqrt{2}}{2}\right)$

2. $\sin^{-1}\left(-\dfrac{\sqrt{2}}{2}\right)$

3. $\cos^{-1}(0)$

4. $\cos^{-1}\left(\dfrac{\sqrt{2}}{2}\right)$

5. $\tan^{-1}(1)$

6. $\tan^{-1}(-1)$

7. $\sin^{-1}\left(-\dfrac{1}{2}\right)$

8. $\cos^{-1}\left(-\dfrac{\sqrt{3}}{2}\right)$

Integrate using substitution.

9. $\displaystyle\int \dfrac{e^t}{1 + e^{2t}}\, dt$

10. $\displaystyle\int \dfrac{-1}{\sqrt{1 - 25x^2}}\, dx$

11. Show that

$$\int \frac{1}{\sqrt{1 - x^2}}\, dx = \sin^{-1} x + C.$$

12. Integrate using the formula in Exercise 11:

$$\int \frac{1}{\sqrt{1 - 49x^2}}\, dx.$$

 If your calculator has an inverse function key, find each of the following.

13. $\sin^{-1}(0.9874)$

14. $\cos^{-1}(-0.3487)$

15. $\cos^{-1}(0.9988)$

16. $\tan^{-1}(2000)$

SYNTHESIS EXERCISE

17. Since

$$\int \frac{1}{1 + x^2}\, dx = \tan^{-1} x + C,$$

it follows that if $u = \tan^{-1} x$,

then

$$\frac{du}{dx} = \frac{1}{1 + x^2}.$$

Use this result and integration by parts to find

$$\int \tan^{-1} x \, dx.$$

CHAPTER SUMMARY AND REVIEW 8

TERMS TO KNOW

Angle, p. 594
Degrees, p. 594
Radians, p. 595
Trigonometric function, p. 596
Circular function, p. 596
Sine function, p. 596, 598

Cosine function, p. 596, 598
Periodic function, p. 602
Tangent function, p. 604
Cotangent function, p. 604
Secant function, p. 604
Cosecant function, p. 604

Trigonometric identity, p. 605
Sum–difference identities, p. 606
Double-angle identity, p. 607
Inverse trigonometric function,
  p. 626

REVIEW EXERCISES

These review exercises are for test preparation. They can also be used as a lengthened practice test. Answers are at the back of the book. The answers also contain bracketed section references, which tell you where to restudy if your answer is incorrect.

Convert to radian measure.

1. $240°$

2. $315°$

Convert to degree measure.

3. $\dfrac{5\pi}{6}$

4. $\dfrac{7\pi}{3}$

Find each of the following.

5. $\sin \dfrac{2\pi}{3}$

6. $\cos \left(-\dfrac{\pi}{2}\right)$

7. $\tan \dfrac{\pi}{4}$

8. $\csc \pi$

Differentiate.

9. $y = \tan (3x^2 - x)$

10. $y = \sin^6 x$

11. $y = \ln (\tan x)$

12. $f(x) = e^{\sin 3x}$

13. $f(x) = \dfrac{1 + \cos 2x}{\sin 2x}$

14. $f(x) = \cos \sqrt{t} + \sqrt{\cos t}$

15. $y = \sqrt{\tan x}$

16. $y = e^x \sin x$

17. $f(x) = \sin 3x \cos 3x$

18. $y = \sin x - x \cos x$

19. *Physical science: Pendulum oscillation.* A pendulum oscillates in such a way that its horizontal position at time t, from its position at rest, is given by

$$y = 3 \cos (3t - 2\pi).$$

Find dy/dt.

20. *Life science: Temperature during an illness.* The temperature of a patient during an 8-day illness is given by

$$T(t) = 99.7° + 5 \sin \dfrac{\pi}{3} t.$$

Find $T'(t)$.

21. Find the area under the graph of $y = \cos x$ on the interval $[0, \pi/6]$.

Integrate using substitution.

22. $\displaystyle\int e^x \cos e^x \, dx$

23. $\displaystyle\int \sin (x - 1) \, dx$

24. $\displaystyle\int \cos 8x \, dx$

Integrate by parts.

25. $\displaystyle\int 3x \cos 2x \, dx$

26. $\displaystyle\int x \sin 3x \, dx$

Integrate.

27. $\displaystyle\int \dfrac{1}{\sqrt{1 - 64t^2}} \, dt$

28. $\displaystyle\int \dfrac{5 - \cos x}{5x - \sin x} \, dx$

SYNTHESIS EXERCISES

29. Find $\dfrac{d^3y}{dx^3}$: $y = e^{\sin x}$.

30. Find f_x and f_{xy}: $f(x, y) = \sin y \cos x$.

EXERCISES FOR THINKING AND WRITING

31. Compare the notions of radian and degree.

32. A graphing calculator has many advantages. Explain a disadvantage of such a calculator when graphing a function like

$$f(x) = \dfrac{\sin x}{x}.$$

33. Suppose you know that

$$\int_a^b \sin x \, dx = 0.$$

Explain the meaning of this result in terms of area.

34. How many maximum values are there of the cosine function? Where do they occur?

CHAPTER TEST

8

1. Convert $120°$ to radian measure. (Leave the answer in terms of π.)

2. Convert $5\pi/3$ to degree measure.

Find each of the following.

3. $\sin \dfrac{\pi}{6}$

4. $\cos \pi$

Differentiate.

5. $y = \cos t$

6. $y = \sin (3x^2 - 5x)$

7. $f(x) = \dfrac{x}{\sin x}$

8. $f(t) = \tan^2 t$

9. $f(x) = \sqrt{\sin x + \cos x}$

10. $y = \dfrac{\sin x + \cos x}{\sin x - \cos x}$

11. *Business: Seasonal sales.* A company has seasonal sales as given by

$$S(t) = 20 \left(1 + \sin \frac{\pi}{8} t \right).$$

Find $S'(t)$.

12. Find the area under $y = \sin x$ on the interval $[0, \pi/6]$.

Integrate using substitution.

13. $\displaystyle\int \sin^9 x \cos x \, dx$

14. $\displaystyle\int \cos 5t \, dt$

15. Integrate by parts: $\displaystyle\int 5x \sin 5x \, dx$.

16. Integrate using substitution: $\displaystyle\int \dfrac{1}{1 + 16t^2} \, dt$.

Integrate.

17. $\displaystyle\int \dfrac{2 - \cos x}{2x - \sin x} \, dx$

18. $\displaystyle\int_0^{\pi/4} \sec^2 x \, dx$

SYNTHESIS EXERCISES

19. Find

$$\frac{dy}{dx} \quad \text{and} \quad \frac{d^2y}{dx^2},$$

if $y = 2e^{\cos x}$.

20. Find f_x and f_{xy}: $f(x, y) = \dfrac{\cos x}{\cos y}$.

21. Use an input–output table or a graph to find the following limit:

$$\lim_{x \to 0} \ln |x + \sin x|.$$

9

DIFFERENTIAL EQUATIONS

A *differential equation* is an equation that involves derivatives or differentials. In this chapter we learn many ways to solve differential equations and we apply these methods to the solution of many interesting problems.

AN APPLICATION

In 1937, eight pheasants were released on Protection Island off the coast of Washington State. Find the limiting value of the pheasant population on the island.

THE MATHEMATICS

A type of *differential equation* used to analyze population growth is

$$\frac{dP}{dt} = kP(L - P),$$

where L is the limiting value. We use this equation to see that the limiting value is 3796 pheasants.

9.1

OBJECTIVES

a) Solve certain differential equations, giving both general and particular solutions.

b) Solve certain differential equations given a condition $f(a) = b$.

c) Verify that a given function is a solution of a given differential equation.

DIFFERENTIAL EQUATIONS*

A *differential equation* is an equation that involves derivatives or differentials. In Chapter 4, we studied one very important differential equation,

$$\frac{dP}{dt} = kP,$$

where P, or $P(t)$, is the population at time t. This equation is a model of uninhibited population growth. Its solution is

$$P = P_0 e^{kt},$$

where the constant P_0 is the size of the initial population; that is, at $t = 0$. As this one equation illustrated, differential equations are rich in application.

Solving Certain Differential Equations

In this chapter we will frequently use the notation y' for a derivative—mainly because it is simple. Thus, if $y = f(x)$, then

$$y' = \frac{dy}{dx} = f'(x).$$

We have already found solutions of certain differential equations when we found antiderivatives or indefinite integrals. The differential equation

$$\frac{dy}{dx} = g(x) \quad \text{or} \quad y' = g(x)$$

has the solution

$$y = \int g(x)\, dx.$$

EXAMPLE 1   Solve $y' = 2x$.

Solution

$$y = \int 2x\, dx = x^2 + C \qquad \qquad ❖$$

---

*To the instructor: Sections 9.1 and 9.2 are a repetition of Section 6.7. They are repeated here to allow flexibility in teaching.

DO EXERCISE 1.

Look again at the solution to Example 1. Note the constant of integration. This solution is called a *general solution* because taking all values of C gives *all* the solutions. Taking specific values of C gives particular solutions. For example, the following are particular solutions of $y' = 2x$:

$$y = x^2 + 3, \qquad y = x^2, \qquad y = x^2 - 3.$$

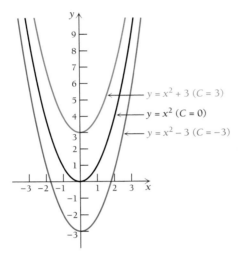

FIGURE 1

The graph in Fig. 1 shows the curves of a few particular solutions. The general solution can be envisioned as the set of all particular solutions, a *family* of curves.

DO EXERCISE 2.

Knowing the value of a function at a particular point may allow us to select a particular solution from the general solution.

EXAMPLE 2   Solve $f'(x) = e^x + 5x - x^{1/2}$, given that $f(0) = 8$.

Solution

a)   First find the general solution:

$$f(x) = \int f'(x)\,dx = e^x + \tfrac{5}{2}x^2 - \tfrac{2}{3}x^{3/2} + C.$$

1. Solve $y' = 3x^2$.

2. Given $y' = 3x^2$.
   a)   Write the general solution.
   b)   Write three particular solutions.

Waves can be represented by differential equations.

3. Solve

$$y' = 2x$$

given that $y = 7$ when $x = 1$.

4. Solve

$$f'(x) = \frac{1}{x} - 2x + x^{1/2}$$

given that $f(1) = 4$.

5. Show that

$$y = 2e^x - 7e^{3x}$$

is a solution of

$$y'' - 4y' + 3y = 0.$$

6. Show that

$$y = xe^{2x}$$

is a solution of

$$\frac{dy}{dx} - 2y = e^{2x}.$$

b) Since $f(0) = 8$, we substitute to find C:

$$8 = e^0 + \tfrac{5}{2} \cdot 0^2 - \tfrac{2}{3} \cdot 0^{3/2} + C$$
$$8 = 1 + C$$
$$7 = C.$$

Thus the particular solution is $f(x) = e^x + \tfrac{5}{2}x^2 - \tfrac{2}{3}x^{3/2} + 7$. ❖

DO EXERCISES 3 AND 4.

Verifying Solutions

To verify that a function is a solution of a differential equation, we find the necessary derivatives and substitute.

EXAMPLE 3 Show that $y = 4e^x + 5e^{3x}$ is a solution of

$$y'' - 4y' + 3y = 0.$$

Solution

a) We first find y' and y'':

$$y' = 4e^x + 15e^{3x},$$
$$y'' = 4e^x + 45e^{3x}.$$

b) Then we substitute in the differential equation, as follows.

$y'' - 4y' + 3y = 0$	
$(4e^x + 45e^{3x}) - 4(4e^x + 15e^{3x}) + 3(4e^x + 5e^{3x})$	0
$4e^x + 45e^{3x} - 16e^x - 60e^{3x} + 12e^x + 15e^{3x}$	
0	0

❖

DO EXERCISES 5 AND 6.

EXERCISE SET **9.1**

Find the general solution and three particular solutions.

1. $y' = 4x^3$

2. $y' = 6x^5$

3. $y' = e^{2x} + x$

4. $y' = e^{3x} - x$

5. $y' = \dfrac{3}{x} - x^2 + x^5$

6. $y' = \dfrac{5}{x} + x^2 - x^4$

Find the particular solution determined by the given condition.

7. $y' = x^2 + 2x - 3;\ y = 4$ when $x = 0$

8. $y' = 3x^2 - x + 5;\ y = 6$ when $x = 0$

9. $f'(x) = x^{2/3} - x;\ f(1) = -6$

10. $f'(x) = x^{2/5} + x; f(1) = -7$

11. Show that $y = x \ln x + 3x - 2$ is a solution of

$$y'' - \frac{1}{x} = 0 \quad \text{for } x > 0.$$

12. Show that $y = x \ln x - 5x + 7$ is a solution of

$$y'' - \frac{1}{x} = 0 \quad \text{for } x > 0.$$

13. Show that $y = e^x + 3xe^x$ is a solution of

$$y'' - 2y' + y = 0.$$

14. Show that $y = -2e^x + xe^x$ is a solution of

$$y'' - 2y' + y = 0.$$

APPLICATIONS

❖ **Business and Economics**

15. *Total cost from marginal cost.* Marginal cost for a certain product is $C'(x) = 2.6 - 0.02x$. Find the total-cost function $C(x)$ and the average cost $A(x)$, assuming fixed costs are $120; that is, $C(0) = 120.

16. *Total revenue from marginal revenue.* Marginal revenue for a certain product is $R'(x) = 300 - 2x$. Find the total-revenue function $R(x)$ assuming $R(0) = 0$.

SYNTHESIS EXERCISE

17. *Total profit from marginal profit.* A firm's marginal profit P as a function of total cost C is given by

$$\frac{dP}{dC} = \frac{-200}{(C + 3)^{3/2}}.$$

a) Find the profit function $P(C)$ if $P = 10 when $C = 61.

b) At what cost will the firm break even $(P = 0)$?

9.2

SEPARATION OF VARIABLES

Consider the differential equation

$$\frac{dy}{dx} = 2xy. \tag{1}$$

We treat dy/dx as a quotient, as we did in Chapter 5. We multiply Eq. (1) by dx and then by $1/y$ to get

$$\frac{dy}{y} = 2x \, dx, \quad y \neq 0. \tag{2}$$

OBJECTIVE

a) Solve certain differential equations using separation of variables.

1. Use separation of variables to solve

$$\frac{dy}{dx} = 3x^2 y.$$

At the 1968 Olympic Games in Mexico City, Bob Beamon made what was believed to be a miracle long jump of 29 ft, $2\frac{1}{2}$ in. Many believed this was due to the altitude, which was 7400 ft. Using differential equations for analysis, M. N. Bearley refuted the altitude theory in "The Long Jump Miracle of Mexico City" (*Mathematics Magazine*, vol. 45, November 1972, pp. 241–246). Bearley argues that the world record jump was a result of Beamon's exceptional speed (9.5 sec in the 100-yd dash) and the fact that he hit the take-off board in perfect position.

We now say that we have *separated the variables*, meaning that all the expressions involving y are on one side and all those involving x are on the other. We then integrate both sides of Eq. (2):

$$\int \frac{dy}{y} = \int 2x \, dx + C$$

$$\ln y = x^2 + C, \qquad y > 0.$$

We use only one constant because any two antiderivatives differ by a constant. Recall that the definition of logarithms says that if $\log_a b = t$, then $b = a^t$. Now, $\ln y = \log_e y = x^2 + C$, so by the definition of logarithms, we have

$$y = e^{x^2 + C}$$
$$= e^{x^2} \cdot e^C.$$

Thus the solution to differential equation (1) is

$$y = C_1 e^{x^2}, \quad \text{where } C_1 = e^C.$$

DO EXERCISE 1.

EXAMPLE 1   Solve

$$3y^2 \frac{dy}{dx} + x = 0, \quad \text{where } y = 5 \text{ when } x = 0.$$

Solution

a)   We first separate the variables:

$$3y^2 \frac{dy}{dx} = -x$$

$$3y^2 \, dy = -x \, dx.$$

We then integrate both sides:

$$\int 3y^2 \, dy = \int -x \, dx + C$$

$$y^3 = -\frac{x^2}{2} + C = C - \frac{x^2}{2}$$

$$y = \sqrt[3]{C - \frac{x^2}{2}}. \qquad \text{Taking the cube root}$$

b) Since $y = 5$ when $x = 0$, we substitute to find C:

$$5 = \sqrt[3]{C - \frac{0^2}{2}} \qquad \text{Substituting 5 for } y \text{ and 0 for } x$$

$$5 = \sqrt[3]{C}$$

$$125 = C. \qquad \text{Cubing both sides}$$

The particular solution is

$$y = \sqrt[3]{125 - \frac{x^2}{2}}.$$

❖

DO EXERCISE 2.

EXAMPLE 2 Solve

$$\frac{dy}{dx} = \frac{x}{y}.$$

Solution We first separate variables:

$$y\frac{dy}{dx} = x$$

$$y \, dy = x \, dx.$$

We then integrate both sides:

$$\int y \, dy = \int x \, dx$$

$$\frac{y^2}{2} = \frac{x^2}{2} + C$$

$$y^2 = x^2 + 2C$$

$$= x^2 + C_1,$$

where $C_1 = 2C$. We make this substitution to simplify the equation. We then obtain the solutions

$$y = \sqrt{x^2 + C_1}$$

and

$$y = -\sqrt{x^2 + C_1}.$$

❖

DO EXERCISE 3.

2. Solve

$$3y^2 \frac{dy}{dx} - 2x = 0,$$

where $y = 7$ when $x = 2$.

3. Solve $\dfrac{dy}{dx} = \dfrac{5}{y}$.

4. Solve $y' = 2x + xy$.

EXAMPLE 3   Solve $y' = x - xy$.

Solution   Before we separate variables, we replace y' by dy/dx:

$$\frac{dy}{dx} = x - xy.$$

Now we separate variables:

$$dy = (x - xy)\ dx$$
$$dy = x(1 - y)\ dx$$
$$\frac{dy}{1 - y} = x\ dx.$$

Then we integrate both sides:

$$\int \frac{dy}{1 - y} = \int x\ dx$$

$$-\ln\ (1 - y) = \frac{x^2}{2} + C \qquad 1 - y > 0$$

$$\ln\ (1 - y) = -\frac{x^2}{2} - C$$

$$1 - y = e^{-x^2/2 - C}$$

$$-y = e^{-x^2/2 - C} - 1$$

$$y = -e^{-x^2/2 - C} + 1$$

$$= -e^{-x^2/2} \cdot e^{-C} + 1.$$

Thus,

$$y = 1 + C_1 e^{-x^2/2}, \quad \text{where } C_1 = -e^{-C}.$$ ❖

DO EXERCISE 4.

An Economic Application: Elasticity

EXAMPLE 4   Suppose for a certain product that the elasticity of demand is 1 for all $p > 0$. That is, $E(p) = 1$ for all $p > 0$. Find the demand function $x = D(p)$. (See Section 4.6.)

Solution   Since $E(p) = 1$ for all $p > 0$,

$$1 = E(p) = -\frac{p\ D'(p)}{D(p)} = -\frac{p}{x} \cdot \frac{dx}{dp}.$$

Then

$$\frac{dx}{dp} = -\frac{x}{p}. \tag{3}$$

Separating the variables, we get

$$\frac{dp}{p} = -\frac{dx}{x}.$$

Now we integrate both sides:

$$\int \frac{dp}{p} = -\int \frac{dx}{x}$$

$\ln p = -\ln x + C.$ $p > 0$ and $dx/dp < 0$, since demand functions are decreasing. So $x > 0$. See Eq. (3).

We can express $C = \ln C_1$, since any real number C is the natural logarithm of some number C_1. Then

$$\ln p = \ln C_1 - \ln x$$
$$= \ln \frac{C_1}{x},$$

so

$$p = \frac{C_1}{x} \quad \text{and} \quad x = \frac{C_1}{p}.$$

This characterizes those demand functions for which the elasticity is always 1.   ❖

DO EXERCISE 5.

A Psychological Application: Reaction to a Stimulus

The Weber–Fechner Law

In psychology, one model of stimulus–response asserts that the rate of change dR/dS of the reaction R with respect to a stimulus S is inversely proportional to the stimulus. That is,

$$\frac{dR}{dS} = \frac{k}{S},$$

where k is some positive constant.

5. *Economics: Elasticity.* Find the demand function, $x = D(p)$, given the elasticity condition

$$E(p) = 3 \quad \text{for all } p > 0.$$

To solve this equation, we first separate the variables:

$$dR = k \cdot \frac{dS}{S}.$$

We then integrate both sides:

$$\int dR = \int k \cdot \frac{dS}{S}$$

$$R = k \ln S + C. \qquad (4)$$

Now suppose we let S_0 be the lowest level of the stimulus that can be detected consistently. This is the *threshold value,* or *detection threshold.* For example, the lowest level of sound that can be consistently detected is the tick of a watch at 20 feet, under very quiet conditions. If S_0 is the lowest level of sound that can be detected, it seems reasonable that the reaction to it would be 0; that is, $R(S_0) = 0$. Substituting this condition in Eq. (4), we get

$$0 = k \ln S_0 + C,$$

or

$$-k \ln S_0 = C.$$

Replacing C in Eq. (4) by $-k \ln S_0$, we get

$$R = k \cdot \ln S - k \cdot \ln S_0$$
$$= k(\ln S - \ln S_0).$$

Using a property of logarithms, we have

$$R = k \cdot \ln \frac{S}{S_0}.$$

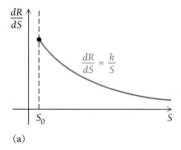

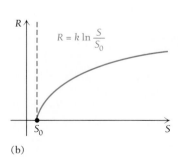

(a) (b)

FIGURE 1

Look at the graphs of dR/dS and R in Fig. 1. Note that as the stimulus gets larger, the rate of change decreases; that is, the reaction becomes smaller as the stimulation you receive gets stronger. For example, suppose you are in a room with one lamp and that lamp has a 50-watt bulb in it. If the bulb were suddenly changed to 100 watts, you would probably be very aware of the difference. That is, your reaction would be strong. If the bulb were then changed to 150 watts, your reaction would not be as great as it was to the change from 50 to 100 watts. A change from a 150- to a 200-watt bulb would cause even less reaction, and so on.

For your interest, here are some other detection thresholds.

Stimulus	Detection threshold
Light	The flame of a candle 30 miles away on a dark night
Taste	Water diluted with sugar in the ratio of 1 teaspoon to 2 gallons
Smell	One drop of perfume diffused into the volume of three average-sized rooms
Touch	The wing of a bee dropped on your cheek at a distance of 1 centimeter (about $\frac{3}{8}$ of an inch)

DO EXERCISE 6.

6. *The Brentano–Stevens Law.* The Weber–Fechner Law has been the subject of great debate among psychologists as to its validity. The model

$$\frac{dR}{dS} = k \cdot \frac{R}{S},$$

where k is a positive constant, has also been conjectured and experimented with. Find the general solution to this equation. (This has also been referred to as the *Power Law of Stimulus–Response.*)

EXERCISE SET 9.2 _____

Solve.

1. $\dfrac{dy}{dx} = 4x^3y$

2. $\dfrac{dy}{dx} = 5x^4y$

3. $3y^2\dfrac{dy}{dx} = 5x$

4. $3y^2\dfrac{dy}{dx} = 7x$

5. $\dfrac{dy}{dx} = \dfrac{2x}{y}$

6. $\dfrac{dy}{dx} = \dfrac{x}{2y}$

7. $\dfrac{dy}{dx} = \dfrac{3}{y}$

8. $\dfrac{dy}{dx} = \dfrac{4}{y}$

9. $y' = 3x + xy,$
     $y = 5$ when $x = 0$

10. $y' = 2x - xy,$
     $y = 9$ when $x = 0$

11. $y' = 5y^{-2},$
     $y = 3$ when $x = 2$

12. $y' = 7y^{-2},$
     $y = 3$ when $x = 1$

13. $\dfrac{dy}{dx} = 3y$

14. $\dfrac{dy}{dx} = 4y$

15. $\dfrac{dP}{dt} = 2P$

16. $\dfrac{dP}{dt} = 4P$

APPLICATIONS

❖ **Business and Economics**

17. *Economics: Utility.* The reaction R, in pleasure units, by a consumer receiving S units of a product can be modeled by the differential equation

$$\frac{dR}{dS} = \frac{k}{S+1},$$

where k is a positive constant.

   a) Use separation of variables to solve the differential equation.

   b) Rewrite the solution in terms of the initial condition $R(0) = 0$.

   c) Explain why the condition $R(0) = 0$ is reasonable.

18. *Domar's capital expansion model* is

$$\frac{dI}{dt} = hkI,$$

where I = investment, h = investment productivity (constant), k = marginal productivity to consume (constant), and t = time.

   a) Use separation of variables to solve the differential equation.

   b) Rewrite the solution in terms of the condition $I_0 = I(0)$.

Elasticity. Find the demand function $p = D(x)$ given the following elasticity conditions.

19. $E(p) = \dfrac{p}{200 - p}$; $x = 190$ when $p = 10$

20. $E(p) = \dfrac{4}{p}$; $x = e$ when $p = 4$

21. $E(p) = n$ for some constant n and all $p > 0$

22. $E(p) = 2$ for all $p > 0$

❖ **Life and Physical Sciences**

23. a) Use separation of variables to solve the differential-equation model of uninhibited growth,

$$\frac{dP}{dt} = kP.$$

   b) Rewrite the solution in terms of the condition $P_0 = P(0)$.

24. *Newton's Law of Cooling.* The temperature T of a cooling object drops at a rate that is proportional to the difference $T - M$, where M is the constant temperature of the surrounding medium. Thus,

$$\frac{dT}{dt} = -k(T - M),$$

where k is a positive constant and t is time.

   a) Solve the differential equation.

   b) Rewrite the solution in terms of the condition $T(0) = 200°$.

9.3

OBJECTIVES

a) State the differential equation

$$\frac{dP}{dt} = kP(L - P)$$

for the inhibited growth model.

(continued)

APPLICATIONS: THE INHIBITED GROWTH MODEL $dP/dt = kP(L - P)$

Recall the model of uninhibited growth

$$\frac{dP}{dt} = kP,$$

with solution (by separation of variables)

$$P = P_0 e^{kt}.$$

A more realistic model of population growth will take into account factors other than the size of the population. For example, there may be some compelling reason why the population P can never grow beyond a limiting value L—perhaps a limitation on food, living space, or other natural resources. In such cases we expect the *growth rate* to lessen as the population size increases, and to approach 0 as P approaches L.

EXAMPLE 1   Buy an ant colony at a pet shop. Place a fixed amount of food and water in the colony (enough to last a long period of time) and close it up. The growth rate of the ants is inhibited by the fixed amount of food and water.   ❖

The simplest model for such an inhibition to express itself is for the growth rate dP/dt to be directly proportional to *both* the population size P and to its remaining possible room for growth $L - P$. Then we have the following.

DEFINITION

The model of inhibited growth is

$$\frac{dP}{dt} = kP(L - P),$$

where $L =$ the *limiting value* and $k > 0$.

Work through the exploratory exercise set on the following page, either on your own or in class with your instructor.

Now let us solve the differential equation

$$\frac{dP}{dt} = kP(L - P).$$

Note that this equation defines P as a function of t implicitly. That is, it is assumed that P is a function of t even though t does not appear on the right-hand side of the equation. We first separate variables:

$$dP = kP(L - P) \, dt$$

$$\frac{dP}{P(L - P)} = k \, dt.$$

b)   State the solution to the inhibited growth model

$$P(t) = \frac{P_0 L}{P_0 + e^{-Lkt}(L - P_0)}.$$

c)   Given the point of inflection, find the limiting value.

d)   Given the values of the constants for the equation in part (b):

   i)   write the equation for $P(t)$ in terms of the given constants;

   ii) sketch a graph of $P(t)$, given the time t at which the point of inflection occurs;

   iii) find t such that $P(t) = M$, for some number M.

e)   Given certain constants and function values, find the value of k in the formula in part (b).

EXPLORATORY EXERCISES: SHIPWRECKED ON AN ISLAND

A ship carrying 100 passengers wrecks on an island, never to be rescued. The population grows over the next 120 years as indicated in the table.

Time, t (in years)	Population, P	Growth rate, $\Delta P/\Delta t$	Time, t (in years)	Population, P	Growth rate, $\Delta P/\Delta t$
0	100	0 (assumed)	70	470	
10	120	2	80	500	
20	150	3	90	520	
30	200		100	530	
40	270		110	535	
50	350		120	538	
60	420				

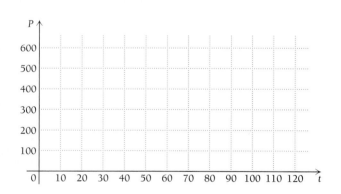

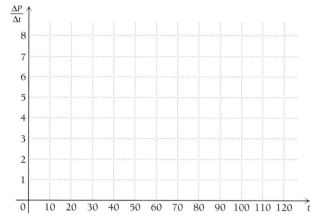

a) Complete the table.

b) Plot P versus t. Use solid dots and connect them with a smooth curve.

c) Plot $\Delta P/\Delta t$ versus t. Use open dots and connect them with a smooth curve.

d) Does P resemble the graph of an exponential function of the type $P(t) = P_0 e^{kt}$?

e) Between what values of t is $\Delta P/\Delta t$ increasing?

f) Between what values of t is $\Delta P/\Delta t$ decreasing?

g) Is there a point of inflection? Where?

h) Does there appear to be a limiting value to the population growth? If so, what is it?

Now we integrate both sides. We use Table 1 to integrate the left side. A similar example appears in Example 1 of Section 5.7:

$$\int \frac{dP}{P(L - P)} = \int k \, dt$$

$$\frac{1}{L} \ln \frac{P}{L - P} = kt + C.$$

To solve for P, we first find an expression for $P/(L - P)$:

$$\ln \frac{P}{L - P} = Lkt + LC$$

$$\frac{P}{L - P} = e^{Lkt + LC} = e^{Lkt} \cdot e^{LC}.$$

Now, letting $C_1 = e^{LC}$, we have

$$\frac{P}{L - P} = C_1 e^{Lkt}.$$

We complete the solution for P as follows, dropping the subscript from C_1 for simplicity:

$$P = Ce^{Lkt}(L - P)$$
$$P = CLe^{Lkt} - PCe^{Lkt}$$
$$P + PCe^{Lkt} = CLe^{Lkt}$$
$$P(1 + Ce^{Lkt}) = CLe^{Lkt}$$
$$P = \frac{CLe^{Lkt}}{1 + Ce^{Lkt}}.$$

THEOREM 1

The *model of inhibited growth*

$$\frac{dP}{dt} = kP(L - P) \qquad (1)$$

has the solution

$$P(t) = \frac{CLe^{Lkt}}{1 + Ce^{Lkt}}. \qquad (2)$$

DO EXERCISE 1.

For each value of the constant C, we get an S-shaped curve. These curves fill up the strip $0 < P < L$ (see Fig. 1).

1. Let $C = 1$, $L = 1$, and $k = 1$. Then Eq. (2) becomes

$$P(t) = \frac{e^t}{1 + e^t}.$$

We want to graph $P(t)$, but to ease computations we multiply by 1 as follows:

$$P(t) = \frac{e^t}{1 + e^t} \cdot \frac{e^{-t}}{e^{-t}}$$

$$= \frac{e^t \cdot e^{-t}}{1 \cdot e^{-t} + e^t \cdot e^{-t}}$$

$$= \frac{1}{e^{-t} + 1}.$$

a) 📱 Complete this table of function values.

t	-3	-2	-1	0	1	2	3
$P(t)$							0.95

For example,

$$P(3) = \frac{1}{e^{-3} + 1} \approx \frac{1}{0.05 + 1}$$

$$= \frac{1}{1.05} \approx 0.95.$$

b) Graph $P(t)$.

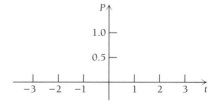

2. Given

$$P = \frac{CLe^{Lkt}}{1 + Ce^{Lkt}} \qquad (1)$$

and

$$C = \frac{P_0}{L - P_0}. \qquad (2)$$

a) Replace C in Eq. (1) by the right-hand side of Eq. (2) and show your work in simplifying to get

$$P = \frac{P_0 L e^{Lkt}}{P_0 e^{Lkt} + (L - P_0)}. \qquad (3)$$

b) Simplify Eq. (3) by multiplying by e^{-Lkt}/e^{-Lkt}.

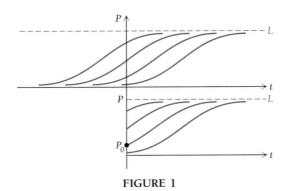

FIGURE 1

If we don't project these curves into the past (that is, if we don't use *negative time*), then the solution curves are confined to the part of the strip for which $t \geqslant 0$. For any initial population P_0, there is a growth curve emanating from $(0, P_0)$, showing the evolution of that population. That is, $P = P_0$ at $t = 0$ is a condition that picks out a unique solution. We can eliminate the constant C and express P in terms of P_0 as follows:

$$P_0 = P(0) = \frac{CLe^{Lk0}}{1 + Ce^{Lk0}} = \frac{CL}{1 + C}. \qquad (3)$$

Solving for C, we get

$$C = \frac{P_0}{L - P_0},$$

$$P(t) = \frac{P_0 L e^{Lkt}}{P_0 e^{Lkt} + (L - P_0)}. \qquad (4)$$

We can simplify this further by multiplying by e^{-Lkt}/e^{-Lkt}.

DO EXERCISE 2.

THEOREM 2

The *model of inhibited growth* (often called the *logistic function*)

$$\frac{dP}{dt} = kP(L - P)$$

has the solution

$$P(t) = \frac{P_0 L}{P_0 + e^{-Lkt}(L - P_0)}.$$

Recall that an inflection point is a point across which a curve changes concavity (see p. 196). The inflection point on the graph of $P(t)$ is important in application. It gives the time and population size at which the *rate* of population growth reaches a maximum and changes from an increasing function to a decreasing function. The following tells us where the inflection point of P occurs.

THEOREM 3

P has an inflection point at that value of t where

$$P = \frac{1}{2}L.$$

The following example is based on an actual experiment.*

EXAMPLE 2   Before 1937 there were no pheasants on Protection Island off the coast of Washington State. In 1937 eight pheasants, two cocks and six hens, were released on the island. The curve in Fig. 2 shows the growth of the pheasant population over the next six years.

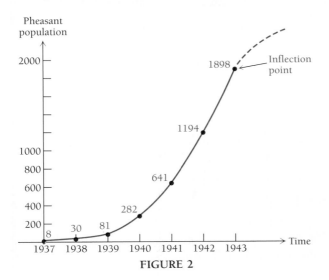

FIGURE 2

---

*A. S. Einarsen, "Some Factors Affecting Ring-Necked Pheasant Population Density," *Murrelet,* **26** (1945): 2–9, 39–44.

A ring-necked pheasant.

3. The following shows the growth of an ant colony with a point of inflection at P_1. What is the limiting value?

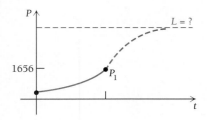

In 1943 the army arrived and started shooting the pheasants. It was theorized on the basis of the data that the population had reached its point of inflection. That is, the growth would stop turning upward and start turning downward as it approached its limiting value. What was the limiting value of the pheasant population?

Solution   At the inflection point,

$$P = \tfrac{1}{2}L.$$

Now $P = 1898$ from the graph, so

$$1898 = \tfrac{1}{2}L$$
$$3796 = L. \qquad \qquad ❖$$

DO EXERCISE 3.

The spread of an epidemic can also be described by the model of inhibited growth.

EXAMPLE 3   *Spread of an epidemic.* In a town whose total population is 2000, the disease *Rottenich* creates an epidemic. The initial number of people infected is 10. The rate of spread of the infection follows the inhibited growth model

$$\frac{dP}{dt} = kP(L - P),$$

where

P = the number of people infected after time t (in weeks),
L = 2000, the total population of the town,
k = 0.003, and
P_0 = the number of people initially infected = 10.

a)   Write the equation for $P(t)$ in terms of the given constants.

b)   Sketch a graph of $P(t)$ given that the point of inflection occurs at $t = 0.882$ week.

c)   At what time t are 1600 people affected?

Solution

a) $P(t) = \dfrac{P_0 L}{P_0 + e^{-Lkt}(L - P_0)} = \dfrac{10 \cdot 2000}{10 + e^{-2000(.003)t}(2000 - 10)}$

$\qquad = \dfrac{20{,}000}{10 + 1990e^{-6t}}$

b) See Fig. 3. We first mark the limiting value 2000 on the vertical axis in the figure and draw a horizontal line through it. Then we mark P_0, or 10, on the vertical axis and mark the inflection point (0.882, 1000). Finally, we draw an S-shaped curve from P_0 through the inflection point and upward, approaching the limiting value.

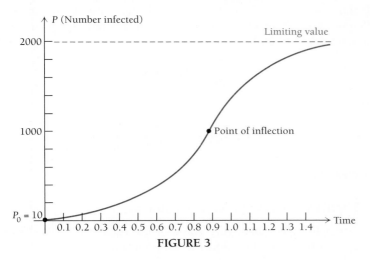

FIGURE 3

c) We solve the following equation for t:

$$1600 = \frac{20,000}{10 + 1990e^{-6t}}$$

$$1600(10 + 1990e^{-6t}) = 20,000$$

$$10 + 1990e^{-6t} = \frac{20,000}{1600} = 12.5$$

$$1990e^{-6t} = 2.5$$

$$e^{-6t} = \frac{2.5}{1990}.$$

We use natural logarithms to solve this equation:

$$\ln e^{-6t} = \ln \frac{2.5}{1990}$$

$$-6t = \ln 2.5 - \ln 1990$$

$$-6t = 0.916291 - 7.595890$$

$$t = \frac{0.916291 - 7.595890}{-6}$$

$$t \approx 1.113 \text{ weeks.}$$

4. *Spread of an epidemic.* A town whose population is 4000 is struck by the disease *Andromedamama*. The initial number of people infected is 20. The rate of spread of the infection follows the inhibited growth model

$$\frac{dP}{dt} = kP(L - P),$$

where

P = the number infected after time t (in months),

L = 4000, the total population of the town,

k = 0.0005, and

P_0 = the number of people initially infected = 20.

a) Write the equation for $P(t)$ in terms of the given constants.

b) Sketch a graph of $P(t)$, given that the point of inflection occurs at t = 2.9 months. Use graph paper.

c) At what time t are 800 people infected?

5. State the solution, $Q(y)$, to

$$\frac{dQ}{dy} = rQ(M - Q).$$

DO EXERCISE 4.

As you will see better after you have done the exercises in this section, and as you have already seen in the preceding example, the differential equation $dP/dt = kP(L - P)$ has many applications. The solution to the equation, however, can be used in all the applications. Of course, different variables or constants are used in different applications. For example, the solution to

$$\frac{dN}{dw} = cN(K - N)$$

is

$$N(w) = \frac{N_0 K}{N_0 + e^{-Kcw}(K - N_0)}.$$

DO EXERCISE 5.

Population Growth: Some Comments

Can a J-shaped curve (the graph of the equation $P = P_0 e^{kt}$) or an S-shaped curve be used to model world population growth? To find out let us look at a log–log graph* of population growth over the last one million years, shown in Fig. 4.

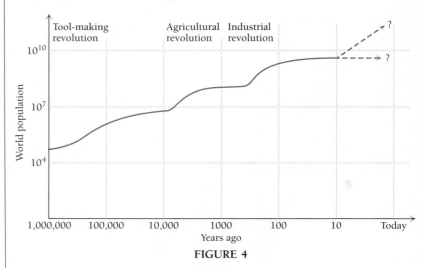

FIGURE 4

---

*The axes are scaled using logarithms of the numbers in question. This condenses the graph.

What happens is that we get a series of S-shaped curves. At three intervals there are surges of growth. These smaller intervals could be modeled by J-shaped curves. Following each surge, there seems to be a leveling off, but an advance in technology causes another surge. One comes after the *tool-making revolution*. Another comes after the *agricultural revolution*, and another after the *industrial revolution*. Judging from the graph, and from the current concerns about availability of food, energy, and other resources, we might speculate that we are in the midst of a leveling-off period when birth rate and death rate are about the same, or will be by the year 2000.

EXERCISE SET 9.3

For the inhibited growth model:

1. $P = 450$ at the point of inflection. What is the limiting value?

2. $P = 670$ at the point of inflection. What is the limiting value?

APPLICATIONS

❖ **Business and Economics**

3. *Business: Advertising.* A company introduced a new product on a trial run in a city. They advertised the product on television and gathered data concerning the percentage P of people who bought the product in relation to the number of times the product was advertised on television. The graph of the data resembled an S-shaped curve, so they hypothesized that it satisfied the inhibited growth model

$$\frac{dP}{dt} = kP(L - P),$$

where

$P =$ the percentage of people who buy the product,

$t =$ the number of times the product is advertised on TV,

$L = 100\%$, or 1, the highest percentage of people who could buy the product,

$k = 0.13$, and

$P_0 = 2\%$, or 0.02, the percentage who bought the product without having seen an ad.

a) Write the equation for $P(t)$ in terms of the given constants.

b) Sketch a graph of $P(t)$ given that the point of inflection (50%) occurs at $t = 30$.

c) How many times t would the company have to advertise its product so that 80% of the people would buy?

❖ **Life and Physical Sciences**

4. *Population growth.* In an experiment* it was found that the rate of growth of yeast cells in a laboratory satisfied the inhibited growth model

$$\frac{dP}{dt} = kP(L - P),$$

---

*G. F. Gause, *The Struggle for Existence* (Baltimore: Williams and Wilkins, 1934).

when handed the 85-g rock? Does this seem reasonable?

115								

where

experiment, you should read it over. The percentage P who respond "yes" to a rock of weight w satisfies the differential equation

$$\frac{dP}{dw} = kP(L - P),$$

where

P = % who respond "yes" to a rock of weight w,

L = 100%, or 1, and

P_{85} = 4%, percent who respond "yes" to the weight of w = 85 g (gram).

Look the graph over. Doesn't it seem reasonable that only 4% would have judged the 85-g rock to be heavier than the 100-g rock? As the weight of the comparison rocks gets to the standard, the percent increases. When the comparison weighs 100 g, the same as the standard, 50% say it is heavier. That is, half would say it is heavier and half would say it isn't. Then as the comparison rock gets heavier and heavier, the percent of "yes" responses increases toward 100%.

a) Write the equation for $P(w)$, the solution to the differential equation, in terms of the given constants. Note that $w = 0$ means 85 g, $w = 1$ means

90 g, $w = 2$ means 95 g, and so on. Note that we have not yet determined k.

b) Use the fact that $P = 50\% = 0.5$ at $w = 3$ (100 g) to find k.

c) Rewrite the equation for $P(w)$ in terms of k and the other constants.

d) At what weight w do 80% respond "yes"?

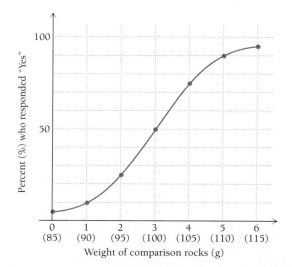

Percent (%) who responded "Yes"

Weight of comparison rocks (g)

9.4

APPLICATIONS: THE GROWTH MODEL $dP/dt = k(L - P)$

OBJECTIVES

a) State the solution to the growth model

$$\frac{dP}{dt} = k(L - P).$$

b) Use a calculator to graph equations of the type

$$P(t) = L(1 - e^{-kt}).$$

(continued)

We have considered two models of growth:

$$\frac{dP}{dt} = kP, \qquad \text{Uninhibited growth model}$$

$$\frac{dP}{dt} = kP(L - P). \qquad \text{Inhibited growth model}$$

A third kind of growth assumes a limited population L, but assumes that the growth rate dP/dt is directly proportional only to its remaining possible room for growth $L - P$, independent of the number who already exist. The model is

$$\frac{dP}{dt} = k(L - P), \qquad \text{where } P = 0 \text{ when } t = 0.$$

This model has not proved suitable to describe biological population growth. Thus we will not consider that here. It is applicable, however, to other areas. Before solving the differential equation, we consider one such application in Margin Exercise 1.

DO EXERCISE 1.

We have considered a problem similar to this before in Exercise 4 of Exercise Set 9.3 where we encountered an S-shaped curve. In Margin Exercise 1, the graph did not resemble an S-shaped curve. In most advertising applications, a curve like that in Margin Exercise 1 is what occurs. Models can vary, of course.

Now let us solve the differential equation

$$\frac{dP}{dt} = k(L - P).$$

We first separate the variables:

$$\frac{dP}{L - P} = k \, dt.$$

Now we integrate both sides:

$$\int \frac{dP}{L - P} = \int k \, dt$$

$$-\ln (L - P) = kt + C$$

$$\ln (L - P) = -kt - C$$

$$L - P = e^{-kt - C}$$

$$= e^{-kt} \cdot e^{-C}.$$

Letting $C_1 = e^{-C}$, we have

$$L - P = C_1 e^{-kt}$$

$$-P = -L + C_1 e^{-kt}$$

$$P = L - C_1 e^{-kt}.$$

c) Given L and a function value for the equation in part (b), find k.

d) Given L and k for the equation in part (b), find a certain function value $P(t)$.

e) Given L and k for the equation in part (b), find t such that $P(t) = M$, for some given number M.

1. *Exploratory exercises: Advertising.* A company introduces a new product on a trial run in a city. It advertises the product on TV and gathers data concerning the percentage P of the people in the city who bought the product after it was advertised a certain number of times.

a) The following table contains the data. Complete the calculations in the table.

t (Number of times ad ran)	P (% who bought product)	$\Delta P/\Delta t$ (% rate of change)
0	0	undefined
10	50	5
20	75	
30	87	
40	94	
50	97	

b) Plot P versus t. Connect the points with a smooth curve.

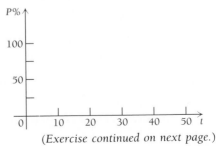

(Exercise continued on next page.)

c) Plot $\Delta P/\Delta t$ versus t. Connect the points with a smooth curve.

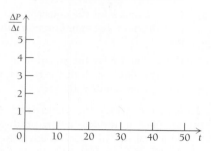

d) Does P resemble an S-shaped curve?

e) Does there seem to be a point of inflection for P?

f) What appears to be the limiting value for P?

g) Does the slope of the graph in part (b) appear to be increasing? decreasing? neither?

2. Let $L = 3$ and $k = 1$, so
$$P(t) = 3(1 - e^{-t}).$$

a) 🖩 Complete this table of function values.

t	0	1	2	3	4	5
$P(t)$			2.6			

For example,
$$P(2) = 3(1 - e^{-2})$$
$$\approx 3(1 - 0.135335)$$
$$\approx 2.6.$$

(*Exercise continued on next page.*)

Now $P = 0$ when $t = 0$, so
$$0 = L - C_1 e^{-k \cdot 0}$$
$$= L - C_1 \cdot 1 = L - C_1.$$

Thus $L = C_1$, and
$$P = L - Le^{-kt}$$
$$= L(1 - e^{-kt}).$$

THEOREM 4

The solution to
$$\frac{dP}{dt} = k(L - P), \qquad P(0) = 0 \tag{1}$$

is
$$P(t) = L(1 - e^{-kt}). \tag{2}$$

DO EXERCISE 2.

As Margin Exercise 2 illustrates, Fig. 1 is a graph of $P(t)$. Note that the graph starts at $(0, 0)$ and increases upward toward the limiting value L. The slope is decreasing and there is no point of inflection.

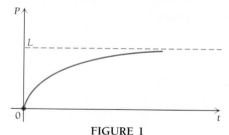

FIGURE 1

Now let us reconsider the problem on advertising in Margin Exercise 1.

EXAMPLE 1   Reread Margin Exercise 1.

a) Suppose we want to "fit" Eq. (2) to these data. We know that $L = 100\%$, or 1, so
$$P(t) = 1 - e^{-kt}.$$

Thus we must determine a value for k. A way to do this is to arbitrarily pick a data point, substitute, and solve for k. We use (10, 50%); that is, $P(10) = 50\%$, or 0.50. Substituting and solving, we have

$$0.5 = 1 - e^{-k \cdot 10}$$
$$-0.5 = -e^{-10k}$$
$$0.5 = e^{-10k}.$$

We use natural logarithms to solve this equation:

$$\ln 0.5 = \ln e^{-10k}$$
$$-0.693147 = -10k \quad \text{}$$
$$0.0693147 = k$$
$$0.07 \approx k.$$

b) We then rewrite $P(t)$ in terms of k:

$$P(t) = 1 - e^{-0.07t}.$$

c) Use the equation in part (b) to find $P(50)$:

$$P(50) = 1 - e^{-0.07 \cdot 50}$$
$$= 1 - e^{-3.5}$$
$$= 1 - 0.030197$$
$$\approx 0.9698.$$

Note that this is close to the data value of 97%. This may not always happen in practice.

d) How many times must the product be advertised so that 99% of the people will buy it? We solve

$$0.99 = 1 - e^{-0.07t}$$

for t and get $t \approx 66$.

DO EXERCISE 3.

b) Graph $P(t)$.

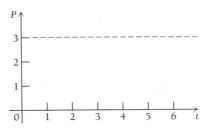

3. a) In another advertising experiment, it was determined that $L = 100\%$, so

$$P(t) = 1 - e^{-kt}.$$

One of the data points was

$$P(10) = 40\%, \quad \text{or } 0.4.$$

Use this value to determine k. Round to the nearest hundredth.

b) Rewrite $P(t)$ in terms of k.

c) Use the equation in part (b) to find $P(60)$.

d) How many times must the product be advertised so that 99% of the people will buy it?

❖

EXERCISE SET	**9.4**

Graph.

1. $P(t) = 4(1 - e^{-t})$ **2.** $P(t) = 2(1 - e^{-t})$ **3.** $P(t) = 1 - e^{-2t}$ **4.** $P(t) = 1 - e^{-3t}$

APPLICATIONS

❖ Business and Economics

5. *Business: Advertising.* In a different advertising experiment, it was determined that $L = 50\%$, or 0.5; that is, no matter how much a company advertised, the percentage of people who bought the product would never reach or exceed 50%.

   a) Express $P(t)$ in terms of $L = 0.5$.

   b) It was determined that $P(10) = 25\%$, or 0.25. Use this to determine k. Round to the nearest hundredth.

   c) Rewrite $P(t)$ in terms of k.

   d) Use the equation in part (c) to find $P(30)$.

   e) How many times must the product be advertised so that 49% of the people will buy it?

6. *Business: Stock growth.* The *growth rate of a certain stock* is modeled by

   $$\frac{dV}{dt} = k(L - V), \qquad V = \$20 \text{ when } t = 0,$$

   where

   $V =$ the value of the stock, per share, after time t (in months),

   $L = \$24.81$, the limiting value of the stock, and

   $k =$ a constant.

   Find the solution to the differential equation in terms of L and k.

❖ Life and Physical Sciences

7. *Acceptance of a new medicine.* A different model for the diffusion of information—for example, awareness of a new medicine by doctors—is the differential equation

   $$\frac{dP}{dt} = k(L - P),$$

   where

   $P =$ the percentage of doctors who are aware of the new medicine after time t (in months),

   $L = 100\%$, or 1, and

   $k =$ a constant.

   a) Write the equation for $P(t)$, the solution to the differential equation, in terms of L and k.

   b) Suppose $P(4) = 50\%$, or 0.5. Use this to determine k. Round to the nearest tenth.

   c) Rewrite $P(t)$ in terms of k.

   d) Use the equation in part (c) to find $P(6)$.

   e) How many months will it take for 90% of the doctors to become aware of the new medicine?

8. *Yield due to the spread of fertilizer.* A farmer is growing soybeans in a field. The more fertilizer he spreads, the greater his yield, up to some limiting value L. A model for this, applied to any given specific farm, is

   $$\frac{dY}{dn} = k(L - Y),$$

   where

   $Y =$ the yield, in bushels per acre, upon spreading n pounds of fertilizer per acre,

   $L = 60$ bushels per acre, and

   $k =$ a constant.

The yield due to fertilizer can be modeled by $Y' = k(L - Y)$.

   a) Write the solution $Y(n)$ in terms of L and k.

b) Suppose $Y(10) = 21$. Use this to determine k. Round to the nearest hundredth.

c) Rewrite $Y(n)$ in terms of k.

d) Use the equation in part (c) to find $Y(5)$.

e) How many pounds of fertilizer, per acre, must be spread to yield 42 bushels per acre?

(*Note:* This model is not completely valid in the sense that there usually is a certain amount of fertilizer that will kill the crop.)

9. *Mass of radioactive material.* Inside nuclear reactors, radioactive material is produced that satisfies

$$\frac{dM}{dt} = k\left(\frac{p}{k} - M\right), \qquad M = 0 \text{ when } t = 0,$$

where

$M =$ the mass of radioactive material inside the reactor after time t (in years),

$p =$ the rate at which the radioactive material is produced, and

$k =$ the annual decay rate of the radioactive material.

Find the solution, $M(t)$, to the differential equation in terms of p and k.

10. *Psychology: Hullian model of learning.* This model is expressed by

$$\frac{dP}{dt} = k(L - P),$$

where

$P =$ the probability of mastery of a certain concept after t learning trials,

$L = 1$, the limiting value of mastery of the learning, and

$k =$ a constant.

a) Write the solution $P(t)$ in terms of L and k.

b) Suppose $P(5) = 0.6$. Use this to determine k. Round to the nearest hundredth.

c) Rewrite $P(t)$ in terms of k.

d) Use the equation in part (c) to find $P(10)$.

e) Find t such that $P(t) = 0.90$. That is, how many trials are necessary in order for the probability of mastery of learning to be 0.90?

11. *Fick's law in biology* is given by

$$\frac{dC}{dt} = \frac{kA}{V}(C_T - C),$$

where

$C =$ the amount of a substance that passes through a tube with cross-sectional area A and volume V at time t,

$k =$ a constant, and

$C_T =$ the total amount of substance.

Find the solution to the differential equation assuming $C(0) = C_0$.

9.5

EXACT EQUATIONS

Consider the differential equation

$$y' = \frac{dy}{dx} = \frac{-3x^2 - y}{x + 1}.$$

Although this equation appears to be a likely candidate for separation of

OBJECTIVES

a) Determine whether or not a given first-order differential equation is exact.

b) Solve an exact first-order differential equation.

1. Check by substitution that

$$y = \frac{C - x^3}{x + 1}$$

satisfies the differential equation

$$y' = \frac{dy}{dx} = \frac{-3x^2 - y}{x + 1}.$$

2. Write the differential equation

$$y' = \frac{dy}{dx} = \frac{-x}{y}$$

in differential form.

variables, a little experimentation will convince you that this cannot be done.

Instead, we try the following approach. We write the equation in *differential form*. This means that we clear the equation of "fractions" by formally multiplying both sides by $(x + 1) \, dx$. This gives us

$$(x + 1) \, dy = (-3x^2 - y) \, dx,$$

or, rewriting,

$$(3x^2 + y) \, dx + (x + 1) \, dy = 0.$$

Now consider the function $F(x, y) = x^3 + xy + y$. Its partial derivatives are $F_x = 3x^2 + y$ and $F_y = x + 1$. The equation above is just

$$F_x \, dx + F_y \, dy = 0.$$

If this last expression looks familiar, it's because we studied the total differential in Chapter 7. We can rewrite this equation as

$$dF = 0,$$

where we have used the fact that

$$dF = F_x \, dx + F_y \, dy.$$

Since $dF = 0$ when $F(x, y) = C$, the last equation suggests that the solution is $F(x, y) = C$. That turns out to be correct. We get

$$x^3 + xy + y = C$$

or, solving for y,

$$y(x + 1) = C - x^3$$

$$y = \frac{C - x^3}{x + 1}.$$

Now that we have the solution, we can check it by direct substitution.

DO EXERCISES 1 AND 2.

Unfortunately, it's not always possible to take a first-order differential equation, write it in differential form, and then find a function F such that the equation becomes $dF = 0$.

DEFINITION

The first-order differential equation

$$P(x, y) \, dx + Q(x, y) \, dy = 0$$

is *exact* if there is a function $F(x, y)$ such that

$$F_x = \frac{\partial F}{\partial x} = P(x, y) \quad \text{and} \quad F_y = \frac{\partial F}{\partial y} = Q(x, y).$$

In other words, $P(x, y)\, dx + Q(x, y)\, dy$, or simply $P\, dx + Q\, dy = 0$, is exact if we can solve it just as we solved the example above. The following theorem tells us when an equation is exact.

THEOREM 5

If $P(x, y)$ and $Q(x, y)$ have continuous partial derivatives for all (x, y) in a rectangle R, then

$$P(x, y)\, dx + Q(x, y)\, dy = 0$$

is *exact* if and only if

$$P_y = Q_x.$$

EXAMPLE 1 Show that the equation

$$2xy\, dx + x^2\, dy = 0$$

is exact.

Solution   Here $P(x, y) = 2xy$ and $Q(x, y) = x^2$. So,

$$\frac{\partial P}{\partial y} = P_y = 2x \quad \text{and} \quad \frac{\partial Q}{\partial x} = Q_x = 2x.$$

Since $P_y = Q_x$, the equation is exact.   ❖

DO EXERCISES 3–5.

EXAMPLE 2 Solve $x\, dx + y\, dy = 0$.

Solution   We know from Margin Exercise 3 that this equation is exact. The problem here is to find $F(x, y)$. Now, $F(x, y)$ must satisfy $F_x = x$ and $F_y = y$. Consider the first equation. We can antidifferentiate it to get $F(x, y) = \frac{1}{2}x^2 + C$. To be correct, however, recall that since F is a function of two variables and since we are integrating with respect to x (and thinking of y as a constant), the constant C above may involve y-terms. We write

$$F(x, y) = \tfrac{1}{2}x^2 + g(y),$$

3. Show that

$$x\, dx + y\, dy = 0$$

is exact.

4. Show that

$$(6xy + 2y)\, dx + (3x^2 + 2x + 1)\, dy = 0$$

is exact.

5. a) "Simplify" the equation

$$2xy\, dx + x^2\, dy = 0$$

by dividing by x.

b) Show that the resulting equation is *not* exact.

6. Solve the exact equation in Example 1.

where g is an unknown function. To find g, recall that F must also satisfy $F_y = y$. Here, $F_y = \partial F/\partial y = g'(y)$. Thus $g'(y) = y$, and, antidifferentiating, we have $g(y) = \frac{1}{2}y^2 + C_1$, where C_1 is a constant. So we find that

$$F(x, y) = \tfrac{1}{2}x^2 + \tfrac{1}{2}y^2 + C_1.$$

The solution to

$$x\,dx + y\,dy = 0,$$

then, is given by $\frac{1}{2}x^2 + \frac{1}{2}y^2 + C_1 = C$, where C is a constant. We can simplify this result by subtracting C_1 from both sides and multiplying by 2. We then get

$$x^2 + y^2 = K,$$

where $K = 2(C - C_1)$ is a constant. The solution curves are a family of concentric circles, each centered at $(0, 0)$.

7. Solve $y\,dx + x\,dy = 0$.

Note, in retrospect, that the constant C_1 was irrelevant and could have been left out. We will do so in the future. ❖

DO EXERCISES 6–8.

EXAMPLE 3   Solve the initial-value problem

$$(2xy^2 - y + 5)\,dx + (2x^2y - x)\,dy = 0, \qquad y(1) = 2.$$

8. Solve the equation in Margin Exercise 4.

Solution   We first solve the equation just as we would if $y(1) = 2$ were not given. After solving the equation, we'll use the initial condition to eliminate the constant of integration.

Here, $P(x, y) = 2xy^2 - y + 5$ and $Q(x, y) = 2x^2y - x$. Testing for exactness, we find that $P_y = 4xy - 1$ and $Q_x = 4xy - 1$. The equation is exact, since $P_y = Q_x$.

To find $F(x, y)$, we recall that $F_x = P = 2xy^2 - y + 5$ and integrate with respect to x. We get

$$F(x, y) = \int P(x, y)\,dx$$

$$= \int (2xy^2 - y + 5)\,dx$$

$$= x^2y^2 - xy + 5x + g(y).$$

So $\partial F/\partial y = F_y = 2x^2y - x + g'(y)$. Since we also must have $F_y = Q = 2x^2y - x$, we see that $g'(y) = 0$. So, $g(y) = C_0$, a constant, and may be taken to be 0. Thus

$$F(x, y) = x^2y^2 - xy + 5x \quad \text{and} \quad x^2y^2 - xy + 5x = C$$

gives the solution.

To evaluate C, note that $y = 2$ when $x = 1$. Direct substitution above yields

$$(1)^2(2)^2 - 1 \cdot 2 + 5 \cdot 1 = C$$

or

$$C = 1 \cdot 4 - 2 + 5$$
$$= 9 - 2$$
$$= 7.$$

Therefore,

$$x^2y^2 - xy + 5x = 7.$$

It isn't necessary to do so, but you may solve the last equation for y by regarding $a = x^2$, $b = -x$, and $c = 5x - 7$ and applying the quadratic formula to $ay^2 + by + c = 0$. ❖

DO EXERCISES 9–11.

9. Use the quadratic formula to solve

$$x^2y^2 - xy + 5x = 7$$

for y.

10. Solve the initial-value problem

$$2xy^3\,dx + 3x^2y^2\,dy = 0, \quad y(1) = 2.$$

11. Solve the initial-value problem

$$(1 + 2y)\,dx + 2x\,dy = 0, \quad y(1) = -1.$$

EXERCISE SET 9.5

Determine whether the given equation is exact.

1. $(2x + 3)\,dx + 6y\,dy = 0$

2. $(2y + 3)\,dx + 5x\,dy = 0$

3. $(6y + 3)\,dx + 6x\,dy = 0$

4. $\sin y\,dx + \cos x\,dy = 0$

5. $\sin y\,dx + x\cos y\,dy = 0$

6. $\sin x\,dx + y\cos x\,dy = 0$

7. $\left(y^3e^x + \dfrac{y}{x}\right)dx + (3y^2e^x + 2y + \ln x)\,dy = 0$

8. $(4y^5x^3 - 6xy^2)\,dx + (5x^4y^4 - 6x^3y)\,dy = 0$

9. Solve the equation in Exercise 1.

10. Solve the equation in Exercise 3.

11. Solve the equation in Exercise 5.

12. Solve the equation in Exercise 7.

Show that each of the following equations is exact. Then solve the initial-value problem.

13. $(3x^2 + 3y)\,dx + (3x + 3y^2)\,dy = 0$, $y(1) = 1$

14. $2xy\,dx + (1 + x^2)\,dy = 0$, $y(1) = \frac{1}{2}$

15. $(\cos y - y\sin x)\,dx + (\cos x - x\sin y)\,dy = 0$, $y(0) = 1$

16. $(x^3 + 2xy)\,dx + (x^2 + y^3)\,dy = 0$, $y(2) = 2$

17. Find the equation of the curve that satisfies the differential equation

$$\frac{dy}{dx} = \frac{y - x}{2y - x}$$

and passes through $(1, 1)$.

APPLICATIONS

❖ **Business and Economics**

18. *Business.* A company finds that if p is the price per unit (in dollars) of its goods and if x represents the quantity

of goods sold, then the elasticity of demand is

$$E(p) = \frac{1 + 3p^2}{1 + p^2}.$$

Use the fact that

$$E = -\frac{p}{x} \cdot \frac{dx}{dp}$$

(see Section 4.6) to set up and solve a differential equation for x, given that $x = 100$ when $p = 2$.

19. *Business.* A company discovers that the price P of one of its products and the number of units u of the product demanded by consumers at that price are related (for a certain range of values) by the equation

$$\frac{dP}{du} = \frac{-3P}{2P + 3u}.$$

If $P = 5$ when $u = 5$, find P when $u = 16$.

20. *Business.* A firm's average cost per unit A (in dollars) of producing x thousands of units of a certain product satisfies the equation

$$\frac{dA}{dx} = \frac{4x - 3A}{3x + 7}.$$

Find A as a function of x if $A(1) = 1$.

SYNTHESIS EXERCISE

21. Show that the substitution $y = vx$ in the differential equation in Exercise 17 changes it to a problem that can be solved by separation of the variables.

9.6

FIRST-ORDER LINEAR EQUATIONS

OBJECTIVE

a) Solve any first-order linear differential equation.

We begin this section with the following definition.

DEFINITION

A **first-order differential equation** is called *linear* if it can be written in the form

$$a_0(x)y' + a_1(x)y = b(x). \tag{1}$$

Note that $a_0(x)$ and $a_1(x)$, which are called the *coefficients* of the equation, are functions of x alone. So too is the function $b(x)$ on the right-hand side of the equation. So, for example, the equation

$$(x^2 + 1)y' - xy = e^x$$

is a first-order linear differential equation with $a_0(x) = x^2 + 1$, $a_1(x) = -x$, and $b(x) = e^x$. On the other hand, $(y')^3(y) - e^y = 0$ is *not* a linear equation.

In this text, we will always assume that the coefficients $a_0(x)$ and $a_1(x)$ and the function $b(x)$ (sometimes called the *forcing function*) are all continuous on some closed interval $[c, d]$. This guarantees that Eq. (1) has a solution on $[c, d]$.

EXAMPLE 1   Solve $xy' - 3x^3y = 0$.

Solution   Comparing this equation with the definition above, we have $a_0(x) = x$, $a_1(x) = -3x^3$, and $b(x) = 0$. (Equations in which $b(x) = 0$ are called *homogeneous*.) Dividing through by x, we have

$$y' - 3x^2y = 0.$$

Using $y' = dy/dx$, we have

$$\frac{dy}{dx} = 3x^2y.$$

We see now that we can separate the variables in this equation. Doing so, we obtain

$$\frac{dy}{y} = 3x^2 \, dx.$$

Integrating both sides, we have

$$\int \frac{dy}{y} = \int 3x^2 \, dx$$

$$\ln y = x^3 + C.$$

Applying the definition of logarithms to $\ln y = \log_e y = x^3 + C$ gives us

$$y = e^{x^3 + C}$$

or

$$y = e^{x^3}e^C$$

or

$$y = C_1 e^{x^3}, \quad \text{where } C_1 = e^C. \qquad \text{❖}$$

DO EXERCISE 1.

It turns out that solving an arbitrary *homogeneous* first-order differential equation is no harder than solving Example 1 or Margin Exercise 1. That is, every homogeneous linear first-order equation can be solved by separating the variables (see Exercise Set 9.6). Unfortunately, though, separation of variables usually doesn't work in the nonhomogeneous

1. Solve $xy' - y = 0$.

case. Nor is such an equation exact. Consider, for example,

$$a_0(x)y' + a_1(x)y = b(x).$$

Dividing through by $a_0(x)$ yields

$$y' + \frac{a_1(x)}{a_0(x)} y = \frac{b(x)}{a_0(x)},$$

where we assume that $a_0(x) \neq 0$. For simplicity, let

$$p(x) = \frac{a_1(x)}{a_0(x)} \quad \text{and} \quad q(x) = \frac{b(x)}{a_0(x)}.$$

Then our equation becomes

$$y' + p(x)y = q(x).$$

This is sometimes called the *standard form* of the equation. Now, letting $y' = dy/dx$ and rearranging gives

$$p(x)y - q(x) + \frac{dy}{dx} = 0$$

or

$$[p(x)y - q(x)] \, dx + dy = 0 \qquad (2)$$

in differential form. Here, to consider exactness, let $P(x, y) = p(x)y - q(x)$ and $Q(x, y) = 1$. Testing for exactness, we find that $\partial P/\partial y = p(x)$ and $\partial Q/\partial x = 0$, so that the equation isn't exact except in the case $p(x) = 0$.

In Margin Exercise 5 of Section 9.5, we saw that exactness could be destroyed by dividing by a term. If this is the case, then we can restore it by multiplying by that term. Here, we can make the equation exact by multiplying by $e^{\int p(x)\,dx}$. We'll check this out in a moment, but first, let's calculate the derivative (with respect to x) of $e^{\int p(x)\,dx}$. Recall that (by the Chain Rule) the derivative of $e^{f(x)}$ is $e^{f(x)} \cdot f'(x)$ for any differentiable function $f(x)$. If $v = e^{\int p(x)\,dx}$, then

$$\frac{dv}{dx} = e^{\int p(x)\,dx} \cdot \frac{d}{dx} \left[\int p(x) \, dx \right].$$

But the derivative of $\int p(x) \, dx$ is just $p(x)$. So,

$$\frac{dv}{dx} = p(x) \cdot e^{\int p(x)\,dx}.$$

In differential form, our original equation was

$$[p(x)y - q(x)]dx + dy = 0.$$

Multiplying by $e^{\int p(x)\,dx}$ gives

$$[p(x)e^{\int p(x)\,dx}y - q(x)e^{\int p(x)\,dx}] \, dx + e^{\int p(x)\,dx} \, dy = 0.$$

The partial derivative with respect to y of the left-hand term is $p(x)e^{\int p(x)\,dx}$. The partial derivative with respect to x of $e^{\int p(x)\,dx}$ is just its derivative with respect to x, and, from above, that's also $p(x)e^{\int p(x)\,dx}$. The new form of the equation is exact!

We'll illustrate with an example.

EXAMPLE 2 Solve $xy' - 3x^3y = 0$.

Solution Dividing through by x, we have $y' - 3x^2y = 0$. Here, $p(x) = -3x^2$ so that $\int p(x)\,dx = -x^3$ (to simplify matters, we omit the constant of integration). So we'll multiply the equation by e^{-x^3}. Doing so yields

$$e^{-x^3}y' - 3x^2e^{-x^3}y = 0 \cdot e^{-x^3}$$

or

$$e^{-x^3}y' - 3x^2e^{-x^3}y = 0.$$

Now the left-hand side of the equation is just the derivative of $e^{-x^3} \cdot y$ (you should check this yourself using the Product Rule). Thus we have

$$(e^{-x^3}y)' = 0.$$

Integrating, we get

$$e^{-x^3}y = C, \quad \text{a constant.}$$

Multiplying both sides by e^{x^3} gives

$$e^{x^3}e^{-x^3}y = Ce^{x^3}$$

or

$$e^0 \cdot y = Ce^{x^3}$$

or

$$y = Ce^{x^3}, \quad \text{since } e^0 = 1.$$

Note that this agrees with the answer we got in Example 1. ❖

DO EXERCISES 2 AND 3.

EXAMPLE 3 Solve $xy' + 2y = 5x^3$.

Solution We divide through by x (valid if we restrict ourselves to positive values of x), obtaining

$$y' + \frac{2}{x}y = 5x^2, \quad x > 0.$$

2. By direct substitution, show that $y = Ce^{x^3}$ is a solution to $xy' - 3x^3y = 0$.

3. Solve $y' + 2xy = 0$ in two different ways.

4. Solve $x^2y' + 4xy = 8x$.

Here,

$$p(x) = \frac{2}{x} \quad \text{and} \quad \int p(x)\, dx = 2\int \frac{dx}{x} = 2\ln x.$$

Note that we can write $\int p(x)\, dx = 2\ln x = \ln x^2$, using a property of logarithms. Now we'll multiply the equation by $e^{\int p(x)\, dx} = e^{\ln x^2}$ (this function is sometimes called the *integrating factor* for the equation). Note that $e^{\ln x^2} = x^2$, by the definition of logarithms. Multiplying through by x^2 gives

$$x^2y' + x^2 \cdot \left(\frac{2}{x}\right)y = 5x^2 \cdot x^2,$$

or

$$x^2y' + 2xy = 5x^4.$$

Now the left-hand side is just the derivative of x^2y. That is, we have

$$(x^2y)' = 5x^4.$$

Integrating, we have $x^2y = x^5 + C$, where C is a constant of integration. So, solving for y, we now have

$$y = x^3 + \frac{C}{x^2} = x^3 + Cx^{-2},$$

5. Solve $y' + 2y = e^{3x}$.

valid for $x > 0$. Note that the solution is undefined at $x = 0$. We knew that our solution might not be valid there, since we had to divide through by x in the first step of the solution. ❖

DO EXERCISES 4 AND 5.

EXAMPLE 4 *Mixing.* A tank initially contains a brine solution consisting of 400 pounds of salt dissolved in 1200 gallons of water (see Fig. 1).

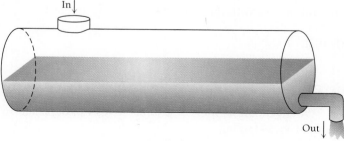

FIGURE 1

Salt water containing 1 pound of salt per gallon of water is pumped into the tank at the rate of 5 gallons per minute. The brine is pumped out at the rate of 5 gallons per minute. Assuming that the mixture is kept uniform by stirring, find the amount of salt in the tank after 5 hours.

Solution   Let $A(t)$ be the amount of salt (in pounds) that is in the tank after t minutes. We are given that $A(0) = 400$ and we want to find $A(300)$. We will find this by setting up and then solving a differential equation for $A(t)$. Note that the expression $A'(t) = dA/dt$ represents the instantaneous rate of change (the rate of increase or decrease) of the amount of salt in the tank at time t.

Since salt water containing 1 pound per gallon is being pumped into the tank at a rate of 5 gallons per minute, we see that salt is entering the tank at the rate of 5 pounds per minute.

The mixture is being pumped out of the tank at the rate of 5 gallons per minute. How much salt does each gallon of the mixture contain? At time t, there are $A(t)$ pounds of salt dissolved in 1200 gallons of water. The concentration of salt (in pounds per gallon) at time t is $A(t)/1200$. Since 5 gallons per minute are being pumped out, this means that

$$\frac{A(t)}{1200} \cdot 5 = \frac{A(t)}{240}$$

pounds per minute are being pumped out.

Now the rate of change of salt in the tank, $dA/dt = A'(t)$, is just the difference between the rate at which salt enters the tank and the rate at which it leaves the tank at time t. That is,

$$\frac{dA}{dt} = (\text{Rate entering}) - (\text{Rate leaving})$$

$$= 5 - \frac{A(t)}{240}.$$

Rearranging, we have

$$A'(t) + \frac{A(t)}{240} = 5,$$

a first-order linear equation. To solve it, note that the integrating factor is $e^{\int (1/240)\, dt} = e^{t/240}$. The equation becomes

$$(e^{t/240} A)' = 5e^{t/240},$$

or, after integrating,

$$e^{t/240} A = 1200e^{t/240} + C$$

6. Show that the equation

$$A'(t) = 5 - \frac{A(t)}{240}$$

from Example 4 can be solved by separating the variables.

7. Find the answer to the question in Example 4 if the solution pumped into the tanks contains only 0.6 pound of salt per gallon.

or

$$A(t) = 1200 + Ce^{-t/240}.$$

Recall that $A(0) = 400$. Letting $t = 0$ and $A = 400$ above, we get

$$400 = 1200 + Ce^{-0/240}$$

or

$$400 = 1200 + Ce^0$$

and

$$C = -800.$$

So

$$A(t) = 1200 - 800e^{-t/240}.$$

In particular,

$$\begin{aligned} A(300) &= 1200 - 800e^{-300/240} \\ &= 1200 - 800e^{-1.25} \\ &= 1200 - 800(0.2865048) \\ &\approx 1200 - 229.204 \\ &\approx 970.796 \text{ pounds.} \end{aligned}$$

❖

DO EXERCISES 6 AND 7.

EXERCISE SET **9.6**

Solve by separating the variables.

1. $y' - 3y = 0$

2. $y' - 2xy = 0$

3. $y' + \dfrac{3}{x} y = 0$

4. $y' - 2xy = 2x$

Show that the given equation is not exact. Then find an integrating factor $m(x)$ for the equation. Don't actually solve the equation.

5. $y' - 3y = e^{5x}$

6. $y' + \dfrac{4}{x} y = \sin x$

7. $y' - \dfrac{5}{x^2} y = x^3$

8. $y' + \dfrac{2x}{x^2 + 1} y = 3x + 2$

Solve by finding an appropriate integrating factor.

9. $y' - 3y = 0$

10. $y' + 4y = 0$

11. $y' - 2xy = 0$

12. $y' + \dfrac{3}{x} y = 0$

13. $y' + 5y = 0, \ y(0) = 7$

14. $xy' + y = 0, \ y(1) = 3$

15. $y' - 2xy = 2x$

16. $y' + \dfrac{3}{x} y = 6x^2 - 4$

17. $y' + \dfrac{2x}{1 + x^2} y = x$

18. $y' - \dfrac{\sin x}{\cos x} y = \dfrac{1}{\cos x}$

19. $x^2 y' + xy = x^4, \ y(2) = 5$

20. $y' + \dfrac{3x^2}{(1 + x^3)} y = 1, \ y(0) = 2$

APPLICATIONS

❖ **Life and Physical Sciences**

21. *Mixing.* A tank initially contains 180 pounds of salt dissolved in 500 gallons of water. Salt water containing 0.5 pound of salt per gallon enters the tank at the rate of 3 gallons per minute. The mixture (kept uniform by stirring) is removed at the same rate. How many pounds of salt are in the tank after an hour? a day?

22. *Mixing.* A 2000-gallon tank initially contains 200 pounds of salt dissolved in 500 gallons of water. Pure water is pumped into the tank at the rate of 5 gallons per minute, while the (well-stirred) mixture is drawn off at the rate of 2 gallons per minute. The process stops when the tank is full. How much salt is left in the tank when the tank is full?

23. *Newton's Law of Cooling.* The temperature T of a cooling object drops at a rate that is proportional to the difference $T - M$, where M is the constant temperature of the surrounding medium. Thus,

$$\frac{dT}{dt} = -k(T - M),$$

where k is a positive constant and t is time.

a) Use the technique of this section to solve the differential equation.

b) If a metal sphere that has been heated to 143°F is observed to cool to 117° in 30 minutes in a room whose temperature is kept constant at 70°, find the temperature of the sphere at any later time t. How long will it take the sphere to cool to 90°?

24. *Revenue growth.* The annual revenue R (in millions of dollars) of a company satisfies the equation

$$\frac{dR}{dt} = 6 - (0.1)R + (0.05)t$$

over a certain period of time, where t is the time in years. If the revenues are $4 million after two years, find the annual revenue after five years.

SYNTHESIS EXERCISES

25. Let $f(x)$ satisfy $f'(x) = p(x)$, where $p(x)$ is continuous on $[a, b]$.

a) Show that if $v = ye^{f(x)}$, then
$v' = e^{f(x)}y' + p(x)e^{f(x)}y.$

b) Then show that under the substitution $v = ye^{f(x)}$, the equation

$$y' + y \cdot p(x) = q(x)$$

becomes

$$v' = e^{-\int p(x)\, dx} \cdot q(x),$$

which may be solved by an integration.

26. In each of the following, show that the given equation is not exact. Then show that the given function $m(x, y)$ is

an integrating factor for the equation, and solve the equation.

a) $3y^4\, dx + 4xy^3\, dy = 0$; $m(x, y) = x^2$

b) $5xy\, dx + 3x^2\, dy = 0$; $m(x, y) = x^3y^2$

c) $(2y - 15xy^2)\, dx + (3x - 20x^2y)\, dy = 0$; $m(x, y) = xy^2$

d) $2xy\, dx + 5x^2\, dy = 0$; $m(x, y) = y^4$

27. Consider the first-order homogeneous equation $a_0y' + a_1y = 0$, where the coefficients are assumed to be constant. Show that $y = Ce^{-a_1x/a_0}$ is the solution to this equation.

28. Use the techniques of this section to solve $dP/dt = k(L - P)$ with the condition $P(0) = 0$.

9.7

OBJECTIVES

a) Solve a linear homogeneous differential equation with constant coefficients.

b) Solve a linear (nonhomogeneous) differential equation with constant coefficients if the forcing function is a polynomial.

HIGHER-ORDER DIFFERENTIAL EQUATIONS

In the preceding section, we saw that we could solve any first-order linear differential equation. Unfortunately, not every higher-order linear differential equation can be solved explicitly. For the moment, we'll consider second-order linear homogeneous equations with constant coefficients, that is, equations of the form

$$a_0 y'' + a_1 y' + a_2 y = 0, \tag{1}$$

where a_0, a_1, and a_2, the coefficients of the equation, are constants.

The corresponding first-order equation was solved explicitly in Exercise 27 of Exercise Set 9.6. The solution is an exponential function. This suggests that the solution to Eq. (1) may be an exponential function, which turns out to be correct. If we let $y = e^{rx}$ (r a constant) in Eq. (1) above, then, since $y' = re^{rx}$ and $y'' = r^2 e^{rx}$, we'll get

$$a_0(r^2 e^{rx}) + a_1(re^{rx}) + a_2(e^{rx}) = 0.$$

Noting that each term in the equation above has the common factor e^{rx}, we write

$$e^{rx}(a_0 r^2 + a_1 r + a_2) = 0.$$

Now e^{rx} is a positive number for all choices of real numbers r and x. Dividing by e^{rx}, we get

$$a_0 r^2 + a_1 r + a_2 = 0.$$

The quadratic formula yields two values of r.

EXAMPLE 1   Solve $y'' - 5y' + 6y = 0$.

Solution   We try $y = e^{rx}$. Then $y' = re^{rx}$ and $y'' = r^2 e^{rx}$. Substituting these quantities into the equation gives

$$r^2 e^{rx} - 5re^{rx} + 6e^{rx} = 0$$

or

$$e^{rx}(r^2 - 5r + 6) = 0$$

or

$$r^2 - 5r + 6 = 0.$$

This last equation factors as $(r - 2)(r - 3) = 0$, and we get solutions, or roots, 2 and 3. This gives us the two solutions e^{2x} and e^{3x}.

It turns out that the general solution to the equation is given by $y = C_1 e^{2x} + C_2 e^{3x}$, where C_1 and C_2 are arbitrary constants. We can verify this by finding y' and y'' and substituting y, y', and y'' into the original equation. We get

$$y = C_1 e^{2x} + C_2 e^{3x},$$
$$y' = 2C_1 e^{2x} + 3C_2 e^{3x},$$
$$y'' = 4C_1 e^{2x} + 9C_2 e^{3x}.$$

Then, substituting, we get

$y'' - 5y' + 6y$

$\quad = (4C_1 e^{2x} + 9C_2 e^{3x}) - 5(2C_1 e^{2x} + 3C_2 e^{3x}) + 6(C_1 e^{2x} + C_2 e^{3x})$

$\quad = 4C_1 e^{2x} + 9C_2 e^{3x} - 10C_1 e^{2x} - 15C_2 e^{3x} + 6C_1 e^{2x} + 6C_2 e^{3x})$

$\quad = 0.$ ❖

DO EXERCISES 1–3.

In Example 1, we used the fact that if $f(x)$ and $g(x)$ are solutions to a linear homogeneous differential equation, then so too is $C_1 f(x) + C_2 g(x)$. We state this important principle in the form of a theorem.

THEOREM 6

If $f_1(x), f_2(x), \ldots, f_n(x)$ are solutions to a linear *homogeneous* differential equation, then so too is

$$\sum_{i=1}^{n} C_i f_i(x) = C_1 f_1(x) + C_2 f_2(x) + \cdots + C_n f_n(x),$$

where $C_1, C_2, \ldots, C_n$ are arbitrary constants.

Note that the equation need not have constant coefficients. Nor must it be a second-order equation. The only requirements are that it be linear and homogeneous. We leave the proof of this theorem for the exercises.

In Example 1, we obtained the expression $y = C_1 e^{2x} + C_2 e^{3x}$ for the solution to $y'' - 5y' + 6y = 0$. This expression is called the *general solution* to the equation. This means that every function of this form is a solution to the equation and that every solution to the equation can be written in this form.

For the equation $y'' - 5y' + 6y = 0$, the general solution contained two arbitrary constants. In all of the equations that we'll consider, the number of constants in the general solution will be equal to the order of the equation.

1. a) By substituting directly, show that $y = e^{2x}$ is a solution to
$$y'' - 5y' + 6y = 0.$$
b) Repeat part (a) for $y = e^{3x}$.
c) Repeat part (a) for $y = 2e^{3x} - 7e^{2x}$.

2. Solve $y'' - 6y' + 8y = 0$.

3. Solve $2y'' - 7y' + 3y = 0$.

4. Solve $y''' + y'' - 2y' = 0$.

5. Find the auxiliary equation of $y'' - 7y' + 12y = 0$.

6. Find the auxiliary equation of $y''' - 6y'' + 11y' - 6y = 0$.

EXAMPLE 2   Find the general solution to $y''' + y'' - 6y' = 0$.

Solution   As in Example 1, we let $y = e^{rx}$. We obtain $y' = re^{rx}$, $y'' = r^2 e^{rx}$, and $y''' = r^3 e^{rx}$. Substituting into the equation, we obtain

$$r^3 e^{rx} + r^2 e^{rx} - 6re^{rx} = 0,$$

or

$$e^{rx}(r^3 + r^2 - 6r) = 0.$$

Since e^{rx} is never 0, we get $r^3 + r^2 - 6r = 0$, or

$$r(r^2 + r - 6) = 0.$$

Factoring yields $r(r + 3)(r - 2) = 0$, so that 0, -3, and 2 are its roots. It is easy to check that e^{-3x}, e^{2x}, and e^{0x}, or 1, are solutions to $y''' + y'' - 6y' = 0$. The general solution is

$$y = C_1 e^{-3x} + C_2 e^{2x} + C_3 \cdot 1,$$

or

$$y = C_1 e^{-3x} + C_2 e^{2x} + C_3. \qquad \diamondsuit$$

DO EXERCISE 4.

In each of Examples 1 and 2, we made the substitution $y = e^{rx}$, reduced the given differential equation to a polynomial equation in the variable r, and then solved this polynomial equation for r. The resulting values of r determined the solution to the differential equation.

But there's no need to repeat the substitution $y = e^{rx}$ and the subsequent algebra each time. We can bypass this work, going directly from the given differential equation to the associated polynomial equation (called the *characteristic equation* or the *auxiliary equation*).

EXAMPLE 3   Find the auxiliary equation of $y'' - 7y' + 10y = 0$.

Solution   Replacing y'' by r^2, y' by r, and $y(= y^{(0)})$ by $r^0 = 1$ gives

$$r^2 - 7r + 10 = 0. \qquad \diamondsuit$$

DO EXERCISES 5 AND 6.

EXAMPLE 4   Solve $y'' - 7y' + 10y = 0$.

Solution   From Example 3, the auxiliary equation is $r^2 - 7r + 10 = 0$. Factoring, we get $(r - 2)(r - 5) = 0$, so that its roots are 2 and 5. The general solution to the differential equation is

$$y = C_1 e^{2x} + C_2 e^{5x}.$$

❖

DO EXERCISE 7.

The technique we have developed works well so long as the auxiliary equation has real roots, all of which are distinct. It must be modified if the real roots are repeated or if the roots are complex.

EXAMPLE 5 Solve $y'' - 2y' + y = 0$.

Solution The auxiliary equation is $r^2 - 2r + 1 = 0$. Factoring yields $(r - 1)^2 = 0$. The roots are 1 and 1, or simply 1. (This is the case of repeated real roots.) Sure enough, $r = 1$ yields the solution $y = e^x$. We might try $y = C_1 e^x + C_2 e^x$ as the general solution. But

$$C_1 e^x + C_2 e^x = (C_1 + C_2)e^x = Ke^x,$$

where $K = C_1 + C_2$. In other words, simply repeating the e^x gets us nowhere.

Let's substitute $y = ve^x$ into the original equation. Then, by the Product Rule,

$$y' = ve^x + v'e^x$$

and

$$y'' = (ve^x + v'e^x) + (v'e^x + v''e^x) = ve^x + 2v'e^x + v''e^x$$

and substituting all of this into

$$y'' - 2y' + y = 0$$

gives

$$(ve^x + 2v'e^x + v''e^x) - 2(ve^x + v'e^x) + ve^x = 0.$$

Simplifying, we get $v''e^x = 0$, or, since e^x is never 0, $v'' = 0$. Integrating twice yields $v = C_1 + C_2 x$. Replacing v in the expression $y = ve^x$ gives

$$y = (C_1 + C_2 x)e^x$$

or

$$y = C_1 e^x + C_2 x e^x.$$

This is the general solution to the original equation and can be checked in the original equation.

❖

DO EXERCISE 8.

7. Solve $y'' - 7y' + 12y = 0$.

8. Solve $y'' - 4y' + 4y = 0$.

9. Solve $y'' - 6y' + 9y = 0$.

10. Solve $4y'' + 4y' + y = 0$.

11. Solve $y''' + 3y'' + 3y' + y = 0$.

The calculations in Example 5 weren't easy. Rather than repeat them each time repeated roots occur, we simply remember that if the root r is repeated exactly once, the corresponding solutions of the differential equation are e^{rx} and xe^{rx}.

EXAMPLE 6 Solve $y'' + 4y' + 4y = 0$.

Solution The auxiliary equation is $r^2 + 4r + 4 = 0$, or $(r + 2)^2 = 0$. So -2 and -2 are the roots. The corresponding solutions to the differential equation are e^{-2x} and xe^{-2x}. The general solution is

$$y = C_1 e^{-2x} + C_2 xe^{-2x}.$$ ❖

DO EXERCISES 9 AND 10.

If a given number, say r, appears as a root three times, we use e^{rx}, xe^{rx}, and $x^2 e^{rx}$ for the three solutions.

EXAMPLE 7 Solve $y''' - 3y'' + 3y' - y = 0$.

Solution The auxiliary equation is $r^3 - 3r^2 + 3r - 1 = 0$, or $(r - 1)^3 = 0$. So the roots are 1, 1, and 1. Three solutions are e^x, xe^x, and $x^2 e^x$. The general solution is

$$y = C_1 e^x + C_2 xe^x + C_3 x^2 e^x.$$ ❖

DO EXERCISE 11.

We have considered the cases in which the roots of the auxiliary equation are real numbers. Consider the equation $y'' + y = 0$. The auxiliary equation is $r^2 + 1 = 0$. The roots of this equation are $r = \pm\sqrt{-1}$, or $r = \pm i$, where i denotes $\sqrt{-1}$ (which doesn't exist as a real number). On the other hand, it's easy to solve $y'' + y = 0$. If $y = \sin x$, for example, then $y' = \cos x$ and $y'' = -\sin x$. In other words, $y = \sin x$ is one solution. A similar calculation (do it!) reveals that $y = \cos x$ is another solution. The general solution is $y = C_1 \sin x + C_2 \cos x$.

This example illustrates the general case. If the roots of the auxiliary equation are the complex numbers $r_1 \pm r_2 i$, then the two corresponding solutions are $e^{r_1 x} \cos (r_2 x)$ and $e^{r_1 x} \sin (r_2 x)$.

EXAMPLE 8 Solve $y'' + 16y = 0$.

Solution The auxiliary equation is $r^2 + 16 = 0$. Solving directly yields $r^2 = -16$, or $r = \pm\sqrt{-16}$, or $r = \pm 4\sqrt{-1} = \pm 4i = 0 \pm 4i$. This is the case where $r_1 = 0$ and $r_2 = 4$. Two solutions are $e^{0x} \sin 4x = \sin 4x$ and $e^{0x} \cos 4x = \cos 4x$. The general solution is

$$y = C_1 \sin 4x + C_2 \cos 4x. \qquad ❖$$

DO EXERCISE 12.

EXAMPLE 9 Solve $y'' - 2y' + 10y = 0$.

Solution The auxiliary equation is $r^2 - 2r + 10 = 0$. Applying the quadratic formula yields

$$r = \frac{-(-2) \pm \sqrt{(-2)^2 - 4(1)(10)}}{2 \cdot 1}$$

$$= \frac{2 \pm \sqrt{4 - 40}}{2} = \frac{2 \pm \sqrt{-36}}{2}$$

$$= \frac{2 \pm 6\sqrt{-1}}{2} = 1 \pm 3i.$$

So $r_1 = 1$ and $r_2 = 3$. The corresponding solutions are $e^x \sin 3x$ and $e^x \cos 3x$ and the general solution is $y = C_1 e^x \sin 3x + C_2 e^x \cos 3x$. ❖

DO EXERCISES 13 AND 14.

If some of the roots are real and some are complex, we simply combine our earlier techniques.

EXAMPLE 10 Solve $y''' - 4y'' + 8y' - 8y = 0$.

Solution The auxiliary equation is $r^3 - 4r^2 + 8r - 8 = 0$. This factors as $(r - 2)(r^2 - 2r + 4) = 0$. The roots turn out to be $r = 2$, $1 \pm \sqrt{3}i$. The corresponding solutions are e^{2x}, $e^x \sin(\sqrt{3}x)$, and $e^x \cos(\sqrt{3}x)$. The general solution is

$$y = C_1 e^{2x} + C_2 e^x \sin(\sqrt{3}x) + C_3 e^x \cos(\sqrt{3}x). \qquad ❖$$

DO EXERCISE 15.

If complex roots are repeated, then we modify the solutions just as we did in the case of repeated real roots.

EXAMPLE 11 Solve $y'''' + 2y'' + y = 0$.

Solution The auxiliary equation is $r^4 + 2r^2 + 1 = 0$, or $(r^2 + 1)^2 = 0$. The roots are $\pm i$ and $\pm i$. Two solutions corresponding to $\pm i$ are $\sin x$ and

12. Solve $y'' + 9y = 0$.

13. Solve $y'' - 4y' + 29y = 0$.

14. Solve $y'' + 2y' + 5y = 0$.

15. Solve $y''' - 2y'' + 8y' = 0$.

16. Solve $y'''' + 8y'' + 16y = 0$.

17. Check by direct calculation that
$y = C_1 \sin x + C_2 \cos x + 5$
is a solution to $y'' + y = 5$ for all
choices of C_1 and C_2.

18. Solve $4y'' + 4y' + y = 3$ (see Margin
Exercise 10).

$\cos x$. Two more corresponding to the second pair $\pm i$ are $x \sin x$ and $x \cos x$. The general solution is

$$y = C_1 \sin x + C_2 \cos x + C_3 x \sin x + C_4 x \cos x. \qquad ❖$$

DO EXERCISE 16.

Nonhomogeneous Equations

So far in this section, the equations considered have been linear homogeneous equations with constant coefficients. We'll relax that now, considering equations of the form

$$a_0 y^{(n)} + a_1 y^{(n-1)} + \cdots + a_{n-1} y' + a_n y = b(x); \qquad (2)$$

that is, we no longer require that the equation be homogeneous. We'll consider only the case in which $b(x)$ is a polynomial.

It turns out to be easy to solve such an equation. To do so, we first find the general solution (call it y_h) to the *associated homogeneous equation,* that is, to Eq. (2) with $b(x) = 0$. Next, we find a *particular solution* (call it y_p) to Eq. (2). Finally, we add y_p to y_h to obtain the general solution to Eq. (2).

EXAMPLE 12   Solve $y'' + y = 5$.

Solution   We must first solve the associated homogeneous equation, $y'' + y = 0$. From above, the general solution, y_h, is

$$y_h = C_1 \sin x + C_2 \cos x.$$

It remains to find a particular solution y_p. To do so, we make an educated guess. Here, we guess that $y_p = C$ must be a constant (more about how to make this "guess" later). Since $y_p' = 0$, we get $y_p'' = 0$, and, since y_p must satisfy $y'' + y = 5$, we find that

$$(0) + (C) = 5,$$

or

$$C = 5.$$

In other words, $y_p = 5$ is a particular solution to $y'' + y = 5$. In retrospect, we see that this is correct.

To complete the problem, we add y_p and y_h, getting

$$y = C_1 \sin x + C_2 \cos x + 5. \qquad ❖$$

DO EXERCISES 17 AND 18.

In Example 12 and Margin Exercise 18, it turned out that y_p was equal to the forcing term $b(x)$. That was a coincidence.

EXAMPLE 13   Solve $y'' - 6y' + 8y = 2x^2 - 1$.

Solution   We first solve $y'' - 6y' + 8y = 0$. From Margin Exercise 2, we know that its solution is $y_h = C_1 e^{4x} + C_2 e^{2x}$. It remains only to find a particular solution y_p. To do this, we try $y_p = Ax^2 + Bx + C$, where A, B, and C are numbers to be determined. Then $y_p' = 2Ax + B$ and $y_p'' = 2A$. Since y_p must satisfy the given equation, we substitute, getting

$$(2A) - 6(2Ax + B) + 8(Ax^2 + Bx + C) = 2x^2 - 1,$$

or

$$2A - 12Ax - 6B + 8Ax^2 + 8Bx + 8C = 2x^2 - 1.$$

Regrouping yields

$$8Ax^2 + (-12A + 8B)x + (2A - 6B + 8C) = 2x^2 - 1.$$

In order for these two expressions to be equal, the coefficients of corresponding terms must be equal. So,

$$8A = 2, \qquad x^2\text{-coefficients are equal}$$
$$8B - 12A = 0, \qquad x\text{-coefficients are equal}$$
$$2A - 6B + 8C = -1. \qquad \text{Constant terms are equal}$$

The second equation gives $B = \frac{3}{2}A$. Since $8A = 2$, we get $A = \frac{1}{4}$ and $B = \frac{3}{2} \cdot \frac{1}{4} = \frac{3}{8}$. Solving the third equation for C yields

$$8C = -2A + 6B - 1 = -2(\tfrac{1}{4}) + 6(\tfrac{3}{8}) - 1,$$

or

$$8C = -\tfrac{1}{2} + \tfrac{18}{8} - 1 = -\tfrac{2}{4} + \tfrac{9}{4} - \tfrac{4}{4} = \tfrac{3}{4},$$

and

$$C = \tfrac{3}{32}.$$

So,

$$y_p = \tfrac{1}{4}x^2 + \tfrac{3}{8}x + \tfrac{3}{32}.$$

The general solution to the given equation is

$$y = y_h + y_p,$$

or

$$y = C_1 e^{4x} + C_2 e^{2x} + \tfrac{1}{4}x^2 + \tfrac{3}{8}x + \tfrac{3}{32}. \qquad ❖$$

DO EXERCISE 19.

19. Solve $y'' - 9y = x^2 + 2x$.

20. Solve $y'' + y' = 7$.

The two examples above show that making the correct guess for the form of the particular solution y_p is the crucial step in solving these equations. If the forcing term is a polynomial, then we will usually let y_p be a polynomial of the same degree. This works in many cases. But consider the equation below.

EXAMPLE 14   Solve $y'' - 2y' = 5$.

Solution   We first solve $y'' - 2y' = 0$. The auxiliary equation is $r^2 - 2r = 0$, or $r(r - 2) = 0$. So the roots are 0 and 2 and

$$y_h = C_1 e^{0 \cdot x} + C_2 e^{2x} = C_1 + C_2 e^{2x}.$$

To complete the problem, we need only find a particular solution y_p. Since the forcing term is a constant, 5, we might try $y_p = C$. Then $y_p' = 0$ and $y_p'' = 0$. Substitution into the original equation gives $(0) - 2(0) = 5$, or $0 = 5$, which is not true. We conclude that $y_p = C$ was the wrong trial solution. The trouble is that 0 was a root of the auxiliary equation. Whenever 0 is a root of the auxiliary equation, we must remember to multiply the trial solution we would generally use by x^m (m is the number of times that 0 occurs as a root of the auxiliary equation). Above, we would ordinarily have used $y_p = C$. Since 0 is a root of the auxiliary equation, we must multiply by x. So we use $y_p = Cx$. Then $y_p' = C$ and $y_p'' = 0$. Substituting, we obtain

$$(0) - 2(C) = 5,$$

or

$$-2C = 5,$$

or

$$C = -\tfrac{5}{2}.$$

Thus $y_p = -\tfrac{5}{2}x$ is a particular solution to $y'' - 2y' = 5$. The general solution is then

$$y = y_h + y_p$$
$$= C_1 + C_2 e^{2x} - \tfrac{5}{2}x. \qquad \diamond$$

DO EXERCISE 20.

EXAMPLE 15   Solve

$$y''' - y'' = 2x - 3,$$
$$y(0) = 1, \qquad y'(0) = 0, \qquad y''(0) = 3.$$

Solution As before, we first find the general solution to the differential equation. Then we use the initial conditions $y(0) = 1$, $y'(0) = 0$, and $y''(0) = 3$ to eliminate the constants.

To solve $y''' - y'' = 2x - 3$, we first solve the associated homogeneous equation $y''' - y'' = 0$. The auxiliary equation is $r^3 - r^2 = 0$, or $r^2(r - 1) = 0$. The roots are 0, 0, and 1. So $y_h = C_1 + C_2 x + C_3 e^x$.

Next, we must find y_p. Since the forcing term is $2x - 3$, a polynomial of degree one, we would ordinarily use $y_p = Ax + B$. In this case, that won't work (as you would discover by trying it), since 0 is a root (here, a double root) of the auxiliary equation. Since 0 occurs twice as a root of $r^3 - r^2 = 0$, we multiply our usual trial solution, $Ax + B$, by x^2, obtaining $y_p = Ax^3 + Bx^2$. Then $y_p' = 3Ax^2 + 2Bx$ and $y_p'' = 6Ax + 2B$. Finally, $y_p''' = 6A$. Substituting into the original equation, we get

$$(6A) - (6Ax + 2B) = 2x - 3,$$

or

$$-6Ax + (6A - 2B) = 2x - 3.$$

Equating corresponding coefficients yields $-6A = 2$ and $6A - 2B = -3$. Solving the first one gives $A = -\frac{1}{3}$. Substituting this into the second equation yields $6(-\frac{1}{3}) - 2B = -3$, or $-2 - 2B = -3$. Solving gives $-2B = -1$, or $B = \frac{1}{2}$. Thus $y_p = -\frac{1}{3}x^3 + \frac{1}{2}x^2$. The general solution is then

$$y = y_h + y_p = C_1 + C_2 x + C_3 e^x - \tfrac{1}{3}x^3 + \tfrac{1}{2}x^2.$$

It remains to use the initial conditions to eliminate C_1, C_2, and C_3. Letting $x = 0$ above, we get

$$y(0) = C_1 + C_2(0) + C_3 e^0 - \tfrac{1}{3}(0)^3 + \tfrac{1}{2}(0)^2,$$

or

$$y(0) = C_1 + C_3.$$

But $y(0) = 1$, so $C_1 + C_3 = 1$. Now, differentiating the solution gives

$$y' = C_2 + C_3 e^x - x^2 + x.$$

Letting $x = 0$, we have

$$y'(0) = C_2 + C_3 e^0 - (0)^2 + 0,$$

or

$$y'(0) = C_2 + C_3.$$

So, since $y'(0) = 0$, we get $C_2 + C_3 = 0$. Finally, differentiating again gives

$$y'' = C_3 e^x - 2x + 1,$$

21. Solve $y''' + 2y'' = 6x - 4$.

22. Solve $y'' - y = -5$, $y(0) = 1$, $y'(0) = 13$.

so that

$$y''(0) = C_3 e^0 - 2(0) + 1 = C_3 + 1.$$

Since $y''(0) = 3$, we have $C_3 + 1 = 3$, or $C_3 = 2$. This enables us to find C_1 and C_2, for $C_2 + C_3 = 0$ gives $C_2 = -C_3 = -2$ and $C_1 + C_3 = 1$ yields $C_1 = 1 - C_3 = 1 - 2 = -1$. So,

$$y = -1 - 2x + 2e^x - \tfrac{1}{3}x^3 + \tfrac{1}{2}x^2.$$

Observe that three initial conditions are needed to eliminate the three constants in the general solution. ❖

DO EXERCISES 21 AND 22.

EXERCISE SET 9.7

Solve.

1. $y'' - 6y' + 5y = 0$
2. $y'' - 8y' + 15y = 0$
3. $y'' - y' - 2y = 0$
4. $y'' + 2y' - 3y = 0$
5. $y'' + 3y' + 2y = 0$
6. $y'' + 5y' + 6y = 0$
7. $2y'' - 5y' + 2y = 0$
8. $2y'' - y' - y = 0$
9. $y'' - 9y = 0$
10. $y'' - 3y = 0$
11. $y'' + 10y' + 25y = 0$
12. $y'' - 8y' + 16y = 0$
13. $4y'' + 12y' + 9y = 0$
14. $y''' + 8y'' + 16y' = 0$
15. $y''' - y'' - 8y' + 12y = 0$
16. $y''' - 7y'' + 11y' - 5y = 0$
17. $y'''' + 3y''' + 3y'' + y' = 0$
18. $y'''' - 4y''' = 0$
19. $y'''' - 5y''' + 4y'' = 0$
20. $y'''' + 3y''' - y' - 3y' = 0$
21. $y'' + 36y = 0$
22. $y'' + 4y' + 13y = 0$
23. $y'' + 8y' + 41y = 0$
24. $y'' + y' + y = 0$
25. $y''' + 2y'' + 5y' = 0$
26. $y''' - 2y'' + 4y' - 8y = 0$
27. $y''' + y'' + 15y' - 17y = 0$
28. $y'''' + 4y'' + 4y = 0$
29. $y'' + y = 7$
30. $y'' - 3y' + 2y = 4$
31. $y'' - 2y' + y = 3$
32. $y'' - 7y' + 10y = 10x - 27$
33. $y'' + 4y' + 4y = 8 - 12x$
34. $y'' - 4y' + 5y = 5x + 1$
35. $y'' + 4y' + 3y = 6x^2 - 4$
36. $y'' + 4y = 8x^2 - 12x$
37. $y'' - 3y' = 4$
38. $y''' + y' = -2$
39. $y''' + y'' = -2$
40. $y'' - 2y' = -4x$
41. $y''' + 4y'' + 20y' = 40x - 12$
42. $y'' - y' = 3x^2 - 8x + 5$
43. $y'' - y' = 0$, $y(0) = 0$, $y'(0) = -1$
44. $y'' + y = 0$, $y(0) = 4$, $y'(0) = 3$
45. $y'' - y = 0$, $y(0) = 1$, $y'(0) = 3$
46. $y'' - 2y' + y = 0$, $y(0) = 1$, $y'(0) = -2$
47. $y''' - y' = 0$, $y(0) = 1$, $y'(0) = 8$, $y''(0) = 4$

48. $y'' - y' - 2y = 2x - 1$, $y(0) = 6$, $y'(0) = 0$

49. $y'' + 9y = 9 - 9x$, $y(0) = 3$, $y'(0) = 2$

50. $y''' + 3y'' = 18x - 18$, $y(0) = 3$, $y'(0) = -1$, $y''(0) = 10$

SYNTHESIS EXERCISES

51. *Commodities.* Over a certain period of time, a trader notes that the price P of gold satisfies the differential equation

$$P'' + 120P' - 575P = 290,$$

where P is measured in dollars per ounce and t is measured in trading days. If the initial price is \$304 per ounce and the initial value of P' is $\frac{1}{10}$, find the price of gold after 50 trading days as well as the highest trading price per ounce over the next 50 trading days.

52. *Population growth.* The estimated number N of a certain bacteria (in thousands) present t hours after the beginning of an experiment satisfies the differential equation

$$N'' - 3N' + 2N = 0.$$

Find the approximate number of bacteria present after 4 hours if $N(0) = 5$ and $N'(0) = 15$.

53. Let y_1 and y_2 both be solutions to the second-order linear homogeneous differential equation

$$a_0(x)y'' + a_1(x)y' + a_2(x)y = 0.$$

(Note that we are *not* assuming that the coefficients are constants.) Show that $C_1 y_1 + C_2 y_2$ is also a solution to the equation, where C_1 and C_2 are arbitrary constants.

54. Show that the conclusion of Exercise 53 holds for the nth-order homogeneous linear equation

$$a_0(x)y^{(n)} + a_1(x)y^{(n-1)} + \cdots + a_{n-1}(x)y' + a_n(x)y = 0.$$

55. a) Let y_p be a solution to

$$a_0(x)y'' + a_1(x)y' + a_2(x)y = b(x),$$

and let y_h be a solution to

$$a_0(x)y'' + a_1(x)y' + a_2(x)y = 0.$$

Show that $y_p + y_h$ is a solution to the nonhomogeneous equation.

b) Let y_1 and y_2 *both* be solutions to the nonhomogeneous equation

$$a_0(x)y'' + a_1(x)y' + a_2(x)y = b(x).$$

Show that $y_1 - y_2$ is a solution to the homogeneous equation

$$a_0(x)y'' + a_1(x)y' + a_2(x)y = 0.$$

c) Use part (b) to show that if y_1 is a given solution to

$$a_0(x)y'' + a_1(x)y' + a_2(x)y = b(x),$$

then every other solution to it is of the form $y_1 + y_2$, where y_2 is a solution to the associated homogeneous equation.

56. You are told that the roots of the auxiliary equation of a certain sixth-order linear homogeneous differential equation with constant coefficients are 2, 2, 3, 3, 3, and 3. Find the general solution to the equation.

57. Repeat Exercise 56 if the roots are 7, 9, $2 \pm 4i$, and $2 \pm 4i$.

COMPUTER-GRAPHING CALCULATOR EXERCISES

58. Plot the graphs of the price of gold, $P(t)$, in Exercise 51 above, and its derivative, $P'(t)$, simultaneously. Does the price of gold peak when the derivative is greatest?

THE CALCULUS EXPLORER:
Derivatives

Use this program to do Exercise 58.

9.8

OBJECTIVES

a) Solve a system of first-order linear homogeneous differential equations.

b) Solve the corresponding non-homogeneous systems if the forcing functions are polynomials.

1. Solve

$$\frac{dx}{dt} = 4x,$$

$$\frac{dy}{dt} = y.$$

SYSTEMS OF DIFFERENTIAL EQUATIONS

In many applications, we are interested in finding the values of two (or more) quantities, say x and y, that are functions of yet another variable t (usually thought of as being time), whose rates of change are given as a system of equations.

EXAMPLE 1 Solve

$$\frac{dx}{dt} = 5x,$$

$$\frac{dy}{dt} = 3y.$$

Solution We want to find x and y in terms of t. In this case, that's particularly easy to do because the equations in the system aren't interrelated. That is, the first equation doesn't involve y and the second one doesn't involve x.

The first equation is a first-order linear equation (with variables separable) whose solution is $x = C_1 e^{5t}$. Similarly, the solution to the second equation is $y = C_2 e^{3t}$.

Note that there will be two constants in the solution. ❖

DO EXERCISE 1.

EXAMPLE 2 Solve

$$\frac{dx}{dt} = y,$$

$$\frac{dy}{dt} = 4x.$$

Solution In this system the variables *are* interrelated, or coupled. That is, x and y both occur in the first equation as well as the second equation. We can "uncouple" them as follows. We differentiate the first equation with respect to t to get

$$\frac{d^2x}{dt^2} = x'' = \frac{dy}{dt}.$$

From the second equation, we have

$$\frac{dy}{dt} = 4x.$$

Combining the equations gives us

$$x'' = \frac{dy}{dt} = 4x,$$

or

$$x'' - 4x = 0.$$

This is a second-order linear homogeneous equation with constant coefficients. Its auxiliary equation is $r^2 - 4 = 0$, or $(r + 2)(r - 2) = 0$. Since 2 and -2 are the roots, we find that $x(t) = C_1 e^{2t} + C_2 e^{-2t}$. Note that since $y = dx/dt$, we know immediately that we have

$$y(t) = 2C_1 e^{2t} - 2C_2 e^{-2t} = \frac{dx}{dt}. \qquad ❖$$

DO EXERCISES 2 AND 3.

We can solve any first-order linear system of the form

$$\frac{dx}{dt} = ax + by,$$

$$\frac{dy}{dt} = cx + dy,$$

where a, b, c, and d are constants, by combining the technique shown in Example 2 with a little algebra. To see how, consider the following example.

EXAMPLE 3   Solve

$$\frac{dx}{dt} = x + 2y,$$

$$\frac{dy}{dt} = 4x - y.$$

Solution   Our objective is to combine the two equations so as to eliminate one of the dependent variables, say y. We write the first equation as $x' = x + 2y$, and then differentiate it to get

$$x'' = x' + 2y'.$$

2. Solve

$$\frac{dx}{dt} = 2y,$$

$$\frac{dy}{dt} = 8x.$$

3. Solve

$$\frac{dx}{dt} = y,$$

$$\frac{dy}{dt} = -4x.$$

4. Check the solutions obtained in Example 3 by direct substitution into the original system of equations.

Now, we'll use the second equation, $y' = 4x - y$, to eliminate y' above. We get

$$x'' = x' + 2(4x - y),$$

or

$$x'' = x' + 8x - 2y.$$

Unfortunately, a y-term remains. But we can eliminate it by using the original version of the first equation. Solving $x' = x + 2y$ for $2y$ yields $2y = x' - x$. Substituting this into the last equation gives us

$$x'' = x' + 8x - (x' - x),$$

or

$$x'' = x' + 8x - x' + x,$$

or

$$x'' - 9x = 0.$$

We have eliminated y and what remains is a second-order linear homogeneous equation in x. Its auxiliary equation is $r^2 - 9 = 0$. The roots here are 3 and -3, so that $x = C_1 e^{3t} + C_2 e^{-3t}$. It still remains to find y. Recall though that in the process of eliminating y, we had obtained $2y = x' - x$. Now using our solution for x gives

$$2y = (3C_1 e^{3t} - 3C_2 e^{-3t}) - (C_1 e^{3t} + C_2 e^{-3t}),$$

or

$$2y = 2C_1 e^{3t} - 4C_2 e^{-3t},$$

or

$$y = C_1 e^{3t} - 2C_2 e^{-3t}.$$

So the solution is

$$x = C_1 e^{3t} + C_2 e^{-3t},$$
$$y = C_1 e^{3t} - 2C_2 e^{-3t}.$$

❖

DO EXERCISES 4 AND 5.

5. Solve

$$\frac{dx}{dt} = 4x + 3y,$$

$$\frac{dy}{dt} = -4x - 4y.$$

EXAMPLE 4 Solve

$$\frac{dx}{dt} = y, \qquad x(0) = 1,$$

$$\frac{dy}{dt} = -x + 2y, \qquad y(0) = 3.$$

Solution   Just as in previous sections, we'll first find the general solution to the problem. Then we'll use the initial conditions to evaluate the constants in the general solution.

Differentiating the first equation gives $x'' = y'$. Using the second equation to replace y', we get $x'' = -x + 2y$. From the first equation, we have $y = x'$. Combining the last two equations gives $x'' = -x + 2x'$, or $x'' - 2x' + x = 0$.

Again, we have a second-order linear equation with constant coefficients. The auxiliary equation is $r^2 - 2r + 1 = 0$, or $(r - 1)^2 = 0$, so that the roots are 1 and 1. Since the roots are repeated, we have

$$x = C_1 e^t + C_2 t e^t.$$

Since $y = dx/dt = x'$, we get

$$y = C_1 e^t + C_2 t e^t + C_2 e^t,$$

or

$$y = (C_1 + C_2) e^t + C_2 t e^t.$$

Letting $t = 0$ in the equation for x gives

$$x(0) = C_1 e^0 + C_2 \cdot 0 \cdot e^0$$
$$= C_1.$$

Since $x(0) = 1$ was one of the initial conditions, we have $x(0) = C_1 = 1$. Letting $t = 0$ in the equation for y gives $y(0) = C_1 + C_2$. Since $y(0) = 3$, we have $C_1 + C_2 = 3$. Since $C_1 = 1$, we must have $C_2 = 2$. So, $x = e^t + 2te^t$ and $y = 3e^t + 2te^t$.

It is easy to check that these solutions satisfy both the original equations and the initial conditions.   ❖

DO EXERCISES 6 AND 7.

If $f(t)$ and $g(t)$ are polynomials in t, then the preceding techniques can be used to solve systems of the form

$$\frac{dx}{dt} = ax + by + f(t),$$

$$\frac{dy}{dt} = cx + dy + g(t),$$

where, as above, a, b, c, and d are constants.

6. Solve

$$\frac{dx}{dt} = 3x + y, \qquad x(0) = 1,$$

$$\frac{dy}{dt} = -x + y, \qquad y(0) = 3.$$

7. Solve

$$\frac{dx}{dt} = 3x - y, \qquad x(0) = 3,$$

$$\frac{dy}{dt} = 2x, \qquad y(0) = 5.$$

8. Solve

$$\frac{dx}{dt} = 4x + y,$$

$$\frac{dy}{dt} = 2x + 3y + 50.$$

EXAMPLE 5   Solve

$$\frac{dx}{dt} = x + y + 4,$$

$$\frac{dy}{dt} = 4x + y + 3.$$

Solution   We differentiate the first equation to obtain $x'' = x' + y'$. Using the second equation to replace y' yields

$$x'' = x' + (4x + y + 3),$$

or

$$x'' = x' + 4x + y + 3.$$

Solving the first equation for y yields $y = x' - x - 4$, so that combining the last two equations gives us

$$x'' = x' + 4x + (x' - x - 4) + 3,$$

or

$$x'' = 2x' + 3x - 1,$$

or

$$x'' - 2x' - 3x = -1.$$

Note that the only difference between this example and our previous ones is that we end up with a nonhomogeneous second-order equation for x rather than a homogeneous one.

Here, the auxiliary equation of the associated homogeneous equation $x'' - 2x' - 3x = 0$ is $r^2 - 2r - 3 = 0$, which factors as $(r - 3)(r + 1) = 0$. The roots are 3 and -1, and $x_h = C_1 e^{3t} + C_2 e^{-t}$. Note that the constant function $x_p = \frac{1}{3}$ is a particular solution, so that

$$x = C_1 e^{3t} + C_2 e^{-t} + \tfrac{1}{3}$$

is the general solution for x. Since $y = x' - x - 4$, we get

$$y = (3C_1 e^{3t} - C_2 e^{-t}) - (C_1 e^{3t} + C_2 e^{-t} + \tfrac{1}{3}) - 4$$
$$y = 3C_1 e^{3t} - C_2 e^{-t} - C_1 e^{3t} - C_2 e^{-t} - \tfrac{1}{3} - \tfrac{12}{3},$$

or

$$y = 2C_1 e^{3t} - 2C_2 e^{-t} - \tfrac{13}{3}.$$

DO EXERCISE 8.

EXAMPLE 6   Tank A contains 2000 pounds of salt dissolved in 1000 gallons of water. Tank B contains 1000 pounds of salt dissolved in 1000 gallons of water. The mixture from tank A is pumped to tank B at the rate of 500 gallons per day, while that from tank B is pumped to tank A at the same rate. Assuming that the mixture in each tank is kept uniform by stirring, how much salt is in each tank after t days?

Solution   We know that the initial concentration of salt in tank A is 2000 lb/1000 gal = 2 lb/gal. The initial concentration in tank B is 1000 lb/1000 gal = 1 lb/gal. As we pump the saltier mixture from tank A to tank B, we'd expect the mixture in tank B to get saltier. At the same time, the mixture being pumped from tank B to tank A should gradually dilute the mixture in tank A. We might guess that the concentrations in each tank might approach an equilibrium concentration.

To see what actually happens, let

$A(t)$ = the number of pounds of salt in tank A after t days, and

$B(t)$ = the number of pounds of salt in tank B after t days.

Note that $A(0) = 2000$ and $B(0) = 1000$. The concentration of salt in tank A after t days is $(A(t)/1000)$ lb/gal, since tank A always contains exactly 1000 gallons of water. Similarly, the concentration in tank B after t days is $B(t)/1000$ lb/gal. To set up our differential equations we observe that

Rate of change of salt in tank A = (Rate of flow in)
$$- \text{(Rate of flow out)}.$$

The rate at which salt flows into tank A is $(500) \cdot B(t)/1000$ pounds per day; that is, the product of the flow into tank A, 500 gallons per day, and the concentration in tank B, $B(t)/1000$ pounds per gallon.

Similarly, the flow out of tank A is $500 \cdot A(t)/1000$. Thus the rate of change of the amount of salt in tank A, dA/dt, is

$$\frac{dA}{dt} = \frac{500B(t)}{1000} - \frac{500A(t)}{1000},$$

or

$$\frac{dA}{dt} = \frac{1}{2}B(t) - \frac{1}{2}A(t).$$

The amount of salt flowing into tank B is just the negative of that flowing into tank A (equivalently, what flows out of tank A goes into tank B and vice versa). So

$$\frac{dB}{dt} = -\frac{dA}{dt} = \frac{1}{2}A(t) - \frac{1}{2}B(t).$$

We have a system of equations, namely,

$$\frac{dA}{dt} = -\frac{1}{2}A + \frac{1}{2}B \quad \text{and} \quad \frac{dB}{dt} = \frac{1}{2}A - \frac{1}{2}B$$

with the initial conditions $A(0) = 2000$ and $B(0) = 1000$.

Differentiating the first equation, $A' = -\frac{1}{2}A + \frac{1}{2}B$, gives $A'' = -\frac{1}{2}A' + \frac{1}{2}B'$. Using the second equation, we get

$$A'' = -\frac{1}{2}A' + \frac{1}{2} \cdot (\frac{1}{2}A - \frac{1}{2}B),$$

or

$$A'' = -\frac{1}{2}A' + \frac{1}{4}A - \frac{1}{4}B.$$

From the first equation, we have $\frac{1}{2}B = A' + \frac{1}{2}A$ or, substituting above,

$$A'' = -\frac{1}{2}A' + \frac{1}{4}A - \frac{1}{4}(2A' + A),$$

or

$$A'' = -A'.$$

The auxiliary equation of $A'' + A' = 0$ is $r^2 + r = 0$, or $r(r + 1) = 0$. Since its roots are 0 and -1, we get

$$A(t) = C_1 + C_2 e^{-t}.$$

Since $B = A + 2A'$ (solving the first equation for B), we get

$$B(t) = (C_1 + C_2 e^{-t}) + 2(-C_2 e^{-t}),$$

or

$$B(t) = C_1 - C_2 e^{-t}.$$

To evaluate the constants, recall that $A(0) = 2000$, whereas if we let $t = 0$ in our formula for $A(t)$, we get

$$A(0) = C_1 + C_2 e^{-0} = C_1 + C_2.$$

So $C_1 + C_2 = 2000$. Similarly, $B(0) = 1000$ while $B(0) = C_1 - C_2 e^{-0} = C_1 - C_2$. Thus $C_1 - C_2 = 1000$. Adding, we get

$$\begin{aligned} C_1 + C_2 &= 2000 \\ \underline{C_1 - C_2} &= \underline{1000} \\ 2C_1 \quad\;\; &= 3000 \end{aligned}$$

or $C_1 = 1500$. From the first equation, $1500 + C_2 = 2000$, or $C_2 = 500$. We have

$$A(t) = 1500 + 500e^{-t},$$
$$B(t) = 1500 - 500e^{-t}.$$

Note that $A(t) + B(t) = 3000$, which is correct, since salt neither enters nor leaves the system.

As for our intuitive feeling that the concentration of salt in each tank should approach an equilibrium concentration, note that if t is a large positive number, then e^{-t} is a very small (positive) number. If you have a calculator handy, you can check that

$$e^{-3} \approx 0.04978707,$$
$$e^{-14} \approx 0.00000083,$$

and

$$e^{-30} \approx 1 \times 10^{-13} \qquad (0.0000000000001).$$

Then we can see that as t increases, the amount of salt in each tank will approach 1500 pounds, as we might have expected. ❖

DO EXERCISES 9 AND 10.

EXAMPLE 7 *Commodities prices.* The prices of two commodities after t days are $P(t)$ and $Q(t)$, respectively. Traders note that over a period of time the prices obey the system

$$\frac{dP}{dt} = 3000 - 5Q$$

and

$$\frac{dQ}{dt} = 20P - 5000.$$

If the initial prices are $P(0) = \$260$ and $Q(0) = \$560$, find the prices at any later time (assume that t is in days and P and Q are in dollars per unit).

Solution Differentiating the first equation yields $P'' = -5Q'$. Since $Q' = 20P - 5000$ (the second equation), we get

$$P'' = -5(20P - 5000),$$

or

$$P'' = -100P + 25{,}000,$$

or

$$P'' + 100P = 25{,}000.$$

We have a nonhomogeneous second-order equation with constant coefficients. Note that the constant function $P_p = 250$ is a particular solution to the equation. Since the auxiliary equation of the associated homo-

9. Find the answer to Example 6 if tank A initially contains 1600 pounds of salt and tank B contains 800 pounds.

10. Find the answer to Example 6 if both tanks initially contain 1100 pounds of salt. (*Warning:* Before computing, stop and *think!*)

geneous equation $(P'' + 100P = 0)$ is $r^2 + 100 = 0$, the roots are $r = \pm 10i$ and we get

$$P_h = C_1 \sin (10t) + C_2 \cos (10t).$$

So

$$P = C_1 \sin (10t) + C_2 \cos (10t) + 250.$$

Solving the first equation for Q gives $5Q = -dP/dt + 3000$, or

$$Q = -\frac{1}{5} \cdot \frac{dP}{dt} + 600.$$

So

$$Q = -\tfrac{1}{5}(10C_1 \cos (10t) - 10C_2 \sin (10t)) + 600$$
$$Q = -2C_1 \cos (10t) + 2C_2 \sin (10t) + 600,$$

or

$$Q = 2C_2 \sin (10t) - 2C_1 \cos (10t) + 600.$$

We were given that $P(0) = 260$. From our solution,

$$P(0) = C_1 \sin (10 \cdot 0) + C_2 \cos (10 \cdot 0) + 250$$
$$P(0) = C_1 \sin (0) + C_2 \cos (0) + 250,$$

or

$$P(0) = C_2 + 250.$$

So $260 = C_2 + 250$, or $C_2 = 10$. Similarly, $Q(0) = 560$, while, from the solution,

$$Q(0) = 2C_2 \sin (10 \cdot 0) - 2C_1 \cos (10 \cdot 0) + 600,$$

or

$$Q(0) = 2C_2 \sin 0 - 2C_1 \cos 0 + 600,$$

or

$$Q(0) = 600 - 2C_1.$$

So $560 = 600 - 2C_1$ and we get $C_1 = 20$. Thus after t days, the prices of the commodities are

$$P(t) = 20 \sin (10t) + 10 \cos (10t) + 250$$

and

$$Q(t) = 20 \sin (10t) - 40 \cos (10t) + 600.$$

The following table gives the values of P and Q after the first few days.

t	P	Q
0	260.00	560.00
1	230.73	622.68
2	272.34	601.94
3	231.78	574.07
4	258.23	641.58
5	254.40	556.15

The prices fluctuate back and forth about the values $250 and $600, respectively. It is *not* true, however, that the greatest price for P coincides with the least for Q (there is a time lag involved). ❖

DO EXERCISE 11.

EXAMPLE 8 *Population growth.* The populations of two species at time t are $X(t)$ and $Y(t)$, respectively. If it is assumed that the species interact according to

$$\frac{dX}{dt} = 3X - 5Y,$$

$$\frac{dY}{dt} = 5X - 3Y - 480,$$

and if $X(0) = 160$ and $Y(0) = 92$, find $X(t)$ and $Y(t)$. Describe what happens to the populations over a long period of time.

Solution The usual sequence of steps (we omit the details) leads to the second-order equation $X'' + 16X = 2400$. The general solution to this is $X = C_1 \sin 4t + C_2 \cos 4t + 150$. Solving for Y in the first equation yields

$$Y = (\tfrac{3}{5}C_1 + \tfrac{4}{5}C_2) \sin 4t + (-\tfrac{4}{5}C_1 + \tfrac{3}{5}C_2) \cos 4t + 90.$$

Using the initial conditions yields $160 = C_2 + 150$, or $C_2 = 10$, and $92 = -\tfrac{4}{5}C_1 + \tfrac{3}{5}C_2 + 90$, or $2 = -\tfrac{4}{5}C_1 + 6$, or $C_1 = 5$. Substitution and simplification result in

$$X = 5 \sin 4t + 10 \cos 4t + 150,$$
$$Y = 11 \sin 4t + 2 \cos 4t + 90.$$

11. If $P(0) = \$24$, $Q(0) = \$38$, and P and Q are related by

$$\frac{dP}{dt} = 160 - 4Q,$$

$$\frac{dQ}{dt} = 16P - 320,$$

find $P(t)$ and $Q(t)$.

The populations fluctuate. If we use the fact that

$$A \sin (\lambda t) + B \cos (\lambda t)$$

always remains between $\pm \sqrt{A^2 + B^2}$, then we see that the X-population stays between $150 \pm \sqrt{5^2 + 10^2}$, or between 139 and 161, while the Y-population remains between 79 and 101 (approximately).   ❖

EXERCISE SET 9.8

Find the general solution to each of the following systems of equations.

1. $x' = 2x,\ y' = 3y$

2. $x' = -x,\ y' = 4y$

3. $x' = y,\ y' = -2x + 3y$

4. $x' = x + 2y,\ y' = x + 2y$

5. $x' = 2x + y,\ y' = 3x + 4y$

6. $x' = 3x - 4y,\ y' = 2x - 3y$

7. $x' = -\frac{1}{2}x + y,\ y' = \frac{1}{2}x$

8. $x' = -x - 3y,\ y' = -y$

9. $x' = 4x + y,\ y' = -x + 2y$

10. $x' = 2x - y,\ y' = 4x - 2y$

11. $x' = 2y,\ y' = -18x$

12. $x' = x - y,\ y' = 5x - y$

13. $x' = 3x - 5y,\ y' = x - y$

14. $x' = -3x + 5y,\ y' = -x + y$

Find the functions that satisfy both the system of equations and the given initial conditions.

15. $x' = 2x - y,\ y' = 2x + 5y,\ x(0) = 3,\ y(0) = -5$

16. $x' = x + y,\ y' = x + y,\ x(0) = 2,\ y(0) = 0$

17. $x' = 2x + 3y,\ y' = -3x + 8y,\ x(0) = 1,\ y(0) = 1$

18. $x' = 2x + 3y,\ y' = -3x + 8y,\ x(0) = 1,\ y(0) = 2$

19. $x' = y,\ y' = -4x,\ x(0) = 1,\ y(0) = 2$

20. $x' = x + 5y,\ y' = -x - 3y,\ x(0) = 10,\ y(0) = -3$

Find the solution to each of the following systems of equations, finding the particular solution if initial conditions are given.

21. $x' = 5x - y + 5,\ y' = 2x + 2y + 2$

22. $x' = 2x + y + 1,\ y' = -2y - 22$

23. $x' = y + 2t + 3,\ y' = -x + 4t - 2$

24. $x' = -x + y + 2t + 2,\ y' = 2x + 2t - 1$

25. $x' = 2x + y + 3,\ y' = 5x - 2y + 12,\ x(0) = 0,\ y(0) = -2$

26. $x' = y + 5,\ y' = -x + 2y + 10,\ x(0) = 1,\ y(0) = -2$

APPLICATIONS

❖ Life and Physical Sciences

27. *Mixing chemicals.* Tank A initially contains 200 pounds of salt dissolved in 500 gallons of water, while tank B contains 100 pounds of salt dissolved in 500 gallons of water. The mixture in tank A is pumped to tank B at the rate of 100 gallons per day and the mixture from B is pumped to A at the same rate. Set up the system of differential equations for $A(t)$ and $B(t)$, the number of pounds of salt in tanks A and B, after t days, respectively. Then solve them. What is the "limiting concentra-

tion" of salt in tank A? in tank B?

28. *Population growth.* The numbers $X(t)$ and $Y(t)$ of two species in a certain area (in hundreds) satisfy the system of equations

$$\frac{dX}{dt} = X + Y - 100, \qquad X(0) = 40,$$

$$\frac{dY}{dt} = -8X - 3Y + 450, \qquad Y(0) = 60,$$

where t is measured in months.

a) Find $X(t)$ and $Y(t)$.

b) Find $X(1)$, $Y(1)$, $X(10)$, and $Y(10)$.

c) What are $X(\infty)$ and $Y(\infty)$?

❖ **Business and Economics**

29. *Commodities prices.* The prices P and Q of two interrelated commodities are determined by the following system of equations:

$$P' = P + 13Q - 400, \qquad P(0) = 49,$$
$$Q' = -2P - Q + 100, \qquad Q(0) = 32.$$

Find $P(t)$ and $Q(t)$.

30. *Projected sales.* A securities analyst finds that the projected sales $X(t)$ and $Y(t)$ of two competing firms (in millions of dollars) are given by

$$X' = 4Y - 100t, \qquad X(0) = 377,$$
$$Y' = -X + 40t + 400, \qquad Y(0) = 12.$$

Find $X(t)$ and $Y(t)$ (where t is in years).

SYNTHESIS EXERCISES

31. Solve

$$x' = -2x + 2y + 2e^t,$$
$$y' = 2x + y.$$

32. Solve

$$\frac{dx}{dt} = -2x + y - z,$$
$$\frac{dy}{dt} = y + z,$$
$$\frac{dz}{dt} = 3z.$$

COMPUTER–GRAPHING CALCULATOR EXERCISES

33. Plot the graph of the solution to Example 7, using

$$x(t) = 20 \sin(10t) + 10 \cos(10t) + 250$$

and

$$y(t) = 20 \sin(10t) - 40 \cos(10t) + 600.$$

a) Is the graph an ellipse?

b) Is the graph traced out in the clockwise or counterclockwise sense?

c) When $x(t)$ (or $P(t)$) reaches its maximum value, is $y(t)$ ($Q(t)$) increasing or decreasing?

d) From the graph, estimate the maximum and minimum values of both $x(t)$ and $y(t)$.

THE CALCULUS EXPLORER:
Power Grapher

Use the program to do Exercise 33.

CHAPTER SUMMARY AND REVIEW

TERMS TO KNOW

Differential equation, p. 633
General solution, p. 635
Separation of variables, p. 638
Elasticity, p. 640
Weber–Fechner Law, p. 641
Inhibited growth model,
$$\frac{dP}{dt} = kP(L - P),\ \text{p. 644}$$

Growth model, $\frac{dP}{dt} = k(L - P)$, p. 656

Differential form, p. 662
Exact first-order differential equation, p. 663
Linear first-order differential equation, p. 666
Forcing function, p. 667
Homogeneous linear first-order differential equation, p. 667
Integrating factor, p. 670

Linear differential equations with constant coefficients, p. 674
Characteristic equation (auxiliary equation), p. 676
Associated homogeneous equation, p. 680
System of first-order linear homogeneous differential equations, p. 686

REVIEW EXERCISES

These review exercises are for test preparation. They can also be used as a lengthened practice test. Answers are at the back of the book. The answers also contain bracketed section references, which tell you where to restudy if your answer is incorrect.

Solve.

1. $\dfrac{dy}{dx} = 6x^2$

2. $\dfrac{dy}{dx} = 6x^2 y$

3. $y' = 5x^2 - x^2 y$

4. $\dfrac{dy}{dx} = 2x^2 - 4x + 3$, given that $y = \dfrac{2}{3}$ when $x = 1$

5. $\dfrac{dv}{dt} = 2v^{-3}$

6. Solve $dy/dx = 4x^3 - 2x$, $y(0) = 1$.

7. Verify that $y = x^2$ is a solution to $x(dy/dx) - 2y = 0$.

8. Graph $P(t) = 1 - e^{-4t}$.

9. *Spread of a rumor.* A group of 8 students at State University start the rumor "All final examinations for the year have been cancelled." The rate at which this rumor

spreads satisfies the equation

$$\frac{dN}{dt} = kN(L - N),$$

where

$N =$ the number of students that have heard this rumor after t minutes,

$L = 1200$, the total enrollment of the university,

$N_0 = 8$, the number of students who started the rumor, and

$k = 0.0002$.

a) Write the equation for $N(t)$.

b) Sketch a graph of $N(t)$ given that the point of inflection occurs at $t = 21$ minutes.

c) After how many minutes t will 800 students have heard the rumor?

10. *Hullian learning model.* The Hullian model of learning is $dP/dt = k(L - P)$, where

$P =$ the probability of mastery of the process of differentiating quotients after t learning trials,

$L = 1$, the limiting value of the probability of mastery of the process, and

$k =$ a constant.

a) Write the solution $P(t)$ in terms of L and k.

b) Suppose that the probability of mastery of differentiating quotients is 0.80 after 10 trials. Use this information to determine k to the nearest hundredth.

c) Rewrite $P(t)$ in terms of k.

d) Find $P(12)$.

e) Using the value of k determined in part (b), find the number of trials required so that $P(t) = 0.95$.

11. For the inhibited growth model, $P = 375$ at the point of inflection. What is the limiting value?

Solve.

12. $2x\,dx + 3y^2\,dy = 0$, $y(0) = 2$

13. $(4x^3 + 6xy)\,dx + (3x^2 - 4y^3)\,dy = 0$, $y(0) = 1$

Solve.

14. $xy' + 4y = 0$, $y(1) = 3$

15. $y' - 2y = e^x$

Solve.

16. $y'' - 7y' + 10y = 0$

17. $y'' - 6y' + 9y = 0$

18. $y'' + 25y = 100$

Solve.

19. $x' = 2x + 6y$, $y' = 4x$

20. $x' = 18y$, $y' = -2x$

21. $x' = -4x + y$, $y' = -4x$

22. $x' = 3x + y$, $y' = x + 3y$, $x(0) = 2$, $y(0) = 0$

23. $x' = y$, $y' = 4x$, $x(0) = 3$, $y(0) = 6$

EXERCISES FOR THINKING AND WRITING

24. Look up the word *Malthusian* in the dictionary. Who was Malthus? What ideas did he espouse and what do his views have to do with this chapter? Were Malthus's predictions for the population of Great Britain accurate? for the population of the earth?

25. How many different applications of differential equations can you find? to business and economics? to the physical sciences and life sciences? to engineering?

Good places to look are advanced textbooks and/or journals in each area.

26. Find all solutions to $y^2 + (y')^2 = -1$.

27. *Miniproject:* Look up some population statistics in your library for **uncontrolled** growth of a population. Plot your data. Do they have the shape of a logistic curve?

CHAPTER TEST 9

1. Solve the differential equation $y' = x^2 + 3x - 5$, given the condition $y = 7$ when $x = 0$.

Solve the following differential equations.

2. $\dfrac{dy}{dx} = 8x^7 y$

3. $\dfrac{dy}{dx} = \dfrac{9}{y}$

4. $\dfrac{dy}{dt} = 6y$

5. For the inhibited growth model, $P = 680$ at the point of inflection. What is the limiting value?

6. *Growth of bacteria.* The growth rate for a certain population of bacteria cells is given by

$$\frac{dP}{dt} = kP(L - P).$$

where

P = the number of cells present after time t (in hours),

$L = 800$, the limiting value,

$k = 0.0009$, and

$P_0 = 20$.

a) Write the solution $P(t)$ in terms of the given constraints.

b) Sketch a graph of $P(t)$ given that the point of inflection occurs at $t = 5$.

c) At what time t are 500 cells present?

7. Graph $P(t) = 5(1 - e^{-t})$.

8. *Sales growth.* The growth rate of a company is modeled by

$$\frac{dS}{dt} = k(L - S), \qquad S = \$0 \text{ when } t = 0;$$

where

S = the total sales, in millions of dollars, in the tth year of operation,

$L = \$20$ million, the limiting value, and

k = a constant.

a) Write the solution $S(t)$ in terms of L and k.

b) Suppose $S(2) = \$4$ million; that is, sales during the second year are $4 million. Use this to determine k. Round to the nearest hundredth.

c) Rewrite $S(t)$ in terms of k.

d) Use the equation in part (c) to find $S(10)$, the sales in the tenth year.

e) In what year will the sales be $15 million?

Solve each of the following.

9. $3x^2 \, dx + 2y \, dy = 0$, $y(0) = 1$

10. $(3x^2 + 4xy) \, dx + (2x^2 - 5y^4) \, dy = 0$, $y(0) = 1$

11. $xy' + 3y = 0$, $y(1) = 4$

12. $y' - 3y = e^{4x}$

13. $y'' - 5y' + 6y = 0$

14. $y'' - 12y' + 36y = 0$

15. $y'' + 36y = 72$

16. $x' = 2x + 5y$, $y' = 3x$

17. $x' = 6y$, $y' = -24x$

18. $x' = -6x + 3y$, $y' = -3x$

19. $x' = -6x + y$, $y' = -2x - 3y$, $x(0) = 0$, $y(0) = -1$

CHAPTER

10

TAYLOR SERIES AND l'HÔPITAL'S RULE

In this chapter we will learn about infinite series of numbers and infinite series of functions as well as how to calculate with these series. These calculations will, in turn, enable us to evaluate limits and to evaluate certain integrals that cannot be calculated easily in any other way.

AN APPLICATION

A corporation deposits $1000 in a bank. The bank may, and does, loan out 90% of the money deposited. That money, in turn, is deposited in another bank, which, in turn, lends out 90% of it. Assuming the process continues indefinitely, find the total economic effect of the initial $1000 deposit.

THE MATHEMATICS

The total effect of the initial $1000 deposit can be modeled as the sum of the infinite series

$$\$1000 + \$1000(0.90) \\ + \$1000(0.90)^2 \\ + \$1000(0.90)^3 + \cdots.$$

This series is a *geometric series* whose sum is given by

$$s = \$1000/[1 - 0.90] = \$10,000.$$

The sum $10,000 is the result of what is referred to in economics as the *multiplier effect*.

10.1

OBJECTIVE

a) Expand a given polynomial in powers of $x - a$ (for any given real number a).

TAYLOR'S FORMULA FOR POLYNOMIALS

The function f given by

$$f(x) = 2x^2 - 16x + 35$$

can also be expressed in the form

$$f(x) = 2(x - 3)^2 - 4(x - 3) + 5.$$

This can be checked by multiplying and simplifying the right-hand side of the second equation. In the second form, $f(x)$ is expressed as a sum of powers of $x - 3$. In general, the formulas for many functions can be written in the form*

$$f(x) = \sum_{n=0}^{\infty} c_n(x - a)^n,$$

where a is a fixed number and $c_1, c_2, \ldots, c_n, \ldots$ is a sequence of real numbers. In this chapter we will study expansions like this. Each is called a *Taylor series expansion*. This tool is useful in many areas of mathematics.

To introduce the subject, we first consider the Taylor expansion for polynomial functions. Recall the following definition from Section 1.5.

DEFINITION

A *polynomial function (in x)* is a function that can be written as

$$f(x) = c_n x^n + c_{n-1} x^{n-1} + \cdots + c_1 x + c_0,$$

where $c_n, c_{n-1}, \ldots, c_1, c_0$ are real numbers, with $c_n \neq 0$, called the *coefficients* of the polynomial and where n is a nonnegative integer called the *degree* of the polynomial.

Thus each of the following is a polynomial function:

$$\begin{aligned} f(x) &= -5 & \text{(a constant function)}, \\ f(x) &= x + 3 & \text{(a linear function)}, \end{aligned}$$

and

$$f(x) = 2x^3 - 4x^2 + x + 1.$$

---

*You may need to review summation notation in Section 5.8.

The polynomial $2x^3 - 4x^2 + x + 1$ has degree 3. The polynomial $x + 3$ has degree 1 (since $x^1 = x$). The polynomial $0 \cdot x^5 + 3x^4$ has degree 4. The polynomial 6, or $6x^0$, has degree 0.

Consider the following example.

EXAMPLE 1   Simplify

$$\frac{x^2 - 3x + 1}{x}, \qquad x \neq 0.$$

Solution   We write this expression in the form

$$\frac{x^2}{x} - \frac{3x}{x} + \frac{1}{x},$$

and, performing the indicated divisions, we obtain

$$x - 3 + \frac{1}{x} \quad \text{for } x \neq 0. \qquad \qquad ❖$$

DO EXERCISE 1.

EXAMPLE 2   Simplify

$$\frac{x^3 - 3x^2 + 3x - 4}{x - 1}, \qquad x \neq 1.$$

Solution   This looks harder. We can write this equation as

$$\frac{x^3}{x - 1} - \frac{3x^2}{x - 1} + \frac{3x}{x - 1} - \frac{4}{x - 1},$$

but this doesn't simplify as in Example 1. Suppose, however, that we know that

$$x^3 - 3x^2 + 3x - 4 = (x - 1)^3 - 3.$$

This is true, although you would not know it offhand. Then we can simplify as follows:

$$\frac{x^3 - 3x^2 + 3x - 4}{x - 1} = \frac{(x - 1)^3 - 3}{x - 1} = \frac{(x - 1)^3}{x - 1} - \frac{3}{x - 1}$$

$$= (x - 1)^2 - \frac{3}{x - 1},$$

whenever $x \neq 1$. ❖

DO EXERCISE 2.

1. Simplify

$$\frac{2(x - 3)^2 + 4(x - 3) + 1}{x - 3}$$

for $x \neq 3$.

2. Use the identity

$$x^3 - 6x^2 + 14x - 8$$
$$= (x - 2)^3 + 2(x - 2) + 4$$

to simplify

$$\frac{x^3 - 6x^2 + 14x - 8}{(x - 2)^2}.$$

In Example 2, the key to the solution was the identity

$$x^3 - 3x^2 + 3x - 4 = (x - 1)^3 - 3.$$

We were given the expansion of $x^3 - 3x^2 + 3x - 4$ in powers of $x - 1$. We want to be able to write down identities of this kind quickly and easily. Suppose that we want to write the polynomial $2x^2 - 16x + 35$ in powers of $x - 3$. That is, we want to find real numbers c_2, c_1, and c_0 so that

$$2x^2 - 16x + 35 = c_2(x - 3)^2 + c_1(x - 3) + c_0.$$

One possible solution is to multiply out the right-hand side of the equation, grouping the x^2-, x-, and constant terms together. Then we can equate corresponding coefficients to obtain three linear equations in the three unknowns c_2, c_1, and c_0, each of which can then be found. But here's another way.

EXAMPLE 3 Let $f(x) = 2x^2 - 16x + 35$. Find c_2, c_1, and c_0 so that

$$f(x) = c_2(x - 3)^2 + c_1(x - 3) + c_0. \tag{1}$$

Solution The equation above must hold for all real numbers x. So it must be true if we let $x = 3$. We obtain

$$\begin{aligned} f(3) &= c_2(3 - 3)^2 + c_1(3 - 3) + c_0 \\ &= c_2 \cdot 0 + c_1 \cdot 0 + c_0 \\ &= c_0. \end{aligned}$$

Since $f(3) = 2(3)^2 - 16(3) + 35 = 18 - 48 + 35 = 5$, we get

$$\begin{aligned} c_0 &= f(3) \\ &= 5. \end{aligned}$$

Next, we apply some calculus. We differentiate both sides of Eq. (1) with respect to x to get

$$\begin{aligned} f'(x) &= 2c_2(x - 3) + c_1 + 0 \\ &= 2c_2(x - 3) + c_1. \end{aligned} \tag{2}$$

Letting $x = 3$, we get

$$f'(3) = 2c_2(3 - 3) + c_1,$$

or

$$c_1 = f'(3).$$

Since $f(x) = 2x^2 - 16x + 35$, then

$$f'(x) = 4x - 16 \quad \text{and} \quad f'(3) = 4(3) - 16 = -4.$$

So we get

$$c_1 = f'(3)$$
$$= -4.$$

To find c_2, we differentiate Eq. (2). We get $f''(x) = 2c_2$, a constant function. Since $f'(x) = 4x - 16$, we also have $f''(x) = 4$, so that $2c_2 = 4$, or $c_2 = 2$. Thus $c_2 = 2$, $c_1 = -4$, $c_0 = 5$, and

$$2x^2 - 16x + 35 = 2(x - 3)^2 - 4(x - 3) + 5. \qquad ❖$$

DO EXERCISE 3.

EXAMPLE 4 Write $f(x) = x^3 - 7x + 3$ in powers of $x - 2$.

Solution We want to write

$$f(x) = c_3(x - 2)^3 + c_2(x - 2)^2 + c_1(x - 2) + c_0. \qquad (3)$$

We let $x = 2$ in Eq. (3) to get

$$f(2) = c_3(2 - 2)^3 + c_2(2 - 2)^2 + c_1(2 - 2) + c_0$$
$$= c_0.$$

Differentiating, we obtain

$$f'(x) = 3c_3(x - 2)^2 + 2c_2(x - 2) + c_1. \qquad (4)$$

Letting $x = 2$, we get

$$f'(2) = c_1.$$

Differentiating Eq. (4) yields

$$f''(x) = 6c_3(x - 2) + 2c_2,$$

and, letting $x = 2$, we get

$$f''(2) = 2c_2.$$

Another differentiation yields

$$f'''(x) = 6c_3.$$

Since this equation is true for all values of x—that is, since $f'''(x)$ is a constant function—we have, in particular,

$$f'''(2) = 6c_3.$$

3. Let $f(x) = x^2 - 4x + 2$. Find c_2, c_1, and c_0 so that

$$f(x) = c_2(x - 3)^2 + c_1(x - 3) + c_0.$$

Before we compute c_3, c_2, c_1, and c_0, note that there seems to be a pattern to the c_n's. That is,

$$c_0 = f(2),$$
$$c_1 = f'(2),$$
$$c_2 = \tfrac{1}{2}f''(2),$$
$$c_3 = \tfrac{1}{6}f'''(2).$$

Returning to the problem, since $f(x) = x^3 - 7x + 3$, we have

$$c_0 = f(2) = 2^3 - 7 \cdot 2 + 3$$
$$= 8 - 14 + 3 = -3.$$

Differentiating, we have $f'(x) = 3x^2 - 7$, so that

$$c_1 = f'(2) = 3(2)^2 - 7$$
$$= 12 - 7 = 5.$$

Also, $f''(x) = 6x,$ so that $f''(2) = 6 \cdot 2 = 12,$ and $c_2 = \tfrac{1}{2}f''(2) = \tfrac{1}{2} \cdot 12 = 6$. Finally, $f'''(x) = 6$, so that $c_3 = \tfrac{1}{6}f'''(2) = \tfrac{1}{6} \cdot 6 = 1$. Thus,

$$x^3 - 7x + 3 = (x-2)^3 + 6(x-2)^2 + 5(x-2) - 3. \qquad ❖$$

Factorial Notation

Before continuing, we introduce some notation. Products like $3 \cdot 2 \cdot 1$ or $6 \cdot 5 \cdot 4 \cdot 3 \cdot 2 \cdot 1$ will occur so often that it is convenient to adopt a notation for them. We define $5 \cdot 4 \cdot 3 \cdot 2 \cdot 1 = 5!$, read "5 factorial."

DEFINITION

We define $n!$, read "n factorial," as follows:

$$n! = n(n-1)(n-2) \cdots 3 \cdot 2 \cdot 1,$$

for any integer n, $n > 1$. For convenience in later notation, we define $1!$ (*one factorial*) to be 1 and $0!$ (*zero factorial*) to be 1.

Thus we have

$$3! = 3 \cdot 2 \cdot 1 = 6,$$
$$5! = 5 \cdot 4 \cdot 3 \cdot 2 \cdot 1 = 120,$$

and

$$6! = 6 \cdot 5 \cdot 4 \cdot 3 \cdot 2 \cdot 1$$
$$= 6 \cdot (5 \cdot 4 \cdot 3 \cdot 2 \cdot 1)$$
$$= 6 \cdot 5! = 6 \cdot 120 = 720.$$

DO EXERCISE 4.

For a moment, look back at Example 4, where we list the coefficients c_3, c_2, c_1, and c_0 in the expansion of $f(x) = x^3 - 7x + 3$ in powers of $x - 2$. Note now that we can rewrite them as follows:

$$c_0 = f(2) = \frac{f(2)}{0!} \qquad \text{since } 0! = 1,$$

$$c_1 = f'(2) = \frac{f'(2)}{1!} \qquad \text{since } 1! = 1,$$

$$c_2 = \frac{1}{2}f''(2) = \frac{f''(2)}{2}$$

$$= \frac{f''(2)}{2!},$$

and

$$c_3 = \frac{1}{6}f'''(2) = \frac{f'''(2)}{6}$$

$$= \frac{f'''(2)}{3!}.$$

In words, c_3 is the value of the third derivative of $f(x)$ at $x = 2$ (since we were expanding in powers of $x - 2$) divided by 3!.

To make the pattern stand out more clearly, recall (Section 2.9) that $f^{(3)}(x)$ is sometimes used for the third derivative of $f(x)$. Thus $f^{(1)}(x) = f'(x)$ and $f^{(6)}(x)$ denotes the sixth derivative of $f(x)$. Finally, we let $f^{(0)}(x) = f(x)$. Then we can rewrite the preceding formulas as follows:

$$c_0 = \frac{f(2)}{0!} = \frac{f^{(0)}(2)}{0!},$$

$$c_1 = \frac{f'(2)}{1!} = \frac{f^{(1)}(2)}{1!},$$

$$c_2 = \frac{f''(2)}{2!} = \frac{f^{(2)}(2)}{2!},$$

and

$$c_3 = \frac{f'''(2)}{3!} = \frac{f^{(3)}(2)}{3!}.$$

Examples 3 and 4 illustrate the following theorem.

4. Compute.

a) $4!$

b) $\dfrac{16}{2!}$

c) $\dfrac{-5}{0!}$

d) $\dfrac{720}{6!}$

TAYLOR SERIES AND l'HÔPITAL'S RULE

THEOREM 1

Taylor's Formula for Polynomials

Let $f(x)$ be a polynomial of degree n and let a be any number. Then $f(x)$ has a unique expansion in powers of $x - a$:

$$f(x) = c_0 + c_1(x - a) + c_2(x - a)^2 + \cdots + c_{n-1}(x - a)^{n-1} + c_n(x - a)^n.$$

Here, the jth coefficient, c_j, can be computed from the formula

$$c_j = \frac{f^{(j)}(a)}{j!}, \qquad j = 0, 1, \ldots, n.$$

EXAMPLE 5   Write $f(x) = x^3 - 7x^2 + 2x + 3$ in powers of $x - 1$.

Solution   We apply Taylor's formula with $a = 1$. Thus,

$$c_0 = \frac{f^{(0)}(1)}{0!} = \frac{(1)^3 - 7(1)^2 + 2 \cdot 1 + 3}{1}$$

$$= 1 - 7 + 2 + 3$$

$$= -1.$$

To compute c_1, we need a formula for $f'(x)$. Differentiating $f(x) = x^3 - 7x^2 + 2x + 3$, we get

$$f^{(1)}(x) = f'(x)$$
$$= 3x^2 - 14x + 2.$$

Hence,

$$c_1 = \frac{f^{(1)}(1)}{1!} = \frac{3(1)^2 - 14 \cdot 1 + 2}{1}$$

$$= -9.$$

Continuing, we have $f^{(2)}(x) = f''(x) = 6x - 14$, so that

$$c_2 = \frac{f^{(2)}(1)}{2!} = \frac{6 \cdot 1 - 14}{2}$$

$$= \frac{-8}{2} = -4.$$

Finally, $f^{(3)}(x) = f'''(x) = 6$, so that

$$c_3 = \frac{f^{(3)}(1)}{3!} = \frac{6}{3!}$$

$$= \frac{6}{6} = 1.$$

Note that $f^{(4)}(x) = 0$ for all x. In particular, c_4 must be 0. Since the same is true for all higher derivatives of $f(x)$, we can stop here—that is, we have found all the nonzero c_j's. So the expansion of $f(x) = x^3 - 7x^2 + 2x + 3$ in powers of $x - 1$ is

$$f(x) = -1 - 9(x - 1) - 4(x - 1)^2 + (x - 1)^3,$$

or, equivalently,

$$f(x) = (x - 1)^3 - 4(x - 1)^2 - 9(x - 1) - 1. \qquad \diamondsuit$$

DO EXERCISES 5 AND 6.

EXAMPLE 6   Expand $f(x) = x^3 - 7x^2 + 2x + 3$ in powers of $x - 2$.

Solution   We have

$$c_0 = \frac{f^{(0)}(2)}{0!} = \frac{f(2)}{0!}$$
$$= (2)^3 - 7(2)^2 + 2 \cdot 2 + 3$$
$$= 8 - 28 + 4 + 3$$
$$= -13.$$

Since $f'(x) = 3x^2 - 14x + 2$ (from Example 5), then

$$c_1 = \frac{f^{(1)}(2)}{1!} = \frac{3 \cdot 2^2 - 14 \cdot 2 + 2}{1}$$
$$= \frac{12 - 28 + 2}{1}$$
$$= -14.$$

Continuing, we have $f^{(2)}(x) = 6x - 14$ and

$$c_2 = \frac{f^{(2)}(2)}{2!} = \frac{6 \cdot 2 - 14}{2} = \frac{-2}{2} = -1.$$

Since $f^{(3)}(x) = 6$,

$$c_3 = \frac{f^{(3)}(2)}{3!} = \frac{6}{6} = 1.$$

So

$$f(x) = (x - 2)^3 - (x - 2)^2 - 14(x - 2) - 13.$$

Note that even though the function being expanded is the same in Examples 5 and 6, the coefficients of the two expansions are different. $\diamondsuit$

DO EXERCISES 7 AND 8.

5. Write $f(x) = 2x^2 - x + 3$ in powers of $x - 1$.

6. Write $f(x) = x^3 - x$ in powers of $x - 3$.

7. Write $f(x) = x^2 - 3x + 1$ in powers of $x - 4$.

8. Write $f(x) = x^2 - 3x + 1$ in powers of $x - 2$.

EXAMPLE 7 Expand $f(x) = x^2 + 3x + 1$ in powers of $x + 1$.

Solution Since $x + 1 = x - (-1)$, we have $a = -1$. So

$$c_0 = \frac{f^{(0)}(-1)}{0!}$$

$$= \frac{(-1)^2 + 3(-1) + 1}{1}$$

$$= 2 - 3 = -1.$$

Next, $f^{(1)}(x) = f'(x) = 2x + 3$, so that

$$c_1 = \frac{f^{(1)}(-1)}{1!}$$

$$= \frac{2(-1) + 3}{1} = 1.$$

Finally, $f^{(2)}(x) = 2$ and

$$c_2 = \frac{f^{(2)}(-1)}{2!} = \frac{2}{2} = 1.$$

So

$$f(x) = 1 \cdot (x + 1)^2 + 1 \cdot (x + 1) - 1$$

or

$$f(x) = (x + 1)^2 + (x + 1) - 1. \tag{5}$$

Note that if we multiply this last expression out, we obtain

$$f(x) = (x^2 + 2x + 1) + (x + 1) - 1$$
$$= x^2 + (2x + x) + (1 + 1 - 1)$$
$$= x^2 + 3x + 1.$$

Our check shows that formula (5) is correct.   ❖

So far, we have regarded Taylor's theorem for polynomials only as a way to write a given polynomial $f(x)$ in powers of $x - a$ (a given). There is a slightly different point of view that will be helpful to us later in the chapter. According to this view, Taylor's theorem says that if $f(x)$ is a polynomial of degree n, and if we know the value of $f(x)$ and its first n derivatives at $x = a$, then we can find the value of $f(x)$ at any point. To illustrate this, we consider the following example.

EXAMPLE 8 Find a polynomial of degree 3 such that $f(5) = 2$, $f'(5) = 0$, $f''(5) = -3$, and $f'''(5) = 12$.

Solution If we use the given data to expand $f(x)$ about $x = 5$ (that is, in powers of $x - 5$), we obtain

$$c_0 = \frac{f^{(0)}(5)}{0!} = \frac{f(5)}{1} = 2,$$

$$c_1 = \frac{f^{(1)}(5)}{1!} = \frac{0}{1} = 0,$$

$$c_2 = \frac{f^{(2)}(5)}{2!} = \frac{-3}{2} = -\frac{3}{2},$$

and

$$c_3 = \frac{f^{(3)}(5)}{3!} = \frac{12}{6} = 2.$$

Thus,

$$f(x) = 2(x - 5)^3 - \tfrac{3}{2}(x - 5)^2 + 0(x - 5) + 2$$
$$= 2(x - 5)^3 - \tfrac{3}{2}(x - 5)^2 + 2.$$

From Theorem 1, we know that $f(x)$ is the *only* polynomial of degree 3 that satisfies the given conditions. ❖

DO EXERCISES 9–11.

9. Find a polynomial $f(x)$ of degree 2 that satisfies $f(1) = 2, f'(1) = -1$, and $f''(1) = 2$.

10. Find a polynomial $f(x)$ of degree 3 such that $f(-1) = 1, f'(-1) = -\frac{1}{2}$, $f''(-1) = 4$, and $f'''(-1) = 18$.

11. Find a polynomial of degree 5 that satisfies $f(0) = 1, f'(0) = 2$, $f''(0) = -1, f^{(3)}(0) = 42$, $f^{(4)}(0) = 12$, and $f^{(5)}(0) = -60$.

EXERCISE SET **10.1**

Simplify.

1. $\dfrac{x^2 - 4x + 3}{x}$, $x \neq 0$

2. $\dfrac{(x + 4)^2 - 3(x + 4) + 1}{x + 4}$, $x \neq -4$

3. $\dfrac{2(x - 3)^2 - 6(x - 3) + 7}{x - 3}$, $x \neq 3$

4. $\dfrac{x^3 - 5x^2 + 7x - 3}{x}$, $x \neq 0$

5. $\dfrac{(x - 2)^3 + 7(x - 2) + 4}{x - 2}$, $x \neq 2$

6. $\dfrac{3(x + 4)^3 + 3(x + 4)^2 + 5(x + 4)}{x + 4}$, $x \neq -4$

Expand each of the given polynomials in powers of $x - a$.

7. $x^2 - 2x + 3$, $a = 2$

8. $x^2 - 2x + 3$, $a = 1$

9. $x^2 - 5x + 7$, $a = 2$

10. $x^2 - 5x + 7$, $a = 1$

11. $x^2 - 5x + 7$, $a = -1$

12. $x^2 - 5x + 7$, $a = 0$

13. $x^2 - 3x + 4$, $a = -2$

14. $x^2 - 3x + 4$, $a = 2$

15. $x^3 - 2x^2 + 4x + 1$, $a = 1$

16. $x^3 - 2x^2 + 4x + 1$, $a = 2$

17. $x^3 - 2x^2 + 4x + 1$, $a = -1$

18. $x^3 + 5$, $a = 1$

19. $2x^2 - 6x + 3$, $a = 3$

20. $x^3 - x - 1$, $a = 2$

21. $3x^4 - 7x^2 + 1$, $a = 1$

22. $x^4 - 5x^3 + 3$, $a = 1$

23. $x^5 - 5x^4 + 5x - 1$, $a = 1$

24. $x^5 - 3x + 1$, $a = 1$

25. Check the answers to Exercises 11 and 13 by multiplying them out and comparing your results with the given polynomial.

26. Repeat Exercise 25 using the solutions to Exercise 15 and 21.

Simplify using Taylor's formula.

27. $\dfrac{x^2 - 5}{x - 1}$, $x \neq 1$
   28. $\dfrac{x^2 - 3x + 1}{x - 2}$, $x \neq 2$
   29. $\dfrac{x^2 + 1}{x + 1}$, $x \neq -1$
   30. $\dfrac{x^2 - 7x - 1}{x + 3}$, $x \neq -3$

SYNTHESIS EXERCISES

31. Calculate $f'(1)$ and $f''(1)$ for

$$f(x) = x^4 - 4x^3 + 6x^2 - 4x + 4.$$

Conclude that $x = 1$ is a critical point for $f(x)$ but that the Second Derivative Test (Section 3.2) fails. Next, write $f(x)$ in powers of $x - 1$. Use your answers to show that $f(x)$ has a minimum point at $x = 1$ and that $x = 1$ is the only maximum or minimum point for (x).

32. Calculate $f'(1)$ and $f''(1)$ for

$$f(x) = x^3 - 3x^2 + 3x - 3.$$

Conclude, as in Exercise 31, that the Second Derivative Test fails. Write $f(x)$ in powers of $x - 1$. Use your answers to conclude that $f(x)$ has no maximum or minimum points and that $f(x)$ has exactly one point of inflection, at $x = 1$.

COMPUTER–GRAPHING CALCULATOR EXERCISE

33. Calculate $10!$, $12!$, $15!$, and $18!$.

10.2

TAYLOR POLYNOMIALS

OBJECTIVES

a) Compute the Taylor polynomial of degree n for $f(x)$ at the point $x = a$.

b) Find a Taylor polynomial and use it to approximate a function value.

Finding Taylor Polynomials

We now extend the ideas discussed in Section 10.1 to a wide class of functions.

DEFINITION

Let $f(x)$ have derivatives of all orders. Then the *Taylor polynomial* $p_n(x)$ *of degree* n *at* $x = a$ is defined to be the polynomial given by

$$p_n(x) = c_0 + c_1(x - a) + \cdots + c_n(x - a)^n$$

$$= \sum_{j=0}^{n} c_j(x - a)^j,$$

where

$$c_j = \frac{f^{(j)}(a)}{j!}, \quad \text{for } j = 0, 1, \ldots, n.$$

Compare this with the formula in Theorem 1 (Section 10.1).

EXAMPLE 1 Find the Taylor polynomial of degree 2 at $x = 0$ if $f(x) = e^x$.

Solution We must apply the definition with $n = 2$ and $a = 0$. Doing so, we obtain $p_2(x) = c_0 + c_1 x + c_2 x^2$, where

$$c_j = \frac{f^{(j)}(0)}{j!}.$$

But $f^{(0)}(x) = f(x) = e^x$, so that

$$c_0 = \frac{f^{(0)}(0)}{0!} = \frac{e^0}{0!} = 1.$$

Since $f^{(1)}(x) = f'(x) = e^x$ and $f^{(2)}(x) = e^x$, we have

$$c_1 = \frac{f^{(1)}(0)}{1!} = \frac{e^0}{1} = 1$$

and

$$c_2 = \frac{f^{(2)}(0)}{2!} = \frac{e^0}{2} = \frac{1}{2}.$$

So

$$p_2(x) = 1 + x + \tfrac{1}{2}x^2. \qquad \qquad ❖$$

DO EXERCISE 1.

EXAMPLE 2 Find the Taylor polynomial of degree 2 at $x = -1$ if $f(x) = x^2 + 3x + 1$.

Solution We have $p_2(x) = c_0 + c_1(x + 1) + c_2(x + 1)^2$, where

$$c_j = \frac{f^{(j)}(-1)}{j!}.$$

Since $f(x) = f^{(0)}(x) = x^2 + 3x + 1$, we get

$$c_0 = \frac{f^{(0)}(-1)}{0!}$$

$$= \frac{(-1)^2 + 3(-1) + 1}{1}$$

$$= \frac{1 - 3 + 1}{1} = -1.$$

1. Find the Taylor polynomial of degree 2 at $x = 0$ if $f(x) = 2e^{-3x}$.

2. a) Find the Taylor polynomial of degree 2 at $x = 4$ if

$$f(x) = x^2 - 3x + 1.$$

b) Compare Example 2 and Margin Exercise 2 with Example 7 and Margin Exercise 7 of Section 10.1.

3. Find the Taylor polynomials of degree 2 and 3 at $x = 0$ if $f(x) = \sin x$.

Similarly, $f'(x) = f^{(1)}(x) = 2x + 3$ and

$$c_1 = \frac{f^{(1)}(-1)}{1!} = \frac{2(-1) + 3}{1} = \frac{-2 + 3}{1} = 1,$$

and since $f''(x) = f^{(2)}(x) = 2$, we have

$$c_2 = \frac{f^{(2)}(-1)}{2!} = \frac{2}{2} = 1.$$

So

$$p_2(x) = -1 + (x + 1) + (x + 1)^2. \qquad \diamond$$

DO EXERCISE 2.

EXAMPLE 3 Find the Taylor polynomials of degrees 2 and 3 at $x = 0$ if $f(x) = \cos x$.

Solution Recalling that $f'(x) = f^{(1)}(x) = -\sin x$ and that $f''(x) = f^{(2)}(x) = -\cos x$, we have

$$c_0 = \frac{f^{(0)}(0)}{0!} = \frac{\cos (0)}{1} = 1,$$

$$c_1 = \frac{f^{(1)}(0)}{1!} = \frac{-\sin (0)}{1} = 0,$$

and

$$c_2 = \frac{f^{(2)}(0)}{2!} = \frac{-\cos (0)}{2!} = \frac{-1}{2} = -\frac{1}{2}.$$

We find that

$$p_2(x) = 1 + 0 \cdot x - \tfrac{1}{2}x^2 = 1 - \tfrac{1}{2}x^2.$$

To compute $p_3(x)$, we need only calculate

$$c_3 = \frac{f^{(3)}(0)}{3!}.$$

Since $f''(x) = -\cos x$, differentiation yields $f'''(x) = f^{(3)}(x) = \sin x$ and $f^{(3)}(0) = \sin (0) = 0$. So

$$c_3 = \frac{f^{(3)}(0)}{3!} = 0,$$

and we have

$$p_3(x) = 1 + 0 \cdot x - \tfrac{1}{2}x^2 + 0 \cdot x^3 = 1 - \tfrac{1}{2}x^2. \qquad \diamond$$

DO EXERCISE 3.

EXAMPLE 4   Compute the Taylor polynomial of degree 2 at $x = 1$ for

$$f(x) = \frac{1}{1 + x^2}.$$

Solution   Since $f^{(0)}(1) = \frac{1}{2}$, we have $c_0 = \frac{1}{2}$. Applying the Quotient Rule, we get

$$f'(x) = \frac{(1 + x^2) \cdot 0 - 1(2x)}{(1 + x^2)^2} = \frac{-2x}{(1 + x^2)^2}.$$

So

$$f'(1) = f^{(1)}(1) = \frac{-2(1)}{(1 + 1^2)^2} = \frac{-2}{4} = -\frac{1}{2}$$

and

$$c_1 = \frac{f^{(1)}(1)}{1!} = -\frac{1}{2}.$$

Finally, differentiating $f'(x)$ yields

$$f''(x) = \frac{(1 + x^2)^2(-2) - (-2x) \cdot 2(1 + x^2) \cdot 2x}{(1 + x^2)^4}$$

$$= \frac{-2(1 + x^2)^2 + 8x^2(1 + x^2)}{(1 + x^2)^4}$$

$$= \frac{-2(1 + x^2) + 8x^2}{(1 + x^2)^3}$$

$$= \frac{6x^2 - 2}{(1 + x^2)^3}.$$

So

$$c_2 = \frac{f^{(2)}(1)}{2!} = \frac{1}{2!} \cdot \frac{6(1)^2 - 2}{(1 + 1^2)^3}$$

$$= \frac{1}{2} \cdot \frac{4}{8} = \frac{1}{4}.$$

Thus,

$$p_2(x) = \frac{1}{2} - \frac{1}{2}(x - 1) + \frac{1}{4}(x - 1)^2.$$

❖

DO EXERCISE 4.

4. Find the Taylor polynomial of degree 2 at $x = 0$ if $f(x) = 1/(1 - x^2)$.

Approximating Using Taylor Polynomials

We turn now to a geometrical or graphical interpretation of Taylor polynomials. Let $f(x) = e^x$, the graph that appears in color in Fig. 1. We'll consider the Taylor polynomials for e^x at $x = 0$.

The Taylor polynomial of degree 0 at $x = 0$ is the constant function $p_0(x) = f(0) = e^0 = 1$. This function is the horizontal (black) line passing through (0, 1) in Fig. 1.

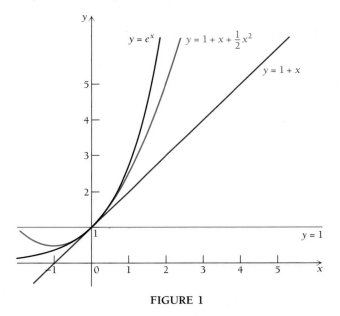

FIGURE 1

The Taylor polynomial of degree 1 at $x = 0$ is $p_1(x) = f(0) + f'(0)x = 1 + x$. This function is also linear and its graph is the dashed line in Fig. 1. Note that it, too, passes through (0, 1). Since the slope of $1 + x$ is 1 and the slope of e^x at (0, 1) is also $e^0 = 1$, we see that $1 + x$ is the tangent line to the graph of $y = e^x$ at (0, 1).

In general, the Taylor polynomial of degree 1 for $f(x)$ at $x = a$ is always the tangent line to the graph of $y = f(x)$ at the point $(a, f(a))$ on the graph.

Next, consider the Taylor polynomial of degree 2 for e^x at $x = 0$. From Example 1, we have $p_2(x) = 1 + x + \frac{1}{2}x^2$. This polynomial is the color dashed curve in Fig. 1. Since $p_2(0) = 1$, it passes through (0, 1). By the way in which it was constructed, $p_2'(0) = 1 = f'(0)$ and $p_2''(0) = 1 = f''(0)$. We might, therefore, think of $p_2(x)$ as being the "tangent quadratic" to the graph of $f(x) = e^x$ at $x = 0$.

Note that although $p_2(x)$ agrees with e^x only at $x = 0$, it is "close to e^x" for values of x that are close to 0. This suggests that we might use the Taylor polynomials of e^x to approximate e^x.

EXAMPLE 5   Use $p_2(x) = 1 + x + \frac{1}{2}x^2$ to calculate an approximate value for $e^{0.1}$.

Solution   We know that

$$p_2(0.1) = 1 + 0.1 + \tfrac{1}{2}(0.1)^2 = 1.105.$$

Thus $e^{0.1}$ is approximately equal to 1.105. A hand-held calculator gives $e^{0.1} = 1.1051709$, so that the error involved here is about 0.0001709. We can obtain a more accurate answer by using Taylor polynomials of higher degree.   ❖

DO EXERCISE 5.

EXAMPLE 6   Use the Taylor polynomial of degree 3 to calculate $\sin\left(\frac{1}{5}\right)$.

Solution   We calculate $p_3(x)$ for $f(x) = \sin x$ at $x = 0$ and then approximate $\sin\left(\frac{1}{5}\right)$ by $p_3\left(\frac{1}{5}\right)$. Since $f'(x) = \cos x, f''(x) = -\sin x,$ and $f'''(x) = -\cos x$, we have

$$c_0 = \frac{f^{(0)}(0)}{0!} = \frac{\sin\ (0)}{1} = 0,$$

$$c_1 = \frac{f^{(1)}(0)}{1!} = \frac{\cos\ (0)}{1} = 1,$$

$$c_2 = \frac{f^{(2)}(0)}{2!} = \frac{-\sin\ (0)}{2} = 0,$$

and

$$c_3 = \frac{f^{(3)}(0)}{3!} = \frac{-\cos\ (0)}{6} = \frac{-1}{6},$$

so that $p_3(x) = x - \frac{1}{6}x^3$. Now

$$p_3(\tfrac{1}{5}) = \tfrac{1}{5} - \tfrac{1}{6}(\tfrac{1}{5})^3 \approx 0.2 - 0.0013333 = 0.1986667.$$

Thus,

$$\sin \tfrac{1}{5} \approx 0.1986667.$$

(Recall that $\approx$ means *approximately equal to*.) A hand-held calculator (make sure that yours is set to handle radians before you compute) yields

5. Find the Taylor polynomial of degree 3 for e^x at $x = 0$ and use it to calculate an approximate value for $e^{0.1}$.

6. Use a Taylor polynomial to approximate $\tan\left(\frac{1}{10}\right)$.

$\sin\left(\frac{1}{5}\right) \approx 0.19866933$.

DO EXERCISE 6.

❖

Find $p_2(x)$ for the given function and the specified value of x.

1. $y = e^{-x}$; $x = 0$

2. $y = \dfrac{1}{x}$; $x = 1$

3. $y = \sin(x^2)$; $x = 0$

4. $y = 5x^2 - 6x + 3$; $x = 0$

5. $f(x) = 2x^3 - 4x^2 + x - 1$; $x = 0$

6. $f(x) = \sqrt{x}$; $x = 1$

7. $f(x) = \sec x$; $x = 0$

8. $f(x) = \ln x$; $x = 1$

9. $y = x \sin x$; $x = 0$

10. $y = x^2 \cos x$; $x = 0$

Find $p_3(x)$ for the given function and the specified value of x.

11. $y = e^{(1/2)x}$; $x = 0$

12. $y = xe^x$; $x = 0$

13. $f(x) = x^3$; $x = 1$

14. $f(x) = \ln(1 + x)$; $x = 0$

15. $y = \ln(1 - x)$; $x = 0$

16. $y = \dfrac{1}{1 - x}$; $x = 0$

Find both $p_4(x)$ and $p_5(x)$ for the given function and the specified value of x.

17. $f(x) = \sin x$; $x = 0$ **18.** $f(x) = e^x$; $x = 0$

19. $y = \cos x$; $x = 0$ **20.** $y = \ln(1 + x)$; $x = 0$

Sketch the graphs of $f(x)$, $p_0(x)$, $p_1(x)$, and $p_2(x)$ near the given value of x for each of the following.

21. $f(x) = \sin x$; $x = 0$ **22.** $f(x) = \cos x$; $x = 0$

23. $f(x) = e^{-x}$; $x = 0$ **24.** $f(x) = \sqrt{x}$; $x = 1$

Use $p_3(x)$ from above to calculate an approximate value for each of the following.

25. $\sin(0.13)$ **26.** $\cos\left(\frac{1}{8}\right)$ **27.** $e^{-0.05}$

28. $\ln(1.05)$ **29.** $\ln(0.95)$ **30.** $\sin\left(\frac{1}{5}\right)$

APPLICATIONS

❖ **Social Sciences**

31. *Psychology: The Hullian model of learning.* It is found that the probability P of mastery of a certain concept after t trials is given by

$$P(t) = 1 - e^{-0.18t}.$$

a) Compute $P(1)$, $P(2)$, and $P(3)$.

b) Find the Taylor polynomial $p_3(t)$ of order 3 for $P(t)$ at $t = 0$.

c) Use the polynomial in (b) to calculate approximate values for $P(1)$, $P(2)$, and $P(3)$.

❖ **Life and Physical Sciences**

32. *Newton's Law of Cooling.* An object heated to 200°F is placed in a room whose temperature is kept constant at 50°F. The temperature of the object after t hours in the room is determined to be given by

$$T(t) = 50 + 150e^{-0.5t}.$$

a) Find the temperature of the object after 1 and 2 hours.

b) Use the Taylor polynomial $p_3(t)$ of degree 3 at $t = 0$ to compute approximate values for $T(1)$ and $T(2)$.

33. *Ecology: Population growth.* The population of fish (in thousands) in a certain lake is given by

$$P(t) = \sqrt{22 + t}$$

over a five-year period beginning at $t = 0$.

a) Calculate the Taylor polynomial $p_2(t)$ of order 2 at $t = 3$ for $P(t)$.

b) Use $p_2(3.5)$ to find an approximate value for $P(3.5)$. Do the same for $P(4)$.

34. *Science: Motion of a spring.* The displacement y of a spring after t seconds is given by

$$y(t) = 6.5 \sin\left(t + \frac{\pi}{4}\right).$$

a) Find the Taylor polynomial $p_2(t)$ for $y(t)$ at $t = \pi/4$.

b) Use $p_2(\pi/3)$ to compute an approximate value for the displacement after $\pi/3$ seconds.

COMPUTER–GRAPHING CALCULATOR EXERCISES

35. Graph

$$f(x) = \sin x \quad \text{and} \quad g(x) = x - \frac{x^3}{3!} + \frac{x^5}{5!}$$

simultaneously on the interval $[-1, 1]$. Repeat the exercise on $[-3, 3]$ and then on $[-5, 5]$. On which interval is the approximation best?

36. Repeat Exercise 35 with

$$f(x) = \sin x \quad \text{and} \quad g(x) = x - \frac{x^3}{3!} + \frac{x^5}{5!} - \frac{x^7}{7!}.$$

 THE CALCULUS EXPLORER:
PowerGrapher

Use this program to do Exercises 35 and 36.

10.3

INFINITE SERIES

An *infinite series* is a special kind of sum

$$a_1 + a_2 + \cdots + a_n + \cdots, \tag{1}$$

where each number of the *sequence* $a_1, a_2, \ldots, a_n$ is a real number. The numbers $a_1, a_2, \ldots, a_n$ are called the *terms* of the infinite series.

Two examples of infinite series are

$$\frac{1}{2} + \frac{1}{4} + \cdots + \frac{1}{2^n} + \cdots \tag{2}$$

and

$$1 + 2 + \cdots + n + \cdots.$$

OBJECTIVES

a) Tell whether a given geometric series converges or diverges by calculating its ratio. If it converges, compute its sum.

b) Tell whether an alternating series converges or diverges. If it converges, approximate its sum by a partial sum and determine the error involved in this approximation.

In the first series, the nth term of the series is $1/2^n$. The fourth term, for instance, is

$$\frac{1}{2^4} = \frac{1}{16}$$

and the first term is

$$\frac{1}{2^1} = \frac{1}{2}.$$

In the second series, the nth term is n.

In ordinary arithmetic, numbers are added two at a time. To add $2 + 4 + 5 + 8$, for example, requires three additions. But an infinite sum like

$$\frac{1}{2} + \frac{1}{4} + \cdots + \frac{1}{2^n} + \cdots$$

would require an infinite number of additions, so it cannot be carried out in a finite length of time. Thus in order to give meaning to an expression like (1) or (2), we use the limit concept.

If the series (1) is given, then

$$s_n = a_1 + a_2 + \cdots + a_n,$$

is called the nth *partial sum* of the series.

DEFINITION

The infinite series (1) *converges to s*, or *has the sum s*, written

$$s = a_1 + a_2 + \cdots + a_n + \cdots,$$

if the limit of the partial sums s_n of the series is s. Here s is a (finite) real number. In other words,

$$s = \lim_{n \to \infty} (a_1 + a_2 + \cdots + a_n)$$

$$= \lim_{n \to \infty} s_n.$$

If the partial sums do not have a limit, then we say that the series *diverges*.

EXAMPLE 1   Find the sum of Eq. (2), that is, of

$$\frac{1}{2} + \frac{1}{4} + \frac{1}{8} + \cdots + \frac{1}{2^n} + \cdots.$$

Solution   Let's consider s_n for a few values of n:

$$s_1 = \tfrac{1}{2} \qquad\qquad = \tfrac{1}{2} = 0.5,$$
$$s_2 = \tfrac{1}{2} + \tfrac{1}{4} \qquad\qquad = \tfrac{3}{4} = 0.75,$$
$$s_3 = \tfrac{1}{2} + \tfrac{1}{4} + \tfrac{1}{8} \qquad\qquad = \tfrac{7}{8} = 0.875,$$
$$s_4 = \tfrac{1}{2} + \tfrac{1}{4} + \tfrac{1}{8} + \tfrac{1}{16} \qquad = \tfrac{15}{16} = 0.9375,$$
$$s_5 = \tfrac{1}{2} + \tfrac{1}{4} + \tfrac{1}{8} + \tfrac{1}{16} + \tfrac{1}{32} = \tfrac{31}{32} = 0.96875.$$

Note that when s_n is written as a fraction, as in the next-to-last column, the numerator is one less than the denominator. Since the denominator is 2^n, this suggests that

$$s_n = \frac{2^n - 1}{2^n}$$

$$= 1 - \frac{1}{2^n}$$

for all n. This formula is true. Note that as n increases, the sequence $1/2^n$ decreases rapidly toward 0. So

$$\lim_{n \to \infty} \left(1 - \frac{1}{2^n}\right) = 1.$$

Then

$$s = \lim_{n \to \infty} s_n = 1. \qquad \diamondsuit$$

DO EXERCISE 1.

EXAMPLE 2   Find the sum of $1 + 1 + \cdots + 1 + \cdots$.

Solution   The nth partial sum of this series is

$$\underbrace{1 + 1 + \cdots + 1}_{n \text{ times}} = n.$$

So the partial sums tend to infinity as n increases. Since $+\infty$ is not a real number, the series diverges. $\qquad \diamondsuit$

DO EXERCISE 2.

A special case in which we can always determine convergence or divergence is that of a geometric series.

1. a)   Find an expression for the nth term a_n of the infinite series

$$\tfrac{2}{3} + \tfrac{2}{9} + \tfrac{2}{27} + \tfrac{2}{81} + \cdots.$$

b)   Calculate the first six partial sums of the series above. (Round your answers to seven decimal places.)

2. a)   Find an expression for the nth term of

$$1 + 3 + 5 + 7 + 9 + \cdots.$$

b)   Calculate the first five partial sums of this series.

c)   Find an expression for the nth partial sum s_n of this series.

d)   Does the series converge?

3. **a)** Find an expression for the nth term a_n of the geometric series
$$2 + \tfrac{2}{5} + \tfrac{2}{25} + \cdots.$$

b) Find the fifth and sixth terms of the series above.

DEFINITION

The infinite series
$$a + ar + ar^2 + \cdots + ar^{n-1} + \cdots \qquad (3)$$
is called a *geometric series with initial term a and ratio r*. Here $a \neq 0$.

The series in Example 1 is a geometric series with initial term $\tfrac{1}{2}$ and ratio $\tfrac{1}{2}$.

If $a + ar + \cdots + ar^{n-1} + \cdots$ is a geometric series, then the ratio of the second term, ar, to the first term, a, is
$$\frac{ar}{a} = r.$$

The ratio of the third term, ar^2, to the second term, ar, is
$$\frac{ar^2}{ar} = r.$$

In general, the ratio of any term of a geometric series to the preceding term is r. In other words, $a_{n+1}/a_n = r$, or $a_{n+1} = ra_n$ for all n.

4. Find the initial term and the ratio of the geometric series
$$\tfrac{1}{5} + \tfrac{1}{15} + \tfrac{1}{45} + \cdots.$$

EXAMPLE 3 Find the fifth and sixth terms of the geometric series $2 + 6 + 18 + \cdots$.

Solution The initial term a is 2. The ratio of the second term to the first is $\tfrac{6}{2} = 3$. So $r = 3$. This means that the nth term of the series is $ar^{n-1} = 2 \cdot 3^{n-1}$. Thus the fifth term is $2 \cdot 3^{5-1} = 2 \cdot 3^4 = 2 \cdot 81 = 162$. The sixth term is three times the fifth term, or $a_6 = 3 \cdot 162 = 486$. ❖

DO EXERCISES 3 AND 4.

The Partial Sums of a Geometric Series

We want to find a formula for the nth partial sum s_n of a geometric series:
$$s_n = a + ar + ar^2 + \cdots + ar^{n-1}.$$

If we multiply both sides of this equation by the ratio r, we get
$$rs_n = ar + ar^2 + \cdots + ar^{n-1} + ar^n.$$

If we multiply the formula for s_n by -1, we have
$$-s_n = -a - ar - ar^2 - \cdots - ar^{n-1}.$$

Adding these expressions, we get

$$rs_n - s_n = ar^n - a,$$

or

$$(r - 1)s_n = a(r^n - 1).$$

To solve this equation for s_n, we can divide both sides by $r - 1$ (if $r \neq 1$), getting the following formula.

THEOREM 2

The nth partial sum of a geometric series is given by

$$s_n = \frac{a(r^n - 1)}{r - 1} \qquad (4)$$

for any $r \neq 1$.

EXAMPLE 4 Find the sum of the first n terms of the geometric series

$$\frac{1}{2} + \frac{1}{4} + \cdots + \frac{1}{2^n} + \cdots .$$

Solution This is a geometric series with $a = \frac{1}{2}$ and $r = \frac{1}{2}$. Using Eq. (4), we have

$$s_n = \frac{\frac{1}{2}\left[\left(\frac{1}{2}\right)^n - 1\right]}{\frac{1}{2} - 1} = \frac{\frac{1}{2}\left[\left(\frac{1}{2}\right)^n - 1\right]}{-\frac{1}{2}}$$

$$= -\left[\left(\frac{1}{2}\right)^n - 1\right]$$

or

$$s_n = 1 - \frac{1}{2^n}.$$

Note that this justifies our use of this formula in Example 1. ❖

DO EXERCISE 5.

The Sum of a Geometric Series

We want to be able to tell when a given geometric series converges and when it diverges. If it converges, we want to be able to find its sum.

5. Find the nth partial sum s_n of the geometric series

$$2 + \frac{2}{3} + \frac{2}{9} + \frac{2}{27} + \cdots .$$

6. Find the sum of the geometric series

$$2 + \tfrac{2}{5} + \tfrac{2}{25} + \cdots .$$

Formula (4) enables us to do this. For if r is a number between -1 and 1, then it follows that r^n gets closer and closer to 0 as n gets larger. (Pick a number r between -1 and 1 and check this by finding larger and larger powers on your calculator.) As r^n gets closer and closer to 0, so does ar^n. Since

$$s_n = \frac{a(r^n - 1)}{r - 1} = \frac{ar^n - a}{r - 1},$$

we see that s_n gets closer to

$$\frac{-a}{r - 1} = \frac{a}{1 - r}$$

as n increases, if r is between -1 and 1.

If $|r|$ is greater than 1, then $|r|^n$ gets larger and larger as n increases. (▣ Check this on your calculator.) In this case, s_n does not have a limit.

If $|r| = 1$, the series also diverges. This case is considered in Exercises 53 and 54.

THEOREM 3

The geometric series $a + ar + ar^2 + \cdots + ar^n + \cdots$ converges to

$$s = \frac{a}{1 - r}$$

if $|r| < 1$, that is, if r is between -1 and 1. Otherwise, it diverges.

---

7. Find the sum of the geometric series

$$\tfrac{1}{5} + \tfrac{1}{15} + \tfrac{1}{45} + \cdots .$$

EXAMPLE 5   Find the sum of

$$2 + \frac{2}{3} + \frac{2}{9} + \frac{2}{27} + \cdots + \frac{2}{3^{n-1}} + \cdots .$$

Solution   This is a geometric series with initial term $a = 2$ and ratio $r = \tfrac{1}{3}$. Since r is between -1 and 1, the series converges and the sum is

$$s = \frac{a}{1 - r}$$

$$= \frac{2}{1 - \tfrac{1}{3}}$$

$$= \frac{2}{\tfrac{2}{3}} = 2 \cdot \frac{3}{2} = 3. \qquad ❖$$

DO EXERCISES 6 AND 7.

EXAMPLE 6 Find the sum of

$$1 - 1 + 1 - 1 + \cdots + (-1)^{n+1} + \cdots .$$

Solution This is a geometric series with initial term $a = 1$ and ratio $r = -1$. Since r is not between -1 and 1, this series diverges. Note that the partial sums of this series are the sequence $1, 0, 1, 0, 1, 0, 1, \ldots$, which has no limit.

DO EXERCISE 8.

EXAMPLE 7 Find the sum of

$$1 - \frac{1}{2} + \frac{1}{4} - \frac{1}{8} + \cdots + \left(-\frac{1}{2}\right)^{n-1} + \cdots .$$

Solution This is a geometric series with $a = 1$ and $r = -\frac{1}{2}$. Thus it converges and its sum is

$$\frac{a}{1 - r} = \frac{1}{1 - (-\frac{1}{2})} = \frac{1}{1 + \frac{1}{2}} = \frac{1}{\frac{3}{2}} = \frac{2}{3}. \qquad ❖$$

DO EXERCISE 9.

EXAMPLE 8 *Economic multiplier.* The United States banking laws require banks to maintain a reserve equivalent to a certain proportion of their outstanding deposits. This enables such banks, when they wish and when they can find borrowers, to loan out a certain proportion of the funds that have been deposited in them. Let us assume that this proportion is 0.90 (or 90%). Now suppose a corporation deposits $1000 in a bank that subsequently is able to loan the maximum amount legally possible, and this loan is redeposited elsewhere, and so on. What is the total effect of the $1000 on the economy?

Solution The total effect can be modeled as the sum of the infinite geometric series

$$\$1000 + \$1000(0.90) + \$1000(0.90)^2 + \$1000(0.90)^3 + \cdots,$$

which is given by

$$s = \frac{\$1000}{1 - 0.90}$$
$$= \$10,000.$$

The sum $10,000 is the result of what is referred to in economics as the *multiplier effect.* ❖

8. Find the sum of

$$1 - 2 + 4 - 8 + \cdots .$$

9. Find the sum of

$$1 - \tfrac{2}{5} + \tfrac{4}{25} - \tfrac{8}{125} + \cdots .$$

10. Find s if the proportion in Example 8 is changed to 0.95 (or 95%).

11. Find s if the proportion in Example 8 is changed to 85%.

12. Write $0.444\overline{4}$ . . . as a fraction.

13. Write $0.459\overline{4594}$ . . . as a fraction.

DO EXERCISES 10 AND 11.

EXAMPLE 9   Write 0.63636363 . . . , or $0.\overline{63}$, as a fraction. (The bar above indicates the portion that repeats.)

Solution   The key to the problem is to note that we can write this decimal as a geometric series. Thus,

$$0.63636363 \ldots = 0.63 + 0.0063 + 0.000063 + \cdots$$

$$= \frac{63}{100} + \frac{63}{10,000} + \frac{63}{1,000,000} + \cdots .$$

The right-hand side is a geometric series with initial term $a = \frac{63}{100} = 0.63$ and ratio $r = \frac{1}{100} = 0.01$. Since the ratio is between -1 and 1, the sum of the series can be expressed as

$$\frac{a}{1-r} = \frac{\frac{63}{100}}{1 - \frac{1}{100}} = \frac{\frac{63}{100}}{\frac{99}{100}}$$

$$= \frac{63}{99} = \frac{7}{11}.$$

Thus,

$$\frac{7}{11} = 0.63636363 \ldots , \quad \text{or } 0.\overline{63}. \qquad \blacklozenge$$

DO EXERCISES 12 AND 13.

Alternating Series

We reconsider the series

$$1 - \frac{1}{2} + \frac{1}{4} - \frac{1}{8} + \cdots + \left(-\frac{1}{2}\right)^{n-1} + \cdots = \frac{2}{3}.$$

The first six partial sums are as follows:

$$
\begin{aligned}
s_1 &= 1 & &= 1 = 1, \\
s_2 &= 1 - \tfrac{1}{2} & &= \tfrac{1}{2} = 0.5, \\
s_3 &= 1 - \tfrac{1}{2} + \tfrac{1}{4} & &= \tfrac{3}{4} = 0.75, \\
s_4 &= 1 - \tfrac{1}{2} + \tfrac{1}{4} - \tfrac{1}{8} & &= \tfrac{5}{8} = 0.625, \\
s_5 &= 1 - \tfrac{1}{2} + \tfrac{1}{4} - \tfrac{1}{8} + \tfrac{1}{16} & &= \tfrac{11}{16} = 0.6875, \\
s_6 &= 1 - \tfrac{1}{2} + \tfrac{1}{4} - \tfrac{1}{8} + \tfrac{1}{16} - \tfrac{1}{32} & &= \tfrac{21}{32} = 0.65625.
\end{aligned}
$$

Note that the odd-numbered partial sums 1, 0.75, 0.6875, . . . are all greater than $\frac{2}{3}$, the sum of the given series. The even-numbered partial sums are all less than $\frac{2}{3}$.

Furthermore, the odd-numbered partial sums 1, 0.75, 0.6875, . . . are decreasing. The even-numbered partial sums 0.5, 0.625, 0.65625, . . . are increasing.

DO EXERCISE 14.

The discussion above and Margin Exercise 14 illustrate the following theorem.

THEOREM 4

The Alternating Series Test

Let

$$a_1 \geqslant a_2 \geqslant \cdots \geqslant a_n \geqslant \cdots$$

be positive numbers with $\lim_{n \to \infty} a_n = 0$. Then the *alternating series*

$$a_1 - a_2 + a_3 - a_4 + \cdots + (-1)^{n-1}a_n + \cdots$$

converges. Furthermore, the odd-numbered partial sums are all greater than s, the sum of the series. In fact, they decrease to s. The even-numbered partial sums are less than s (and increase to s).

EXAMPLE 10   Determine whether the series

$$1 - \frac{1}{2} + \frac{1}{3} - \frac{1}{4} + \cdots + (-1)^{n-1}\frac{1}{n} + \cdots$$

converges or diverges.

Solution   The given series is an alternating series with $a_n = 1/n$ for all positive integers n. Since these numbers decrease to 0, the series converges.   ❖

DO EXERCISE 15.

EXAMPLE 11   Show that

$$1 - \frac{1}{8} + \frac{1}{27} + \cdots + (-1)^{n-1}\frac{1}{n^3} + \cdots$$

14. Calculate the first six partial sums of

$$2 - \frac{2}{3} + \frac{2}{9} - \frac{2}{27} + \frac{2}{81} - \cdots .$$

15. Determine whether or not

$$1 - \frac{1}{4} + \frac{1}{9} - \frac{1}{16} + \cdots$$
$$+ (-1)^{n+1}\frac{1}{n^2} + \cdots$$

converges or diverges.

16. Use the third and fourth partial sums to estimate the sum of

$$1 - \frac{1}{16} + \frac{1}{81} - \frac{1}{256} + \cdots$$

$$+ (-1)^{n+1} \frac{1}{n^4} + \cdots.$$

converges, and use the fourth and fifth partial sums to estimate the sum s of the series.

Solution The series is an alternating series with $a_n = 1/n^3$ for each positive integer n. Since these numbers decrease to 0, the series converges. The fourth partial sum is

$$1 - \tfrac{1}{8} + \tfrac{1}{27} - \tfrac{1}{64} \approx 0.896412.$$

The fifth partial sum is

$$1 - \tfrac{1}{8} + \tfrac{1}{27} - \tfrac{1}{64} + \tfrac{1}{125} \approx 0.904412.$$

According to the test, the sum s of the series is between these two numbers. Thus,

$$0.896412 < s < 0.904412. \qquad \diamondsuit$$

DO EXERCISE 16.

In Example 11, we saw that the sum s of the series

$$1 - \frac{1}{8} + \frac{1}{27} + \cdots + (-1)^{n-1} \frac{1}{n^3} + \cdots$$

was between 0.896412 and 0.904412, the fourth and fifth partial sums of the series. The difference between the fifth partial sum and the fourth partial sum of this series is the fifth term of the series, $\frac{1}{125}$. If we want to use 0.896412 as an approximate value for s, then the error involved is no more than $\frac{1}{125} = 0.008$. This fact is so important that we state it separately.

THEOREM 5

If $a_1 \geqslant a_2 \geqslant \cdots \geqslant a_n \geqslant \cdots$ are positive numbers with $\lim_{n \to \infty} a_n = 0$ and if s is the sum of the (convergent) alternating series

$$a_1 - a_2 + a_3 + \cdots + (-1)^{n-1} a_n + \cdots,$$

then the absolute value of the difference between s and the nth partial sum is less than or equal to a_{n+1}. That is,

$$|s - s_n| \leqslant a_{n+1}.$$

EXAMPLE 12 Let

$$s = 1 - \frac{1}{7} + \frac{1}{49} + \cdots + \frac{(-1)^{n-1}}{7^{n-1}} + \cdots.$$

Find the error involved in approximating s by the fourth partial sum

$$s_4 = 1 - \tfrac{1}{7} + \tfrac{1}{49} - \tfrac{1}{343}$$
$$\approx 0.87463557.$$

Check by finding s explicitly.

Solution By the alternating series test, we know that the series converges. According to Theorem 5 above, the error involved in approximating s by 0.87463557 is less than $1/7^4 = \tfrac{1}{2401} \approx 0.0004165$. Since the given series happens to be a geometric series with $a = 1$ and $r = -\tfrac{1}{7}$, we see that the sum is

$$s = \frac{a}{1-r} = \frac{1}{1-(-\tfrac{1}{7})} = \frac{1}{\tfrac{8}{7}} = \frac{7}{8} = 0.875.$$

Thus the error involved in approximating s by 0.87463557 is 0.00036443, so that our error estimate above is correct. ❖

DO EXERCISE 17.

It is important to keep in mind that the alternating-series test applies only if the series is indeed alternating.

EXAMPLE 13 Use the alternating-series test to determine whether

$$1 + \frac{1}{2} + \frac{1}{3} + \cdots + \frac{1}{n} + \cdots$$

converges or diverges.

Solution This series is not an alternating series! Thus the alternating series test cannot be applied. Contrary to what you might expect, this series *diverges*, as we will show in Exercise Set 10.3. ❖

17. a) Use the third and fourth partial sums to estimate the sum of

$$1 - \tfrac{1}{9} + \tfrac{1}{81} - \tfrac{1}{729} + \cdots.$$

b) Then check your answer by finding the sum of this geometric series.

EXERCISE SET 10.3

Find an expression for the nth term a_n of the series.

1. $\tfrac{1}{5} + \tfrac{1}{25} + \tfrac{1}{125} + \cdots$
2. $1 + 4 + 9 + 16 + 25 + \cdots$
3. $\tfrac{1}{2} + \tfrac{2}{3} + \tfrac{3}{4} + \tfrac{4}{5} + \tfrac{5}{6} + \cdots$
4. $1 - \tfrac{1}{2} + \tfrac{1}{3} - \tfrac{1}{4} + \cdots$
5. $\tfrac{2}{3} + \tfrac{4}{9} + \tfrac{8}{27} + \tfrac{16}{81} + \tfrac{32}{243} + \cdots$
6. $\tfrac{1}{4} - \tfrac{1}{16} + \tfrac{1}{64} - \tfrac{1}{256} + \cdots$
7. $1 - 2 + 3 - 4 + \cdots$
8. $1 + \tfrac{1}{6} + \tfrac{1}{36} + \tfrac{1}{216} + \cdots$

Find the initial term, the ratio, and the fifteenth partial sum for the geometric series.

9. $1 + \tfrac{1}{3} + \tfrac{1}{9} + \cdots$
10. $1 + \tfrac{1}{5} + \tfrac{1}{25} + \cdots$

11. $1 - \frac{1}{3} + \frac{1}{9} - \cdots$

12. $1 - \frac{1}{5} + \frac{1}{25} - \cdots$

13. $\frac{1}{8} - \frac{1}{12} + \frac{1}{18} - \cdots$

14. $\frac{1}{2} + \frac{2}{3} + \frac{8}{9} + \cdots$

15. $\frac{5}{3} - \frac{2}{3} + \frac{4}{15} - \cdots$

16. $\frac{1}{3} - \frac{1}{2} + \frac{3}{4} - \frac{9}{8} + \cdots$

Determine whether the geometric series converges or diverges. If it converges, find its sum.

17. $1 + \frac{1}{8} + \frac{1}{64} + \cdots$

18. $1 + 3 + 9 + 27 + \cdots$

19. $1 + \frac{1}{9} + \frac{1}{81} + \frac{1}{729} + \cdots$

20. $1 - 2 + 4 - 8 + \cdots$

21. $1 - \frac{2}{3} + \frac{4}{9} - \cdots$

22. $\frac{1}{2} - \frac{1}{2} + \frac{1}{2} - \frac{1}{2} + \cdots$

23. $\frac{1}{4} + \frac{1}{4} + \frac{1}{4} + \cdots$

24. $\frac{1}{2} - \frac{1}{6} + \frac{1}{18} - \cdots$

Write the decimal as a rational fraction (the bar above indicates the portion that repeats).

25. $0.7777\overline{7}$

26. $0.41414\overline{1}$

27. $2.3838\overline{38}$

28. $0.03666\overline{6}$

29. $0.142857\overline{142857}$

30. $0.1234567890\overline{1234567890}$

Determine whether the infinite series converges or diverges.

31. $\frac{1}{7} + \frac{2}{7} + \frac{4}{7} + \frac{8}{7} + \cdots$

32. $1 - \frac{1}{3} + \frac{1}{5} - \frac{1}{7} + \cdots$

33. $1 - \frac{9}{10} + \frac{81}{100} - \frac{729}{1000} + \cdots$

34. $1 - \frac{1}{9} + \frac{1}{25} - \frac{1}{49} + \frac{1}{81} - \cdots$

35. $1 - \frac{1}{4} + \frac{1}{7} - \frac{1}{10} + \cdots$

36. $\frac{1}{9} - \frac{1}{6} + \frac{1}{4} - \frac{3}{8} + \cdots$

37. $1 - \frac{3}{4} + \frac{9}{16} - \frac{27}{64} + \cdots$

38. $-\frac{1}{5} + \frac{1}{6} - \frac{1}{7} + \frac{1}{8} - \cdots$

39. $\frac{1}{5} - \frac{1}{7} + \frac{1}{9} - \frac{1}{11} + \cdots$

40. $\frac{1}{36} + \frac{1}{24} + \frac{1}{16} + \frac{3}{32} + \cdots$

Use the fourth and fifth partial sums of the (convergent) series to estimate the sum of the series.

41. $1 - \frac{1}{3} + \frac{1}{9} - \frac{1}{27} + \cdots$

42. $1 - \frac{1}{6} + \frac{1}{36} - \frac{1}{216} + \cdots$

43. $1 - \frac{1}{16} + \frac{1}{81} - \frac{1}{256} + \cdots + (-1)^{n+1} \cdot \frac{1}{n^4} + \cdots$

44. $\frac{1}{2} - \frac{1}{8} + \frac{1}{24} - \frac{1}{64} + \cdots + (-1)^{n-1} \cdot \frac{1}{n2^n} + \cdots$

45. $\frac{1}{9} - \frac{1}{25} + \frac{1}{49} - \frac{1}{81} + \cdots + (-1)^{n+1} \cdot \frac{1}{(2n+1)^2} + \cdots$

46. $-\frac{1}{4} + \frac{1}{16} - \frac{1}{36} + \frac{1}{64} - \frac{1}{100} + \cdots$

APPLICATIONS

❖ **Business and Economics**

47. *Economics.* The government makes an $8,000,000,000 expenditure for a new type of aircraft. If 75% of this gets spent again, and 75% of that gets spent, and so on, what is the total effect on the economy?

48. *Finance.* A certain *perpetuity* is to pay $100 each year (starting today). If the current interest rate is i, then the present value of a payment of $100 n years from today is defined to be $100(1 + i)^{-n}$, so that the *total* present value P of the payments is given by

$$P = 100 + \frac{100}{1 + i} + \frac{100}{(1 + i)^2} + \cdots.$$

a)   Find the present value of the perpetuity if $i = 0.10$ (10% interest).

b)   What happens to the present value if the interest rate drops from 10% to 8%?

49. *Investments.* A certain stock pays a dividend of $5 per year at present. If the dividend grows 2% each year and if the interest rate is 12.5%, then the present value V of the next twelve payments (beginning today) is given by

$$V = 5 + 5\left(\frac{1.02}{1.125}\right) + 5\left(\frac{1.02}{1.125}\right)^2 + \cdots + 5\left(\frac{1.02}{1.125}\right)^{11}.$$

Find the sum.

❖ **Life and Physical Sciences**

50. *Medicine.* A patient is to receive 1 mg (milligram) of a certain drug once every day. By the time the next dose is administered, it is expected that 50% of the drug will have been eliminated from the patient's system.

a)   Find the amount of the drug in the patient's system immediately after the first four doses have been administered.

b)   If we assume that the pattern in part (a) holds, how much of the drug will be present after the fifteenth dose is administered?

c) When will the amount of the drug in the patient's system exceed 3 mg?

51. *Medicine.* If, in Exercise 50, the daily dosage is 5 mg and the percentage eliminated daily is 60%, find each of the following.

a) The amount of the drug in the patient's system immediately after the first, second, and third doses have been administered

b) The amount of the drug in the patient's system immediately after the ninth dose has been administered

c) The "limiting value" of the amount of the drug in

the patient's system, that is, the sum of the geometric series in question

52. *Chemistry.* A factory deposits waste material containing 1 kg (kilogram) of a certain radioactive isotope on a slag heap each day. It is known that after one day, 3% of the isotope will decay.

a) How much of the isotope will be present after the first, second, and third deposits?

b) How much of the isotope will be present after the fiftieth deposit?

c) Find the limiting value of the amount of isotope in the slag heap.

SYNTHESIS EXERCISES

53. By computing the nth partial sum s_n, show that a geometric series with initial term $a \neq 0$ and ratio $r = 1$ diverges.

54. By computing the nth partial sum s_n, show that a geometric series with initial term $a \neq 0$ and ratio $r = -1$ diverges.

55. A patient is able to eliminate p $(0 < p < 1)$ of the concentration of a certain drug between successive doses. If the daily dosage of the drug is d mg, find the limiting value of the amount of the drug in the patient's system.

56. A perpetuity pays $\$r$ annually beginning today. If the interest rate is i, then the present value of the annuity, denoted P, is given by

$$P = r + \frac{r}{1 + i} + \frac{r}{(1 + i)^2} + \cdots.$$

Find the sum of the series.

57. Find the value of the perpetuity in Exercise 56 if the first payment is to be made one year from today.

58. We consider the harmonic series

$$1 + \frac{1}{2} + \frac{1}{3} + \frac{1}{4} + \frac{1}{5} + \cdots + \frac{1}{n} + \cdots.$$

Note that

$$\frac{1}{3} + \frac{1}{4} > \frac{1}{4} + \frac{1}{4} = \frac{1}{2},$$

that

$$\frac{1}{5} + \frac{1}{6} + \frac{1}{7} + \frac{1}{8} > \frac{1}{8} + \frac{1}{8} + \frac{1}{8} + \frac{1}{8} = \frac{1}{2},$$

and so forth.

a) Show that

$$\frac{1}{2^n + 1} + \frac{1}{2^n + 2} + \cdots + \frac{1}{2^{n+1}} > \frac{1}{2}.$$

b) Use the result from part (a) to show that $s_4 > 2$, $s_8 > \frac{5}{2}$, $s_{16} > 3$; then show that $s_{2^{n+1}} > (n + 3)/2$ for all integers n.

c) Conclude that the harmonic series diverges.

10.4

OBJECTIVES

a) Compute the Taylor series expansion for a given function.

b) Use Taylor series and the alternating-series test to approximate certain integrals.

TAYLOR SERIES

In Section 10.2, we saw that we can approximate many functions by using an appropriately constructed polynomial, called the Taylor polynomial (of degree n) of $f(x)$ at $x = a$. Although exact equality occurs only at $x = a$, the approximation is excellent near $x = a$ and the accuracy can be improved by using a polynomial of increased degree. This raises the question of whether we can obtain exact equality by taking the limit (as n tends to infinity) of the Taylor polynomial of degree n. When this occurs, we say that $f(x)$ is represented by its Taylor series at $x = a$.

DEFINITION

If $f(x)$ has derivatives of all orders at $x = a$, then *the Taylor series of $f(x)$ at $x = a$* is the series

$$\sum_{j=0}^{\infty} c_j(x - a)^j,$$

where

$$c_j = \frac{f^{(j)}(a)}{j!} \quad \text{for } j = 0, 1, \ldots.$$

Note that the nth partial sum of the series is given by

$$p_n(x) = \sum_{j=0}^{n} c_j(x - a)^j,$$

the Taylor polynomial of $f(x)$ (of degree n) at $x = a$. We say that $f(x)$ is *represented by its Taylor series at x* if

$$f(x) = \sum_{j=0}^{\infty} c_j(x - a)^j$$

$$= \lim_{n \to \infty} \sum_{j=0}^{n} c_j(x - a)^j$$

$$= \lim_{n \to \infty} p_n(x).$$

EXAMPLE 1   Compute the Taylor series of $f(x) = e^x$ at $x = 0$.

Solution   In this problem, $a = 0$. We must compute the coefficients, given by $c_j = f^{(j)}(0)/j!$. Since $f(x) = e^x$, the required derivatives are easy to compute in this case. We have $f^{(j)}(x) = e^x$ for all j. So $f^{(j)}(0) = e^0 = 1$ for all values of j and $c_j = f^{(j)}(0)/j! = 1/j!$ for all j. Thus the Taylor series for e^x at $x = 0$ is

$$1 + x + \frac{x^2}{2!} + \frac{x^3}{3!} + \cdots + \frac{x^n}{n!} + \cdots = \sum_{j=0}^{\infty} \frac{x^j}{j!}. \qquad \text{❖}$$

DO EXERCISE 1.

EXAMPLE 2   Compute the Taylor series of $f(x) = \sin x$ at $x = 0$.

Solution   We have $f'(x) = \cos x$ and $f''(x) = f^{(2)}(x) = -\sin x$. Differentiating again, we obtain $f^{(3)}(x) = -\cos x$, and, differentiating again, $f^{(4)}(x) = \sin x$. Note that $f^{(4)}(x) = f(x)$, the function with which we started. This means that $f^{(5)}(x) = f'(x) = \cos x$, that $f^{(6)}(x) = f^{(2)}(x) = -\sin x$, and so forth, repeating every four terms. If we let $x = 0$, we get

$$f(0) = \sin (0) = 0,$$
$$f'(0) = \cos (0) = 1,$$
$$f''(0) = f^{(2)}(0) = -\sin (0) = 0,$$
$$f^{(3)}(0) = -\cos (0) = -1,$$
$$f^{(4)}(0) = \sin (0) = 0,$$
$$f^{(5)}(0) = f^{(1)}(0) = 1,$$

and so on. In other words, the derivatives, when evaluated at $x = 0$, produce the sequence $0, 1, 0, -1, 0, 1, -1, 0, 1, \ldots$, in which the four numbers $0, 1, 0, -1$ (in that order) are continually repeated.

It is helpful to be able to describe this sequence concisely. The even-numbered terms are all equal to 0. This is equivalent to $f^{(j)}(0) = 0$ for $j = 2k$, or $f^{(2k)}(0) = 0$ for all k. If we let $j = 2k + 1$, then, as k takes on the values $0, 1, 2, 3, \ldots$, j takes on the values $1, 3, 5, 7, \ldots$. Using this, we can write $f^{(j)}(0) = (-1)^k$ for $j = 2k + 1$, or $f^{(2k+1)}(0) = (-1)^k$. Note that as k takes on the values $0, 1, 2, \ldots$, the quantity $(-1)^k$ is equal to $1, -1, 1, -1, \ldots$.

Finally, since $c_j = f^{(j)}(0)/j!$, we see that for j even (that is, $j = 2k$), we have $c_{2k} = 0$, whereas for j odd ($j = 2k + 1$), we have

$$c_{2k+1} = (-1)^k/(2k+1)!.$$

Thus the Taylor series for $\sin x$ at $x = 0$ is

$$\sum_{k=0}^{\infty} \frac{(-1)^k x^{2k+1}}{(2k+1)!},$$

or

$$x - \frac{x^3}{3!} + \frac{x^5}{5!} - \frac{x^7}{7!} + \frac{x^9}{9!} - \cdots. \qquad \text{❖}$$

1. Compute the Taylor series of

$$f(x) = e^{-x} \quad \text{at } x = 0.$$

2. Compute the Taylor series of

$$f(x) = \cos x \quad \text{at } x = 0.$$

DO EXERCISE 2.

EXAMPLE 3   Compute the Taylor series of $f(x) = 2x^2 - 3x + 5$ at $x = 0$.

Solution   We have

$$f(0) = 2(0)^2 - 3(0) + 5 = 5.$$

Since $f'(x) = 4x - 3$, we have

$$f'(0) = 4(0) - 3 = -3.$$

Next, $f^{(2)}(x) = 4$, so that $f^{(2)}(0) = 4$. Observe that $f^{(3)}(x)$ and all higher-order derivatives of $f(x)$ are equal to 0. So

$$c_0 = \frac{f(0)}{0!} = \frac{5}{1} = 5,$$

$$c_1 = \frac{f^{(1)}(0)}{1!} = \frac{-3}{1} = -3,$$

and

3. Compute the Taylor series of

$$f(x) = 7x^3 - 5x^2 + 2x + 4$$

at $x = 0$.

$$c_2 = \frac{f^{(2)}(0)}{2!} = \frac{4}{2} = 2,$$

with all other coefficients equal to 0. Thus the Taylor series for $f(x) = 2x^2 - 3x + 5$ at $x = 0$ is $5 - 3x + 2x^2 + 0 \cdot x^3 + 0 \cdot x^4 + \cdots$, or $5 - 3x + 2x^2$, which is what we started with.   ❖

DO EXERCISE 3.

EXAMPLE 4   Compute the Taylor series of $f(x) = \ln x$ at $x = 1$.

Solution   We have $f'(x) = 1/x = x^{-1}$. So, differentiating again, we get

$$f^{(2)}(x) = f''(x) = (-1)x^{-2} = -1/x^2.$$

Continuing, we have

$$f^{(3)}(x) = (-1)\,(-2x^{-3}) = 2x^{-3} = \frac{2}{x^3},$$

$$f^{(4)}(x) = 2(-3x^{-4}) = -(3!)x^{-4},$$

and

$$f^{(5)}(x) = -(3!)\,(-4x^{-5}) = (4!)x^{-5} = \frac{4!}{x^5}.$$

To see the pattern, consider the following:

$$f^{(1)}(x) = (0!)x^{-1},$$
$$f^{(2)}(x) = -(1!)x^{-2},$$
$$f^{(3)}(x) = (2!)x^{-3},$$
$$f^{(4)}(x) = -(3!)x^{-4},$$
$$f^{(5)}(x) = (4!)x^{-5}.$$

We see that $f^{(j)}(x) = (-1)^{j+1}(j-1)!x^{-j}$ [here, the $(-1)^{j+1}$ term takes care of the alternating signs of the derivatives]. We then have

$$f^{(j)}(1) = (-1)^{j+1}(j-1)!(1)^{-j} = (-1)^{j+1}(j-1)! \quad \text{for } j > 0,$$

so that

$$c_j = \frac{f^{(j)}(1)}{j!} = \frac{(-1)^{j+1}(j-1)!}{j!} = \frac{(-1)^{j+1}(j-1)!}{j(j-1)!}$$
$$= \frac{(-1)^{j+1}}{j} \quad \text{for } j > 0.$$

Recalling that

$$c_0 = \frac{f^{(0)}(1)}{0!} = \frac{\ln(1)}{1} = 0,$$

we see that the Taylor series of $\ln x$ at $x = 1$ is

$$(x-1) - \tfrac{1}{2}(x-1)^2 + \tfrac{1}{3}(x-1)^3 - \tfrac{1}{4}(x-1)^4 + \cdots$$
$$= \sum_{j=1}^{\infty} \frac{(-1)^{j+1}(x-1)^j}{j}. \qquad \clubsuit$$

EXAMPLE 5 Compute the Taylor series of $f(x) = 1/(1-x)$ at $x = 0$.

Solution Writing $f(x) = (1-x)^{-1}$, we see that

$$f'(x) = -1(1-x)^{-2}(-1) = (1-x)^{-2}.$$

So

$$f^{(2)}(x) = f''(x) = -2(1-x)^{-3}(-1) = 2(1-x)^{-3}$$

and

$$f^{(3)}(x) = 2(-3)(1-x)^{-4}(-1) = 3!(1-x)^{-4}.$$

We see that

$$f^{(j)}(x) = (j!)(1-x)^{-(j+1)} \quad \text{for all } j > 0.$$

4. Compute the Taylor series of

$$f(x) = 1/(1 + x^2) \quad \text{at } x = 0.$$

Thus,

$$f^{(j)}(0) = (j!)(1 - 0)^{-(j + 1)} = (j!)(1)^{-(j + 1)} = j!.$$

We get

$$c_j = \frac{f^{(j)}(0)}{j!} = \frac{j!}{j!} = 1 \quad \text{for all } j \geq 0,$$

and the Taylor series for $1/(1 - x)$ at $x = 0$ is

$$1 + x + x^2 + \cdots + x^n + \cdots = \sum_{j=0}^{\infty} x^j.$$

Note that this is a geometric series.   ❖

DO EXERCISE 4.

In Example 5, we showed that the Taylor series for $1/(1 - x)$ at $x = 0$ is given by $1 + x + x^2 + \cdots + x^n + \cdots$. In Section 10.3, we showed that the geometric series $1 + r + r^2 + \cdots + r^n + \cdots$ converges to $1/(1 - r)$ if $-1 < r < 1$, and diverges otherwise. This shows that the Taylor series for $f(x) = 1/(1 - x)$ *is equal to* $f(x)$ for values of x between -1 and 1, that is,

$$\frac{1}{1 - x} = 1 + x + x^2 + \cdots + x^n + \cdots, \qquad -1 < x < 1,$$

and that this formula makes no sense for other values of x.

It turns out that this example illustrates the general case. The series $\sum_{j=0}^{\infty} c_j(x - a)^j$ always converges at $x = a$, since all the terms except the first are 0 in this case. Also, there is a nonnegative number R so that the series converges on the interval $(a - R, a + R)$ and diverges if $|x - a| > R$ (where R may be 0 or $+\infty$). If R is positive and finite, then the series converges on the interval shown in Fig. 1 and diverges for $x > a + R$ and $x < a - R$ (convergence at the endpoints $a + R$ and $a - R$ must be decided on a case-by-case basis). If $R = 0$, then the series converges only at $x = a$ (where it always converges) and diverges for all other values of x. If $R = +\infty$, then the series converges for all real numbers x. Fortunately, for a large class of functions, it turns out that R (the *radius of convergence*) is positive and $f(x)$ is equal to its Taylor series on $(a - R, a + R)$, the *interval of convergence* of the series.

FIGURE 1

If $f(x) = \sum_{j=0}^{\infty} c_j(x - a)^j$ on $(a - R, a + R)$ for R positive, then this formula can be differentiated or integrated termwise within the given interval.

We can also make appropriate substitutions. These techniques may make the computation of Taylor series much less tedious. We illustrate with several examples.

EXAMPLE 6 Compute the Taylor series of $f(x) = 1/(1 - x)^2$ at $x = 0$.

Solution We use

$$(1 - x)^{-1} = \frac{1}{1 - x} = 1 + x + x^2 + \cdots, \qquad \text{valid for } -1 < x < 1,$$

from Example 5. Differentiating $(1 - x)^{-1}$ gives

$$(-1)(1 - x)^2(-1) = (1 - x)^{-2}.$$

Differentiating the series $1 + x + x^2 + x^3 + \cdots$ term by term gives $0 + 1 + 2x + 3x^2 + \cdots$. Since the formula with which we started was correct for all x between -1 and 1, so too is the new formula. That is,

$$(1 - x)^{-2} = \frac{1}{(1 - x)^2} = 1 + 2x + 3x^2 + \cdots,$$

for all x with $-1 < x < 1$.   ❖

DO EXERCISES 5 AND 6.

EXAMPLE 7 Compute the Taylor series of $f(x) = 1/(1 + x^2)$ at $x = 0$.

Solution We could compute the Taylor series directly, but those calculations are lengthy (see Margin Exercise 4). For a shortcut, we use

$$\frac{1}{1 - t} = 1 + t + t^2 + t^3 + \cdots, \qquad -1 < t < 1.$$

Letting $t = -x^2$, we get

$$\frac{1}{1 - (-x^2)} = 1 + (-x^2) + (-x^2)^2 + (-x^2)^3 + \cdots$$

or

$$\frac{1}{1 + x^2} = 1 - x^2 + x^4 - x^6 + \cdots.$$

5. Compute the Taylor series of
$$f(x) = 1/(1 + x)^2 \quad \text{at } x = 0.$$

6. Compute the Taylor series of
$$f(x) = 1/(1 - x)^3 \quad \text{at } x = 0.$$

7. Compute the Taylor series of

$$f(x) = 1/(1 - x^3) \quad \text{at } x = 0.$$

This formula is valid for $-1 < x^2 < 1$ or, in other words, for $-1 < x < 1$.   ❖

DO EXERCISE 7.

EXAMPLE 8   Compute the Taylor series of $f(x) = e^{-x^2}$ at $x = 0$.

Solution   We use the fact that e^x is equal to its Taylor series at $x = 0$ for *all* values of x. That is,

$$e^t = 1 + t + \frac{t^2}{2!} + \frac{t^3}{3!} + \cdots \quad \text{for all } t.$$

Letting $t = -x^2$, we get

8. Compute the Taylor series of

$$f(x) = \sin(x^2) \quad \text{at } x = 0.$$

$$e^{-x^2} = 1 - x^2 + \frac{x^4}{2!} - \frac{x^6}{3!} + \frac{x^8}{4!} - \frac{x^{10}}{5!} + \cdots \quad \text{for all } x.$$ ❖

DO EXERCISES 8 AND 9.

EXAMPLE 9   Compute the Taylor series of $f(x) = \ln(1 + x)$ at $x = 0$.

Solution   Letting $x = -t$ in the series for $1/(1 - x)$ yields

$$\frac{1}{1 + t} = 1 - t + t^2 - t^3 + \cdots, \quad -1 < t < 1.$$

Integrating both sides from 0 to x, we obtain

9. Compute the Taylor series of

$$f(x) = e^{-5x} \quad \text{at } x = 0.$$

$$\int_0^x \frac{1}{1 + t} \, dt = \int_0^x (1 - t + t^2 - t^3 + \cdots) \, dt$$

$$= \int_0^x dt - \int_0^x t \, dt + \int_0^x t^2 \, dt - \int_0^x t^3 \, dt + \cdots,$$

or

$$[\ln(1 + t)]_0^x = [t]_0^x - \left[\frac{t^2}{2}\right]_0^x + \left[\frac{t^3}{3}\right]_0^x - \left[\frac{t^4}{4}\right]_0^x + \cdots,$$

or

$$\ln(1 + x) - \ln(1 + 0) = (x - 0) - \left(\frac{x^2}{2} - 0\right)$$

$$+ \left(\frac{x^3}{3} - 0\right) - \left(\frac{x^4}{4} - 0\right) + \cdots.$$

Thus,

$$\ln (1 + x) = x - \frac{x^2}{2} + \frac{x^3}{3} - \frac{x^4}{4} + \cdots, \qquad -1 < x < 1. \qquad ❖$$

DO EXERCISE 10.

EXAMPLE 10 Compute the Taylor series of $f(x) = \tan^{-1} x$ at $x = 0$.

Solution Recall that the derivative of $\tan^{-1} x$ is $1/(1 + x^2)$ (see Section 8.4). In Example 7, we found that the Taylor series at $x = 0$ of $1/(1 + x^2)$ is

$$\frac{1}{1 + x^2} = 1 - x^2 + x^4 - x^6 + \cdots, \qquad -1 < x < 1.$$

Replacing x by t and integrating from $t = 0$ to $t = x$ yields

$$\int_0^x \frac{1}{1 + t^2} \, dt = \int_0^x \{1 - t^2 + t^4 - t^6 + \cdots\} \, dt, \qquad -1 < x < 1,$$

or

$$[\tan^{-1} t]_0^x = \left[t - \frac{t^3}{3} + \frac{t^5}{5} - \frac{t^7}{7} + \cdots \right]_0^x, \qquad -1 < x < 1$$

and

$$\tan^{-1} x - \tan^{-1} 0 = \left(x - \frac{x^3}{3} + \frac{x^5}{5} - \frac{x^7}{7} + \cdots \right) - 0.$$

Since $\tan^{-1} 0 = 0$, we have

$$\tan^{-1} x = x - \frac{x^3}{3} + \frac{x^5}{5} - \frac{x^7}{7} + \cdots, \qquad -1 < x < 1. \qquad ❖$$

EXAMPLE 11 Use Taylor series to estimate

$$\int_0^{1/2} \frac{1}{1 + x^3} \, dx$$

with an error of less than 0.001.

Solution We use a combination of several techniques from this chapter. First, we compute the Taylor series for $1/(1 + x^3)$ at $x = 0$ by letting $t = x^3$ in the expression for $1/(1 + t)$ to get

$$\frac{1}{1 + x^3} = 1 - x^3 + x^6 - x^9 + \cdots, \qquad -1 < x < 1.$$

10. Compute the Taylor series of
$$f(x) = \ln (1 - x) \quad \text{at } x = 0.$$

11. Use a Taylor series to compute

$$\int_0^{1/2} \frac{1}{1+x^4}\, dx$$

with an error of less than 0.001.

Integrating both sides from 0 to $\frac{1}{2}$ yields

$$\int_0^{1/2} \frac{1}{1+x^3}\, dx = \left[x - \frac{x^4}{4} + \frac{x^7}{7} - \frac{x^{10}}{10} + \cdots \right]_0^{1/2},$$

or

$$\int_0^{1/2} \frac{1}{1+x^3}\, dx = \frac{1}{2} - \frac{1}{4}\left(\frac{1}{2}\right)^4 + \frac{1}{7}\left(\frac{1}{2}\right)^7 - \frac{1}{10}\left(\frac{1}{2}\right)^{10} + \cdots.$$

Before proceeding, note that the series on the right is an alternating series. Thus if we were to approximate the integral by using, say, $\frac{1}{2} - \frac{1}{4}(\frac{1}{2})^4$, the resulting approximation would be smaller than the actual value of the integral, and the absolute value of error would be less than the next term,

$$\frac{1}{7}\left(\frac{1}{2}\right)^7 = \frac{1}{7\cdot 2^7} = \frac{1}{7\cdot 128} = \frac{1}{896}.$$

If we use the sum of the first three terms to approximate the integral, then we will have an overestimate of the integral, and the error will be less than

$$\frac{1}{10}\cdot\left(\frac{1}{2}\right)^{10} = \frac{1}{10\cdot 2^{10}} = \frac{1}{10,240}.$$

Since we want an error of less than $0.001 = \frac{1}{1000}$, and since $\frac{1}{896} > \frac{1}{1000}$, we must use three terms. We get

$$\int_0^{1/2} \frac{1}{1+x^3}\, dx \approx \frac{1}{2} - \frac{1}{4}\left(\frac{1}{2}\right)^4 + \frac{1}{7}\left(\frac{1}{2}\right)^7$$

$$= \frac{1}{2} - \frac{1}{64} + \frac{1}{896},$$

or

$$\int_0^{1/2} \frac{1}{1+x^3}\, dx \approx 0.485491$$

with an error of less than $1/10,000 = 0.0001$.   ❖

DO EXERCISE 11.

EXAMPLE 12   Compute the Taylor series expansion of $f(x) = x^2 e^x$ at $x = 0$.

Solution   We could calculate the coefficients of the Taylor series of $x^2 e^x$ by directly computing the derivatives of $f(x)$ (see Exercise Set 10.4).

There is, however, an easier way. From Example 1, we have

$$e^x = 1 + x + \frac{1}{2}x^2 + \frac{1}{3!}x^3 + \cdots ,$$

valid for all values of x. Simply multiplying both sides of this equation by x^2 yields

$$x^2 e^x = x^2 \left(1 + x + \frac{1}{2}x^2 + \frac{1}{3!}x^3 + \cdots \right)$$

$$= x^2 + x^3 + \frac{1}{2}x^4 + \frac{1}{3!}x^5 + \cdots ,$$

valid for all real numbers x.   ❖

DO EXERCISES 12 AND 13.

EXAMPLE 13   Calculate the Taylor series expansion of $f(x)$ at $x = 0$, where $f(x)$ is given by

$$f(x) = \begin{cases} \dfrac{\sin x}{x}, & x \neq 0, \\ 1, & x = 0. \end{cases}$$

Solution   The formula $(\sin x)/x$ makes sense only for $x \neq 0$. From Example 2, we get

$$\sin x = x - \frac{1}{3!}x^3 + \frac{1}{5!}x^5 - \frac{1}{7!}x^7 + \cdots ,$$

valid for all real numbers x. If x is not equal to 0, then we can divide both sides of this equation by x, getting

$$\frac{\sin x}{x} = 1 - \frac{1}{3!}x^2 + \frac{1}{5!}x^4 - \frac{1}{7!}x^6 + \cdots ,$$

valid for all x (except $x = 0$, of course). Note that as x approaches 0 in the formula above, the right-hand side will approach 1. That is,

$$\lim_{x \to 0} \frac{\sin x}{x} = 1,$$

a fact that we will use in Section 10.5. For the present, note that we have shown that

$$f(x) = 1 - \frac{x^2}{3!} + \frac{x^4}{5!} - \frac{x^6}{7!} + \cdots$$

for *all* values of x.   ❖

DO EXERCISE 14.

12. Compute the Taylor series of
$$f(x) = x^2 \cos x \quad \text{at } x = 0.$$

13. Compute the Taylor series of
$$f(x) = x \tan^{-1} x \quad \text{at } x = 0.$$

14. Compute the Taylor series of $f(x)$ at $x = 0$ if
$$f(x) = \begin{cases} \dfrac{e^x - 1}{x}, & x \neq 0, \\ 1, & x = 0 \end{cases}.$$

EXERCISE SET 10.4

Compute the Taylor series expansion of $f(x)$ at the prescribed value of x by directly calculating the coefficients.

1. $f(x) = e^{4x}$; $x = 0$

2. $f(x) = \sin(5x)$; $x = 0$

3. $f(x) = e^{1-x}$; $x = 1$

4. $f(x) = \cos(x - 2)$; $x = 2$

5. $f(x) = x^3 - 1$; $x = -1$

6. $f(x) = x^2 - 5x + 1$; $x = 3$

7. $f(x) = \dfrac{1}{1 - 2x}$; $x = 0$

8. $f(x) = \dfrac{1}{2 - x}$; $x = 0$

Using any method you wish, compute the Taylor series expansion of $f(x)$ for the prescribed value of x.

9. $f(x) = e^{6x}$; $x = 0$

10. $f(x) = \cos(x^2)$; $x = 0$

11. $f(x) = \dfrac{1}{1 - x^5}$; $x = 0$

12. $f(x) = \dfrac{1}{1 - bx}$ $(b \neq 0)$; $x = 0$

13. $f(x) = \ln(1 + 4x)$; $x = 0$

14. $f(x) = \dfrac{1}{(1 - bx)^2}$ $(b \neq 0)$; $x = 0$

15. $f(x) = \dfrac{1}{2 + x}$; $x = -1$ $\left(\text{Hint: } \dfrac{1}{2 + x} = \dfrac{1}{1 + (1 + x)}. \right)$

16. $f(x) = \dfrac{1}{(2 + x)^2}$; $x = -1$

17. $f(x) = \ln(2 + x)$; $x = -1$

18. $f(x) = \dfrac{1}{(1 + x)^3}$; $x = 0$

19. $f(x) = \dfrac{5x^4}{1 - x^5}$; $x = 0$ (See Exercise 11 above.)

20. $f(x) = \ln(1 - x^5)$; $x = 0$

21. $f(x) = \begin{cases} \dfrac{\cos x - 1}{x^2}, & x \neq 0, \\ -\dfrac{1}{2}, & x = 0 \end{cases}$

22. $f(x) = x^3 e^{-x}$; $x = 0$

23. $f(x) = x^5 \sin(x^3)$; $x = 0$

24. $f(x) = \begin{cases} \dfrac{\sin(x^3)}{x}, & x \neq 0, \\ 0, & x = 0 \end{cases}$

Use the Taylor series and the alternating series test to find an approximate value for each of the following integrals (use three terms of the series and give an error estimate).

25. $\displaystyle\int_0^1 \cos(x^2)\, dx$ (See Exercise 10 above.)

26. $\displaystyle\int_0^1 x^5 \sin(x^3)\, dx$ (See Exercise 23 above.)

27. $\displaystyle\int_0^1 e^{-x^3}\, dx$ **28.** $\displaystyle\int_0^{0.1} \ln(1 + x)\, dx$

29. $\displaystyle\int_0^{0.5} x^5 \sin(x^3)\, dx$

30. $\displaystyle\int_0^{0.25} \cos(x^2)\, dx$

SYNTHESIS EXERCISE

31. *Probability theory.* Let

$$F(t) = \int_0^t e^{(-x^2)/2}\, dx.$$

a) Find the Taylor series expansion of $F(t)$ at $x = 0$.

b) Use your answer to part (a) to calculate

$$P(0 \leq t \leq 0.1) = \dfrac{1}{\sqrt{2\pi}} F(0.1)$$

for the normal distribution.

c)   Check your answer to (b) in Table 2.

What we have given here is a method for constructing that table.

32. Repeat Exercise 35 of Section 10.2.
33. Repeat Exercise 36 of Section 10.2.

THE CALCULUS EXPLORER:
Taylor Series

Use this program to do Exercises 32 and 33.

10.5

INDETERMINATE FORMS AND l'HÔPITAL'S RULE

We will lay the groundwork for this section with three preliminary examples. You may wish to review the definition of continuous function given in Section 2.1.

EXAMPLE 1   Evaluate

$$\lim_{x \to 0} \frac{x^2 + 1}{x + 3}.$$

Solution   Let $f(x) = x^2 + 1$ and $g(x) = x + 3$. Then $f(x)$ and $g(x)$ are continuous functions for all x. It follows that

$$\frac{f(x)}{g(x)} = \frac{x^2 + 1}{x + 3}$$

is continuous except when $g(x) = 0$ (that is, when $x = -3$). So we can evaluate the given limit as follows:

$$\lim_{x \to 0} \frac{x^2 + 1}{x + 3} = \frac{\lim_{x \to 0} (x^2 + 1)}{\lim_{x \to 0} (x + 3)} = \frac{(0)^2 + 1}{0 + 3} = \frac{1}{3}.$$ ❖

DO EXERCISE 1.

OBJECTIVE

a)   Determine whether or not a given limit expression is an indeterminate form. If it is, use a version of l'Hôpital's Rule to evaluate it.

1. Evaluate

$$\lim_{x \to 3} \frac{x + 1}{x^2 + 2}.$$

Evaluate.

2. $\lim\limits_{x \to 0} \dfrac{1 - \cos x}{x^2 + 2}$

3. $\lim\limits_{x \to 1} \dfrac{\ln x}{x}$

EXAMPLE 2   Evaluate

$$\lim_{x \to 0} \frac{\sin x}{x + 3}.$$

Solution   Let $f(x) = \sin x$ and $g(x) = x + 3$. Then $f(x)$ and $g(x)$ are continuous functions. It follows that $f(x)/g(x)$ is continuous except at $x = -3$. So, just as in Example 1, we have

$$\lim_{x \to 0} \frac{\sin x}{x + 3} = \frac{\lim\limits_{x \to 0} \sin x}{\lim\limits_{x \to 0} x + 3}$$

$$= \frac{\sin (0)}{0 + 3} = \frac{0}{3} = 0.$$

Note that the fact that the *numerator* of the resulting fraction is 0 does not affect the validity of the steps that were used.   ❖

DO EXERCISES 2 AND 3.

EXAMPLE 3   Evaluate

$$\lim_{x \to 1} \frac{x^2}{x - 1}.$$

Solution   Let $f(x) = x^2$ and $g(x) = x - 1$. Then, just as before, $f(x)$ and $g(x)$ are continuous functions. Also, just as before,

$$\frac{f(x)}{g(x)} = \frac{x^2}{x - 1}$$

is continuous for all values of x such that $g(x) \neq 0$ (in this case, except for $x = 1$). But this is exactly the point at which we wish to evaluate the limit. We cannot use the techniques of Examples 1 and 2 since the expression $1/0$ is meaningless.

To try to understand what is happening, we compute a few values of $x^2/(x - 1)$ for choices of x near 1 (but different from 1). For $x = 1.1$, we obtain

$$\frac{(1.1)^2}{1.1 - 1} = \frac{1.21}{0.1} = 12.1.$$

For $x = 1.01$, we get

$$\frac{(1.01)^2}{1.01 - 1} = \frac{1.0201}{0.01} = 102.01.$$

For $x = 1.001$, we get

$$\frac{(1.001)^2}{1.001 - 1} = \frac{1.002001}{0.001}$$

$$= 1002.001.$$

Note that as we take values of x that are closer and closer to 1, the numerator of $x^2/(x - 1)$ gets closer and closer to 1 while the denominator gets closer and closer to 0, so that the resulting quotients are larger and larger positive numbers.

The values of x used above were all larger than 1. If we had used values of x smaller than 1 (and getting closer and closer to 1), the quotients would be negative numbers with large absolute values.

In any event, the quotients do not get closer to a real number. In this case, we say that the limit *does not exist.* ❖

DO EXERCISES 4 AND 5.

In investigating

$$\lim_{x \to a} \frac{f(x)}{g(x)},$$

where $f(x)$ and $g(x)$ are continuous functions, we have seen that we can compute the limit by substitution so long as $g(a) \neq 0$. If the denominator approaches 0 as x approaches a, while the numerator approaches a nonzero number, then the limit does not exist. This leaves unresolved only the case in which both the numerator and the denominator are 0 at $x = a$. This is the most interesting case.

The Indeterminate Form 0/0

EXAMPLE 4 Evaluate

$$\lim_{x \to 0} \frac{x^2 + 3x}{5x}.$$

Solution This limit is of the form

$$\lim_{x \to a} \frac{f(x)}{g(x)},$$

where $a = 0$, $f(x) = x^2 + 3x$, and $g(x) = 5x$. Substitution in the numerator yields $f(0) = (0)^2 + 3(0) = 0$. Substitution in the denominator yields $g(0) = 5(0) = 0$. Since substitution of $x = a$ in both numerator and

4. Compute the ratio $x^2/(x - 1)$ for $x = 0.93$, $x = 0.991$, and $x = 0.999$.

5. Evaluate

$$\lim_{x \to -1} \frac{x}{x + 1}.$$

6. Evaluate

$$\lim_{x \to 0} \frac{x^2 - 7x}{x}.$$

denominator yields 0, we say that

$$\lim_{x \to 0} \frac{x^2 + 3x}{5x}$$

is an indeterminate form of the type $\frac{0}{0}$.

In this case, we cannot evaluate the limit by direct substitution. Note, however, that we can factor an x out of $x^2 + 3x$, obtaining

$$\lim_{x \to 0} \frac{x^2 + 3x}{5x} = \lim_{x \to 0} \frac{x(x + 3)}{5x}.$$

If x is not zero,

$$\frac{x(x + 3)}{5x} = \frac{x + 3}{5}.$$

Thus we have

$$\lim_{x \to 0} \frac{x^2 + 3x}{5x} = \lim_{x \to 0} \frac{x + 3}{5}.$$

Note that the limit on the right-hand side of the last equation is *not* an indeterminate form. In fact, direct substitution yields

$$\lim_{x \to 0} \frac{x + 3}{5} = \frac{(0) + 3}{5} = \frac{3}{5}.$$

So

$$\lim_{x \to 0} \frac{x^2 + 3x}{5x} = \frac{3}{5}.$$ ❖

DO EXERCISE 6.

EXAMPLE 5 Evaluate

$$\lim_{x \to -1} \frac{x^2 + 4x + 3}{x + 1}.$$

Solution Substitution in the numerator yields

$$(-1)^2 + 4(-1) + 3 = 1 - 4 + 3 = 0.$$

Since substitution of $x = -1$ in the denominator also yields 0, this is an indeterminate form. We can write the numerator in the form $x^2 + 4x + 3 = (x + 1)(x + 3)$. For values of x that are not equal to -1, we have

$$\frac{x + 4x + 3}{x + 1} = \frac{(x + 1)(x + 3)}{x + 1} = x + 3.$$

So

$$\lim_{x \to -1} \frac{x^2 + 4x + 3}{x + 1} = \lim_{x \to -1} (x + 3) = (-1) + 3 = 2.$$

❖

DO EXERCISE 7.

EXAMPLE 6   Evaluate

$$\lim_{x \to 0} \frac{\sin x}{x}.$$

Solution   Both the numerator and the denominator are continuous functions of x, but substitution of $x = 0$ yields sin $(0) = 0$ in the numerator and 0 in the denominator. This limit is an indeterminate form. In this case, we recall from Section 10.4 that $f(x) = \sin x$ can be represented as

$$\sin x = x - \frac{x^3}{3!} + \frac{x^5}{5!} - \frac{x^7}{7!} + \cdots .$$

This formula is valid for all values of x. Dividing both sides by x, we see that if x is not equal to 0, then

$$\frac{\sin x}{x} = 1 - \frac{x^2}{3!} + \frac{x^4}{5!} - \frac{x^6}{7!} + \cdots .$$

So

$$\lim_{x \to 0} \frac{\sin x}{x} = \lim_{x \to 0} \left(1 - \frac{x^2}{3!} + \frac{x^4}{5!} - \frac{x^6}{7!} + \cdots \right)$$

$$= 1 - \frac{(0)^2}{3!} + \frac{(0)^4}{5!} - \frac{(0)^6}{7!} + \cdots = 1.$$

❖

DO EXERCISE 8.

So far, we have evaluated several indeterminate forms using algebraic techniques that are often tedious and time-consuming. Fortunately, there is an easier way to do many of these problems.

Consider

$$\lim_{x \to a} \frac{f(x)}{g(x)},$$

where we assume that $f(a) = 0$ and $g(a) = 0$, so that this limit is an indeterminate form. Suppose that $f(x)$ and $g(x)$ are both represented by their Taylor series expansion (at $x = a$) on an interval containing $x = a$.

7. Evaluate

$$\lim_{x \to -3} \frac{x^2 + 4x + 3}{x + 3}.$$

8. Evaluate

$$\lim_{x \to 0} \frac{1 - \cos x}{x^2}.$$

Then

$$f(x) = f(a) + f'(a) (x - a) + \frac{f''(a)}{2!} (x - a)^2 + \cdots .$$

Since $f(a) = 0$, we have

$$f(x) = f'(a) (x - a) + \frac{f''(a)}{2!} (x - a)^2 + \frac{f'''(a)}{3!} (x - a)^3 + \cdots .$$

Factoring the common term $x - a$ from the right-hand side yields

$$f(x) = (x - a) \left[f'(a) + \frac{f''(a)}{2!} (x - a) + \frac{f'''(a)}{3!} (x - a)^2 + \cdots \right].$$

Likewise,

$$g(x) = (x - a) \left[g'(a) + \frac{g''(a)}{2!} (x - a) + \frac{g'''(a)}{3!} (x - a)^2 + \cdots \right],$$

since $g(a) = 0$. So, for values of x that are not equal to a, we have

$$\frac{f(x)}{g(x)} = \frac{(x - a) \left[f'(a) + \frac{f''(a)}{2!} (x - a) + \cdots \right]}{(x - a) \left[g'(a) + \frac{g''(a)}{2!} (x - a) + \cdots \right]}$$

$$= \frac{f'(a) + \frac{f''(a)}{2!} (x - a) + \cdots}{g'(a) + \frac{g''(a)}{2!} (x - a) + \cdots}.$$

Note that substituting $x = a$ in the right-hand side yields $f'(a)/g'(a)$. This suggests the following rule.

THEOREM 6

l'Hôpital's Rule

If

$$\lim_{x \to a} \frac{f(x)}{g(x)}$$

is an indeterminate form of the type $\frac{0}{0}$ and if

$$\lim_{x \to a} \frac{f'(x)}{g'(x)}$$

exists, then

$$\lim_{x \to a} \frac{f(x)}{g(x)} = \lim_{x \to a} \frac{f'(x)}{g'(x)}.$$

EXAMPLE 7   Use l'Hôpital's Rule to evaluate

$$\lim_{x \to 0} \frac{x^2 + 3x}{5x}.$$

Solution   This is an indeterminate form of the type $\frac{0}{0}$, with $f(x) = x^2 + 3x$ and $g(x) = 5x$. Here, $f'(x) = 2x + 3$ and $g'(x) = 5$. So

$$\lim_{x \to 0} \frac{x^2 + 3x}{5x} = \lim_{x \to 0} \frac{2x + 3}{5}$$

$$= \frac{2(0) + 3}{5} = \frac{3}{5}.$$

This is the correct answer, as we saw in Example 4.   ❖

DO EXERCISES 9 AND 10.

EXAMPLE 8   Evaluate

$$\lim_{x \to 0} \frac{e^{2x} - 1}{x}.$$

Solution   Substitution of $x = 0$ in $f(x) = e^{2x} - 1$ yields

$$f(0) = e^{2(0)} - 1 = e^0 - 1 = 1 - 1 = 0.$$

Since the same substitution into $g(x) = x$ also yields 0, this is an indeterminate form. Since $f'(x) = 2e^{2x}$ and $g'(x) = 1$, we have

$$\lim_{x \to 0} \frac{e^{2x} - 1}{x} = \lim_{x \to 0} \frac{2e^{2x}}{1} = \frac{2e^{2(0)}}{1} = 2.$$ ❖

DO EXERCISE 11.

EXAMPLE 9   Evaluate

$$\lim_{x \to 0} \frac{1 - \cos x}{x^2}.$$

Solution   This is an indeterminate form of the type $\frac{0}{0}$. Since $f(x) = 1 - \cos x$ and $g(x) = x^2$, we have $f'(x) = \sin x$ and $g'(x) = 2x$. Now, both $\sin x$ and $2x$ tend to 0 as x approaches 0. In other words,

$$\lim_{x \to 0} \frac{f'(x)}{g'(x)} = \lim_{x \to 0} \frac{\sin x}{2x}$$

is *another* indeterminate form. But we may apply l'Hôpital's Rule again and again, stopping as soon as we obtain a limit that is not an indeter-

9. Use l'Hôpital's Rule to evaluate

$$\lim_{x \to -1} \frac{x^2 + 4x + 3}{x + 1}.$$

10. Use l'Hôpital's Rule to evaluate

$$\lim_{x \to -3} \frac{x^2 + 4x + 3}{x + 3}.$$

11. Use l'Hôpital's Rule to evaluate

$$\lim_{x \to 0} \frac{\sin x}{x}.$$

12. Use l'Hôpital's Rule to evaluate

$$\lim_{x \to 1} \frac{1 + \ln x - x}{(x-1)^2}.$$

13. Use l'Hôpital's Rule, if possible, to evaluate

$$\lim_{x \to 2} \frac{e^x}{x-1}.$$

minate form. In this case, we obtain

$$\lim_{x \to 0} \frac{f''(x)}{g''(x)} = \lim_{x \to 0} \frac{\cos x}{2} = \frac{\cos (0)}{2} = \frac{1}{2}.$$

Thus,

$$\lim_{x \to 0} \frac{f(x)}{g(x)} = \lim_{x \to 0} \frac{f'(x)}{g'(x)} = \lim_{x \to 0} \frac{f''(x)}{g''(x)} = \frac{1}{2}.$$

DO EXERCISE 12.

EXAMPLE 10 Use l'Hôpital's Rule to evaluate

$$\lim_{x \to 0} \frac{e^x}{\sin x}.$$

Solution As x approaches 0, substitution shows that the numerator approaches $e^0 = 1$, while the denominator approaches sin (0) = 0. This is *not* an indeterminate form of the type $\frac{0}{0}$ and l'Hôpital's Rule does *not* apply. In fact, this limit is similar to the limit in Example 3. So

$$\lim_{x \to 0} \frac{e^x}{\sin x} \quad \text{does not exist.}$$

DO EXERCISE 13.

EXAMPLE 11 Evaluate

$$\lim_{x \to 0} \frac{\sin x}{x^2}.$$

Solution This is an indeterminate form of the type $\frac{0}{0}$. Here, $f(x) = \sin x$ and $g(x) = x^2$, so that $f'(x) = \cos x$ and $g'(x) = 2x$. But

$$\lim_{x \to 0} \frac{\cos x}{2x}$$

does not exist since as x approaches 0, the numerator approaches cos (0) = 1 while the denominator approaches 0. L'Hôpital's Rule, as stated earlier, does not apply. It turns out, though, that if the numerator $f'(x)$ has a nonzero limit and the denominator $g'(x)$ approaches 0 as x approaches a, then the original limit,

$$\lim_{x \to a} \frac{f(x)}{g(x)},$$

does not exist either. Thus,

$$\lim_{x \to 0} \frac{\sin x}{x^2}.$$

does not exist.   ❖

DO EXERCISE 14.

The Indeterminate Form ∞/∞

In applications, one encounters limits of the form

$$\lim_{x \to a} \frac{f(x)}{g(x)},$$

where both $f(x)$ and $g(x)$ tend to infinity as x approaches a. In this case, the following version of l'Hôpital's Rule applies.

THEOREM 7

If $f(x)$ and $g(x)$ are continuous and if $f(x)$ and $g(x)$ approach infinity as x approaches a, then

$$\lim_{x \to a} \frac{f(x)}{g(x)} = \lim_{x \to a} \frac{f'(x)}{g'(x)}$$

if this latter limit exists. If $f'(x)/g'(x)$ tends to infinity as x approaches a, then so does $f(x)/g(x)$.

EXAMPLE 12   Evaluate

$$\lim_{x \to +\infty} \frac{2x - 3}{x + 1}.$$

Solution   Both numerator and denominator tend to infinity as x does. Applying l'Hôpital's Rule, we have

$$\lim_{x \to +\infty} \frac{2x - 3}{x + 1} = \lim_{x \to +\infty} \frac{2}{1} = 2.$$

Thus,

$$\lim_{x \to +\infty} \frac{2x - 3}{x + 1} = 2.$$

                                                                               ❖

DO EXERCISE 15.

14. Evaluate

$$\lim_{x \to 0} \frac{e^x - 1}{x^3}.$$

15. Evaluate

$$\lim_{x \to +\infty} \frac{5x - 1}{4 - 3x}.$$

16. Evaluate

$$\lim_{x \to +\infty} \frac{7e^{3x}}{5x}.$$

EXAMPLE 13 Evaluate

$$\lim_{x \to +\infty} \frac{e^x}{x}.$$

Solution Again, both numerator and denominator tend to infinity as x does. By l'Hôpital's Rule,

$$\lim_{x \to +\infty} \frac{e^x}{x} = \lim_{x \to +\infty} \frac{e^x}{1} = +\infty.$$

That is, the limit does not exist. ❖

DO EXERCISE 16.

EXAMPLE 14 Evaluate

$$\lim_{x \to +\infty} \frac{3x^2 - 5x + 1}{x^2 - 3x + 7}.$$

Solution One application of l'Hôpital's Rule yields

$$\lim_{x \to +\infty} \frac{3x^2 - 5x + 1}{x^2 - 3x + 7} = \lim_{x \to +\infty} \frac{6x - 5}{2x - 3},$$

which is again an indeterminate form of the type ∞/∞. A second application, however, gives

$$\lim_{x \to +\infty} \frac{6x - 5}{2x - 3} = \lim_{x \to +\infty} \frac{6}{2} = 3.$$

Thus,

$$\lim_{x \to +\infty} \frac{3x^2 - 5x + 1}{x^2 - 3x + 7} = 3.$$ ❖

DO EXERCISE 17.

17. Evaluate

$$\lim_{x \to +\infty} \frac{3x^2 - 5x + 1}{x^2 + 1}.$$

EXERCISE SET 10.5

Find the limit, if possible, *without* using l'Hôpital's Rule.

1. $\lim_{x \to 1} \dfrac{1}{x}$

2. $\lim_{x \to 2} \dfrac{x - 2}{x^2}$

3. $\lim_{x \to -1} \dfrac{3}{x + 1}$

4. $\lim_{x \to -5} \dfrac{7}{x + 5}$

5. $\lim_{x \to -1} \dfrac{x+1}{3}$

6. $\lim_{x \to -5} \dfrac{x+3}{7}$

7. $\lim_{x \to 0} \dfrac{7x+3}{x^2}$

8. $\lim_{x \to 0} \dfrac{x^2}{7x+3}$

9. $\lim_{x \to 0} \dfrac{7x+3x^2}{x}$

10. $\lim_{x \to -1} \dfrac{x+1}{x^2+6x+5}$

11. $\lim_{x \to 2} \dfrac{x-2}{x^2-4}$

12. $\lim_{x \to 1} \dfrac{x-1}{x^3-1}$

13. $\lim_{x \to +\infty} \dfrac{5}{x}$

14. $\lim_{x \to 1} \dfrac{x^3}{x-1}$

15. $\lim_{x \to +\infty} \dfrac{2x+1}{x-6}$

16. $\lim_{x \to 0} \dfrac{7+3x}{x}$

Evaluate the limit, if it exists, using l'Hôpital's Rule, if it applies.

17. $\lim_{x \to 1} \dfrac{x-1}{x^3-1}$

18. $\lim_{x \to 2} \dfrac{x-2}{x^2-4}$

19. $\lim_{x \to 0} \dfrac{1-e^x}{x}$

20. $\lim_{x \to 0} \dfrac{\sin(ax)}{x}$

21. $\lim_{x \to -1} \dfrac{x+1}{x^3+3x^2-x-3}$

22. $\lim_{x \to 1} \dfrac{\ln x}{x-1}$

23. $\lim_{x \to \pi} \dfrac{\sin x}{x-\pi}$

24. $\lim_{x \to 0} \dfrac{x \cos x}{\sin(2x)}$

25. $\lim_{x \to 0} \dfrac{e^x}{\cos x}$

26. $\lim_{x \to +\infty} \dfrac{2x+1}{x-6}$

27. $\lim_{x \to 1} \dfrac{\ln x - x + 1}{x-1}$

28. $\lim_{x \to 0} \dfrac{e^x - x - 1}{x^2}$

29. $\lim_{x \to +\infty} \dfrac{x^2}{7}$

30. $\lim_{x \to +\infty} \dfrac{x^2}{e^x}$

31. $\lim_{x \to 0} \dfrac{x^2-2x+2}{3x+4}$

32. $\lim_{x \to 1} \dfrac{\ln x}{(x-1)^3}$

33. $\lim_{x \to +\infty} \dfrac{x^2-5x+1}{3x^2-7x+2}$

34. $\lim_{x \to +\infty} \dfrac{x^3}{x^3-1}$

35. $\lim_{x \to +\infty} \dfrac{e^{5x}}{x^3}$

36. $\lim_{x \to +\infty} \dfrac{\tan^{-1} x}{x}$

37. $\lim_{x \to 0} \dfrac{e^{5x}}{x^3}$

38. $\lim_{x \to 0} \dfrac{e^{3x}-1}{x}$

SYNTHESIS EXERCISES

APPLICATIONS

❖ **Life and Physical Sciences**

39. *Chemistry.* In a certain chemical reaction, the amount of the product at time t is given by

$$x(t) = ab \left\{ \dfrac{r - pe^{kt}}{pa - be^{kt}} \right\},$$

where $a, b, k, p,$ and r are positive constants. Use l'Hôpital's Rule to find the limiting amount of the product present, that is, to evaluate

$$\lim_{t \to +\infty} x(t).$$

40. *Physics.* The velocity v of a parachutist t seconds after ejection from a plane is given by

$$v(t) = a \left[\dfrac{e^{bt} - e^{-bt}}{e^{bt} + e^{-bt}} \right],$$

where a and b are positive constants. Determine the limiting velocity of the parachutist,

$$\lim_{t \to +\infty} v(t).$$

41. If f is continuous on $(-\infty, +\infty)$, find

$$\lim_{x \to 0} \dfrac{1}{x} \int_0^x f(t)\, dt.$$

42. Plot the graph of $(\sin x)/x$ on $[-3, 3]$. Then use the graph to estimate $\lim_{x \to 0} (\sin x)/x$ (don't forget radian mode).

43. Graph $(1 - e^x)/x$ on $[-1, 1]$ and estimate

$$\lim_{x \to 0} (1 - e^x)/x.$$

44. Solve Exercise 17 above graphically.

45. Redo Exercise 42, leaving your calculator in degree

mode to estimate $\lim_{x \to 0} (\sin x)/x$ when x is measured in degrees.

THE CALCULUS EXPLORER:
PowerGrapher   Limit Problems

Use these programs to investigate Problems 42–46.

CHAPTER SUMMARY AND REVIEW

10

TERMS TO KNOW

Polynomial function, p. 702
n factorial, p. 706
Taylor's formula for polynomials, p. 708
Taylor polynomial of degree n, p. 712
Infinite series, p. 719
Convergent infinite series, p. 720

Divergent infinite series, p. 720
Sum of an infinite series, p. 720
Geometric series, p. 722
Initial term, ratio, of a geometric series, p. 722
nth partial sum of a geometric series, p. 723
Alternating series, p. 727
Taylor series of $f(x)$, p. 732

$f(x)$ is represented by its Taylor series, p. 732
Radius of convergence of a Taylor series, p. 736
Interval of convergence of a Taylor series, p. 736
Indeterminate form (%), p. 745
l'Hôpital's Rule, p. 748
Indeterminate form (∞/∞), p. 751

REVIEW EXERCISES

These review exercises are for test preparation. They can also be used as a lengthened practice test. Answers are at the back of the book. The answers also contain bracketed section references, which tell you where to restudy if your answer is incorrect.

1. Expand $x^2 - 5x + 3$ in powers of $x - 1$.

Find $p_2(x)$ and $p_3(x)$ for the given function and the specified value of x.

2. $y = x^3$; $x = -1$

3. $y = e^{x^2}$; $x = 0$

4. $y = \sin (x - 2)$; $x = 2$

Use a Taylor series to estimate the expressions in Exercises 5 and 6.

5. $\sin (0.05)$

6. $e^{0.05}$

7. Find the sum of $\frac{8}{5} - \frac{6}{5} + \frac{9}{10} - \frac{27}{40} + \frac{81}{160} - \frac{243}{640} + \cdots$.

8. Find the sum of $1 - \frac{3}{2} + \frac{9}{4} - \frac{27}{8} + \cdots$.

9. Evaluate the sum of $\frac{1}{3} + \frac{1}{5} + \frac{3}{25} + \cdots + (\frac{1}{3})(\frac{3}{5})^{10}$.

10. Write $1.0641\overline{41}$ as a rational fraction.

Determine whether the given series converges or diverges. If it converges, estimate its sum with an error of less than 0.01.

11. $1 - \dfrac{1}{3 \cdot 5} + \dfrac{1}{5 \cdot 5^2} - \dfrac{1}{7 \cdot 5^3} + \cdots = \displaystyle\sum_{i=0}^{\infty} \dfrac{(-1)^i}{[2i + 1] \cdot 5^i}$

12. $1 - 3 + 5 - 7 + 9 - 11 + \cdots$

13. $\dfrac{2}{5} + \dfrac{3}{6} + \dfrac{4}{7} + \dfrac{5}{8} + \dfrac{6}{9} + \cdots$

14. $\dfrac{2}{5} - \dfrac{3}{6} + \dfrac{4}{7} - \dfrac{5}{8} + \dfrac{6}{9} - \cdots$

15. $1 - \dfrac{1}{12} + \dfrac{1}{144} - \dfrac{1}{1728} + \dfrac{1}{12^4} - \dfrac{1}{12^5} + \cdots$

Expand each of the following functions in a Taylor series using the prescribed value of x.

16. $y = x^3 \sin(x)$; $x = 0$ **17.** $y = e^{(x + 1)^2}$; $x = -1$

18. $y = (1 - 3x)^{-1}$; $x = 0$

19. Use a Taylor series to evaluate
$$\lim_{x \to 0} \frac{\sin(x^3)}{x^3}.$$

Evaluate, where possible.

20. $\displaystyle\lim_{x \to 0} \dfrac{1 - \cos x}{3x \sin x}$ **21.** $\displaystyle\lim_{x \to 0} \dfrac{1 - \cos x}{2e^x}$

22. $\displaystyle\lim_{x \to \pi} \dfrac{\sin x}{x - \pi}$

23. $\displaystyle\lim_{x \to +\infty} \dfrac{x^{2n}}{e^{5x}}$, where n is a positive integer

24. $\displaystyle\lim_{x \to +\infty} \dfrac{x^2 - 3x + 5}{3x^2 - 7x + 1}$

SYNTHESIS EXERCISES

25. *Business: Value of a stock.* A certain stock pays a dividend of $6 per year. Assuming that the dividend remains constant forever and that interest rates are constant at 10%, so that the present value of the dividend to be paid n years from today is $6(1.10)^{-n}$, find the sum of the present value of all future dividend payments, assuming that the first payment is one year from today.

26. Find a function $f(x)$, if there is one, such that $f(0) = 0$ and $f^{(n)}(0) = 1/n^2$ for all positive integers n.

EXERCISES FOR THINKING AND WRITING

27. Is the function $f(x) = |x|$ represented by some Taylor series, say, on $[-1, 1]$? Explain your assertion. What about $g(x) = x|x|$?

28. A function $f(x)$ is defined for all real numbers by $f(0) = 1$ and $f'(x) = 2f(x)$. Without doing any computation, find $f'(0)$. Then differentiate both sides of $f'(x) = 2f(x)$ to find $f''(0)$. Find the Taylor series for $f(x)$ (in powers of x). Does this suggest a new technique for solving differential equations?

29. In Example 4 of Section 9.8, we were given the system of equations $dx/dt = y$ and $dy/dt = -x + 2y$ and initial conditions $x(0) = 1$ and $y(0) = 3$. Find $x'(0)$ and $y'(0)$ without doing any computation. Can you find $x''(0)$ and $y''(0)$? Does this suggest a different method of solution of that problem?

CHAPTER TEST

10

Find $p_2(x)$ and $p_3(x)$ for the given function and the specified value of x.

1. $y = x^3$; $x = -2$ **2.** $y = e^{x^3}$; $x = 0$

3. $y = \sin(x - 1)$; $x = 1$

Use a Taylor polynomial of degree 3 to find an approximate value for each of the following.

4. $\sin(0.04)$ **5.** $e^{0.07}$

6. Find the sum of the infinite series

$$\frac{3}{5} - \frac{2}{5} + \frac{4}{15} - \frac{8}{45} + \frac{16}{135} - \frac{32}{405} + \cdots.$$

7. Evaluate

$$\frac{1}{2} + \frac{1}{7} + \frac{2}{49} + \cdots + \frac{1}{2}\left(\frac{2}{7}\right)^{12}.$$

8. Write $2.431313\overline{31}$ as a rational fraction.

Determine whether the given infinite series converges or diverges. If it converges, estimate its sum with an error of less than 0.01.

9. $1 - \dfrac{1}{3 \cdot 4} + \dfrac{1}{5 \cdot 4^2} - \dfrac{1}{7 \cdot 4^3} + \dfrac{1}{9 \cdot 4^4} - \cdots$

10. $1 - 2 + 4 - 8 + 16 - 32 + \cdots$

11. $\dfrac{2}{3} - \dfrac{3}{4} + \dfrac{4}{5} - \dfrac{5}{6} + \dfrac{6}{7} - \dfrac{7}{8} + \cdots$

12. $1 - \dfrac{1}{16} + \dfrac{1}{256} - \dfrac{1}{4^6} + \dfrac{1}{4^8} - \cdots$

Expand each of the following functions in a Taylor series using the prescribed value of x.

13. $y = x^2 \sin x$; $x = 0$ **14.** $y = e^{(x-1)^2}$; $x = 1$

15. $y = \dfrac{1}{1 - 5x}$; $x = 0$

16. Use a Taylor series to evaluate

$$\lim_{x \to 0} \frac{1 - \cos(x^2)}{x^4}.$$

Evaluate each of the following limits.

17. $\displaystyle\lim_{x \to 0} \frac{x \sin x}{\cos x - 1}$ **18.** $\displaystyle\lim_{x \to 1} \frac{3x \ln x}{e^x}$

19. $\displaystyle\lim_{x \to \pi/2} \frac{\cos x}{x - \pi/2}$

20. $\displaystyle\lim_{x \to +\infty} \frac{x^n}{e^x}$ (n is a positive integer)

21. $\displaystyle\lim_{x \to +\infty} \frac{x^3 - 5x + 6}{2x^3 - 7x^2 + 1}$

SYNTHESIS EXERCISES

22. A certain stock pays a dividend of $2 per year. Assuming that the dividend remains constant forever and that interest rates are constant at 10% so that the present value of the dividend payment n years from today is $2(1.10)^{-n}$, find the sum of the present value of all future dividend payments, assuming that the first payment is one year from today.

23. In Problem 22, if the dividend payment is $D and the interest rate is i, find the present value of all future dividend payments.

24. Find a function $f(x)$ (if there is one) such that $f(0) = 0$ and $f^{(n)}(0) = 1/n$ for all positive integers n.

11

NUMERICAL

TECHNIQUES

AN APPLICATION

An investor buys an annuity (a sequence of future payments) for $2700. The annuity will return $1000 per year for three years, beginning one year from today. Find the yield rate on this investment.

THE MATHEMATICS

If i is the yield rate, then it is more convenient to first find $v = (1 + i)^{-1}$. It turns out that elementary algebra will lead us to the equation

$$v^3 + v^2 + v - 2.7 = 0,$$

an equation that is difficult to solve exactly. Using Newton's method, a technique for finding approximate solutions to such equations, we will obtain $i \approx 0.0546$, or 5.46%.

Many interesting problems lead to the solution of an algebraic equation, to the evaluation of a definite integral, or to the solution of a differential equation. Unfortunately, it may be difficult (or impossible) to find the exact solution in one of these situations. This chapter deals with finding approximate solutions to the problems described above. The techniques are used to solve several interesting applied problems.

11.1

SOLVING EQUATIONS NUMERICALLY: ITERATION

The solutions of many equations that arise in practical applications cannot be found exactly using standard algebraic techniques. An example of an equation of this sort is $\cos x = x$, which we consider below. We use the following general procedure.

To solve a given equation, we start by making an *initial approximation*, or *initial guess* (denoted x_1), of the solution to the equation. Then we use our initial guess x_1 to construct a new value x_2 (which we expect to be closer to the solution than x_1 is). Continuing, we use x_2 to construct a new value x_3, and so forth. In theory, we construct an infinite sequence $x_1, x_2, \ldots, x_n, \ldots$, which converges to a solution, say $\bar{x}$, of the given equation. In practice, we must stop the process after a finite number of steps N, using the approximate solution x_N instead of the actual solution $\bar{x}$.

EXAMPLE 1   Solve $\cos x = x$.

Solution   The graphs of $y = \cos x$ and $y = x$ are plotted in Fig. 1. They appear to intersect at one point only. The x-coordinate of that point, $\bar{x}$, is the solution to our equation. Judging from the graph, $x_1 = \frac{1}{2} = 0.5000000$ is not an unreasonable initial guess.

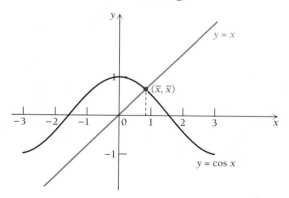

FIGURE 1

We'll construct x_2 using what is called the *method of iteration*. In this case, it means that $x_2 = \cos(x_1) = \cos(0.5)$. Computing (using a calculator in radian mode), we get $x_2 \approx 0.8775826$. We continue by computing $x_3 = \cos(x_2) = \cos(0.8775826)$. We get $x_3 \approx 0.6390125$.

Then, with $x_4 = \cos(x_3)$, we find that $x_4 \approx 0.8026851$. We continue with $x_{n+1} = \cos(x_n)$, for all n. We list a few values (check them on your calculator):

$$x_5 \approx 0.6947780,$$
$$x_6 \approx 0.7681958,$$
$$x_7 \approx 0.7191654,$$
$$x_8 \approx 0.7523558,$$
$$x_9 \approx 0.7300811,$$
$$x_{10} \approx 0.7451203,$$
$$\cdot$$
$$\cdot$$
$$\cdot$$
$$x_{17} \approx 0.7387045,$$
$$x_{18} \approx 0.7393415,$$
$$x_{19} \approx 0.7389124,$$
$$\cdot$$
$$\cdot$$
$$\cdot$$
$$x_{43} \approx 0.7390851,$$
$$x_{44} \approx 0.7390851,$$

with the readings unchanged after that. We conclude that $\bar{x} \approx 0.7390851$. As a check, we substitute this value into the original equation to get

$$\cos(\bar{x}) \approx 0.7390851 \approx \bar{x}. \qquad \diamond$$

DO EXERCISE 1.

We will apply the method of iteration only to equations of the form $g(x) = x$, where $g(x)$ is a continuous function. Remember that $x_{n+1} = g(x_n)$ for all n. If the sequence $x_1, x_2, \ldots x_n, \ldots$, has a limit, say $\bar{x}$, then

$$\bar{x} = \lim_{x \to \infty} x_n.$$

For continuous functions g,

$$g(\bar{x}) = \lim_{n \to \infty} g(x_n)$$
$$= \lim_{n \to \infty} x_{n+1} = \bar{x},$$

or

$$g(\bar{x}) = \bar{x}.$$

1. Using $x_1 = 1.8$ and $x_{n+1} = \sqrt{x_n}$ for $n = 1, 2, \ldots$, solve $\sqrt{x} = x$.

2. Try to solve $x^2 = x$ using $x_1 = 1.1$.

Thus if the sequence of iterates $x_1, x_2, \ldots x_n, \ldots$, has a limit $\bar{x}$ and if g is continuous, then $\bar{x}$ is a solution to $g(x) = x$. The catch, of course, is that the sequence of iterates may not converge.

EXAMPLE 2 Solve $x^2 = x$ using $x_1 = 1.5$.

Solution Here, $g(x) = x^2$. So $x_{n+1} = x_n^2$ for $n = 1, 2, \ldots$. Using $x_1 = 1.5$, we obtain $x_2 = (1.5)^2 = 2.25$, $x_3 = 5.0625$, $x_4 \approx 25.62891$, and so on. The sequence of iterates diverges. Note that $x = 0$ and $x = 1$ are the only solutions to $x^2 = x$. ❖

3. Try to solve $x^2 = x$ using each of the following.

a) $x_1 = \frac{1}{3}$

b) $x_1 = -\frac{1}{2}$

c) $x_1 = \frac{2}{3}$

DO EXERCISES 2 AND 3.

EXAMPLE 3 Solve $x^2 - 5x + 6 = 0$.

Solution We write $x^2 + 6 = 5x$, or $\frac{1}{5}x^2 + \frac{6}{5} = x$. Letting $g(x) = \frac{1}{5}x^2 + \frac{6}{5}$, we have $g(x) = x$.

The original equation can be factored as $(x - 2)(x - 3) = 0$, so that $\bar{x} = 2$ and $\bar{x} = 3$ are the solutions. Alternatively, we could have used the quadratic formula to obtain the solutions.

Ignoring the fact that we can easily find the solutions in this case, we start with $x_1 = 1$. Then

$$x_2 = g(x_1) = \tfrac{1}{5}(1)^2 + \tfrac{6}{5} = 1.4.$$

4. What happens in Example 3 if we use $x_1 = 2.9$?

Continuing, we have

$$x_3 = g(x_2) = \tfrac{1}{5}(1.4)^2 + \tfrac{6}{5} = 1.592$$

and

$$x_4 = g(x_3) = \tfrac{1}{5}(1.592)^2 + \tfrac{6}{5} \approx 1.7068928.$$

Similarly, $x_5 \approx 1.782697$, $x_6 \approx 1.835601$, $x_7 \approx 1.873887$, and $x_8 \approx 1.902290$. We find that $x_{14} \approx 1.976536$ and, after a few more iterations, $x_{20} \approx 1.99398$. It seems clear that these iterates have the limit $\bar{x} = 2$.

Now, what happens if we start with the initial guess $x_1 = 3.2$? Then $x_2 = \frac{1}{5}(3.2)^2 + \frac{6}{5} = 3.248$. Continuing, we have $x_3 = g(x_2) = \frac{1}{5}(3.248)^2 + \frac{6}{5} \approx 3.30990$, a bad sign, since $x_3 - x_2$ is larger than $x_2 - x_1$. Continuing, we find that $x_4 \approx 3.391089$, $x_5 \approx 3.499896$, $x_6 \approx 3.649855$, and $x_7 \approx 3.864288$, with the sequence of iterates diverging. ❖

5. Solve $x^2 + 1 = 4x$ using $x_1 = 2$.

DO EXERCISES 4 AND 5.

Examples 2 and 3 and Margin Exercises 2, 3, and 4 show that the sequence of iterates may diverge for one choice of x_1 and converge for another choice. The following theorem clarifies the situation.

THEOREM 1

If $g(x)$ and $g'(x)$ are continuous and $|g'(x)| < 1$ on an interval containing *both* x_1 and an $\bar{x}$ satisfying $g(\bar{x}) = \bar{x}$, and if we put

$$x_{n+1} = g(x_n) \quad \text{for } n = 1, 2, \ldots,$$

then the sequence $x_1, x_2, \ldots, x_n, \ldots$, converges to $\bar{x}$.

6. Solve $x^2 + 1 = 4x$ using $x_1 = 2$ and $g(x) = \sqrt{4x - 1}$.

We reconsider Examples 2 and 3. In Example 2, we had $g(x) = x^2$, so that $g'(x) = 2x$. If x is between $-\frac{1}{2}$ and $\frac{1}{2}$, then $g'(x) = 2x$ will be between -1 and 1. Since the interval contains $\bar{x} = 0$, the theorem guarantees us that if we pick x_1 between $-\frac{1}{2}$ and $\frac{1}{2}$ and define $x_{n+1} = x_n^2$ for $n = 1, 2, \ldots$, then the iterates will converge to $\bar{x} = 0$.

In Example 3, we had $g(x) = \frac{1}{5}x^2 + \frac{6}{5}$. Thus $g'(x) = \frac{2}{5}x$. In order to guarantee that $-1 < \frac{2}{5}x < 1$, we must choose x so that $-\frac{5}{2} < x < \frac{5}{2}$ (multiplying the original inequality by $\frac{5}{2}$). This interval contains $\bar{x} = 2$. It follows that if we select x_1 between $-\frac{5}{2}$ and $\frac{5}{2}$, then the iterates will converge to 2. For this function, we can show that the iterates will converge to 2 if we select any x_1 between -3 and 3.

EXAMPLE 4   Use iteration to solve $x^2 - 5x + 6 = 0$.

Solution   Yes! This is the same equation that appeared in Example 3 above. But there may be many ways to rewrite $x^2 - 5x + 6 = 0$ in the form $g(x) = x$. For instance, let's write $x^2 = 5x - 6$. Then, taking the principal square roots of both sides (assuming that x and $5x - 6$ are nonnegative), we get $x = \sqrt{5x - 6}$. Let $g(x) = \sqrt{5x - 6}$. If we define

$$x_{n+1} = g(x_n) = \sqrt{5x_n - 6},$$

then we obtain a new and different iteration formula for this equation. If we pick $x_1 = 4$, for example, then

$$x_2 = \sqrt{5x_1 - 6} = \sqrt{5(4) - 6} = \sqrt{14} \approx 3.741657.$$

Continuing, we get $x_3 = \sqrt{5x_2 - 6} \approx \sqrt{12.708287} \approx 3.564868$, $x_4 \approx 3.438654$, $x_5 \approx 3.345635$, $x_6 \approx 3.275389$, . . . , $x_{24} \approx 3.008502$, $x_{25} \approx 3.007077$, $x_{26} \approx 3.005892$, and so forth. Theorem 1 can be applied to show that the iterates converge to $\bar{x} = 3$, although the convergence is very slow. Examples 3 and 4 show that in solving an equation by iteration, some experimentation may be required to determine a workable iteration formula.   ❖

DO EXERCISE 6.

EXERCISE SET 11.1

Solve the given equation by iteration, using the indicated initial guess. Check your answers.

1. $x = \frac{1}{3}x^2 + \frac{1}{3}$; $x_1 = 0$

2. $x = \frac{1}{3}x^2 + \frac{1}{3}$; $x_1 = 3$

3. $x = \sin x + \frac{1}{2}$; $x_1 = 1$
(x in radians)

4. $x = \frac{1}{2}\cos x$; $x_1 = 1$

5. $x = 2\cos x$; $x_1 = 1$

6. $x = \sqrt{3x - 1}$; $x_1 = 2$

7. $x = \frac{1}{3}x^3 + 1$; $x_1 = 0$

8. $x = \sqrt[3]{3x - 3}$; $x_1 = 0$

Use iteration to find the solution to each of the following. Check your answers.

9. $x - e^{-x} = 0$

10. $x^5 - 5x + 5 = 0$

11. $3\cos x - 7x = 0$

12. $\cos x - 2x^2 = 0$

SYNTHESIS EXERCISE

13. Define $x_{n+1} = \frac{1}{3}x_n^2 + \frac{1}{3}$ for all n, as in Exercise 2 above.

   a) Show that if $x_n > 3$, then $x_{n+1} > \frac{10}{3} > 3$. Conclude that if $x_1 > 3$, then all the succeeding x_n's are greater than $\frac{10}{3}$.

   b) By writing

   $$\frac{x_{n+1} + 1}{x_n} = \frac{1}{3}x_n + \frac{1}{3x_n},$$

   show that if $x_n > \frac{10}{3}$, then

   $$\frac{x_{n+1}}{x_n} > \frac{10}{9}.$$

   c) By combining parts (a) and (b), show that if $x_1 > 3$, then $x_2 > \frac{10}{9}$ and $x_{k+2} > (\frac{10}{9})^k \cdot x_2$, which in turn shows that the iterates in Exercise 2 diverge.

COMPUTER–GRAPHING CALCULATOR EXERCISES

14. On the same graph, graph $f(x) = x$ and $g(x) = \frac{1}{2}\cos x$. Then find the point of intersection, $(\bar{x}, \bar{x})$, of the two curves. Compare your answer with your answer to Exercise 4 above. (**Note: Make sure that you are in radian mode before graphing.**)

15. Repeat Exercise 14, graphing both $f(x) = x$ and $g(x) = \sin(x) + \frac{1}{2}$. Compare your answer with your answer to Exercise 3 above.

THE CALCULUS EXPLORER:
PowerGrapher

Do Exercises 14 and 15 using these programs.

11.2

NEWTON'S METHOD

We introduce Newton's method, another technique for solving equations numerically. This method is fairly easy to apply and usually yields more rapid convergence than the method of iteration. To apply this method, we write the given equation in the form $f(x) = 0$. Let x_1 be our initial approximation, as shown in Fig. 1.

OBJECTIVE

a) Solve equations using Newton's method.

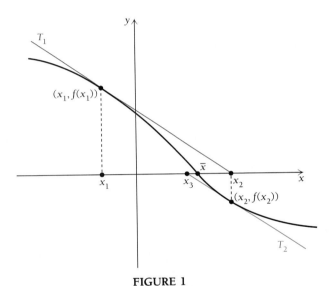

FIGURE 1

Assuming that $f'(x_1) \neq 0$, the tangent line T_1 to the graph of $y = f(x)$ at the point $(x_1, f(x_1))$ will intersect the x-axis in exactly one point. We call this point $(x_2, 0)$. The x-coordinate of this point, x_2, will be our second approximation. Next we locate $(x_2, f(x_2))$ on the graph of $y = f(x)$ and construct the tangent line T_2 to the graph at this new point. Then the tangent line T_2 will intersect the x-axis at $(x_3, 0)$ (assuming that $f'(x_2) \neq 0$). The x-coordinate of this point, x_3, is our third approximation to the solution $\bar{x}$. Continuing, we can generate as many approximations as we need.

We now have a geometric method for constructing x_{n+1} from x_n. To obtain a formula relating x_n and x_{n+1}, we return to the tangent line T_1.

The slope of T_1 is $f'(x_1)$, the value of the derivative of $f(x)$ at $x = x_1$. Now T_1 passes through the point $(x_1, f(x_1))$. Applying the point–slope formula from Section 1.4, we see that the equation of T_1 is

$$y - f(x_1) = f'(x_1)\ (x - x_1).$$

The point $(x_2, 0)$ is on T_1 and so its coordinates must satisfy the equation of T_1. Thus, setting $x = x_2$ and $y = 0$, we have

$$0 - f(x_1) = f'(x_1)\ (x_2 - x_1),$$

or

$$-f(x_1) = f'(x_1)\ (x_2 - x_1).$$

Dividing both sides by $f'(x_1)$ (recall that $f'(x_1) \neq 0$), we get

$$\frac{-f(x_1)}{f'(x_1)} = x_2 - x_1.$$

Solving for x_2, we have

$$x_2 = x_1 - \frac{f(x_1)}{f'(x_1)}.$$

In general, we have the following:

$$x_{n+1} = x_n - \frac{f(x_n)}{f'(x_n)}. \tag{1}$$

EXAMPLE 1   Use Newton's method with $x_1 = 1.5$ to solve $x^2 = x$.

Solution   To apply Newton's method, we must write the equation in the form $f(x) = 0$. So we write $x^2 - x = 0$. If we put $f(x) = x^2 - x$, then the equation takes the form $f(x) = 0$ and we can apply formula (1). Since $f(x) = x^2 - x$, we get $f'(x) = 2x - 1$. So formula (1) becomes

$$x_{n+1} = x_n - \frac{f(x_n)}{f'(x_n)}$$

$$= x_n - \frac{x_n^2 - x_n}{2x_n - 1}.$$

If $n = 1$, we get

$$x_2 = x_1 - \frac{x_1^2 - x_1}{2x_1 - 1}$$

and, with $x_1 = 1.5$,

$$x_2 = 1.5 - \frac{(1.5)^2 - 1.5}{2(1.5) - 1}$$

$$= 1.5 - \frac{2.25 - 1.5}{3 - 1}$$

$$= 1.5 - \frac{0.75}{2} = 1.5 - 0.375$$

$$= 1.125.$$

Then

$$x_3 = x_2 - \frac{x_2^2 - x_2}{2x_2 - 1}$$

$$= 1.125 - \frac{(1.125)^2 - (1.125)}{2(1.125) - 1}$$

$$= 1.125 - \frac{1.265625 - 1.125}{2.25 - 1}$$

$$= 1.125 - \frac{0.140625}{1.25}$$

or

$$x_3 = 1.125 - 0.1125 = 1.0125.$$

Then

$$x_4 = 1.0125 - \frac{(1.0125)^2 - 1.0125}{2(1.0125) - 1}$$

$$\approx 1.0125 - 0.1234756$$

$$\approx 1.000152,$$

and

$$x_5 \approx 1.000152 - \frac{(1.000152)^2 - (1.000152)}{2(1.000152) - 1}$$

$$\approx 1.0000000.$$

With $x_5 = 1$ (correct to seven decimal places), we get $x_6 = 1, \ldots$, and we see that 1 is a solution to $f(x) = 0$.

Note the graph of $f(x) = x^2 - x$ in Fig. 2. It's a parabola and the initial guess x_1, the tangent line T_1, and x_2 are all shown. If we pick x_1 greater than 0.5, then the iterates given by Newton's method will con-

1. Use Newton's method with $x_1 = \frac{1}{3}$ to solve $x^2 = x$.

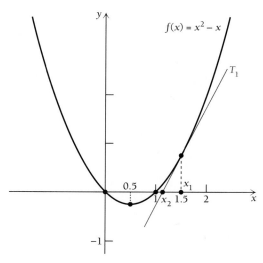

FIGURE 2

verge to 1. If x_1 is less than 0.5, then the iterates will converge to 0. Note that we may not choose $x_1 = 0.5$ since $f'(x_1) = 2x_1 - 1 = 2(0.5) - 1 = 0$ and the tangent line at $x = 0.5$ is horizontal. ❖

DO EXERCISE 1.

EXAMPLE 2 Solve $\cos x = x$.

Solution To put the equation in the correct form, we write $x - \cos x = 0$ and let $f(x) = x - \cos x$. Then $f'(x) = 1 + \sin x$ and formula (1) becomes

$$x_{n+1} = x_n - \frac{x_n - \cos (x_n)}{1 + \sin (x_n)}.$$

If $x_1 = 0.5$, then

$$x_2 = x_1 - \frac{x_1 - \cos (x_1)}{1 + \sin (x_1)} = 0.5 - \frac{(0.5) - \cos (0.5)}{1 + \sin (0.5)}$$

$$= 0.5 - \frac{0.5 - (0.8775826)}{1 + 0.4794255} = 0.5 - \frac{-0.3775826}{1.4794255}$$

$$\approx 0.5 + 0.2552224$$

$$\approx 0.7552224.$$

So,

$$x_3 \approx 0.7552224 - \frac{(0.7552224) - \cos (0.7552224)}{1 + \sin (0.7552224)}$$

$$\approx 0.7552224 - \frac{0.0271033}{1.685451}$$

$$\approx 0.7391417.$$

2. Use Newton's method to solve $\cos x = 2x$. Check your answer.

Continuing, we have

$$x_4 \approx 0.7391417 - \frac{(0.7391417) - \cos (0.7391417)}{1 + \sin (0.7391417)}$$

$$\approx 0.7391417 - \frac{0.000094672}{1.6736538}$$

$$\approx 0.7391417 - 0.0000565661$$

$$\approx 0.7390851,$$

which is correct to seven decimal places, as we saw in Example 1 of Section 11.1.

Be sure to note the difference in the speed of convergence of the two methods. Newton's method generally gives much faster convergence than simple iteration does. ❖

DO EXERCISE 2.

Newton's method doesn't always work. One equation on which it must fail is $x^2 + 1 = 0$, because the equation has no real solutions.

Even if the equation has a solution, though, the iterates generated by Newton's method may not converge to it. Consider the graph in Fig. 3. The function has no particular formula—call it $f(x)$.

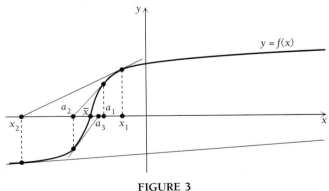

FIGURE 3

The equation $f(x) = 0$ has exactly one solution, $\bar{x}$. But if x_1 is chosen as shown, then x_2 will be farther from $\bar{x}$ than x_1 is. In fact, it appears from the graph that successive estimates will get worse and worse. Note that if the initial guess had been a_1 instead of x_1, the successive approximations would seem to converge rapidly to $\bar{x}$.

The following theorem gives us a way to make a correct choice.

THEOREM 2

Newton's Method

If there exists an interval containing a solution $\bar{x}$ to $f(x) = 0$ on which $f(x)$, $f'(x)$, and $f''(x)$ are continuous, $f'(x) \neq 0$, and

$$\left| \frac{f(x)f''(x)}{(f'(x))^2} \right| < 1,$$

and if x_1 is chosen in this interval, then the sequence given by

$$x_{n+1} = x_n - \frac{f(x_n)}{f'(x_n)}$$

converges to $\bar{x}$.

We omit the proof. To see how to apply it, consider the following example.

EXAMPLE 3   Use Newton's method to solve $x^2 - 3 = 0$.

Solution   We set $f(x) = x^2 - 3$. Then

$$f(1) = (1)^2 - 3 = -2 \quad \text{and} \quad f(2) = (2)^2 - 3 = 1,$$

so that the equation $f(x) = 0$ has a solution $\bar{x}$ in the interval from 1 to 2.

In this case, $f'(x) = 2x$ and $f''(x) = 2$ so that

$$\frac{f(x)f''(x)}{(f'(x))^2} = \frac{(x^2 - 3)(2)}{(2x)^2}$$

$$= \frac{2x^2 - 6}{4x^2}$$

$$= \frac{1}{2} - \frac{3}{2x^2}.$$

This expression is always less than 1 for x between 1 and 2. To keep this ratio between -1 and 1, as the theorem requires, we need only choose x so that

$$-1 < \frac{2x^2 - 6}{4x^2}$$

or

$$-4x^2 < 2x^2 - 6.$$

Rearranging terms, we get

$$6 < 2x^2 + 4x^2,$$

or

$$6 < 6x^2,$$

which is satisfied if x is greater than 1 (or less than -1). Now $f'(x) = 2x$ is 0 only when $x = 0$. So, if we choose x_1 to be greater than 1, then all the requirements of the theorem are met and the iterates will converge to the solution $\bar{x}$, which lies between 1 and 2.

If $x_1 = 2$, for example, then we have

$$\begin{aligned} x_{n+1} &= x_n - \frac{x_n^2 - 3}{2x_n} \\ &= \frac{2x_n^2}{2x_n} - \frac{x_n^2 - 3}{2x_n} \\ &= \frac{2x_n^2 - x_n^2 + 3}{2x_n} \\ &= \frac{x_n^2 + 3}{2x_n}. \end{aligned}$$

So

$$x_2 = \frac{(2)^2 + 3}{2 \cdot 2} = \frac{4 + 3}{4} = 1.75.$$

Then

$$x_3 = \frac{(1.75)^2 + 3}{2 \cdot (1.75)} = \frac{3.0625 + 3}{3.5},$$

or

$$x_3 \approx 1.732143.$$

Continuing, we have

$$x_4 \approx \frac{(1.732143)^2 + 3}{2 \cdot (1.732143)} \approx 1.732051.$$

3. Using Newton's method with $x_1 = -2$, solve $x^3 - 3x + 3 = 0$. Check your answer.

Another iteration yields $x_5 \approx x_4$. If we put $\bar{x} = 1.732051$, then $(\bar{x})^2 \approx 3$.

For this particular function, the iterates given by Newton's method will converge for any initial choice x_1 except $x_1 = 0$ (the *one* value of x for which $f'(x) = 0$). If $x_1 > 0$, the iterates will converge to $\sqrt{3}$, whereas if $x_1 < 0$, they will converge to $-\sqrt{3}$ (see Fig. 4).

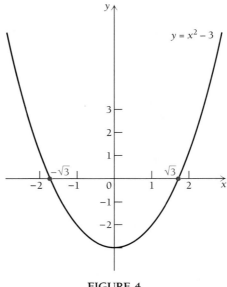

FIGURE 4

4. Use Newton's method to solve $x^3 = 10$. Check your answer.

DO EXERCISES 3 AND 4.

Use the indicated choice of x_1 and Newton's method to solve the given equation. Check your answers.

1. $x = \frac{1}{3}x^2 + \frac{1}{3}$; $x_1 = 0$
2. $x = \frac{1}{3}x^2 + \frac{1}{3}$; $x_1 = 3$
3. $x = \sin x + \frac{1}{2}$ (x in radians); $x_1 = 1$
4. $x = 2 \cos x$; $x_1 = 1$
5. $x^3 - 3x + 3 = 0$; $x_1 = -3$
6. $x^4 - 4x + 1 = 0$; $x_1 = 0$

Use Newton's method to find all real solutions to the given equation.

7. $x^3 = e^{-x}$
8. $x^4 - x - 2 = 0$
9. $2x^3 + 4x^2 - 6x - 1 = 0$
10. $x^4 + 4x^2 + 1 = 0$
11. $x^7 - 7x + 8 = 0$
12. $3 \cos x - 7x = 0$
13. $\cos x - 2x^2 = 0$
14. $x + \ln x = 0$

SYNTHESIS EXERCISES

15. a) Show that applying Newton's method to

$$x^2 - a = 0$$

yields the iteration scheme

$$x_{n+1} = \frac{x_n}{2} + \frac{a}{2x_n},$$

provided $x_n \neq 0$.

b) Use the formula above to approximate $\sqrt{19}$.

16. a) Sketch the graph of $y = f(x) = \tan^{-1} x$ to show that the equation $\tan^{-1} x = 0$ has only the solution $\bar{x} = 0$.

b) Calculate

$$\left| \frac{f(x)f''(x)}{(f'(x))^2} \right|.$$

c) Use Newton's method and $x_1 = 4$ to try to find the solution.

d) Repeat part (c) using $x_1 = 0.5$.

e) Explain the different results in parts (c) and (d) by referring to Theorem 2 and part (b) above.

17. a) Use Newton's method to show that the iterates

$$x_{n+1} = \frac{k-1}{k} x_n + \frac{a}{kx_n^{k-1}} \qquad (x_1, a > 0)$$

converge to $a^{1/k}$.

b) Use the algorithm above to calculate $\sqrt[10]{20}$.

COMPUTER–GRAPHING CALCULATOR EXERCISES

18. Graph the functions $f(x) = x$ and $g(x) = \frac{1}{3}x^2 + \frac{1}{3}$ simultaneously. Then find their point of intersection, $(\bar{x}, \bar{x})$. Compare this result with your answer to Exercise 1 above.

19. Graph $f(x) = x^3$ and $g(x) = e^{-x}$ simultaneously. Plot the point of intersection, $(\bar{x}, \bar{x}^3)$, and compare the result to your answer to Exercise 7 above.

20. Solve the equation in Exercise 12 graphically. (Don't forget to use radian mode!)

 CALCULUS EXPLORER:
PowerGrapher

Do Exercises 18–20 using these programs.

11.3

AN APPLICATION: YIELD RATE (INTERNAL RATE OF RETURN)

Which is the better deal: a payment of $1000 immediately or a payment of $1100 one year from today? (It's understood here that inflation and taxes are *not* to be taken into account in our analysis.) Consider the following. If we deposit $1000 in a savings account (or money-market fund) at 12%

OBJECTIVES

a) Set up the equation for the yield rate for a (finite) sequence of investments and returns on those investments.

b) Solve the equations resulting from (a).

1. If $8000 is deposited in a savings account at $8\frac{1}{4}$% interest compounded annually, find its accumulated value after three years.

interest compounded annually, then, at the end of one year, we will have $1000 plus 12% of $1000, that is, $1120. In this situation (12% interest), we would be better off taking the $1000 now and depositing it at 12% interest, since, at the end of one year, we would have $1120 instead of $1100. On the other hand, if the interest rate is 8%, then our $1000 would earn $(0.08) \times (\$1000) = \$80$ interest. At the end of one year, we would have only $1080. In this case (8% interest), we would be better off taking $1100 at the end of one year.

A little thought shows that if the interest rate i is greater than 10%, then we should take the $1000 immediately. If i is less than 10%, we should take the $1100 at the end of the year. If i equals 10%, then the choices are equivalent.

If the interest rate is 8%, that is, if $i = 0.08$, then $1000 will accumulate to $1080. A deposit of $P will accumulate to $P + \$(0.08)P = \$(1.08)P$. In general, $P deposited at a rate of interest i (compounded annually) will accumulate to $P(1 + i)$ at the end of one year and $P(1 + i)^n$ at the end of n years.

EXAMPLE 1   If $1400 is deposited in a money-market fund at $11\frac{1}{2}$% interest (compounded annually), find its accumulated value after two years.

Solution   Here, $P = 1400$, $i = 0.115$, and $n = 2$. If A denotes the accumulated value, then

$$A = (1400)(1 + 0.115)^2$$
$$= (1400)(1.115)^2$$
$$= \$1740.52. \qquad \diamondsuit$$

DO EXERCISE 1.

To find the *present value* of an amount $A payable n years from today at a rate of interest i, we solve the equation $P(1 + i)^n = A$ for P. In other words, we find the amount P, which, if deposited at a rate of interest i, would accumulate to A in n years. Dividing both sides by $(1 + i)^n$, we have

$$P = \frac{A}{(1 + i)^n},$$

or

$$P = A(1 + i)^{-n}.$$

2. Find the present value of $1100, payable one year from today, at a rate of interest of 12%.

DEFINITION

The *present value* $\$P$ of a payment of $\$A$ to be made n years from today at a rate of interest i (compounded annually) is

$$P = A(1 + i)^{-n}.$$

Unless otherwise indicated, assume in this section that interest is compounded annually.

EXAMPLE 2   Find the present value of $1100, payable one year from today, at a rate of interest of 8%.

Solution   Here, $A = 1100$, $i = 0.08$, and $n = 1$. Thus

$$P = (1100)(1 + 0.08)^{-1} = (1100)(1.08)^{-1}$$
$$= (1100)(0.9259259) \approx \$1018.52.$$

This means that $1018.52, invested at 8%, will yield $1100 after one year.   ❖

DO EXERCISE 2.

If we revisit the problem of the present payment of $1000 versus the payment of $1100 one year from today, we see, in light of Example 2 and Margin Exercise 2, that the present value of the future payment of $1100 one year from today will be less than $1000 if the rate of interest is greater than 10% and more than $1000 if the rate of interest is less than 10%. Although the actual numbers are different, the results of the analysis are identical.

We define the *present value* of a series of future payments as the *sum of the present values of the payments.*

EXAMPLE 3   An annuity will pay $1000 at the end of each year for the next three years. Find the present value of the annuity if the interest rate is 8%.

Solution   The present value of the first payment (one year from today) is $(1000)(1.08)^{-1}$. The present value of the second payment is $(1000)(1.08)^{-2}$, and the present value of the third payment is $(1000)(1.08)^{-3}$. The present value P of the annuity is the sum

$$P = (1000)(1.08)^{-1} + (1000)(1.08)^{-2} + (1000)(1.08)^{-3}$$
$$= 925.93 + 857.34 + 793.83 = \$2577.10.$$ ❖

3. An annuity will pay $250 at the end of each year for the next four years. Find the present value of the annuity if the interest rate is 12%.

DO EXERCISE 3.

In the preceding problems, we were given the dollar amount of a series of future payments and the interest rate and were asked to find the *present value* of the future payments. In practice, we may be given the dollar amount of the future payments and the present value of the future payments and asked to find an interest rate i. In this case, we can call i the *yield rate,* or the *internal rate of return,* of the initial investment.

EXAMPLE 4   An investor makes two investments of $1000 each. The first returns $1050 at the end of one year, while the second returns $1210 at the end of two years. Find the yield rate i.

Solution   Note what happens if we consider the investments separately. The first investment of $1000 returns $1050 at the end of one year. Solving the equation

$$1000 = 1050(1 + i_1)^{-1}$$

for i_1 gives us $i_1 = 5\%$. The second investment of $1000 returns $1210 at the end of two years. Solving

$$1000 = 1210(1 + i_2)^{-2}$$

for i_2 gives us $i_2 = 10\%$. We have two interest rates, each coming as a part of the investment. What we want is one interest rate that represents the entire investment. This is the *yield rate.*

We define the yield rate i as the value of i at which the sum of the present value of the returns equals the sum of the present value of the investments. In this case, the present value of the investments is $2000. The present value of the returns is $(1050)(1 + i)^{-1}$ and $(1210)(1 + i)^{-2}$, respectively. Therefore, the value of i that we are looking for is the solution to

$$2000 = (1050)(1 + i)^{-1} + (1210)(1 + i)^{-2}.$$

To simplify the equation, let

$$v = \frac{1}{1 + i} = (1 + i)^{-1}.$$

Then we get

$$2000 = 1050v + 1210v^2,$$

a quadratic equation in v. Writing it in the form

$$1210v^2 + 1050v - 2000 = 0$$

and applying the quadratic formula gives

$$v = \frac{-1050 \pm \sqrt{(1050)^2 - 4(1210)(-2000)}}{2(1210)}$$

$$= \frac{-1050 \pm \sqrt{1,102,500 + 9,680,000}}{2420}$$

$$= \frac{-1050 \pm 3283.67}{2420}.$$

We get $v_1 \approx 0.923005$ and $v_2 \approx -1.790773$. Note that even if i is the worst case possible (-100%, or $i = -1$, at which point we lose all our money), $1 + i$ is a nonnegative number. In this situation, then, negative values of v are irrelevant and v_2 can be discarded. To complete the problem, observe that since

$$v = \frac{1}{1 + i},$$

we can take reciprocals to get

$$\frac{1}{v} = 1 + i.$$

Then,

$$i = \frac{1}{v} - 1 = \frac{1}{v} - \frac{v}{v}$$

$$= \frac{1 - v}{v}.$$

Here,

$$i \approx \frac{1 - 0.923005}{0.923005} \approx 0.083418.$$

The yield rate of this investment "package" is approximately 8.34%. We now have one interest rate that represents the entire investment.   ❖

DO EXERCISE 4.

EXAMPLE 5   An investor buys an annuity for $2700. The annuity will return $1000 per year for three years, beginning one year from today. Find the yield rate on this investment.

Solution   The present value of the three payments, $2700, is given by

$$2700 = 1000(1 + i)^{-1} + 1000(1 + i)^{-2} + 1000(1 + i)^{-3}.$$

4. An investor invests a total of $4000 in two projects. The first returns $2000 at the end of one year, while the second returns $2500 at the end of two years. Find the investor's yield rate on this package.

Does this yield rate conform to the notion of yield rate introduced in this section?

Letting

$$v = \frac{1}{1 + i} = (1 + i)^{-1}$$

(v is called *the discount factor*), we see that the equation to be solved is

$$2700 = 1000v + 1000v^2 + 1000v^3,$$

or, if we rearrange and divide by 1000,

$$v^3 + v^2 + v - 2.7 = 0.$$

To solve this, we'll use Newton's method. First, though, observe that since i is generally a small positive number, $v = 1/(1 + i)$ will be close to 1 (and usually less than 1). So, $v_1 = 1$ is a good first approximation.

Newton's method yields

$$v_{j+1} = v_j - \frac{f(v_j)}{f'(v_j)}, \qquad j = 1, 2, \ldots .$$

Since $f(v) = v^3 + v^2 + v - 2.7$, differentiation yields $f'(v) = 3v^2 + 2v + 1$. Thus,

$$v_{j+1} = v_j - \frac{v_j^3 + v_j^2 + v_j - 2.7}{3v_j^2 + 2v_j + 1}.$$

If we let $v_1 = 1$, we have

$$v_2 = 1 - \frac{(1)^3 + (1)^2 + 1 - 2.7}{3(1)^2 + 2(1) + 1}$$

$$= 1 - \frac{1 + 1 + 1 - 2.7}{3 + 2 + 1}$$

$$= 1 - \frac{0.3}{6} = 1 - 0.05 = 0.95.$$

Using $v_2 = 0.95$ gives us

$$v_3 = 0.95 - \frac{(0.95)^3 + (0.95)^2 + (0.95) - 2.7}{3(0.95)^2 + 2(0.95) + 1}$$

or

$$v_3 = 0.95 - \frac{0.857375 + 0.9025 + 0.95 - 2.7}{2.7075 + 1.90 + 1}$$

$$= 0.95 - \frac{0.009875}{5.6075}$$

$$= 0.95 - 0.00176,$$

or

$$v_3 \approx 0.948239.$$

Using this value for v yields

$$i = \frac{1 - v}{v} \approx \frac{1 - 0.948239}{0.948239} \approx 0.0545865.$$

Therefore, the yield rate is approximately 5.46% on this annuity.   ❖

DO EXERCISE 5.

It is interesting to compare the results of Examples 3 and 5. The series of payments is identical. Note that as the price (present value) of the annuity increased from $2577.10 to $2700, the yield rate *dropped* from 8% to roughly 5.46%. This illustrates the general principle that as interest rates rise, the prices of annuities (and bonds) fall, while as interest rates fall, the prices of annuities (and bonds) rise.

Note too that in Example 4, an initial investment with return payments after one and two years led to a quadratic equation in v, whereas in Example 5 an investment with return payments in one, two, and three years led to a cubic equation. The general situation of an initial investment followed by return payments at the end of the year for a period of n years will lead to a polynomial equation of degree n. If payments are made at times other than the end of the year, fractional exponents may be required to solve the resulting equation (see Exercise Set 11.3). In any case, a technique such as Newton's method (interpolation in tables is another way) is generally needed.

The Method of Bisection

We revisit the equation $f(x) = x^3 - 3x + 3 = 0$, which we solved using Newton's method (in Margin Exercise 3 of Section 11.2). We note that the solution $\bar{x}$ to $f(x) = 0$ lies between -2 and -3. In fact,

$$f(-3) = (-3)^3 - 3(-3) + 3 = -15$$

and

$$f(-2) = (-2)^3 - 3(-2) + 3 = 1,$$

so that, by continuity, a solution to $f(x) = 0$ must lie between -2 and -3. Now, we bisect the interval from -3 to -2. The midpoint is -2.5. Then

$$f(-2.5) = (-2.5)^3 - 3(-2.5) + 3 = -5.125.$$

5. An investor pays $5000 for three annual payments of $2000 per year beginning one year from today. Find the yield rate on this annuity.

Since $f(-2.5)$ is negative, while $f(-2)$ is positive, there must be a root between -2.5 and -2. (There *may* also be a root, or even several roots, between -3 and -2.5, but the argument *guarantees* one between -2.5 and -2.) We repeat the procedure. Bisecting the interval yields -2.25. We compute

$$f(-2.25) = (-2.25)^3 - 3(-2.25) + 3 = -1.640625,$$

which is negative. Arguing as before, we see that a solution to $f(x) = 0$ must lie between -2.25 and -2. Bisecting again, we arrive at -2.125. Calculating, we now have

$$f(-2.125) = (-2.125)^3 - 3(-2.125) + 3 \approx -0.22070313.$$

A solution lies between -2.125 and -2. Note that had we wanted to stop at this step, we could have used the midpoint of the interval, -2.0625, for $\bar{x}$ with the assurance that the error involved here is less than one-half the length of the interval (in this case, 0.0625).

The method of bisection is simple, but usually much too laborious for hand calculations. It is, however, ideal for high-speed computing machinery since it is easy to program and gives an error estimate at each stage of the calculation.

EXERCISE SET 11.3

APPLICATIONS

❖ **Business and Economics**

Accumulated value. Find the accumulated value of an initial amount P for n years at a rate of interest i (compounded annually).

1. $P = 500$, $n = 7$, $i = 10\%$
2. $P = 500$, $n = 8$, $i = 10\%$
3. $P = 1000$, $n = 30$, $i = 12\%$
4. $P = 1000$, $n = 30$, $i = 10\%$
5. $P = 1000$, $n = 30$, $i = 8\%$
6. $P = 1000$, $n = 30$, $i = 6\%$

Present value. Find the present value of a payment of $\$A$ to be made n years from today at a rate of interest i (compounded annually).

7. $A = \$1000$, $n = 3$, $i = 10\%$
8. $A = \$1000$, $n = 5$, $i = 10\%$
9. $A = \$1000$, $n = 20$, $i = 10\%$
10. $A = \$1000$, $n = 10$, $i = 12\%$
11. $A = \$1000$, $n = 10$, $i = 9\%$
12. $A = \$1000$, $n = 10$, $i = 6\%$
13. $A = \$10,062.66$, $n = 30$, $i = 8\%$
14. $A = \$974.36$, $n = 7$, $i = 10\%$

Annuity. An annuity pays $\$R$ per year for n years beginning one year from today. Find the present value of the annuity if the rate of interest is i (compounded annually).

15. $R = \$1000$, $n = 2$, $i = 12\%$

16. $R = \$1000$, $n = 2$, $i = 9\%$

17. $R = \$1000$, $n = 2$, $i = 6\%$

18. $R = \$1000$, $n = 2$, $i = 19\%$

19. $R = \$200$, $n = 5$, $i = 10\%$

20. $R = \$5000$, $n = 6$, $i = 11\frac{1}{2}\%$

21. *Yield rate.* An investor makes two different investments of $5000, and receives a payment of $6000 after one year on the first and $5800 after two years on the second. Find the equation that determines the yield rate i (in terms of the discount factor $v = (1 + i)^{-1}$) for this pair of investments. Then calculate i.

22. *Yield rate.* An investor makes two investments, one of $3000, the second of $5000. The first returns $3500 after one year, the second $6500 after four years. Find the equation that determines the yield rate and use Newton's method to solve it.

23. *Yield rate.* An investor invests a total of $9000 in two projects, receiving return payments of $4000 after two years and $7000 after five years. Find the internal rate of return on the investment.

24. *Yield rate.* An investor pays $3000 for four annual payments of $1000, beginning one year from today. Find the yield rate of this annuity. Check your answer.

25. *Yield rate.* An investor must choose between two annuities. The first pays $1000 a year for three years and sells for $2400. The second pays $800 a year for five years and sells for $2900. Which should the investor buy?

SYNTHESIS EXERCISES

26. *Yield rate.* An investor makes an initial investment of $2000 in a company, receives a payment of $4300 one year later, and makes another investment of $2310 in the company two years after the initial investment. Set up the equation for the investor's yield rate and show that the equation has two solutions, that is, that the yield rate is not well defined.

27. *Yield rate.* Repeat Exercise 26 given a second investment of $2500 (instead of $2310).

28. *Yield rate.* An investor makes an initial investment of P and receives payment of $R_1, R_2, \ldots, R_k$ dollars after $1, 2, \ldots, k$ years. Find the equation that determines the investor's yield rate and use calculus to show that there is *exactly one* positive number v that solves the equation.

COMPUTER–GRAPHING CALCULATOR EXERCISE

29. Solve the yield rate equation $f(v) = 0$ (for v) in Exercise 21 by sketching the graph of $f(x)$ and finding all positive x such that $f(x) = 0$.

THE CALCULUS EXPLORER:
PowerGrapher

Use the program to do Exercise 29.

11.4

a) Find a numerical estimate for
$\int_a^b f(x)\,dx$ using each of the
following.

 i) **Riemann sums**

 ii) **The Trapezoidal Rule**

 iii) **Simpson's Rule**

NUMERICAL INTEGRATION

In Chapters 5 and 6, we saw that many interesting problems in business, medicine, probability, and the physical sciences lead to the evaluation of definite integrals, that is, expressions of the form

$$\int_a^b f(x)\,dx.$$

In many cases, $f(x)$ has an antiderivative that we can find and evaluate. Thus, for example,

$$\int_0^5 3x^2\,dx = [x^3]_0^5 = 5^3 - 0^3 = 125.$$

In many other cases, though, this won't be so. As we saw in Chapter 6, this happens with the function $f(x) = e^{-(1/2)x^2}$, a function that occurs naturally in many probability applications, but has no elementary antiderivative.

In cases where $f(x)$ has no elementary antiderivative, we may have to evaluate $\int_a^b f(x)\,dx$ approximately. In this section, we will consider several approximation schemes for estimating integrals.

Let $f(x)$ be continuous on $[a, b]$ and consider $\int_a^b f(x)\,dx$. If we subdivide $[a, b]$ into n equal pieces (see Fig. 1) of length Δx, then $\Delta x = (b - a)/n$. Let $a = x_1$ and choose $x_2, x_3, \ldots, x_n, x_{n+1}$ so that $x_{i+1} - x_i = \Delta x$ for $i = 1, 2, \ldots, n$, that is, so that the points are equally spaced. Just as in Section 5.8, we see that the area of the first rectangle is $f(x_1) \cdot \Delta x$, and so forth, so that the sum of the areas of the n inscribed

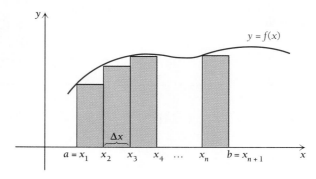

FIGURE 1

rectangles indicated in Fig. 1 is

$$\sum_{i=1}^{n} f(x_i)\, \Delta x.$$

In Section 5.8, we saw that

$$\int_{a}^{b} f(x)\, dx = \lim_{n \to \infty} \sum_{i=1}^{n} f(x_i)\, \Delta x.$$

For large values of n, then, we expect that $\sum_{i=1}^{n} f(x_i)\, \Delta x$ can yield a good approximation to $\int_{a}^{b} f(x)\, dx$.

EXAMPLE 1   Approximate $\int_{0}^{1}(1 + x_2)\, dx$ using $\sum_{i=1}^{n} f(x_i)\, \Delta x$ with $n = 5$ and compare your approximation with the exact answer.

Solution   The integral represents the area shown in Fig. 2.

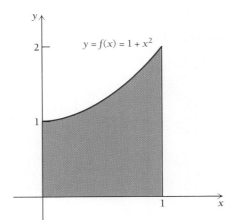

FIGURE 2

We can find its exact value as follows:

$$\int_{0}^{1}(1 + x^2)\, dx = \left[x + \frac{1}{3}x^3 \right]_{0}^{1}$$

$$= \left(1 + \frac{1}{3}(1)^3 \right) - \left(0 + \frac{1}{3}(0)^3 \right)$$

$$= 1 + \frac{1}{3} = \frac{4}{3} \approx 1.333333.$$

If $n = 5$, then $\Delta x = (1 - 0)/5 = \frac{1}{5} = 0.2$. The points of subdivision are

1. Approximate $\int_0^1 (1 + x)\, dx$ using $\sum_{i=1}^{n} f(x_i)\, \Delta x$ with $n = 4$, and compare your approximation with the exact answer.

$x_1 = 0$, $x_2 = 0.2$, $x_3 = 0.4$, $x_4 = 0.6$, $x_5 = 0.8$, and $x_6 = 1$. Since $f(x) = 1 + x^2$, we have

$$f(x_1) = f(0) = 1 + (0)^2 = 1,$$
$$f(x_2) = f(0.2) = 1 + (0.2)^2 = 1.04,$$
$$f(x_3) = f(0.4) = 1 + (0.4)^2 = 1.16,$$
$$f(x_4) = f(0.6) = 1 + (0.6)^2 = 1.36,$$
$$f(x_5) = f(0.8) = 1 + (0.8)^2 = 1.64.$$

Therefore,

$$\sum_{i=1}^{5} f(x_i)\, \Delta x = f(x_1)\, \Delta x + f(x_2)\, \Delta x + \cdots + f(x_5)\, \Delta x$$
$$= [f(x_1) + f(x_2) + \cdots + f(x_5)]\, \Delta x$$
$$= [1 + (1.04) + (1.16) + (1.36) + (1.64)] \cdot \tfrac{1}{5}$$
$$= [6.2](0.2)$$
$$= 1.24.$$

This sum equals the sum of the areas of the rectangles shown in Fig. 3.

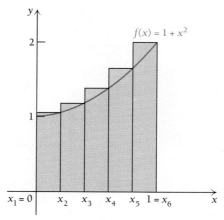

FIGURE 3

The error involved in our approximation is about $(1.333333) - (1.24) \approx 0.09$. ❖

DO EXERCISE 1.

In Example 1, we used the values of $f(x)$ at the left-hand endpoints of the subintervals, $f(x_1), f(x_2), \ldots, f(x_5)$, for the altitudes of the approximating rectangles. If, instead, we had used the values of $f(x)$ at the right-hand endpoints of the subintervals, $f(x_2), f(x_3), \ldots, f(x_6)$, we would have obtained the rectangles shown in Fig. 4.

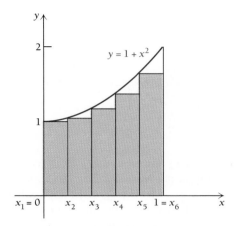

$x_1 = 0 \quad x_2 \quad x_3 \quad x_4 \quad x_5 \quad 1 = x_6$

$y = 1 + x^2$

FIGURE 4

EXAMPLE 2 Approximate $\int_0^1 (1 + x^2) \, dx$ by using $\sum_{i=1}^{n} f(x_{i+1}) \, \Delta x$ with $n = 5$, as shown above. Compare your result with the exact answer.

Solution We have

$$\sum_{i=1}^{5} f(x_{i+1}) \, \Delta x = [f(x_2) + f(x_3) + \cdots + f(x_6)] \, \Delta x$$
$$= [(1.04) + (1.16) + (1.36) + (1.64) + 2] \cdot \tfrac{1}{5}$$
$$= [7.2](0.2) = 1.44.$$

The error involved is about $(1.44) - (1.333333) \approx 0.106667.$ ❖

DO EXERCISE 2.

In each of Examples 1 and 2, we approximated $\int_0^1 (1 + x^2) \, dx$ by using the sum of areas of rectangles. In Example 1, we used the values of $f(x)$ at the left-hand endpoints for the altitudes of the rectangles, whereas in Example 2 we used the right-hand endpoints. Both cases are examples of *approximation by Riemann sums*. In the general case, we pick any number c_1 in $[x_1, x_2]$ and use $f(c_1)$ for the altitude of the first rectangle (see Fig. 5).

2. Approximate $\int_0^1 (1 + x) \, dx$ using $\sum_{i=1}^{n} f(x_{i+1}) \, \Delta x$ with $n = 4$, and compare your approximation with the exact answer.

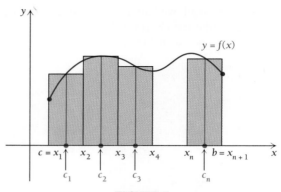

FIGURE 5

We pick c_2 between x_2 and x_3 and use $f(c_2)\,\Delta x$ for the area of the second rectangle. Continuing, we pick $c_3, c_4, \ldots, c_n$ so that c_i is between x_i and x_{i+1} for $i = 1, 2, \ldots, n$. Then we can approximate $\int_a^b f(x)\,dx$ by the sum $\sum_{i=1}^n f(c_i)\,\Delta x$. This sum is the sum of the areas of the rectangles shown in Fig. 5. Any sum of the form

$$\sum_{i=1}^n f(c_1)\,\Delta x$$

is called a *Riemann sum for* $\int_a^b f(x)\,dx$.

The sums $\sum_{i=1}^n f(x_i)\,\Delta x$ (using the left-hand endpoints of the subintervals for c_i) and $\sum_{i=1}^n f(x_{i+1})\,\Delta x$ (using the right-hand endpoints) are both special cases of Riemann sums. We might use the midpoints of the subintervals or any other convenient choice of $c_1, c_2, \ldots, c_n$.

In general, though, Riemann sums are not an efficient way to approximate integrals. For example, suppose that we were to approximate $\int_0^1 (1 + x^2)\,dx$ using left-hand endpoints and $n = 100$ (think of the computations required!). It turns out that we would obtain

$$\sum_{i=1}^{100} f(x_i)\,\Delta x = 1.32835.$$

Note that even with 100 subintervals, the error (which is about $1.333333 - 1.32835 \approx 0.005$) is rather large.

The Trapezoidal Rule

There is an easy and obvious way of improving on approximation by Riemann sums. As before, we partition $[a, b]$ into n equal subintervals. Instead of using rectangles, however, we use trapezoids, as shown in Fig. 6.

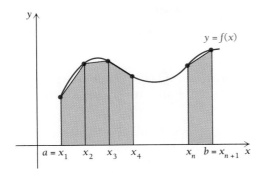

FIGURE 6

Note that the sum of the areas of the trapezoids in the figure, although not equal to $\int_a^b f(x)\ dx$, certainly appears to yield a better approximation than the inscribed rectangles would.

Note that the area A of a trapezoid whose base is b and whose altitudes are h_1 and h_2 (see the figure) is

$$\left(\begin{array}{c}\text{Area of}\\ \text{the rectangle}\end{array}\right) + \left(\begin{array}{c}\text{Area of}\\ \text{the triangle}\end{array}\right) = bh_1 + \frac{1}{2}\,b(h_2 - h_1).$$

Then

$$\begin{aligned}
A &= bh_1 + \tfrac{1}{2}b(h_2 - h_1)\\
&= b[h_1 + \tfrac{1}{2}(h_2 - h_1)]\\
&= b[h_1 + \tfrac{1}{2}h_2 - \tfrac{1}{2}h_1]\\
&= b[\tfrac{1}{2}h_1 + \tfrac{1}{2}h_2]\\
&= \tfrac{1}{2}b[h_1 + h_2].
\end{aligned}$$

Applying these calculations to the situation shown in Fig. 6, we see that each of the trapezoids will have base equal to Δx. The first trapezoid has altitudes $f(x_1)$ and $f(x_2)$, the values of $f(x)$ at the right- and left-hand endpoints of its base. Thus the area of the first trapezoid will be $\frac{1}{2}(f(x_1) + f(x_2)) \cdot \Delta x$. The area of the second trapezoid will be $\frac{1}{2}(f(x_2) + f(x_3)) \cdot \Delta x$. The area of the last ($n$th) trapezoid will be $\frac{1}{2}(f(x_n) + f(x_{n+1})) \cdot \Delta x$. Let T_n denote the sum of the areas of all n trapezoids. Then,

$$\begin{aligned}
T_n &= \tfrac{1}{2}(f(x_1) + f(x_2)) \cdot \Delta x + \tfrac{1}{2}(f(x_2) + f(x_3)) \cdot \Delta x\\
&\quad + \cdots + \tfrac{1}{2}(f(x_n) + f(x_{n+1})) \cdot \Delta x\\
&= \tfrac{1}{2}[(f(x_1) + f(x_2)) + (f(x_2) + f(x_3))\\
&\quad + \cdots + (f(x_n) + f(x_{n+1}))] \cdot \Delta x\\
&= \tfrac{1}{2}[f(x_1) + 2f(x_2) + 2f(x_3) + \cdots + 2(f(x_n) + f(x_{n+1}))] \cdot \Delta x\\
&= [\tfrac{1}{2}f(a) + f(x_2) + f(x_3) + \cdots + f(x_n) + \tfrac{1}{2}f(b)] \cdot \Delta x.
\end{aligned}$$

3. Approximate $\int_0^1 (1 + x) \, dx$ using the Trapezoidal Rule with $n = 4$, that is, by T_4. Compare your result with the exact answer.

Note that the endpoints a and b of the interval receive "special treatment" in the formula since they occur only once as endpoints of subintervals, whereas the other points occur twice.

EXAMPLE 3 Approximate $\int_0^1 (1 + x^2) \, dx$ using the Trapezoidal Rule with $n = 5$, that is, T_5. Compare your answer with the exact value of the integral.

Solution As in the solution to Example 1, we have $x_1 = 0$, $x_2 = 0.2$, $x_3 = 0.4$, $x_4 = 0.6$, $x_5 = 0.8$, and $x_6 = 1$. Also,

$$f(x_1) = 1 + (x_1)^2 = 1 + (0)^2 \quad = 1.00,$$
$$f(x_2) = 1 + (x_2)^2 = 1 + (0.2)^2 = 1.04,$$
$$f(x_3) = 1 + (x_3)^2 = 1 + (0.4)^2 = 1.16,$$
$$f(x_4) = 1 + (x_4)^2 = 1 + (0.6)^2 = 1.36,$$
$$f(x_5) = 1 + (x_5)^2 = 1 + (0.8)^2 = 1.64,$$

and

$$f(x_6) = 1 + (x_6)^2 = 1 + (1)^2 = 2.00.$$

Thus,

$$
\begin{aligned}
T_5 &= [\tfrac{1}{2} f(x_1) + f(x_2) + \cdots + f(x_5) + \tfrac{1}{2} f(x_6)] \cdot \Delta x \\
&= [\tfrac{1}{2}(1) + (1.04) + (1.16) + (1.36) + (1.64) + \tfrac{1}{2}(2)] \cdot \tfrac{1}{5} \\
&= [6.70] (0.2) \\
&= 1.34.
\end{aligned}
$$

The error involved is about $(1.34) - (1.333333) = 0.006667$. Note that this compares with the error we get with $n = 100$ using the rectangle method (Riemann sums) with left-hand endpoints. ❖

DO EXERCISE 3.

In Margin Exercise 3, the Trapezoidal Rule is exact. This is always the case if $f(x) = c_0 + c_1 x$ is a polynomial of degree 1. In fact, we have the following theorem.

THEOREM 3

The Trapezoidal Rule

If $f(x)$ and its first and second derivatives are continuous on $[a, b]$, and if

$$T_n = [\tfrac{1}{2} f(x_1) + f(x_2) + \cdots + f(x_n) + \tfrac{1}{2} f(x_{n+1})] \cdot \Delta x,$$

then

$$\int_a^b f(x)\, dx \approx T_n$$

and the error involved in this approximation is

$$E = \frac{b-a}{12} f''(c)\, (\Delta x)^2,$$

where $a < c < b$.

In particular, if $f(x) = c_0 + c_1 x$, then $f''(x) = 0$ and E must be 0.

If we let $K = [(b-a)/12]f''(c)$ and observe that K is a constant, we see that the error is $K(\Delta x)^2$. The error is proportional to $1/n^2$.

EXAMPLE 4 Use the error estimate in Theorem 3 to calculate the error in approximating $\int_0^1 (1 + x^2)\, dx$ by T_5 in Example 3.

Solution We have $a = 0, b = 1, \Delta x = \frac{1}{5}$, and $f(x) = 1 + x^2$. So $f'(x) = 2x$ and $f''(x) = 2$. Thus $f''(c) = 2$, independent of c. The error E is then given by

$$\frac{b-a}{12} f''(c)\, (\Delta x)^2 = \frac{1-0}{12} \cdot 2 \cdot \left(\frac{1}{5}\right)^2$$

$$= \frac{1}{12} \cdot 2 \cdot \frac{1}{25} = \frac{1}{150} \approx 0.006667.$$

From Example 3, we know that this is correct.   ❖

DO EXERCISE 4.

Margin Exercise 3 and Example 3 suggest that the Trapezoidal Rule is simply the average of the Riemann sums corresponding to the right- and left-hand endpoints, and that is correct.

EXAMPLE 5 Use T_{10} to approximate $\int_1^2 (1/x)\, dx$. Estimate the error involved.

Solution The points of subdivision are $x_1 = 1, x_2 = 1.1, x_3 = 1.2, \ldots,$ $x_{10} = 1.9$, and $x_{11} = 2.0$. Also, $\Delta x = 0.1$. Applying the Trapezoidal Rule gives us

$$T_{10} = \left[\frac{1}{2} \cdot \frac{1}{1} + \frac{1}{1.1} + \frac{1}{1.2} + \cdots + \frac{1}{1.9} + \frac{1}{2} \cdot \frac{1}{2.0}\right](0.1)$$

$$= [6.9377140](0.1)$$

$$= 0.69377140.$$

4. Use the error estimate in Theorem 3 to calculate the error in approximating $\int_0^1 (1 + x)\, dx$ by T_4 in Margin Exercise 3.

5. Use T_{10} to approximate $\int_0^1 x^3 \, dx$. Estimate the error involved.

According to Theorem 3, the error, call it E, is given by

$$E = \frac{b - a}{12} \cdot f''(c) \cdot (\Delta x)^2,$$

where $a < c < b$. Here, $a = 1$, $b = 2$, and $\Delta x = 0.1$. Since $f(x) = 1/x = x^{-1}$, we have $f'(x) = -1x^{-2}$ and $f''(x) = (-1)(-2)x^{-3} = 2/x^3$. So we get

$$E = \frac{2 - 1}{12} \cdot \frac{2}{c^3} \cdot (0.1)^2 = \frac{1}{12} \cdot \frac{2}{c^3} \cdot (0.01) = \frac{1}{600} \cdot \frac{1}{c^3},$$

where $1 < c < 2$. Since $1 < c < 2$, we have $1 < c^3 < 8$ and $\frac{1}{8} < 1/c^3 < 1$. Thus, $E < \frac{1}{600} \approx 0.0016667$. In fact,

$$\int_1^2 \frac{1}{x} \, dx = [\ln x]_1^2 = \ln 2 - \ln 1 = \ln 2.$$

So, to seven decimal places,

$$\int_1^2 \frac{1}{x} \, dx = \ln 2$$

$$\approx 0.6931472.$$

The actual error is about 0.0006242, which, of course, is less than 0.00166667, the error estimate obtained above. ❖

DO EXERCISE 5.

Simpson's Rule

In approximating $\int_a^b f(x) \, dx$ by Riemann sums, we partitioned $[a, b]$ into n equal subintervals and used $f(c_i)$ to approximate $f(x)$ on the ith subinterval. In approximating the same integral using the Trapezoidal Rule, we approximated $f(x)$ on the ith subinterval by means of the line segment joining $(x_i, f(x_i))$ to $(x_{i+1}, f(x_{i+1}))$. This line segment is the top of the ith trapezoid in Fig. 6.

To obtain Simpson's Rule, we partition $[a, b]$ into n equal subintervals and fit a parabola to $f(x)$ on each subinterval. Since it takes three points to uniquely determine a parabola, we use $(x_i, f(x_i))$, $(x_{i+1}, f(x_{i+1}))$ and the point on $y = f(x)$ corresponding to the midpoint of the subinterval,

$$\left(\frac{x_i + x_{i+1}}{2}, f\left(\frac{x_i + x_{i+1}}{2} \right) \right),$$

as shown in Fig. 7. (The parabolas here are P_1, P_2, and P_3.)

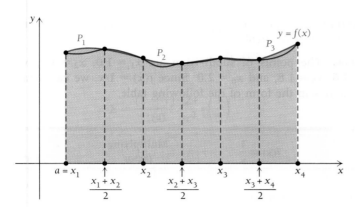

FIGURE 7

Calculating the area under each parabolic arch, summing the areas, and simplifying yields

$$S_n = \frac{1}{6}\left[f(x_1) + 4f\left(\frac{x_1 + x_2}{2}\right) + 2f(x_2) + 4f\left(\frac{x_2 + x_3}{2}\right) + 2f(x_3)\right.$$

$$\left. + \cdots + 2f(x_n) + 4f\left(\frac{x_n + x_{n+1}}{2}\right) + f(x_{n+1})\right] \cdot \Delta x.$$

We reiterate this result and the error estimate in the following theorem.

THEOREM 4

Simpson's Rule

If $f(x)$ and its first four derivatives are continuous on $[a, b]$, and if

$$S_n = \frac{1}{6}\left[f(x_1) + 4f\left(\frac{x_1 + x_2}{2}\right) + 2f(x_2)\right.$$

$$\left. + \cdots + 4f\left(\frac{x_n + x_{n+1}}{2}\right) + f(x_{n+1})\right] \cdot \Delta x,$$

then

$$\int_a^b f(x)\, dx \approx S_n,$$

and the error involved is

$$\frac{(b - a)}{180} \cdot f^{(4)}(c) \cdot (\Delta x)^4,$$

where $a < c < b$.

$$x_1 = 0.5,$$
$$x_2 = \cos(x_1) = \cos(0.5) \approx 0.8775826,$$

794   NUMERICAL TECHNIQUES

1. Using $x_1 = 4$ and $x_{n+1} = \sqrt{5x_n - 6}$, compute x_2 and x_3 by iteration and x_4 by Aitken acceleration. Then compute x_5 and x_6 (from x_4) by iteration and use them to compute x_7 by Aitken's method. Compare your results with those of Example 4 in Section 11.1.

and

$$x_3 = \cos{(x_2)} \approx 0.63901249.$$

Applying Aitken acceleration, we obtain a new estimate, call it x_4 (this is *not* the same x_4 as in Example 1 of Section 11.1), as follows:

$$x_4 = \frac{x_1 x_3 - (x_2)^2}{x_3 - 2x_2 + x_1}$$

$$\approx \frac{(0.5)(0.63901249) - (0.8775826)^2}{(0.63901249) - 2(0.8775826) + 0.5}$$

$$\approx 0.7313852.$$

Of course, we don't have to stop here. Using x_4 and iteration, we get

$$x_5 = \cos{(x_4)} \approx 0.7442499$$

and

$$x_6 = \cos{(x_5)} \approx 0.7355962.$$

Using Aitken acceleration to generate x_7, we get

$$x_7 = \frac{x_4 x_6 - (x_5)^2}{x_6 - 2x_5 + x_4}$$

$$\approx \frac{(0.7313852)(0.7355962) - (0.7442499)^2}{(0.7355962) - 2(0.7442499) + (0.7313852)}$$

$$\approx 0.7390763.$$

Compare this with the results of Example 1 of Section 11.1.   ❖

DO EXERCISE 1.

We obtained Aitken's acceleration formula above because we were able to estimate the error in successive applications of the method of iteration. The technique used above, called *extrapolation*, can be used in numerical integration, too.

Suppose that we would like to evaluate the integral $I = \int_a^b f(x)\ dx$ numerically. We might fix n, partition $[a, b]$ into n subintervals of equal length, and use the Trapezoidal Rule to obtain an estimate T_n for I. According to Theorem 3, the error here, $|I - T_n|$, is less than $K(\Delta x)^2$, where K is a constant and $\Delta x = (b - a)/n$ is the length of each subinterval. The key here is going to be this error estimate. To see what happens, suppose, for a moment, that we repeat the process above using twice as many points. We obtain a new estimate, T_{2n}, for I. How much better is our new estimate? Since we doubled the number of subintervals, it follows that the length of each of the new subintervals is exactly one-half the length of those used the first time. In other words, we simply replaced Δx by $\Delta x/2$. Our new error estimate will look like

$$K\left(\frac{\Delta x}{2}\right)^2 = K\frac{(\Delta x)^2}{4} = \frac{1}{4}K(\Delta x)^2.$$

This suggests that doubling the number of points used should improve the accuracy by a factor of 4. But it gives us a lot more than that.

If we pretend for a moment that our error estimates are exact, then we get

$$\tfrac{1}{4}(I - T_n) = I - T_{2n},$$

or

$$I - T_n = 4(I - T_{2n}).$$

Thus,

$$I - T_n = 4I - 4T_{2n}.$$

Solving for I, we get

$$3I = 4T_{2n} - T_n,$$

or

$$I = \frac{4T_{2n} - T_n}{3}.$$

Of course, the error estimates are only approximations, so that we write

$$I \approx \frac{4T_{2n} - T_n}{3}.$$

EXAMPLE 2   Find an approximate value for $\int_0^1 x^4 \, dx$.

Solution   We know that the exact value of the integral is $\frac{1}{5}$. Using the Trapezoidal Rule with $n = 4$ and $f(x) = x^4$, we have

$$\begin{aligned}
T_4 &= [\tfrac{1}{2}f(0) + f(\tfrac{1}{4}) + f(\tfrac{1}{2}) + f(\tfrac{3}{4}) + \tfrac{1}{2}f(1)] \cdot \tfrac{1}{4} \\
&= [\tfrac{1}{2}(0)^4 + (\tfrac{1}{4})^4 + (\tfrac{1}{2})^4 + (\tfrac{3}{4})^4 + \tfrac{1}{2}(1)^4] \cdot \tfrac{1}{4} \\
&= [0 + \tfrac{1}{256} + \tfrac{1}{16} + \tfrac{81}{256} + \tfrac{1}{2}] \cdot \tfrac{1}{4} \\
&= [0.8828125] \cdot \tfrac{1}{4} = 0.220703125.
\end{aligned}$$

Similarly,

$$\begin{aligned}
T_8 &= [\tfrac{1}{2}(0)^4 + (\tfrac{1}{8})^4 + (\tfrac{2}{8})^4 + \cdots + (\tfrac{7}{8})^4 + \tfrac{1}{2}(1)^4] \cdot \tfrac{1}{8} \\
&\approx [1.6416016] \cdot \tfrac{1}{8} \approx 0.2052002.
\end{aligned}$$

2. Approximate $\int_1^2 (1/x)\, dx$ using T_5. Use T_5 and the value of T_{10} from Example 5 of Section 11.4 to compute a new approximation to the integral.

Using the formula above, we get

$$I \approx \frac{4T_8 - T_4}{3}$$

$$= \frac{4(0.2052002) - (0.220703125)}{3}$$

$$= 0.2000325521.$$

We see that this is an excellent approximation. ❖

DO EXERCISE 2.

Compare the answer to Margin Exercise 2 to that in Example 6 of Section 11.4. The answers are the same. This is no coincidence. The answer that we have obtained from T_n and T_{2n} by extrapolation (or, as it is usually called, *Romberg integration*) is S_n. In other words,

$$S_n = \frac{4T_{2n} - T_n}{3}.$$

This will be proved in Exercise Set 11.5. In the meantime, note that we can use our earlier argument, along with the error estimate from Theorem 4, to obtain even better numerical estimates.

If $I = \int_a^b f(x)\, dx$ and S_n is the Simpson's Rule approximation to I, then the absolute error $|I - S_n|$ is less than $M(\Delta x)^4$, where M is a constant and Δx is the length of the subintervals used. Here, if we double the number of points used, then we halve the length of each subinterval and, replacing Δx by $\Delta x/2$ in our error estimate, we get

$$M\left(\frac{\Delta x}{2}\right)^4 = M \cdot \frac{(\Delta x)^4}{16}.$$

Doubling the number of points should improve the accuracy by a factor of 16. Thus,

$$I - S_{2n} \approx \tfrac{1}{16}(I - S_n)$$

or

$$16I - 16S_{2n} \approx I - S_n$$

or

$$15I \approx 16S_{2n} - S_n.$$

Solving, we get

$$I \approx \frac{16S_{2n} - S_n}{15}.$$

Define the Romberg sum R_n by

$$R_n = \frac{16S_{2n} - S_n}{15}$$

for all positive integers n.

EXAMPLE 3   Estimate

$$\int_0^1 \frac{x}{1 + x^2}\, dx$$

by S_2, S_4, and the Romberg sum

$$R_2 = \frac{16S_4 - S_2}{15}.$$

Solution   Here,

$$f(x) = \frac{x}{1 + x^2}.$$

Thus,

$$
\begin{aligned}
S_2 &= \tfrac{1}{6} \cdot [f(0) + 4f(\tfrac{1}{4}) + 2f(\tfrac{1}{2}) + 4f(\tfrac{3}{4}) + f(1)] \cdot \Delta x \\
&= \tfrac{1}{6} \cdot [0 + 4(\tfrac{4}{17}) + 2(\tfrac{2}{5}) + 4(\tfrac{12}{25}) + \tfrac{1}{2}] \cdot \tfrac{1}{2} \\
&\approx \tfrac{1}{6} \cdot [4.161176] \cdot \tfrac{1}{2} \\
&\approx 0.3467647.
\end{aligned}
$$

Similarly,

$$
\begin{aligned}
S_4 &= \tfrac{1}{6}[f(0) + 4f(\tfrac{1}{8}) + 2f(\tfrac{1}{4}) + \cdots + 4f(\tfrac{7}{8}) + f(1)] \cdot \Delta x \\
&= \tfrac{1}{6} \cdot [0 + 4 \cdot \tfrac{8}{65} + 2 \cdot \tfrac{4}{17} + \cdots + 4 \cdot \tfrac{56}{113} + \tfrac{1}{2}] \cdot \tfrac{1}{4} \\
&\approx \tfrac{1}{24} \cdot [8.3180181] \\
&\approx 0.3465841.
\end{aligned}
$$

Now,

$$
\begin{aligned}
R_2 &= \frac{16S_4 - S_2}{15} \\
&\approx \frac{16(0.3465841) - (0.3467647)}{15} \\
&\approx \frac{5.1985809}{15} \\
&\approx 0.34657206.
\end{aligned}
$$

3. Estimate

$$\int_0^1 \frac{1}{1+x}\, dx$$

using S_2, S_4, and the Romberg sum

$$R_2 = \frac{16 S_4 - S_2}{15}.$$

To get an idea of the accuracy, we write

$$I = \int_0^1 \frac{x}{1+x_2}\, dx$$

$$= \left[\frac{1}{2} \ln\,(1+x^2) \right]_0^1$$

$$= \frac{1}{2} \ln\, 2.$$

So, using a calculator, we get $I = \frac{1}{2} \ln\, 2 \approx 0.34657359$.   ❖

DO EXERCISE 3.

EXERCISE SET 11.5

Use the initial guess x_1 and iteration to compute x_2 and x_3; then calculate x_4 using Aitken's method.

1. $x = \frac{1}{3}x^2 + \frac{1}{3}$; $x_1 = 0$

2. $x = \frac{1}{3}x^2 + \frac{1}{3}$; $x_1 = \frac{1}{2}$

3. $x = \sin x + \frac{1}{2}$ (x in radians); $x_1 = 1$

4. $x = \frac{1}{2}\cos x$; $x_1 = 1$

5. $x = \sqrt[3]{3x - 3}$; $x_1 = -1$

6. $x = \sqrt{3x - 1}$; $x_1 = 2$

Calculate T_2, T_4, and T_8 for each of the following integrals. Then use Romberg integration to calculate S_2 and S_4. Finally, use Romberg integration to calculate R_2 from S_2 and S_4.

7. $\int_0^1 x^2\, dx$

8. $\int_0^1 x^4\, dx$

9. $\int_0^1 \frac{1}{1+x^2}\, dx$

10. $\int_0^1 \frac{1}{1+x^6}\, dx$

SYNTHESIS EXERCISES

11. Let

$$G = \frac{x_n x_{n+2} - (x_{n+1})^2}{x_{n+2} - 2x_{n+1} + x_n}$$

for a fixed positive integer n. Also, let

$$x_n = \bar{x} + K^{n-1} e_1, \tag{1}$$

$$x_{n+1} = \bar{x} + K^n e_1, \tag{2}$$

and

$$x_{n+2} = \bar{x} + K^{n+1} e_1, \tag{3}$$

where K, $\bar{x}$, and e_1 are constants. Substitute Eqs. (1), (2), and (3) into the expression for G, expand, and simplify to obtain $G = \bar{x}$.

12. Let $f(x)$ be a continuous function on $[a, b]$, and let T_n and S_n be as in Sections 11.4 and 11.5. Show that

$$\frac{4T_{2n} - T_n}{3} = S_n \quad \text{for all } n.$$

COMPUTER–GRAPHING CALCULATOR EXERCISES

THE CALCULUS EXPLORER:
Integral Evaluation

13. The exact value of the integral in Exercise 9 is $\pi/4$. Use Simpson's Rule, the Trapezoidal Rule, and Romberg integration with $n = 8$ to approximate the integral. Which method gives the most accurate estimate?

11.6

EULER'S METHOD

In Chapter 9, we saw that many differential equations can be solved directly using various methods that depend on the form of the equations. Unfortunately, many applications give rise to differential equations that cannot be solved directly. In these cases, it is necessary to have methods for generating accurate, but approximate, solutions to the given equation.

We first need to introduce some notation. In Example 3 of Section 9.2, we solved the differential equation

$$y' = x - xy.$$

Note that this equation involves variables x and y, with y being a function of x. It can then be thought of as a function of two variables and can be written

$$y' = x - xy = f(x, y).$$

The solution of the equation is

$$y = 1 + C_1 e^{-x^2/2}.$$

Now we know that y is a function of x. Though we have not done so in the past, we can express this as $y = y(x)$. With the initial condition $y(0) = 2$, the solution can be written as

$$y = 1 + e^{-x^2/2}.$$

OBJECTIVES

a) Find an approximate solution to a first-order differential equation with initial condition using either Euler's method or the three-term method.

b) Find an approximate solution to a system of two first-order differential equations with initial conditions using Euler's method.

1. a) By separating the variables, solve $y' = -2xy$, $y(0) = 1$ explicitly.

b) Using Euler's method with $n = 5$ on $[0, 1]$, find an approximate solution to $y' = -2xy$, $y(0) = 1$.

c) Repeat part (b) with a step size of 0.1.

Next,

$$y_2 = y_1 + f(x_1, y_1)\, \Delta x$$
$$= 2.4 + y_1 \cdot (0.2)$$
$$= 2.4 + (2.4) \cdot (0.2)$$
$$= 2.88.$$

Continuing, we have

$$y_3 = 2.88 + (2.88)(0.2)$$
$$= 3.456,$$
$$y_4 = 3.456 + (3.456)(0.2)$$
$$= 4.1472,$$

and

$$y_5 = 4.1472 + (4.1472)(0.2)$$
$$= 4.97664.$$

If these points were plotted against the graph of the exact solution $y = 2e^x = g(x)$, then we would find that on the left-hand side of the interval, the approximations aren't so bad, but when we reach $x_5 = 1$, the approximate value, $y_5 = 4.97664$, bears little resemblance to the actual value,

$$g(1) = 2e^1 = 2e \approx 5.4365637.$$

The great defect in Euler's method (sometimes called the *tangent-line method*) is that it isn't very accurate. ❖

DO EXERCISE 1.

There are many ways to alter Euler's method so as to improve its accuracy. To obtain one such improvement, return to Fig. 1. There we obtained our initial approximations by use of the tangent line T_1 to the graph of the exact solution $y = g(x)$. To obtain better accuracy, we could use the "tangent quadratic" to the graph of $y = g(x)$. Suppose for the moment that $y = g(x)$ is represented by its Taylor series at $x = x_0$. Then,

$$g(x) = g(x_0) + g'(x_0)(x - x_0) + \frac{g''(x_0)}{2!}(x - x_0)^2 + \cdots.$$

We want to use this formula to compute $y_1 = g(x_1)$. Setting $x = x_1$ and ignoring all terms after the third one, we obtain

$$y_1 = g(x_1) \approx g(x_0) + g'(x_0)(x_1 - x_0) + \frac{g''(x_0)}{2!}(x_1 - x_0)^2.$$

Now $x_1 - x_0 = \Delta x$, the step size. So,

$$y_1 \approx g(x_0) + g'(x_0) \cdot \Delta x + \frac{g''(x_0)}{2!} (\Delta x)^2.$$

But $g(x_0) = y_0$ and $g'(x_0)$ is just $y'(x_0) = f(x_0, y_0)$, which is given. Substituting, we get

$$y_1 \approx y_0 + f(x_0, y_0) \Delta x + \tfrac{1}{2}g''(x_0) (\Delta x)^2.$$

Every quantity on the right-hand side above is given except Δx, the step size, which we are free to choose, and $g''(x_0)$. The last quantity may or may not be easy to compute. We illustrate with two examples.

EXAMPLE 2   Compute $g''(x_0) = y''(x_0)$ for the initial-value problem

$$y' = xy^2,$$
$$y(1) = 2.$$

Solution   In this problem, $f(x, y) = xy^2$, $x_0 = 1$, and $y_0 = 2$. Thus,

$$y'(x_0) = f(x_0, y_0) = f(1, 2) = 1 \cdot (2)^2 = 4.$$

To get y'', we must differentiate $y' = g(x, y) = xy^2$ with respect to x. Using the Product Rule, we obtain

$$y'' = x(y^2)' + y^2(x)'$$
$$= x(y^2)' + y^2,$$

and, applying the Chain Rule, we get

$$y'' = x(2yy') + y^2.$$

So $y'' = 2xyy' + y^2$. Replacing y' by xy^2 gives

$$y'' = 2xy(xy^2) + y^2,$$

or

$$y'' = 2x^2y^3 + y^2.$$

Even a very simple choice of $f(x, y)$ results in a fairly complicated expression for y''.   ❖

DO EXERCISE 2.

When y'' is relatively simple, however, this method (sometimes called the *three-term method*) yields superior results. The general iteration formula will be as follows:

$$y_{n+1} = y_n + f(x_n, y_n) \cdot \Delta x + \tfrac{1}{2}y''(x_n, y_n) (\Delta x)^2.$$

2. Compute y'' for the initial-value problem $y' = y \sin(xy)$, $y(1) = 1$.

3. Use the three-term method to find an approximate solution to $y' = -2xy$, $y(0) = 1$, on the interval $[0, 1]$ using a step size of 0.2.

EXAMPLE 3 Apply the three-term method to $y' = y$, $y(0) = 2$, on the interval $[0, 1]$ with $n = 5$.

Solution This is the problem from Example 1 given earlier. To apply the method, we must compute y''. Here, since $y' = y$, we get $y'' = y'$ by differentiation. Combining the two equations, we have $y'' = y$. Since $f(x, y) = y$ and $y'' = g'' = y$, the iteration formula becomes

$$y_{n+1} = y_n + y_n \cdot (\Delta x) + \tfrac{1}{2} y_n \cdot (\Delta x)^2,$$

or

$$y_{n+1} = y_n(1 + \Delta x + \tfrac{1}{2}(\Delta x)^2).$$

Here, $\Delta x = 0.2$. So,

$$y_{n+1} = y_n(1 + (0.2) + \tfrac{1}{2}(0.2)^2)$$
$$= y_n(1 + 0.2 + 0.02)$$
$$= 1.22 y_n.$$

Using $y(x_0) = y(0) = 2$, we find that

$$y_1 = y(x_1) = 1.22 y_0$$
$$= 1.22 \cdot (2)$$
$$= 2.44.$$

Next, we have

$$y_2 = 1.22 y_1 = 1.22(2.44),$$

or

$$y_2 = 2.9768.$$

Similarly,

$$y_3 = 1.22(2.9768) = 3.631696,$$
$$y_4 = 1.22(3.631696) \approx 4.4306691,$$

and

$$y_5 \approx 1.22(4.4306691) \approx 5.4054163. \qquad ❖$$

DO EXERCISE 3.

We compare the result of using Euler's method with the three-term method on the initial-value problem $y' = y$, $y(0) = 2$, in the table below. The exact values in the right-hand column were calculated using $y = 2e^x$.

	Euler	Three-term method	Exact
$y_0 = y(0)$	2	2.	2
$y_1 = y(0.2)$	2.4	2.44	2.4428055
$y_2 = y(0.4)$	2.88	2.9768	2.9836494
$y_3 = y(0.6)$	3.456	3.631696	3.6442376
$y_4 = y(0.8)$	4.1472	4.4306691	4.4510819
$y_5 = y(1)$	4.97664	5.4054163	5.4365637

DO EXERCISE 4.

As expected, the three-term method yields greater accuracy (although more computation is involved) than Euler's method.

We can also apply these techniques to systems of first-order differential equations. They could be used if, for instance, we have

$$x' = f(x, y, t) \quad \text{and} \quad y' = g(x, y, t)$$

with the variables x and y being thought of as functions of the third variable t. If the initial conditions are

$$x(t_0) = x_0,$$
$$y(t_0) = y_0,$$

then Euler's method becomes

$$x_{n+1} = x_n + f(x_n, y_n, t_n) \cdot \Delta t$$

and

$$y_{n+1} = y_n + g(x_n, y_n, t_n) \cdot \Delta t$$

for all n.

EXAMPLE 4   Use Euler's method to find approximate values for $x(t)$ and $y(t)$ on $[0, 1]$, where x and y satisfy

$$x' = x - 5y,$$
$$y' = 2x - 6y,$$

with

$$x(0) = 6,$$
$$y(0) = 3$$

using $\Delta t = 0.2$.

4. Compare the results of Margin Exercises 1 and 3 with the corresponding values of the exact solution $y = e^{-x^2}$ as in the table at the left.

5. a) Show that $x(t) = 5e^{-t} + e^{-4t}$ and $y(t) = 2e^{-t} + e^{-4t}$ are the exact solutions to the initial-value problem in Example 4.

b) Use part (a) to compute the exact values of $y(0.4)$, $x(0.6)$, and $y(1)$.

6. Use Euler's method to find two approximate values for $t = 0.2$ and $t = 0.4$ for the system

$$x' = x - 2y,$$
$$y' = -2x + y,$$

with initial conditions $x(0) = 2$, $y(0) = 0$. Use $\Delta t = 0.2$.

Solution Here, $f(x, y, t) = x - 5y$ and $g(x, y, t) = 2x - 6y$. Our formulas become

$$x_{n+1} = x_n + (x_n - 5y_n) \cdot \Delta t$$

and

$$y_{n+1} = y_n + (2x_n - 6y_n) \cdot \Delta t,$$

or, if we use $\Delta t = 0.2$,

$$x_{n+1} = x_n + (x_n - 5y_n) \cdot (0.2)$$

and

$$y_{n+1} = y_n + (2x_n - 6y_n) \cdot (0.2).$$

Simplification yields

$$x_{n+1} = 1.2x_n - y_n$$

and

$$y_{n+1} = 0.4x_n - 0.2y_n.$$

Using $x_0 = x(0) = 6$ and $y_0 = y(0) = 3$, we find that

$$x_1 = 1.2x_0 - y_0$$
$$= (1.2)(6) - 3 = 4.2,$$

whereas

$$y_1 = 0.4x_0 - 0.2y_0$$
$$= (0.4)(6) - (0.2)(3) = 1.8.$$

Continuing, we get

$$x_2 = (1.2)x_1 - y_1 = (1.2)(4.2) - (1.8) = 3.24,$$
$$y_2 = (0.4)x_1 - (0.2)y_1 = (0.4)(4.2) - (0.2)(1.8) = 1.32,$$
$$x_3 = (1.2)(3.24) - (1.32) = 2.568,$$
$$y_3 = (0.4)(3.24) - (0.2)(1.32) = 1.032,$$
$$x_4 = (1.2)(2.568) - (1.032) = 2.0496,$$
$$y_4 = (0.4)(2.568) - (0.2)(1.032) = 0.8208,$$

and

$$x_5 = (1.2)(2.0496) - (0.8208) = 1.63872,$$
$$y_5 = (0.4)(2.0496) - (0.2)(0.8208) = 0.65568. \qquad \diamondsuit$$

DO EXERCISES 5 AND 6.

EXERCISE SET ## 11.6

Find the solution to the initial-value problem using Euler's method on [0, 1] with the indicated step size Δx.

1. $y' = -1$, $y(0) = 1$, $\Delta x = 0.1$
2. $y' = 2x$, $y(0) = -1$, $\Delta x = 0.1$
3. $y' = 2x$, $y(0) = 0$, $\Delta x = 0.25$
4. $y' = 3x^2 - 1$, $y(0) = 1$, $\Delta x = 0.2$
5. $y' = 2xy$, $y(0) = 1$, $\Delta x = 0.2$
6. $y' = -y$, $y(0) = 1$, $\Delta x = 0.25$
7. $y' = x\sqrt{y}$, $y(0) = 1$, $\Delta x = 0.2$

Find the solutions to the initial-value problem using the three-term method on [0, 1] with the indicated step size Δx.

8. $y' = -1$, $y(0) = 1$, $\Delta x = 0.1$
9. $y' = 2x$, $y(0) = -1$, $\Delta x = 0.1$
10. $y' = 2x$, $y(0) = 0$, $\Delta x = 0.25$
11. $y' = 3x^2 - 1$, $y(0) = 1$, $\Delta x = 0.2$
12. $y' = 2xy$, $y(0) = 1$, $\Delta x = 0.2$
13. $y' = -y$, $y(0) = 1$, $\Delta x = 0.25$
14. $y' = x\sqrt{y}$, $y(0) = 1$, $\Delta x = 0.2$

For each of the following initial-value problems, find the exact solution $y = g(x)$. Then make a table comparing the exact values with the approximate values obtained using Euler's method and the three-term method.

15. $y' = -1$, $y(0) = 1$
16. $y' = 2x$, $y(0) = -1$
17. $y' = 2xy$, $y(0) = 1$
18. $y' = -y$, $y(0) = 1$

Use Euler's method with the indicated value of Δt to approximate the solution to the given system of differential equations on [0, 1]. Compare these with the exact solutions, where possible.

19. $x'(t) = x + y$,
    $y'(t) = 3x - y$,
    $x(0) = 3$, $y(0) = -1$, $\Delta t = 0.2$
20. $x'(t) = x + y$,
    $y'(t) = 4x - 2y$,
    $x(0) = 3$, $y(0) = -2$, $\Delta t = 0.25$
21. $x'(t) = y$,
    $y'(t) = -2x + 3y$,
    $x(0) = 0$, $y(0) = 1$, $\Delta t = 0.2$
22. $x'(t) = -3x + 4y$,
    $y'(t) = -2x + 3y$,
    $x(0) = 4$, $y(0) = 3$, $\Delta t = 0.1$
23. $x' = x^3$,
    $y' = -y$,
    $x(0) = 0$, $y(0) = 1$, $\Delta t = 0.1$
24. $x' = x + ty$,
    $y' = xy$,
    $x(0) = 1$, $y(0) = 1$, $\Delta t = 0.2$

APPLICATIONS

❖ **Business and Economics**

25. *Accumulated present value of a continuous cash flow.* If money flows continuously into an investment at the rate $R(t)$ (in thousands of dollars), then the *accumulated present value* $V(t)$ satisfies the initial-value problem (see Exercise Set 6.3)

$$V' = R(t)e^{-kt}, \qquad V(0) = 0,$$

where k is the current interest rate. Using step size $\Delta t = 0.2$, calculate $V(1)$ using Euler's method if $R(t) = 1/(1 + t^2)$ and $k = 0.08$.

26. *Accumulated present value of dividends.* Suppose that $d(t)$ represents the instantaneous dividend payment of a stock at time t. Then $d(t)e^{-mt}$ is the present value of that payment, where m is the current interest rate. Then the accumulated value $D_p(t)$ of the dividends at time t satisfies the initial-value problem

$$D_p'(t) = d(t)e^{-mt}, \qquad D(0) = 0.$$

Use step size $\Delta t = 1$ to estimate $D_p(10)$ if $m = 0.08$ and $d(t) = \$10$ using (a) Euler's method and (b) the three-term method.

27. *Commodities prices.* The price (in dollars) of two commodities at time t (in weeks) are $P(t)$ and $Q(t)$, respectively. Given that the prices satisfy the system

$$P' = -0.5P + 0.5Q + e^{-t^2},$$
$$Q' = 0.25P - 0.25Q - 3e^{-t^2},$$
$$P(0) = 225,$$
$$Q(0) = 100,$$

use Euler's method with step size $\Delta t = 1$ to estimate $P(10)$ and $Q(10)$.

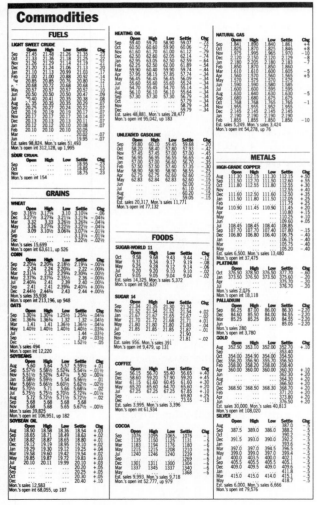

Commodities listing.

❖ **Life and Physical Sciences**

28. *Biology: Growth of a culture.* In an experiment (see Exercise 3 of Exercise Set 9.3), it was found that the rate of growth of yeast cells in a laboratory satisfied the initial-value problem

$$\frac{dP}{dt} = 0.0008P(700 - P), \qquad P(0) = 10,$$

where $P(t)$ is the number of cells present after time t (in hours). Using step size 1, calculate $P(5)$ and $P(8)$ using (a) Euler's method; (b) the three-term method; and (c) the exact solution.

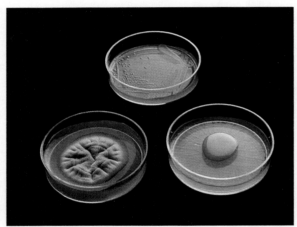

Yeast cells growing in a Petri dish.

29. *Fick's law in biology* is given by

$$\frac{dC}{dt} = \frac{kA}{V}(C_T - C),$$

where $C(t)$ is the amount of a substance that passes through a tube with cross-sectional area A and volume V at time t (see Exercise 11 of Exercise Set 9.4), k is a constant, and C_T is the total amount of substance. Assuming that $k = 1$, $A = 3$, $V = 20$, $C_T = 1000$, and $C(0) = C_0 = 2$, use step size $\Delta t = 1$ to estimate $C(5)$ using (a) Euler's method; (b) the three-term method; and (c) the exact solution.

30. *Population.* The population of two species satisfies the system

$$\frac{dx}{dt} = 0.10x - 1.5y,$$

$$\frac{dy}{dt} = 0.015x + 0.3y,$$

$$x(0) = 1600,$$

$$y(0) = 75,$$

where $x(t)$ and $y(t)$ are the populations of the two species at time t (in years), respectively. Use Euler's method with step size $\Delta t = 1$ to estimate $x(5)$ and $y(5)$.

31. Let $F(x)$ be defined by

$$F(x) = \int_0^x f(t)\, dt$$

where $f(t)$ is continuous on $[0, 10]$.

a) Show that $F(x)$ satisfies the initial-value problem

$$F'(x) = f(x), \qquad F(0) = 0$$

for all x in $[0, 10]$.

b) Use Euler's method with step size $\Delta x = 1$ to estimate $F(10)$.

Predator–prey. Suppose that $x(t)$ and $y(t)$ denote the number of hares and the number of foxes, respectively, present at time t in a certain enclosed wilderness area. If we assume that plenty of food is provided for the hares, then it seems reasonable that their birth rate should be a (large) constant that is independent of time. Their death rate, on the other hand, ought to be proportional to the number of encounters between foxes and hares. Since the rate of change of the hare population will be the difference between the birth and death rates, we assume that

$$\frac{dx}{dt} = ax - bxy,$$

where a and b are positive constants.

For the foxes, the birth rate tends to increase with the supply of food (hares). Here, we might take the birth rate to be proportional to xy. As the fox population increases, there will be shortages of food and the death rate of the foxes will be proportional to the number of foxes present. We get

$$\frac{dy}{dt} = cxy - dy,$$

where c and d are positive constants. Thus we have the

(nonlinear) system of equations

$$\frac{dx}{dt} = ax - bxy,$$

$$\frac{dy}{dx} = cxy - dy.$$

We will omit the derivation but we can show that these equations lead to the equation

$$\frac{dy}{dx} = \frac{cxy - dy}{ax - bxy} = \frac{y(cx - d)}{x(a - by)}.$$

32. a) Use separation of variables to solve the equation in the predator–prey discussion above.

b) Evaluate the constant of integration in part (a) in terms of a, b, c, d, and the initial populations $x_0 = x(0)$ and $y_0 = y(0)$.

Although the equation relating $x(t)$ and $y(t)$ that we obtained in the predator–prey discussion above cannot generally be solved for y or x as a function of the other variable, the graph of the relation can be plotted, as shown here.

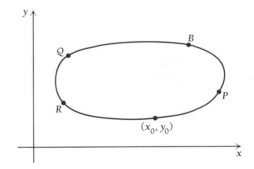

Note that if we start at (x_0, y_0) and proceed counterclockwise around the figure, then the number of hares increases at first, as does the number of foxes. This results in a dramatic increase in the number of foxes (starting at P). Eventually, the foxes begin to devour the hares and we have a long period in the cycle during which the hare population drops steadily, while the fox population remains constant (arc BQ). Finally, there are so few hares that the foxes begin to starve (at Q) and their population dwindles. As the number of foxes declines, the hare population begins to grow (R), and the cycle begins anew.

33. In a certain forest, the population of hares and foxes (in thousands) is determined by the system

$$\frac{dx}{dt} = 0.75x - 0.04xy, \qquad x(0) = 60,$$

$$\frac{dy}{dt} = 0.004xy - 0.25y, \qquad y(0) = 20.$$

Use Euler's method with step size $\Delta t = 1$ to find an approximation for **(a)** $x(5)$ and $y(5)$ and **(b)** $x(10)$ and $y(10)$.

CHAPTER SUMMARY AND REVIEW

11

TERMS TO KNOW

Initial approximation, p. 758
Method of iteration, p. 758
Sequence of iterates, p. 760
Iteration formula, p. 761
Newton's formula, p. 764
Present value, p. 773
Yield rate, p. 774

Internal rate of return, p. 774
Discount factor, p. 776
Method of Bisection, p. 777
Approximation by Riemann sums, p. 783
Trapezoidal Rule, p. 784

Simpson's Rule, p. 789
Extrapolation, p. 792
Aitken acceleration, p. 793
Romberg integration, p. 796
Euler's method, p. 800
Three-term method, p. 803

REVIEW EXERCISES

These review exercises are for test preparation. They can also be used as a lengthened practice test. Answers are at the back of the book. The answers also contain bracketed section references, which tell you where to restudy if your answer is incorrect.

Use iteration and the suggested initial guesses to solve each of the given equations.

1. $x = \sqrt{5x - 6}$; $x_1 = 2.5$
2. $x = (3x - 2)^{1/3}$; $x_1 = 2$
3. $x = ((2x + 5)/9)^{1/4}$; $x_1 = 1$
4. $2x = \cos(x)$; $x_1 = 0.8$
5. $x = 2\cos(x)$; $x_1 = 0.8$

Use Newton's method to find all solutions in $[0, 2]$ to each of the following equations.

6. $x^3 - 4x + 1 = 0$ 7. $5x^4 - x - 3 = 0$
8. $x^8 - 4x + 2 = 0$

9. Find the equilibrium points for the supply and demand functions $S(x) = x^3 + 6x + 2$ and $D(x) = 10x^{-2}$, $x > 0$.

10. *Finance: Present value of a future payment.* If interest is compounded at 8.75% annually, find the present value of a payment of $100,000 to be made twenty years from today.

11. *Finance: Yield rates.* An investor makes an initial investment of $4000 followed by an additional $1000 one

year later. The investment returns $7000 four years after the original investment was made.

a) Set up the equation for the yield rate of this investment.

b) Find the investor's yield rate.

12. *Finance: Yield rates.* An investor must choose between two investment packages. The first requires an initial investment of $1000 and returns $1750 after three years. The second requires an initial investment of $1000, with another $1000 after one year, and returns $9500 twelve years after the initial investment was made. Find the respective yield rates.

13. Let

$$f(x) = \begin{bmatrix} \dfrac{1 - \cos\,(x)}{x}, & \text{if } x > 0, \\ 0, & \text{if } x = 0 \end{bmatrix}.$$

Use the Trapezoidal Rule with $n = 2$, 4, and 8 to approximate $\int_0^1 f(x)\,dx$.

14. Use Simpson's Rule to calculate S_2 and S_4 for Exercise 13.

15. Use the initial guess $x_1 = 0.8$ and iteration to obtain x_2 and x_3 for the equation $x = 2\cos\,(x)$. Then use Aitken

16. Repeat Exercise 15 with $2x = \cos\,(x)$.

17. Use Romberg integration and the results of Exercise 14 to find an approximate value for $\int_0^1 f(x)\,dx$, where $f(x)$ is as in Exercises 13 and 14.

18. Use Euler's method with step size $\Delta x = 0.2$ to find an approximate solution to $y' = -y$ with $y(0) = 3$ on the interval $[0, 1]$.

19. Use the three-term method with $\Delta x = 0.2$ to find an approximate solution to Exercise 18.

20. a) Separate the variables to find the exact solution to the equation in Exercise 18.

b) Compare the exact values obtained above with the approximate values obtained in your solution to Exercises 18 and 19.

21. Use Euler's method with step size $\Delta t = 0.2$ on the interval $[0, 1]$ to approximate a solution to

$$\frac{dx}{dt} = -4x + y, \qquad x(0) = 3,$$

$$\frac{dy}{dt} = -3x, \qquad y(0) = 1.$$

acceleration to calculate x_4, etc. Compare your answer here with your answer to Exercise 5 above.

SYNTHESIS EXERCISES

22. Using the techniques of Section 9.8, find the exact solution to the system of differential equations in Exercise 21.

23. Use the exact solution obtained in Exercise 22 to compute x- and y-values at $t = 0, 0.2, 0.4, 0.6, 0.8$, and 1. Then compare these results with the approximations obtained in Exercise 21.

EXERCISES FOR THINKING AND WRITING

24. Both $g(x) = \tan x$ and $g'(x) = \sec^2\,(x)$ are continuous on $[-\pi/4, \pi/4]$ and $\bar{x} = 0$ satisfies $g(\bar{x}) = \bar{x}$. Yet if $x_1 = 0.5$ and $x_{n+1} = g(x_n) = \tan\,(x_n)$ for all $n \geqslant 1$, the sequence of iterates seems to diverge. Does this contradict Theorem 1?

25. How many different financial situations can you think of in which the ideas of *present value* and/or *internal rate of return* play a role?

26. A certain stock pays a dividend of d dollars per year. If the value of the stock is solely determined as the sum of the present values of its future dividend payments and the rate of interest is i, use the formula in Exercise 22 of

the Chapter Test to find the value of one share of this stock.

27. Use Theorems 3 and 4 to show that the Trapezoidal Rule and Simpson's Rule both give the exact value of $\int_a^b f(x)\,dx$ when $f(x)$ is a polynomial of degree one. What happens if $f(x)$ is a polynomial of degree two? of degree three?

28. Does Romberg integration give the exact value of $\int_a^b f(x)\,dx$ when $f(x)$ is a polynomial of degree one? of degree two? Why? What happens if $f(x)$ is a polynomial of degree three?

CHAPTER TEST

11

Use iteration and the suggested initial guesses to solve each of the given equations.

1. $x = \sqrt{3x - 2}; \; x_1 = 1.5$

2. $x = \sqrt[3]{3x - 2}; \; x_1 = 0$

3. $x = \sqrt[4]{\dfrac{x + 3}{6}}; \; x_1 = 1$

4. $x = \frac{2}{5} \cos x; \; x_1 = 0.75$

5. $x = \frac{5}{2} \cos x; \; x_1 = 0.75$

Use Newton's method to find all solutions in $[0, 2]$ of each of the given equations.

6. $2x^3 - 5x + 1 = 0$ 7. $6x^4 - x - 3 = 0$

8. $x^9 - 4x + 2 = 0$

9. Find the equilibrium point for the supply and demand functions

$$S(x) = x^3 + 6x + 1 \quad \text{and} \quad D(x) = \frac{12}{x^2}.$$

10. *Finance: Present value of a future payment.* If interest is compounded annually at 9.25%, find the present value of a payment of $100,000 to be made twenty years from today.

11. *Finance: Yield rates.* An investor makes an initial investment of $3000 followed by an additional $1000 one year later. The investment returns $6000 four years after the original investment was made.

    a) Set up the equation for the yield rate for this investment.

    b) Find the investor's yield rate.

12. *Finance: Yield rates.* An investor must choose between two investment packages. The first requires an initial investment of $1000 and returns $1500 after three years. The second requires an initial investment of $1000, another $1000 after one year, and returns $10,000 12 years after the initial investment was made. Find the respective yield rates.

13. Let

$$f(x) = \begin{cases} \dfrac{\sin x}{x}, & \text{if } x > 0, \\ 1, & \text{if } x = 0. \end{cases}$$

Use the Trapezoidal Rule with $n = 2$, 4, and 8 to approximate

$$\int_0^1 f(x) \, dx.$$

14. Use Simpson's Rule to calculate S_2 and S_4 for Problem 13.

15. Use the initial guess $x_1 = 0.75$ and iteration to obtain x_2 and x_3 for $x = \frac{5}{2} \cos x$. Then use Aitken acceleration to calculate x_4, etc. Compare your answer here with your answer to Problem 5 above.

16. Repeat Problem 15 with $x = \frac{2}{5} \cos x$.

17. Use Romberg integration and the results of Problem 14 to find an approximate value for

$$\int_0^1 f(x) \, dx.$$

18. Use Euler's method with step size $\Delta x = 0.2$ to find an approximate solution to $y' = -y^2$ with $y(0) = 1$ on the interval $[0, 1]$.

19. Use the three-term method with $\Delta x = 0.2$ to find an approximate solution to Problem 18.

20. a) Separate the variables to find the exact solution to Problem 18.

    b) Compare the exact values with the approximate values obtained in Problems 18 and 19.

21. Use Euler's method with step size $\Delta t = 0.2$ on the interval $[0, 1]$ to approximate a solution to

$$\frac{dx}{dt} = -3x + 2y, \qquad x(0) = 1,$$

$$\frac{dy}{dt} = -x, \qquad y(0) = 2.$$

SYNTHESIS EXERCISES

22. *Business: Present value of an annuity.* Show that the present value V of a series of (equal) payments of R dollars per year for n years (starting one year from today) at a rate of interest i is

$$V = Rv \cdot \frac{1 - v^n}{1 - v},$$

where $v = (1 + i)^{-1}$.

23. Show that Aitken's formula can be written in the form

$$x_n - \frac{(x_{n+1} - x_n)^2}{x_{n+2} - 2x_{n+1} + x_n}.$$

90. Use Newton's Method to find the solution of $x^3 - 6x + 4 = 0$ that lies between 0 and 1.

91. Find the present value of a payment of $1000 to be made 8 years from today at a rate of interest of 10%.

92. An investor pays $7000 for three annual payments of $3000 each, beginning one year from today. Find the yield rate of this investment and check your answer.

93. Use the Trapezoidal Rule with $n = 5$ to estimate

$$\int_0^1 \frac{dx}{x + 2}.$$

Compare your value with the exact answer.

94. Use Simpson's Rule with $n = 5$ to estimate

$$\int_0^1 \frac{dx}{x + 2}.$$

95. a) Use Simpson's Rule with $n = 2$ and $n = 4$ to estimate

$$I = \int_0^1 x^8 \, dx.$$

b) Then use S_2 and S_4 to compute R_2.

TABLE 1

INTEGRATION FORMULAS

(Whenever $\ln X$ is used, it is assumed that $X > 0$.)

1. $\int x^n \, dx = \dfrac{x^{n+1}}{n+1} + C, \, n \neq -1$

2. $\int \dfrac{dx}{x} = \ln x + C$

3. $\int u \, dv = uv - \int v \, du$

4. $\int e^x \, dx = e^x + C$

5. $\int e^{ax} \, dx = \dfrac{1}{a} \cdot e^{ax} + C$

6. $\int x e^{ax} \, dx = \dfrac{1}{a^2} \cdot e^{ax}(ax - 1) + C$

7. $\int x^n e^{ax} \, dx = \dfrac{x^n e^{ax}}{a} - \dfrac{n}{a} \int x^{n-1} e^{ax} \, dx + C$

8. $\int \ln x \, dx - x \ln x - x + C$

9. $\int (\ln x)^n \, dx = x(\ln x)^n - n \int (\ln x)^{n-1} \, dx + C, \, n \neq -1$

10. $\int x^n \ln x \, dx = x^{n+1} \left[\dfrac{\ln x}{n+1} - \dfrac{1}{(n+1)^2} \right] + C, \, n \neq -1$

11. $\int a^x \, dx = \dfrac{a^x}{\ln a} + C, \, a > 0, \, a \neq 1$

12. $\int \dfrac{1}{\sqrt{x^2 + a^2}} \, dx = \ln \left(x + \sqrt{x^2 + a^2} \right) + C$

13. $\int \dfrac{1}{\sqrt{x^2 - a^2}} \, dx = \ln \left(x + \sqrt{x^2 - a^2} \right) + C$

14. $\int \dfrac{1}{x^2 - a^2} \, dx = \dfrac{1}{2a} \ln \left(\dfrac{x - a}{x + a} \right) + C$

15. $\int \dfrac{1}{a^2 - x^2} \, dx = \dfrac{1}{2a} \ln \left(\dfrac{a + x}{a - x} \right) + C$

16. $\int \dfrac{1}{x \sqrt{a^2 + x^2}} \, dx = -\dfrac{1}{a} \ln \left(\dfrac{a + \sqrt{a^2 + x^2}}{x} \right) + C$

17. $\int \dfrac{1}{x \sqrt{a^2 - x^2}} \, dx = -\dfrac{1}{a} \ln \left(\dfrac{a + \sqrt{a^2 - x^2}}{x} \right) + C, \, 0 < x < a$

18. $\int \dfrac{x}{ax + b} \, dx = \dfrac{b}{a^2} + \dfrac{x}{a} - \dfrac{b}{a^2} \ln (ax + b) + C$

19. $\displaystyle \int \frac{x}{(ax + b)^2} \, dx = \frac{b}{a^2(ax + b)} + \frac{1}{a^2} \ln \, (ax + b) + C$

20. $\displaystyle \int \frac{1}{x(ax + b)} \, dx = \frac{1}{b} \ln\!\left(\frac{x}{ax + b}\right) + C$

21. $\displaystyle \int \frac{1}{x(ax + b)^2} \, dx = \frac{1}{b(ax + b)} + \frac{1}{b^2} \ln\!\left(\frac{x}{ax + b}\right) + C$

22. $\displaystyle \int \sqrt{x^2 \pm a^2} \, dx = \tfrac{1}{2}[x\sqrt{x^2 \pm a^2} \pm a^2 \ln \, (x + \sqrt{x^2 \pm a^2})] + C$

23. $\displaystyle \int \sin x \, dx = -\cos x + C, \int \cos x \, dx = \sin x + C$

24. $\displaystyle \int \sec^2 x \, dx = \tan x + C, \int \csc^2 x \, dx = -\cot x + C$

25. $\displaystyle \int \tan x \sec x \, dx = \sec x + C, \int \cot x \csc x = -\csc x + C$

26. $\displaystyle \int \sec x \, dx = \ln \, |\sec x + \tan x| + C, \int \csc x \, dx = -\ln \, |\csc x + \cot x| + C$

27. $\displaystyle \int \frac{1}{1 + x^2} \, dx = \tan^{-1} x + C$

28. $\displaystyle \int \frac{1}{\sqrt{1 - x^2}} \, dx = \sin^{-1} x + C, \int \frac{-1}{\sqrt{1 - x^2}} \, dx = \cos^{-1} x + C$

29. $\displaystyle \int x\sqrt{a + bx} \, dx = \frac{2}{15b^3} \, (3bx - 2a)(a + bx)^{3/2} + C$

30. $\displaystyle \int x^2\sqrt{a + bx} \, dx = \frac{2}{105b^3} \, (15b^2x^2 - 12abx + 8a^2)(a + bx)^{3/2} + C$

31. $\displaystyle \int \frac{x \, dx}{\sqrt{a + bx}} = \frac{2}{3b^2} \, (bx - 2a)\sqrt{a + bx} + C$

32. $\displaystyle \int \frac{x^2 \, dx}{\sqrt{a + bx}} = \frac{2}{15b^3} \, (3b^2x^2 - 4abx + 8a^2)\sqrt{a + bx} + C$

33. $\displaystyle \int \sin^n x \, dx = -\frac{1}{n} \, \sin^{n-1} x \cos x + \frac{n - 1}{n} \int \sin^{n-2} x \, dx + C$

34. $\displaystyle \int \cos^n x \, dx = \frac{1}{n} \, \cos^{n-1} x \sin x + \frac{n - 1}{n} \int \cos^{n-2} x \, dx + C$

35. $\displaystyle \int e^{ax} \sin x \, dx = \frac{e^{ax}}{a^2 + b^2} \, (a \sin bx - b \cos bx) + C$

36. $\displaystyle \int e^{ax} \cos x \, dx = \frac{e^{ax}}{a^2 + b^2} \, (a \cos bx + b \sin bx) + C$

TABLE 2

| | | AREAS FOR A STANDARD NORMAL DISTRIBUTION | | | | | | | | |

Entries in the table represent area under the curve between $t = 0$ and a positive value of t. Because of the symmetry of the curve, area under the curve between $t = 0$ and a negative value of t would be found in a similar manner.

Area = Probability
$= P(0 \leq x \leq t)$
$= \int_0^t \frac{1}{\sqrt{2\pi}} e^{-x^2/2}\, dx$

t	0.00	0.01	0.02	0.03	0.04	0.05	0.06	0.07	0.08	0.09
0.0	.0000	.0040	.0080	.0120	.0160	.0199	.0239	.0279	.0319	.0359
0.1	.0398	.0438	.0478	.0517	.0557	.0596	.0636	.0675	.0714	.0753
0.2	.0793	.0832	.0871	.0910	.0948	.0987	.1026	.1064	.1103	.1141
0.3	.1179	.1217	.1255	.1293	.1331	.1368	.1406	.1443	.1480	.1517
0.4	.1554	.1591	.1628	.1664	.1700	.1736	.1772	.1808	.1844	.1879
0.5	.1915	.1950	.1985	.2019	.2054	.2088	.2123	.2157	.2190	.2224
0.6	.2257	.2291	.2324	.2357	.2389	.2422	.2454	.2486	.2517	.2549
0.7	.2580	.2611	.2642	.2673	.2704	.2734	.2764	.2794	.2823	.2852
0.8	.2881	.2910	.2939	.2967	.2995	.3023	.3051	.3078	.3106	.3133
0.9	.3159	.3186	.3212	.3238	.3264	.3289	.3315	.3340	.3365	.3389
1.0	.3413	.3438	.3461	.3485	.3508	.3531	.3554	.3577	.3599	.3621
1.1	.3643	.3665	.3686	.3708	.3729	.3749	.3770	.3790	.3810	.3830
1.2	.3849	.3869	.3888	.3907	.3925	.3944	.3962	.3980	.3997	.4015
1.3	.4032	.4049	.4066	.4082	.4099	.4115	.4131	.4147	.4162	.4177
1.4	.4192	.4207	.4222	.4236	.4251	.4265	.4279	.4292	.4306	.4319
1.5	.4332	.4345	.4357	.4370	.4382	.4394	.4406	.4418	.4429	.4441
1.6	.4452	.4463	.4474	.4484	.4495	.4505	.4515	.4525	.4535	.4545
1.7	.4554	.4564	.4573	.4582	.4591	.4599	.4608	.4616	.4625	.4633
1.8	.4641	.4649	.4656	.4664	.4671	.4678	.4686	.4693	.4699	.4706
1.9	.4713	.4719	.4726	.4732	.4738	.4744	.4750	.4756	.4761	.4767
2.0	.4772	.4778	.4783	.4788	.4793	.4798	.4803	.4808	.4812	.4817
2.1	.4821	.4826	.4830	.4834	.4838	.4842	.4846	.4850	.4854	.4857
2.2	.4861	.4864	.4868	.4871	.4875	.4878	.4881	.4884	.4887	.4890
2.3	.4893	.4896	.4898	.4901	.4904	.4906	.4909	.4911	.4913	.4916
2.4	.4918	.4920	.4922	.4925	.4927	.4929	.4931	.4932	.4934	.4936
2.5	.4938	.4940	.4941	.4943	.4945	.4946	.4948	.4949	.4951	.4952
2.6	.4953	.4955	.4956	.4957	.4959	.4960	.4961	.4962	.4963	.4964
2.7	.4965	.4966	.4967	.4968	.4969	.4970	.4971	.4972	.4973	.4974
2.8	.4974	.4975	.4976	.4977	.4977	.4978	.4979	.4979	.4980	.4981
2.9	.4981	.4982	.4982	.4983	.4984	.4984	.4985	.4985	.4986	.4986
3.0	.4987	.4987	.4987	.4988	.4988	.4989	.4989	.4989	.4990	.4990

PHOTO CREDITS

Note: All non-linear curves in the answer section have been computer-generated utilizing Mathematica®, a software-based symbolic math processor. The curves are actually composed of tightly packed laser exposures based on Mathematica's polygonal data output. Ordered pairs are sampled at a rate optimized to produce smooth, plotter-like curves on high-resolution imagesetters.

CHAPTER 1

Margin Exercises, Section 1.1, pp. 2–9

1. $3 \cdot 3 \cdot 3 \cdot 3$, or 81   2. $(-3)(-3)$, or 9
3. $1.02 \times 1.02 \times 1.02$, or 1.061208   4. $\frac{1}{4} \cdot \frac{1}{4}$, or $\frac{1}{16}$
5. 625   6. -625 7. 1   8. $5t$ 9. 1   10. m 11. $\frac{1}{4}$
12. 1   13. $\frac{1}{2 \cdot 2 \cdot 2 \cdot 2}$, or $\frac{1}{16}$ 14. $\frac{1}{10 \cdot 10}$, or $\frac{1}{100}$, or 0.01
15. 64   16. $\frac{1}{t^7}$ 17. $\frac{1}{e^t}$ 18. $\frac{1}{M}$ 19. $\frac{1}{(x+1)^2}$
20. t^9 21. t^{-3} 22. $50e^{-13}$ 23. t^{-6} 24. $24b^3$
25. x^4 26. x^{-4} 27. 1   28. e^{2-k} 29. e^{12}
30. e^2 31. x^{-12} 32. e^+ 33. e^{3x} 34. $25x^6y^{10}$
35. $\frac{1}{256} x^{20}y^{24}z^{-8}$, or $\frac{x^{20}y^{24}}{256z^8}$ 36. $2x + 14$ 37. $P - Pi$
38. $x^2 + 3x - 28$ 39. $a^2 - 2ab + b^2$ 40. $a^2 - b^2$
41. $x^2 - 2xh + h^2$ 42. $9x^2 + 6xt + t^2$ 43. $25t^2 - m^2$
44. $P(1 - i)$ 45. $(x + 5y)^2$ 46. $4(x + 5)(x + 2)$
47. $(5c - d)(5c + d)$ 48. $(4y + 7)(3y - 2)$ 49. 1.01
50. \$1188.10   51. \$1378.84

Exercise Set 1.1, p. 9

1. $5 \cdot 5 \cdot 5$, or 125   3. $(-7)(-7)$, or 49   5. 1.0201
7. $\frac{1}{16}$ 9. 1   11. t 13. 1   15. $\frac{1}{3^2}$, or $\frac{1}{9}$ 17. 8
19. 0.1   21. $\frac{1}{e^b}$ 23. $\frac{1}{b}$ 25. x^5 27. x^{-6}, or $\frac{1}{x^6}$

29. $35x^5$ 31. x^4 33. 1   35. x^3 37. x^{-3}, or $\frac{1}{x^3}$
39. 1   41. e^{t-4} 43. t^{14} 45. t^2 47. a^6b^5
49. t^{-6}, or $\frac{1}{t^6}$ 51. e^{+x} 53. $8x^6y^{12}$
55. $\frac{1}{81} x^8y^{20}z^{-16}$, or $\frac{x^8y^{20}}{81z^{16}}$ 57. $9x^{-16}y^{14}z^4$, or $\frac{9y^{14}z^4}{x^{16}}$
59. $5x - 35$ 61. $x - xt$ 63. $x^2 - 7x + 10$
65. $a^3 - b^3$ 67. $2x^2 + 3x - 5$ 69. $a^2 - 4$
71. $25x^2 - 4$ 73. $a^2 - 2ah + h^2$ 75. $25x^2 + 10xt + t^2$
77. $5x^5 + 30x^3 + 45x$ 79. $a^3 + 3a^2b + 3ab^2 + b^3$
81. $x^3 - 15x^2 + 75x - 125$ 83. $x(1 - t)$
85. $(x + 3y)^2$ 87. $(x - 5)(x + 3)$ 89. $(x - 5)(x + 4)$
91. $(7x - t)(7x + t)$ 93. $4(3t - 2m)(3t + 2m)$
95. $ab(a + 4b)(a - 4b)$
97. $(a^4 + b^4)(a^2 + b^2)(a + b)(a - b)$
99. $10x(a + 2b)(a - 2b)$
101. $2(1 + 4x^2)(1 + 2x)(1 - 2x)$ 103. $(9x - 1)(x + 2)$
105. $(x + 2)(x^2 - 2x + 4)$
107. $(y - 4t)(y^2 + 4yt + 16t^2)$ 109. (a) 0.81;
(b) 0.0801;   (c) 0.008001   111. (a) \$1160;
(b) \$1166.40;   (c) \$1169.86;   (d) \$1173.47, assuming 365 days in a year;   (e) \$1173.51   113. \$353.52

Margin Exercises, Section 1.2, pp. 13–18

1. $\frac{56}{9}$ 2. \$725   3. $0, -2, \frac{3}{2}$ 4. $-4, 3$ 5. $0, -1, 1$
6. $x < \frac{11}{5}$ 7. $x \geqslant \frac{20}{17}$ 8. More than 475 suits

9. (a) $(-1, 3)$; **(b)** $(1, 4)$ **10. (a)** $(-1, 4)$; **(b)** $\left(-\frac{1}{4}, \frac{1}{4}\right)$
11. (a) $[-1, 4]$; **(b)** $(-1, 4]$; **(c)** $[-1, 4)$; **(d)** $(-1, 4)$
12. (a) $(-\sqrt{2}, \sqrt{2})$; **(b)** $[0, 1)$; **(c)** $(-6.7, -4.2]$;
(d) $[3, 7\frac{1}{2}]$ **13. (a)** $(-\infty, 5]$; **(b)** $(4, \infty)$; **(c)** $(-\infty, 4.8)$;
(d) $[3, \infty)$ **14. (a)** $[8, \infty)$; **(b)** $(-\infty, -7)$; **(c)** $(10, \infty)$;
(d) $(-\infty, -0.78]$

6.

Inputs	Outputs
5	$\frac{1}{5}$
$-\frac{2}{3}$	$-\frac{3}{2}$
$\frac{1}{4}$	4
$\dfrac{1}{a}$	a
k	$\dfrac{1}{k}$
$1 + t$	$\dfrac{1}{1 + t}$

Exercise Set 1.2, p. 18

1. $\frac{7}{4}$ **3.** -8 **5.** 120 **7.** 200 **9.** $0, -3, \frac{4}{5}$ **11.** $0, 2$
13. $0, 3$ **15.** $0, 7$ **17.** $0, \frac{1}{3}, -\frac{1}{3}$ **19.** 1 **21.** $x \geqslant -\frac{4}{5}$
23. $x > -\frac{1}{12}$ **25.** $x > -\frac{4}{7}$ **27.** $x \leqslant -3$ **29.** $x > \frac{2}{3}$
31. $x < -\frac{2}{5}$ **33.** $2 < x < 4$ **35.** $\frac{3}{2} \leqslant x \leqslant \frac{11}{2}$
37. $-1 \leqslant x \leqslant \frac{14}{5}$ **39.** $(0, 5)$ **41.** $[-9, -4)$
43. $[x, x + h]$ **45.** (p, ∞) **47.** $[-3, 3]$
49. $[-14, -11)$ **51.** $(-\infty, -4]$ **53.** \$650
55. More than 7000 units   **57.** 480 lb   **59.** 810,000
61. $60\% \leqslant x < 100\%$

Margin Exercises, Section 1.3, pp. 21–33

1.

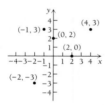

2 (a) Yes; **(b)** no

7. $f(5) = \frac{1}{5}, f(-2) = -\frac{1}{2}, f(\frac{1}{4}) = 4, f\left(\dfrac{1}{a}\right) = a,$

$f(k) = \dfrac{1}{k}, f(1 + t) = \dfrac{1}{1 + t}, f(x + h) = \dfrac{1}{x + h}$

8. $t(5) = 30, t(-5) = 20, t(x + h) = x + h + x^2 + 2xh + h^2$
9. (a) All real numbers except 3, since an input of 3 would result in division by 0;

(b) $f(5) = \frac{1}{2}, f(4) = 1, f(2.5) = -2, f(x + h) = \dfrac{1}{x + h - 3}$

10. Same as Margin Exercise 3, only labeled $f(x) = -2x + 1$
11. Same as Margin Exercise 4, only labeled $g(x) = x^2 - 3$
12. (a) No; **(b)** no; **(c)** yes; **(d)** yes
13.

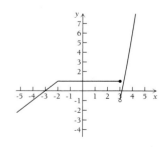

3.

4.

5.

Exercise Set 1.3, p. 34

1. **(a)**

Inputs	Outputs
4.1	11.2
4.01	11.02
4.001	11.002
4	11

(b) $f(5) = 13, f(-1) = 1, f(k) = 2k + 3, f(1 + t) = 2t + 5$, $f(x + h) = 2x + 2h + 3$
3. $g(-1) = -2, g(0) = -3, g(1) = -2, g(5) = 22$, $g(u) = u^2 - 3, g(a + h) = a^2 + 2ah + h^2 - 3$, $g(1 - h) = h^2 - 2h - 2$
5. **(a)** $f(4) = 1, f(-2) = 25, f(0) = 9, f(a) = a^2 - 6a + 9$, $f(t + 1) = t^2 - 4t + 4, f(t + 3) = t^2$, $f(x + h) = x^2 + 2xh + h^2 - 6x - 6h + 9$;
(b) Take an input, square it, subtract 6 times the input, add 9.

7.

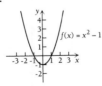

9.

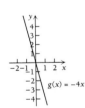

11.

13.

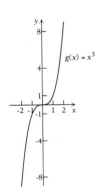

15. Yes   **17.** Yes   **19.** No   **21.** No   **23.** Yes
25. Yes   **27.** **(a)**

(b) no   **29.** $f(x + h) = x^2 + 2xh + h^2 - 3x - 3h$
31.

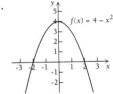

33.

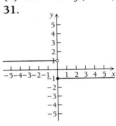

35. $R(10) = \$2050, R(100) = \$20,050$
37. $y = 5$; a function   **39.** $y = \pm\sqrt{x}$; not a function
41.

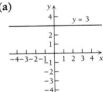

43.

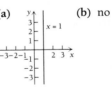

Margin Exercises, Section 1.4, pp. 37–49

1. **(a)**

(b) yes   **2.** **(a)**

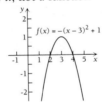

(b) no

3. **(a)**

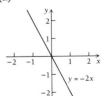

(b) yes; **(c)** -2

4. (a) *A, B;* **(b)** *C, D, E;* **(c)** *A;* **(d)** *E*
5. (a) $T = \frac{1}{36}h$; **(b)** 4.5
6. (a)

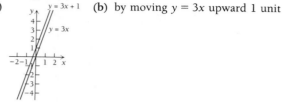

(b) by moving $y = 3x$ upward 1 unit

7. $m = -\frac{2}{3}$, y-intercept: (0, 2)   **8.** $y = -4x + 1$
9. $y + 7 = -4(x - 2)$, or $y = -4x + 1$ **10.** 2   **11.** $\frac{1}{8}$
12. $-\frac{17}{2}$ **13.** 0   **14.** 0   **15.** No slope
16. (a)

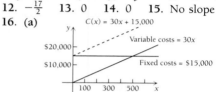

(b) $C(100) = \$18,000$, $C(400) = \$27,000$;
(c) $C(400) - C(100) = \$9000$
17. (a), (b)

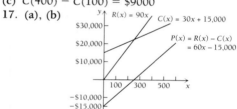

(c) break-even is 250

Exercise Set 1.4, p. 49

1.

3.

5. $m = -3$, y-intercept: (0, 0)

7. $m = 0.5$, y-intercept: (0, 0)

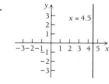

9. $m = -2$, y-intercept: (0, 3)

11. $m = -1$, y-intercept: (0, -2)

13. $m = -2$, y-intercept: (0, 2)   **15.** $m = -1$, y-intercept: $(0, -\frac{5}{2})$ **17.** $y + 5 = -5(x - 1)$, or $y = -5x$
19. $y - 3 = -2(x - 2)$, or $y = -2x + 7$ **21.** $y = \frac{1}{2}x - 6$
23. $y = 3$ **25.** $\frac{3}{2}$ **27.** $\frac{1}{2}$ **29.** No slope   **31.** 0
33. 3   **35.** 2   **37.** $y - 1 = \frac{3}{2}(x + 2)$, or
$y + 2 = \frac{3}{2}(x + 4)$, or $y = \frac{3}{2}x + 4$ **39.** $y + 4 = \frac{1}{2}(x - 2)$, or
$y + 3 = \frac{1}{2}(x - 4)$, or $y = \frac{1}{2}x - 5$ **41.** $x = 3$ **43.** $y = 3$
45. $y = 3x$ **47.** $y = 2x + 3$
49. (a) $A = P + 14\%P = P + 0.14P = 1.14P$;
(b) \$114; **(c)** \$240   **51. (a)** $C(x) = 20x + 100,000$;
(b) $R(x) = 45x$; **(c)** $P(x) = R(x) - C(x) = 25x - 100,000$;
(d) \$3,650,000, a profit; **(e)** 4000
53. (a) $C(x) = x + 250$; **(b)** $R(x) = 10x$. The student
charges \$10 per lawn. **(c)** 28
55. (a) $R = 4.17T$; **(b)** $R = 25.02$ **57. (a)** $B = 0.025W$;
(b) $B = 2.5\%W$. The weight of the brain is 2.5% of the
body weight. **(c)** 3 lb   **59. (a)** $D(0°) = 115$ ft,
$D(-20°) = 75$ ft, $D(10°) = 135$ ft, $D(32°) = 179$ ft;
(b)

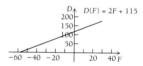

(c) Temperatures below $-57.5°$ would yield a negative
stopping distance, which has no meaning here. For
temperatures above 32°, there would be no ice.
61. (a) 200.69 cm; **(b)** 195.23 cm   **63. (a)** $A(0) = 19.7$,
$A(1) = 19.78$, $A(10) = 20.5$, $A(30) = 22.1$, $A(40) = 22.9$;
(b) 23.38; **(c)**

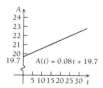

Margin Exercises, Section 1.5, pp. 55–67

1.

 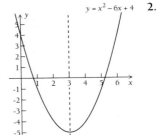

2.

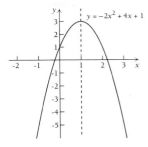

3. $\dfrac{-1 \pm \sqrt{22}}{3}$

4.

5. **(a)** All real numbers except -2; **(b)** all real numbers except $-5, 1$; **(c)** all real numbers except 5

6.

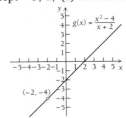

7.

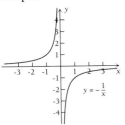

8. 160,000

9.

10.

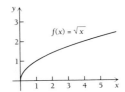

11. $[-\frac{3}{2}, \infty)$ **12.** $t^{3/4}$ **13.** $y^{1/5}$ **14.** $x^{-2/5}$ **15.** $t^{-1/3}$
16. x^3 **17.** $x^{7/2}$ **18.** $\sqrt[7]{y}$ **19.** $\sqrt{x^3}$

20. $\dfrac{1}{\sqrt{t^3}}$ **21.** $\dfrac{1}{\sqrt{b}}$ **22.** 32   **23.** 9   **24.** ($12.50, 250)

Exercise Set 1.5, p. 68

1.

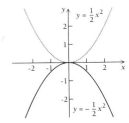

3.

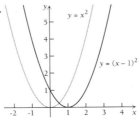

5.

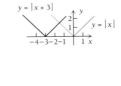

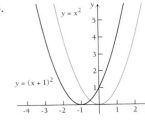

9.

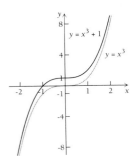

11.

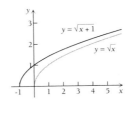

13.

15.

17.

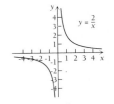

19.

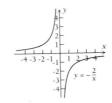

21.

x	-2	-1	$-\frac{1}{2}$	$\frac{1}{2}$	1	2
y	$\frac{1}{4}$	1	4	4	1	$\frac{1}{4}$

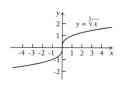

23.

25.

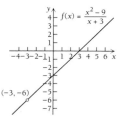

27.

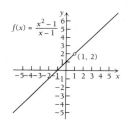

29. $1 \pm \sqrt{3}$ **31.** $-3 \pm \sqrt{10}$ **33.** $\dfrac{1 \pm \sqrt{2}}{2}$

35. $\dfrac{-4 \pm \sqrt{10}}{3}$ **37.** $x^{3/2}$ **39.** $a^{3/5}$

41. $t^{1/7}$ **43.** $t^{-4/3}$ **45.** $t^{-1/2}$ **47.** $(x^2 + 7)^{-1/2}$

49. $\sqrt[5]{x}$ **51.** $\sqrt[3]{y^2}$

53. $\dfrac{1}{\sqrt[5]{t^2}}$ **55.** $\dfrac{1}{\sqrt[3]{b}}$ **57.** $\dfrac{1}{\sqrt[6]{e^{17}}}$ **59.** $\dfrac{1}{\sqrt{x^2 - 3}}$

61. 27   **63.** 16   **65.** 8   **67.** All real numbers except 5

69. All real numbers except 2, 3   **71.** $[-\frac{4}{5}, \infty)$

73. ($4, 2)   **75.** ($50, 500)   **77.** ($5, 1)   **79.** ($1, 4)

81. ($2, 3)   **83.** $2.27   **85.** 84, 220, 364

87. $-2, 0, 2$

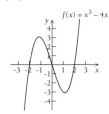

89. -1.2543

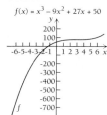

Margin Exercises, Section 1.6, pp. 72–76

1. $3.71924 \approx 3{:}43.2$ **2.** 1993

3. (a) $A = 0.5t + 18.5$; (b) 52¢

4. (a) $A = \frac{131}{300}x^2 - 39\frac{7}{10}x + 1039\frac{1}{3}$,

$A = 0.437x^2 - 39.7x + 1039.333$; (b) 516

Exercise Set 1.6, p. 76

1. (b) Linear; (c) $C = 32x + 9998$; (d) $10,126;

$10,318; (e) $C = 29.5x + 10,000$; (f) $10,118; $10,295

3. (b) Constant, linear; (c) $S = -20t + 100,330$; (d) Let

$S = b$, where b is the average of sales totals: $100,296.25.

5. (b) Quadratic; (c) $t = 0.0000033N^2 + 0.00233N$;

(d) 172, 408, 586

Summary and Review: Chapter 1, p. 79

1. [1.1] $(2.01)(2.01)$ **2.** [1.1] y^3 **3.** [1.1] y^{-7}

4. [1.1] $243t^{-10}m^{20}$ **5.** [1.1] $9x^2 - 6xt + t^2$ **6.** [1.1]

$9x^2 - t^2$ **7.** [1.1] $12x^2 + 11xt - 5t^2$ **8.** [1.1]

$(x - 8)(x + 6)$ **9.** [1.1] $(5x - 4t)(5x + 4t)$ **10.** [1.1]

$(a - 4b)^2$ **11.** [1.1] $(3x - 1)(7x - 4)$ **12.** [1.1]

$1212.75   **13.** [1.1] $5017.60   **14.** [1.2] $x \geqslant 1$

15. [1.2] 1   **16.** [1.2] $\frac{5}{4}, -\frac{5}{4}$ **17.** [1.2] $0, 3, -\frac{5}{2}$

18. [1.2] $(-6, 1]$ **19.** [1.2] $(0, \infty)$ **20.** [1.3] 13

21. [1.3] $2h^2 + 3h + 4$ **22.** [1.3] 3   **23.** [1.3] 36

24. [1.3] $h^2 - 2h + 1$ **25.** [1.3] 9

26. [1.3]   **27.** [1.3]

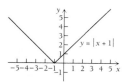

28. [1.5]

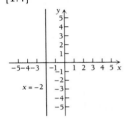

29. [1.3] No   **30.** [1.3] Yes

1. $\dfrac{1}{e^k}$ **2.** e^{-13}, or $\dfrac{1}{e^{13}}$ **3.** $x^2 + 2xh + h^2$
4. $(5x - t)(5x + t)$ **5.** \$920   **6.** $x > -4$ **7.** (a) 5;
(b) $x^2 + 2xh + h^2 - 4$ **8.** $m = -3$; y-intercept: $(0, 2)$
9. $y + 5 = \frac{1}{4}(x - 8)$, or $y = \frac{1}{4}x - 7$ **10.** $m = 6$
11. $F = \frac{2}{3}W$ **12.** (a) $C(x) = 0.5x + 10{,}000$;
(b) $R(x) = 1.3x$;   (c) $P(x) = 0.8x - 10{,}000$;   (d) 12,500
13. (\$3, 16)

31. [1.3] Yes

32. [1.4]

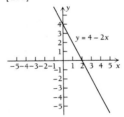

14.   **15.** $t^{-1/2}$ **16.** $\dfrac{1}{\sqrt[5]{t^3}}$

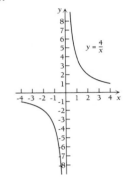

33. [1.4]

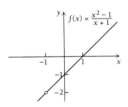

34. [1.4] $y = -\frac{7}{11}x + \frac{6}{11}$, or $7x + 11y = 6$
35. [1.4] $y = 8x + 7$
36. [1.4] $m = -\frac{1}{6}$, y-intercept: $(0, 3)$ **37.** [1.4] 72 lb
38. [1.4] (a) $R(x) = 28x$;   (b) $C(x) = 16x + 80{,}000$;
(c) $P(x) = 12x - 80{,}000$;   (d) \$16,000 profit;   (e) 6667
39. [1.5] $\sqrt[6]{y}$ **40.** [1.5] $x^{3/20}$ **41.** [1.5] 125
42. [1.5] $(-\infty, 3]$
43. [1.5] All real numbers except 1 and -1
44. [1.5] (\$3, 27)   **45.** [1.6] $y = -\frac{11}{6}x + 10$
46. [1.6] $y = x^2 - 2x + 5$
47. [1.3]   **48.** [1.5] $\frac{1}{32}$

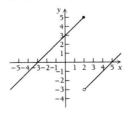

17.

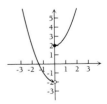

18. All real numbers except 2, -7 **19.** $[-2, \infty)$
20. $y = 4x - 1$ **21.** $y = -4.5x^2 + 17.5x - 8$ **22.** $[c, d)$
23.

24. (a) 525;   (b) \$4.31

CHAPTER 2

Margin Exercises, Section 2.1, pp. 85-94

1. (a) 14, 16.4, 16.7, 16.97, 16.997; 17.003, 17.03, 17.3, 18.2, 20; **(b)** 17;
(c)

(d) -4; **(e)** 5; **(f)** -1

2. (a) $-0.5, -1, -10, -100, -1000; 1000, 100, 10, 1.25,$ 1; **(b)** does not exist;
(c)

(d) does not exist; **(e)** -0.5; **(f)** 1

3. (a) 4; **(b)** -3; **(c)** does not exist; **(d)** 0; **(e)** 0; **(f)** 0; **(g)** 3; **(h)** 3; **(i)** 3
4. (a)

(b) does not exist; **(c)** 2

5. (a)

(b) 4; **(c)** 4; **(d)** 4; **(e)** 4;

(f) 4; **(g)** no; **(h)** yes

6. (a) 2; **(b)** 3; **(c)** yes; **(d)** no   **7. (a)** Yes; **(b)** yes; **(c)** no; **(d)** no   **8. (a)** No, yes; **(b)** yes, no   **9. (a)** Yes, 17; **(b)** yes, 17; *(c)* yes; **(d)** yes   **10. (a)** No; **(b)** yes   **11.** No   **12.** No   **13.** x is continuous by C2, and $\sqrt[3]{x}$ is continuous by C2; 7 is continuous by C1, and x^2 is continuous by C2, so $7x^2$ is continuous by C3. Then $\sqrt[3]{x} - 7x^2$ is continuous by C3. Now x is continuous by C2, and 2 is continuous by C1, so $x - 2$ is continuous by C3. Thus $\dfrac{\sqrt[3]{x} - 7x^2}{x - 2}$ is continuous by C3 and C4, except when $x = 2$.

Exercise Set 2.1, p. 94

1. No   **3.** Yes   **5. (a)** -1, 2, does not exist; **(b)** -1; **(c)** no; **(d)** 3; **(e)** 3; **(f)** yes   **7. (a)** 2; **(b)** 2; **(c)** yes; **(d)** 0; **(e)** 0; **(f)** yes   **9.** No, yes, no, yes
11. 29¢, 52¢, does not exist   **13.** 75¢
15. \$20, \$28, does not exist   **17.** No, yes
19. 100, 30, does not exist   **21.** No, yes   **23. (a)** The "hole" may or may not appear, depending on the software package or the graphing calculator. **(b)** 0.1667

Margin Exercises, Section 2.2, pp. 97–103

1. 53   **2.** $\sqrt{8}$ **3.** -8 **4.** $\frac{1}{6}$ **5. (a)** $2x + 1$, $2x + 0.7$, $2x + 0.4$, $2x + 0.1$, $2x + 0.01$, $2x + 0.001$; **(b)** $2x$
6. (a) 4.00, 4.50, 5.14, 6.00, 7.20, 9.00, 12.00, 18.00, 36.00, 54.00, 108.00; **(b)** ∞; **(c)** ∞ **7. (a)** 3.25, 2.25, 2.0625, 2.025, 2.005, 2.0005; **(b)** 2   **8.** $\frac{2}{3}$ **9.** 0   **10.** ∞

Exercise Set 2.2, p. 103

1. -2 **3.** Does not exist   **5.** 11   **7.** -10 **9.** $-\frac{5}{2}$
11. 5   **13.** 2   **15.** 3   **17.** Does not exist   **19.** $\frac{2}{7}$
21. $\frac{5}{4}$ **23.** $\frac{1}{2}$ **25.** $6x^2$ **27.** $\dfrac{-2}{x^3}$ **29.** $\frac{2}{5}$ **31.** 5
33. $\frac{1}{2}$ **35.** $\frac{2}{3}$ **37.** 0   **39.** ∞ **41.** 0   **43.** ∞
45. (a) 1; **(b)** 1; **(c)** 1; **(d)** 1; **(e)** yes, no
47. (a) \$800, \$736, \$677.12, \$622.95, \$573.11; **(b)** \$5656.12; **(c)** \$10,000   **49.** \$13 per unit. This means that as more and more units are produced, the average cost gets closer and closer to \$13 per unit.
51. Does not exist   **53.** $\frac{3}{2}$ **55.** $-\infty$ **57. (a)** 1, -1; **(b)** They are not defined because the radicand in the numerator is negative. **(c)** 0, 0

Margin Exercises, Section 2.3, pp. 107–112

1. (a) 20 suits/hr, 35 suits/hr, 9 suits/hr, 36 suits/hr;
(b) 11 A.M. to 12 P.M.; (c) workers were anticipating the lunch break; answers may vary; (d) 10 A.M. to 11 A.M.;
(e) workers were fatigued or took longer breaks than they should have; answers may vary; (f) 25 suits/hr
2. (a) 21; (b) 7; (c) 28; (d) 21 **3.** (a) $\frac{1}{2}$; (b) $\frac{1}{2}$; (c) $\frac{1}{2}$

4.

x	h	$x + h$	$f(x)$	$f(x + h)$	$f(x + h) - f(x)$	$\dfrac{f(x + h) - f(x)}{h}$
3	2	5	36	100	64	32
3	1	4	36	64	28	28
3	0.1	3.1	36	38.44	2.44	24.4
3	0.01	3.01	36	36.2404	0.2404	24.04
3	0.001	3.001	36	36.024004	0.024004	24.004

5. (a) $f(x + h) = 4x^2 + 8xh + 4h^2$;
(b) $f(x + h) - f(x) = 8xh + 4h^2$;
(c) $\dfrac{f(x + h) - f(x)}{h} = 4(2x + h)$;
(d) 36, 40, 44, 47.6, 47.96, 47.996
6. (a) $4(3x^2 + 3xh + h^2)$; (b) 28, 45.64, 47.7604, 47.976004
7. (a) $\dfrac{-1}{x(x + h)}$;
(b) -0.1; -0.1667, -0.2381, -0.2488, -0.2499

Exercise Set 2.3, p. 113

1. (a) $7(2x + h)$; (b) 70, 63, 56.7, 56.07
3. (a) $-7(2x + h)$; (b) -70, -63, -56.7, -56.07
5. (a) $7(3x^2 + 3xh + h^2)$; (b) 532, 427, 344.47, 336.8407
7. (a) $\dfrac{-5}{x(x + h)}$; (b) -0.2083, -0.25, -0.3049, -0.3117
9. (a) -2; (b) all -2 **11.** (a) $2x + h - 1$; (b) 9, 8, 7.1, 7.01 **13.** (a) 70, 39, 29, 23 pleasure units/unit of product; (b) The more you get, the less pleasure you get from each additional unit. **15.** (a) $300,093.99;

(b) $299,100; (c) $993.99; (d) $993.99
17. (a) $4770.67 million/yr; (b) $2235 million/yr
19. (a) 1.25 words/min, 1.25 words/min, 0.625 words/min, 0 words/min, 0 words/min; (b) You have reached a saturation point; you cannot memorize any more. **21.** (a) 144 ft; (b) 400 ft; (c) 128 ft/sec
23. (a) 125 million people/yr for each; (b) no; (c) A: 290 million people/yr, -40 million people/yr, -50 million people/yr, 300 million people/yr; B: 125 million people/yr in all intervals; (d) A **25.** $\approx 44{,}182$ marriages/yr

27. $2ax + b + ah$ **29.** $\dfrac{1}{\sqrt{x + h} + \sqrt{x}}$

31. $\dfrac{-2x - h}{x^2(x + h)^2}$ **33.** $\dfrac{1}{(x + 1)(x + 1 + h)}$

Margin Exercises, Section 2.4, pp. 117–127

1. (a) L_2, L_3, L_4, L_6; (b) $-2, -1, 0, 1, 2$; (c) $m(x) = 2x$
2. Drawing left to student **3.** $f'(5) = 10$
4. $f'(x) = 8x, f'(5) = 40 =$ the slope of the tangent line at $(5, f(5))$, or $(5, 100)$ **5.** $f'(x) = 12x^2, f'(-5) = 300$,

$f'(0) = 0$ **6. (a)** $f'(x) = \dfrac{-1}{x^2}$;

(b) $f'(-10) = -\dfrac{1}{100} = -0.01, f'(-2) = -\dfrac{1}{4} = -0.25$;

(c) $y = -0.01x - 0.2$; **(d)** $y = -0.25x - 1$

7. x_2, x_4, x_5, x_6

Exercise Set 2.4, p. 127

1. $f'(x) = 10x, f'(-2) = -20, f'(-1) = -10, f'(0) = 0,$
$f'(1) = 10, f'(2) = 20$ **3.** $f'(x) = -10x, f'(-2) = 20,$
$f'(-1) = 10, f'(0) = 0, f'(1) = -10, f'(2) = -20$
5. $f'(x) = 15x^2, f'(-2) = 60, f'(-1) = 15, f'(0) = 0,$
$f'(1) = 15, f'(2) = 60$ **7.** $f'(x) = 2$, all 2
9. $f'(x) = -4$, all -4 **11.** $f'(x) = 2x + 1, f'(-2) = -3,$
$f'(-1) = -1, f'(0) = 1, f'(1) = 3, f'(2) = 5$
13. $f'(x) = \dfrac{-4}{x^2}; f'(-2) = -1; f'(-1) = -4;$

$f'(0)$ does not exist; $f'(1) = -4; f'(2) = -1$
15. $f'(x) = m$, all m **17.** $y = 6x - 9; y = -2x - 1;$
$y = 20x - 100$ **19.** $y = -5x + 10; y = -5x - 10;$
$y = -0.0005x + 0.1$ **21.** $y = 2x + 5; y = 4;$
$y = -10x + 29$ **23.** $x_0, x_3, x_4, x_6, x_{12}$

25. 1, 2, 3, 4, and so on **27.** $\dfrac{-2}{x^3}$ **29.** $\dfrac{1}{(1+x)^2}$

31. (a) **(b)** $v = 21.545t + 3.455;$

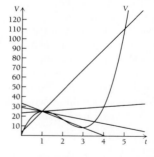

(c) \$21.545 million per year; **(d)** $v = 1.666667t + 23.33333,$
\$1.67 million per year; $v = -8.169875t + 33.16988,$
$-\$8.17$ million per year; $v = -4.002602t + 29.0026,$
$-\$4$ million per year; **(e)** 0

Margin Exercises, Section 2.5, pp. 130–138

1. (a) $3x^2$; **(b)** $3x^2$; **(c)** 48 **2.** $\dfrac{dy}{dx} = 6x^5$

3. $\dfrac{dy}{dx} = -7x^{-8}$, or $-\dfrac{7}{x^8}$ **4.** $\dfrac{dy}{dx} = \dfrac{1}{3}x^{-2/3}$, or $\dfrac{1}{3\sqrt[3]{x^2}}$

5. $\dfrac{dy}{dx} = -\dfrac{1}{4}x^{-5/4}$, or $\dfrac{-1}{4\sqrt[4]{x^5}}$ **6.** $g'(x) = 0$

7. $\dfrac{dy}{dx} = 100x^{19}$ **8.** $\dfrac{dy}{dx} = \dfrac{3}{x^2}$

9. $\dfrac{dy}{dx} = -4x^{-1/2}$, or $\dfrac{-4}{\sqrt{x}}$ **10.** $\dfrac{dy}{dx} = x^{5.25}$

11. $\dfrac{dy}{dx} = -\dfrac{1}{4}$ **12.** $\dfrac{dy}{dx} = 28x^3 + 12x$

13. $\dfrac{dy}{dx} = 30x - \dfrac{4}{x^2} + \dfrac{1}{2\sqrt{x}}$ **14.** $(2, \frac{8}{3})$

15. $(2 + \sqrt{3}, \frac{8}{3} + \sqrt{3}), (2 - \sqrt{3}, \frac{8}{3} - \sqrt{3})$

Exercise Set 2.5, p. 138

1. $7x^6$ **3.** 0 **5.** $600x^{149}$ **7.** $3x^2 + 6x$

9. $\dfrac{4}{\sqrt{x}}$ **11.** $0.07x^{-0.93}$ **13.** $\dfrac{2}{5\sqrt[5]{x}}$ **15.** $\dfrac{-3}{x^4}$

17. $6x - 8$ **19.** $\dfrac{1}{4\sqrt[4]{x^3}} + \dfrac{1}{x^2}$ **21.** $1.6x^{1.5}$

23. $\dfrac{-5}{x^2} - 1$ **25.** 4 **27.** 4 **29.** x^3

31. $-0.02x - 0.5$
33. $-2x^{-5/3} + \frac{3}{4}x^{-1/4} + \frac{6}{5}x^{1/5} - 24x^{-4}$

35. $-\dfrac{2}{x^2} - \dfrac{1}{2}$ **37.** $-16x^{-2} + 24x^{-4} - 4x^{-5}$

39. $\frac{1}{2}x^{-1/2} + \frac{1}{3}x^{-2/3} - \frac{1}{4}x^{-3/4} + \frac{1}{5}x^{-4/5}$ **41.** $y = 10x - 15;$
$y = x + 3; y = -2x + 1$ **43.** $(0, 0)$ **45.** $(0, 0)$
47. $(\frac{5}{6}, \frac{23}{12})$ **49.** $(-25, 76.25)$ **51.** There are none.
53. The tangent is horizontal at all points on the graph.
55. $(\frac{5}{3}, \frac{148}{27}), (-1, -4)$
57. $(\sqrt{3}, 2 - 2\sqrt{3}), (-\sqrt{3}, 2 + 2\sqrt{3})$ **59.** $(\frac{19}{2}, \frac{399}{4})$
61. $(60, 150)$ **63.** $(-2 + \sqrt{3}, \frac{4}{3} - \sqrt{3}),$
$(-2 - \sqrt{3}, \frac{4}{3} + \sqrt{3})$
65. $(0, -4), \left(\sqrt{\frac{2}{3}}, -\frac{40}{9}\right), \left(-\sqrt{\frac{2}{3}}, -\frac{40}{9}\right)$

67. $2x - 1$ **69.** $2x + 1$ **71.** $3x^2 - \dfrac{1}{x^2}$

73. $-192x^2$ **75.** $\dfrac{2}{3\sqrt[3]{x^2}}$ **77.** $3x^2 + 6x + 3$

79. $\dfrac{F(x + h) - F(x)}{h} =$

$\dfrac{[f(x + h) - g(x + h)] - [f(x) - g(x)]}{h} =$

$$\frac{f(x+h)-f(x)}{h}-\frac{g(x+h)-g(x)}{h}.$$ As $h \to 0$, the

two terms on the right approach $f'(x)$ and $g'(x)$, respectively, so their difference approaches $f'(x) - g'(x)$. Thus, $F'(x) = f'(x) - g'(x)$.

81. (a)

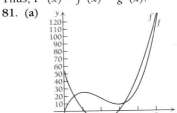

(b) (1.0836, 25.1029), (2.9503, 8.62466)

Margin Exercises, Section 2.6, pp. 141–145

1. (a) 70 mi/hr; **(b)** 100 mi/hr **2. (a)** $v(t) = 32t$;
(b) $v(2) = 64$ ft/sec; **(c)** $v(10) = 320$ ft/sec **3.** $a(t) = 32$ ft/sec^2 **4. (a)** $V'(s) = 3s^2$; **(b)** $V'(10) = 300$ ft^2
5. (a) $P'(t) = 9700 + 20{,}000t$; **(b)** 308,500; 109,700 bacteria/hr; **(c)** 428,200; 129,700 bacteria/hr
6. (a) $P(x) = 40x - 0.5x^2 - 3$; **(b)** $R(40) = \$1200$, $C(40) = \$403$, $P(40) = \$797$; **(c)** $R'(x) = 50 - x$, $C'(x) = 10$, $P'(x) = 40 - x$; **(d)** $R'(40) = \$10$ per unit, $C'(40) = \$10$ per unit, $P'(40) = \$0$ per unit; **(e)** no

Exercise Set 2.6, p. 146

1. (a) $v(t) = 3t^2 + 1$; **(b)** $a(t) = 6t$; **(c)** $v(4) = 49$ ft/sec, $a(4) = 24$ ft/sec^2 **3. (a)** $P(x) = -0.001x^2 + 3.8x - 60$;
(b) $R(100) = \$500$, $C(100) = \$190$, $P(100) = \$310$;
(c) $R'(x) = 5$, $C'(x) = 0.002x + 1.2$, $P'(x) = -0.002x + 3.8$; **(d)** $R'(100) = \$5$ per unit, $C'(100) = \$1.40$ per unit, $P'(100) = \$3.60$ per unit
5. (a) $N'(a) = -2a + 300$; **(b)** 2906; **(c)** 280 units/thousand dollars **7. (a)** 630 units per month, 980 units per month, 1430 units per month, 630 units per month; **(b)** $M'(t) = -4t + 100$; **(c)** 80, 60, 0, -80
9. (a) $\dfrac{dD}{dp} = -\dfrac{1}{2\sqrt{p}}$; **(b)** 95; **(c)** $-\frac{1}{10}$ **11.** $\dfrac{dC}{dr} = 2\pi$
13. (a) $T'(t) = -0.2t + 1.2$; **(b)** $100.175°$;
(c) 0.9 degrees/day
15. $\dfrac{dT}{dW} = 1.31W^{0.31}$ **17.** $\dfrac{dR}{dQ} = kQ - Q^2$
19. $A'(t) = 0.08$

Margin Exercises, Section 2.7, pp. 150–154

1. $f'(x) = 54x^{17}$
2. $f'(x) = (9x^3 + 4x^2 + 10)(-14x + 4x^3) + (27x^2 + 8x)(-7x^2 + x^4)$
3. $f'(x) = (3x^4 + \sqrt{x})(2x - 6) + (12x^3 + \frac{1}{2}x^{-1/2})(x^2 - 6x)$
4. $f'(x) = 4x^3$ **5.** $f'(x) = \dfrac{3x^2 - 5}{x^6}$
6. $f'(x) = \dfrac{-x^4 + 3x^2 + 2x}{(x^3 + 1)^2}$
7. (a) $R(p) = p(200 - p) = 200p - p^2$;
(b) $R'(p) = 200 - 2p$

Exercise Set 2.7, p. 154

1. $11x^{10}$ **3.** $\dfrac{1}{x^2}$ **5.** $3x^2$

7. $(8x^5 - 3x^2 + 20)\left(32x^3 - \dfrac{3}{2\sqrt{x}}\right) + (40x^4 - 6x)(8x^4 - 3\sqrt{x})$ **9.** $300 - 2x$
11. $(4\sqrt{x} - 6)(3x^2 - 2) + (2x^{-1/2})(x^3 - 2x + 4)$
13. $2x + 6$ **15.** $6x^5 - 32x^3 + 32x$
17. $(5x^{-3})(4x^3 - 15x^2 + 10) - 15x^{-4}(x^4 - 5x^3 + 10x - 2)$, or $5 - 100x^{-3} + 30x^{-4}$
19. $\left(x + \dfrac{2}{x}\right)(2x) + \left(1 - \dfrac{2}{x^2}\right)(x^2 - 3)$, or $3x^2 + \dfrac{6}{x^2} - 1$
21. $\dfrac{300}{(300 - x)^2}$ **23.** $\dfrac{17}{(2x + 5)^2}$
25. $\dfrac{-x^4 - 3x^2 - 2x}{(x^3 - 1)^2}$ **27.** $\dfrac{1}{(1 - x)^2}$
29. $\dfrac{2}{(x + 1)^2}$ **31.** $\dfrac{-1}{(x - 3)^2}$ **33.** $\dfrac{-2x^2 + 6x + 2}{(x^2 + 1)^2}$
35. $\dfrac{-18x + 35}{x^8}$ **37.** $\dfrac{4\sqrt{x} - \frac{1}{2}x^{-1/2}(4x + 3)}{x}$, or $\dfrac{4x - 3}{2x^{3/2}}$
39. $y = 2$; $y = \frac{1}{2}x + 2$
41. (a) $D'(p) = \dfrac{-2978}{(10p + 11)^2}$; **(b)** $-\dfrac{2978}{2601}$
43. (a) $R(p) = p(400 - p) = 400p - p^2$;
(b) $R'(p) = 400 - 2p$ **45. (a)** $R(p) = 4000 + 3p$;
(b) $R'(p) = 3$
47. $A'(x) = \dfrac{xC'(x) - C(x)}{x^2}$ **49.** $\dfrac{5x^3 - 30x^2\sqrt{x}}{2\sqrt{x}(\sqrt{x} - 5)^2}$
51. $\dfrac{-3(1 + 2v)}{(1 + v + v^2)^2}$

53. $\dfrac{2t^3 - t^2 + 1}{(1 - t + t^2 - t^3)^2}$ **55.** $\dfrac{5x^3 + 15x^2 + 2}{2x\sqrt{x}}$

57. $[x(9x^2 + 6) + (3x^3 + 6x - 2)](3x^4 + 7) +$
$12x^4(3x^3 + 6x - 2)$

59. $\dfrac{6t^2(t^5 + 3)}{(t^3 + 1)^2} + \dfrac{5t^4(t^3 - 1)}{t^3 + 1}$

61.
$\dfrac{\{(x^7 - 2x^6 + 9)[(2x^2 + 3)(12x^2 - 7) + 4x(4x^3 - 7x + 2)] \\ \quad - (7x^6 - 12x^5)(2x^2 + 3)(4x^3 - 7x + 2)\}}{(x^7 - 2x^6 + 9)^2}$

63. $(-1.1547, 3.0792), (1.1547, -3.0792)$

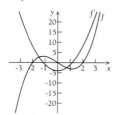

65. No points at which the tangent line is horizontal.

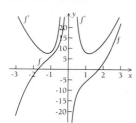

67. $(-0.2, -0.75), (0.2, 0.75)$

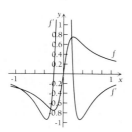

Margin Exercises, Section 2.8, pp. 157–163

1. $20x(1 + x^2)^9$ **2.** $\dfrac{-x}{\sqrt{1 - x^2}}$

3. $-2x(1 + x^2)(1 + 3x^2)$ **4.** $2(x - 4)^4(6 - x)^2(21 - 4x)$

5. $\left(\dfrac{x + 5}{x - 4}\right)^{-2/3} \cdot \dfrac{-3}{(x - 4)^2}$

6. $3(x^2 - 1), 9x^2 - 1$ **7.** $4\sqrt[3]{x} + 5, \sqrt[3]{4x + 5}$

8. $2u, 3x^2, 6x^2(x^3 + 2)$ **9.** (a) $f(x) = \sqrt[3]{x}$,

$g(x) = x^2 + 1$; (b) $f(x) = \dfrac{1}{x}, g(x) = (x + 5)^4$

Exercise Set 2.8, p. 163

1. $-55(1 - x)^{54}$ **3.** $\dfrac{4}{\sqrt{1 + 8x}}$ **5.** $\dfrac{3x}{\sqrt{3x^2 - 4}}$

7. $-240x(3x^2 - 6)^{-41}$

9. $\sqrt{2x + 3} + \dfrac{x}{\sqrt{2x + 3}}$, or $\dfrac{3(x + 1)}{\sqrt{2x + 3}}$

11. $2x\sqrt{x - 1} + \dfrac{x^2}{2\sqrt{x - 1}}$, or $\dfrac{5x^2 - 4x}{2\sqrt{x - 1}}$ **13.** $\dfrac{-6}{(3x + 8)^3}$

15. $(1 + x^3)^2(-3x^2 - 12x^5)$, or $-3x^2(1 + x^3)^2(1 + 4x^3)$

17. $4x - 400$ **19.** $2(x + 6)^9(x - 5)^3(7x - 13)$

21. $4(x - 4)^7(3 - x)^3(10 - 3x)$ **23.** $4(2x - 3)^2(3 - 8x)$

25. $\left(\dfrac{1 - x}{1 + x}\right)^{-1/2} \cdot \dfrac{-1}{(x + 1)^2}$ **27.** $\left(\dfrac{3x - 1}{5x + 2}\right)^3 \cdot \dfrac{44}{(5x + 2)^2}$

29. $\frac{1}{3}(x^4 + 3x^2)^{-2/3}(4x^3 + 6x)$

31. $100(2x^3 - 3x^2 + 4x + 1)^{99}(6x^2 - 6x + 4)$

33. $\left(\dfrac{2x + 3}{5x - 1}\right)^{-5} \cdot \dfrac{68}{(5x - 1)^2}$ **35.** $\left(\dfrac{x^2 + 1}{x^2 - 1}\right)^{-1/2} \cdot \dfrac{-2x}{(x^2 - 1)^2}$

37. $\dfrac{(2x + 3)^3(-6x - 61)}{(3x - 2)^6}$

39. $16(2x + 1)^{-1/3}(3x - 4)^{5/4} + 45(2x + 1)^{2/3}(3x - 4)^{1/4}$,
or $(2x + 1)^{-1/3}(3x - 4)^{1/4}(138x - 19)$

41. $\frac{1}{2}u^{-1/2}, 2x, \dfrac{x}{\sqrt{x^2 - 1}}$

43. $50u^{49}, 12x^2 - 4x, 50(4x^3 - 2x^2)^{49}(12x^2 - 4x)$

45. $2u + 1, 3x^2 - 2, (2x^3 - 4x + 1)(3x^2 - 2)$

47. $y = \frac{5}{4}x + \frac{3}{4}$

49. (a) $\dfrac{2x - 3x^2}{(1 + x)^6}$; (b) $\dfrac{2x - 3x^2}{(1 + x)^6}$; (c) same

51. $12x^2 - 12x + 5, 6x^2 + 3$ **53.** $\dfrac{16}{x^2} - 1, \dfrac{2}{4x^2 - 1}$

55. $x^4 - 2x^2 + 2, x^4 + 2x^2$

57. $f(x) = x^5, g(x) = 3x^2 - 7$

59. $f(x) = \dfrac{x + 1}{x - 1}, g(x) = x^3$ **61.** $C'(x) = \dfrac{1500x^2}{\sqrt{x^3 + 2}}$

63. $P'(x) = \dfrac{2000x}{\sqrt{x^2 + 3}} - \dfrac{1500x^2}{\sqrt{x^3 + 2}}$ **65.** $\$3000(1 + i)^2$

67. (a) $D(t) = \dfrac{80{,}000}{1.6t + 9}$; **(b)** $D'(t) = \dfrac{-128{,}000}{(1.6t + 9)^2}$;

(c) $\dfrac{-128{,}000}{28{,}561}$, or about -4.48 **69.** $\dfrac{x^2 - 2}{\sqrt[3]{x^3 - 6x + 1}^2}$

71. $\dfrac{x - 2}{2(x - 1)^{3/2}}$ **73.** $\dfrac{-4(1 + 2v)^3}{v^5}$

75. $\dfrac{1}{\sqrt{1 - x^2}(1 - x)}$, or $\dfrac{\sqrt{1 - x^2}}{(1 - x^2)(1 - x)}$

77. $3\left(\dfrac{x^2 - x - 1}{x^2 + 1}\right)^2 \cdot \dfrac{x^2 + 4x - 1}{(x^2 + 1)^2}$, or

$\dfrac{3(x^2 - x - 1)^2(x^2 + 4x - 1)}{(x^2 + 1)^4}$ **79.** $\dfrac{1}{\sqrt{t}(1 + \sqrt{t})^2}$

81. $(-2.14476, -7.728)$, $(2.14476, 7.728)$

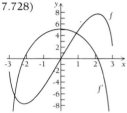

Margin Exercises, Section 2.9, pp. 166–168

1. $f'(x) = 12x^5 - 5x^4$, $f''(x) = 60x^4 - 20x^3$,
$f'''(x) = 240x^3 - 60x^2$, $f^{(4)}(x) = 720x^2 - 120x$,
$f^{(5)}(x) = 1440x - 120$, $f^{(6)}(x) = 1440$

2. (a) $\dfrac{dy}{dx} = 7x^6 - 3x^2$; **(b)** $\dfrac{d^2y}{dx^2} = 42x^5 - 6x$;

(c) $\dfrac{d^3y}{dx^3} = 210x^4 - 6$; **(d)** $\dfrac{d^4y}{dx^4} = 840x^3$ **3.** $\dfrac{d^2y}{dx^2} = \dfrac{4}{x^3}$

4. $y' = 60(x^2 - 12x)^{29}(x - 6)$;
$y'' = 60(x^2 - 12x)^{28}[59x^2 - 708x + 2088]$
5. $a(t) = 12t^2$

Exercise Set 2.9, p. 168

1. 0 **3.** $-\dfrac{2}{x^3}$ **5.** $-\dfrac{3}{16}x^{-7/4}$ **7.** $12x^2 + \dfrac{8}{x^3}$ **9.** $\dfrac{12}{x^5}$

11. $n(n - 1)x^{n - 2}$ **13.** $12x^2 - 2$

15. $-\dfrac{1}{4}(x - 1)^{-3/2}$, or $\dfrac{-1}{4\sqrt{(x - 1)^3}}$ **17.** $2a$

19. $86(x^2 - 8x)^{41}(85x^2 - 680x + 1344)$

21. $200x^2(x^4 - 4x^2)^{48}(199x^4 - 798x^2 + 792)$
23. $-\dfrac{2}{9}x^{-4/3}$ **25.** $-\dfrac{3}{16}(x - 8)^{-5/4}$ **27.** $6x^{-4} + 24x^{-5}$
29. 24 **31.** $720x$ **33.** $20(x^2 - 5)^8[19x^2 - 5]$
35. $a(t) = 6t + 2$ **37.** $P''(t) = 200{,}000$
39. $y' = -x^{-2} - 2x^{-3}$, $y'' = 2x^{-3} + 6x^{-4}$,
$y''' = -6x^{-4} - 24x^{-5}$

41. $y' = \dfrac{1 + 2x^2}{\sqrt{1 + x^2}}$, $y'' = \dfrac{2x^3 + 3x}{(1 + x^2)^{3/2}}$, $y''' = \dfrac{3}{(1 + x^2)^{5/2}}$

43. $y' = \dfrac{11}{(2x + 3)^2}$, $y'' = \dfrac{-44}{(2x + 3)^3}$, $y''' = \dfrac{264}{(2x + 3)^4}$

45. $y' = \dfrac{x - 2}{2(x - 1)^{3/2}}$, $y'' = \dfrac{4 - x}{4(x - 1)^{5/2}}$, $y''' = \dfrac{3(x - 6)}{8(x - 1)^{7/2}}$

47. $\dfrac{2}{(x - 1)^3}$ **49.** $\dfrac{3}{(x + 2)^2}$, $\dfrac{-6}{(x + 2)^3}$, $\dfrac{18}{(x + 2)^4}$, $\dfrac{-72}{(x + 2)^5}$,

$\dfrac{360}{(x + 2)^6}$ **51.**

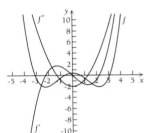

53.

Summary and Review: Chapter 2, p. 170

1. [2.1], [2.2] -4 **2.** [2.1], [2.2] 10
3. [2.1], [2.2] -10 **4.** [2.2] 5 **5.** [2.2] 0
6. [2.1] No   **7.** [2.1] Yes   **8.** [2.1] -4 **9.** [2.1] -4
10. [2.1] Yes   **11.** [2.1] Does not exist   **12.** [2.1] -2
13. [2.1] No   **14.** [2.3] 1 **15.** [2.3] -3
16. [2.3] $4x + 2h$ **17.** [2.4] $y = x - 1$ **18.** [2.5] $(4, 5)$
19. [2.5] $(5, -108)$ **20.** [2.5] $20x^4$ **21.** [2.5] $x^{-2/3}$
22. [2.5] $64x^{-9}$ **23.** [2.5] $6x^{-3/5}$

24. [2.5] $0.7x^6 - 12x^3 - 3x^2$ **25.** [2.5] $x^5 + 32x^3 - 5$

26. [2.7] $2x$ **27.** [2.7] $\dfrac{-x^2 + 16x + 8}{(8 - x)^2}$

28. [2.8] $2(5 - x)(2x - 1)^4(26 - 7x)$

29. [2.8] $35x^4(x^5 - 2)^6$

30. [2.8] $3x^2(4x + 3)^{-1/4} + 2x(4x + 3)^{3/4}$, or $(4x + 3)^{-1/4}(11x^2 + 6x)$

31. [2.9] $240x^{-6}$ **32.** [2.9] $840x^3$ **33.** [2.6]
(a) $1 + 4t^3$; (b) $12t^2$; (c) $33, 48$ **34.** [2.6]
(a) $-8x^2 + 47x + 10$; (b) $\$800, \$3050, -\$2250$;
(c) $40, 16x - 7, -16x + 47$;
(d) $\$40$ per unit, $\$313$ per unit, $-\$273$ per unit
35. [2.6] (a) $100t$; (b) $30{,}000$; (c) 2000 per yr
36. [2.8] $4x^2 - 4x + 6, -2x^2 - 9$ **37.** [2.2] 0

38. [2.8] $\dfrac{-9x^4 - 4x^3 + 9x + 2}{2\sqrt{1 + 3x}(1 + x^3)^2}$

Test: Chapter 2, p. 172

1. Yes **2.** No **3.** Does not exist **4.** 1 **5.** No
6. 3 **7.** 3 **8.** Yes **9.** 6 **10.** $\frac{1}{2}$ **11.** Does not exist
12. 4 **13.** ∞ **14.** $3(2x + h)$ **15.** $y = \frac{3}{4}x + 2$

16. $(0, 0), (2, -4)$ **17.** $84x^{83}$ **18.** $\dfrac{5}{\sqrt{x}}$ **19.** $\dfrac{10}{x^2}$

20. $\frac{5}{4}x^{1/4}$, or $\frac{5}{4}\sqrt[4]{x}$ **21.** $-x + 0.61$

22. $x^2 - 2x + 2$ **23.** $\dfrac{-6x + 20}{x^5}$

24. $\dfrac{5}{(5 - x)^2}$ **25.** $(x + 3)^3(7 - x)^4(-9x + 13)$

26. $-5(x^5 - 4x^3 + x)^{-6}(5x^4 - 12x^2 + 1)$

27. $\sqrt{x^2 + 5} + \dfrac{x^2}{\sqrt{x^2 + 5}}$, or $\dfrac{2x^2 + 5}{\sqrt{x^2 + 5}}$

28. $24x$ **29.** (a) $P(x) = -0.001x^2 + 48.8x - 60$;
(b) $R(10) = \$500, C(10) = \$72.10, P(10) = \$427.90$;
(c) $R'(x) = 50, C'(x) = 0.002x + 1.2, P'(x) = -0.002x + 48.8$;
(d) $R'(10) = \$50$ per unit,
$C'(10) = \$1.22$ per unit, $P'(10) = \$48.78$

30. (a) $\dfrac{dM}{dt} = -0.003t^2 + 0.2t$; (b) 9; (c) 1.7 words/min

31. $x^3 + x^6, x^3 + 3x^4 + 3x^5 + x^6$ **32.** 12

33. $-2\left(\dfrac{1 + 3x}{1 - 3x}\right)^{1/3} + \left(\dfrac{1 - 3x}{1 + 3x}\right)^{2/3}$

CHAPTER 3

Margin Exercises, Section 3.1, pp. 178–189

1. (a) $[a, b], [c, d]$; (b) $[a, b), (c, d)$; (c) $[b, c], [d, e]$;
(d) $(b, c), (d, e)$ **2.** (a) $x_1, x_3, x_5, x_6, x_8, x_{10}$; (b) x_4, x_7, x_9; (c) all but x_2 **3.** Answers will vary, but the graph must have at least one critical point.

4. It is not possible. **5.** Relative maximum at $(-3, 15)$; relative minimum at $(1, -17)$

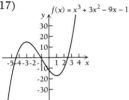

6. Relative minimum at $(0, 0)$

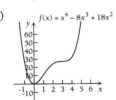

7. Relative minimum at $(0, 0)$

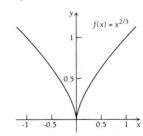

Exercise Set 3.1, p. 189

1. Relative minimum at $(2, 1)$

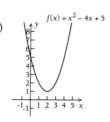

3. Relative maximum at $(\frac{1}{2}, \frac{21}{4})$

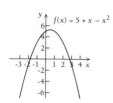

5. Relative minimum at $(-1, -2)$

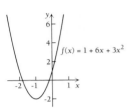

7. Relative minimum at $(1, 1)$; relative maximum at $(-\frac{1}{3}, \frac{59}{27})$

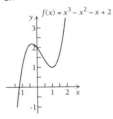

9. Relative maximum at $(-1, 8)$; relative minimum at $(1, 4)$

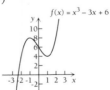

11. Relative minimum at $(0, 0)$; relative maximum at $(1, 1)$

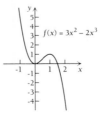

13. None

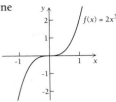

15. Relative maximum at $(0, 10)$; relative minimum at $(4, -22)$

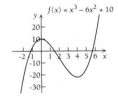

17. Relative maximum at $(\frac{3}{4}, \frac{27}{256})$

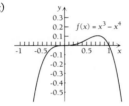

19. Relative minima at $(-2, -13)$ and $(2, -13)$; relative maximum at $(0, 3)$

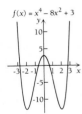

21. Relative maximum at $(0, 1)$

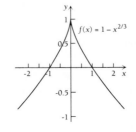

23. Relative minimum at $(0, -8)$

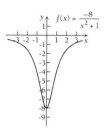

25. Relative minimum at $(-1, -2)$;   **27.** None
relative maximum at $(1, 2)$

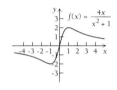

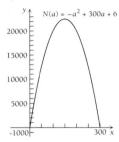

29. Relative maximum at $(150, 22{,}506)$

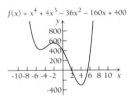

31. 5   **33.** Relative minima at $(-5, 425)$ and $(4, -304)$;
relative maximum at $(-2, 560)$

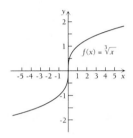

Margin Exercises, Section 3.2, pp. 191–204

1. (a) $[b, c]$, $[d, e]$; (b) $[b, c]$, $[d, e]$; (c) $[a, b]$, $[c, d]$;
(d) $[a, b]$, $[c, d]$ **2.** (a) $(0, b]$; (b) $[a, 0)$ **3.** Relative
maximum at $(-1, 4)$; relative minimum at $(3, -28)$

4. Relative maximum at $(-1, 2)$; relative minimum at
$(1, -2)$ **5.** R, T, V   **6.** Relative maximum at $(-1, \frac{13}{6})$;
relative minimum at $(2, -\frac{7}{3})$

7. Relative minima at $(-1, 0)$ and $(1, 0)$; relative
maximum at $(0, 2)$

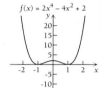

Exercise Set 3.2, p. 204

1. Relative maximum at $(0, 2)$

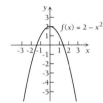

3. Relative minimum at $(-\frac{1}{2}, -\frac{5}{4})$

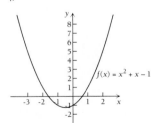

5. Relative maximum at $(\frac{3}{8}, -\frac{7}{16})$

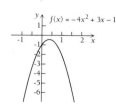

7. Relative maximum at $(-2, 72)$; relative minimum at $(3, -53)$

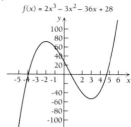

$f(x) = 2x^3 - 3x^2 - 36x + 28$

9. Relative maximum at $(-\frac{1}{2}, 1)$; relative minimum at $(\frac{1}{2}, -\frac{1}{3})$

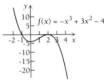

$f(x) = \frac{8}{3}x^3 - 2x + \frac{1}{3}$

11. Relative minimum at $(0, -4)$; relative maximum at $(2, 0)$

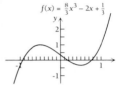

$f(x) = -x^3 + 3x^2 - 4$

13. Relative minima at $(0, 0)$ and $(3, -27)$; relative maximum at $(1, 5)$

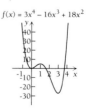

$f(x) = 3x^4 - 16x^3 + 18x^2$

15. Relative minimum at $(-1, 0)$

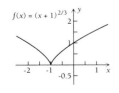

$f(x) = (x + 1)^{2/3}$

17. Relative minima at $(-\sqrt{3}, -9)$ and $(\sqrt{3}, -9)$; relative maximum at $(0, 0)$

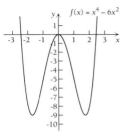

$f(x) = x^4 - 6x^2$

19. Relative maximum at $(-\frac{2}{3}, \frac{121}{27})$; relative minimum at $(2, -5)$

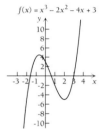

$f(x) = x^3 - 2x^2 - 4x + 3$

21. Relative minimum at $(-1, -1)$

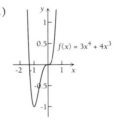

$f(x) = 3x^4 + 4x^3$

23. Relative maximum at $(-5, 400)$; relative minimum at $(9, -972)$

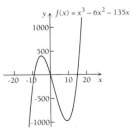

$f(x) = x^3 - 6x^2 - 135x$

25. Relative minimum at $(-1, -\frac{1}{2})$; relative maximum at $(1, \frac{1}{2})$

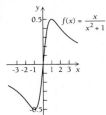

27. Relative maximum at $(0, 3)$

29. None

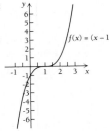

31. Relative minima at $(0, 0)$ and $(1, 0)$; relative maximum at $(\frac{1}{2}, \frac{1}{16})$

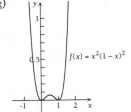

33. Relative minimum at $(-2, -64)$; relative maximum at $(2, 64)$

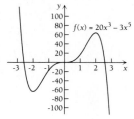

35. Relative minimum at $(-\sqrt{2}, -2)$; relative maximum at $(\sqrt{2}, 2)$

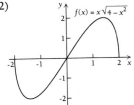

37. None

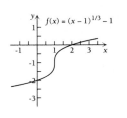

39.

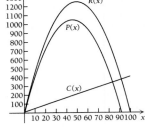

41. An object of radius $13\frac{1}{3}$ mm   **43.** $g = h'$
45. Relative maximum at $(1, 1)$; relative minimum at $(0, 0)$

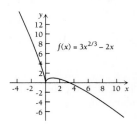

47. Relative minimum at $(1, -1)$

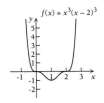

49. Relative minimum at $(\frac{1}{4}, -\frac{1}{4})$

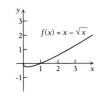

Margin Exercises, Section 3.3, pp. 210–221

1. $x = -7, x = 4$ **2.** $x = -2, x = 0, x = 3$
3. (b) and (c) **4.** $y = \frac{1}{2}$ **5.** $y = 3$ **6.** $y = 3x - 1$
7. $y = 5x$ **8.** $(0, 0), (-5, 0), (3, 0)$
9. $(0, 0), (-3, 0), (1, 0)$

9.
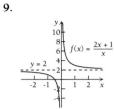

11.
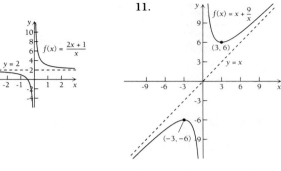

10.
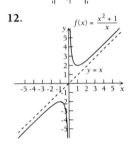

11.
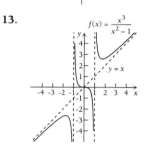

12.

13.
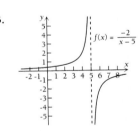

13.
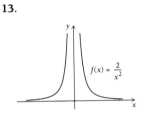

15.
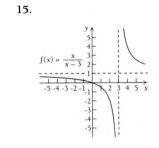

Exercise Set 3.3, p. 222

1.
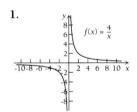

3.

17.
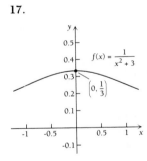

19.
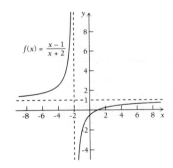

5.
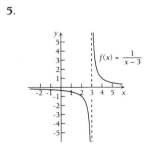

7.

21.
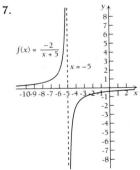

23.
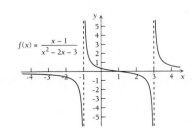

25.

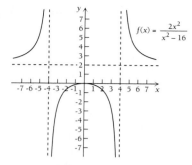

$f(x) = \dfrac{2x^2}{x^2 - 16}$

27. (a) \$50, \$37.24, \$32.64, \$26.37; (b) Maximum = 50 at $x = 0$; (c)

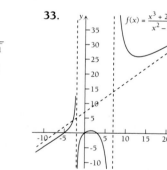

$V(t) = 50 - \dfrac{25t^2}{(t + 2)^2}$

(d) 25; (e) yes, \$25

29. (a) \$480, \$600, \$2400, \$4800; (b) ∞; (c)

$C(p) = \dfrac{48{,}000}{100 - p}$

(d) no

31.

$f(x) = \dfrac{x}{\sqrt{x^2 + 1}}$

33.

$f(x) = \dfrac{x^3 + 2x^2 - 15x}{x^2 - 5x - 14}$

35.

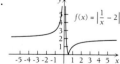

$f(x) = \left|\dfrac{1}{x} - 2\right|$

Margin Exercises, Section 3.4, pp. 224–233

1. (a) Maximum at c_2; minimum at b;
(b) maximum at c_1; minimum at c_2
2. Maximum $= \frac{59}{27}$ at $x = -\frac{1}{3}$; minimum $= -8$ at $x = -2$

A-20

3. Maximum = 4 at $x = 2$; minimum = 1 at $x = -1$ and $x = 1$ **4.** Minimum = -4 at $x = 2$
5. Minimum = -4 at $x = 2$; maximum = 0 at $x = 0$ and $x = 4$ **6.** None
7. Maximum = 8 at $x = 2$; minimum = -8 at $x = -2$; the endpoints
8. Minimum $= \dfrac{10}{\sqrt{10}} + \sqrt{10}$, or $2\sqrt{10}$, at $x = \dfrac{1}{\sqrt{10}}$

Exercise Set 3.4, p. 233

1. (a) 41 mph; (b) 80 mph; (c) 13.5 mpg; (d) 16.5 mpg; (e) about 22%
3. Maximum $= 5\frac{1}{4}$ at $x = \frac{1}{2}$; minimum = 3 at $x = 2$
5. Maximum = 4 at $x = 2$; minimum = 1 at $x = 1$
7. Maximum $= \frac{59}{27}$ at $x = -\frac{1}{3}$; minimum = 1 at $x = -1$
9. Maximum = 1 at $x = 1$; minimum = -5 at $x = -1$
11. Maximum = 15 at $x = -2$; minimum = -13 at $x = 5$
13. Maximum = minimum = -5 for all x in $[-1, 1]$
15. Maximum = 4 at $x = -1$; minimum = -12 at $x = 3$
17. Maximum $= 3\frac{1}{5}$ at $x = -\frac{1}{5}$; minimum = -48 at $x = 3$
19. Minimum = -4 at $x = 2$; maximum = 50 at $x = 5$
21. Minimum = -110 at $x = -5$; maximum = 2 at $x = -1$
23. Maximum = 513 at $x = -8$; minimum = -511 at $x = 8$
25. Maximum = 17 at $x = 1$; minimum = -15 at $x = -3$
27. Maximum = 32 at $x = -2$; minimum $= -\frac{27}{16}$ at $x = \frac{3}{2}$
29. Maximum = 13 at $x = -2$ and $x = 2$; minimum = 4 at $x = -1$ and $x = 1$
31. Minimum = -5 at $x = -3$; maximum = -1 at $x = 5$
33. Minimum = 2 at $x = 1$; maximum $= 20\frac{1}{20}$ at $x = 20$
35. Maximum $= \frac{4}{5}$ at $x = -2$ and $x = 2$; minimum = 0 at $x = 0$
37. Minimum = -1 at $x = -2$; maximum = 3 at $x = 26$
39. Maximum = 1225 at $x = 35$
41. Minimum = 200 at $x = 10$
43. Maximum $= \frac{1}{3}$ at $x = \frac{1}{2}$ **45.** Maximum $= \frac{289}{4}$ at $x = \frac{17}{2}$
47. Maximum $= 2\sqrt{3}$ at $x = -\sqrt{3}$; minimum $= -2\sqrt{3}$ at $x = \sqrt{3}$
49. Maximum = 5700 at $x = 2400$
51. Minimum $= -55\frac{1}{3}$ at $x = 1$
53. Maximum = 2000 at $x = 20$; minimum = 0 at $x = 0$ and $x = 30$
55. Minimum = 24 at $x = 6$
57. Minimum = 108 at $x = 6$
59. Maximum = 3 at $x = -1$; minimum $= -\frac{3}{8}$ at $x = \frac{1}{2}$
61. Maximum = 2 at $x = 8$; minimum = 0 at $x = 0$
63. None

65. Maximum = -1 at $x = 1$; minimum = -5 at $x = -1$
67. None
69. Minimum = 0 at $x = 0$; maximum = 1 at $x = -1$ and
$x = 1$ **71.** None
73. Maximum = $-\frac{10}{3} + 2\sqrt{3}$ at $x = 2 - \sqrt{3}$;
minimum = $-\frac{10}{3} - 2\sqrt{3}$ at $x = 2 + \sqrt{3}$
75. Minimum = -1 at $x = -1$ and $x = 1$
77. Maximum = 1430 units per month at $t = 25$ years of
service   **79.** Maximum = \$1500 at $x = 3$
81. 61.64 mph   **83.** Maximum = $3\sqrt{6}$ at $x = 3$;
minimum = -2 at $x = -2$ **85.** 7

Margin Exercises, Section 3.5, pp. 237–250

1. **(a)** y: 20, 16, 13.5, 12, 10, 8, 6.8, 0; A: 0, 64, 87.75,
96, 100, 96, 89.76, 0;   **(b)**

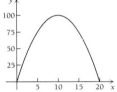

(c) yes;   **(d)** maximum = 100 ft^2 at $x = 10$ ft
2. 25 ft by 25 ft; 625 ft^2 **3.** **(a)** h: 4, 3.5, 3, 2.5, 2, 1.7,
1.5, 1, 0.6, 0.5, 0; V:0, 3.5, 12, 22.5, 32, 35.972, 37.5, 36,
27.744, 24.5, 0;
(b)

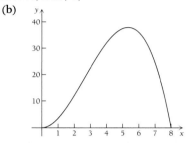

(c) about 38; between 5 and 6, maybe 5.2
4. Maximum volume = $74\frac{2}{27}$ in^3; dimensions $6\frac{2}{3}$ in. by
$6\frac{2}{3}$ in. by $1\frac{2}{3}$ in.   **5.** Minimum area = 300 cm^2;
dimensions 10 cm by 10 cm by 5 cm
6. **(a)** $R(x) = x(200 - x) = 200x - x^2$;
(b) $P(x) = -x^2 + 192x - 5000$;   **(c)** 96;   **(d)** \$4216;
(e) $200 - 96$, or \$104 per unit   **7.** 40¢

8.

x	$\dfrac{2500}{x}$	$\dfrac{x}{2}$	$10 \cdot \dfrac{x}{2}$	$20 + 9x$	$(20 + 9x)\dfrac{2500}{x}$	$10 \cdot \dfrac{x}{2} + (20 + 9x)\dfrac{2500}{x}$
2500	1	1250	\$12,500	\$22,520	\$22,520	\$35,020
1250	2	625	\$6250	\$11,270	\$22,540	\$28,790
500	5	250	\$2500	\$4520	\$22,600	\$25,100
250	10	125	\$1250	\$2270	\$22,700	\$23,950
167	15	84	\$840	\$1523	\$22,845	\$23,685
125	20	63	\$630	\$1145	\$22,900	\$23,530
100	25	50	\$500	\$920	\$23,000	\$23,500
90	28	45	\$450	\$830	\$23,240	\$23,690
50	50	25	\$250	\$470	\$23,500	\$23,750

Lot size $\approx$ 100 to minimize cost; 25 orders

9. 15 times per year at lot size 40   **10.** 19 times per year at lot size 31

Exercise Set 3.5, p. 250

1. 25 and 25; maximum product $= 625$ **3.** No; $Q = x(50 - x)$ has no minimum   **5.** 2 and -2; minimum product $= -4$ **7.** $x = \frac{1}{2}$, $y = \sqrt{\frac{1}{2}}$; maximum $= \frac{1}{4}$
9. $x = 10$, $y = 10$; minimum $= 200$ **11.** $x = 2$, $y = \frac{32}{3}$; maximum $= \frac{64}{3}$ **13.** 30 yd by 60 yd; maximum area $= 1800$ yd^2 **15.** 13.5 ft by 13.5 ft; 182.25 ft^2
17. 20 in. by 20 in. by 5 in.; maximum $= 2000$ in^3
19. 5 in. by 5 in. by 2.5 in.; minimum $= 75$ in^2 **21.** 46 units; maximum profit $= \$1048$ **23.** 70 units; maximum profit $= \$19$ **25.** Approximately 1667 units; maximum profit $\approx \$5481$ **27. (a)** $R(x) = 150x - 0.5x^2$;
(b) $P(x) = -0.75x^2 + 150x - 4000$; **(c)** 100; **(d)** \$3500;
(e) \$100   **29.** \$5.75, 72,500 (Will the stadium hold that many?)   **31.** 25   **33.** 4 ft by 4 ft by 20 ft   **35.** 9%
37. Reorder 5 times per year; lot size $= 20$ **39.** Reorder 12 times per year; lot size $= 30$ **41.** Reorder about 13 times per year; lot size $= 28$ **43.** 14 in. by 14 in. by 28 in.
45. $x = y = \dfrac{24}{4 + \pi}$ ft   **47.** $\sqrt[3]{\dfrac{1}{10}}$ **49.** Reorder

$\dfrac{Q}{\sqrt{\dfrac{2bQ}{a}}}$ times per year; lot size $= \sqrt{\dfrac{2bQ}{a}}$

51. Minimum at $x = \dfrac{24\pi}{\pi + 4} \approx 10.55$ in., $24 - x = \dfrac{96}{\pi + 4}$
≈ 13.45 in. There is no maximum if the string is to be cut. One would interpret the maximum to be at the endpoint, with the string uncut and used to form a circle.
53. S should be about 4.25 miles downshore from A.

55. (a) $C'(x) = 8 + \dfrac{3x^2}{100}$; $A(x) = 8 + \dfrac{20}{x} + \dfrac{x^2}{100}$;

(c) $A'(x) = \dfrac{x}{50} - \dfrac{20}{x^2}$;

(d) minimum $= 11$ at $x_0 = 10$, $C'(10) = 11$;
(e) $A(10) = 11$, $C'(10) = 11$ **57.** Minimum $= 6 - 4\sqrt{2}$
at $x = 2 - \sqrt{2}$ and $y = -1 + \sqrt{2}$

Margin Exercises, Section 3.6, pp. 257–262

1. $\Delta y = -0.59$ **2.** $\Delta y = 19$ **3.** 8.1875; 8.185354 on calculator   **4. (a)** $\Delta C = \$4.11$, $C'(5) = \$4.10$;

(b) $\Delta C = \$6.01$, $C'(100) = \$6$
5. (a) $dy = (3x^2 - 10x - 4)dx$; **(b)** $dy = -0.7$

Exercise Set 3.6, p. 262

1. $\Delta y = 0.0401$; $f'(x)\,\Delta x = 0.04$ **3.** 0.2816, 0.28
5. -0.556, -1 **7.** 6, 6   **9.** $\Delta C = \$2.01$, $C'(70) = \$2$
11. $\Delta R = \$2$, $R'(70) = \$2$
13. (a) $P(x) = -0.01x^2 + 1.4x - 30$; **(b)** $\Delta P = -\$0.01$,
$P'(70) = 0$ **15.** 4.375   **17.** 10.1   **19.** 2.167
21. $dy = 9x^2(2x^3 + 1)^{1/2}\,dx$ **23.** $dy = \frac{1}{5}(x + 27)^{-4/5}\,dx$
25. $dy = (4x^3 - 6x^2 + 10x + 3)\,dx$ **27.** $dy = 3.1$

29. -0.01 **31.** 100   **33.** 2.512 cm^3 **35.** $\dfrac{5}{\pi} \approx 1.6$ ft

Margin Exercises, Section 3.7, pp. 264–268

1. $\dfrac{dy}{dx} = \dfrac{2}{5y^4}$ **2. (a)** $\dfrac{dy}{dx} = \dfrac{-3x^2 - 2xy^4}{3y^2 + 4x^2y^3}$; **(b)** 0

3. $\dfrac{dp}{dx} = \dfrac{-\sqrt{p}}{50}$ **4.** 24π ft^2/sec ≈ 75 ft^2/sec

5. (a) $\dfrac{dR}{dt} = \$0$ per day; **(b)** $\dfrac{dC}{dt} = \$64$ per day;

(c) $\dfrac{dP}{dt} = -\$64$ per day

Exercise Set 3.7, p. 268

1. $\dfrac{1 - y}{x + 2}$; $-\dfrac{1}{9}$ **3.** $-\dfrac{x}{y}$; $-\dfrac{1}{\sqrt{3}}$

5. $\dfrac{6x^2 - 2xy}{x^2 - 3y^2}$; $-\dfrac{36}{23}$ **7.** $-\dfrac{y}{x}$ **9.** $\dfrac{x}{y}$

11. $\dfrac{3x^2}{5y^4}$ **13.** $\dfrac{-3xy^2 - 2y}{3x + 4x^2y}$

15. $\dfrac{dp}{dx} = \dfrac{-2}{2p + 1}$ **17.** $\dfrac{dp}{dx} = -\dfrac{p + 4}{x + 3}$

19. $-\frac{3}{4}$ **21.** \$400 per day, \$80 per day, \$320 per day
23. \$16 per day, \$8 per day, \$8 per day
25. 0.1728π cm^3/day ≈ 0.54 cm^3/day

27. (a) $1000R \cdot \dfrac{dR}{dt}$; **(b)** -0.01125 mm^3/min

29. 65 mph   **31.** $-\dfrac{\sqrt{y}}{\sqrt{x}}$ **33.** $\dfrac{2}{3y^2(x + 1)^2}$

35. $-\dfrac{9}{4}\sqrt[3]{y}\sqrt{x}$ **37.** $\dfrac{dy}{dx} = \dfrac{1+y}{2-x}, \dfrac{d^2y}{dx^2} = \dfrac{2+2y}{(2-x)^2}$

39. $\dfrac{dy}{dx} = \dfrac{x}{y}, \dfrac{d^2y}{dx^2} = \dfrac{y^2 - x^2}{y^3}$ **41.**

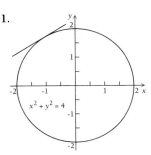

Summary and Review: Chapter 3, p. 271

1. [3.1], [3.2] Relative maximum at $(-1, 4)$

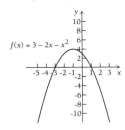

2. [3.1], [3.2] Relative maximum at $(0, 3)$; relative minima at $(-1, 2)$ and $(1, 2)$

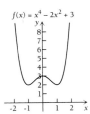

3. [3.1], [3.2] Relative maximum at $(-1, 4)$; relative minimum at $(1, -4)$

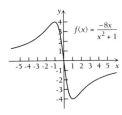

4. [3.1], [3.2] None

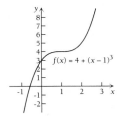

5. [3.1], [3.2] Relative maximum at $(-1, 4)$; relative minimum at $(\frac{1}{3}, \frac{76}{27})$

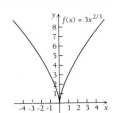

6. [3.1], [3.2] Relative minimum at $(0, 0)$

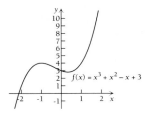

7. [3.1], [3.2] Relative maximum at $(-1, 17)$; relative minimum at $(2, -10)$

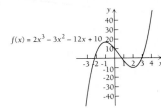

8. [3.1], [3.2] Relative maximum at $(-1, 4)$; relative minimum at $(1, 0)$

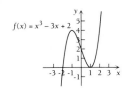

9. [3.3]

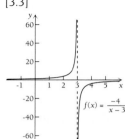

10. [3.3]

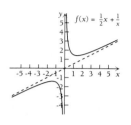

11. [3.3]

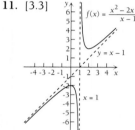

12. [3.3]

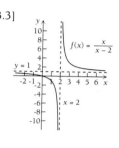

13. [3.4] None   **14.** [3.4] Maximum = 32 at $x = -2$; minimum = -17 at $x = 5$ **15.** [3.4] Maximum = 19 at $x = -1$; minimum = -35 at $x = 2$
16. [3.4] Maximum = -2 at $x = 1$; minimum = -12 at $x = -1$
17. [3.4] Minimum = 3 at $x = -1$ **18.** [3.4] None
19. [3.4] Maximum = 163 at $x = -3$; minimum = -1 at $x = -1$
20. [3.4] Minimum = 10 at $x = 1$
21. [3.4] Maximum = 256 at $x = 4$; minimum = 0 at $x = 0$ and $x = 5$
22. [3.4] Maximum = $\frac{53}{4}$ at $x = \frac{5}{2}$ **23.** [3.5] 30, 30
24. [3.5] $Q = -1$ at $x = -1$, $y = -1$
25. [3.5] 30 units; $451 profit
26. [3.5] 10 ft × 10 ft × 25 ft; $1500
27. [3.5] 12 times; 30 bicycles
28. [3.6] $\Delta y = -10.875$; $f'(x)\,\Delta x = -13$
29. [3.6] **(a)** $dy = (3x^2 - 1)\,dx$; **(b)** $dy = 0.11$

30. [3.6] 8.3125   **31.** [3.7] $\dfrac{2x^2 + 3y}{-2y^2 - 3x}; \dfrac{4}{5}$

32. [3.7] $-1\frac{3}{4}$ ft/sec
33. [3.7] $600 per day, $450 per day, $150 per day
34. [3.4] Minimum = 0 at $x = 3$

35. [3.7] $\dfrac{3x^5 - 4x^3 - 12xy^2}{12x^2y + 4y^3 - 3y^5}$

36. [3.1], [3.2] Relative minimum at $(-9, -9477)$; relative maximum at $(0, 0)$; relative minimum at $(15, -37{,}125)$

Test: Chapter 3, p. 272

1. Relative minimum at $(2, -9)$

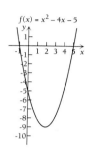

2. Relative minima at $(-1, -1)$ and $(1, -1)$; relative maximum at $(0, 1)$

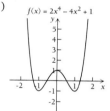

3. Relative minimum at $(2, -4)$

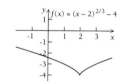

4. Relative maximum at $(0, 4)$

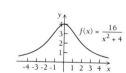

5. Relative maximum at $(-1, 2)$; relative minimum at $(\frac{1}{3}, \frac{22}{27})$

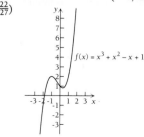

6. Relative minimum at $(-1, 2)$; relative maximum at $(1, 6)$

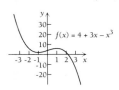
$f(x) = 4 + 3x - x^3$

7.   None

$f(x) = (x + 2)^3$

8. Relative minimum at $\left(-\dfrac{3}{\sqrt{2}}, -\dfrac{9}{2}\right)$;

relative maximum at $\left(\dfrac{3}{\sqrt{2}}, \dfrac{9}{2}\right)$

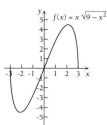
$f(x) = x\sqrt{9 - x^2}$

9.

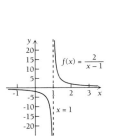
$f(x) = \dfrac{2}{x - 1}$

10.

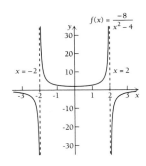
$f(x) = \dfrac{-8}{x^2 - 4}$

11.

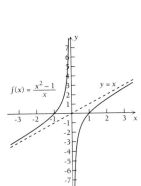
$f(x) = \dfrac{x^2 - 1}{x}$

12.

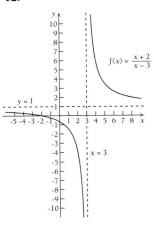
$f(x) = \dfrac{x + 2}{x - 3}$

13. Maximum $= 9$ at $x = 3$

14. Maximum $= 2$ at $x = -1$; minimum $= -1$ at $x = -2$

15. Maximum $= 28.49$ at $x = 4.3$

16. Maximum $= 7$ at $x = -1$; minimum $= 3$ at $x = 1$

17. None   **18.** Minimum $= -\frac{13}{12}$ at $x = \frac{1}{6}$

19. Minimum $= 48$ at $x = 4$ **20.** 4 and -4

21. $x = 5$, $y = -5$; minimum $= 50$

22. Maximum profit $= \$24,980$; 500 units

23. 40 in. by 40 in. by 10 in.; maximum volume $=$ 16,000 in^3 **24.** 35 times at lot size 35

25. $\Delta y = 1.01, f'(x)\, \Delta x = 1$ **26.** 10.2

27. (a) $\dfrac{x\, dx}{\sqrt{x^2 + 3}}$; (b) 0.0075593

28. $-\dfrac{x^2}{y^2}, -\dfrac{1}{4}$ **29.** -0.96 ft/sec

30. Maximum $= \frac{1}{3} \cdot 2^{2/3}$ at $x = \sqrt[3]{2}$; minimum $= 0$ at $x = 0$

31. (a) $A(x) = \dfrac{C(x)}{x} = 100 + \dfrac{100}{\sqrt{x}} + \dfrac{\sqrt{x}}{100}$;

(b) minimum $= 102$ at $x = 10,000$

CHAPTER 4

Margin Exercises, Section 4.1, pp. 277–287

1. 8, 8.574188, 8.815241, 8.821353, 8.824411, 8.824962; $2^\pi \approx 8.82$

2. (a)

x	0	$\frac{1}{2}$	1	2	-1	-2
3^x	1	1.7	3	9	$\frac{1}{3}$	$\frac{1}{9}$

(b)

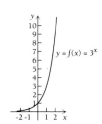
$y = f(x) = 3^x$

3. (a)

x	0	$\frac{1}{2}$	1	2	-1	-2
$\left(\frac{1}{3}\right)^x$	1	0.6	$\frac{1}{3}$	$\frac{1}{9}$	3	9

(b)

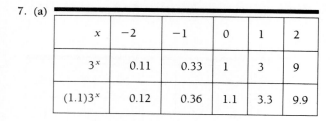

$y = f(x) = \left(\frac{1}{3}\right)^x$

4. (a) 0.8284, 0.7568, 0.7369, 0.7083, 0.7007, 0.6934;
(b) 0.7   **5. (a)** 1.4641, 1.2643, 1.2113, 1.1372, 1.1177, 1.09928; **(b)** 1.1

6. (a)

x	-3	-2	-1	0	1	2	3
2^x	0.125	0.25	0.5	1	2	4	8
$(0.7)2^x$	0.09	0.18	0.35	0.7	1.4	2.8	5.6

(b) See Fig. 6 in text.

7. (a)

x	-2	-1	0	1	2
3^x	0.11	0.33	1	3	9
$(1.1)3^x$	0.12	0.36	1.1	3.3	9.9

(b) See Fig. 7 in text.

8. $2, $2.25, $2.370370, $2.441406, $2.613035, $2.692597, $2.714567, $2.718121, $2.718869   **9.** $6e^x$

10. $e^x(x^3 + 3x^2)$, or $x^2e^x(x + 3)$ **11.** $\dfrac{e^x(x - 2)}{x^3}$

12. $-4e^{-4x}$ **13.** $(3x^2 + 8)e^{x^3 + 8x}$ **14.** $\dfrac{xe^{\sqrt{x^2 + 5}}}{\sqrt{x^2 + 5}}$

15.

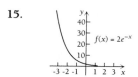

$f(x) = 2e^{-x}$

16. (a)

x	0	$\frac{1}{2}$	1	2	3	4
$f(x)$	0	0.39	0.63	0.86	0.95	0.98

(b)

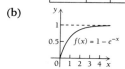

$f(x) = 1 - e^{-x}$

Exercise Set 4.1, p. 287

1.

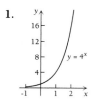

$y = 4^x$

3.

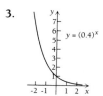

$y = (0.4)^x$

5.

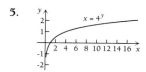

$x = 4^y$

7. $3e^{3x}$ **9.** $-10e^{-2x}$

11. e^{-x} **13.** $-7e^x$ **15.** e^{2x} **17.** $x^3e^x(x + 4)$

19. $\dfrac{e^x(x - 4)}{x^5}$ **21.** $(-2x + 7)e^{-x^2 + 7x}$ **23.** $-xe^{-x^2/2}$

25. $\dfrac{e^{\sqrt{x - 7}}}{2\sqrt{x - 7}}$ **27.** $\dfrac{e^x}{2\sqrt{e^x - 1}}$

29. $(1 - 2x)e^{-2x} - e^{-x} + 3x^2$ **31.** e^{-x} **33.** ke^{-kx}

35.

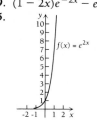

$f(x) = e^{2x}$

37.

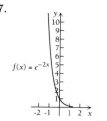

$f(x) = e^{-2x}$

39.

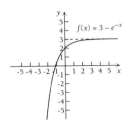

41. $y = x + 1$ **43.** (a) $C'(t) = 50e^{-t}$; (b) \$50 million;
(c) \$0.916 million
45. (a) Approximately 335, 280, 173; (b)

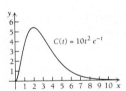

(c) $D'(p) = 1.44e^{-0.003p}$
47. (a) 0 ppm, 3.7 ppm, 5.4 ppm, 4.48 ppm, 0.05 ppm;
(b)

(c) $C'(t) = e^{-t}(20t - 10t^2)$; (d) $C \approx 5.4$ ppm at $t = 2$ hr
49. $15(e^{3x} + 1)^4 e^{3x}$ **51.** $-e^{-t} - 3e^{3t}$
53. $\dfrac{e^x(x-1)^2}{(x^2+1)^2}$ **55.** $\dfrac{e^{\sqrt{x}}}{2\sqrt{x}} + \dfrac{1}{2}\sqrt{e^x}$
57. $\dfrac{1}{2}e^{x/2}\left[\dfrac{x}{\sqrt{x-1}}\right]$ **59.** $\dfrac{4}{(e^x + e^{-x})^2}$
61. 2, 2.25, 2.48832, 2.59374, 2.71692
63. Maximum $= 4e^{-2} \approx 0.54$ at $x = 2$ **65.** Relative
minimum at $(0, 0)$; relative maximum at $(2, 4e^{-2})$

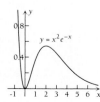

Margin Exercises, Section 4.2, pp. 290–305

1. (a) $b^T = P$; (b) $9^{1/2} = 3$; (c) $10^3 = 1000$;
(d) $10^{-1} = 0.1$
2. (a) $\log_e T = k$; (b) $\log_{16} 2 = \frac{1}{4}$; (c) $\log_{10} 10{,}000 = 4$;
(d) $\log_{10} 0.001 = -3$ **3.**

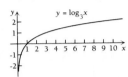

4. (a) 1000; (b) 3; (c) 0.699; (d) ≈ 5.000 **5.** 0.602
6. 1 **7.** -0.398 **8.** 0.398 **9.** -0.699 **10.** $\frac{3}{2}$
11. 1.699 **12.** 1.204 **13.** (a) 4.4977; (b) -0.0419;
(c) 1.8954; (d) 0.8954; (e) -0.1046; (f) -1.1046;
(g) -2.1046 **14.** 2.3025 **15.** -0.9163 **16.** 0.9163
17. 2.7724 **18.** 2.6094 **19.** $\frac{1}{2}$ **20.** -1.6094
21. 0.693147 **22.** 2.995732 **23.** 4.605170
24. -2.599375 **25.** 0.076961 **26.** -0.000100
27. $t \approx 4.4$ **28.** $t \approx 44$

29. (a)

x	0.5	1	2	3	4
$\ln x$	-0.7	0	0.7	1.1	1.4

(b) same as text figure
30. $\dfrac{5}{x}$ **31.** $x^2(1 + 3\ln x) + 4$ **32.** $\dfrac{1 - 2\ln x}{x^3}$ **33.** $\dfrac{1}{x}$
34. $\dfrac{6x}{3x^2 + 4}$ **35.** $\dfrac{1}{x \ln 5x}$ **36.** $\dfrac{4x^5 + 2}{x(x^5 - 2)}$ **37.** (a) 500;
(b) $N'(a) = \dfrac{200}{a}$; (c) minimum $= 500$ at $a = 1$
38. 87 days

Exercise Set 4.2, p. 305

1. $2^3 = 8$ **3.** $8^{1/3} = 2$ **5.** $a^J = K$ **7.** $b^v = T$
9. $\log_e b = M$ **11.** $\log_{10} 100 = 2$ **13.** $\log_{10} 0.1 = -1$
15. $\log_M V = p$ **17.** 2.708 **19.** 0.51 **21.** -1.609
23. $\frac{3}{2}$ **25.** 2.609 **27.** 3.218 **29.** 2.9957 **31.** 0.2231
33. -1.3863 **35.** 2.3863 **37.** 4 **39.** 2.7726
41. 8.681690 **43.** -4.006334 **45.** 0.631272
47. -3.973898 **49.** 6.809039 **51.** -4.509860

53. $t \approx 4.6$ 55. $t \approx 4.1$ 57. $t \approx 2.3$ 59. $t \approx 141$

61. $-\dfrac{6}{x}$ 63. $x^3(1 + 4 \ln x) - x$ 65. $\dfrac{1 - 4 \ln x}{x^5}$

67. $\dfrac{1}{x}$ 69. $\dfrac{10x}{5x^2 - 7}$ 71. $\dfrac{1}{x \ln 4x}$ 73. $\dfrac{x^2 + 7}{x(x^2 - 7)}$

75. $e^x\left(\dfrac{1}{x} + \ln x\right)$ 77. $\dfrac{e^x}{e^x + 1}$ 79. $\dfrac{2 \ln x}{x}$

81. (a) 1000; (b) $N'(a) = \dfrac{200}{a}$, $N'(10) = 20$;

(c) minimum = 1000 at $a = 1$ 83. 58 days
85. (a) \$58.69, \$78.00; (b) $\$63.80e^{-1.1t}$; (c) 2.7
87. (a) 18.1%, 69.9%; (b) $P'(t) = 0.2e^{-0.2t}$; (c) 11.5
89. (a) 68%; (b) 36%; (c) 3.6%; (d) 5%;

(e) $-\dfrac{20}{t + 1}$;   (f) maximum = 68% at $t = 0$

91. (a) 2.37 ft/sec; (b) 3.37 ft/sec;

(c) $v'(p) = \dfrac{0.37}{p}$, $v'(p)$ = the acceleration of the population

93. $\dfrac{-4(\ln x)^{-5}}{x}$ 95. $\dfrac{15t^2}{t^3 + 1}$ 97. $\dfrac{4[\ln (x + 5)]^3}{x + 5}$

99. $\dfrac{5t^4 - 3t^2 + 6t}{(t^3 + 3)(t^2 - 1)}$ 101. $\dfrac{24x + 25}{8x^2 + 5x}$

103. $\dfrac{2(1 - \ln t^2)}{t^3}$ 105. $x^n \ln x$ 107. $\dfrac{1}{\sqrt{1 + t^2}}$

109. $\dfrac{3}{x \ln x}$ 111. 1   113. $\sqrt[e]{e} \approx 1.444667861$;
$\sqrt[e]{e} > \sqrt[x]{x}$ for any $x > 0$ such that $x \ne e$

115. $t = \dfrac{\ln P - \ln P_0}{k}$ 117. ∞

119. Let $a = \ln x$; then $e^a = x$, so $\log x = \log e^a = a \log e$.
Then $a = \ln x = \dfrac{\log x}{\log e} \approx \dfrac{\log x}{0.4343} \approx 2.3026 \log x$.

121. $-\dfrac{1}{2e}$

Margin Exercises, Section 4.3, pp. 309–319

1. (a) $\dfrac{dy}{dx} = 20e^{4x}$; (b) $\dfrac{dy}{dx} = 4y$ 2. (a) $N(t) = ce^{kt}$;

(b) $f(t) = ce^{kt}$
3. (a) Should be about $\frac{1}{8}$ in., or 0.125 in.;
(b) 0.032, 0.064, 0.128; (c) about 2 mi

4. (a) $P(t) = P_0e^{0.13t}$; (b) \$1138.83; (c) 5.3 yr
5. 2%, 34.7; 6.9%, 10; 14%, 5.0; 4.6%, 15; 1%, 69.3
6. (a) $P(t) = 241e^{0.009t}$; (b) 268 million;
(c) 77 yr (2063)
7. (a) $C(t) = 80e^{kt}$; (b) $k = 0.124822$;
(c) $C(t) = 80e^{0.124822t}$; (d) \$2636 thousand
8. (a) 96; (b) 286, 1005, 1719, 1971, 1997;

(c) $N'(t) = \dfrac{23{,}880e^{-0.6t}}{(1 + 19.9e^{-0.6t})^2}$;   (d)

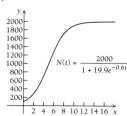

Exercise Set 4.3, p. 320

1. $Q(t) = Q_0e^{kt}$ 3. (a) $P(t) = P_0e^{0.09t}$;
(b) \$1094.17, \$1197.22; (c) 7.7 yr   5. 6.9%
7. (a) $N(t) = 50e^{0.1t}$; (b) 369; (c) 6.9 yr   9. 6.9 yr (1996)
11. (a) $k = 0.161602$; $V(t) = \$84{,}000e^{0.161602t}$;
(b) \$271,279,000; (c) 4.3 yr; (d) 58 yr
13. (a) $k = 0.061248$, $P(t) = \$100e^{0.061248t}$;
(b) \$628.04; (c) 11.3 yr (1978)
15. $\approx \$202{,}000{,}000{,}000{,}000$ 17. $k = 0.16778$, or
16.778%; $\approx \$2{,}718{,}267$; $\approx \$4{,}496{,}659$
19. (a) 2%; (b) 3.8%, 7.0%, 21.6%, 50.2%, 93.1%,

98.01%; (c) $P'(t) = \dfrac{6.37e^{-0.13t}}{(1 + 49e^{-0.13t})^2}$; (d)

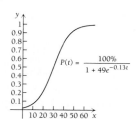

21. 19.8 yr   23. (a) $P(t) = 209e^{0.01t}$; (b) 312 million;
(c) 69.3 yr   25. 0.20%   27. (a) $k = 0.024464$,
$P(t) = 107e^{0.024464t}$; (b) 158 thousand   29. (a) 400,
520, 1214, 2059, 2396, 2478;

(b) $P'(t) = \dfrac{4200e^{-0.32t}}{(1 + 5.25e^{-0.32t})^2}$;

(c)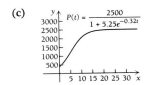

31. **(a)** 0%, 33.0%, 55.1%, 69.9%, 86.5%, 99.2%, 99.8%;
(b) $P'(t) = 0.4e^{-0.4t}$;
(c)

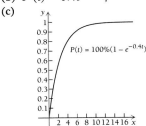

33. 15.03%   **35.** 9%   **37.** $T_3 = \dfrac{\ln 3}{k}$

39. Answers depend on particular data.

41. $\dfrac{\ln 2}{24} \approx 2.9\%$ per hr

Margin Exercises, Section 4.4, pp. 324–333

1. See Exercises 35 and 37 of Exercise Set 4.1.
2. **(a)** $N(t) = N_0 e^{-0.14t}$; **(b)** 246.6 g; **(c)** 5 days
3. 30 yr   **4.** 5.3%   **5.** 13,356 yr   **6.** $2231.30
7. $24,261.23   **8.** **(a)** $a = 130°$; **(b)** $k = 0.01$; **(c)** 188°;
(d) 256 min   **9.** Answers will vary. "Theoretically" it is
never possible for the temperature of the water to be the
same as the room temperature.   **10.** 10:00 A.M.

Exercise Set 4.4, p. 333

1. $826.49   **3.** $22,973.57   **5.** **(a)** $40,000; **(b)** $5413
7. $27,194.82   **9.** 7.2 days   **11.** 23% per min
13. 10.1 g   **15.** 19,109 yr   **17.** 7604 yr
19. **(a)** $A = A_0 e^{-kt}$; **(b)** 9 hr   **21.** **(a)** 0.8%;
(b) 78.7% W_0 **23.** **(a)** $25\%I_0$, $6\%I_0$, $1.5\%I_0$;
(b) $0.00008\%I_0$ **25.** **(a)** 25°; **(b)** $k = 0.05$; **(c)** 84.2°;
(d) 32 min   **27.** 7 P.M.   **29.** **(a)** $k = 0.0162649$,
$P(t) = 453,000e^{-0.162649t}$, where t = the number of years
since 1970; **(b)** 327,207 **(c)** 801 yr   **31.** 233 million,
173 million   **33.** **(a)** 11 watts; **(b)** 173 days; **(c)** 402 days;
(d) 50 watts   **35.** ($166.16, 292)

Margin Exercises, Section 4.5, pp. 337–339

1. $e^{6.9315}$ **2.** $e^{0.6931x}$ **3.** $(\ln 5)5^x$ **4.** $(\ln 4)4^x$
5. $(\ln 4.3)(4.3)^x$ **6.** $\dfrac{1}{\ln 2} \cdot \dfrac{1}{x}$ **7.** $-\dfrac{7}{x \ln 10}$
8. $x^5\left(\dfrac{1}{\ln 10} + 6 \log x\right)$

Exercise Set 4.5, p. 340

1. $e^{6.4378}$ **3.** $e^{12.238}$ **5.** $e^{k \cdot \ln 4}$ **7.** $e^{kT \cdot \ln 8}$
9. $(\ln 6)6^x$ **11.** $(\ln 10)10^x$ **13.** $(6.2)^x[x \ln 6.2 + 1]$
15. $10^x x^2[x \ln 10 + 3]$
17. $\dfrac{1}{\ln 4} \cdot \dfrac{1}{x}$ **19.** $\dfrac{2}{\ln 10} \cdot \dfrac{1}{x}$ **21.** $\dfrac{1}{\ln 10} \cdot \dfrac{1}{x}$
23. $x^2\left[\dfrac{1}{\ln 8} + 3 \log_8 x\right]$
25. $-250,000(\ln 4)\left(\frac{1}{4}\right)^t$ **27.** 5
29. **(a)** $I = 10^7 \cdot I_0$; **(b)** $I = 10^8 \cdot I_0$; **(c)** the intensity in
(b) is 10 times that in part (a);
(d) $\dfrac{dI}{dR} = (I_0 \cdot \ln 10) \cdot 10^R$ **31.** $\dfrac{1}{\ln 10} \cdot \dfrac{1}{I}$ **33.** $\dfrac{m}{\ln 10} \cdot \dfrac{1}{x}$
35. $2(\ln 3)3^{2x}$ **37.** $(\ln x + 1)x^x$
39. $x^{e^x}e^x\left(\ln x + \dfrac{1}{x}\right)$ **41.** $\dfrac{1}{\ln a} \cdot \dfrac{f'(x)}{f(x)}$

Margin Exercises, Section 4.6, pp. 345–347

1. **(a)** 245; **(b)** $E(p) = \dfrac{p}{300 - p}$;
(c) $E(100) = \frac{1}{2}$, $E(200) = 2$; **(d)** $p = \$150$;
(e) $R(p) = pD(p) = 300p - p^2$; **(f)** $p = \$150$
2. **(a)** $E(55) = \frac{11}{49}$, an increase in price will result in an
increase in revenue; **(b)** $E(200) = 2$, an increase in price
will result in a decrease in revenue.

Exercise Set 4.6, p. 347

1. **(a)** $E(p) = \dfrac{p}{400 - p}$;
(b) $E(125) = \frac{5}{11}$, inelastic; **(c)** $p = \$200$
3. **(a)** $E(p) = \dfrac{p}{50 - p}$; **(b)** $E(46) = \frac{23}{2}$, elastic;

(c) $p = \$25$ **5. (a)** $E(p) = 1$, for all $p > 0$;
(b) $E(50) = 1$, unit elasticity; **(c)** Total
revenue $= R(p) = 400$, for all $p > 0$. It has 400 as a
maximum for all $p > 0$.

7. (a) $E(p) = \dfrac{p}{1000 - 2p}$; **(b)** $E(400) = 2$, elastic;
(c) $p = \$\frac{1000}{3} \approx \$333.33$

9. (a) $E(p) = \dfrac{p}{4}$; **(b)** $E(10) = \frac{5}{2}$, elastic; **(c)** $p = \$4$

11. (a) $E(p) = \dfrac{2p}{p + 3}$; **(b)** $E(1) = \frac{1}{2}$, inelastic; **(c)** $p = \$3$

13. (a) $E(p) = \dfrac{25p}{967 - 25p}$; **(b)** $p = 19.34¢$;
(c) $p > 19.34¢$; **(d)** $p < 19.34¢$;
(e) $p = 19.34¢$; **(f)** decrease

15. (a) $E(p) = \dfrac{3p^3}{2(200 - p^3)}$; **(b)** $\frac{81}{346} \approx 0.234$;

(c) increases it   **17. (a)** $E(p) = n$; **(b)** no, E is a
constant n; **(c)** only when $n = 1$ **19.** $E(p) = -pL'(p)$

Summary and Review: Chapter 4, p. 350

1. [4.2] $\dfrac{1}{x}$ **2.** [4.1] e^x **3.** [4.2] $\dfrac{4x^3}{x^4 + 5}$ **4.** [4.1] $\dfrac{e^{2\sqrt{x}}}{\sqrt{x}}$

5. [4.2] $\dfrac{6}{x}$ **6.** [4.1] $4e^{4x} + 4x^3$ **7.** [4.2] $\dfrac{1 - 3\ln x}{x^4}$

8. [4.1], [4.2] $e^{x^2}\left(\dfrac{1}{x} + 2x \ln 4x\right)$

9. [4.1], [4.2] $4e^{4x} - \dfrac{1}{x}$

10. [4.2] $8x^7 - \dfrac{8}{x}$ **11.** [4.1], [4.2] $\dfrac{1 - x}{e^x}$

12. [4.2] 6.9300   **13.** [4.2] -3.2698 **14.** [4.2] 8.7601
15. [4.2] 3.2698   **16.** [4.2] 2.54995   **17.** [4.2]
-3.6602 **18.** [4.3] $Q(t) = Q_0 e^{kt}$ **19.** [4.3] 4.3%
20. [4.3] 8.7 yr   **21.** [4.3] **(a)** $k = 0.050991$,
$C(t) = \$4.65e^{0.050991t}$; **(b)** $\$27.70, \$46.13$ **22.** [4.3]
(a) $N(t) = 60e^{0.12t}$; **(b)** 199; **(c)** 5.8 yr   **23.** [4.4] 5.3 yr
24. [4.4] 18.2% per day   **25.** [4.4] **(a)** $A(t) = 800e^{-0.07t}$;
(b) 197 grams; **(c)** 9.9 days   **26.** [4.3] **(a)** 0.5034,
0.7534, 0.9698, 0.9991, 0.9999; **(b)** $p'(t) = 0.7e^{-0.7t}$;

(c)

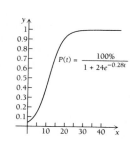

27. [4.4] $\$34,735.26$ **28.** [4.5] $(\ln 3)3^x$

29. [4.5] $\dfrac{1}{\ln 15} \cdot \dfrac{1}{x}$ **30.** [4.6] **(a)** $E(p) = \dfrac{2p}{p + 4}$;
(b) $E(\$1) = \frac{2}{5}$, inelastic; **(c)** $E(\$12) = \frac{3}{2}$, elastic;
(d) decrease; **(e)** $p = \$4$

31. [4.1] $\dfrac{-8}{(e^{2x} - e^{-2x})^2}$ **32.** [4.2] $-\dfrac{1}{1024e}$

Test: Chapter 4, p. 351

1. e^x **2.** $\dfrac{1}{x}$ **3.** $-2xe^{-x^2}$ **4.** $\dfrac{1}{x}$ **5.** $e^x - 15x^2$

6. $3e^x\left(\dfrac{1}{x} + \ln x\right)$ **7.** $\dfrac{e^x - 3x^2}{e^x - x^3}$

8. $\dfrac{\dfrac{1}{x} - \ln x}{e^x}$, or $\dfrac{1 - x \ln x}{xe^x}$

9. 1.0674   **10.** 0.5554   **11.** 0.4057   **12.** $M(t) = M_0 e^{kt}$
13. 17.3% per hr   **14.** 10 yr   **15. (a)** 0.0330326;
$C(t) = \$0.54e^{0.0330326t}$; **(b)** 3.79, $\$5.28$
16. (a) $A(t) = 3e^{-0.1t}$; **(b)** 1.1 cc; **(c)** 6.9 hr
17. 63 days   **18.** 0.000069% per yr   **19. (a)** 4%;
(b) 5.2%, 14.5%, 40.7%, 73.5%, 91.8%, 99.5%, 99.9%;

(c) $P'(t) = \dfrac{6.72e^{-0.28t}}{(1 + 24e^{-0.28t})^2}$; **(d)**

20. $\$28,503.49$ **21.** $(\ln 20)20^x$

22. $\dfrac{1}{\ln 20} \cdot \dfrac{1}{x}$ **23. (a)** $E(p) = \dfrac{p}{5}$;

(b) $E(\$3) = \frac{3}{5}$, inelastic; **(c)** $E(\$18) = \frac{18}{5}$, elastic;
(d) increase; **(e)** $p = \$5$ **24.** $(\ln x)^2$
25. Maximum $= 256e^4 \approx 4.69$ at $x = 4$; minimum $= 0$ at
$x = 0$

CHAPTER 5

Margin Exercises, Section 5.1, pp. 355–361

1. $y = 7x$, $y = 7x - \frac{1}{2}$, $y = 7x + C$ (answers may vary)
2. $y = -2x$, $y = -2x + 27$, $y = -2x + C$
(answers may vary)

3. $\dfrac{x^2}{2} + C$ **4.** $\dfrac{x^4}{4} + C$ **5.** $e^x + C$ **6.** $\ln x + C$

7. $\dfrac{x^4}{4} + C$ **8.** $e^x + C$ **9.** $\frac{8}{3}e^{3x} + C$

10. $\ln x + C$ **11.** $\frac{7}{5}x^5 + x^2 + C$ **12.** $3e^{2x} - \frac{5}{7}x^{7/5} + C$
13. $5 \ln x - 7x - \frac{1}{5}x^{-5} + C$
14.

15. $f(x) = \dfrac{x^3}{3} + \dfrac{23}{3}$ **16.** $g(x) = x^2 - 4x + 13$

17. $C(x) = \frac{1}{3}x^3 + \frac{5}{2}x^2 + 35$ **18.** $s(t) = t^4 + 13$
19. $s(t) = 2t^4 - 6t^2 + 7t + 8$

Exercise Set 5.1, p. 362

1. $\dfrac{x^7}{7} + C$ **3.** $2x + C$ **5.** $\frac{4}{5}x^{5/4} + C$

7. $\dfrac{x^3}{3} + \dfrac{x^2}{2} - x + C$ **9.** $\dfrac{t^3}{3} - t^2 + 3t + C$ **11.** $\frac{5}{8}e^{8x} + C$

13. $\dfrac{x^4}{4} - \dfrac{7}{15}x^{15/7} + C$ **15.** $1000 \ln x + C$ **17.** $-x^{-1} + C$

19. $\frac{2}{3}x^{3/2} + C$ **21.** $-18x^{1/3} + C$ **23.** $-4e^{-2x} + C$

25. $\frac{1}{3}x^3 - x^{3/2} - 3x^{-1/3} + C$ **27.** $f(x) = \dfrac{x^2}{2} - 3x + 13$

29. $f(x) = \dfrac{x^3}{3} - 4x + 7$ **31.** $C(x) = \dfrac{x^4}{4} - x^2 + 100$

33. (a) $R(x) = \dfrac{x^3}{3} - 3x$;

(b) If you sell no products, you make no money.

35. $D(p) = \dfrac{4000}{p} + 3$ **37. (a)** $E(t) = 30t - 5t^2 + 32$;

(b) $E(3) = 77\%$, $E(5) = 57\%$
39. $s(t) = t^3 + 4$ **41.** $v(t) = 2t^2 + 20$
43. $s(t) = -\frac{1}{3}t^3 + 3t^2 + 6t + 10$
45. $s(t) = -16t^2 + v_0 t + s_0$ **47.** $\frac{1}{4}$ mi
49. (a) $M(t) = -0.001t^3 + 0.1t^2$; **(b)** 6

51. $f(t) = \dfrac{t^{\sqrt{3}+1}}{\sqrt{3}+1} + 8$ **53.** $\dfrac{x^6}{6} - \dfrac{2}{5}x^5 + \dfrac{1}{4}x^4 + C$

55. $\frac{2}{5}t^{5/2} + 4t^{3/2} + 18\sqrt{t} + C$

57. $\dfrac{t^4}{4} + t^3 + \dfrac{3}{2}t^2 + t + C$, or $\dfrac{(t+1)^4}{4} + C$ **59.** $\dfrac{b}{a}e^{ax} + C$

61. $\frac{12}{7}x^{7/3} + C$ **63.** $\dfrac{t^3}{3} - t^2 + 4t + C$

Margin Exercises, Section 5.2, pp. 365–374

1. (a) $A(x) = 3x$; **(b)** $A(1) = 3$, $A(2) = 6$, $A(5) = 15$;
(c)

(d) $A(x)$ is an antiderivative of $f(x)$ **2. (a)** $C(x) = 50x$;
(b) $A(x) = 50x$, same;
(c) As x increases, the area over $[0, x]$ increases.
Also, as x increases, total cost increases.

3. (a) $5000 + 4000 + 3000 + 2000$, or $\$14,000$;
(b) 14,000, same

4. (a) $A(x) = \frac{3}{2}x^2$; **(b)** $A(1) = \frac{3}{2}$; **(c)** $A(2) = 6$;
(d) $A(3.5) = 18.375$; **(e)**

(f) $A(x)$ is an antiderivative of $f(x)$
5. (a) $C(x) = -0.05x^2 + 50x$; **(b)** $A(x) = -0.05x^2 + 50x$;
(c) $12,000; less than in Margin Exercise 3   **6.** $5\frac{1}{3}$
7. $13\frac{1}{2}$ **8. (a)** Yes; **(b)** $4100

Exercise Set 5.2, p. 374

1. 8   **3.** 8   **5.** $41\frac{2}{3}$ **7.** $\frac{1}{4}$ **9.** $10\frac{2}{3}$
11. $e^3 - 1 \approx 19.086$ **13.** $\ln 3 \approx 1.0986$ **15.** 51
17. An antiderivative, velocity
19. An antiderivative, energy used in time t
21. An antiderivative, total revenue
23. An antiderivative, amount of drug in blood
25. (a) $C(x) = 100x - 0.1x^2$; **(b)** $R(x) = 100x + 0.1x^2$;
(c) $P(x) = 0.2x^2$; **(d)** $P(1000) = 200,000
27. (a) $s(t) = t^3 + t^2$; **(b)** 150   **29.** $3\frac{1}{2}$ **31.** $359\frac{7}{15}$
33. $6\frac{3}{4}$

Margin Exercises, Section 5.3, pp. 377–385

1. $\int_a^b 2x \, dx = b^2 - a^2$ **2.** $\int_a^b e^x \, dx = e^b - e^a$

3. 8   **4.** $1 - \dfrac{1}{e^2}$ **5.** $\frac{2}{3}$ **6.** $e + e^3 - 3$ **7.** $5\frac{1}{3}$

8. $13\frac{1}{2}$ **9.** $\ln 7$ **10.** $-\frac{1}{3}b^{-3} + \frac{1}{3}$ **11.** 75   **12.** $17\frac{1}{3}$
13. (a) $595,991.60; **(b)** $k = 7.3$, so it is the 8th day;
(c) $4,185,966.53

Exercise Set 5.3, p. 385

1. $\frac{1}{6}$ **3.** $\frac{4}{15}$ **5.** $e^b - e^a$ **7.** $b^3 - a^3$

9. $\dfrac{e^2}{2} + \dfrac{1}{2}$ **11.** $\frac{2}{3}$ **13.** $\frac{5}{34}$ **15.** 4   **17.** $9\frac{5}{6}$

19. 12   **21.** $e^5 - \dfrac{1}{e}$ **23.** $7\frac{1}{3}$ **25.** $17\frac{1}{3}$ **27. (a)** $2948.26;

(b) $2913.90; **(c)** $k = 6.9$, so it will be on the 7th day
29. $3600   **31.** $68,676,000   **33.** 90
35. (a) 2,253,169; **(b)** 21.8 days   **37.** 9   **39.** $\frac{307}{6}$

41. $\frac{15}{4}$ **43.** 8   **45.** 12   **47.** See Exercise Set 1.5,
Exercise 87 for the graph. The area is 4.   **49.** See
Exercise Set 1.5, Exercise 89 for the graph. The area is
529.356.

Margin Exercises, Section 5.4, pp. 389–395

1. 4   **2.** 4   **3. (a)** $-\frac{1}{6}$; **(b)** $\frac{5}{6}$; **(c)** $\frac{4}{6}$ **4. (a)** -4;
(b) 4;   **(c)** 0   **5. (a)** 0; **(b)** $A - A$, or 0
6. (a) Negative; **(b)** $A - 2A$, or $-A$ **7. (a)** Positive;
(b) $3A - A$, or $2A$ **8. (a)** Negative; **(b)** $-3A$ **9.** $\frac{1}{6}$
10. (a) $P(k) = 5k^2 + 10k$; **(b)** $P(5) = 175$

Exercise Set 5.4, p. 395

1. $\frac{1}{4}$ **3.** $\frac{9}{2}$ **5.** $\frac{125}{6}$ **7.** $\frac{9}{2}$ **9.** $\frac{1}{6}$ **11.** $\frac{125}{3}$ **13.** $\frac{32}{3}$
15. 3   **17.** 128   **19.** 9   **21. (a)** B; **(b)** 2

23. $\dfrac{(e-1)^2}{e}$ **25.** $\frac{2}{3}$ **27.** 16   **29.** 96

31. (a)

$c = 1.4594$; **(c)** 64.5239; **(d)** 17.683

(b) $a = -1.8623$, $b = 0$,

Margin Exercises, Section 5.5, pp. 399–402

1. $dy = (12x + 1) \, dx$ **2.** $du = dx$ **3.** $dy = 3x^2 e^{x^3} dx$
4. $e^{x^3} + C$ **5.** $\ln(5 + x^2) + C$

6. $-\dfrac{1}{3 + x^2} + C$ **7.** $\dfrac{(\ln x)^2}{2} + C$ **8.** $\frac{1}{3}e^{x^3} + C$

9. $\frac{1}{5}e^{5x} + C$ **10.** $\frac{1}{0.02}e^{0.02x} + C$, or $50e^{0.02x} + C$
11. $-e^{-x} + C$ **12.** $\frac{1}{80}(x^4 + 5)^{20} + C$

13. $\displaystyle\int \dfrac{\ln x \, dx}{x} = \dfrac{(\ln x)^2}{2} + C$, so $\displaystyle\int_1^e \dfrac{\ln x \, dx}{x} =$

$\left[\dfrac{(\ln x)^2}{2}\right]_1^e = \dfrac{(\ln e)^2}{2} - \dfrac{(\ln 1)^2}{2} = \dfrac{1}{2} - 0 = \dfrac{1}{2}$

Exercise Set 5.5, p. 403

1. $\ln(7 + x^3) + C$ **3.** $\frac{1}{4}e^{4x} + C$ **5.** $2e^{x/2} + C$

7. $\frac{1}{4}e^{x^4} + C$ 9. $-\frac{1}{3}e^{-t^3} + C$ 11. $\frac{(\ln 4x)^2}{2} + C$

13. $\ln(1 + x) + C$ 15. $-\ln(4 - x) + C$

17. $\frac{1}{24}(t^3 - 1)^8 + C$ 19. $\frac{1}{8}(x^4 + x^3 + x^2)^8 + C$

21. $\ln(4 + e^x) + C$ 23. $\frac{1}{4}(\ln x^2)^2 + C$, or $(\ln x)^2 + C$

25. $\ln(\ln x) + C$ 27. $\frac{2}{3a}(ax + b)^{3/2} + C$

29. $\frac{b}{a}e^{ax} + C$ 31. $\frac{-1}{4(1 + x^3)^4} + C$

33. $-\frac{21}{8}(4 - x^2)^{4/3} + C$ 35. $e - 1$ 37. $\frac{21}{4}$

39. $\ln 4 - \ln 2 = \ln \frac{4}{2} = \ln 2$ 41. $\ln 19$

43. $1 - \frac{1}{e^b}$ 45. $1 - \frac{1}{e^{mb}}$ 47. $\frac{208}{3}$ 49. $\frac{1640}{6561}$ 51. $\frac{315}{8}$

53. $D(p) = 2000\sqrt{25 - p^2} + 5000$

55. (a) $P(T) = 2000T^2 + [-250,000(e^{-0.1T} - 1)] -$ 250,000, or $2000T^2 - 250,000e^{-0.1T}$;
(b) $P(10) = \$108,030$ 57. $\frac{128}{3}$

59. $-\frac{5}{12}(1 - 4x^2)^{3/2} + C$ 61. $-\frac{1}{3}e^{-x^3} + C$

63. $-e^{1/t} + C$ 65. $-\frac{1}{3}(\ln x)^{-3} + C$

67. $\frac{2}{9}(x^3 + 1)^{3/2} + C$

69. $\frac{3}{4}(x^2 - 6x)^{2/3} + C$ 71. $\frac{1}{8}[\ln(t^4 + 8)]^2 + C$

73. $x + \frac{9}{x + 3} + C$ 75. $t - \ln(t - 4) + C$

77. $-\ln(1 + e^{-x}) + C$

79. $\frac{1}{n + 1}(\ln x)^{n + 1} + C$ 81. $\ln[\ln(\ln x)] + C$

83. $\frac{9(7x^2 + 9)^{n + 1}}{14(n + 1)} + C$

Margin Exercises, Section 5.6, pp. 406–411

1. $xe^{3x} - \frac{1}{3}e^{3x} + C$ 2. $x \ln 5x - x + C$

3. $\frac{x^2}{2} \ln 4x - \frac{x^2}{4} + C$ 4. $\frac{2}{3}x(x + 3)^{3/2} - \frac{4}{15}(x + 3)^{5/2} + C$

5. $\ln 32 - \frac{3}{4}$ 6. (a) $e^x(x^2 - 2x + 2) + C$;
(b) $17e^5 - 2 \approx 2521.02$
7. $e^x(x^4 - 4x^3 + 12x^2 - 24x + 24) + C$

Exercise Set 5.6 p. 412

1. $xe^{5x} - \frac{1}{5}e^{5x} + C$ 3. $\frac{1}{2}x^6 + C$ 5. $\frac{x}{2}e^{2x} - \frac{1}{4}e^{2x} + C$

7. $-\frac{x}{2}e^{-2x} - \frac{1}{4}e^{-2x} + C$ 9. $\frac{x^3}{3} \ln x - \frac{x^3}{9} + C$

11. $\frac{x^2}{2} \ln x^2 - \frac{x^2}{2} + C$, or $x^2 \ln x - \frac{x^2}{2} + C$

13. $(x + 3) \ln(x + 3) - x + C$. Let $u = \ln(x + 3)$, $dv = dx$, and choose $v = x + 3$ for an antiderivative of v.

15. $\left(\frac{x^2}{2} + 2x\right) \ln x - \frac{x^2}{4} - 2x + C$

17. $\left(\frac{x^2}{2} - x\right) \ln x - \frac{x^2}{4} + x + C$

19. $\frac{2}{3}x(x + 2)^{3/2} - \frac{4}{15}(x + 2)^{5/2} + C$

21. $\frac{x^4}{4} \ln 2x - \frac{x^4}{16} + C$ 23. $x^2e^x - 2xe^x + 2e^x + C$

25. $\frac{1}{2}x^2e^{2x} + \frac{1}{4}e^{2x} - \frac{1}{2}xe^{2x} + C$

27. $e^{-2x}(-\frac{1}{2}x^3 - \frac{3}{4}x^2 - \frac{3}{4}x - \frac{3}{8}) + C$

29. $e^{3x}[\frac{1}{3}(x^4 + 1) - \frac{4}{9}x^3 + \frac{4}{9}x^2 - \frac{8}{27}x + \frac{8}{81}] + C$

31. $\frac{8}{3} \ln 2 - \frac{7}{9}$ 33. $9 \ln 9 - 5 \ln 5 - 4$

35. 1 37. $C(x) = \frac{8}{3}x(x + 3)^{3/2} - \frac{16}{15}(x + 3)^{5/2}$

39. (a) $10[e^{-T}(-T - 1) + 1]$; (b) ≈ 9.084

41. $\frac{2}{9}x^{3/2}[3 \ln x - 2] + C$

43. $\frac{e^t}{t + 1} + C$ 45. $2\sqrt{x} \ln x - 4\sqrt{x} + C$

47. $(13t^2 - 48)[\frac{5}{16}(4t + 7)^{4/5}] - t[\frac{325}{288}(4t + 7)^{9/5}] + \frac{1625}{16,128}(4t + 7)^{14/5} + C$

49. Let $u = x^n$ and $dv = e^x\, dx$. Then $du = nx^{n-1}\, dx$ and $v = e^x$. Then use integration by parts.   51. $355,986$

Margin Exercises, Section 5.7, pp. 414–416

1. $\frac{1}{7}\ln\left(\frac{x}{2x + 7}\right) + C$ 2. $\frac{15}{4} - \frac{3x}{2} - \frac{15}{4}\ln(5 - 2x) + C$

3. $\frac{5}{4(5 - 2x)} + \frac{1}{4}\ln(5 - 2x) + C$

4. $\frac{5}{2}y\sqrt{y^2 - \frac{4}{25}} - \frac{2}{5}\ln\left(y + \sqrt{y^2 - \frac{4}{25}}\right) + C$

5. $\frac{1}{7}\ln\left(\frac{7 + x}{7 - x}\right) + C$

6. $x(\ln x)^2 - 2(x \ln x - x) + C$

Exercise Set 5.7, p. 416

1. $\frac{1}{9}e^{-3x}(-3x - 1) + C$ 3. $\frac{5^x}{\ln 5} + C$

5. $\frac{1}{8}\ln\left(\frac{4 + x}{4 - x}\right) + C$ 7. $5 - x - 5 \ln(5 - x) + C$

9. $\dfrac{1}{5(5-x)} + \dfrac{1}{25}\ln\left(\dfrac{x}{5-x}\right) + C$

11. $(\ln 3)x + x\ln x - x + C$

13. $\dfrac{x^4 e^{5x}}{5} - \dfrac{4}{25}x^3 e^{5x} + \dfrac{12}{125}x^2 e^{5x} - \dfrac{24}{3125}e^{5x}(5x-1) + C$

15. $x^4(\frac{1}{4}\ln x - \frac{1}{16}) + C$ 17. $\ln(x + \sqrt{x^2+7}) + C$

19. $\dfrac{2}{5-7x} + \dfrac{2}{5}\ln\left(\dfrac{x}{5-7x}\right) + C$ 21. $-\dfrac{5}{4}\ln\left(\dfrac{x-\frac{1}{2}}{x+\frac{1}{2}}\right) + C$

23. $m\sqrt{m^2+4} + 4\ln(m + \sqrt{m^2+4}) + C$

25. $\dfrac{5\ln x}{2x^2} + \dfrac{5}{4x^2} + C$

27. $x^3 e^x - 3x^2 e^x + 6xe^x - 6e^x + C$

29. $S(p) = 100\left[\dfrac{20}{20-p} + \ln(20-p)\right]$

31. $-4\ln\left(\dfrac{x}{3x-2}\right) + C$

33. $\dfrac{1}{-2(x-2)} + \dfrac{1}{4}\ln\left(\dfrac{x}{x-2}\right) + C$

35. $\dfrac{-3}{(e^{-x}-3)} + \ln(e^{-x}-3) + C$

Margin Exercises, Section 5.8, pp. 418–427

1. $\displaystyle\sum_{i=1}^{6} i^2$ 2. $\displaystyle\sum_{i=1}^{4} e^i$ 3. $\displaystyle\sum_{i=1}^{38} P(x_i)\,\Delta x$

4. $4^1 + 4^2 + 4^3$, or 84   5. $e^1 + 2e^2 + 3e^3 + 4e^4 + 5e^5$

6. $t(x_1)\,\Delta x + t(x_2)\,\Delta x + \cdots + t(x_{20})\,\Delta x$ 7. 35,937,500

8. \$11,250   9. (a) 3.0667; (b) 2.4500;

(c) $\displaystyle\int_1^7 \dfrac{1}{x}\,dx = \ln 7 \approx 1.9459$ 10. 2

11. (a) $\frac{95}{3} \approx 31.7°$; (b) $-10°$; (c) 46.25°   12. \$257

Exercise Set 5.8, p. 427

1. (a) 1.4914; (b) 0.8571   3. 0   5. $e-1$

7. $\frac{4}{3}$ 9. 13   11. $\dfrac{1}{n+1}$

13. (a) 90 words per minute; (b) 96 words per minute at $t = 1$; (c) 82 words per minute; the typist tires as time increases.   15. 252 million   17. (a) 100 after 10 hr; (b) $33\frac{1}{3}$ 19. 47.5

Summary and Review: Chapter 5, p. 430

1. [5.1] $\frac{8}{5}x^5 + C$ 2. [5.1] $3e^x + 2x + C$
3. [5.1] $t^3 + \frac{7}{2}t^2 + \ln t + C$ 4. [5.2], [5.3] 9
5. [5.2], [5.3] $\frac{62}{3}$ 6. [5.2] Total number of words written in t minutes; an antiderivative   7. [5.2] Total amount of sales at the end of t days; an antiderivative
8. [5.3] $\frac{1}{6}(b^6 - a^6)$ 9. [5.4] $-\frac{2}{5}$ 10. [5.3] $e - \frac{1}{2}$
11. [5.3] $3\ln 3$ 12. [5.4] 0   13. [5.4] Negative
14. [5.4] Positive   15. [5.4] $\frac{27}{2}$ 16. [5.5] $\frac{1}{4}e^{x^4} + C$
17. [5.5] $\ln(4t^6 + 3) + C$ 18. [5.5] $\frac{1}{4}(\ln 4x)^2 + C$
19. [5.5] $-\frac{2}{3}e^{-3x} + C$ 20. [5.6] $e^{3x}(x - \frac{1}{3}) + C$
21. [5.6] $x\ln x^7 - 7x + C$

22. [5.6] $x^3 \ln x - \dfrac{x^3}{3} + C$ 23. [5.7] $\dfrac{1}{14}\ln\left(\dfrac{7+x}{7-x}\right) + C$

24. [5.7] $\frac{1}{5}x^2 e^{5x} - \frac{2}{25}xe^{5x} + \frac{2}{125}e^{5x} + C$

25. [5.7] $\dfrac{1}{49} + \dfrac{x}{7} - \dfrac{1}{49}\ln(7x+1) + C$

26. [5.7] $\ln(x + \sqrt{x^2 - 36}) + C$

27. [5.7] $x^7\left(\dfrac{\ln x}{7} - \dfrac{1}{49}\right) + C$

28. [5.7] $\frac{1}{64}e^{8x}(8x - 1) + C$ 29. [5.8] 3.667

30. [5.8] $\frac{1}{2}(11 - e^2)$ 31. [5.3] 80   32. [5.5] $e^{12} - 1$, or about \$162,754

33. [5.6] $e^{0.1x}(10x^3 - 300x^2 + 6000x - 60,000) + C$
34. [5.5] $\ln(4t^3 + 7) + C$
35. [5.7] $\frac{2}{75}(5x - 8)\sqrt{4 + 5x} + C$
36. [5.5] $e^{x^5} + C$ 37. [5.5] $\ln(x + 9) + C$
38. [5.5] $\frac{1}{96}(t^8 + 3)^{12} + C$ 39. [5.6] $x\ln 7x - x + C$

40. [5.6] $\dfrac{x^2}{2}\ln 8x - \dfrac{x^2}{4} + C$

41. [5.5], [5.7] $\frac{1}{10}[\ln(t^5 + 3)]^2 + C$
42. [5.5] $-\frac{1}{2}\ln(1 + 2e^{-x}) + C$
43. [5.5], [5.6], [5.7] $(\ln\sqrt{x})^2 + C$

44. [5.6], [5.7] $\dfrac{x^{92}}{92}(\ln x - \dfrac{1}{92}) + C$

45. [5.6] $(x - 3)\ln(x - 3) - (x - 4)\ln(x - 4) + C$

46. [5.5] $-\dfrac{1}{3(\ln x)^3} + C$

Test: Chapter 5, p. 432

1. $x + C$ 2. $200x^5 + C$ 3. $e^x + \ln x + \frac{8}{11}x^{11/8} + C$
4. $\frac{1}{6}$ 5. $4\ln 3$ 6. An antiderivative, total number of

words typed in t minutes   **7.** 12

8. $-\frac{1}{2}\left(\frac{1}{e^2} - 1\right)$ **9.** $\ln b - \ln a$, or $\ln \frac{b}{a}$ **10.** Positive

11. $\ln (x + 8) + C$ **12.** $-2e^{-0.5x} + C$

13. $\frac{1}{40}(t^4 + 1)^{10} + C$

14. $\frac{x}{5}e^{5x} - \frac{e^{5x}}{25} + C$ **15.** $\frac{x^4}{4} \ln x^4 - \frac{x^4}{4} + C$

16. $\frac{2^x}{\ln 2} + C$ **17.** $\frac{1}{7} \ln \left(\frac{x}{7 - x}\right) + C$ **18.** 6   **19.** $\frac{1}{3}$

20. 95   **21.** $49,000   **22.** 94 words

23. $\frac{1}{10} \ln \left(\frac{x}{10 - x}\right) + C$,

24. $e^x(x^5 - 5x^4 + 20x^3 - 60x^2 + 120x - 120) + C$

25. $\frac{1}{6}e^{x^6} + C$ **26.** $\frac{2}{3}x^{3/2} \ln x - \frac{4}{9}x^{3/2} + C$

27. $\frac{1}{15}(3x^2 - 8)(x^2 + 4)^{3/2} + C$

28. $\frac{1}{16} \ln \left(\frac{8 + x}{8 - x}\right) + C$

29. $e^{0.1x}(10x^4 - 400x^3 + 12,000x^2 - 240,000x + 2,400,000) + C$

30. $\frac{x^2}{2} \ln 13x - \frac{x^2}{4} + C$

31. $\frac{(\ln x)^4}{4} - \frac{4}{3}(\ln x)^3 + 5 \ln x + C$

32. $(x + 3) \ln (x + 3) - (x + 5) \ln (x + 5) + C$

33. $\frac{3}{10}(8x^3 + 10)(5x - 4)^{2/3} - \frac{108}{125}x^2(5x - 4)^{5/3} + \frac{81}{625}x(5x - 4)^{8/3} - \frac{243}{34,375}(5x - 4)^{11/3} + C$

CHAPTER 6

Margin Exercises, Section 6.1, pp. 438–443

1. $53.33   **2.** $1.17   **3.** (a) $(2, \$16)$; (b) $18.67; (c) $7.33

Exercise Set 6.1, p. 443

1. (a) $(6, \$5)$; (b) $15; (c) $9   **3.** (a) $(1, \$9)$; (b) $3.33; (c) $1.67   **5.** (a) $(3, \$9)$; (b) $36; (c) $18
7. (a) $(50, \$500)$; (b) $12,500; (c) $6250
9. (a) $(2, \$3)$; (b) $2; (c) $0.35   **11.** (a) $(5, \$0.61)$;

(b) $86.36; (c) $2.45   **13.**

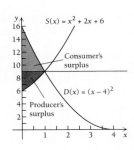

Margin Exercises, Section 6.2, pp. 444–450

1. $1161.83   **2.** $29,786.72   **3.** $215.41
4. 112,891.1 million barrels   **5.** After 27 yr (2017)
6. $10,369.61   **7.** $67,032.00   **8.** $15,563.96

Exercise Set 6.2, p. 451

1. $131   **3.** $5610.71   **5.** $340,754.12   **7.** $949.94
9. $259.37   **11.** $8264.94   **13.** $29,676.18
15. $17,802.91   **17.** $511,471.15
19. 3,674,343,000 tons   **21.** After 38 yr (2028)
23. 19.994 lb   **25.** $2,383,120   **27.** $125.30
29.

Margin Exercises, Section 6.3, pp. 453–457

1. $\frac{2}{3}, \frac{9}{10}, \frac{99}{100}, \frac{199}{200}$ **2.** $\frac{1}{2}$ **3.** ∞ **4.** Convergent, 1
5. Divergent   **6.** $36,000

Exercise Set 6.3, p. 457

1. Convergent, $\frac{1}{3}$ **3.** Divergent   **5.** Convergent, 1
7. Convergent, $\frac{1}{2}$ **9.** Divergent   **11.** Convergent, 5
13. Divergent   **15.** Divergent   **17.** Divergent
19. Convergent, 1   **21.** 1   **23.** $45,000   **25.** $6250

27. \$4500 **29.** 33,333 lb **31.** Divergent
33. Convergent, 2 **35.** Convergent, $\frac{1}{2}$

37. $\dfrac{1}{k^2}$; the total dose of the drug

39.

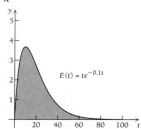

$E(t) = te^{-0.1t}$

Margin Exercises, Section 6.4, pp. 460–472

1. (a) $\frac{7}{20}$; (b) $\frac{6}{20}$, or $\frac{3}{10}$; (c) $\frac{4}{20}$, or $\frac{1}{5}$; (d) 0 **2.** (a) $\frac{1}{2}$; (b) $\frac{1}{8}$;
(c) $\frac{1}{4}$; (d) $\frac{1}{8}$ **3.** [15, 25] **4.** [0, ∞) **5.** $\frac{2}{3}$

6. $\displaystyle\int_1^2 \frac{2}{3}x\,dx = \left[\frac{2}{3}\cdot\frac{1}{2}x^2\right]_1^2 = \frac{1}{3}(2^2 - 1) = 1$

7. (a) $\displaystyle\int_3^8 \frac{24}{5t^2}\,dt = \frac{24}{5}\left[\frac{t^{-1}}{-1}\right]_3^8 = -\frac{24}{5}[t^{-1}]_3^8 =$

$-\dfrac{24}{5}\left(\dfrac{1}{8} - \dfrac{1}{3}\right) = 1;$

(b) $\frac{16}{25}$; (c) $\frac{2}{5}$ **8.** $\frac{3}{26}$; $f(x) = \frac{3}{26}x^2$ **9.** 4; $f(x) = 4x^3$
10. $\frac{1}{4}$ **11.** $\frac{1}{4}$

12. $\displaystyle\int_0^\infty ke^{-kx}\,dx = \lim_{b\to\infty}\int_0^b ke^{-kx}\,dx = \lim_{b\to\infty}\left[\frac{k}{-k}e^{-kx}\right]_0^b =$

$\lim_{b\to\infty}[-e^{-kx}]_0^b = \lim_{b\to\infty}[-e^{-kb} - (-e^{-k(0)})] =$

$\lim_{b\to\infty}\left[-\dfrac{1}{e^{kb}} + 1\right] = 0 + 1 = 1$

13. 0.3297

Exercise Set 6.4, p. 472

1. $\displaystyle\int_0^1 2x\,dx = [x^2]_0^1 = 1^2 - 0^2 = 1$

3. $\displaystyle\int_4^7 \frac{1}{3}\,dx = \left[\frac{1}{3}x\right]_4^7 = \frac{1}{3}(7 - 4) = 1$

5. $\displaystyle\int_1^3 \frac{3}{26}x^2\,dx = \left[\frac{3}{26}\cdot\frac{x^3}{3}\right]_1^3 = \frac{1}{26}(3^3 - 1^3) = 1$

7. $\displaystyle\int_1^e \frac{1}{x}\,dx = [\ln x]_1^e = \ln e - \ln 1 = 1 - 0 = 1$

9. $\displaystyle\int_{-1}^1 \frac{3}{2}x^2\,dx = \left[\frac{3}{2}\cdot\frac{1}{3}x^3\right]_{-1}^1 = \frac{1}{2}(1^3 - (-1)^3) =$

$\frac{1}{2}(1 + 1) = 1$

11. $\displaystyle\int_0^\infty 3e^{-3x}\,dx = \lim_{b\to\infty}\int_0^b 3e^{-3x}\,dx =$

$\lim_{b\to\infty}\left[\frac{3}{-3}e^{-3x}\right]_0^b = \lim_{b\to\infty}[-e^{-3x}]_0^b =$

$\lim_{b\to\infty}[-e^{-3b} - (-e^{-3\cdot0})] - \lim_{b\to\infty}\left(1 - \dfrac{1}{e^{3b}}\right) = 1$
13. $\frac{1}{4}$; $f(x) = \frac{1}{4}x$ **15.** $\frac{3}{2}$; $f(x) = \frac{3}{2}x^2$ **17.** $\frac{1}{5}$; $f(x) = \frac{1}{5}$

19. $\frac{1}{2}$; $f(x) = \dfrac{2 - x}{2}$ **21.** $\dfrac{1}{\ln 3}$; $f(x) = \dfrac{1}{x\ln 3}$

23. $\dfrac{1}{e^3 - 1}$; $f(x) = \dfrac{e^x}{e^3 - 1}$ **25.** $\frac{8}{25}$ **27.** $\frac{1}{2}$

29. 0.3297 **31.** 0.99995 **33.** 0.3935 **35.** 0.9502
37. $b = \sqrt[4]{4}$, or $\sqrt{2}$

Margin Exercises, Section 6.5, pp. 477–485

1. $E(x) = \frac{2}{3}$, $E(x^2) = \frac{1}{2}$ **2.** $\mu = \frac{2}{3}$, $\sigma^2 = \frac{1}{18}$, $\sigma = \sqrt{\frac{1}{18}} = \frac{1}{3}\sqrt{\frac{1}{2}}$
3. (a) 0.4850; (b) 0.4608; (c) 0.9559; (d) 0.2720;
(e) 0.1102; (f) 0.0307 **4.** 0.1210

Exercise Set 6.5, p. 485

1. $\mu = E(x) = \frac{7}{2}$, $E(x^2) = 13$, $\sigma^2 = \frac{3}{4}$, $\sigma = \frac{1}{2}\sqrt{3}$
3. $\mu = E(x) = 2$, $E(x^2) = \frac{9}{2}$, $\sigma^2 = \frac{1}{2}$, $\sigma = \sqrt{\frac{1}{2}}$
5. $\mu = E(x) = \frac{14}{9}$, $E(x^2) = \frac{5}{2}$, $\sigma^2 = \frac{13}{162}$, $\sigma = \sqrt{\frac{13}{162}} = \frac{1}{9}\sqrt{\frac{13}{2}}$
7. $\mu = E(x) = -\frac{5}{4}$, $E(x^2) = \frac{11}{5}$, $\sigma^2 = \frac{51}{80}$, $\sigma = \sqrt{\frac{51}{80}} = \frac{1}{4}\sqrt{\frac{51}{5}}$
9. $\mu = E(x) = \dfrac{2}{\ln 3}$, $E(x^2) = \dfrac{4}{\ln 3}$, $\sigma^2 = \dfrac{4\ln 3 - 4}{(\ln 3)^2}$,

$\sigma = \dfrac{2}{\ln 3}\sqrt{\ln 3 - 1}$ **11.** 0.4964

13. 0.3665 **15.** 0.6442 **17.** 0.0078 **19.** 0.1716
21. 0.0013 **23.** (a) 0.6826; (b) 68.26% **25.** 0.2898
27. 0.4514 **29.** 0.62% **31.** 0.1115

33. $\mu = E(x) = \dfrac{a + b}{2}$, $E(x^2) = \dfrac{b^3 - a^3}{3(b - a)}$,

or $\dfrac{b^2 + ba + a^2}{3}$, $\sigma^2 = \dfrac{(b - a)^2}{12}$, $\sigma = \dfrac{b - a}{2\sqrt{3}}$

35. $\sqrt{2}$ **37.** $\dfrac{\ln 2}{k}$ **39.** 5.801 oz

Margin Exercises, Section 6.6, pp. 489–490

1. $\dfrac{\pi}{3}$ 2. $\dfrac{\pi}{2}(e^2 - e^{-4})$

Exercise Set 6.6, p. 490

1. $\dfrac{9\pi}{2}$ 3. $\dfrac{7\pi}{3}$ 5. $\dfrac{\pi}{2}(e^{10} - e^{-4})$ 7. $\dfrac{2\pi}{3}$ 9. $\pi \ln 3$

11. 32π 13. $\dfrac{32\pi}{5}$ 15. 56π 17. $\dfrac{32\pi}{3}$ 19. $2\pi e^3$

21. (a) Divergent; (b) π

Margin Exercises, Section 6.7, pp. 492–500

1. $y = x^3 + C$ 2. (a) $y = x^3 + C$; (b) $y = x^3 - 7$,
$y = x^3 + \frac{1}{2}$, $y = x^3$. Answers may vary.   3. $y = x^2 + 6$
4. $f(x) = \ln x - x^2 + \frac{2}{3}x^{3/2} + \frac{13}{3}$ 5. $y' = 2e^x - 21e^{3x}$,
$y'' = 2e^x - 63e^{3x}$. Then

$$y'' - 4y' + 3y = 0$$

$$\begin{array}{c|c} 2e^x - 63e^{3x} - 4(2e^x - 21e^{3x}) + 3(2e^x - 7e^{3x}) & 0 \\ 2e^x - 63e^{3x} - 8e^x + 84e^{3x} + 6e^x - 21e^{3x} & \\ & 0 \end{array}$$

6. $\dfrac{dy}{dx} = 2xe^{2x} + e^{2x}$. Then

$$\dfrac{dy}{dx} - 2y = e^{2x}$$

$$\begin{array}{c|c} (2xe^{2x} + e^{2x}) - 2(xe^{2x}) & e^{2x} \\ \quad\quad\quad e^{2x} & \end{array}$$

7. $y = C_1 e^{x^3}$, where $C_1 = e^C$ 8. $y = \sqrt[3]{x^2 + 339}$
9. $y = \sqrt{10x + C_1}$, $y = -\sqrt{10x + C_1}$, where $C_1 = 2C$
10. $y = -2 + C_1 e^{x^2/2}$, where $C_1 = e^C$
11. $x = \dfrac{C_1}{p^3}$ 12. $R = C_1 \cdot S^k$, where $C_1 = e^C$

Exercise Set 6.7, p. 500

1. $y = x^4 + C$; $y = x^4 + 3$, $y = x^4$, $y = x^4 - 796$; answers
may vary   3. $y = \frac{1}{2}e^{2x} + \frac{1}{2}x^2 + C$; $y = \frac{1}{2}e^{2x} + \frac{1}{2}x^2 - 5$,
$y = \frac{1}{2}e^{2x} + \frac{1}{2}x^2 + 7$; $y = \frac{1}{2}e^{2x} + \frac{1}{2}x^2$; answers may vary
5. $y = 3 \ln x - \frac{1}{3}x^3 + \frac{1}{6}x^6 + C$; $y = 3 \ln x - \frac{1}{3}x^3 + \frac{1}{6}x^6 - 15$,
$y = 3 \ln x - \frac{1}{3}x^3 + \frac{1}{6}x^6 - 7$, $y = 3 \ln x - \frac{1}{3}x^3 + \frac{1}{6}x^6$;
answers may vary   7. $y = \frac{1}{3}x^3 + x^2 - 3x + 4$
9. $y = \frac{3}{5}x^{5/3} - \frac{1}{2}x^2 - \frac{61}{10}$

11. $y'' = \dfrac{1}{x}$. Then

$$y'' - \dfrac{1}{x} = 0$$

$$\begin{array}{c|c} \dfrac{1}{x} - \dfrac{1}{x} & 0 \\ 0 & \end{array}$$

13. $y' = 4e^x + 3xe^x$, $y'' = 7e^x + 3xe^x$. Then

$$y'' - 2y' + y = 0$$

$$\begin{array}{c|c} (7e^x + 3xe^x) - 2(4e^x + 3xe^x) + (e^x + 3xe^x) & 0 \\ 7e^x + 3xe^x - 8e^x - 6xe^x + e^x + 3xe^x & \\ 0 & \end{array}$$

15. $y = C_1 e^{x^4}$, where $C_1 = e^C$ 17. $y = \sqrt[3]{\frac{5}{2}x^2 + C}$
19. $y = \sqrt{2x^2 + C_1}$, $y = -\sqrt{2x^2 + C_1}$, where $C_1 = 2C$
21. $y = \sqrt{6x + C_1}$, $y = -\sqrt{6x + C_1}$, where $C_1 = 2C$
23. $y = -3 + 8e^{x/2}$ 25. $y = \sqrt[3]{15x - 3}$
27. $y = C_1 e^{3x}$, where $C_1 = e^C$ 29. $P = C_1 e^{2t}$, where
$C_1 = e^C$ 31. $f(x) = \ln x - 2x^2 + \frac{2}{3}x^{3/2} + 9$

33. $C(x) = 2.6x - 0.01x^2 + 120$, $A(x) = 2.6 - 0.01x + \dfrac{120}{x}$

35. (a) $P(C) = \dfrac{400}{(C + 3)^{1/2}} - 40$; (b) \$97

37. (a) $R = k \cdot \ln (S + 1) + C$; (b) $R = k \cdot \ln (S + 1)$;
(c) no units, no pleasure from them

39. $x = 200 - p$ 41. $x = \dfrac{C_1}{p^n}$

Summary and Review: Chapter 6, p. 502

1. [6.1] (16, \$2)   2. [6.1] \$18.67   3. [6.1] \$5.33
4. [6.2] \$83,560.54   5. [6.2] \$251.50   6. [6.2] 282,732
thousand tons   7. [6.2] 183 yr   8. [6.2] \$247.88
9. [6.2] \$44,517   10. [6.2], [6.3] \$53,333
11. [6.3] Convergent, 1   12. [6.3] Divergent

13. [6.3] Convergent, $\frac{1}{2}$ 14. [6.4] $\frac{8}{3}$; $f(x) = \dfrac{8}{3x^3}$

15. [6.4] 0.6   16. [6.5] $\frac{3}{5}$ 17. [6.5] $\frac{3}{4}$ 18. [6.5] $\frac{3}{4}$

19. [6.5] $\frac{3}{80}$ 20. [6.5] $\dfrac{\sqrt{15}}{20}$ 21. [6.5] 0.4678

22. [6.5] 0.8827   23. [6.5] 0.1002   24. [6.5] 0.5000

25. [6.5] 0.0918   26. [6.6] $\dfrac{127\pi}{7}$ 27. [6.6] $\dfrac{\pi}{6}$

28. [6.7] $y = C_1 e^{x^{11}}$, where $C_1 = e^C$
29. [6.7] $y = \pm\sqrt{4x + C_1}$, where $C_1 = 2C$
30. [6.7] $y = 5e^{4x}$ **31.** [6.7] $v = \sqrt[3]{15t + 19}$
32. [6.7] $y = \pm\sqrt{3x^2 + C_1}$, where $C_1 = 2C$
33. [6.8] $y = 8 - C_1 e^{-x^2/2}$, where $C_1 = e^{-C}$
34. [6.8] $x = 100 - p$ **35.** [6.7] $V = \$36.37 - \$6.37e^{-kt}$
36. [6.4] $c = \sqrt[9]{4.5}$ **37.** [6.3] Convergent, $-\frac{1}{5}$
38. [6.3] Convergent, 3

Test: Chapter 6, p. 504

1. (16, \$3) **2.** \$45 **3.** \$22.50 **4.** \$29,192
5. \$344.66 **6.** 37,121,870 thousand tons **7.** 39 yr
8. \$2465.97 **9.** \$30,717.71 **10.** \$34,545.45
11. Convergent, $\frac{1}{4}$ **12.** Divergent
13. $\frac{1}{4}$; $f(x) = \dfrac{x^3}{4}$ **14.** 0.8647 **15.** $E(x) = \frac{13}{6}$
16. $E(x^2) = 5$ **17.** $\mu = \frac{13}{6}$ **18.** $\sigma^2 = \frac{11}{36}$
19. $\sigma = \dfrac{\sqrt{11}}{6}$ **20.** 0.4332
21. 0.4420 **22.** 0.9071 **23.** 0.4207
24. $\pi \ln 5$ **25.** $\dfrac{5\pi}{2}$ **26.** $y = C_1 e^{x^8}$, where $C_1 = e^C$
27. $y = \sqrt{18x + C_1}$, $y = -\sqrt{18x + C_1}$, where $C_1 = 2C$
28. $y = 11e^{6t}$ **29.** $y = 5 - C_1 e^{-x^3/3}$, where $C_1 = e^{-C}$
30. $v = \pm\sqrt[4]{8t + C_1}$, where $C_1 = 4C$
31. $y = C_1 e^{4x + x^2/2}$, where $C_1 = e^C$
32. $x = \dfrac{C_1}{p^4}$ **33.** (a) $V(t) = 36(1 - e^{-kt})$;
(b) $k = 0.12$; (c) $V(t) = 36(1 - e^{-0.12t})$;
(d) $V(12) \approx \$27.47$; (e) $t \approx 14.9$ months **34.** $b = \sqrt[6]{6}$
35. Convergent, $-\frac{1}{4}$

CHAPTER 7

Margin Exercises, Section 7.1, pp. 509–512

1. (a) 128; the profit from selling 14 items of the first product and 12 of the second is \$128; (b) 48; the profit from selling none of the first product and 8 items of the second is \$48 **2.** 7.2, 94.5, 0.6 **3.** 2000 **4.** (a) 352;

(b) 15.4 **5.** \$125,656.83 **6.** (a) 4; (b) 4
7.

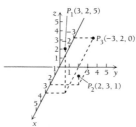

Exercise Set 7.1, p. 516

1. $f(0, -2) = 0$, $f(2, 3) = -8$, $f(10, -5) = 200$
3. $f(0, -2) = 1$, $f(-2, 1) = -13\frac{8}{9}$, $f(2, 1) = 23$
5. $f(e, 2) = \ln e + 2^3 = 1 + 8 = 9$, $f(e^2, 4) = 66$, $f(e^3, 5) = 128$ **7.** $f(-1, 2, 3) = 6$, $f(2, -1, 3) = 12$
9. \$151,571.66 **11.** 7.7% **13.** 244.7 mph **15.** $0°$
17. $-22°$ **19.**

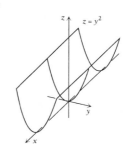

$z = y^2$

21.

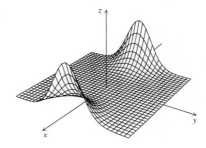

$z = (x^4 - 16x^2)e^{-y^2}$

23.

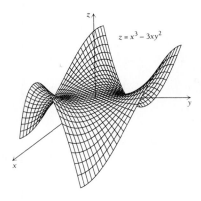

$z = x^3 - 3xy^2$

Margin Exercises, Section 7.2, pp. 518–523

1. (a) $f(x, 4) = -x^2 - 15$; (b) $-2x$ **2.** $\dfrac{\partial f}{\partial x} = -2x$

3. (a) $\dfrac{\partial z}{\partial x} = 6xy + 15x^2$; (b) $\dfrac{\partial z}{\partial y} = 3x^2$

4. (a) $\dfrac{\partial t}{\partial x} = y + z + 2x$; (b) $\dfrac{\partial t}{\partial y} = x + 3y^2$; (c) $\dfrac{\partial t}{\partial z} = x$

5. (a) $f_x = 9x^2y + 2y$; (b) $f_x(-4, 1) = 146$;
(c) $f_y = 3x^3 + 2x$; (d) $f_y(2, 6) = 28$

6. $f_x = \dfrac{1}{x} + ye^x$, $f_y = \dfrac{1}{y} + e^x$

7. (a) 108,000 units; (b) $\dfrac{\partial p}{\partial x} = 600\left(\dfrac{y}{x}\right)^{1/4}$,

$\dfrac{\partial p}{\partial y} = 200\left(\dfrac{x}{y}\right)^{3/4}$; (c) 1000, 43.2

Exercise Set 7.2, p. 523

1. $\dfrac{\partial z}{\partial x} = 2 - 3y$, $\dfrac{\partial z}{\partial y} = -3x$, $\dfrac{\partial z}{\partial x}\Big|_{(-2, -3)} = 11$, $\dfrac{\partial z}{\partial y}\Big|_{(0, -5)} = 0$

3. $\dfrac{\partial z}{\partial x} = 6x - 2y$, $\dfrac{\partial z}{\partial y} = -2x + 1$, $\dfrac{\partial z}{\partial x}\Big|_{(-2, -3)} = -6$,

$\dfrac{\partial z}{\partial y}\Big|_{(0, -5)} = 1$ **5.** $f_x = 2$, $f_y = -3$, $f_x(-2, 4) = 2$,

$f_y(4, -3) = -3$ **7.** $f_x = \dfrac{x}{\sqrt{x^2 + y^2}}$, $f_y = \dfrac{y}{\sqrt{x^2 + y^2}}$,

$f_x(-2, 1) = \dfrac{-2}{\sqrt{5}}$, $f_y(-3, -2) = \dfrac{-2}{\sqrt{13}}$

9. $f_x = 2e^{2x + 3y}$, $f_y = 3e^{2x + 3y}$ **11.** $f_x = ye^{xy}$, $f_y = xe^{xy}$

13. $f_x = \dfrac{y}{x + y}$, $f_y = \dfrac{y}{x + y} + \ln(x + y)$

15. $f_x = 1 + \ln xy$, $f_y = \dfrac{x}{y}$ **17.** $f_x = \dfrac{1}{y} + \dfrac{y}{x^2}$, $f_y = -\dfrac{x}{y^2} - \dfrac{1}{x}$

19. $f_x = 12(2x + y - 5)$, $f_y = 6(2x + y - 5)$

21. $\dfrac{\partial f}{\partial b} = 12m + 6b - 30$, $\dfrac{\partial f}{\partial m} = 28m + 12b - 64$

23. $f_x = 3y - 2\lambda$, $f_y = 3x - \lambda$, $f_\lambda = -(2x + y - 8)$
25. $f_x = 2x - 10\lambda$, $f_y = 2y - 2\lambda$, $f_\lambda = -(10x + 2y - 4)$

27. (a) 3,888,064 units; (b) $\dfrac{\partial p}{\partial x} = 1117.8\left(\dfrac{y}{x}\right)^{0.379}$

$\dfrac{\partial p}{\partial y} = 682.2\left(\dfrac{x}{y}\right)^{0.621}$; (c) 965.8, 866.8

29. (a) 99.6°; (b) 117.8°; (c) 121.3°; (d) 97.5°;
(e) $1.09(T - 58)$; (f) $1.98 - 1.09(1 - H)$

31. $f_x = \dfrac{-4xt^2}{(x^2 - t^2)^2}$, $f_t = \dfrac{4x^2t}{(x^2 - t^2)^2}$

33. $f_x = \dfrac{1}{\sqrt{x}(1 + 2\sqrt{t})}$, $f_t = \dfrac{-1 - 2\sqrt{x}}{\sqrt{t}(1 + 2\sqrt{t})^2}$

35. $f_x = 4x^{-1/3} - 2x^{-3/4}t^{1/2} + 6x^{-3/2}t^{3/2}$,
$f_t = -4x^{1/4}t^{-1/2} - 18x^{-1/2}t^{1/2}$

Margin Exercises, Section 7.3, pp. 525–527

1. (a) $\dfrac{\partial z}{\partial y} = 6xy + 2x$; (b) $\dfrac{\partial}{\partial x}\left(\dfrac{\partial z}{\partial y}\right) = 6y + 2$;

(c) $\dfrac{\partial}{\partial y}\left(\dfrac{\partial z}{\partial y}\right) = 6x$

2. (a) $f_y = 6xy + 2x$; (b) $f_{yx} = 6y + 2$; (c) $f_{yy} = 6x$

3. $\dfrac{\partial^2 f}{\partial x^2} = 2$, $\dfrac{\partial^2 f}{\partial y\,\partial x} = 6y + 2 + \dfrac{1}{y}$,

$\dfrac{\partial^2 f}{\partial x\,\partial y} = 6y + 2 + \dfrac{1}{y}$, $\dfrac{\partial^2 f}{\partial y^2} = 6x - \dfrac{x}{y^2}$

Exercise Set 7.3, p. 527

1. $\dfrac{\partial^2 f}{\partial x^2} = 6$, $\dfrac{\partial^2 f}{\partial y\,\partial x} = \dfrac{\partial^2 f}{\partial x\,\partial y} = -1$, $\dfrac{\partial^2 f}{\partial y^2} = 0$

3. $\dfrac{\partial^2 f}{\partial x^2} = 0$, $\dfrac{\partial^2 f}{\partial y\,\partial x} = \dfrac{\partial^2 f}{\partial x\,\partial y} = 3$, $\dfrac{\partial^2 f}{\partial y^2} = 0$

5. $\dfrac{\partial^2 f}{\partial x^2} = 20x^3 y^4 + 6xy^2$, $\dfrac{\partial^2 f}{\partial y\,\partial x} = \dfrac{\partial^2 f}{\partial x\,\partial y} =$

$20x^4 y^3 + 6x^2 y$, $\dfrac{\partial^2 f}{\partial y^2} = 12x^5 y^2 + 2x^3$

7. $f_{xx} = 0$, $f_{yx} = 0$, $f_{xy} = 0$, $f_{yy} = 0$
9. $f_{xx} = 4y^2 e^{2xy}$, $f_{yx} = f_{xy} = 4xye^{2xy} + 2e^{2xy}$, $f_{yy} = 4x^2 e^{2xy}$
11. $f_{xx} = 0$, $f_{yx} = f_{xy} = 0$, $f_{yy} = e^y$

13. $f_{xx} = -\dfrac{y}{x^2}$, $f_{yx} = f_{xy} = \dfrac{1}{x}$, $f_{yy} = 0$

15. $f_{xx} = \dfrac{-6y}{x^4}$, $f_{yx} = f_{xy} = \dfrac{2(y^3 - x^3)}{x^3 y^3}$, $f_{yy} = \dfrac{6x}{y^4}$

17. $\dfrac{\partial^2 f}{\partial x^2} = \dfrac{2y^2 - 2x^2}{(x^2 + y^2)^2}$, $\dfrac{\partial^2 f}{\partial y^2} = \dfrac{2x^2 - 2y^2}{(x^2 + y^2)^2}$, so the sum is 0

19. (a) $-y$; (b) x; (c) $f_{yx}(0, 0) = 1$, $f_{xy}(0, 0) = -1$; so $f_{yx}(0, 0) \neq f_{xy}(0, 0)$

Margin Exercises, Section 7.4, pp. 532–536

1. Minimum $= -9$ at $(-3, 3)$
2. Maximum $= \frac{1}{108}$ at $(\frac{1}{6}, \frac{1}{6})$
3. 5 thousand of the $15 calculator; 7 thousand of the $20 calculator

Exercise Set 7.4, p. 536

1. Minimum $= -\frac{1}{3}$ at $(-\frac{1}{3}, \frac{2}{3})$
3. Maximum $= \frac{4}{27}$ at $(\frac{2}{3}, \frac{2}{3})$
5. Minimum $= -1$ at $(1, 1)$
7. Minimum $= -7$ at $(1, -2)$
9. Minimum $= -5$ at $(-1, 2)$ **11.** None
13. 6 thousand of the $17 radio and 5 thousand of the $21 radio
15. Maximum of $P = 35$ (million dollars) when $a = 10$ (million dollars) and $p = \$3$
17. The dimensions of the bottom are 8 ft by 8 ft, and the height is 5 ft.
19. (a) $R(p_1, p_2) = 64p_1 - 4p_1^2 - 4p_1 p_2 + 56p_2 - 4p_2^2$;
(b) $p_1 = 6(\$60)$, $p_2 = 4(\$40)$;
(c) $q_1 = 32$ (hundreds), $q_2 = 28$ (hundreds); (d) $\$304,000$
21. None; saddle point at $(0, 0)$
23. Minimum $= 0$ at $(0, 0)$; saddle points at $(2, 1)$ and $(-2, 1)$

Margin Exercises, Section 7.5, pp. 539–545

1. $dz = e^y\,dx + xe^y\,dy$ **2.** $\dfrac{1}{x}\,dx + \dfrac{1}{y}\,dy$ **3.** $9\,dx + 12\,dy$

4. -0.66 **5.** $dw = \left(-\dfrac{1}{x^2} - y\right) dx - x\,dy + 2\,dz$

6. $\dfrac{x\,dx + y\,dy + z\,dz}{(x^2 + y^2 + z^2)^{1/2}}$

7. $2(x_1\,dx_1 + x_2\,dx_2 + x_3\,dx_3 + x_4\,dx_4)$
8. $dw = -2\,dx + 3\,dy - 2\,dz$ **9.** 0.3 **10.** $\Delta z = 0.7947$
11. 0.78 **12.** $\Delta w = 0.2226$, $dw = 0.22$
13. $\ln([1.01][0.97]) \approx -0.02$ **14.** 0.04
15. 28.875 psi

Exercise Set 7.5, p. 545

1. $dz = 2x\,dx + 8y\,dy$ **3.** $e^{4y}\,dx + 4xe^{4y}\,dy$

5. $\left(4x^3 + \dfrac{1}{y}\right) dx + \left(3y^2 - \dfrac{x}{y^2}\right) dy$

7. $2x\,dx - 2y\,dy + 6z\,dz$
9. $yz^3[yz\,dx + 2xz\,dy + 4xy\,dz]$
11. $dw = 2x_1\,dx_1 + x_3\,dx_2 + x_2\,dx_3 - 2x_4\,dx_4$
13. $\Delta z = 0.5805$, $dz = 0.58$
15. $\Delta z = 0.010553$, $dz = 0.0105$
17. $\Delta w = 0.0801$, $dw = 0.08$
19. $\Delta w = -0.062608$, $dw = -0.06$ **21.** 167.66
23. 4.09 **25.** 4.53025 sq ft   **27.** 39.06 sq ft
29. (a) $P(500, 100) = -50{,}000 + 300(100) + 1000(500) - 3(100)^2 - (500)^2 + (100)(500) = \$-50{,}000 + 30{,}000 + 500{,}000 - 30{,}000 - 250{,}000 + 50{,}000 = \$250{,}000$; (b) $\$300$ **31.** Estimated: 28,800; exact: 28,374   **33.** (a) $R = NT$; (c) $dR \approx -\$500$
35. 0.00125

Margin Exercises, Section 7.6, pp. 548–554

1. (a) $10.1 billion; (b) $9.5 billion; (c) differ by $0.6 billion   **2.** (a) $y = 68.7x + 191.5$; (b) 1,291,000; 1,703,000   **3.** Same answer
4. (a) $y = 225e^{0.179130x}$; (b) 3,953,000; 11,579,000

Exercise Set 7.6, p. 554

1. (a) $y = 2.29x + 33.67$; (b) $68.02, $81.76
3. (a) $y = 202.50x + 275.33$; (b) $2,908 million; $6,958

million   **5. (a)** $y = 1.068421x - 1.236842$; **(b)** 85.3
7. (a) $y = -0.005881x + 15.46388$; **(b)** $3:43.5$; $3:42.1$;
(c) According to the regression line, the record should
have been $3:47.3$, so Cram beat the regression-line
prediction.   **9. (a)** $y = 33.27948e^{0.061289x}$; **(b)** $83.45,
$120.54; **(c)** This model predicts higher prices.

Margin Exercises, Section 7.7, pp. 560–563

1. (a) $A(x, y) = xy$, subject to $x + y = 50$;
(b) maximum $= 625$ at $(25, 25)$ **2.** Maximum $= 125$ at
$(2.5, 10)$ **3.** $r \approx 1.66$ in., $h \approx 3.33$ in.; surface area is
about 52.05 in^2

Exercise Set 7.7, p. 564

1. Maximum $= 8$ at $(2, 4)$ **3.** Maximum $= -16$ at $(2, 4)$
5. Minimum $= 20$ at $(4, 2)$
7. Minimum $= -96$ at $(8, -12)$
9. Minimum $= \frac{3}{2}$ at $(1, \frac{1}{2}, -\frac{1}{2})$ **11.** 35 and 35
13. 3 and -3 **15.** $9\frac{3}{4}$ in., $9\frac{3}{4}$ in.; $95\frac{1}{16}$ in^2; no

17. $r = \sqrt[3]{\dfrac{27}{2\pi}} \approx 1.6$ ft; $h = 2 \cdot r \approx 3.2$ ft;

minimum surface area ≈ 48.3 ft^2
19. Maximum of $S = 800$ at $L = 20$, $M = 60$
21. (a) $C(x, y, z) = 7xy + 6yz + 6xz$;
(b) $x = 60$ ft, $y = 60$ ft, $z = 70$ ft; $75,600
23. 10,000 on A, 100 on B
25. Minimum $= -\frac{155}{128}$ at $(-\frac{7}{16}, -\frac{3}{4})$

27. Maximum $= \dfrac{1}{27}$ at $\left(\dfrac{1}{\sqrt{3}}, \dfrac{1}{\sqrt{3}}, \dfrac{1}{\sqrt{3}}\right)$

and $\left(-\dfrac{1}{\sqrt{3}}, -\dfrac{1}{\sqrt{3}}, -\dfrac{1}{\sqrt{3}}\right)$

29. Maximum $= 2$ at $\left(\dfrac{1}{2}, \dfrac{1}{2}, \dfrac{1}{2}, \dfrac{1}{2}\right)$ **31.** $\lambda = \dfrac{p_x}{c_1} = \dfrac{p_y}{c_2}$

Margin Exercises, Section 7.8, pp. 570–576

1. $4y$ **2.** $\frac{1}{2}y^2$ **3.** $\frac{1}{3}x$ **4.** $\frac{3}{2}x + 3x^2$ **5.** $2rx^2 + 2y$
6. $\frac{1}{6}$ **7.** $\frac{25}{6}$ **8.** $\frac{1}{6}$; they are equal   **9.** $\frac{25}{6}$; equal
10. (a) $4x^2$; **(b)** $\frac{224}{3}$ **11.** 72   **12.** $\frac{3}{2}(e - 1)$ **13.** $\frac{7}{6}$
14. $\frac{5}{6}$

Exercise Set 7.8, p. 576

1. $\frac{1}{3}y$ **3.** $\frac{8}{3}$ **5.** $\frac{4}{3}$ **7.** 1   **9.** 12   **11.** $\frac{1}{3} \ln 2$
13. 2   **15.** $8 \ln 2$ **17.** $f_{av} = 1$

Margin Exercises, Section 7.9, pp. 578–587

1. $1 - x^3$ **2.** $8x$ **3.** $\frac{3}{4}$ **4.** 4
5. (a) 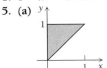   **(b)** $\frac{3}{4}$ **6. (a)**   **(b)** 4

7. (a) 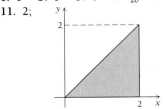   **(b)** 28   **8.** $\frac{3}{4}$ **9.** $\frac{1}{2}(e^2 - 1)$
10. $\frac{3}{2}$ **11.** $\frac{3}{8}$ **12.** $\frac{27}{32}$
13. $\frac{8}{3}$ **14.** $\frac{3}{4}$ **15.** 4

Exercise Set 7.9, p. 588

1. 1   **3.** 0   **5.** 6   **7.** $\frac{3}{20}$ **9.** 4
11. 2;   **13.** $\frac{1}{2}$;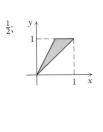

15. $\frac{4}{15}$ **17.** $\frac{375}{4}$ **19.** $\frac{14}{3}$

21. (a) $\displaystyle\int_1^{e^2}\left[\int_{\ln y}^2 dx\right] dy$; **(b)** $e^2 - 3$ **23.** 1   **25.** $\frac{5}{12}$

27. 39   **29.** $\frac{13}{240}$ **31.** $\frac{1}{8}$

Summary and Review: Chapter 7, p. 590

1. [7.2] $3y^3$ **2.** [7.2] $e^y + 9xy^2 + 2$ **3.** [7.3] $9y^2$
4. [7.3] $9y^2$ **5.** [7.3] 0 **6.** [7.3] $e^y + 18xy$

7. [7.2] $2x \ln y$ **8.** [7.2] $\dfrac{x^2}{y} + 4y^3$ **9.** [7.3] $\dfrac{2x}{y}$

10. [7.3] $\dfrac{2x}{y}$ **11.** [7.3] $2 \ln y$ **12.** [7.3] $-\dfrac{x^2}{y^2} + 12y^2$

13. [7.4] Minimum $= -\frac{549}{20}$ at $(5, \frac{27}{2})$
14. [7.4] Minimum $= -4$ at $(0, -2)$
15. [7.4] Maximum $= \frac{45}{4}$ at $(\frac{3}{2}, -3)$
16. [7.4] Minimum $= 27$ at $(-1, 2)$
17. [7.5] $dz = 10x\, dx - 2e^{2y}\, dy$
18. [7.5] $4xy\, dx + 2x^2\, dy$
19. [7.5] 1.1 ft^2 **20.** [7.5] 0.72 ft^3
21. [7.6] **(a)** $y = 2.5x + 21.67$; **(b)** $\$46.67, \$74.17$
22. [7.6] **(a)** $y = 0.6x + 6.7$; **(b)** 9.1 million
23. [7.7] Minimum $= \frac{125}{4}$ at $(-\frac{3}{2}, -3)$
24. [7.7] Maximum $= 300$ at $(5, 10)$ **25.** [7.9] $\frac{97}{30}$
26. [7.9] $\frac{1}{60}$ **27.** [7.8] 0 **28.** [7.8] 2 **29.** [7.9] 16
30. [7.9] $\frac{8}{3}$ **31.** [7.9] 0 **32.** [7.7] The cylindrical container

Test: Chapter 7, p. 592

1. $e^{-1} - 2$ **2.** $e^x + 6x^2y$ **3.** $2x^3 + 1$ **4.** $e^x + 12xy$
5. $6x^2$ **6.** $6x^2$ **7.** 0 **8.** Minimum $= -\frac{7}{16}$ at $(\frac{3}{4}, \frac{1}{2})$
9. None **10.** **(a)** $y = \frac{9}{2}x + \frac{17}{3}$; **(b)** $\$23.67$ million
11. Maximum $= -19$ at $(4, 5)$ **12.** 1
13. $dz = 3x^2\, dx + e^y\, dy$ **14.** $dz = 3y^2\, dx + 6xy\, dy$
15. $dV = 0.59$ ft^3 **16.** $\frac{4}{3}$ **17.** $\frac{3}{20}$
18. $\$400,000$ for labor; $\$200,000$ for capital
19. $f_x = \dfrac{-x^4 + 4xt + 6x^2 t}{(x^3 + 2t)^2}, \quad f_t = \dfrac{-2x^2(x+1)}{(x^3 + 2t)^2}$

CHAPTER 8

Margin Exercises, Section 8.1, pp. 595–607

1. **(a)** I; **(b)** III; **(c)** IV; **(d)** III; **(e)** I; **(f)** IV; **(g)** I

2. **(a)** $\frac{3}{4}\pi$; **(b)** $\frac{7}{4}\pi$; **(c)** $-\dfrac{\pi}{2}$; **(d)** 4π; **(e)** $-\dfrac{5\pi}{4}$;

(f) $-\dfrac{7\pi}{4}$; **(g)** $\dfrac{9\pi}{4}$; **(h)** $\dfrac{8\pi}{3}$

3. **(a)** $60°$; **(b)** $135°$; **(c)** $450°$; **(d)** $1800°$; **(e)** $-210°$;
(f) $54,000°$; **(g)** $-48,600°$; **(h)** $1125°$

4. **(a)** $-\dfrac{\sqrt{3}}{2}$; **(b)** $\frac{1}{2}$; **(c)** $-\dfrac{\sqrt{3}}{2}$; **(d)** $\frac{1}{2}$

5. **(a)** 1; **(b)** $\sqrt{3}$; **(c)** 0; **(d)** $\sqrt{2}$; **(e)** 2; **(f)** 1

6. **(a)** $\sqrt{3}$; **(b)** undefined; **(c)** $\dfrac{\sqrt{3}}{3}$; **(d)** 2;

(e) undefined; **(f)** $\sqrt{2}$

7. $1 + \cot^2 t = \csc^2 t$ **8.** $\dfrac{\sqrt{6} - \sqrt{2}}{4}$

9. $\dfrac{\sqrt{2} + \sqrt{6}}{4}$ **10.** $2 \sin u \cos u$

Exercise Set 8.1, p. 607

1. I **3.** III **5.** $\dfrac{\pi}{6}$ **7.** $\dfrac{\pi}{3}$ **9.** $\dfrac{5\pi}{12}$ **11.** $270°$

13. $-45°$ **15.** $1440°$ **17.** $\left(\dfrac{180}{\pi}\right)^{\circ} \approx 57.3°$ **19.** $\dfrac{\sqrt{3}}{2}$

21. 0 **23.** $\frac{1}{2}$ **25.** Undefined **27.** $\dfrac{\sqrt{3}}{3}$ **29.** -1

31. $\tan(u + v) = \dfrac{\sin(u+v)}{\cos(u+v)}$. Then use identities (6)
and (4). **33.** 0.5210 **35.** -0.8632 **37.** -0.3327
39. Undefined **41.** $\$0, \$0.9$ thousand, $\$3.5$ thousand, $\$7$
thousand, $\$14$ thousand, $\$0$, $\$7$ thousand

Margin Exercises, Section 8.2, pp. 611–618

1. $\cot x = \dfrac{\cos x}{\sin x}$,

so $\dfrac{d}{dx} \cot x = [\sin x\,(-\sin x) - \cos x\,(\cos x)] \div \sin^2 x =$

$\dfrac{-(\sin^2 x + \cos^2 x)}{\sin^2 x} = -\dfrac{1}{\sin^2 x} = -\csc^2 x$ **2.** $\dfrac{3 \sin^2 x}{\cos^4 x}$

3. $(4x^3 + 6x) \cos(x^4 + 3x^2)$
4. $3x^2 \cos x^3 \cos(e^{x^2}) - 2xe^{x^2} \sin x^3 \sin(e^{x^2})$

5. **(a)** Amplitude $= 4$, period $= \pi$, phase shift $= \dfrac{\pi}{4}$;

(b)

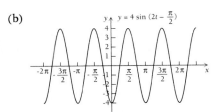

(c) $8 \cos \left(2t - \dfrac{\pi}{2} \right)$

6. $\dfrac{dy}{dx} = 0.03594\pi \cos 1.198\pi x$

7. $\dfrac{dL}{dT} = 2\pi A f \cos \left(2\pi f T - \dfrac{d}{\omega} \right)$

8. **(a)** $L = (a^{2/3} + b^{2/3})^{3/2}$; **(b)** 27

Exercise Set 8.2, p. 618

1. $\sec x = \dfrac{1}{\cos x}$, so $\dfrac{d}{dx} \sec x = \dfrac{\cos x \cdot 0 - (-\sin x) \cdot 1}{\cos^2 x} =$
$\dfrac{\sin x}{\cos^2 x} = \dfrac{\sin x}{\cos x} \cdot \dfrac{1}{\cos x} = \tan x \sec x$ **3.** $x \cos x + \sin x$

5. $e^x(\cos x + \sin x)$ **7.** $\dfrac{x \cos x - \sin x}{x^2}$

9. $2 \sin x \cos x$ **11.** $\cos^2 x - \sin^2 x$ **13.** $\dfrac{1}{1 + \cos x}$

15. $\dfrac{2 \sin x}{\cos^3 x}$ **17.** $\dfrac{-\sin x}{2\sqrt{1 + \cos x}}$ **19.** $-x^2 \sin x$

21. $\cos x \cdot e^{\sin x}$ **23.** $-\sin x$

25. $(2x + 3x^2) \cos (x^2 + x^3)$

27. $(4x^3 - 5x^4) \sin (x^5 - x^4)$ **29.** $-\frac{1}{2}x^{-1/2} \cdot \sin \sqrt{x}$

31. $-\sin x \cdot \cos (\cos x)$ **33.** $\dfrac{-2 \sin 4x}{\sqrt{\cos 4x}}$

35. $\frac{2}{3}[\csc^2 (5 - 2x)^{1/3}](5 - 2x)^{-2/3}$

37. $-12 (\tan 3x)(\sec 3x)^2$ **39.** $\cot x$

41. $40{,}000(\cos t - \sin t)$ **43.** $\dfrac{3\pi}{8} \cos \dfrac{\pi}{8}t$

45. $-\dfrac{1000\pi}{9} \left[\sin \dfrac{\pi}{45}(t - 10) \right]$

47. **(a)** Amplitude $= 5$, period $= \dfrac{\pi}{2}$, phase shift $= -\dfrac{\pi}{4}$;

(b) $20 \cos (4t + \pi)$ **49.** $16\sqrt{2}$ **51.** $90°$

53. $f_x = -ye^x$, $f_y = 2 \cos 2y - e^x$, $f_{xx} = -ye^x$,
$f_{yx} = f_{xy} = -e^x$, $f_{yy} = -4 \sin 2y$

55. $f_x = -2 \sin (2x + 3y)$, $f_y = -3 \cdot \sin (2x + 3y)$,
$f_{xx} = -4 \cos (2x + 3y)$, $f_{yx} = f_{xy} = -6 \cos (2x + 3y)$,
$f_{yy} = -9 \cos (2x + 3y)$

57. $f_x = 5x^3y \cdot \sec^2 (5xy) + 3x^2 \cdot \tan (5xy)$,
$f_y = 5x^4 \cdot \sec^2 (5xy)$, $f_{xx} = 50x^3y^2 \cdot \sec^2 (5xy) \cdot \tan (5xy) +$
$30x^2y \sec^2 (5xy) + 6x \tan (5xy)$,
$f_{yx} = f_{xy} = 50x^4y \cdot \sec^2 (5xy) \cdot \tan (5xy) + 20x^3 \cdot \sec^2 (5xy)$,
$f_{yy} = 50x^5 \cdot \sec^2 (5xy) \cdot \tan (5xy)$

59. $\dfrac{\cos y}{1 + x \sin y}$

61.

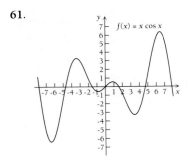

63.

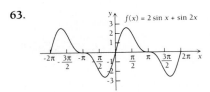

65.

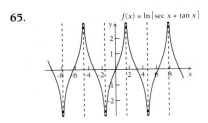

67.

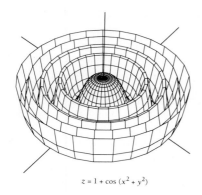

$z = 1 + \cos (x^2 + y^2)$

Margin Exercises, Section 8.3, pp. 621–623

1. 2   **2.** $\dfrac{(\sin x)^4}{4} + C$ **3.** $\frac{1}{4} \sin 4x + C$

4. $\ln |x^2 + \cos x| + C$ **5.** $\sin x - x \cos x + C$
6. $-\ln |\csc x + \cot x| + C$

Exercise Set 8.3, p. 623

1. $\frac{1}{2}$ **3.** 0   **5.** 1   **7.** $\dfrac{\sin^5 x}{5} + C$ **9.** $\dfrac{\cos^3 x}{3} + C$

11. $\sin (x + 3) + C$ **13.** $-\frac{1}{2} \cos 2x + C$
15. $\frac{1}{2} \sin x^2 + C$ **17.** $-\cos (e^x) + C$
19. $-\ln |\cos x| + C$, or $\ln |\sec x| + C$
21. $\frac{1}{4}x \sin 4x + \frac{1}{16} \cos 4x + C$
23. $3x \sin x + 3 \cos x + C$
25. $2x \sin x - (x^2 - 2) \cos x + C$ **27.** $\tan x - x + C$
29. \$84 thousand   **31.** $\frac{1}{2}x[\sin (\ln x) - \cos (\ln x)] + C$
33. $\dfrac{e^x}{2}(\cos x + \sin x) + C$

35. $-x^3 \cos x + 3x^2 \sin x + 6x \cos x - 6 \sin x + C$
37. $\frac{1}{7} \tan 7x + C$ **39.** $\tan x + \sec x + C$
41. $-\csc u + C$ **43.** $\cos x - \ln |\csc x + \cot x| + C$
45. $x + 2 \ln |\sec x + \tan x| + \tan x + C$
47. $\dfrac{\sqrt{2}}{2}$ **49.** π

Margin Exercises, Section 8.4, pp. 626–628

1. $\dfrac{\pi}{6}$ **2.** 0   **3.** $-\dfrac{\pi}{3}$ **4.** $\dfrac{\pi}{3}$ **5.** $\dfrac{\pi}{6}$ **6.** $\dfrac{2\pi}{3}$
7. $\frac{1}{5} \tan^{-1} (5x) + C$ **8.** $\frac{1}{2} \cos^{-1} (2x) + C$

Exercise Set 8.4, p. 629

1. $\dfrac{\pi}{4}$ **3.** $\dfrac{\pi}{2}$ **5.** $\dfrac{\pi}{4}$ **7.** $-\dfrac{\pi}{6}$ **9.** $\tan^{-1} (e^t) + C$
11. Use substitution. Let $u = \sin^{-1} x$; then $x = \sin u$ and

$dx = \cos u \, du$. Then the integral $\displaystyle\int \dfrac{1}{\sqrt{1 - x^2}} \, dx =$

$\displaystyle\int \dfrac{1}{\sqrt{1 - \sin^2 u}} \cos u \, du = \int \dfrac{1}{\cos u} \cdot \cos u \, du =$

$\displaystyle\int du = u + C = \sin^{-1} x + C.$ **13.** 1.412 radians

15. 0.049 radian   **17.** $x \tan^{-1} x - \frac{1}{2} \ln (1 + x^2) + C$

Summary and Review: Chapter 8, p. 630

1. [8.1] $\dfrac{4\pi}{3}$ **2.** [8.1] $\dfrac{7\pi}{4}$ **3.** [8.1] 150°   **4.** [8.1] 420°

5. [8.1] $\dfrac{\sqrt{3}}{2}$ **6.** [8.1] 0   **7.** [8.1] 1

8. [8.1] Undefined
9. [8.2] $(6x - 1) \sec^2 (3x^2 - x)$ **10.** [8.2] $6 \sin^5 x \cos x$

11. [8.2] $\dfrac{1}{\cos x \sin x}$ **12.** [8.2] $3e^{\sin 3x} \cos 3x$

13. [8.2] $\dfrac{-2(1 + \cos 2x)}{\sin^2 2x}$ **14.** [8.2] $-\dfrac{\sin \sqrt{t}}{2\sqrt{t}} - \dfrac{\sin t}{2\sqrt{\cos t}}$

15. [8.2] $\dfrac{\sec^2 x}{2\sqrt{\tan x}}$ **16.** [8.2] $e^x(\cos x + \sin x)$

17. [8.2] $3(\cos^2 3x - \sin^2 3x)$ **18.** [8.2] $x \sin x$

19. [8.2] $-9 \sin (3t - 2\pi)$ **20.** [8.2] $\dfrac{5\pi}{3} \cos \dfrac{\pi}{3}t$

21. [8.3] $\frac{1}{2}$ **22.** [8.3] $\sin e^x + C$
23. [8.3] $-\cos (x + 1) + C$ **24.** [8.3] $\frac{1}{8} \sin 8x + C$
25. [8.3] $\frac{3}{2}x \sin 2x + \frac{3}{4} \cos 2x + C$
26. [8.3] $-\frac{1}{3}x \cos 3x + \frac{1}{9} \sin 3x + C$
27. [8.4] $\frac{1}{8} \sin^{-1} 8t + C$ **28.** [8.3] $\ln |5x - \sin x| + C$
29. [8.2] $e^{\sin x} \cos x(\cos^2 x - 3 \sin x - 1)$
30. [8.2] $f_x = -\sin y \sin x, f_{xy} = -\sin x \cos y$

Test: Chapter 8, p. 631

1. $\dfrac{2\pi}{3}$ **2.** 300°   **3.** $\frac{1}{2}$ **4.** -1 **5.** $-\sin t$

6. $(6x - 5) \cos (3x^2 - 5x)$ 7. $\dfrac{\sin x - x \cdot \cos x}{\sin^2 x}$

8. $2 \tan t \cdot \sec^2 t$ 9. $\dfrac{\cos x - \sin x}{2\sqrt{\sin x + \cos x}}$

10. $\dfrac{-2}{1 - 2 \sin x \cos x}$ 11. $S'(t) = \dfrac{5\pi}{2} \cos \dfrac{\pi}{8}t$

12. $1 - \dfrac{\sqrt{3}}{2}$ 13. $\frac{1}{10}(\sin x)^{10} + C$ 14. $\frac{1}{5} \sin 5t + C$

15. $\frac{1}{5} \sin 5x - x \cos 5x + C$ 16. $\frac{1}{4} \tan^{-1} 4t + C$

17. $\ln |2x - \sin x| + C$ 18. 1

19. $\dfrac{dy}{dx} = -2 (\sin x)e^{\cos x}$, $\dfrac{d^2 y}{dx^2} = 2e^{\cos x}(\sin^2 x - \cos x)$

20. $f_x = -\dfrac{\sin x}{\cos y}$, $f_{xy} = -\dfrac{\sin x \sin y}{\cos^2 y}$

21. Does not exist; $-\infty$

CHAPTER 9

Margin Exercises, Section 9.1, pp. 635–636

1. $y = x^3 + C$ 2. (a) $y = x^3 + C$; (b) $y = x^3 - 7$,
$y = x^3 + \frac{1}{2}$, $y = x^3$. Answers may vary.   3. $y = x^2 + 6$
4. $f(x) = \ln x - x^2 + \frac{2}{3}x^{3/2} + \frac{13}{3}$
5. $y' = 2e^x - 21e^{3x}$, $y'' = 2e^x - 63e^{3x}$. Then

$$y'' - 4y' + 3y = 0$$

$$\begin{array}{c|c} 2e^x - 63e^{3x} - 4(2e^x - 21e^{3x}) + 3(2e^x - 7e^{3x}) & 0 \\ 2e^x - 63e^{3x} - 8e^x + 84e^{3x} + 6e^x - 21e^{3x} & \\ & 0 \end{array}$$

6. $\dfrac{dy}{dx} = 2xe^{2x} + e^{2x}$. Then

$$\dfrac{dy}{dx} - 2y = e^{2x}$$

$$\begin{array}{c|c} (2xe^{2x} + e^{2x}) - 2(xe^{2x}) & e^{2x} \\ e^{2x} & \end{array}$$

Exercise Set 9.1, p. 636

1. $y = x^4 - C$; $y = x^4 + 3$, $y = x^4$, $y = x^4 - 796$; answers
may vary.

3. $y = \frac{1}{2}e^{2x} + \frac{1}{2}x^2 + C$; $y = \frac{1}{2}e^{2x} + \frac{1}{2}x^2 - 5$,
$y = \frac{1}{2}e^{2x} + \frac{1}{2}x^2 + 7$; $y = \frac{1}{2}e^{2x} + \frac{1}{2}x^2$; answers may vary.
5. $y = 3 \ln x - \frac{1}{3}x^3 + \frac{1}{6}x^6 + C$;
$y = 3 \ln x - \frac{1}{3}x^3 + \frac{1}{6}x^6 - .15$, $y = 3 \ln x - \frac{1}{3}x^3 + \frac{1}{6}x^6 - 7$,
$y = 3 \ln x - \frac{1}{3}x^3 + \frac{1}{6}x^6$; answers may vary.
7. $y = \frac{1}{3}x^3 + x^2 - 3x + 4$ 9. $y = \frac{3}{5}x^{5/3} - \frac{1}{2}x^2 - \frac{61}{10}$

11. $y'' = \dfrac{1}{x}$. Then

$$y'' - \dfrac{1}{x} = 0$$

$$\begin{array}{c|c} \dfrac{1}{x} - \dfrac{1}{x} & 0 \\ 0 & \end{array}$$

13. $y' = 4e^x + 3xe^x$, $y'' = 7e^x + 3xe^x$. Then

$$y'' - 2y' + y = 0$$

$$\begin{array}{c|c} (7e^x + 3xe^x) - 2(4e^x + 3xe^x) + (e^x + 3xe^x) & 0 \\ 7e^x + 3xe^x - 8e^x - 6xe^x + e^x + 3xe^x & \\ 0 & \end{array}$$

15. $C(x) = 2.6x - 0.01x^2 + 120$, $A(x) = 2.6 - 0.01x + \dfrac{120}{x}$

17. (a) $P(C) = \dfrac{400}{(C + 3)^{1/2}} - 40$; (b) \$97

Margin Exercises, Section 9.2, pp. 638–643

1. $y = C_1 e^{x^3}$, where $C_1 = e^C$ 2. $y = \sqrt[3]{x^2 + 339}$
3. $y = \sqrt{10x + C_1}$, $y = -\sqrt{10x + C_1}$, where $C_1 = 2C$
4. $y = -2 + C_1 e^{x^2/2}$, where $C_1 = e^C$ 5. $x = \dfrac{C_1}{p^3}$
6. $R = C_1 \cdot S^k$, where $C_1 = e^C$

Exercise Set 9.2, p. 643

1. $y = C_1 e^{x^4}$, where $C_1 = e^C$ 3. $y = \sqrt[3]{\dfrac{5}{2}x^2 + C}$
5. $y = \sqrt{2x^2 + C_1}$, $y = -\sqrt{2x^2 + C_1}$, where $C_1 = 2C$
7. $y = \sqrt{6x + C_1}$, $y = -\sqrt{6x + C_1}$, where $C_1 = 2C$
9. $y = -3 + 8e^{x^2/2}$ 11. $y = \sqrt[3]{15x - 3}$
13. $y = C_1 e^{3x}$, where $C_1 = e^C$

15. $P = C_1 e^{2t}$, where $C_1 = e^C$

17. (a) $R = k \cdot \ln (S + 1) + C$; (b) $R = k \cdot \ln (S + 1)$;
(c) no units, no pleasure from them

19. $x = 200 - p$ 21. $x = \dfrac{C_1}{p^n}$

23. (a) $P = C_1 e^{kt}$ where $C_1 = e^C$; (b) $P = P_0 e^{kt}$

Margin Exercises, Section 9.3, pp. 647–652

1. (a)

t	-3	-2	-1	0	1	2	3
$P(t)$	0.05	0.12	0.27	0.5	0.73	0.88	0.95

(b)

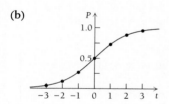

2. (a) $P = \dfrac{\dfrac{P_0}{L - P_0} \cdot L e^{Lkt}}{1 + \dfrac{P_0}{L - P_0} e^{Lkt}} = \dfrac{P_0 L e^{Lkt}}{(L - P_0) + P_0 e^{Lkt}}$;

(b) $\dfrac{P_0 L e^{Lkt}}{(L - P_0) + P_0 e^{Lkt}} \cdot \dfrac{e^{-Lkt}}{e^{-Lkt}} = \dfrac{P_0 L}{P_0 + e^{-Lkt}(L - P_0)}$
(add exponents)

3. 3312   4. (a) $P(t) = \dfrac{80{,}000}{20 + 3980e^{-2t}}$;

(b)

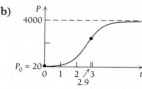

(c) $t \approx 1.95$ mo   5. $Q(y) = \dfrac{Q_0 M}{Q_0 + e^{-Mry}(M - Q_0)}$

Exercise Set 9.3, p. 653

1. 900   3. (a) $P(t) = \dfrac{0.02}{0.02 + 0.98e^{-0.13t}}$;

(b)

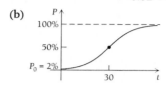

(c) $t = 40.6 \approx 41$ 5. (a) $N(t) = \dfrac{4800}{6 + 794e^{-800kt}}$;

(b) $k = 0.0005$; (c) $t \approx 17.08$ min

7. (a) $P(w) = \dfrac{0.04}{0.04 + 0.96e^{-kw}}$; (b) $k = 1.06$;

(c) $P(w) = \dfrac{0.04}{0.04 + 0.96e^{-1.06w}}$; (d) $w \approx 4.3$ (about 106 g)

Margin Exercises, Section 9.4, pp. 657–659

1. (a) Undefined, 5%, 2.5%, 1.2%, 0.7%, 0.3°;

(b)   (c)

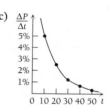

(d) no; (e) no; (f) 100%; (g) decreasing

2. (a)

t	0	1	2	3	4	5
$P(t)$	0	1.9	2.6	2.85	2.95	2.98

(b)

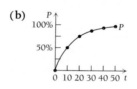

3. (a) $k = 0.05$; (b) $P(t) = 1 - e^{-0.05t}$;
(c) $P(60) = 0.9502$, or 95.02%; (d) $t \approx 92$

Exercise Set 9.4, p. 659

1. **3.**

5. (a) $P(t) = 0.5(1 - e^{-kt})$; (b) $k = 0.07$;
(c) $P(t) = 0.5(1 - e^{-0.07t})$; (d) $P(30) = 0.439$, or 43.9%;
(e) $t \approx 56$
7. (a) $P(t) = 1 - e^{-kt}$; (b) $k = 0.2$; (c) $P(t) = 1 - e^{-0.2t}$;
(d) $P(6) = 0.6998$, or 69.98%; (e) 11.5 mo
9. $M(t) = \dfrac{P}{k}(1 - e^{-kt})$
11. $C(t) = C_T - (C_T - C_0)e^{-(kA/v)t}$

Margin Exercises, Section 9.5, pp. 662–665

1. It checks. **2.** $x\,dx + y\,dy = 0$ **3.** $P_y = 0 = Q_x$
4. $P_y = 6x - 2 = Q_x$ **5.** (a) $2y\,dx + x\,dy = 0$;
(b) $2 = P_y \neq Q_x = 1$ **6.** $x^2 y = C$ **7.** $xy = C$
8. $(3x^2 - 2x + 1)y = C$ **9.** $y = \dfrac{x \pm \sqrt{x^2 - 4x^2(5x - 7)}}{2x^2}$
10. $x^2 y^3 = 8$ **11.** $y = \dfrac{-(x+1)}{2x}$

Exercise Set 9.5, p. 665

1. Exact **3.** Exact **5.** Exact **7.** Exact
9. $x^2 + 3x + 3y^2 = C$ **11.** $x \sin y = C$
13. $x^3 + 3xy + y^3 = 5$ **15.** $x \cos y + y \cos x = 1$
17. $x^2 - 2xy + 2y^2 = 1$ **19.** Exact, $P^2 + 3Pu = 100$,
$P = 2$ when $u = 16$

Margin Exercises, Section 9.6, pp. 667–672

1. $y = C_1 x$ **2.** It checks. **3.** $y = C_1 e^{-x^2}$
4. $y = 2 + Cx^{-4}$, $x > 0$ **5.** $y = \frac{1}{5}e^{3x} + Ce^{-2x}$
6. $A(t) = 1200 - Ce^{-t/240}$
7. $A(t) = 720 - 320e^{-t/240}$; $A(300) \approx 628.32$ lb

Exercise Set 9.6, p. 672

1. $y = Ce^{3x}$ **3.** $y = Cx^{-3}$ **5.** $m(x) = e^{-3x}$
7. $m(x) = e^{5/x}$ **9.** $y = Ce^{3x}$ **11.** $y = Ce^{x^2}$
13. $y = 7e^{-5x}$ **15.** $y = Ce^{x^2} - 1$
17. $y = \dfrac{2x^2 + x^4 + 4C}{4(1 + x^2)}$ **19.** $y = \dfrac{x^3}{4} + \dfrac{6}{x}$
21. (a) 201.16 lb; (b) 249.99 lb
23. (a) $T(t) = M + Ce^{-kt}$; (b) 88.21 min

Margin Exercises, Section 9.7, pp. 675–684

1. They all check. **2.** $y = C_1 e^{4x} + C_2 e^{2x}$
3. $y = C_1 e^{(1/2)x} + C_2 e^{3x}$ **4.** $y = C_1 e^x + C_2 e^{-2x} + C_3$
5. $r^2 - 7r + 12 = 0$ **6.** $r^3 - 6r^2 + 11r - 6 = 0$
7. $y = C_1 e^{3x} + C_2 e^{4x}$ **8.** $y = C_1 e^{2x} + C_2 xe^{2x}$
9. $y = C_1 e^{3x} + C_2 xe^{3x}$ **10.** $y = C_1 e^{-(1/2)x} + C_2 xe^{-(1/2)x}$
11. $y = C_1 e^{-x} + C_2 xe^{-x} + C_3 x^2 e^{-x}$
12. $y = C_1 \sin 3x + C_2 \cos 3x$
13. $y = C_1 e^{2x} \sin 5x + C_2 e^{2x} \cos 5x$
14. $y = C_1 e^{-x} \sin 2x + C_2 e^{-x} \cos 2x$
15. $y = C_1 + C_2 e^x \sin(\sqrt{7}x) + C_3 e^x \cos(\sqrt{7}x)$
16. $y = C_1 \sin 2x + C_2 \cos 2x + C_3 x \sin 2x + C_4 x \cos 2x$
17. It checks. **18.** $y = C_1 e^{-(1/2)x} + C_2 xe^{-(1/2)x} + 3$
19. $y = C_1 e^{3x} + C_2 e^{-3x} - \frac{1}{9}x^2 - \frac{2}{9}x - \frac{2}{81}$
20. $y = C_1 + C_2 e^{-x} + 7x$
21. $y = C_1 + C_2 x + C_3 e^{-2x} + \frac{1}{2}x^3 - \frac{7}{4}x^2$
22. $y = \frac{9}{2}e^x - \frac{17}{2}e^{-x} + 5$

Exercise Set 9.7, p. 684

1. $y = C_1 e^x + C_2 e^{5x}$ **3.** $y = C_1 e^{-x} + C_2 e^{2x}$
5. $y = C_1 e^{-x} + C_2 e^{-2x}$ **7.** $y = C_1 e^{(1/2)x} + C_2 e^{2x}$
9. $y = C_1 e^{3x} + C_2 e^{-3x}$ **11.** $y = C_1 e^{-5x} + C_2 xe^{-5x}$
13. $y = C_1 e^{-(3/2)x} + C_2 xe^{-(3/2)x}$
15. $y = C_1 e^{2x} + C_2 xe^{2x} + C_3 e^{-3x}$
17. $y = C_1 e^{-x} + C_2 xe^{-x} + C_3 x^2 e^{-x} + C_4$
19. $y = C_1 e^x + C_2 e^{4x} + C_3 + C_4 x$
21. $y = C_1 \sin 6x + C_2 \cos 6x$
23. $y = C_1 e^{-4x} \sin 5x + C_2 e^{-4x} \cos 5x$
25. $y = C_1 e^{-x} \sin 2x + C_2 e^{-x} \cos 2x + C_3$
27. $y = C_1 e^x + C_2 e^{-x} \sin 4x + C_3 e^{-x} \cos 4x$
29. $y = C_1 \sin x + C_2 \cos x + 7$
31. $y = C_1 e^x + C_2 xe^x + 3$
33. $y = C_1 e^{-2x} + C_2 xe^{-2x} - 3x + 5$

35. $y = C_1 e^{-x} + C_2 e^{-3x} + 2x^2 - \frac{16}{3}x + \frac{40}{9}$
37. $y = C_1 + C_2 e^{3x} - \frac{4}{3}x$
39. $y = C_1 + C_2 x + C_3 e^{-x} - x^2$
41. $y = C_1 + C_2 e^{-2x} \sin 4x + C_3 e^{-2x} \cos 4x + x^2 - x$
43. $y = 1 - e^x$ 45. $y = 2e^x - e^{-x}$
47. $y = 6e^x - 2e^{-x} - 3$
49. $y = \sin 3x + 2 \cos 3x - x + 1$
57. $y = C_1 e^{7x} + C_2 e^{9x} + C_3 e^{2x} \sin 4x + C_4 e^{2x} \cos 4x + C_5 x e^{2x} \sin 4x + C_6 x e^{2x} \cos 4x$

Margin Exercises, Section 9.8, pp. 686–695

1. $x = C_1 e^{4t}, y = C_2 e^t$
2. $x = C_1 e^{4t} + C_2 e^{-4t}, y = 2C_1 e^{4t} - 2C_2 e^{-4t}$
3. $x = C_1 \sin 2t + C_2 \cos 2t, y = -2C_2 \sin 2t + 2C_1 \cos 2t$
4. They check.
5. $x = C_1 e^{2t} + C_2 e^{-2t}, y = -\frac{2}{3}C_1 e^{2t} - 2C_2 e^{-2t}$
6. $x = e^{2t} + 4te^{2t}, y = 3e^{2t} - 4te^{2t}$
7. $x = 2e^t + e^{2t}, y = 4e^t + e^{2t}$
8. $x = C_1 e^{2t} + C_2 e^{5t} + 5, y = -2C_1 e^{2t} + C_2 e^{5t} - 20$
9. $A(t) = 1200 + 400e^{-t}, B(t) = 1200 - 400e^{-t}$
10. $A(t) = 1100, B(t) = 1100$
11. $P(t) = \sin 8t + 4 \cos 8t + 20,$
$Q(t) = 40 - 2 \cos 8t + 8 \sin 8t$

Exercise Set 9.8, p. 696

1. $x = C_1 e^{2t}, y = C_2 e^{3t}$
3. $x = C_1 e^t + C_2 e^{2t}, y = C_1 e^t + 2C_2 e^{2t}$
5. $x = C_1 e^{5t} + C_2 e^t, y = 3C_1 e^{5t} - C_2 e^t$
7. $x = C_1 e^{(1/2)t} + C_2 e^{-t}, y = C_1 e^{(1/2)t} - \frac{1}{2}C_2 e^{-t}$
9. $x = C_1 e^{3t} + C_2 t e^{3t}, y = (C_2 - C_1)e^{3t} - C_2 t e^{3t}$
11. $x = C_1 \sin 6t + C_2 \cos 6t, y = -3C_2 \sin 6t + 3C_1 \cos 6t$
13. $x = C_1 e^t \sin t + C_2 e^t \cos t,$
$y = (\frac{2}{5}C_1 + \frac{1}{5}C_2)e^t \sin t + (-\frac{1}{5}C_1 + \frac{2}{5}C_2)e^t \cos t$
15. $x = e^{3t} + 2e^{4t}, y = -e^{3t} - 4e^{4t}$ 17. $x = e^{5t}, y = e^{5t}$
19. $x = \sin 2t + \cos 2t, y = -2 \sin 2t + 2 \cos 2t$
21. $x = C_1 e^{3t} - C_2 e^{4t} - 1, y = 2C_1 e^{3t} + C_2 e^{4t}$
23. $x = C_1 \sin t + C_2 \cos t + 4t,$
$y = -C_2 \sin t + C_1 \cos t - 2t + 1$
25. $x = e^{3t} + e^{-3t} - 2, y = e^{3t} - 5e^{-3t} + 1$
27. $A' = -\frac{1}{5}A + \frac{1}{5}B, B' = \frac{1}{5}A - \frac{1}{5}B, A(0) = 200,$
$B(0) = 100;$
$A(t) = 150 + 50e^{-(2/5)t}, B(t) = 150 - 50e^{-(2/5)t};$ Limiting
concentration in both is 0.3 lb/gal
29. $P(t) = 13 \sin 5t + 13 \cos 5t + 36,$
$Q(t) = -6 \sin 5t + 4 \cos 5t + 28$

31. $x = C_1 e^{2t} + C_2 e^{-3t}, y = 2C_1 e^{2t} - \frac{1}{2}C_2 e^{-3t} - e^t$
33. (a) Yes; (b) counterclockwise; (c) increasing;
(d) x min ≈ 257.5, x max ≈ 262.5; y min ≈ 557.5,
y max ≈ 642.5

Summary and Review: Chapter 9, p. 698

1. [9.1] $y = 2x^3 + C$ 2. [9.2] $y = C_1 e^{2x^3}$, where $C_1 = e^C$
3. [9.2] $y = 5 - C_1 e^{-x^3/3}$, where $C_1 = e^{-C}$
4. [9.1] $y = \frac{2}{3}x^3 - 2x^2 + 3x - 1$
5. [9.2] $y = \pm (8t + 4C)^{1/4}$ 6. [9.1] $y = x^4 - x^2 + 1$
7. [9.1] $x(2x) - 2(x^2) = 0$

8. [9.3] 9. [9.3] (a) $N(t) = \dfrac{9600}{8 + 1192e^{-0.24t}}$;
(b)

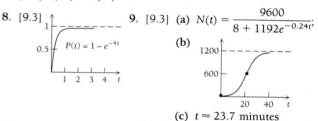

$P(t) = 1 - e^{-4t}$

(c) $t \approx 23.7$ minutes
10. [9.4] (a) $P(t) = 1 - e^{-kt}$; (b) $k \approx 0.16$;
(c) $P(t) = 1 - e^{-0.16t}$; (d) $P(12) \approx 0.85$; (e) $t \approx 18.7$
11. [9.4] 750 12. [9.5] $x^2 + y^3 = 8$
13. [9.5] $x^4 + 3x^2 y - y^4 = -1$
14. [9.6] $y = 3x^{-4}, x > 0$
15. [9.6] $y = -e^{3x} + Ce^{2x}$ 16. [9.7] $y = C_1 e^{2x} + C_2 e^{5x}$
17. [9.7] $y = C_1 e^{3x} + C_2 x e^{3x}$
18. [9.7] $y = C_1 \sin (5x) + C_2 \cos (5x) + 4$
19. [9.8] $x = C_1 e^{6t} + C_2 e^{-4t}, y = \frac{2}{3}C_1 e^{6t} - C_2 e^{-4t}$
20. [9.8] $x = C_1 \sin (6t) + C_2 \cos (6t),$
$y = -\frac{1}{3} \sin (6t) + \frac{1}{3} \cos (6t)$
21. [9.8] $x = C_1 e^{-2t} + C_2 t e^{-2t},$
$y = (2C_1 + C_2)e^{-2t} + 2C_2 t e^{-2t}$
22. [9.8] $x = e^{4t} + e^{2t}, y = e^{4t} - e^{2t}$
23. [9.8] $x = 3e^{2t}, y = 6e^{2t}$

Test: Chapter 9, p. 699

1. $y = \frac{1}{3}x^3 + \frac{3}{2}x^2 - 5x + 7$ 2. $y = C_1 e^{x^8}$, where $C_1 = e^C$
3. $y = \sqrt{18x + C_1}, y = -\sqrt{18x + C_1}$, where $C_1 = 2C$
4. $y = C_1 e^{6t}$, where $C_1 = e^C$. 5. 1360
6. (a) $P(t) = \dfrac{16,000}{20 + 780e^{-0.72t}}$;

(b)
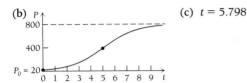
(c) $t = 5.798$

7.

$P(t) = 5(1 - e^{-t})$

8. (a) $S(t) = 20(1 - e^{-kt})$; **(b)** $k = 0.11$;
(c) $S(t) = 20(1 - e^{-0.11t})$; **(d)** 13.343 million;
(e) $t = 12.6$ yr, or the 13th yr　**9.** $x^3 + y^2 = 1$

10. $x^3 + 2x^2y - y^5 = -1$　**11.** $y = \dfrac{4}{x^3}, x > 0$

12. $y = e^{4x} + Ce^{3x}$　**13.** $y = C_1 e^{2x} + C_2 e^{3x}$
14. $y = C_1 e^{6x} + C_2 x e^{6x}$
15. $y = C_1 \sin 6x + C_2 \cos 6x + 2$
16. $x = C_1 e^{5t} + C_2 e^{-3t}, y = \frac{3}{5}C_1 e^{5t} - C_2 e^{-3t}$
17. $x = C_1 \sin 12t + C_2 \cos 12t$,
$y = -2C_2 \sin 12t + 2C_1 \cos 12t$
18. $x = C_1 e^{-3t} + C_2 t e^{-3t}, y = (C_1 + \frac{1}{3}C_2)e^{-3t} + C_2 t e^{-3t}$
19. $x = e^{-5t} - e^{-4t}, y = e^{-5t} - 2e^{-4t}$

CHAPTER 10

Margin Exercises, Section 10.1, pp. 703–711

1. $2(x - 3) + 4 + \dfrac{1}{x - 3}$　**2.** $x - 2 + \dfrac{2}{x - 2} + \dfrac{4}{(x - 2)^2}$
3. $(x - 3)^2 + 2(x - 3) - 1$
4. (a) 24; **(b)** 8; **(c)** -5; **(d)** 1
5. $2(x - 1)^2 + 3(x - 1) + 4$
6. $(x - 3)^3 + 9(x - 3)^2 + 26(x - 3) + 24$
7. $(x - 4)^2 + 5(x - 4) + 5$　**8.** $(x - 2)^2 + (x - 2) - 1$
9. $(x - 1)^2 - (x - 1) + 2$
10. $3(x + 1)^3 + 2(x + 1)^2 - \frac{1}{2}(x + 1) + 1$
11. $-\frac{1}{2}x^5 + \frac{1}{2}x^4 + 7x^3 - \frac{1}{2}x^2 + 2x + 1$

Exercise Set 10.1, p. 711

1. $x - 4 + \dfrac{3}{x}$　**3.** $2x - 12 + \dfrac{7}{x - 3}$

5. $(x - 2)^2 + 7 + \dfrac{4}{x - 2}$　**7.** $(x - 2)^2 + 2(x - 2) + 3$
9. $(x - 2)^2 - (x - 2) + 1$
11. $(x + 1)^2 - 7(x + 1) + 13$
13. $(x + 2)^2 - 7(x + 2) + 14$
15. $(x - 1)^3 + (x - 1)^2 + 3(x - 1) + 4$
17. $(x + 1)^3 - 5(x + 1)^2 + 11(x + 1) - 6$
19. $2(x - 3)^2 + 6(x - 3) + 3$
21. $3(x - 1)^4 + 12(x - 1)^3 + 11(x - 1)^2 - 2(x - 1) - 3$
23. $(x - 1)^5 - 10(x - 1)^3 - 20(x - 1)^2 - 10(x - 1)$

27. $x + 1 - \dfrac{4}{x - 1}$　**29.** $x - 1 + \dfrac{2}{x + 1}$

31. $(x - 1)^4 + 3$
33. (a) 3,628,800; **(b)** 479,001,500; **(c)** 1,307,674,368,000;
(d) $6.402373705 \times 10^{15}$

Margin Exercises, Section 10.2, pp. 713–718

1. $2 - 6x + 9x^2$
2. (a) $(x - 4)^2 + 5(x - 4) + 5$; **(b)** they are equal.
3. $p_2(x) = x, p_3(x) = x - \frac{1}{6}x^3$　**4.** $1 + x^2$
5. (a) $1 + x + \frac{1}{2}x^2 + \frac{1}{6}x^3$; **(b)** 1.105167　**6.** 0.100333

Exercise Set 10.2, p. 718

1. $1 - x + \frac{1}{2}x^2$　**3.** x^2　**5.** $-4x^2 + x - 1$　**7.** $1 + \frac{1}{2}x^2$
9. x^2　**11.** $1 + \frac{1}{2}x + \frac{1}{8}x^2 + \frac{1}{48}x^3$
13. $1 + 3(x - 1) + 3(x - 1)^2 + (x - 1)^3$
15. $-x - \frac{1}{2}x^2 - \frac{1}{3}x^3$
17. $p_4(x) = x - \frac{1}{6}x^3; p_5(x) = x - \frac{1}{6}x^3 + \frac{1}{120}x^5$
19. $p_4(x) = 1 - \frac{1}{2}x^2 + \frac{1}{24}x^4 = p_5(x)$

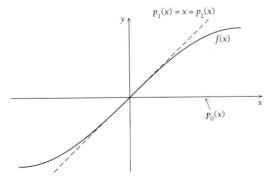

21. $p_0(x) = 0; p_1(x) = x = p_2(x)$

23. $p_0(x) = 1$, $p_1(x) = 1 - x$, $p_2(x) = 1 - x + \frac{1}{2}x^2$

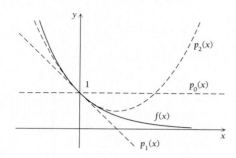

25. 0.1296338   **27.** 0.9512292   **29.** -0.0512917
31. (a) $P(1) \approx 0.16473$, $P(2) \approx 0.30232$, $P(3) \approx 0.41725$;
(b) $p_3(t) = 0.18t - 0.0162t^2 + 0.000972t^3$;
(c) $p_3(1) = 0.164772$, $p_3(2) = 0.302976$, $p_3(3) = 0.420444$
33. (a) $p_2(t) = 5 + 0.1(t - 3) - 0.001(t - 3)^2$;
(b) $P(3.5) \approx 5.04975$, $P(4) \approx 5.099$
35. On $[-1, 1]$

Margin Exercises, Section 10.3, pp. 721–729

1. (a) $a_n = \dfrac{2}{3^n}$; (b) $s_1 = \frac{2}{3} \approx 0.667$, $s_2 = \frac{8}{9} \approx 0.889$,

$s_3 = \frac{26}{27} \approx 0.963$, $s_4 = \frac{80}{81} \approx 0.988$, $s_5 = \frac{242}{243} \approx 0.996$,

$s_6 = \frac{728}{729} \approx 0.999$
2. (a) $a_n = 2n - 1$;
(b) $s_1 = 1$, $s_2 = 4$, $s_3 = 9$, $s_4 = 16$, $s_5 = 25$;
(c) $s_n = n^2$; (d) no, it diverges.

3. (a) $a_n = \dfrac{2}{5^{n-1}} = 2\left(\dfrac{1}{5}\right)^{n-1}$; (b) $a_5 = \frac{2}{625}$, $a_6 = \frac{2}{3125}$

4. $a = \frac{1}{5}$, $r = \frac{1}{3}$ **5.** $s_n = 3[1 - (\frac{1}{3})^n]$ **6.** $\frac{5}{2}$ **7.** $\frac{3}{10}$
8. Diverges ($r = -2$)   **9.** $\frac{5}{7}$ **10.** \$20,000
11. \$6,666.67   **12.** $\frac{4}{9}$ **13.** $\frac{17}{37} (= \frac{459}{999})$
14. 2, 1.3333, 1.5556, 1.4815, 1.5062, and 1.4979
15. Converges
16. $0.9498457 \approx s_3 \geqslant s \geqslant s_4 \approx 0.9459394$
17. (a) $0.8998629 \leqslant s \leqslant 0.9012346$; (b) $\frac{9}{10}$

Exercise Set 10.3, p. 729

1. $a_n = 5^{-n} = \left(\dfrac{1}{5}\right)^n$ **3.** $\dfrac{n}{n+1}$ **5.** $\left(\dfrac{2}{3}\right)^n$
7. $n(-1)^{n+1} = n(-1)^{n-1}$

9. $a = 1$, $r = \dfrac{1}{3}$, $s_{15} = \dfrac{3}{2}\left[1 - \left(\dfrac{1}{3}\right)^{15}\right]$

11. $a = 1$, $r = -\dfrac{1}{3}$, $s_{15} = \dfrac{3}{4}\left[1 + \left(\dfrac{1}{3}\right)^{15}\right]$

13. $a = \dfrac{1}{8}$, $r = -\dfrac{2}{3}$, $s_{15} = \dfrac{3}{40}\left[1 + \left(\dfrac{2}{3}\right)^{15}\right]$

15. $a = \dfrac{5}{3}$, $r = -\dfrac{2}{5}$, $s_{15} = \dfrac{25}{21}\left[1 + \left(\dfrac{2}{5}\right)^{15}\right]$

17. Converges; $s = \frac{8}{7}$ **19.** Converges; $s = \frac{9}{8}$
21. Converges; $s = \frac{3}{5}$ **23.** Diverges   **25.** $\frac{7}{9}$
27. $2 + \frac{38}{99} = \frac{236}{99}$ **29.** $\frac{1}{7}$ **31.** Diverges   **33.** Converges
35. Converges   **37.** Converges   **39.** Converges
41. $0.7407407 \leqslant s \leqslant 0.7530864$
43. $0.9459394 \leqslant s \leqslant 0.9475394$
45. $0.0791736 \leqslant s \leqslant 0.0874381$ **47.** 32 billion dollars
49. \$37.04   **51.** (a) 5 mg, 7 mg, 9.2 mg; (b) 12.29 mg;
(c) 12.5 mg

Margin Exercises, Section 10.4, pp. 733–741

1. $1 - x + \dfrac{x^2}{2!} - \dfrac{x^3}{3!} + \cdots = \displaystyle\sum_{j=0}^{\infty} \dfrac{(-1)^j x^j}{j!}$ **2.** $\displaystyle\sum_{j=0}^{\infty} \dfrac{(-1)^j x^{2j}}{(2j)!}$

3. $4 + 2x - 5x^2 + 7x^3$ **4.** $\displaystyle\sum_{j=0}^{\infty} (-1)^j x^{2j}$

5. $\displaystyle\sum_{j=0}^{\infty} (-1)^j (j + 1) x^j$, $|x| < 1$

6. $1 + 3x + 6x^2 + 10x^3 + \cdots$, $|x| < 1$

7. $1 + x^3 + x^6 + \cdots = \displaystyle\sum_{j=0}^{\infty} x^{3j}$, $|x| < 1$

8. $x^2 - \dfrac{x^6}{3!} + \dfrac{x^{10}}{5!} - \dfrac{x^{14}}{7!} + \cdots = \displaystyle\sum_{j=0}^{\infty} (-1)^j \dfrac{x^{4j+2}}{(2j+1)!}$, all x

9. $e^{-5x} = 1 - 5x + \dfrac{25}{2}x^2 - \dfrac{125}{3!}x^3 + \cdots$, all x

10. $\ln(1 - x) = -x - \dfrac{x^2}{2} - \dfrac{x^3}{3} - \dfrac{x^4}{4} \cdots = \displaystyle\sum_{j=0}^{\infty} -\dfrac{x^{j+1}}{j+1}$

11. 0.494   **12.** $x^2 \cos x = x^2 - \dfrac{1}{2!}x^4 + \dfrac{1}{4!}x^6 - \dfrac{1}{6!}x^8 + \cdots$

13. $x \tan^{-1} x = x^2 - \dfrac{x^4}{3} + \dfrac{x^6}{5} - \dfrac{x^8}{7} + \cdots$, $|x| \leqslant 1$

14. $f(x) = 1 + \dfrac{1}{2!}x + \dfrac{1}{3!}x^2 + \dfrac{1}{4!}x^3 + \cdots$

Exercise Set 10.4, p. 742

1. $1 + 4x + 8x^2 + \frac{32}{3}x^3 + \frac{4^4}{4!}x^4 + \cdots$

3. $1 - (x - 1) + \frac{1}{2!}(x - 1)^2 - \frac{1}{3!}(x - 1)^3 + \cdots$

5. $-2 + 3(x + 1) - 3(x + 1)^2 + (x + 1)^3$

7. $1 + 2x + 4x^2 + 8x^3 + \cdots$

9. $e^{6x} = 1 + (6x) + \frac{1}{2!}(6x)^2 + \frac{1}{3!}(6x)^3 + \cdots$

11. $\frac{1}{1 - x^5} = 1 + x^5 + x^{10} + x^{15} + \cdots, |x| < 1$

13. $\ln(1 + 4x) = 4x - 8x^2 + \frac{64}{3}x^3 - 64x^4 + \cdots, |x| < \frac{1}{4}$

15. $\frac{1}{2 + x} = 1 - (x + 1) + (x + 1)^2 - (x + 1)^3 + \cdots,$
$-2 < x < 0$

17. $\ln(2 + x) = (x + 1) - \frac{1}{2}(x + 1)^2 + \frac{1}{3}(x + 1)^3 - \frac{1}{4}(x + 1)^4 + \cdots, -2 < x < 0$

19. $\frac{5x^4}{1 - x^5} = 5x^4 + 5x^9 + 5x^{14} + \cdots, |x| < 1$

21. $f(x) = -\frac{1}{2} + \frac{1}{4!}x^2 - \frac{1}{6!}x^4 + \cdots,$ all x

23. $x^5 \sin(x^3) = x^8 - \frac{1}{3!}x^{14} + \frac{1}{5!}x^{20} - \frac{1}{7!}x^{26} + \cdots,$ all x

25. $\int_0^1 \cos(x^2)\,dx = 1 - \frac{1}{10} + \frac{1}{216} - \frac{1}{9360} + \cdots \approx 0.90463$

(error $< \frac{1}{9360} \le 0.0002$)   27. 0.82143 (error $< \frac{1}{60} < 0.02$)
29. 0.0002167 (error $< 10^{-13}$)
31. (a) $F(t) = t - \frac{1}{6}t^3 + \frac{1}{40}t^5 - \frac{1}{336}t^7 + \cdots;$
(b) $P(0 \le t \le 0.1) \approx 0.0398278;$ (c) it checks.
33. They are almost identical on $[-1, 1]$ and on $[-3, 3]$, but are quite different on $[3, 5]$ (and on $[-5, -3]$).

Margin Exercises, Section 10.5, pp. 743–752

1. $\frac{4}{11}$ 2. 0   3. 0   4. $-12.356, -109.12,$ and -998.00
5. Does not exist   6. -7 7. -2 8. $\frac{1}{2}$ 9. 2
10. -2 11. 1   12. $-\frac{1}{2}$
13. e^2 (l'Hôpital's Rule doesn't apply)   14. Does not exist
15. $-\frac{5}{3}$ 16. Does not exist   17. 3

Exercise 10.5, p. 752

1. 1   3. Does not exist   5. 0   7. Does not exist
9. 7   11. $\frac{1}{4}$ 13. 0   15. 2   17. $\frac{1}{3}$ 19. -1
21. $-\frac{1}{4}$ 23. -1 25. 1   27. 0   29. Does not exist
31. $\frac{1}{2}$ 33. $\frac{1}{3}$ 35. Does not exist   37. Does not exist
39. ap 41. $f(0)$

43. (a)

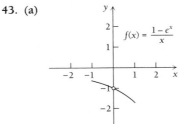

(b) $\lim\limits_{x \to 0} \dfrac{1 - e^x}{x} = -1$

45. (a)

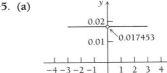

(b) $\pi/180$

Summary and Review: Chapter 10, p. 754

1. [10.1] $x^2 - 5x + 3 = (x - 1)^2 - 3(x - 1) - 1$
2. [10.2] $p_2(x) = -3(x + 1)^2 + 3(x + 1) - 1;$
$p_3(x) = (x + 1)^3 - 3(x + 1)^2 + 3(x + 1) - 1$
3. [10.2] $p_2(x) = p_3(x) = 1 + x^2$
4. [10.2] $p_2(x) = x - 2;\ p_3(x) = (x - 2) - \frac{1}{6}(x - 2)^3$
5. [10.2] 0.049979   6. [10.2] 1.051271   7. [10.3] $\frac{32}{35}$
8. [10.3] Diverges   9. [10.3] $\frac{5}{6}[1 - (\frac{3}{5})^{11}]$ 10. [10.3] $\frac{2107}{1980}$
11. [10.3] Converges; $s \approx s_3 \approx 0.941333$
12. [10.3] Diverges   13. [10.3] Diverges
14. [10.3] Diverges
15. [10.3] Converges; $s = \frac{12}{13} \approx 0.9230769$

16. [10.4] $y = x^4 - \dfrac{x^6}{3!} + \dfrac{x^8}{5!} - \dfrac{x^{10}}{7!} + \cdots,$ for all x

17. [10.4] $y = 1 + (x+1)^2 + \dfrac{(x+1)^4}{2!} + \dfrac{(x+1)^6}{3!} + \cdots$,

for all x

18. [10.4] $y = 1 + 3x + 9x^2 + 27x^3 + \cdots$, for $|x| < \frac{1}{3}$

19. [10.4] 1   **20.** [10.5] $\frac{1}{6}$

21. [10.5] 0 (not an indeterminate form)   **22.** [10.5] -1

23. [10.5] 0   **24.** [10.5] $\frac{1}{3}$ **25.** [10.3] $50

26. [10.4] $\displaystyle\sum_{n=1}^{\infty} \dfrac{x^n}{n^2 \cdot n!}$

27. [10.4] No, it's not differentiable. No.

28. [10.4] **(a)** $f'(0) = 2f(0) = 2$;
(b) $f''(0) = 2f'(0) = 4f(0) = 4$;

(c) $f(x) = \displaystyle\sum_{n=0}^{\infty} \dfrac{2^n}{n!} x^n$

29. **(a)** $x'(0) = 3$, $y'(0) = 5$; **(b)** $x''(0) = y'(0) = 5$,
$y''(0) = -x'(0) + 2y'(0) = 7$; **(c)** yes, Taylor series

Test: Chapter 10, p. 756

1. $p_2(x) = -6(x+2)^2 + 12(x+2) - 8$;
$p_3(x) = (x+2)^3 - 6(x+2)^2 + 12(x+2) - 8$

2. $p_2(x) = 1$; $p_3(x) = 1 + x^3$

3. $p_2(x) = x - 1$; $p_3(x) = (x-1) - \frac{1}{6}(x-1)^3$

4. 0.039989   **5.** 1.072507   **6.** $\frac{9}{25}$ **7.** $\frac{7}{10}[1 - (\frac{2}{7})^{13}]$

8. $\frac{2407}{990}$ **9.** Converges; $s \approx s_3 \approx 0.929167$ **10.** Diverges

11. Diverges   **12.** Converges; $s = \frac{16}{17} \approx 0.941176$

13. $y = x^3 - \dfrac{x^5}{3!} + \dfrac{x^7}{5!} - \dfrac{x^9}{7!} + \cdots$, all x

14. $y = 1 + (x-1)^2 + \dfrac{(x-1)^4}{2!} + \dfrac{(x-1)^6}{3!} + \cdots$, all x

15. $y = 1 + 5x + 25x^2 + 125x^3 + \cdots$, $|x| < \frac{1}{5}$ **16.** $\frac{1}{2}$

17. -2 **18.** 0 (not indeterminate)   **19.** -1 **20.** 0

21. $\frac{1}{2}$ **22.** $20   **23.** $\dfrac{D}{i}$ **24.** $\displaystyle\sum_{n=1}^{\infty} \dfrac{x^n}{n \cdot n!}$

CHAPTER 11

Margin Exercises, Section 11.1, pp. 759–761

1. 1   **2.** The iterates diverge.   **3.** $\bar{x} = 0$ (all 3 cases)

4. 2   **5.** $2 - \sqrt{3} \approx 0.267949$

6. $2 + \sqrt{3} \approx 3.732051$

Exercise Set 11.1, p. 762

1. $\bar{x} = \frac{3}{2} - \frac{1}{2}\sqrt{5} \approx 0.381966$ **3.** 1.497300

5. They diverge.   **7.** They diverge.   **9.** 0.567143

11. 0.395489   **13.** $\bar{x} \approx 1.49787234$

Margin Exercises, Section 11.2, pp. 766–770

1. $x_2 = -\dfrac{1}{3}$, $x_3 = -\dfrac{1}{15}$, $x_4 = \dfrac{-1}{255}$, $x_5 = \dfrac{-1}{65,535}$,
$x_6 \approx x_7 \approx 0 = \bar{x}$ **2.** 0.450184   **3.** -2.103803
4. 2.154435

Exercise Set 11.2, p. 770

1. $\bar{x} = \frac{3}{2} - \frac{1}{2}\sqrt{5} \approx 0.381966$ **3.** 1.497300

5. -2.103803 **7.** 0.772914

9. $\bar{x}_1 \approx -0.152368$; $\bar{x}_2 \approx 1.109643$; $\bar{x}_3 \approx -2.957275$

11. -1.518662 **13.** 0.6345599

15. **(b)** $\sqrt{19} \approx 4.358899$ **17.** **(b)** $\sqrt[10]{20} \approx 1.349283$

19. $\bar{x} \approx 0.7723404255$

Margin Exercises, Section 11.3, pp. 772–777

1. $10,147.84   **2.** $982.14   **3.** $759.34

4. $i \approx 7.9156\%$ **5.** $i \approx 9.7011\%$

Exercise Set 11.3, p. 778

1. $974.36   **3.** $29,959.92   **5.** $10,062.66

7. $751.31   **9.** $148.64

11. $422.41   **13.** $1000   **15.** $1690.05

17. $1833.39   **19.** $758.16

21. **(a)** $5800v^2 + 6000v - 10000 = 0$; **(b)** $i \approx 11.853\%$

23. $i \approx 5.3445\%$

25. The first one (12.044% against 11.775%)

27. **(a)** $2500v^2 - 4300v + 2000 = 0$; **(b)** no real roots

29. $v \approx 0.8941489362$; $i \approx 11.8382\%$

Margin Exercises, Section 11.4, pp. 782–791

1. Approx. = 1.375; exact = 1.5

2. Approx. = 1.625; exact = 1.5

3. $T_4 = 1.5$; exact = 1.5   **4.** Error = 0 since $f'' = 0$

5. $T_{10} = 0.2525$; $E \le \frac{1}{200} = 0.005$ **6.** $S_5 = 0.25$; $E = 0$

Exercise Set 11.4, p. 791

1. (a) 1.9; (b) 2.1   3. (a) 0.833732; (b) 0.733732
5. $T_{10} = 2$ 7. $T_5 = 0.7837315$
9. $T_{10} = 0.170825$; $E \leq 0.01667$
11. $T_{10} \approx 0.385878$; $|E| < 0.000833$ 13. $S_5 = 2$
15. 0.7853982   17. 0.25; $E = 0$
19. 0.3862934; $E \leq 0.0000533$
21. Simpson's Rule

Margin Exercises, Section 11.5, pp. 794–798

1. $x_2 \approx 3.741657$; $x_3 \approx 3.564868$; $x_4 \approx 3.181632$;
$x_5 \approx 3.147723$; $x_6 \approx 3.120675$; $x_7 \approx 3.014054$
2. $T_5 \approx 0.695635$; $T_{10} \approx 0.6937714$; $I \approx 0.6931502$
3. $S_2 \approx 0.6932539$; $S_4 \approx 0.6931545$; $R_2 \approx 0.6931479$

Exercise Set 11.5, p. 798

1. $x_2 = \frac{1}{3}$, $x_3 = \frac{10}{27}$, $x_4 = \frac{3}{8}$
3. $x_2 \approx 1.341471$, $x_3 \approx 1.47382$, $x_4 \approx 1.55758$
5. $x_2 \approx -1.817121$, $x_3 \approx -2.036927$, $x_4 \approx -2.117815$
7. $T_2 = \frac{3}{8} = 0.375$; $T_4 = \frac{11}{32} = 0.34375$;
$T_8 = \frac{43}{128} \approx 0.3359375$; $S_2 = \frac{1}{3}$; $S_4 = \frac{1}{3}$; $R_2 = \frac{1}{3}$
9. $T_2 = 0.775$; $T_4 \approx 0.7827941$; $T_8 \approx 0.7847471$;
$S_2 \approx 0.78539216$; $S_4 \approx 0.7853981$; $R_2 \approx 0.7853985$
13. Romberg integration

Margin Exercises, Section 11.6, pp. 802–806

1. (a) $y = g(x) = e^{-x^2}$; (b) $y_0 = 1$, $y_1 = 1$, $y_2 = 0.92$,
$y_3 = 0.7728$, $y_4 = 0.587328$, $y_5 \approx 0.399383$; (c) $y_0 = 1$,
$y_1 = 1$, $y_2 = 0.98$, $y_3 = 0.9408$, $y_4 = 0.884352$,
$y_5 \approx 0.813604$, $y_6 \approx 0.7322435$, $y_7 \approx 0.644374$,
$y_8 \approx 0.554162$, $y_9 \approx 0.465496$, $y_{10} \approx 0.381707$
2. $y'' = y^2 \cos(xy) + xy^2 \sin(xy) \cos(xy) + y \sin^2(xy)$
3. (a) $y'' = y(4x^2 - 2)$;
(b) $y_{n+1} = y_n[0.96 - (0.4)x_n + (0.08)x_n^2]$; (c) $y_0 = 1$,
$y_1 = 0.96$, $y_2 = 0.847872$, $y_3 = 0.689150$, $y_4 \approx 0.5160358$,
$y_5 \approx 0.356684$

4.

	Euler	Three-term	Exact
$y_0 = y(0)$	1	1	1
$y_1 = y(0.2)$	1	0.96	0.960789
$y_2 = y(0.4)$	0.92	0.847872	0.852144
$y_3 = y(0.6)$	0.7728	0.689150	0.697676
$y_4 = y(0.8)$	0.587328	0.516036	0.527292
$y_5 = y(1)$	0.399383	0.356684	0.367879

5. (b) $y(0.4) \approx 1.5425366$; $x(0.6) \approx 2.8347761$;
$y(1) \approx 0.7540745$
6. $x_1 = 2.4$, $y_1 = -0.8$, $x_2 = 3.2$, $y_2 = -1.92$

Exercise Set 11.6, p. 807

1. 1, 0.9, 0.8, 0.7, 0.6, 0.5, 0.4, 0.3, 0.2, 0.1, and 0
3. 0, 0, 0.125, 0.375, and 0.75
5. 1, 1, 1.08, 1.2528, 1.553472, and 2.050583
7. 1, 1, 1.04, 1.121584, 1.248670, and 1.427460
9. -1, -0.99, -0.96, -0.91, -0.84, -0.75, -0.64,
-0.51, -0.36, -0.19, and 0
11. 1, 0.8, 0.648, 0.592, 0.68, and 0.96
13. 1, 0.78125, 0.6103516, 0.476837, and 0.372529
15. $y = g(x) = 1 - x$. Both methods yield the exact values
17. $y = g(x) = e^{x^2}$. For Euler, see number 5 above.
Three-term gives 1, 1.04, 1.168218, 1.4167056, 1.854184,
and 2.616625.
19. $x = 2e^{2t} + e^{-2t}$, $y = 2e^{2t} - 3e^{-2t}$ are the exact
solutions. Euler gives $x_0 = 3$, $y_0 = -1$, $x_1 = 3.4$, $y_1 = 1$,
$x_2 = 4.28$, $y_2 = 2.84$, $x_3 = 5.704$, $y_3 = 4.84$, $x_4 = 7.8128$,
$y_4 = 7.2944$, $x_5 = 10.83424$, and $y_5 = 10.5232$.
21. $x = e^{2t} - e^t$, $y = 2e^{2t} - e^t$ are the exact solutions.
Euler gives $x_0 = 0$, $y_0 = 1$, $x_1 = 0.2$, $y_1 = 1.6$, $x_2 = 0.52$,
$y_2 = 2.48$, $x_3 = 1.016$, $y_3 = 3.76$, $x_4 = 1.768$, $y_4 = 5.6096$,
$x_5 = 2.88992$, and $y_5 = 8.26816$.
23. $x_0 = x_1 = \cdots = x_{10} = 0$. $y_0 = 1$, $y_1 = 0.9$, $y_2 = 0.81$,
$y_3 = 0.729$, $y_4 = 0.6561$, $y_5 = 0.59049$, $y_6 = 0.531441$,
$y_7 = 0.478297$, $y_8 = 0.430467$, $y_9 = 0.387420$, and
$y_{10} = 0.348678$ 25. $810.80

27. (a) 557.18; **(b)** 527.09; **(c)** 528.58
29. $x(5) \approx 860.94$; $y(5) \approx 491.42$

31. (b) $F(10) = \sum_{i=0}^{9} f(i)$
33. (a) $x(5) \approx 54.7273$, $y(5) \approx 17.3924$; **(b)** $x(10) \approx$
72.204, $y(10) \approx 18.782$

Summary and Review: Chapter 11, p. 810

1. [11.1] 3　**2.** [11.1] 1　**3.** [11.1] 0.9346927
4. [11.1] 0.4501836　**5.** [11.1] Diverges
6. [11.2] $\bar{x}_1 \approx 0.2541017$; $\bar{x}_2 \approx 1.8608058$
7. [11.2] 0.9423132
8. [11.2] $\bar{x}_1 \approx 0.5009922$; $\bar{x}_2 \approx 1.1202946$
9. [11.2] $\bar{x}_E \approx 1.0356017$　**10.** [11.3] \$18,681.63
11. [11.3] **(a)** $7000v^+ = 4000 + 1000v$;
(b) $i \approx 9.2407275\%$
12. [11.3] $i_1 \approx 20.51\%$, $i_2 \approx 14.49\%$
13. [11.4] $T_2 \approx 0.2373419$, $T_4 \approx 0.2391956$,
$T_8 \approx 0.2396578$
14. [11.4] $S_2 \approx 0.2398134$, $S_4 \approx 0.2398118$
15. [11.5] $x_2 \approx 1.3934134$, $x_3 \approx 0.3529083$,
$x_4 \approx 1.0155184$, $x_5 \approx 1.0543590$, $x_6 \approx 0.9875705$,
$x_7 \approx 1.0298004$, $x_8 \approx 1.0299800$, $x_9 \approx 1.0296720$,
$x_{10} \approx 1.0298665$
16. [11.5] $x_2 \approx 0.3483533$, $x_3 \approx 0.4699680$,
$x_4 \approx 0.4441681$, $x_5 \approx 0.45148424$, $x_6 \approx 0.4499003$,
$x_7 \approx 0.4501571$, $x_8 \approx 0.45018938$, $x_9 \approx 0.45018236$,
$x_{10} \approx 0.450180624$　**17.** [11.5] $R_2 \approx 0.2398117$
18. [11.6] $y(0) = 3$, $y(0.2) = 2.4$, $y(0.4) = 1.92$,
$y(0.6) = 1.536$, $y(0.8) = 1.2288$, and $y(1) = 0.99304$
19. [11.6] $y(0) = 3$, $y(0.2) = 2.46$, $y(0.4) = 2.0172$,
$y(0.6) = 1.654104$, $y(0.8) \approx 1.3563653$, $y(1) \approx 1.1122195$
20. [11.6] **(a)** $y = 3e^{-x}$; **(b)** $y(0) = 3$, $y(0.2) = 2.4561923$,
$y(0.4) = 2.0109601$, $y(0.6) = 1.6464349$,
$y(0.8) = 1.34798769$, $y(1) = 1.1036383$
21. [11.6] $x_0 = 3$, $y_0 = 1$, $x_1 = 0.8$, $y_1 = -0.8$, $x_2 = 0$,
$y_2 = -1.28$, $x_3 = -0.256$, $y_3 = -1.28$, $x_4 = -0.3072$,
$y_4 = -1.1264$, $x_5 = -0.28672$, and $y_5 = -0.94208$
22. [9.8] $x = 4e^{-3t} - e^{-t}$, $y = 4e^{-3t} - 3e^{-t}$
23. [11.6] $x_0 = x(0) = 3$, $y_0 = y(0) = 1$,
$x(0.2) \approx 1.376516$, $y(0.2) \approx -0.260946$,
$x(0.4) \approx 0.534457$, $y(0.4) \approx -0.806183$,
$x(0.6) \approx 0.112384$, $y(0.6) \approx -0.985239$,
$x(0.8) \approx -0.086457$, $y(0.8) \approx -0.985115$,
$x(1) \approx -0.168731$, $y(1) \approx -0.90449$

Test: Chapter 11, p. 812

1. 2　**2.** -2　**3.** 0.8977725　**4.** 0.372559
5. The iterates diverge.
6. $\bar{x}_1 \approx 0.2033642$; $\bar{x}_2 \approx 1.469617$　**7.** 0.8977725
8. $\bar{x}_1 \approx 0.50049263$; $\bar{x}_2 \approx 1.1027159$
9. $x_E \approx 1.1366384$　**10.** \$17,043.98
11. (a) $6000v^+ = 3000 + 1000v$; **(b)** $i \approx 11.386791\%$
12. $i_1 \approx 14.47\%$; $i_2 \approx 14.99\%$
13. $T_2 \approx 0.9397933$; $T_4 \approx 0.94451352$; $T_8 \approx 0.94569087$
14. $S_2 \approx 0.94608694$; $S_4 \approx 0.94608331$
15. $x_{10} \approx 1.1105271$　**16.** $x_{10} \approx 0.3725596$
17. $R_2 \approx 0.9460837$　**18.** $y(0) = 1$, $y(0.2) = 0.8$,
$y(0.4) = 0.672$, $y(0.6) = 0.581683$, $y(0.8) = 0.514012$,
and $y(1) = 0.461170$
19. $y(0) = 1$, $y(0.2) = 0.84$, $y(0.4) = 0.722588$,
$y(0.6) = 0.633253$, $y(0.8) = 0.563209$, and
$y(1) = 0.506914$　**20.** $y = (1 + x)^{-1}$
21. $x_0 = 1$, $y_0 = 2$, $x_1 = 1.2$, $y_1 = 1.8$, $x_2 = 1.2$,
$y_2 = 1.56$, $x_3 = 1.104$, $y_3 = 1.32$, $x_4 = 0.9696$,
$y_4 = 1.0992$, $x_5 = 0.82752$, and $y_5 = 0.90528$

Cumulative Review, p. 815

1. $y = -4x - 27$　**2.** $x^2 + 2xh + h^2 - 5$
3. (a)　　　　　　**(b)** 3; **(c)** -3; **(d)** no

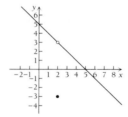

4. -8　**5.** 3　**6.** Does not exist　**7.** 4　**8.** $3x$
9. 0　**10.** -9　**11.** $2x - 7$　**12.** $\frac{1}{4}x^{-3/4}$
13. $-6x^{-7}$　**14.** $(x - 3)5(x + 1)^4 + (x + 1)^5$, or

$2(x + 1)^4(3x - 7)$　**15.** $\dfrac{5 - 2x^3}{x^6}$　**16.** $\dfrac{1 - x}{e^x}$

17. $\dfrac{2x}{x^2 + 5}$　**18.** 1　**19.** $3e^{3x} + 2x$

20. $e^{-x}[2x \cos (x^2 + 3) - \sin (x^2 + 3)]$　**21.** $-\tan x$

22. $2(\sec^2 x)(\tan x)$　**23.** $3xy^2 + \dfrac{y}{x}$

24. $y = x - 2$ **25.** Relative maximum at $(-1, 3)$; relative minimum at $(1, -1)$

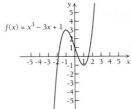

26. Relative maxima at $(-1, -2)$ and $(1, -2)$; relative minimum at $(0, -3)$

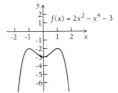

27. Relative minimum at $(-1, -4)$; relative maximum at $(1, 4)$

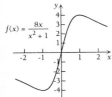

28. Relative maximum at $(0, -2)$. For graph, see Example 7 in Section 3.3.   **29.** Minimum $= -7$ at $x = 1$
30. None   **31.** Maximum $= 6\frac{2}{3}$ at $x = -1$; minimum $= 4\frac{1}{3}$ at $x = -2$ **32.** 15   **33.** 30 times; lot size 15
34. $\Delta y = 3.03$, $f'(x)\, \Delta x = 3$ **35. (a)** $N(t) = 8000e^{0.1t}$;
(b) 17,804; **(c)** 6.9 yr

36. (a) $E(p) = \dfrac{p}{12 - p}$; **(b)** $E(2) = \frac{1}{5}$, inelastic;

(c) $E(9) = 3$; elastic; **(d)** increase; **(e)** $p = \$6$

37. $\frac{1}{2}x^6 + C$ **38.** $3 - \dfrac{2}{e}$

39. $\dfrac{7}{9(7 - 3x)} + \dfrac{1}{9} \ln (7 - 3x) + C$ **40.** $\frac{1}{4}e^{x^4} + C$

41. $\left(\dfrac{x^2}{2} + 3x\right) \ln x - \dfrac{x^2}{4} - 3x + C$

42. $-\cos (x^2 + 3) + C$ **43.** $\sin (e^x) + C$
44. $\frac{232}{3}$ **45.** \$22,679.49   **46.** \$49,444.55
47. Convergent, $\frac{1}{4374}$ **48.** $\frac{3}{2} \ln 3$ **49.** 0.6826
50. $(169, \$7)$; \$751.33   **51.** $-\dfrac{\pi}{2}\left(\dfrac{1}{e^{10}} - 1\right)$

52. $y = C_1 e^{x^2/2}$, where $C_1 = e^C$
53. (a) $P(t) = 1 - e^{-kt}$; **(b)** $k = 0.12$;
(c) $P(t) = 1 - e^{0.12t}$; **(d)** $P(20) = 0.9093$, or 90.93%
54. $8xy^3 + 3$ **55.** $e^y + 24x^2y$ **56.** None
57. Maximum $= 4$ at $(3, 3)$ **58.** $3(e^3 - 1)$ **59.** $\dfrac{\sqrt{2}}{2}$

60. $\frac{1}{2}$ **61.** 0   **62.** $\frac{1}{8} \ln |\sec 8x + \tan 8x| + C$
63. $\frac{1}{3} \tan^{-1} (3x) + C$

64. $[-\csc^2 (\sin (x^3))][\cos (x^3)][3x^2]$ **65.** $\dfrac{1}{\sqrt{1 - x^2}}$

66. $dw = 15x^4y^2\, dx + 6x^5y\, dy$
67. $dw = 7e^x[\sin (yz)\, dx + z \cos (yz)\, dy + y \cos (yz)\, dz]$
68. 13.81 cubic feet   **69.** $\dfrac{1}{12}$ **70.** $\dfrac{2}{3}$ **71.** 0

72. $xy - 3x + y^3 = C$ **73.** $y = Ce^{-\sin x}$
74. $y = C_1 + C_2 \sin 2x + C_3 \cos 2x$
75. $y = C_1 e^{7x} + C_2 x e^{7x}$ **76.** $y = 5 + C_1 e^{-x} + C_2 e^{-3x}$
77. $x = C_1 e^t + C_2 e^{-t}$; $y = 5C_1 e^t + 3C_2 e^{-t}$
78. $5(x - 1)^3 + 2(x - 1)^2 - 3(x - 1) - 5$

79. $\sin (0.07) \approx (0.07) - \dfrac{(0.07)^3}{6} = 0.069942833$

80. $\dfrac{7}{40} = 0.175$ **81.** $\dfrac{679}{999}$ **82.** Converges

83. $x^5 - \dfrac{x^{13}}{3!} + \dfrac{x^{21}}{5!} - \dfrac{x^{29}}{7!} + \cdots = \displaystyle\sum_{k=0}^{\infty} \dfrac{(-1)^k x^{8k+5}}{(2k + 1)!}$

84. 0.32180075   **85.** Does not exist.   **86.** -2 **87.** 4
88. 1   **89.** 0.3725595   **90.** 0.7320508
91. \$466.51   **92.** 13.7%   **93.** $T_5 \approx 0.4059274$;
exact $= \ln (1.5) \approx 0.4054651$ **94.** $s_5 \approx 0.4054653$
95. (a) $s_2 \approx 0.1173604$; $s_4 \approx 0.1115501$;

$I = \dfrac{1}{9} \approx 0.1111111$; **(b)** $R_2 \approx 0.1111627$

INDEX

SUMMARY OF IMPORTANT FORMULAS

18. $\displaystyle\int k\,dx = kx + C$

19. $\displaystyle\int x^r\,dx = \frac{x^{r+1}}{r+1} + C, \quad r \neq -1$

$\displaystyle\int (r+1)x^r\,dx = x^{r+1} + C, \quad r \neq -1$

20. $\displaystyle\int \frac{dx}{x} = \ln x + C, \quad x > 0;$

$\displaystyle\int \frac{dx}{x} = \ln|x| + C, \quad x < 0;$

21. $\displaystyle\int e^x\,dx = e^x + C$

22. $\displaystyle\int k\,f(x)\,dx = k\int f(x)\,dx$

23. $\displaystyle\int [f(x) \pm g(x)]\,dx = \int f(x)\,dx \pm \int g(x)\,dx$

24. $\displaystyle\int u\,dv = uv - \int v\,du$

Further integration formulas occur in Table 1 at the back of the book.